Historical Changes in Large River Fish Assemblages of the Americas

Funding for the publication of this book was provided by

Aquatic Monitoring & Bioassessment Branch
U.S. Environmental Protection Agency
Corvallis, Oregon

Ohio River Valley Water Sanitation Commission (ORSANCO)
Cincinnati, Ohio

Region 3
U.S. Forest Service
Albuquerque, New Mexico

Rocky Mountain Research Station
Flagstaff, Arizona

Water Resources Division
U.S. Geological Survey
Boise, Idaho

Water Quality Section
American Fisheries Society

Historical Changes in Large River Fish Assemblages of the Americas

Edited by

John N. Rinne

*Rocky Mountain Research Station, U.S. Forest Service,
Flagstaff, Arizona 86001, USA*

Robert M. Hughes

*Department of Fisheries and Wildlife, Oregon State University,
Corvallis, Oregon 97331, USA*

Bob Calamusso

Tonto National Forest, U.S. Forest Service, Phoenix, Arizona 85006, USA

American Fisheries Society Symposium 45

Proceedings of the AFS Symposium
Changes in Fish Community Structures in Large USA Rivers
Held in Phoenix, Arizona, USA
21 August 2001

American Fisheries Society
Bethesda, Maryland
2005

The American Fisheries Society Symposium series is a registered serial. Suggested citation formats follow.

Entire book

Rinne J. N., R. M. Hughes, and B. Calamusso, editors. 2005. Historical changes in large river fish assemblages of the Americas. American Fisheries Society, Symposium 45, Bethesda, Maryland.

Chapter within the book

Mueller, G. A., P. C. Marsh, and W. L. Minckley. 2005. A legacy of change: the lower Colorado River, Arizona-California-Nevada, USA and Sonora-Baja Calfornia Norte, Mexico. Pages 139–156 *in* J. N. Rinne, R. M. Hughes, and B. Calamusso, editors. Historical changes in large river fish assemblages of the Americas. American Fisheries Society, Symposium 45, Bethesda, Maryland.

Printed in the United States of America on acid-free paper.

Library of Congress Control Number 2005925668
ISBN 1-888569-72-7
ISSN 0892-2284

American Fisheries Society Web site address: www.fisheries.org

American Fisheries Society
5410 Grosvenor Lane, Suite 110
Bethesda, Maryland 20814-2199
USA

Contents

Preface

In 2000, John Rinne was in the thick of Verde and Gila River data comparisons and began wondering if other American rivers were experiencing similar losses of native riverine fish species, or if the Southwest United States cornered the market. The forum of a national meeting the next summer in nearby Phoenix seemed like a good place to answer that question. So he and Bob Calamusso decided to organize a session entitled *Changes in Fish Community Structure in Large USA Rivers,* which attracted 13 presenters.

In the discussion period that followed the presentations, a few presenters encouraged publication of the talks in a book. Bob Hughes volunteered to solicit manuscripts from other riverine fish biologists and help submit a book proposal to the American Fisheries Society (AFS) by the following spring. He also requested, and obtained, early sponsorship for the effort from the AFS Water Quality Section. Aaron Lerner of AFS had the proposal reviewed and it was approved for publication in the AFS Symposium Series in September 2002. From then on, John, Bob, and Bob Calamusso tried to find time between full-time jobs and long field seasons to bird-dog authors and reviewers for manuscripts. In the two years that followed, several authors completed review and revised drafts on time, while others were delayed by more important events in their lives or were dropped from the publication. Only eight of the presenters completed a chapter, but out of 29 authors originally expressing sincere interest, 3 were dropped and two were added. We thank all the authors for their persistence and scholarship, the many reviewers for suggested improvements, the sponsors for sharing the publication costs of the book, and Julie Frandsen for the copyediting that improved the book's readability.

Our objective in producing this book, as its title indicates, is to document recent and historical changes in the fish assemblages of large American rivers, and to determine patterns in those changes. We believe such information should aid river management in general, and stimulate similar studies elsewhere. We hope you, the reader, will find all our contributions valuable in appreciating what riverine fishes we have lost, as well as the efforts of many (including several of the authors) in rehabilitating and protecting the remaining ones.

Reviewers

Arthur Benke
University of Alabama
Tuscaloosa, Alabama, USA

Robert Bettaso
Arizona Department of Game and Fish
Phoenix, Arizona, USA

Scott Bonar
University of Arizona
Tucson, Arizona, USA

Anne Brasher
U.S. Geological Survey
Salt Lake City, Utah, USA

Lee Bridges
Indiana Department of Environmental Management
Mooresville, Indiana, USA

Larry Brown
U.S. Geological Survey
Sacramento, California, USA

James Chandler
Idaho Power Company
Boise, Idaho, USA

John Chick
Illinois Natural History Survey
Brighton, Illinois, USA

Doug Chipersak
Department of Fisheries and Oceans
Winnipeg, Manitoba, Canada

Andy Clark
Arizona Game & Fish Department
Kingman, Arizona, USA

Robert Clarkson
U.S. Bureau of Reclamation
Phoenix, Arizona, USA

Walt Courtenay
U.S. Geological Survey
Gainesville, Florida, USA

Robert Dubey
New Mexico State University
Las Cruces, New Mexico, USA

Erich Emery
ORSANCO
Cincinnati, Ohio, USA

Dave Galat
University of Missouri
Columbia, Missouri, USA

James Gammon
DePau University
Greencastle, Indiana, USA

Fran Gelwick
Texas A & M University
College Station, Texas, USA

Carter Gilbert
Florida Museum of Natural History
Gainesville, Florida, USA

Hugo Godinho
Pontifical Catholic University of Minas Gerais
Belo Horizonte, Minas Gerais, Brazil

Beth Goldwitz
Platt River Whooping Crane Trust
Wood River, Nebraska, USA

Steve Gutreuter
U.S. Geological Survey
La Crosse, Wisconsin, USA

Clark Hubbs
University of Texas
Austin, Texas, USA

Brian Ickes
U.S. Geological Survey
Onalaska, Wisconsin, USA
James Karr
University of Washington
Seattle, Washington, USA

Philip Kaufmann
U.S. Environmental Protection Agency
Corvallis, Oregon, USA

Gordon Linam
Texas Parks & Wildlife Department
San Marcos, Texas, USA

George Low
Department of Fisheries and Oceans
Hay River, Northwest Territories, Canada

John Lyons
Wisconsin Department of Natural Resources
Monona, Wisconsin, USA

Terry R. Maret
U.S. Geological Survey
Boise, Idaho, USA

Peter McCart
Aquatic Environments Ltd.
Spruce View, Alberta, Canada

Frank McCormick
U.S. Forest Service
Olympia, Washington, USA

Stephan McIninch
Virginia Commonwealth University
Richmond, Virginia, USA

Michael Meador
U.S. Geological Survey
Reston, Virginia, USA

Maurice Mettee
Geological Survey of Alabama
Tuscaloosa, Alabama, USA

Dennis Miller
Western New Mexico University
Silver City, New Mexico, USA

Charles Minckley
U.S. Fish and Wildlife Service
Kingman, Arizona, USA

Joseph Nelson
University of Alberta
Edmonton, Alberta, Canada

Steven Newhouse
Indiana Department of Environmental Management
Indianapolis, Indiana, USA

Mark Oswood
Institute of Arctic Biology
Fairbanks, Alaska, USA

William Pearson
University of Louisville
Louisville, Kentucky, USA

Tadeusz Penczak
University of Lodz
Lodz, Poland

John Peterka
Rochert, Minnesota, USA

Edward Peters
University of Nebraska
Lincoln, Nebraska, USA

David Peterson
U.S. Geological Survey
Cheyenne, Wyoming, USA

John Pitlo
Iowa Department of Natural Resources
Bellevue, Iowa, USA

LeRoy Poff
Colorado State University
Fort Collins, Colorado, USA

Michael Porter
U.S. Bureau of Reclamation
Albuquerque, New Mexico, USA

Ronald Preston
Barnesville, Ohio, USA

David Propst
New Mexico Department of Game & Fish
Santa Fe, New Mexico, USA

J. W. Rachlin
Lehman College
Bronx, New York, USA

Michael Saiki
U.S. Geological Survey
Dixon, California, USA

Dan Sallee
Illinois Department of Natural Resources
Sterling, Illinois, USA

Randall Sanders
Ohio Department of Natural Resources
Columbus, Ohio, USA

Gilmar Santos
Pontifical Catholic University of Minas Gerais
Belo Horizonte, Minas Gerais, Brazil

Isaac Schlosser
University of North Dakota
Grand Forks, North Dakota, USA

Harold Schramm
Mississippi State University
Mississippi State, Mississippi, USA

Gary Scoppettone
U.S. Geological Survey
Reno, Nevada, USA

Patrice Simon
Department of Fisheries & Oceans
Ottawa, Ontario, Canada

Thomas Simon
U.S. Fish & Wildlife Service
Bloomington, Indiana, USA

Clarence Smith
Colorado Springs, Colorado, USA

Gerald Smith
University of Michigan
Ann Arbor, Michigan, USA

William Smith-Vaniz
U.S. Geological Survey
Gainesville, Florida, USA

Blaine Snyder
Tetra Tech
York, Pennsylvania, USA

Michael Stewart
Troy State University
Troy, Alabama, USA

Jeffrey Thomas
Ohio River Valley Water Sanitation Commission
Cincinnati, Ohio, USA

William Tonn
University of Alberta
Edmonton, Alberta, Canada

Vince Travnichek
Missouri Department of Conservation
Columbia, Missouri, USA

Joel Trexler
Florida International University
Miami, Florida, USA

Bruce Vondracek
University of Minnesota
St. Paul, Minnesota, USA

John Waldman
Hudson River Foundation
New York, New York, USA

Lizhu Wang
Institute for Fisheries Research
Ann Arbor, Michigan, USA

Melvin Warren
U.S. Forest Service
Oxford, Mississippi, USA

Steve Wente
U.S. Geological Survey
Mounds View, Minnesota, USA

Gwen White
Indiana Department of Natural Resources
Indianapolis, Indiana, USA

Jack Williams
Southern Oregon University
Ashland, Oregon, USA

Kirk Winemiller
Texas A & M University
College Station, Texas, USA

Luis Zambrano
Universidad Nacional de Mexico
Mexico City, Mexico

Symbols and Abbreviations

The following symbols and abbreviations may be found in this book without definition. Also undefined are standard mathematical and statistical symbols given in most dictionaries.

A	ampere
AC	alternating current
Bq	becquerel
C	coulomb
°C	degrees Celsius
cal	calorie
cd	candela
cm	centimeter
Co.	Company
Corp.	Corporation
cov	covariance
DC	direct current; District of Columbia
D	dextro (as a prefix)
d	day
d	dextrorotatory
df	degrees of freedom
dL	deciliter
E	east
E	expected value
e	base of natural logarithm (2.71828…)
e.g.	(exempli gratia) for example
eq	equivalent
et al.	(et alii) and others
etc.	et cetera
eV	electron volt
F	filial generation; Farad
°F	degrees Fahrenheit
fc	footcandle (0.0929 lx)
ft	foot (30.5 cm)
ft^3/s	cubic feet per second (0.0283 m^3/s)
g	gram
G	giga (10^9, as a prefix)
gal	gallon (3.79 L)
Gy	gray
h	hour
ha	hectare (2.47 acres)
hp	horsepower (746 W)
Hz	hertz
in	inch (2.54 cm)
Inc.	Incorporated
i.e.	(id est) that is
IU	international unit
J	joule
K	Kelvin (degrees above absolute zero)
k	kilo (10^3, as a prefix)
kg	kilogram
km	kilometer
l	levorotatory
L	levo (as a prefix)
L	liter (0.264 gal, 1.06 qt)
lb	pound (0.454 kg, 454g)
lm	lumen
log	logarithm
Ltd.	Limited
M	mega (10^6, as a prefix); molar (as a suffix or by itself)
m	meter (as a suffix or by itself); milli (10^{23}, as a prefix)
mi	mile (1.61 km)
min	minute
mol	mole
N	normal (for chemistry); north (for geography); newton
N	sample size
NS	not significant
n	ploidy; nanno (10^{29}, as a prefix)
o	ortho (as a chemical prefix)
oz	ounce (28.4 g)
P	probability
p	para (as a chemical prefix)
p	pico (10^{212}, as a prefix)
Pa	pascal
pH	negative log of hydrogen ion activity
ppm	parts per million
qt	quart (0.946 L)
R	multiple correlation or regression coefficient
r	simple correlation or regression coefficient
rad	radian
S	siemens (for electrical conductance); south (for geography)
SD	standard deviation
SE	standard error

s	second
T	tesla
tris	tris(hydroxymethyl)-aminomethane (a buffer)
UK	United Kingdom
U.S.	United States (adjective)
USA	United States of America (noun)
V	volt
V, Var	variance (population)
var	variance (sample)
W	watt (for power); west (for geography)
Wb	weber
yd	yard (0.914 m, 91.4 cm)
α	probability of type I error (false rejection of null hypothesis)
β	probability of type II error (false acceptance of null hypothesis)
Ω	ohm
μ	micro (10^{26}, as a prefix)
′	minute (angular)
″	second (angular)
°	degree (temperature as a prefix, angular as a suffix)
%	per cent (per hundred)
‰	per mille (per thousand)

American Fisheries Society Symposium 45:1–12, 2005

Introduction to Historical Changes in Large River Fish Assemblages of the Americas

Robert M. Hughes*
Department of Fisheries and Wildlife, Oregon State University, Corvallis, Oregon 97331, USA

John N. Rinne
Rocky Mountain Research Station, U.S. Forest Service, Flagstaff, Arizona 86001, USA

Bob Calamusso
Tonto National Forest, U.S. Forest Service, Phoenix, Arizona 85006, USA

Abstract.—This book's objective is to document historical changes in the fish assemblages of large American rivers, and to determine patterns in and rationale for those changes. In this chapter, we review pertinent literature on large rivers and fish assemblages worldwide and briefly introduce the chapters. We expect that the information contained in this book will aid river management in general, and stimulate similar historical fish assemblage studies elsewhere. There will never be a better time to learn and understand what has been changed and to reverse or slow undesirable changes.

Introduction

Two recent books have attempted to capture the changing character of large rivers. Dodge (1989) focused on commercial fisheries, and thus on yield, standing stocks, and production of fishes valued for harvest. However, many of the factors affecting fisheries also influence entire fish assemblages, and some chapters in Dodge (1989) depicted fish assemblages. Benke and Cushing (2005) described the ecosystem character of major rivers in North America, and included considerable information on native and alien fish species in each chapter. Although they did not focus on fish assemblages, most of the issues that limit riverine ecosystems certainly affect their fish assemblages. Excellent texts on general river ecology have also been written (Whitton 1975; Davies and Walker 1986), both included chapters on fish assemblages. Hocutt and Wiley (1986) reviewed the ichthyogeograpy of North American freshwater fish by basin, but did not emphasize large rivers. The chapter authors in our book were asked to focus on historical changes in fish assemblages because historical fish assemblage data are available for many rivers, because fish assemblages are excellent indicators of changing ecological conditions in large rivers, and because the public often cares about fish (Schiemer 2000). We concur with Fauth et al. (1996) in this book, restricting use of the word "community" to *all* species occurring in the same place or ecosystem at some time, and use the word "assemblage" for a phylogenetically related subset of a community.

* Corresponding author: hughes.bob@epa.gov

Historical Changes in Fish Assemblages and Species

Several authors have previously employed historical data to assess changes in, and threats to, fish assemblages in large water bodies. Smith (1968) documented dramatic reductions in 11 native salmonids in the Laurentian Great Lakes from dominant and commercially fished in 1900, to vulnerable (three species), threatened (two species), endangered (two species), or extinct (four species) at present. Major causes were over harvest and alien species allowed access by lock construction. In the Maumee and Illinois rivers, Karr et al. (1985) determined that three large-river fish species have been extirpated since 1850, probably because of water pollution. Since 1910, Columbia River anadromous fish runs have declined by 75% following construction of a series of 14 main-stem dams plus another 11 on its major tributary, the Snake River (Ebel et al. 1989). Construction of four lower Snake River dams reduced salmon spawning by 90% since 1960 (Kareiva et al. 2000). Lelek (1989) reported marked reductions in Rhine River lamprey, sturgeon, herring, minnow, and sculpin species as a result of dams and water pollution over the past 200 years. Oberdorff and Hughes (1992) documented sturgeon and herring extirpations as well as species losses in salmonids and lampreys from the Seine River, mostly because of dams, over harvest, and sedimentation. Clearly, deliberate human actions have altered fish assemblages in those waters.

Similar historical declines have been documented for stream fish species. In Illinois, 34% of native fish species have been decimated or extirpated (Smith 1971). Karr et al. (1985) observed that species declines and extirpations in small and medium-sized tributaries to the Maumee and Illinois rivers ranged from 70% to 84%. Moyle and Williams (1990) found that 63% of native California fish species were vulnerable, threatened, endangered, or extinct. Rinne and Minckley (1991) reported 80% of the native fishes in the Southwest are threatened or endangered. Since 1960, 52% of the stream fish species in the Missouri River basin of Wyoming have declined in abundance (Patton et al. 1998). In the southeastern United States, 28% of the fish taxa are considered vulnerable, threatened, or endangered (Warren et al. 2000). Williams et al. (1989) concluded that 27% of fish species in the United States are vulnerable, threatened or endangered. Over the past 100 years, 40 fish taxa in North America have become extinct as a result of physical habitat change and alien species (Miller et al. 1989). Such declines are not simply threats to naturally rare, endemic fish species; Nehlsen et al. (1991) reported that 23% of native anadromous Pacific salmonids are vulnerable or at risk of extinction, despite once supporting commercial harvest. Extinction rates for North American aquatic amphibians, crayfishes, mollusks, and fishes are five times those of temperate terrestrial fauna and similar to those of tropical forests (Ricciardi and Rasmussen 1999), indicating widespread systemic changes.

Current Fish Assemblage Status

The concern does not just lie with individual species, but with entire fish assemblages as well. Judy et al. (1984) used a probability survey to estimate that 81% of U.S. rivers were adversely affected. Roth et al. (1999) employed an index of fish assemblage integrity and a probability survey to find that 54% of all stream lengths in Maryland were in fair to very poor condition, 20% were in good condition, and the remainder were too small to be rated. In the mid-Atlantic highlands of the eastern United States, McCormick et al. (2001) used an index of fish assemblage integrity and a probability survey to determine that 52% of all mapped wadeable stream lengths were in poor or marginal condition, with 27% in good or excellent condition (the remainder were fishless). A similar assessment of all mapped wadeable stream and nonwadeable river lengths in the five mid-Atlantic states indicated that 73% were in poor or marginal condition, with 21% in good condition (USEPA 2003). Hughes et al. (2004) concluded that 45% of the mapped wadeable stream lengths in the Oregon and Washington Coast Range were impaired, based on a probability survey and an index of fish and amphibian assemblage integrity. Such extensive and substantial impairments of fish assemblages indicate fundamental ecosystem alterations.

Causes for Fish Assemblage Declines

Social Causes

The ultimate causes of the declines in fish assemblages of large rivers have been reported on at length and will only be summarized briefly here. Those causes are the same as those for environmental and social degradation in general: excessive human population and consumption stimulated by ethical and economic principles that are not based on ecological sustainability (Ehrlich and Holdren 1971; Daly 1973; Nash 1989; Hughes 1996; Czech et al. 2004). Those social values and policies generate human disturbances across landscapes and riverscapes of such magnitude that fish species often cannot maintain viable population sizes. Thus, the major stressor-producing disturbances fall into two broad categories: land use and inchannel.

Land use.—Land use changes are often visible from high altitudes and outside the Earth's atmosphere. Vitousek et al. (1986) reported that humans use 40% of the global terrestrial net primary production, which has substantial effects on rivers draining intensively farmed, mined, or urbanized basins. Bryce et al. (1999) and Malmqvist and Rundle (2002) list land uses as major human activities leading to degraded rivers. For example, Omernik (1977) determined that land use accounted for 25% to 67% of the variance in stream nutrient concentrations across the conterminous United States. Land use accounted for 20% of the variance in water quality of central Michigan rivers (Johnson et al. 1997). Turner and Rabalais (2003) found that sediment and nitrogen concentrations doubled in the lower Mississippi River following land clearing for agriculture.

How do land use and water quality changes affect fish assemblages? Frequently, land use represents a substantial amount of the variability in fish assemblage integrity. Judy et al. (1984) estimated that fish assemblages in 29% of all U.S. rivers were adversely affected by agriculture. Harding et al. (1998) found that historical catchment agriculture accounted for 53% of the variability in observed fish species richness in western North Carolina streams. Percent catchment agriculture predicted 31–50% of the variability in eastern Wisconsin and eastern Michigan streams (Roth et al. 1996; Fitzpatrick et al. 2001), and 56% of the variability in eastern Idaho rivers (Mebane et al. 2003). In Oregon rivers, urban and agricultural land use was associated with 45% of the variability in fish index of biotic integrity (IBI) scores (Mebane et al. 2003). Those land uses also represented 37% of the variability in fish IBI scores in streams of Oregon's Willamette Valley (Bryce and Hughes 2002; Van Sickle et al. 2004). Percent catchment urbanized was associated with 34–68% of the variability of IBI scores in southern Ontario, eastern Maryland, southeastern Wisconsin, and eastern West Virginia streams (Steedman 1988; Klauda et al. 1998; Wang et al. 2000; Snyder et al. 2003). Catchment land use was associated with 27% of the variability in fish assemblage metrics in highly disturbed Michigan and Wisconsin streams (Wang et al. in press). Richter et al. (1997) determined that agricultural and municipal land uses were the chief sources of stressors to imperiled aquatic species. Although most of the preceding studies involved streams, the hierarchical nature of rivers and their tributary connections suggest that those effects are transmitted downriver, especially in highly disturbed basins (Wang et al. in press).

Inchannel Causes

There are many inchannel means of impairing fish assemblages. Karr and Dudley (1981) and Karr et al. (1986) presented a conceptual model of factors governing the condition of lotic ecosystems: water quality, flow regime, habitat structure, energy sources, and biotic interactions. Of these, Welcomme (1979) and Junk et al. (1989) felt that the flood pulse was the principal driving force for the major biota in large rivers. Minshall (1988) regarded alterations in flood magnitude, frequency, duration, and especially predictability and amount of change as key disturbances. Stanford et al. (1996) considered flow regulation the most pervasive anthropogenic change on rivers world-wide. Poff et al. (1997) argued that flow regime was the master variable, often regulating the others. Allan and Flecker (1993) added alien species and overexploitation to the list, but these can be subsumed by the biotic interactions of Karr et al. (1986). In a survey of conservation biologists, Richter et al. (1997) determined that degraded water quality (increased sediments and nutrients), alien species, and altered flow regimes were the chief threats to imperiled aquatic species.

Although land use affects peak and low flows, water management structures such as dams, levees, and diversions directly modify flow regimes. Dynesius and Nilsson (1994) found that 77% of the flows from the 139 largest rivers in the world's north temperate regions were strongly or moderately affected by dams and diversions. The NRC (1992) reported that more than 85% of lotic systems in the conterminous United States are artificially controlled by channel modifications, with one million river kilometers replaced by reservoirs (Echeverria et al. 1989). Graf (2001) reported that there are over 80,000 dams in the United States large enough to meet the specifications of the Dam Safety Act, and that they can store the entire nation's annual runoff. He added that 80% of conterminous U.S. river length is negatively affected by humans, with only 0.3% officially listed as wild and scenic, 25% of which occurs in the Pacific Northwest. Benke (1990) reported that only 42 high-quality, freely-flowing river segments greater than 200 km long occurred in the conterminous United States. Of these, only 2 were large and only 6 of the 42 rivers (all six of which were low-gradient southeastern U.S. rivers) flowed undammed to the sea. Globally, 54% of runoff is appropriated by humans (Gleick 2000) and in arid regions, water extractions often cause rivers to cease flowing to the sea or to their confluences with receiving rivers (Malmqvist and Rundle 2002).

Altered flow regimes remove the natural dynamism of rivers by which rivers create and maintain diverse habitats necessary for biological diversity, productivity, connectivity, and sustainability. The single strand, straightened, armored, flow-regulated systems free of snags that most of us see and think of when the words "large river" are mentioned are vastly different from what many rivers once were. Of course, some rivers always had constrained channels or reaches. Many rivers, however, once incorporated wide annually-flooding floodplains, meandering braided channels, off-channel sloughs and ponds, sinuous plan forms, highly variable thalweg depths, diverse vegetated and unvegetated banks and bars, islands, and log jams (Leopold 1968; Minckley and Rinne 1982; Sedell and Froggatt 1984; Triska 1984; Sparks 1995; Jungwirth et al. 2002, 2003; Gregory et al. 2003; Hohensinner et al. 2004; Blew and MacCoy in press). All these features typically changed at biennially. This incredible heterogeneity was driven by natural continental variation in flow regimes and geology, as well as natural disturbance regimes that maintained connectivity upstream and downstream and between river channels and riparian/floodplain systems (Ward 1998).

Floodplains, temporarily flooded bars, and complex shorelines are critical sites for many riverine fish species to reproduce, rear, or both. Johnson et al. (1995) proposed that if floodplain waters are included, species richness is highest in large, versus medium-sized, rivers. In natural floodplain rivers, such as the Parana, fish richness increases downriver from 57 to 91 species in its major tributaries to 100 in the lower main stem (Agostinho et al. 2000), as opposed to the decline in species richness observed in lower reaches of many large altered rivers (where floodplains have been eliminated). When floodplains no longer flood, channel complexity is lost, alluvial groundwater recharge is reduced, and sediments and nutrients are flushed down river rather than deposited and recycled in floodplain wetlands and water bodies, or converted to fish biomass (Jungwirth 1998). Continuous steep banks and u-shaped or steep trapezoidal channels typical of engineered rivers offer no refuge against floods, dam releases, piscivores, or low winter temperatures when swimming capacities are reduced (Baras and Lucas 2001).

Modified flow regimes also eliminate signals for upstream and downstream fish migrations, and they break annual wetland flooding or flood timing, thereby hindering spawning, rearing, and escape to refugia (Junk et al. 1989; Welcomme 1992, 1995; Sparks 1995; Pringle et al. 2000; Baras and Lucas 2001). Stabilized flows often favor alien fish species (Meffe 1984; Minckley and Deacon 1991; Fuller et al. 1999; Marchetti et al. 2004a, 2004b). Highly fluctuating flows, produced by diel changes in hydropower demand, favor generalist species over native rheophilic species (Richards 1976; Bain et al. 1988; Travnichek and Maceina 1994; Penczak and Kruk 2000; Brown and Ford 2002; Aarts et al. 2004), or shallow water species and life stages that concentrate at stream margins in slow water (Bain et al. 1988; Stanford et al. 1996; Baras and Lucas 2001). During annual floods, many riverine species lay buoyant eggs that drift downstream to floodplains or shallow bars where the embryos develop and the young rear. Altered flows, reduced channel complexity, and reservoirs remove the turbid water and structural concealment that allow these species to complete their life histories (Baras and Lucas 2001; Dieterman and Galat 2004; Quist et al. 2004; Tieman et al. 2004). Baras and Lucas (2001) list examples of 40 fish species that migrate 100–1000s of kilometers in rivers, but potamodromous and diadromous fish migrations are limited or eliminated by dams (Agostinho et al. 1994, 2000; Pringle et al. 2000; Baras and Lucas 2001). Even where fish passes exist, the altered flow characteristics of reservoirs interrupt movements of adult and larval fish (Agostinho et al. 2002).

Reservoirs also affect water quality. Reservoir releases alter thermal signals, eliminating or reducing densities of coldwater species from low head dams (Richards 1976; Lessard and Hayes 2003) and warmwater species from high-head, hypolimnetic-release dams in temperate regions (Carlson and Muth 1989). Dams also eliminate the upstream movements of nutrients (Pringle 1997; Baras and Lucas 2001) and the downstream movement of water, sediments, nutrients, and large wood debris (Ligon et al. 1995). Thus, river regulation alters most important ecological processes of rivers, thereby reducing channel complexity, habitat diversity, and native and specialist fish species diversity, and it is strongly associated with the proliferation of many alien species (Fuller et al. 1999; Marchetti et al. 2004a, 2004b).

Alien species.—The introduction of alien species has been deemed a major stressor on native fish assemblages. Rahel (2002) determined that most of the homogenization of the fish fauna in the conterminous United States resulted from deliberately introducing alien, cosmopolitan sport and food species. Rinne (2004a) reported that over 100 species have been introduced into southwestern U.S. rivers and streams; over half have become established. Moyle et al. (1986) called this the "Frankenstein Effect" because such attempts to improve on nature can have disastrous results (Miller et al. 1989; Minckley and Deacon 1991), especially when accompanied by physical and chemical habitat change (Moyle and Light 1996; Marchetti et al. 2004a, 2004b; Rinne 2004b).

Conceptual Models of Riverine Structure and Processes

Increased knowledge of how natural rivers function and how their performances are altered by humans has led to several very useful conceptual models. One of the earliest models was that of river zonation, whereby continental European rivers were divided into four zones with their indicator fish species (brown trout *Salmo trutta*, grayling *Thymallus thymallus*, barbel *Barbus barbus*, bream *Abramis brama*; Huet 1959). But this model proved difficult to apply because of the restricted ranges of the species, the absence of distinct zones, and general problems with single indicator species. Later, Vannote et al. (1980) proposed the river continuum concept, which was based on a linear model of forested streams changing continuously as they widened, and they emphasized the downstream flux of materials. Their model seemed poorly applicable to large floodplain rivers and those passing through nonforested ecoregions. Increased knowledge of the differing life history and species requirements of fishes has increased and focused attention on maintaining connectivity for processes and fish among upstream, downstream, and floodplain habitat patches—especially at the aquatic terrestrial interface (Amoros and Roux 1988; Gregory et al. 1991; Schlosser 1991). Ward and Stanford (1983a, 1995) introduced the serial discontinuity concept to clarify how dams alter the river continuum depending on their location. They also stressed the importance of intermediate disturbance for supporting high levels of biological diversity (Ward and Stanford 1983b). This led to the flood pulse concept of Junk et al. (1989). Bayley (1995) and Sparks (1995) stressed that significant alteration from natural flood pulses (such as result from flood prevention) was a serious ecosystem disturbance. Fausch et al. (2002) coined the riverscape model (from riverine landscapes, Ward 1998), to emphasize the blending of landscape ecology and riverine fish ecology. The riverscape concept stresses the physical and biological variability of rivers and the importance of hydrological connectivity in channels and between channels and catchments—all of which are essential to life cycle completion for many fish species.

Chapter Overview

The following chapters employ those conceptual models and trace the recent history of 27 American rivers and their fish assemblages. Although each river is different, the chapters can be placed into five groups based on their fish faunas (Figure 1).

1. Systems naturally characterized by anadromous salmonids and flowing to the Arctic or Pacific oceans (Mackenzie, Slave, Snake, Willamette, Sacramento);
2. Desert rivers once characterized by warm-water potamodromous fishes and flowing to the Gulf of California or Gulf of Mexico (Virgin, Verde, Gila, upper Colo rado, lower Colorado, Grande, Nazas);
3. Mid-continent rivers characterized by cool-water potamodromous fishes and flowing to Hudson's Bay (Red) or the Gulf of Mexico (Platte, Missouri, upper Mississippi, Wisconsin, Wabash, Patoka, Ohio);
4. Rivers with cool water anadromous fishes and draining to the Atlantic Ocean (Sus quehanna, Hudson, St. Johns); and
5. Subtropical rivers supporting warmwater potamodromous faunas (Alabama, Kissimmee, Velhas) connecting to the Gulf of Mexico or Atlantic Ocean.

Acknowledgments

We thank Aaron Lerner for approving publication of this book, Julie Frandsen for copyediting, Joseph

Figure 1. Locations of rivers described in this book.

Tomelleri for the cover artwork, and the American Fisheries Society for printing. This manuscript was partially funded by the USEPA through a National Center for Environmental Research (NCER) STAR Program grant R-829498–01 to Oregon State University, and benefited from reviews by Art Benke, Erich Emery, Jim Gammon, Tad Penczak, and LeRoy Poff.

References

Aarts, B. G. W., F. W. B. Van den Brink, and P. H. Nienhuis. 2004. Habitat loss as the main cause of the slow recovery of fish faunas of regulated large rivers in Europe: the transversal floodplain gradient. River Research and Applications 20:3–23.

Agostinho, A. A., L. C. Gomes, D. R. Fernandez, and H. I. Suzuki. 2002. Efficiency of fish ladders for neotropical ichthyofauna. River Research and Applications 18:299–306.

Agostinho, A. A., H. F. Julio, Jr., and M. Petrere, Jr. 1994. Itaipu reservoir (Brasil): impacts of the impoundment on the fish fauna and fisheries. Pages 171–184 *in* I. G. Cowx, editor. Rehabilitation of inland fisheries. Fishing News Books, Oxford, England.

Agostinho, A. A., S. M. Thomaz, C. V. Minte-Vera, and K. O. Winemiller. 2000. Biodiversity in the high Parana River floodplain. Pages 89–118 *in* B. Gopal, W. J. Junk, and J. A. Davis, editors. Biodiversity in wetlands: assessment, function and conservation. Backhuys Publishers, Leiden, The Netherlands.

Allan, J. D., and A. S. Flecker. 1993. Biodiversity conservation in running waters. BioScience 43:32–43.

Amoros, C., and A. L. Roux. 1988. Interaction between water bodies within the flood plains of large rivers: function and development of connectivity. Munsterische Geographische Arbeiten 29:125–130.

Bain, M. B., J. T. Finn, and H. E. Booke. 1988. Streamflow regulation and fish community structure. Ecology 69:382–392.

Baras, E., and M. C. Lucas. 2001. Impacts of man's modifications of river hydrology on the migration of freshwater fishes: a mechanistic perspective. Ecohydrology and Hydrobiology 1:291–304.

Bayley, P. B. 1995. Understanding large river-floodplain ecosystems. BioScience 45:153–158.

Benke, A. C. 1990. A perspective on America's vanishing streams. Journal of the North American Benthological Society 9:77–88.

Benke, A. C., and C. E. Cushing. 2005. Rivers of North America. Elsevier, Amsterdam, The Netherlands.

Blew, D., and D. MacCoy. In press. Historical perspective on the urbanization of the lower Boise River, Idaho, USA. In L. R. Brown, R. M. Hughes, R. Gray, and M. R. Meador, editors. Effects of urbanization on stream ecosystems. American Fisheries Society, Symposium 47, Bethesda, Maryland.

Brown, L. R., and T. Ford. 2002. Effects of flow on the fish communities of a regulated California river: implications for managing native fishes. River Research and Applications 18:331–342.

Bryce, S. A., and R. M. Hughes. 2002. Variable assemblage response to multiple disturbance gradients: case studies in Oregon and Appalachia, USA. Pages 539–560 *in* T. P. Simon, editor. Biological response signatures: indicator patterns using aquatic communities. CRC Press, Boca Raton, Florida.

Bryce, S. A., D. P. Larsen, R.M. Hughes, and P. R. Kaufmann. 1999. Assessing relative risks to aquatic ecosystems: a mid-Appalachian case study. Journal of the American Water Resources Association 35:23–36.

Carlson, C. A., and R. Muth. 1989. The Colorado River: lifeline of the American Southwest. Pages 220–239 *in* D. P. Dodge, editor. Proceedings of the international large river symposium (LARS). Canadian Special Publication of Fisheries and Aquatic Sciences 106, Department of Fisheries and Oceans, Ottawa.

Czech, B., P. L. Angermeier, H. E. Daly, E. P. Pister, and R. M. Hughes. 2004. Fish conservation, sustainable fisheries, and economic growth: no more fish stories. Fisheries 29(8):36–37.

Daly, H. E. 1973. The steady-state economy: toward a political economy of biophysical equilibrium and moral growth. Pages 149–174 *in* H. E. Daly, editor. Toward a steady-state economy. W. H. Freeman, San Francisco.

Davies, B. R., and K. F. Walker. 1986. The ecology of river systems. Dr. W. Junk Publishers, Dordrecht, Netherlands.

Dieterman, D. J., and D. L. Galat. 2004. Large-scale factors associated with sicklefin chub distribution in the Missouri and lower Yellowstone rivers. Transactions of the American Fisheries Society 133:577–587.

Dodge, D. P. 1989. Proceedings of the international large river symposium (LARS). Canadian Special Publication of Fisheries and Aquatic Sciences 106, Department of Fisheries and Oceans, Ottawa.

Dynesius, M., and C. Nilsson. 1994. Fragmentation and flow regulation of river systems in the northern third of the world. Science 266:753–762.

Ebel, J. W., C. D. Becker, J. W. Mullan, and H. L. Raymond. 1989. The Columbia River—toward holistic understanding. Pages 205–219 *in* D. P. Dodge, editor. Proceedings of the international large river symposium (LARS). Canadian Special Publication of Fisheries and Aquatic Sciences 106, Department of Fisheries and Oceans, Ottawa.

Echeverria, J. D., P. Barrow, and R. Roos-Collins. 1989. Rivers at risk: the concerned citizen's guide to hydropower. Island Press, Covelo, California.

Ehrlich, P. R., and J. P. Holdren. 1971. Impact of population growth. Science 171:1212–1217.

Fausch, K. D., C. E. Torgersen, C. V. Baxter, and H. W. Li. 2002. Landscapes to riverscapes: bridging the gap between research and conservation of stream fishes. BioScience 52:483–498.

Fauth, J. E., J. Bernardo, M. Camara, W. J. Resetarits, Jr., J. Van Buskirk, and S. A. McCollum. 1996. Simplifying the jargon of community ecology: a conceptual approach. The American Naturalist 147:282–286.

Fitzpatrick, F. A., B. C. Scudder, B. N. Lenz, and D. J. Sullivan. 2001. Effects of multi-scale environmental characteristics on agricultural stream biota in eastern Wisconsin. Journal of the American Water Resources Association 37:1489–1507.

Fuller, P. L., L. G. Nico, and J. D. Williams. 1999. Nonindigenous fishes introduced into inland waters of the United States. American Fisheries Society, Bethesda, Maryland.

Gleick, P. H. 2000. The world's water 2000–2001. Island Press, Covelo, California.

Graf, W. L. 2001. Damage control: restoring the physical integrity of America's rivers. Annals of the Association of American Geographers 91:1–27.

Gregory, S. V., K. L. Boyer, and A. M. Gurnell. 2003. The ecology and management of wood in rivers. American Fisheries Society, Bethesda, Maryland.

Gregory, S. V., F. J. Swanson, W. A. McKee, and K. W. Cummins. 1991. An ecosystem perspective of riparian zones. BioScience 41:540–551.

Harding, J. S., E. F. Benfield, P. V. Bolstad, G. S. Helfman, and E. B. B. Jones, III. 1998. Stream biodiversity: the ghost of land use past. Proceedings of the National Academy of Science 95:14843–14847.

Hocutt, C. H., and E. O. Wiley. 1986. The zoogeography of North American freshwater fishes. Wiley, New York.

Hohensinner, S., H. Habersack, M. Jungwirth, and G. Zauner. 2004. Reconstruction of the characteristics of a natural alluvial river-floodplain system and hydromorphological changes following human modifications: the Danube River (1812–1991). River Research and Applications 20:25–41.

Huet, M. 1959. Profiles and biology of western European streams as related to fish management. Transactions of the American Fisheries Society 88:155–163.

Hughes, R.M. 1996. Do we need institutional change? Pages 559–568 *in* D.J. Stouder, P.A. Bisson, and R.J. Naiman, editors. Pacific salmon and their ecosystems. Chapman and Hall, New York.

Hughes, R. M., S. Howlin, and P. R. Kaufmann. 2004. A biointegrity index (IBI) for coldwater streams of western Oregon and Washington. Transactions of the American Fisheries Society 133:1497–1515.

Johnson, B. L., W. B. Richardson, and T. J. Naimo. 1995. Past, present, and future concepts in large river ecology. BioScience 45:134–141.

Johnson, L. B., C. Richards, G. E. Host, and J. W. Arthur. 1997. Landscape influences on water chemistry in Midwestern stream ecosystems. Freshwater Biology 37:193–208.

Judy, R. D., P. N. Seeley, T. M. Murray, S. C. Svirsky, M. R. Whitworth, and L. S. Ischinger. 1984. 1982 national fisheries survey volume 1 technical report: initial findings. U. S. Fish and Wildlife Service, FWS/OBS-84/06, Fort Collins, Colorado.

Jungwirth, M. 1998. River continuum and fish migration—going beyond the longitudinal river corridor in understanding ecological integrity. Pages 19–32 *in* M. Jungwirth, S. Schmutz, and S. Weiss, editors. Fish migration and fish bypasses. Fishing News Books, Oxford, England.

Jungwirth, M., G. Haidvogl, O. Moog, S. Muhar, and S. Schmutz. 2003. Angewandte fischökologie an flieBgewassern. Facultas Verlags und Buchhandels, Vienna.

Jungwirth, M., S. Muhar, and S. Schmutz. 2002. Re-establishing and assessing ecological integrity in riverine landscapes. Freshwater Biology 47:867–887.

Junk, W. J., P. B. Bayley, and R. E. Sparks. 1989. The flood pulse concept in river-floodplain systems. Pages 110–127 *in* D. P. Dodge, editor. Proceedings of the international large river symposium (LARS). Canadian Special Publication of Fisheries and Aquatic Sciences 106, Department of Fisheries and Oceans, Ottawa.

Kareiva, P., M. Marvier, and M. McClure. 2000. Recovery and management options for spring/summer Chinook salmon in the Columbia River basin. Science 290:977–979.

Karr, J. R., and D. R. Dudley. 1981. Ecological perspective on water quality goals. Environmental Management 5:55–68.

Karr, J. R., K. D. Fausch, P. L. Angermeier, P. R. Yant, and I. J. Schlosser. 1986. Assessing biological integrity in running waters: a method and its rationale. Illinois Natural History Survey, Special Publication 5, Champaign, Illinois.

Karr, J. R., L. A. Toth, and D. R. Dudley. 1985. Fish communities of midwestern rivers: a history of degradation. BioScience 35:90–95.

Klauda, R., P. Kazyak, S. Stranko, M. Southerland, N. Roth, and J. Chaillou. 1998. Maryland Biological Stream Survey: a state agency program to assess the impact of anthropogenic stresses on stream habitat quality and biota. Environmental Monitoring and Assessment 51:299–316.

Lelek, A. 1989. The Rhine River and some of its tributaries under human impact in the last two centuries. Pages 469–487 *in* D. P. Dodge, editor. Proceedings of the international large river symposium (LARS). Canadian Special Publication of Fisheries and Aquatic Sciences 106, Department of Fisheries and Oceans, Ottawa.

Leopold, A. 1968. A Sand County almanac and sketches here and there. Oxford University Press, Oxford, England.

Lessard, J. L., and D. B. Hayes. 2003. Effects of elevated water temperature on fish and macroinvertebrate communities below small dams. River Research and Applications 19:721–732.

Ligon, F. K., W. E. Dietrich, and W. J. Trush. 1995. Downstream ecological effects of dams. BioScience 45:183–192.

Malmqvist, B., and S. Rundle. 2002. Threats to running water ecosystems of the world. Environmental Conservation 29(2):134–153.

Marchetti, M. P., T. Light, P. B. Moyle, and J. H. Viers. 2004a. Fish invasions in California watersheds: testing hypotheses using landscape patterns. Ecological Applications 14:1507–1525.

Marchetti, M. P., P. B. Moyle, and R. Levine. 2004b. Alien fishes in California watersheds: characteristics of successful and failed invaders. Ecological Applications 14:587–596.

McCormick, F. H., R. M. Hughes, P. R. Kaufmann, D. V. Peck, J. L. Stoddard, and A. T. Herlihy. 2001. Development of an index of biotic integrity for the mid-Atlantic highlands region. Transactions of the American Fisheries Society 130:857–877.

Mebane, C. A., T. R. Maret, and R. M. Hughes. 2003. An index of biological integrity (IBI) for Pacific Northwest Rivers. Transactions of the American Fisheries Society 132:239–261.

Meffe, G. K. 1984. Effects of abiotic disturbance on coexistence of predator and prey fish species. Ecology 65:1525–1534.

Miller, R. R., J. D. Williams, and J. E. Williams. 1989. Extinctions of North American fishes during the past century. Fisheries 14(6):22–38.

Minckley, W. L., and J. E. Deacon. 1991. Battle against extinction: native fish management in the American West. University of Arizona Press, Tucson.

Minckley, W. L., and J. N. Rinne. 1982. Large woody debris in hot-desert streams: a historical review. Desert Plants 7(3):142–153.

Minshall, G. W. 1988. Stream ecosystem theory: a global perspective. Journal of the North American Benthological Society 7.263–288.

Moyle, P. B., H. W. Li, and B. A. Barton. 1986. The Frankenstein effect: impact of introduced fishes on native fishes of North America. Pages 415–426 *in* R. Stroud, editor. The role of fish culture in fisheries management. American Fisheries Society, Bethesda, Maryland.

Moyle, P. B., and T. Light. 1996. Biological invasions of fresh water: empirical rules and assembly theory. Biological Conservation 78:149–161.

Moyle, P. B., and J. E. Williams. 1990. Biodiversity loss in the temperate zone: decline of the native fish fauna of California. Conservation Biology 4:275–284.

Nash, R. F. 1989. The rights of nature: a history of

environmental ethics. University of Wisconsin Press, Madison.

Nehlsen, W., J. E. Williams, and J. A. Lichatowich. 1991. Pacific salmon at the crossroads: stocks at risk from California, Oregon, Idaho, and Washington. Fisheries 16(2):4–21.

NRC (National Research Council). 1992. Restoration of aquatic systems: science, technology, and public policy. National Academy Press, Washington, D.C.

Oberdorff, T., and R. M. Hughes. 1992. Modification of an index of biotic integrity based on fish assemblages to characterize rivers of the Seine basin, France. Hydrobiologia 228:117–130.

Omernik, J. M. 1977. Nonpoint source-stream nutrient level relationships: a nationwide study. U.S. Environmental Protection Agency, EPA600/3–77-105, Corvallis, Oregon.

Patton, T. M., F. J. Rahel, and W. A. Hubert. 1998. Using historical data to assess changes in Wyoming's fish fauna. Conservation Biology 12:1120–1128.

Penczak, T., and A. Kruk. 2000. Threatened obligatory riverine fishes in human-modified Polish rivers. Ecology of Freshwater Fish 9:109–117.

Poff, N. L., J. D. Allan, M. B. Bain, J. R. Karr, K. L. Prestegaard, B. D. Richter, R. E. Sparks, and J. C. Stromberg. 1997. The natural flow regime. BioScience 47:769–784.

Pringle, C. M. 1997. Exploring how disturbance is transmitted upstream: going against the flow. Journal of the North American Benthological Society 16:425–438.

Pringle, C. M., M. C. Freeman, and B. J. Freeman. 2000. Regional effects of hydrologic alterations on riverine macrobiota in the new world: tropical-temperate comparisons. BioScience 50:807–823.

Quist, M. C., W. A. Hubert, and F. J. Rahel. 2004. Relations among habitat characteristics, exotic species, and turbid-river cyprinids in the Missouri River drainage of Wyoming. Transactions of the American Fisheries Society 133:727–742.

Rahel, F. J. 2002. Homogenization of freshwater faunas. Annual Review of Ecology and Systematics 33:291–315.

Ricciardi, A., and J. B. Rasmussen. 1999. Extinction rates of North American freshwater fauna. Conservation Biology 13:1220–1222.

Richards, J. S. 1976. Changes in fish species composition in the Au Sable River, Michigan from the 1920s to 1972. Transactions of the American Fisheries Society 105:32–40.

Richter, B. D., D. P. Braun, M. A. Mendelson, and L. L. Master. 1997. Threats to imperiled freshwater fauna. Conservation Biology 11:1081–1093.

Rinne, J. N. 2004a. Native and introduced fishes: their status, threats, and conservation. Pages 193–213 *in* M. B. Baker, P. F. Ffolliott, L. F. Debano, and D. G. Neary, editors. Riparian areas of the southwestern United States: history, ecology and management. Lewis, Boca Raton, Florida.

Rinne, J. N. 2004b. Fish habitats: conservation and management implications. Pages 277–297 *in* M. B. Baker, P. F. Ffolliott, L. F. Debano, and D. G. Neary, editors. Riparian areas of the southwestern United States: history, ecology and management. Lewis, Boca Raton, Florida.

Rinne, J. N., and W. L. Minckley. 1991. Native fishes in arid lands, a dwindling natural resource of the desert Southwest. Rocky Mountain Forest and Range Experiment Station, General Technical Report 206, Fort Collins, Colorado.

Roth, N. E., J. D. Allan, and D. E. Erickson. 1996. Landscape influences on stream biotic integrity assessed at multiple scales. Landscape Ecology 11:141–156.

Roth, N. E., M. T. Southerland, G. Mercurio, J. C. Chaillou, D. G. Heimbuch, and J. C. Seibel. 1999. State of the streams: 1995–1997 Maryland Biological Stream Survey results. Maryland Department of Natural Resources, Annapolis.

Schiemer, F. 2000. Fish as indicators for the assessment of the ecological integrity of large rivers. Hydrobiologia 423:271–278.

Schlosser, I. J. 1991. Stream fish ecology: a landscape perspective. BioScience 41:704–712.

Sedell, J. R., and J. L. Froggatt. 1984. Importance of streamside forests to large rivers: the isolation of the Willamette River, Oregon, USA, from its floodplain by snagging and streamside forest removal. Internationale Vereinigung fuer Theoretische und Angewandte Limnologie Verhandlungen 22:1828–1834.

Smith, P. W. 1971. Illinois streams: a classification based on their fishes and an analysis of factors responsible for disappearance of native species. Illinois Natural History Survey, Biological Notes 76, Champaign.

Smith, S. H. 1968. Species succession and fishery exploitation in the Great Lakes. Journal of the Fisheries Research Board of Canada 25:667–693.

Snyder, C. D., J. A. Young, R. Villella, and D. P. Lemarie. 2003. Influences of upland and riparian land use patterns on stream biotic integrity. Landscape Ecology 18:647–664.

Sparks, R. E. 1995. Need for ecosystem management of large rivers and their floodplains. BioScience 45:168–182.

Stanford, J. A., J. V. Ward, W. J. Liss, C. A. Frissell, R. N. Williams, J. A. Lichatowich, and C. C. Coutant. 1996. A general protocol for restoration of regulated rivers. Regulated Rivers Research and Management 12:391–413.

Steedman, R. J. 1988. Modification and assessment of an index of biotic integrity to quantify stream quality in southern Ontario. Canadian Journal of Fisheries and Aquatic Sciences 45:492–501.

Tieman, J. S., D. P. Gillette, M. L. Wildhaber, and D. R. Edds. 2004. Effects of lowhead dams on riffle-dwelling fishes and macroinvertebrates in a midwestern river. Transactions of the American Fisheries Society 133:705–717.

Travnichek, V. H., and M. J. Maceina. 1994. Comparison of flow regulation effects on fish assemblages in shallow and deep water habitats in the Tallapoosa River, Alabama. Journal of Freshwater Ecology 9:207–216.

Triska, F. J. 1984. Role of wood debris in modifying channel geomorphology and riparian areas of a large lowland river under pristine conditions: a case history. Internationale Vereinigung fuer Theoretische und Angewandte Limnologie Verhandlungen 22:1876–1892.

Turner, R. E., and N. N. Rabalais. 2003. Linking landscape and water quality in the Mississippi River basin for 200 years. BioScience 53:563–572.

USEPA. 2003. Mid-Atlantic integrated assessment (MAIA) state of the flowing waters report. U.S. Environmental Protection Agency, Corvallis, Oregon.

Vannote, R. L., G. W. Minshall, K. W. Cummins, J. R. Sedell, and C. E. Cushing. 1980. The river continuum concept. Canadian Journal of Fisheries and Aquatic Sciences 37:130–137.

Van Sickle, J., J. Baker, A. Herlihy, P. Bayley, S. Gregory, P. Haggerty, L. Ashkenas, and J. Li. 2004. Projecting the biological condition of streams under alternative scenarios of human land use. Ecological Applications 14:368–380.

Vitousek, P. M., P. R. Ehrlich, A. H. Ehrlich, and P. A. Matson. 1986. Human appropriation of the products of photosynthesis. BioScience 36:368–373.

Wang, L., J. Lyons, P. Kanehl, R. Bannerman, and E. Emmons. 2000. Watershed urbanization and changes in fish communities in southeastern Wisconsin streams. Journal of the American Water Resources Association 36:1173–1189.

Wang, L., P. W. Seelbach, and J. Lyons. In press. Effects of levels of human disturbance on the influence of catchment, riparian, and reach scale factors on fish assemblages. In R. M. Hughes, L. Wang, and P. W. Seelbach, editors. Influences of landscapes on stream habitats and biological assemblages. American Fisheries Society Symposium, Bethesda, Maryland.

Ward, J. V. 1998. Riverine landscapes: biodiversity patterns, disturbance regimes, and aquatic conservation. Biological Conservation 83:269–278.

Ward, J. V., and J. A. Stanford. 1983a. The serial discontinuity concept of lotic ecosystems. Pages 29–42 *in* T. D. Fontaine, III and S. M. Bartell, editors. Dynamics of lotic ecosystems. Ann Arbor Science, Ann Arbor, Michigan.

Ward, J. V., and J. A. Stanford. 1983b. The intermediate disturbance hypothesis: an explanation for biotic diversity patterns in lotic ecosystems. Pages 347–356 *in* T. D. Fontaine, III and S. M. Bartell, editors. Dynamics of lotic ecosystems. Ann Arbor Science, Ann Arbor, Michigan.

Ward, J. V., and J. A. Stanford. 1995. The serial discontinuity concept: extending the model to floodplain rivers. Regulated Rivers Research and Management 10:159–168.

Warren, M. L., B. M. Burr, S. J. Walsh, H. L. Bart, Jr., R. C. Cashner, D. A. Etnier, B. J. Freeman, B. R. Kuhajda, R. L. Mayden, H. W. Robison, S. T. Ross, and W. C. Starnes. 2000. Diversity, distribution, and conservation status of the native freshwater fishes of the southern United States. Fisheries 25(10):7–31.

Welcomme, R. L. 1979. Fisheries ecology of floodplain rivers. Longman, London.

Welcomme, R. L. 1992. River conservation future prospects. Pages 453–462 *in* P. J. Boon, P. Calow, and G. E. Petts, editors. River conservation and management. Wiley, Chichester, England.

Welcomme, R. L. 1995. Relationships between fisheries and the integrity of river systems. Regulated Rivers Research and Management 11:121–136.

Whitton, B. A. 1975. River ecology. University of California Press. Berkeley.

Williams, J. E., J. E. Johnson, D. A. Hendrickson, S. Contreras-Balderas, J. D. Williams, M. Navarro-Mendoza, D. E. McAllister, and J. E. Deacon. 1989. Fishes of North America endangered, threatened, or of special concern: 1989. Fisheries 14(6):2–20.

American Fisheries Society Symposium 45:13–21, 2005

Stability, Change, and Species Composition of Fish Assemblages in the Lower Mackenzie River: A Pristine Large River

Ross F. Tallman*, Kimberly L. Howland, and Sam Stephenson

Central and Arctic Region, Department of Fisheries and Oceans, Winnipeg, Manitoba, R3T 2N6, Canada

Abstract.—The Mackenzie River is the second longest river in North America and drains 1.8 × 10^6 km^2 of Arctic and sub-Arctic Canada. Thirty-eight fish species have been recorded in the lower Mackenzie River. These species represent a unique mixture of fishes from the Beringian and Agassisian refugia. Many of the species important for subsistence and commercial fisheries in the lower Mackenzie River have complex life cycles and undertake long migrations to spawn, rear, and overwinter. The lower Mackenzie River is a relatively pristine environment with no dams or major industry, a low human population, and species only lightly harvested. This explains why the species composition is relatively stable. However, recently, the effects of climate change may be starting to influence the species composition in terms of greater frequency of rare species such as Pacific salmon. Moreover, a major gas pipeline proposed for the lower Mackenzie River region will probably disturb the fish assemblage structure.

Introduction

The Mackenzie River is the largest north-flowing river in the Americas, and the fourth largest discharging into the Arctic Ocean. Only the Yenesei, Lena, and Ob in Russia are larger (Todd 1970). The Mackenzie River system drains 1.787 million km^2 and annually carries 333 km^2 of water and 118 million metric tons (mt) of suspended sediment to the Arctic Ocean (Brunskill 1986). It extends from 54°N to 69°N and 103°W to 140°W. Much of the basin lies in permafrost, permanently frozen ground, which forms when the mean annual temperature is below 0°C (Rosenberg and Barton 1986). With temperatures below freezing much of the year, great quantities of water are retained as ice (Brunskill 1986). The area around the Mackenzie has a climate of a dry subpolar or polar desert. Hence, mean annual runoff is low, but flow at spring ice breakup can be very high.

The Mackenzie River system includes eight major rivers, three major lakes, and three major delta areas (Figure 1). The lower Mackenzie River flows from the western basin of Great Slave Lake. It is joined by the Liard River from the west, the Great Bear River from the east, and as it bends sharply westward by the Arctic Red and Peel rivers on its southern side. The system empties into the Beaufort Sea through the Mackenzie Delta. Only the Lena River Delta is larger among the Arctic river deltas. Great Bear and Great Slave lakes are Canada's third and fourth largest lakes, respectively (Rosenberg and Barton 1986). Great Bear Lake is the largest freshwater lake situated entirely in Canada.

Man's interaction with the lower Mackenzie River and the fish therein may have first occurred 25,000 years ago when the ancestral American Indians are thought to have moved from Asia across the Bering land bridge. However, the configuration of the Mackenzie basin is relatively recent (Craig and Fyles 1960; Mayewski et al. 1981). During the Wisconsin glaciation, four-fifths of the Mackenzie

*Corresponding author: TallmanR@DFO-MPO.GC.CA

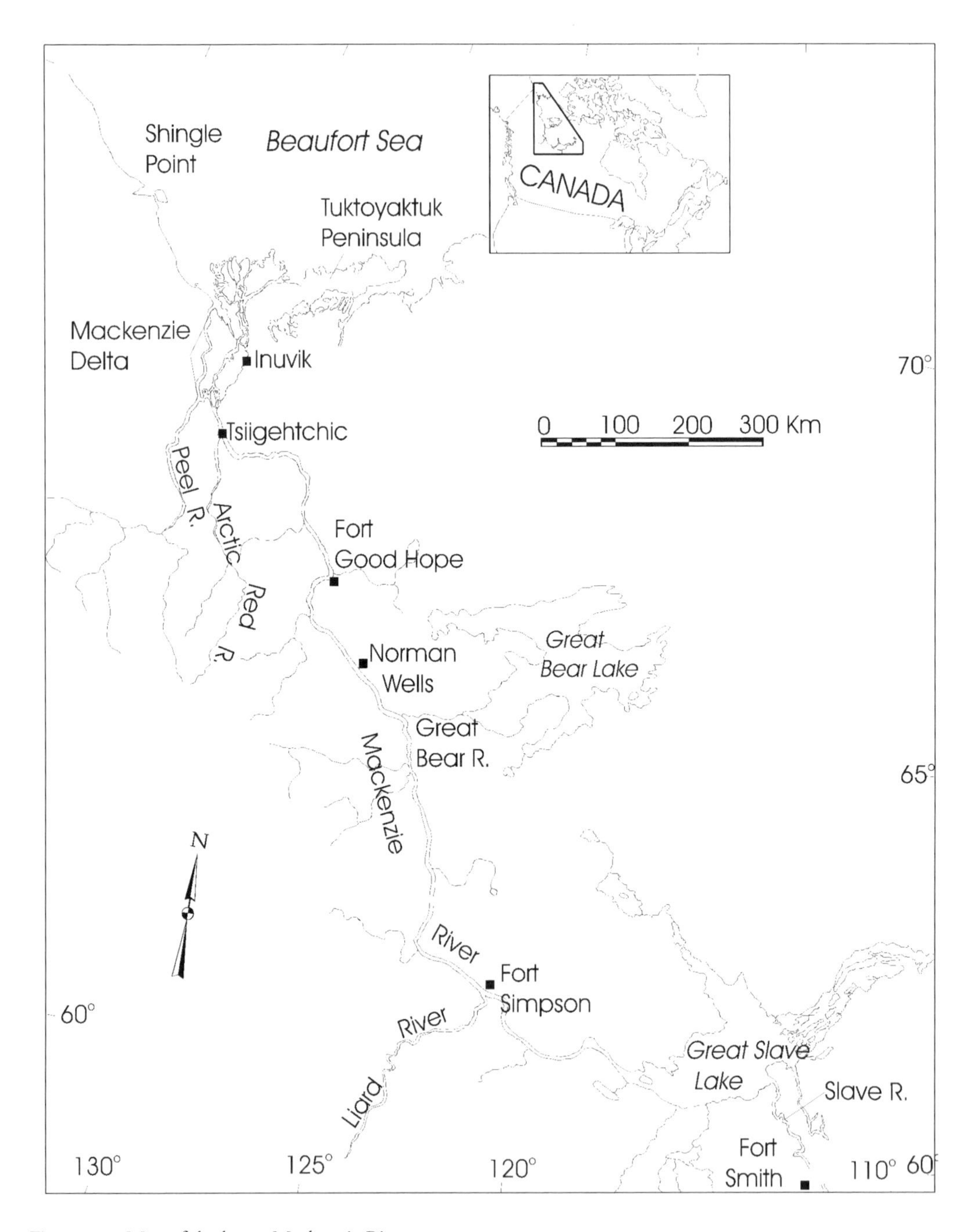

Figure 1.—Map of the lower Mackenzie River.

basin was covered by ice. Deglaciation began about 12,000–14,000 BP and was complete by 8,000 BP (Prest 1969). Therefore, man's activities probably date from 7,000 to 8,000 BP.

In any case, the intensity of man's activities on the fish would have been quite limited until the time of the fur trade when Fort Good Hope was established in 1804 for the Northwest Company. From this point on, dog teams became important to transport trappers and furs to and from the trapping areas and the trading post. To feed both the trappers and the dogs, fish, such as broad whitefish *Coregonus nasus* from the Mackenzie River, were desirable. The level of fishing probably continued to increase until the 1960s and 1970s when snowmobiles began to replace dog teams (Treble and Tallman 1997).

During the late 19th and early 20th centuries, geologists and surveyors explored and mapped the Mackenzie basin (Millar and Fedirchuk 1975). The discovery of oil at Norman Wells in the 1920s led to human population increases. Construction of the Alaska, Mackenzie, Dempster, and Liard highways, along with regular airplane flights, has made much of the lower Mackenzie basin accessible. The economy is mixed between those pursuing traditional ways of life (e.g., trappers, hunters, and fishermen) and those engaged in more modern pursuits (e.g., mining engineers, office workers, and researchers).

Half the Mackenzie River is in the Northwest Territories, with about 10% of the basin's human population (Rosenberg and Barton 1986), but there are only about 17,000 people living along the entire length of the river. Current population of the Northwest Territories is near 20,000. The lower Mackenzie River region has a population of near 8,000, with most in the Inuvik area.

Three aboriginal land claims have been settled along the lower Mackenzie (see Figure 1), and other claims are in progress. The Inuvialuit, Gwich'in, and Sahtu claims will be important in altering the developments along the river in the future.

Compared to most large rivers, the lower Mackenzie River is nearly pristine. The only industrial development along the river is oil drilling based at Norman Wells and this has minimal impact on the river itself. Almost all the communities along the river originated because they were at good fishing sites. People along the river remain heavily dependent upon fish. Scientific studies of the fish assemblage structure in the river date back to the 1960s. Only McCart (1986) and Bodaly et al. (1989) are widely available. The rest of the information has been published in industry and government reports or remains in Department of Fisheries and Oceans files.

Recently, there is concern that climate change is altering the composition of species in the lower Mackenzie River. Climate change will be greatest in the Arctic, and much of the fauna are present only because they are adaptated to the Arctic climate (Peilou 1991).

Development of a natural gas pipeline along the lower Mackenzie River corridor is imminent. The corridor will result in considerable habitat disruption and also provide increased access to fish for human populations. Seismic exploration and barge traffic will increase and landing sites may experience erosion. Development of large camps along the river are likely to deplete fish stocks. There is an urgent need to characterize the present fauna and determine the extent of natural fluctuations in the system. Thus, a description of the fauna and examination of changes in the species composition is timely.

Thirty-eight species of fish have been reported from the lower Mackenzie River, its tributaries, and delta (Scott and Crossman 1973, Shotton 1973, Stein et al. 1973, Johnson 1975, Chang-Kue and Cameron 1980, Corkum and McCart 1981, McLeod and O'Neill 1983, McCart 1986, Bodaly et al. 1989, Howland 1997, Tallman and Reist 1997, Treble and Tallman 1997, Howland et al. 2001, Stephenson 2003) (Table 1). This compares to 36 species in the Athabasca, 31 in the Peace, 26 in the Liard, and 32 in the Slave River (McCart 1986; Tallman et al. 2004). The lower Mackenzie River has relatively low species richness compared to other large rivers of the world (e.g., the Niger River—149 species [Greenwood et al. 1966], the Amazon—1,300+ species [Lowe-McConnell 1986], the Parana River—550+ species [Bonetto 1986]), but supports more species than most rivers of the same latitude.

Methods

A history of studies on the lower Mackenzie River affords a rare opportunity to examine changes in fish assemblage structure of a relatively undisturbed large river over a 30+ year period. The various studies that have been conducted used overlapping, but not identical protocols, due to spatial, temporal, and sampling-gear differences; therefore, care must be taken to limit comparisons to data collected under a standardized protocol.

To examine whether or not there have been changes in the Mackenzie fish assemblage, we ini-

Table 1.—List of scientific names and common names for fish species collected in the lower Mackenzie River (MR), Mackenzie River Delta (MD), Great Bear River (GB), Peel River (PR), and Arctic Red River (ARR).

Family/generic name	Common name	Location
Petromyzontidae		
Lampetra camtschatica	Arctic lamprey[a,b,c,d,e,f]	MR, MD, GB
Salmonidae		
Oncorhychus keta	chum salmon[a,f,g,l,m,n]	MR, MD, PR, ARR
O. nerka	sockeye salmon[l,n]	MR, ARR
O. kisutch	coho salmon[n,o]	MD, GB
O. gorbuscha	pink salmon[o]	MD
O. tshawytscha	Chinook salmon[n,p]	MD, MR
Salvelinus namaycush	lake trout[a,b,c,d,e,f,l]	MR, MD, GB
S. malma *	Dolly Varden[a,c,d,e,f, q]	MR, GB
S. confluentus	bull trout[n]	GB (Reist et al. Arctic vol. 55, no. 1 2002)
Coregonus clupeaformis	lake whitefish [a,b,c,d,e,f,g,h,i,j,k,l,m,n,r]	MR, MD, GB, PR, ARR
C. nasus	broad whitefish [a,b,c,d,e,f,g,h,i,j,k,l,m,n,r]	MR, MD, GB, PR, ARR
		MR, MD, GB, PR, ARR
C. laurettae	Bering cisco[f]	MR
C. artedi	lake cisco[a,b,c,d,e,f]	MR, MD, GB
C. autumnalis	Arctic cisco [a,b,c,d,e,f,g,h,i,j,k,l,r]	MR, MD, GB, PR, ARR
C. sardinella	least cisco [a,b,c,d,e,f,g,h,i,j,k,l,m,n,r]	MR, MD, GB, PR, ARR
Stenodus leucichthys	inconnu [a,b,c,d,e,f,g,h,i,j,k,l,m,n,r]	MR, MD, GB, PR, ARR
Prosopium cylindraceum	round whitefish [a,b,c,d,f,h,q]	MR, MD, GB
Prosopium williamsoni *	mountain whitefish [a,b,c,d,f,h]	MR, MD, GB
Thymallus arcticus	Arctic grayling	
Esocidae		
Esox lucius	northern pike [a,d,e,f,k,l,m,n,r]	MR, MD, GB, PR, ARR
Hiodontidae		
Hiodon alosoides	goldeye[a,b,c,d,e,f]	MR,MD,GB
Cyprinidae		
Couesius plumbeus	lake chub[a,b,c,d,e,f]	MR, MD, GB
Platygobio gracilis	flathead chub[a,b,c,d,e,f,l,n]	MR, MD, GB, ARR
Notropis atherinoides	emerald shiner[a,d,f]	MR, MD, GB
N. hudsonius	spottail shiner[a,d,f,g]	MR, MD, GB
Chrosomus eos	northern redbelly dace[b,f]	MR, MD
Phoxinus neogaeus	finescale dace[b,f]	MR, MD
Rhinichthys cataractae *	longnose dace[b,e,f]	MR,MD,GB
Maragariscus margarita	pearl dace	
Catostomidae		
Catostomus catostomus	longnose sucker[a,b,c,d,f,l,m,n,r]	MR, MD, GB, PR, ARR
C. commersonii	white sucker[a,b,c,d,f,l,m,n]	MR, MD,GB, PR, ARR
Gadidae		
Lota lota	burbot [a,b,c,d,e,f,g,k,l,m,n,r]	MR, MD, GB, PR, ARR
Gasterosteidae		
Culaea inconstans	brook stickleback[b,d,f]	MR, MD
Pungitius pungitius	ninespine stickleback[a,b,c,d,e,f,n]	MR, MD, GB
Percopsidae		
Percopsis omiscomaycus	trout-perch[a,b,c,d,e,f]	MR, MD, GB
Percidae		
Sander vitreus	walleye[a,b,c,f,l,m,n,r]	MR, MD, GB, PR, ARR
Cottidae		
Cottus cognatus	slimy sculpin[a,b,c,d,e,f]	MR, MD, GB
Cottus ricei	spoonhead sculpin[b,f]	PR, MD
Myoxocephalus quadricornis	fourhorn sculpin	

Table 1.—Continued.

Family/generic name	Common name	Location
Osmeridae		
Hypomesus olidus	pond smelt[g,n]	MR, MD
Osmerus mordax	rainbow smelt[g,n]	MD

[a] Chang-Kue and Cameron (1980); [b] Corkum and McCart (1981); [c] Johnson (1975); [d] McCart (1986); [e] Shotton (1973); [f] Stein et al. (1973); [g] Scott and Crossman (1973); [h] Bodaly et al (1989); [i] Tallman and Reist (1997); [j] Treble and Tallman (1997); [k] Howland et al. (2001); [l] Tallman (unpublished data); [m] Toyne (2001); [n] Stephenson (unpublished data); [o] Babaluk et al. (2000); [p] McLeod and O'Neil (1983); [q] Stephenson (2003); [r] Howland (1997); [s] Reist et al. (2002)

tially compared total species diversity and composition of fishes captured by various research programs from prior to 1970 to the present.

Results

Long-Term Change in the Fish Assemblage: the Evidence

With the exception of the 1990s and 2000 sampling periods, species diversity appears to have remained stable over time, ranging from 30 to 38 species, with slight variations between time periods (Table 2). The species composition has also varied to some extent between time periods. For example, prior to the 1980s, Pacific salmon species such as sockeye, coho, and Chinook salmon were not seen but have become more common in recent decades.

The reduced fauna from the 1990s onward reflects the sampling focus more than the actual species list. Studies have been funded mainly by aboriginal resource management boards and have focused on harvestable species rather than the total fauna (Tallman and Reist 1997; Stevenson 2003). This is evident in that the fauna compositions are lacking the smaller species such as lake chub, emerald shiners, spottail shiners, longnose dace, pearl dace, northern redbelly dace, finescale dace, trout-perch, and some members of the stickleback (Gasterosteidae) and sculpin (Cottidae) families. All of these species are generally 3 in or less in length and thus must be sampled using nonfishery gear such as seines.

There is little evidence that fishing pressure or other anthropogenic activities have altered the fish fauna except in a few isolated cases. For example, Howland et al. (2001) reported that northern pike suffered local depletions with fishing activities in the Mackenzie River Delta. However, both Howland et al. (2001) and Treble and Tallman (1997) found no change in the abundance of migratory species such as inconnu and broad whitefish.

Evidence for Change in Species Composition or Abundance

In order to consider the possibility of a major faunal shift, we decided to limit analysis to the species that could be caught within fishermen's gill nets. We also decided not to include the Pacific salmon species that had not been typical in the area because there were questions about whether they may have been missed in earlier studies due to lack of expertise in species identification.

From Table 3, there are only slight differences in the ranking of species throughout the time periods. In general, the coregonid species were the most dominant fishes with broad whitefish, inconnu, and Arctic cisco being most dominant.

We calculated the Spearman rank correlation (van Tongeren 1987; Legendre and Legendre 1998) between the various time periods (Table 4). The values are uniformly high throughout the comparisons. The 2000 period had the lowest correlation to the other time periods (Table 4).

It is interesting that the most recent period appears to have the lowest correlation. Whether this is really indicative of change is difficult to determine. However, there has been evidence of increased occurrence of north Pacific species (salmon, Table 2) and of freshwater species moving northward (e.g., walleye in Tuktoyaktuk Harbor in 1999 [Stevenson 2003]).

Table 2.—Abundance of species in samples. The number of x's represent the relative abundance in catches (x = <1%, xx = 1–10%, xxx = >10% of the catches).

Family and species	<1970 (a,b)	1970s (c,d,e)	1980s (f,g,h)	1990s (i,j)	2000s (k,l,m)
Petromyzontidae					
Lampetra camtschatica	x	x	x	x	x
Hiodontidae					
Hiodon alosoides	x	x	x	x	x
Esocidae					
Esox lucius	x	xx	xx	xx	xx
Salmonidae					
Oncorhynchus keta	x	x	xx	x	x
O. nerka				x	
O. tshawytscha			x		x
O. kisutch				x	x
O. gorbuscha			x		
Salvelinus malma	x	x	x	x	x
S. namaycush	x	x	x	x	x
Coregonus artedii	x	x	xx		x
C. clupeaformis	x	xx	x	xxx	xxx
C. autumnalis	x	xx	xx	xx	xx
C. sardinella	x	x	x	xx	xx
C. nasus	x	xx	xx	xxx	xxx
Prosopium cylindraceum	x	x	x	x	x
P. williamsoni			x		
Stenodus leucichthys	x	x	x	xxx	xxx
Thymallus arcticus	x	x	x	x	x
Cyprinidae					
Couesius plumbeus	x	x	x		x
Platygobio gracilis	x	xx	xx	x	x
Notropis atherinoides	x	x	xx		
N. hudsonius	x	x	x		x
Rhinichthys cataractae			x		
Margariscus margarita			x		
Chrosomus eos		x	x		
Phoxinus neogaeus	x	x	x		
Catostomidae					
Catostomus catostomus	x	xx	xx	xx	x
C. commersonii	x	x	x	x	x
Percopsidae					
Percopsis omiscomaycus	x	x	x		x
Gadidae					
Lota lota	x	x	x	xx	x
Gasterosteidae					
Pungitius pungitius	x	x	x		x
Culaea inconstans		x	x		x
Cottidae					
Cottus cognatus	x	x	x		x
C. ricei	x	x	x		
Myoxocephalus quadricornis	x		x		
Percidae					
Sander vitreus	x	x	x	xx	xx
Osmeridae					
Hypomesus olidus	x	x	x		x
Osmerus mordax	x	x	x		x

a. McPhail and Lindsey (1967); b. Scott and Crossman (1973); c. Hatfield et al. (1972a); d. Hatfield et al. (1972b); e. Stein et al (1973); f. McCart (1986); g. Tallman and Howland (unpublished data); h. Babaluk et al. (2000); i. McLeod and O'Neil (1983); j. Stephenson (2003); k. Stephenson (unpublished data); l. Black and Fehr (2001); m. Toyne and Tallman (unpublished data)

Table 3.—Ranking of species regularly observed in fishermen's nets.

Family and species	<1970 (a,b)	1970s (c,d,e)	1980s (f,g,h)	1990s (i,j)	2000s (k,l,m)
Esocidae					
Esox lucius	11	10	7	8	6
Salmonidae					
Oncorhynchus keta	16	14	16	12	15
Salvelinus malma	10	11	13	14	10
S. namaycush	13	12	14	13	16
Coregonus artedii	14	15	8	16	13
C. clupeaformis	3	3	3	4	5
C. autumnalis	4	4	4	3	1
C. sardinella	6	5	5	5	4
C. nasus	1	1	1	1	3
Stenodus leucichthys	2	2	2	2	2
Thymallus arcticus	12	13	12	11	14
Cyprinidae					
Platygobio gracilis	9	7	11	10	9
Catostomidae					
Catostomus catostomus	5	6	6	7	7
C. commersonii	15	16	15	15	11
Gadidae					
Lota lota	7	8	10	9	12
Percidae					
Sander vitreus	8	9	9	6	8

a. McPhail and Lindsey (1967); b. Scott and Crossman (1973); c. Hatfield et al. (1972a); d. Hatfield et al. (1972b); e. Stein et al (1973); f. McCart (1986); g. Tallman and Howland (unpublished data); h. Babaluk et al. (2000); i. McLeod and O'Neill (1983); j. Stephenson (2003); k. Stephenson (unpublished data); l. Black and Fehr (2001); m. Toyne and Tallman (unpublished data).

Conclusions

In general, the harvestable fish stocks are thought to be in good shape due to an overall reduction in fishing as fewer people are practicing traditional subsistence fishing practices. There are fewer mouths to feed because people are also moving away from the use of dog teams.

It is difficult to be sure about the status of the smaller species as few studies since the 1980s have sampled for them. Since little has changed environmentally other than possible warming effects, it is unlikely that the species composition has altered over time, at least not downwards. Almost all of the smaller species are widespread in the temperate zone and therefore would only be enhanced by a longer growing season. However, cyprinids and sculpins have often been used as indicators of ecosystem health in other studies and may give the first indication of change either through disappearance or arrival of

Table 4.—Spearman rank correlation among the five time period noted in Table 3.

	<1970	1970s	1980s	1990s
1970s	0.974			
1980s	0.885	0.865		
1990s	0.907	0.926	0.882	
2000s	0.844	0.847	0.871	0.874

species or increased abundance (Mebane et al. 2003).

Our analysis suggests that there is little change in the faunal composition along the Mackenzie, at least for the harvestable species. There is evidence that some more temperate species are gradually invading the system as the effects of climate change become manifest. According to our data, changes have occurred only in rare species that were present or absent in various years.

The Mackenzie River remains one of the most pristine rivers in the world. It has the lowest human population of any major river, and it appears that the fish assemblage has remained stable for the last 40 years.

References

Babaluk, J. A., J. D. Reist, J. D. Johnson, L. Johnson. 2000. First records of sockeye (*Oncorhynchus nerka*) and pink salmon (*O. gorbuscha*) from Banks Island and other records of Pacific salmon in Northwest Territories, Canada. Arctic 53(2):161–164.

Black, S., and A. Fehr. 2001. Natural history of the western Arctic: a supplement to Canada's western Arctic: including the Dempster Highway. Western Arctic handbook. Yellowknife, Northwest Territories.

Bodaly, R. A., J. D. Reist, and D. Rosenberg. 1989. Fish and fisheries of the Mackenzie and Churchill river basins, northern Canada. Pages 128–144 *in* D. P. Dodge, editor. Proceedings of the international large river symposium (LARS) Canadian Special Publication of Fisheries and Aquatic Sciences 106.

Brunskill, G. J. 1986. Environmental features of the Mackenzie system, Pages 435–472 *in* B. R Davies, and K. F. Walker, editors. The ecology of river systems. Dr. W. Junk Publishers, Dordrecht, The Netherlands.

Chang-Kue, K. T. J., and R. A. Cameron. 1980. A survey of fish resources of the Great Bear River, N.W.T., 974. Canada Department of Fisheries and Environment, MS 1510, Winnipeg, Manitoba.

Corkum, L., and P. McCart. 1981. A review of the fisheries of the Mackenzie Delta and nearshore Beaufort Sea. Canada Miscellaneous Report Fisheries Aquatic Sciences 1613.

Craig, B. G., and J. G. Fyles. 1960. Pleistocene geology of Arctic Canada. Geological Survey of Canada Paper 60–12.

Hatfield, C. R., J. N. Stein, M. R. Falk, and C. S. Jessop. 1972b. Fish resources of the Mackenzie River Valley. Interim Report 1, Volume II. Fisheries and Marine Service of Environment Canada, Winnipeg, Manitoba.

Howland, K. L. 1997. Migration patterns of freshwater and anadramous inconnu, *Stenodus leucichthys*, within the Mackenzie River system. Master's thesis. University of Alberta, Edmonton.

Howland, K. L., M. Treble, and R. F. Tallman. 2001 A biological analysis and population assessment of northern pike, inconnu and lake whitefish from the Mackenzie River delta exploratory fishery, 1989–1993. Canadian Technical Report of Fisheries and Aquatic Sciences 2330.

Johnson, L. 1975. Distribution of fish species in the Great Bear Lake with reference to zooplankton, benthic invertebrates and ecological conditions. Journal Fisheries Research Board Canada 32:1959–2005.

Mayewski, P. A., G. H. Denton, and T. J. Hughes. 1981. Late Wisconsin ice sheets in North America. Pages 67–178 *in* G. H. Denton and T. J. Hughes, editors. Last great ice sheets. Wiley Interscience, Toronto.

McCart, P. J. 1986. Fish and fisheries of the Mackenzie system. Pages 493–516 *in* B.R. Davies, and K. F. Walker, editors. The ecology of river systems. Dr. W. Junk Publishers, Dordrecht, The Netherlands.

McLeod, C. L., and J. P. O'Neill. 1983. Major range extensions of anadromous salmonids and first record of chinook salmon in the Mackenzie River drainage. Canadian Journal of Zoology 61:2183–2184.

Mebane, C. A., T. R. Maret, and R. M. Hughes. 2003. An index of biological integrity (IBI) for Pacific Northwest rivers. Transactions of the American Fisheries Society 132:239–261.

Millar, J. K. and G. F. Fedirchuk. 1975. Mackenzie River archeological survey. Environmental Social Committee, Northern Pipelines, Task Force on Northern Oil Development, Report 74–47. Information Canada, Cat. R57–22/1974, Ottawa.

Peilou, E. C. 1991. After the ice age. University of Chicago Press, Chicago.

Prest, V. K. 1969. Retreat of Wisconsin and recent

ice in North America. Government of Canada, Geological Survey Canada Map 1235A, Ottawa.

Reist, J. D., G. Low, J. D. Johnson and D. McDowell. 2002. Range extension of bull trout, *Salvelinus confluentus*, to the central Northwest Territories, with notes on identification and distribution of Dolly Varden, *Salvelinus malma*, in the western Canadian Arctic. Arctic 58(1):70–76.

Rosenberg, D. M., and D. R. Barton. 1986. The Mackenzie River system. Pages 425–434 *in* B. R. Davies, and K. F. Walker, editors. The ecology of river systems. Dr. W. Junk Publishers, Dordrecht, The Netherlands.

Scott, W. B., and E. J. Crossman. 1973. Freshwater fishes of Canada. Fisheries Research Board of Canada Bulletin 184, Ottaswa.

Shotton, R. 1973. Towards an environmental impact assessment of the portion of the Mackenzie gas pipeline from Alaska to Alberta. Appendix II. Fish base data. Government of Alberta, Environmental Protection Board, Interim Report 3, Edmonton, Alberta.

Stein, J. N, C. S. Jessop, T. R. Porter and K. T. J. Chang-kue. 1973. Fish resouces of the Mackenzie River Valley. Interim Report II. Fisheries and Marine Service of Environment Canada, Winnipeg.

Stephenson, S. A. 2003. Summary results of the 1999–2001 lower Mackenzie River index netting program (unpublished manuscript). Department of Fisheries and Oceans, Winnipeg, Manitoba.

Tallman, R. F. and J. D. Reist. 1997. The proceedings of the broad whitefish workshop: the biology, traditional knowledge and scientific management of broad whitefish, *Coregonus nasus* (Pallas), in the lower Mackenzie River. Canadian Fisheries and Aquatic Sciences Technical Report 2193.

Todd, D. K. 1970. The water encyclopedia. A compendium of useful information on water resources. Water Information Center Inc., Port Washington, New York.

Toyne, M. 2001. Comparison of growth, age-at-maturity, and fecundity for broad whitefish (*Coregonus nasus*) in the lower Mackenzie Delta, NWT and evaluation of the Peel River monitoring program. M.Sc thesis. University of Manitoba, Winnipeg.

Treble, M. and R. F. Tallman. 1997. An assessment of the exploratory fishery and investigation of the population structure of broad whitefish from the Mackenzie River Delta, 1989–1993. Canadian Fisheries Aquatic Sciences Technical Report 2180.

Van Tongeren, O. F. R. 1987. Cluster analysis. Pages 174–212 *in* R. H. G. Jongman, C. J. F. ter Braak and O. F. R. Van Tongeren, editors. Data analysis in community and landscape ecology. Pudoc, Wageningen, Netherlands.

American Fisheries Society Symposium 45:23–39, 2005

Composition and Changes to the Fish Assemblage in a Large Sub-Arctic Drainage: The Lower Slave River

Ross F. Tallman*, Kimberly L. Howland, and G. Low
Central and Arctic Region, Department of Fisheries and Oceans, Winnipeg, Manitoba, R3T 2N6, Canada

W. M. Tonn
Department of Biological Sciences, University of Alberta, Edmonton, Alberta, T6E 2G4, Canada

A. Little
Golder Associates, 1712 106 Avenue, Edmonton, Alberta T5S 1H9, Canada

Abstract.—The Slave River is the largest tributary to Great Slave Lake and the second largest river flowing northward in North America. There are no dams or major industrial developments on the lower Slave River, but further upstream in its Peace and Athabasca tributaries there are numerous pulp mills and a large hydroelectric project (Bennett Dam). These developments appear to have had limited effects on the Slave River fish fauna. The most significant concern is the reduced flood-pulse due to flow regulation, which is hypothesized to have affected spawning success in some species. The other major human impact is from commercial fishing on Great Slave Lake. Migratory species, such as inconnu, have been extirpated from some tributaries due to overfishing. In the Slave River, however, the impact of fishing on inconnu and other species appears to have been less severe. Although the number of age-groups has decreased within some species, the species composition appears to have remained stable. There is little evidence of species introductions into the system, but some rare species, such as chum salmon *Oncorhynchus keta*, may be extirpated.

Introduction

Large rivers are probably one of the most geologically stable freshwater habitats. Small streams and lakes may dry up or disappear with glaciation or climate change, but large rivers persist in geological time scales—sometimes changing course, but rarely disappearing altogether (Lindsey and McPhail 1986; Pielou 1991). For example, the Yukon and Mackenzie river courses have persisted at least 10,000 years and probably up to 100,000 years (Lindsey and McPhail 1986). However, many rivers throughout the world have been substantially altered by human activities: for example, the most complete inventory of large reservoirs available documents 25,410 dams that have been erected on the world's major river systems (ICOLD 1998), while St. Louis et al. (2000) estimate that the global surface area of all reservoirs, which turns many rivers into chains of man-made lakes, is 1.5 million km^2. Rivers serve as waste disposal media for industry and city sewers, thus altering water quality (Davies and Walker 1986). Fisheries and alien species have directly altered the fish species composition in many systems (Scott and Crossman 1973; Hughes et al. 2005). Forestry operations have changed catchments, increasing runoff and inputs of solutes and sediments and altering flow regimes either directly or through road construction (Hartman and Scriv-

* Corresponding author: TallmanR@DFO-MPO.GC.CA

ener 1990; Baxter et al. 1999). However, it has been difficult in many cases to directly link river alterations to loss (or gain) of fish species. An alternative hypothesis is that river fish assemblages are constantly changing even in the pristine state, and could be more ephemeral than one might imagine. Such changes should ideally be assessed relative to temporal patterns of fish assemblages in unimpaired reference rivers, which may or may not change naturally through time. It is increasingly difficult, however, to find large rivers that can serve as such references.

In contrast to the situation in many large rivers, the lower Slave River has been only lightly touched by human activities. It is a large river with only two small villages (Fort Smith, population 2,653 and Fort Resolution, population 562) and no industrial development. Thus, it presents an ideal situation to observe natural fluctuations in fish species composition over moderate time periods. Despite its relatively remote location, there have been a series of studies since the 1970s on the fishes in the river, although only one (Little et al. 1998) has been published in the primary literature. The rest of the information has been published in industry and government reports or remains in Department of Fisheries and Oceans files. Much of this material has been synthesized and reexamined in this chapter.

Study Area

Description of the Slave River

The Slave River is, by far, the largest tributary of Great Slave Lake (Figure 1). With a length of 434 km, it receives waters from the Peace and Athabasca rivers via the Peace-Athabasca Delta and drains 606,000 km^2. The Slave River has an average annual discharge of 110 km^3 (Brunskill 1986), making it the largest northward flowing river in North America next to the Mackenzie. Throughout its course, the Slave River is turbid and carries a high load of dissolved solids (Brunskill 1986). Below the Rapids of the Drowned at Fort Smith, NT, the last in a series of rapids that serve as a natural biogeographic barrier for some fishes (Nelson and Paetz 1972), the river flows unimpeded for approximately 320 km to the Slave River Delta on Great Slave Lake (Tallman et al. 1996).

The lower Slave River from Fort Smith to the delta is a relatively homogeneous system with a maximum channel width of approximately 3 km, fast-flowing water, and steep riverbanks (Vanderburgh and Smith 1988). The banks are up to 35 m high resulting in very narrow littoral zones for establishing aquatic vegetation (Vanderburgh and Smith 1988). This, in turn, has consequences for the diversity of the fish fauna.

The Slave River Delta is located midway along the south shore of Great Slave Lake, approximately 13 km northeast of Fort Resolution (61°10'N, 113°40'W), where it covers an area of approximately 640 km^2 (NRBS 1996a). The delta has a wide diversity of habitat types compared with the main stem of the river, and over 200 species of vertebrates are residents or visitors to this important area. There are four main channels, ranging in depth from 5 to 32 m, and numerous small and shallow side channels, as well as isolated perched lakes created during spring floods. The depth and permanency of these channels and lakes varies annually and seasonally. Landforms range from large mud flats on the outer edges of the Delta to cut-banks ranging from 0.25 to 3 m (English 1979). Shoreline habitat ranges from gently sloping banks with heavy vegetation to steeper banks with narrower littoral zones and sparse vegetation.

The Salt River, the largest tributary on the Slave River, is located 25 km downstream of Fort Smith. It is a narrow, meandering river with an average maximum depth of 1–2 m and a maximum width of approximately 60 m. The Salt River has a greater abundance of aquatic vegetation than the Slave River (Little et al. 1998) and appears to provide an important refuge and nursery area where juvenile fish can feed and mature (Little 1997).

Human Activities that Could Disturb the Slave River

Commercial and Subsistence Fishing.—Over the last 45 years, more than 90% of the annual commercial fish harvest for the Northwest Territories has been from Great Slave Lake (McCart 1986, C. Day, Department of Fisheries and Oceans,

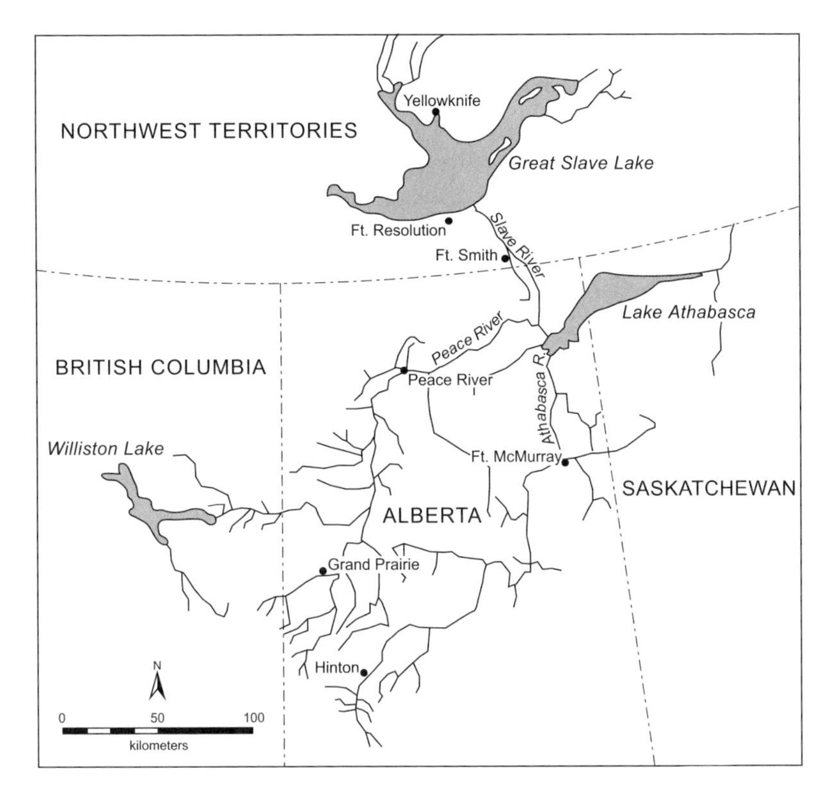

Figure 1.— The Slave River basin, showing locations of Bennett Dam at Williston Lake and major pulp mills at Hinton, Peace River, and Grand Prairie.

personal communication). This commercial fishery has operated continuously since 1945 and, until the mid 1960s, lake whitefish *Coregonus clupeaformis* averaged three-fifths of the catch, with lake trout *Salvelinus namaycush*, inconnu *Stenodus leucichthys*, and northern pike *Esox lucius* making up most of the rest. Subsequently, species such as lake trout and inconnu have declined and the proportion of the catch as lake whitefish has increased to about 80% (Chris Day, Department of Fisheries and Oceans [DFO], personal communication). Overall catches average about 1,200 tons per year. (Chris Day, personal communication). The commercial fishery in the lake directly effects migratory large fish species in the Slave River. Species, such as inconnu and lake whitefish, come into the river to spawn, but spend a large portion of their lives in Great Slave Lake and are thus vulnerable to commercial fishing. Members of the river and delta assemblage that use Great Slave Lake, such as lake trout, inconnu, lake whitefish, northern pike, white sucker *Catostomus commersonii*, and burbot *Lota lota* may also be caught in the commercial nets on the lake. The potential effects of such exploitation are a) a reduction in average size and age as larger older fish are eliminated; b) declining catch per unit of effort (CPUE), as numbers decline in the age-classes susceptible to the gear; c) increased growth rate in the younger age-groups; d) a reduction in the age-at-sexual-maturity. In time, heavily exploited populations may be comprised almost entirely of juveniles and first-time spawners (Ricker 1975).

Fishing on the Slave River itself is mainly for aboriginal subsistence. Fishermen use both large and small mesh gill nets. The intensity of this harvest has declined more recently because fewer people use fish for their own sustenance and to maintain dog teams. Presently, there is only one full time (harvesting throughout the fishing season) subsistence fisherman on the Slave River. The mortality and subsequent population responses to such a fishery are of the same nature as from a commercial fishery (see above). The number of fish killed can be of similar magnitude. While the overall commercial fishery catch is large, it is dispersed over many populations (for example, up to 50 stocks of lake whitefish within the lake itself [DFO, unpublished data]). On the other hand, subsistence fishing takes place in the river when the fish are highly concentrated. Also, the river has a high silt load so that during the daytime, at least, the nets are much less visible to the fish. Yet, the possible impact of subsistence fishing on assemblage structure is often overlooked.

The W.A.C. Bennett Dam.—The Peace Canyon project is a hydroelectric development that harnesses the turbulent water of the upper Peace River, one of the two major sources of water to the Peace-Athabasca Delta and thus to the Slave River. The W.A.C. Bennett Dam at the head of the Peace River Canyon, completed in 1967, is one of the world's largest earth fill structures, stretching 2 km across the head of the canyon and measuring 183 m in height. Behind the dam is British Columbia's largest reservoir, Williston Lake, which covers an area of 166,000 h (Figure 1).

The Bennett Dam has had a variety of significant impacts on the Peace River and Peace-Athabasca Delta (NRBS 1996a). Although more than 1,500 km downstream from the dam, the people of Fort Resolution claim that since the dam was built, the flood-pulse cycle of the Slave River delta has been disrupted. Reports from the Northern Rivers Basin Study also suggest that the Slave River Delta experiences lower peak flows, which would consequently reduce the amount of sediment delivered to the Delta. However, the Slave River Delta itself is a dynamic and complex hydrologic system; given its distance from the Bennett Dam and the presence of intervening water bodies, it is difficult to link specific temporal changes in the Delta to the onset of flow regulation by the Bennett dam (NRBS 1996a). Data from Davies (1981) and other water surveys suggest that flooding has been generally less intense since the creation of the dam, but it is not conclusive. Annual peak flows in the years preceding dam completion averaged about 2,000 m^3/s higher than those afterwards (Figure 2).

Aside from these concerns about structures that are relatively far removed geographically, there are no dams that directly affect the lower Slave River. However, the hydroelectric potential of the rapids near Fort Smith caught the attention of the Alberta government and spawned the Slave River Hydro Feasibility Study of the early 1980s, which generated some of the historic reports on the fishes of the Slave River system (e.g., Tripp et al. 1981; McLeod et al. 1985).

Pulp Mills on Peace and Athabasca Rivers.—South of the Northwest Territorial boundary, there are three pulp mills on the Peace River —two in Alberta and one in British Columbia. The mill in British Columbia makes bleached chemi-thermomechanical pulp. The two mills in Alberta are both bleached kraft pulp mills. All three mills started operation between 1973 and 1990.

There are five mills on the Athabasca River, all in Alberta. Of these, two make bleached kraft pulp, two bleached chemi-thermomechanical pulp, and one thermomechanical pulp and de-inked paper. These mills began operation between 1957 and 1993.

The bleached kraft process uses strong chemicals and heat for wood processing, followed by chemical bleaching (traditionally molecular chlorine, now chlorine dioxide, oxygen or hydrogen peroxide that result in lower concentrations of chlorinated compounds). Bleached chemithermomechanical mills use chemicals (usually sodium sulfite), heat and grinding to process wood. Thermo-mechanical mills use grinding and heat.

The principal contaminants of kraft mills are organochlorines (chlorinated dimethylsulphones, chlorinated aromatics [which include dioxins/furans]) (NRBSa 1996). Chemi-thermo plants produce compounds like terpenes and aromatics. Mills have been switching to improved technologies since the late 1980s, and new regulations came into affect in 1993. This has resulted in near elimination of dioxins and furans, and reduced chlorine use.

Nutrient loading from pulp mills is substantial, creating high biochemical oxygen demand, more so in the Athabasca than the Peace. The majority of the annual phosphorus load (from point sources) is contributed by pulp mills (6–17% of the total load from point sources on the Athabasca) (NRBS 1996a).

Oil Sands on the Athabasca River.—Of the two oil sands operations in Fort McMurray, only one (Suncor Inc.) continuously discharges utility waste-

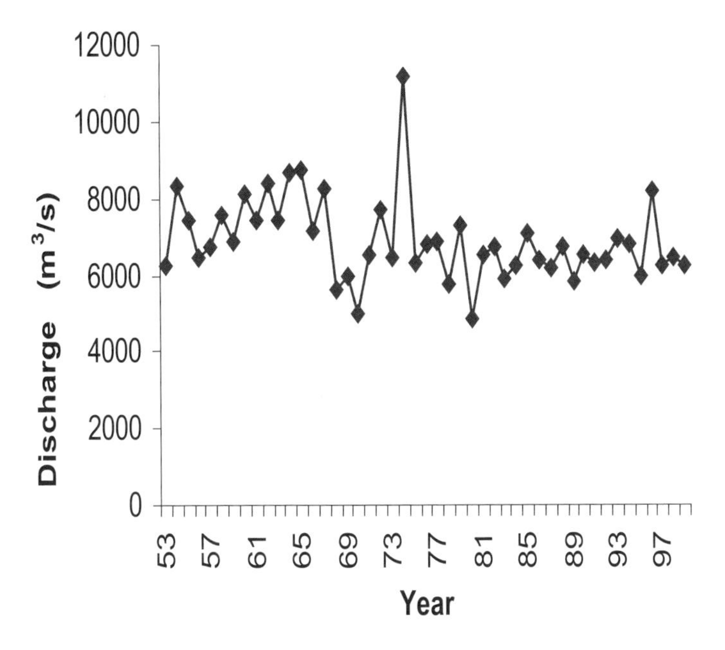

Figure 2.—Peak annual discharge at the Slave River delta (1953–1999). Dam closure occurred in 1967. (Data from Davies 1981 and unpublished Water Survey of Canada data.)

water (not wastewater from oil extraction processes). To ameliorate contamination, the water is treated before release and the resulting effluent is low in nutrients. However, the effluent does contain hydrocarbons and other odor-causing substances. Tailings ponds from both operations seep small amounts, but are being moved and/or upgraded at the present time. Both oil sands and pulp mills emit sulfur dioxide (NRBS 1996a).

Developments that could affect the fish assemblage in the lower Slave River are located considerable distances away, aside from the Great Slave Lake fishery. The Northern River Basin Study (NRBS) final report indicates that the fish populations in the lower Slave River were healthy, although there was some evidence that contaminants had entered the system (NRBS 1996b). Although, it seems contaminants are exported downstream, it is unlikely that there have been any serious changes generated from industrial activity.

Fish Species Composition

Thirty-one fish species have been reported from the Slave River, its tributaries and delta (Falk and Dalkhe 1975; Tripp et al. 1981; McLeod and O'Neill 1983; McLeod et al. 1985; McCart 1986, Tallman et al. 1996a; Little et al. 1998, Tallman and Low, unpublished) (Table 1). This compares to 36 species in the Athabasca, 31 in the Peace, 26 in the Liard, and 38 in the lower Mackenzie River (McCart 1986; Tallman et al. 2005). The Slave has relatively low species richness compared to other large rivers of the world (e.g., the Niger River – 149 spp (Greenwood et al. 1966), the Amazon – 1,300+ spp. (Lowe-McConnell 1986), the Parana River – 550+ spp. (Bonetto 1986)) but is similar in richness to rivers of the same latitude.

Seasonal Variation

Change in Abundance and Composition over the Year

Seasonal movements of fish cause the fish assemblage composition to change spatially and temporally. As species move from one area to another, their exposure to different habitats, prey assemblages, and water quality changes. Extensive movements of fish in large river systems are not uncommon and add complexity to both the sampling and assessment of

Table 1.—List of scientific names, and common names for fish species collected in the lower Slave River (SR), Slave River Delta (SRD) and Salt River (SALT), 1975–2000.

Family /species name	Years	Common name	Location
Petromyzontidae			
Lampetra camtschatica	78–80,83,84,94,95	Arctic lamprey [a,b,c,e]	SR/ SRD
Salmonidae			
Oncorhynchus keta	78–80	chum salmon[a]	SRD
O. nerka	95	sockeye salmon[e]	SR
O. tshawytscha	75–85,95	Chinook salmon[d,e]	SR
O. mykiss	83,84	rainbow trout[b,e]	SR/SRD
Salvelinus namaycush	83,84,94.95,00	lake trout[b,c,e,f]	SRD
*S. malma**	75–85	Dolly Varden[d]	SR
Coregonus clupeaformis	78–80,83,84,94,95,00	lake whitefish[a,b,c,e,f]	SR/SRD/SALT
C. sardinella	95	least cisco[e]	SR
C. artedii	78–80,83,84,95,	lake cisco[a,b,e]	SRD
Coregonus spp.	78–80,83,84	small cisco[a,b]	SRD
Stenodus leucichthys	78–80,83,84,94,95,00	inconnu[a,b,c,e,f]	SR/SRD
Prosopium cylindraceum	78–80,83,84	round whitefish[a,b]	SRD
*P. williamsoni**	75–85	mountain whitefish[d]	SR
Thymallus arcticus		Arctic grayling	
Esocidae			
Esox lucius	78–80,83,84,94,95,00	northern pike[a,b,c,e,f]	SR/SRD/SALT
Hiodontidae			
Hiodon alosoides	78–80,83,84,94,95,00	goldeye[a,b,c,e,f]	SR/SRD/SALT
Cyprinidae			
Couesius plumbeus	78–80,94,95,00	lake chub[a,c,e,f]	SRD
Platygobio gracilis	78–80,83,84,94,95,00	flathead chub[a,b,c,e,f]	SR/SRD/SALT
Notropis atherinoides	78–80,94,95,00	emerald shiner[a,c,e,f]	SR/SRD/SALT
N. hudsonius	78–80,94,95	spottail shiner[a,c,e]	SR/SRD/SALT
Rhinichthys cataractae	75–85	longnose dace[d]	SR
Margariscus margarita	78–80	pearl dace[a]	SR/SRD
Catostomidae			
Catostomus catostomus	78–80,83,84,94,95,00	longnose sucker[a,b,c,e,f]	SR/SRD/SALT
C. commersonii	78–80,83,84,94,95,00	white sucker[a,b,c,e,f]	SR/SALT
Gadidae			
Lota lota	78–80,83,84,94,95,00	burbot[a,b,c,e,f]	SR/SRD/SALT
Gasterosteidae			
Pungitius pungitius	78–80,94,95	ninespine stickleback[a,c,e]	SL/SRD/SALT
Percopsidae			
Percopsis omiscomaycus	78–80,84,94,95,00	trout-perch[a,b,c,e,f]	SR/SRD/SALT
Percidae			
Sander vitreus	78–80,83,84,94,95,00	walleye[a,b,c,e,f]	SR/SRD/SALT
Perca flavescens	78–80,94,95	yellow perch[a,c,e]	SL/SRD
Cottidae			
Cottus cognatus	78–80	slimy sculpin[a]	SRD
C. ricei	78–80,94,95	spoonhead sculpin[a,c,e]	SR/SRD/SALT

[a] Tripp et al. (1981)
[b] McLeod et al. (1985)
[c] Tallman et al. (1996a)
[d] McCart (1986)
[e] Little et al. (1998)
[f] Tallman and Low, unpublished.

fish assemblages. Studies must incorporate such movements into their designs and syntheses need to incorporate studies from different locations and/or times. For example, for the lower Slave River, McLeod et al. (1985) focused only on the fall period near Fort Smith and, therefore, cannot provide a complete picture of movement patterns in this area. Tripp et al. (1981) provides useful information on the pattern of seasonal movement in the Slave River Delta. Tallman et al. (1996d) and Little et al. (1998) sampled throughout the open-water season in the Delta, in the main stem near Fort Smith, in the lower Salt River, and also in the winter near Fort Smith. From these and other sources, the data on changes in abundance and species composition in the Delta and lower Slave River can be blended to give a more detailed picture of activities of fishes in the lower Slave River across seasons.

Spring

In late May after ice break-up, the fish assemblage in the Slave River was dominated by flathead chub, walleye, and goldeye, and moderate numbers of longnose sucker, and lake whitefish (Tallman et al. 1996a; Little 1997). Absent near Fort Smith were inconnu, ciscoes *Coregonus artedi*, white sucker, and adult burbot, (Tallman et al. 1996a; Little 1997).

Goldeye were also abundant in the Slave River Delta in the spring 1980, although not in 1979 (Tripp et al. 1981). Flathead chub were generally abundant in the Delta throughout the month of May as were young-of-the-year burbot (Tripp et al. 1981). Juvenile pike, occurred in the Delta but were also present in certain areas near Fort Smith (Tripp et al. 1981; Tallman et al. 1996). In 1979, adult pike gradually moved into the Delta over the course of the open-water season presumably to feed and overwinter in Great Slave Lake or the Delta after spring spawning near Fort Smith (Tripp et al. 1981). In 1980, however, large numbers of pike in spawning condition were captured in the Delta, demonstrating that in some years, spawning probably occurs there (Tripp et al. 1981). Tripp et al. (1981) also found an increase in the abundance of lake whitefish in the Slave River Delta after ice break-up but the numerical abundance was much lower than at Fort Smith. They attributed this to the general downstream movement by lake whitefish that stay and overwinter in the Slave River after spawning the previous fall. Similarly, Little et al. (1998) noted a spring peak in catches of mature lake whitefish near Fort Smith, and suggested that these might be fall spawning fish that had overwintered on the spawning grounds prior to migrating back to Great Slave Lake.

We suspect that the high catches of flathead chub, goldeye, and walleye were due to spawning aggregations. If so, we would place the spawning of these species in early spring, shortly after river breakup. According to Scott and Crossman (1973), details of the spawning habits of flathead chub are unknown, but the available information indicates that spawning takes place in summer. Olund and Cross (1961) reported collections of males and females in spawning condition, taken in the Milk River in August 1955. However, McPhail and Lindsey (1970) reported the capture of females with large ovaries of almost free eggs, and one spent female in the Mackenzie River at 64°N, on June 27. Olund and Cross (1961) suggested that spawning occurred when water levels receded to the seasonal low during midsummer. However, the seasonal low in the Slave River would be in fall. The nature of the Slave River may encourage spring spawning.

The biology of goldeye is well studied in Canada. In most locales, spawning occurs in the spring from May to the first week of June starting just after the ice breaks and continuing over a period of 3–6 weeks (Battle and Sprules 1960; Kennedy and Sprules 1967; McPhail and Lindsey 1970; Pankhurst et al. 1986).

According to Scott and Crossman (1973), walleye spawning occurs in spring or early summer (early April in southwestern Ontario to the end of June in the far north) depending on the latitude and water temperature. Normally, spawning begins shortly after the ice breaks up at temperatures ranging from 5.6°C to 11.1°C. Spawning grounds are commonly in rocky areas in white water below impassable falls in rivers. Tripp et al. (1981) found two major seasonal migrations of walleye through the Delta. The first was an upstream spawning migration starting in mid-April before spring breakup and continuing through most of May. The second migration was a downstream migration to Great Slave Lake that started in late August and continued until freeze-up, and possibly later. Based on these criteria, we suspect that the Rapids of the Drowned is the spawning area for walleye in the lower Slave River. On the other hand, Tripp et al. (1981) thought that the Salt River might be an important spawning area.

Since longnose sucker spawn in the spring as soon as the water temperature exceeds 5°C (Harris 1962; Geen et al. 1966) we suspect that spawning occurred around mid-May and what we observed was the tail end of the aggregation. According to Geen et al. (1966), spawning takes place at depths of 152–279 mm, with a current of 30–45 cm per second over gravel substrate of 50–100 mm in diameter.

Tallman et al. (1996) and Little et al. (1998) found that the relatively high numbers of lake whitefish during spring were due to the presence of juveniles in the Salt River area. Thus, the Salt River may serve as a nursery area for this species. Whitefish may also be attracted away from Great Slave Lake to benefit from feeding/growing opportunities in the more rapidly warming waters of the river.

The absence of inconnu at this time is not surprising. In other studies, several species only used rivers for reproduction in the fall. Burbot were never very susceptible to the fishing gear employed in these studies. White sucker were only captured in the Salt River area (Little 1997).

Northern pike is an early spring spawner—probably so early here that there was no distinct evidence of a spawning aggregation. Spawning takes place immediately after the ice melts in April to early May when water temperatures range from 4.4°C to 11.1°C (Scott and Crossman 1973). Northern pike also tend to spawn in small tributaries and therefore may have been congregating elsewhere in the river system.

Summer-Fall

In the early part of June, the abundance of the spring spawning species tapered off but were still generally high. Lake whitefish and northern pike juveniles were particularly abundant at this time (Tallman et al. 1996a; Little 1997). The pike may have been attracted by concentrations of flathead chub and juvenile whitefish. All nine major species (flathead chub, goldeye, burbot, inconnu, lake whitefish, walleye, northern pike, and longnose and white sucker) were present near Fort Smith over the course of June and July, but no major aggregations appeared to occur (Tallman et al. 1996a; Little 1997). In the Delta, Tripp et al. (1981) recorded peaks in abundance of longnose sucker in late June (1979) and early July. Tripp et al. (1981) thought that these fish were postspawners from a spring spawning aggregation upstream. In contrast, Little et al. (1998) proposed that longnose suckers spawned in the Delta during spring.

In the latter part of August, fall-spawning species such as inconnu and lake whitefish begin to increase in abundance. Inconnu first appeared in the Slave River near the beginning of August (Tallman et al. 1996b; Howland et al. 2000). The run peaked between September 1 and September 15, and again between October 1 and 15, and was estimated to have ended in the latter part of October. By October 21, most inconnu seemed to have left the Slave River (Tallman et al. 1996b; Howland et al. 2000). However, due to ice formation during this period, sampling was limited.

Distribution and Abundance of Major Species

Little information exists on the biology of inconnu in Canada (Scott and Crossman 1973). Upstream runs into Great Slave Lake tributaries are protracted throughout the summer. Spawning is thought to occur in the early fall only every 2, 3, or 4 years (Scott and Crossman 1973). However, Howland et al. (2000) placed spawning in the Slave River in mid-October and found that a male tagged on the spawning grounds returned to the Slave River the following season. Tripp et al. (1981) found two peaks of inconnu abundance in the fall in the Slave River Delta—one as part of the upstream spawning run in late August and early September, and the other as a concerted downstream run out of the system in late October. Similarly, McLeod et al. (1985, Appendix B) found two peaks of inconnu abundance in the Delta in late August–early September and in the third week of October in 1983 and 1984. In contrast, McLeod et al. (1985, Appendix B) captured relatively few inconnu below the Rapids of the Drowned near Fort Smith during mid-September in 1983 and 1984, but found them to be abundant in the mid-lower Slave River during the first 2 weeks of October. Regardless, radio telemetry indicates several areas near Fort Smith are used as inconnu spawning grounds in the fall (McLeod et al. 1985; Howland et al. 2000).

A transient peak in abundance of lake whitefish in the Slave River Delta has been observed to occur in late August (Tripp et al. 1981; McLeod et al. 1985), presumably from migrating prespawners. Tallman et al. (1996a) and Little (1997) recorded that lake whitefish became abundant in the Slave River during August and that the CPUE steadily climbed to a peak in the first 2 weeks of October. Similarly, McLeod et al. (1985), sampling from the second week in September onward, found that abundance below Rapids of the Drowned increased steadily to a peak in the first 2 weeks of October. According to Rawson (1947), spawning occurred in and around Great Slave Lake from late September to October. Therefore, the peak abundance in the Slave River is presumably a spawning run. Lawler (1965) suggested that spawning in Lake Erie was delayed until the temperature dropped to 7.8°C and below. This agrees well with what was observed in the Slave River by Tallman et al. (1996a) and Little (1997). Tripp et al. (1981) proposed that the base of the rapids on the Slave River near Fort Smith was likely a major spawning area. They found that after October 10, water temperatures were 7.5°C and most of the males were running milt. In 1994, both sexes were ready to spawn by mid-October (Lange and Tallman 2002). By the beginning of November, lake whitefish appeared to have left the Slave River. However, due to ice formation, few nets were set during this period. Floy tagging results of Tripp et al. (1981) suggest that after spawning most adults return to Great Slave Lake. Nonetheless, a substantial proportion of adults and juveniles apparently overwinter in the Slave River (Tripp et al. 1981). Lake whitefish spawning usually occurs in shallow water at depth of less than 8 m, but spawning has been reported in deeper waters by Koelz (1929). It often takes place over a hard or stoney bottom, but sometimes over sand. The Slave River has ample locations with this type of habitat. In northern waters, individuals may spawn only every second year (Scott and Crossman 1973).

Burbot are only occasionally captured in gill nets, the main gear used to sample Slave River fish (Tripp et al. 1981; McLeod et al. 1985; Tallman et al. 1996b). It is unclear if this reflects their low abundance, depths occupied, a lack of movement, or other factors that limit catchability. Tallman et al. (1996d), however, found that radio-tagged burbot were relatively sedentary.

Goldeye is undoubtably the most common fish in the Slave River system. Postspawning peak catches occur between August 1 and August 15, and between October 15 and October 31 (McLeod et al. 1985; Tallman et al. 1996a; Little 1997). The peak in the late fall is possibly a preparatory phase for overwintering. Reductions of goldeye from environmental changes would probably affect the entire Slave River community, including birds and mammals. On the other hand, Sandheinrich and Atchison (1986) show that anthropogenic changes, such as those of the Missouri River, could result in greater habitat for goldeye.

Longnose sucker were present in low numbers throughout the summer and fall (Little 1997). However, Tripp et al. (1981) recorded a peak of abundance in the Slave River Delta in late September (1979) and late October (1980), which they attributed to downstream movements to overwintering areas by fish that stayed upstream during the summer. White sucker only inhabited the Salt River, and their abundance there was relatively low (Tallman et al. 1996a; Little et al. 1998).

Northern pike, walleye, and flathead chub are significant members of the Slave River fish assemblage and were generally present in moderate abundance throughout the summer and fall (Tallman et al. 1996a; Little 1997). Walleye appeared to aggregate in the channel near Fort Smith in October in preparation for overwintering (Tallman et al. 1996d). Juvenile pike were particularly abundant in the Salt River, and this tributary is likely a spawning area for the species.

The abundance of flathead chub may be important to monitor in the future. It is specialized for systems, such as the Slave River, characterized by high turbidity, wide seasonal fluctuations in flow and a wide channel that is in a constant state of change (Pflieger and Grace 1987). Historically, the Slave Delta showed cycles of flooding and drying of about four years (Tom Unka, Fort Resolution Native Band Environmental Council, personal communication). Since the construction of the Bennett Dam, these cycles have been altered (English et al. 1997). Flathead chub declined severely after extensive alterations restricted the Missouri River, reduced turbidity and sediment load, and modified the natural-flow regimen (Pflieger and Grace 1987).

Minimal information is available regarding the species composition of the lower Slave River during the ice-covered winter season. Limited gill netting in December 1994 and 1995 produced most of the major species except inconnu (Alison Little, University of Alberta and Fernand Saurette, Department of Fisheries and Oceans, Winnipeg, personal communication). However, CPUE was low, suggesting that most species use only restricted portions of the river for overwintering or most individuals overwinter in Great Slave Lake.

Methods

Statistical Analysis

To determine if species composition changed over the time period, we calculated a similarity value developed by Legendre and Legendre (1998). This statistic compares the frequency distribution of one population to another and determines the similarity between them. A score near one indicates a high degree of similarity while a score close to 0 indicates that the distributions differ. The statistic is calculated as follows:

$$r = 1/(n-1)\sum_{i=1}^{n}\sum_{j=1}^{n}\left[\left(X_{ij} - \overline{X}/S_x\right)\right]\left[\left(Y_{ij} - \overline{Y}\right)/S_y\right]$$

where x and y are variables measured at locations I, and j and n are the number of elements in the distance matrices, and s_x and s_y are standard deviations for variables x and y.

We calculated the Spearman rank correlation coefficient as a nonparametric alternative comparison between the fish assemblage structure described by the different studies. The Spearmen rank correlation coefficient, r_s, is the ordinary correlation coefficient between the ranked values X_1 and X_2. It can be calculated using the equation (Snedecor and Cochran 1980):

$$r_s = 1 - 6\sum D^2/\left[n\left(n^2-1\right)\right]$$

where D represents the sum of the comparisons of the ranks for each species.

Results and Discussion

Evidence for Long-Term Change in the Fish Assemblage in the Lower Slave River

To examine whether or not there have been changes in the Slave River fish assemblage, we initially compared total species richness and composition of fishes captured by various research programs from 1971 to the present (Table 2). McLeod and O'Neill (1983) only recorded the occurrence of Chinook salmon in the Liard River and not the Slave. Falk and Dalkhe (1975) report catches in streams on the south side of Great Slave Lake from Hay River up to, but not including the Slave River. McLeod et al. (1985) do not report catches of Dolly Varden, longnose dace, or mountain whitefish, and therefore it is difficult to determine the validity of McCart's (1986) report that they occur in the lower Slave River. With the exception of the most recent sampling period (Tallman and Low, unpublished data), species richness has remained stable over time, ranging from 22 to 28 species, with slight variations between time periods. The species composition has also varied to some extent between times. For example, Dolly Varden, mountain whitefish, and longnose dace were only captured during the studies conducted in the mid-1980s (Table 2). Chum salmon were absent from catches in studies conducted after 1985, while round whitefish, pearl dace, and slimy sculpin, were absent from catches in studies conducted after 1990 (Table 2). However, two previously unreported species, the Chinook and sockeye salmon, appeared in catches in the mid-1990s (Table 2). McLeod et al. (1985) sampling only in the fall of 1983 and 1984 found 18 species, including two rainbow trout and lake trout not reported by Tripp et al. (1981). Discounting Dolly Varden, mountain whitefish, and longnose dace, 25 species were present in the early 1980s.

The above comparisons of species richness and composition are confounded by the fact that each of the studies emphasized somewhat different methods for capture and focused on different portions of the Slave River. For example, Little et al. (1998) used mainly gill nets with occasional hand seining. Tripp et al. (1981) used a combination of gill nets, traps, and hand seining. McLeod et al. (1985) used gill nets and hand seining. Tripp et al. (1981) focused their sampling in the Delta with some work in the Slave River channel. McLeod et al. (1985) sampled only in the fall and distributed their effort evenly across the Delta, mid-lower river and near Fort Smith. Little et al. (1998) sampled extensively near Fort Smith in the Slave and Salt rivers and spent about 20% of their time sampling in the Slave River Delta.

When we examined gill-net catches only from the three major studies, species richness ranged from 13 to 16 species in total (Table 3), suggesting that there has only been minor, if any changes in diversity over time. The differences in richness and composition between times was due mainly to the presence or absence of rare species. For example, Tallman et al. (1996a) and Little et al. (1998) did not capture the small bodied ciscoes captured by Tripp et al. (1981) and McLeod et al. (1985). These fish appeared in the Slave River delta during the latter part of October. At this time in 1994 and 1995, the river and Delta were in transition from open water to ice cover and sampling could not be undertaken. If Little et al. (1998) could have achieved the same level of coverage of the Delta as Tripp et al. (1981), it seems probable they would have caught the ciscoes. McLeod et al. (1985) captured a couple of specimens identified as round whitefish that neither Tripp et al. (1981), nor Tallman et al. (1996a) and Little et al. (1998) caught in gill nets. Similarly, while McLeod et al. (1985) failed to capture gill-net trout-perch and yellow perch, Tripp et al. (1981) and Tallman et al. (1996a) and Little et al. (1998) caught a few specimens of both species. Tallman et al. (1996a) and Little et al. (1998) caught one Chinook and one sockeye salmon, neither of which were captured in the previous studies (Table 3).

To compare relative abundance estimates between the 1970s and 2000, we used the data from gill-net catches of only the more commonly captured species: lake whitefish, inconnu, goldeye, northern pike, flathead chub, longnose sucker, white sucker, burbot, and walleye. All data were converted to the percentage of the total catch made up by each species. Catch per unit of effort data for the Slave River from Tripp et al. (1981) were directly converted. For McLeod et al. (1985), we calculated an average percentage of total catch over 1983 and 1984 by combining the data for mid-river and Slave River from both years (see Tables 4.3 and 4.4 in McLeod et al. 1985) and dividing by number of sites represented (four in total). Since other species were caught in this study, we then adjusted the totals to sum to 100%. To calculate average percentage catch for Tallman et al. 1996a and Little et al. 1998, we used CPUE averaged over the entire sampling period (1994–1995, see Tables 2–10 in Tallman et al. 1996d).

Based on our calculations, there have been fluctuations in the relative abundance of common species between 1978 and 2000 (Table 4). Lake whitefish has remained relatively stable at between 5% and 10% of the catch. However, inconnu apparently increased up to 1995. Historically, the subsistence fishery in the area was much more intense than it is presently (Jalkotsky 1976; Bodden 1980, Tripp et al. 1981, MacDonald and Smith 1993; McLeod et al. 1985). Moreover, over the last few years, the Department of Fisheries and Oceans (DFO) has progressively extended a spring conservation zone for inconnu in some of the nearshore areas along the south shore of Great Slave Lake (C. Day, DFO, Winnipeg stock assessment biologist). The combination of much lower subsistence fishing in the river and more protection in Great Slave Lake may be contributing to the increased abundance of inconnu in the Slave River. Goldeye has fluctuated greatly in percentage of the catch, but overall they have been and continue to be the most common species in the Slave River. Northern pike is also abundant, but their rela-

Table 2.—Abundance of species in samples. The number of x's represent the relative abundance in catches (x = <1%, xx = 1–10%, xxx = >10% of the catches).

Family and species	<1975 (a,b,c)	1976–1985 (d)	Mid-1980s (e,f,g)	Mid-1990s (h,i,j)	2000 (k)
Petromyzontidae					
Lampetra camtschatica	x	x	x	x	
Hiodontidae					
Hiodon alosoides	x	xx	xx	xxx	xxx
Esocidae					
Esox lucius	x	xx	xx	xx	xx
Salmonidae					
Oncorhynchus keta	x	x			
O. nerka				x	
O. tshawytsha					x
O. mykiss			x	x	
Salvelinus malma			x		
S. namaycush	x		x	x	
Coregonus artedii	x	xx	xx	x	x
C. sardinella				x	
C. clupeaformis	x	x	x	xx	xx
Prosopium cylindraceum	x	x	x		
P. williamsoni			x		
Stenodus leucichthys	x	x	x	x	xx
Thymallus arcticus		x	x		
Cyprinidae					
Couesius plumbeus	x	xx	x	x	x
Platygobio gracilis	x	xx	xx	x	x
Notropis atherinoides	x	xx	xx	x	x
N. hudsonius	x	x	x	x	
Rhinichthys cataractae			x		
Margariscus margarita	x	x	x		
Catostomidae					
Catostomus catostomus	x	xx	xx	x	xx
C. commersonii	x	x	x	x	x
Percopsidae					
Percopsis omiscomaycus	x	xx	xx	x	x
Gadidae					
Lota lota	x	xx	x	x	x
Gasterosteidae					
Pungitius pungitius	x		x	x	
Cottidae					
Cottus cognatus	x	x	x		
C. ricei	x	x	x	x	
Percidae					
Perca flavescens	x	x	x	x	
Sander vitreus	x	xx	xx	xx	xx

[a] Scott and Crossman (1973); [b] Falk and Dahlke (1975); [c] Keleher (1972); [d] Tripp et al. (1981); [e] McLeod and O'Neil (1983); [f] McLeod et al. (1985); [g] McCart (1986); [h] Tallman et al. (1996a); [i] Tallman et al. (1996d); [j] Tallman and Low (unpublished data); [k] Little et al. (1998)

tive abundance has decreased since the late 1970s. However, Tripp et al. (1981) sampled more heavily in the lower end of the Slave River channel where the more sluggish flows and increased aquatic vegetation favor northern pike. Flathead chub represented a small component of the gillnet catch in all of these studies, ranging from about 1% to 5%. White sucker was also present throughout, but in low abun-

Table 3.—List of fish species collected by gillnet in the lower Slave River, Slave River Delta, and Salt River, 1981–1995 by various studies. X = captured; – = not captured.

Family /species name	Tripp et al. (1981)	McLeod et al. (1985)	Tallman et al.(1996a) /Little et al. (1998)
Salmonidae			
Oncorhynchus nerka	–	–	X
O. tshawytsha	–	–	X
O. mykiss	–	X	X
Salvelinus namaycush	–	X	X
Coregonus clupeaformis	X	X	X
C. artedii	X	X	X
Coregonus spp.	X	X	–
Stenodus leucichthys	X	X	X
Prosopium cylindraceum	–	X	–
Esocidae			
Esox lucius	X	X	X
Hiodontidae			
Hiodon alosoides	X	X	X
Cyprinidae			
Platygobio gracilis	X	X	X
Catostomidae			
Catostomus catostomus	X	X	X
C. commersonii	X	X	X
Gadidae			
Lota lota	X	X	X
Percopsidae			
Percopsis omiscomaycus	X	X	X
Percidae			
Sander vitreus	X	X	X
Perca flavescens	X	–	X
Total	n = 13	n = 15	n = 16

dance, contributing less than 1%. In contrast to inconnu, the longnose sucker appears to have decreased in relative abundance with time. The relative contribution of this species to total catch dropped from almost 9% to about 1%. Similarly, burbot is also less abundant now compared with the late 1970s. Walleye is another important fish in the system. Its relative contribution to the catch has generally fluctuated between 15% and 25%; the rather low value (5%) found by McLeod et al. (1985) is probably a result of their fall sampling, thereby missing the spawning migration. As mentioned previously, assemblage composition changes dramatically with season in relation to the influxes of spawners of different species. Thus, McLeod et al.'s (1985) results may not be completely comparable to the other studies listed in Table 4. If one compares Tripp et al. (1981) and Tallman et al. (1996a) only, one can see that with the exceptions of inconnu, burbot, and longnose sucker, the assemblage composition in the Slave River has changed very little over the last 30 years.

A similarity matrix (Van Tongeren 1987; Legendre and Legendre 1998) shows no obvious trends, but data from 1978 to 1980 and 1983 to 1984 have the lowest similarity (Tables 4 and 5). Spearman's rank correlation (Van Tongeren 1987; Legendre and Legendre 1998) has the 1978 to 1980 period with the lowest correlation to the other time periods (Table 6). The correlations are relatively high for all comparisons.

Our results suggest that there has been little change in fish assemblage structure over time. According to our data changes have occurred only in rare species that were present or absent in various years.

This conclusion is based on an analysis of the

Table 4.—Percentage (and rank) of the gill-net catch for each species recorded by Tripp et al. (1981), McLeod et al. (1985), Tallman et al. (1996), and Tallman and Low (unpublished) for the Slave River Channel.

Species	1978–1980[a]	1983–1984[b]	1994–1995[c]	2000[d]
lake whitefish	5.3% (6)	9.3% (3)	5.6% (5)	10.3% (4)
inconnu	1.3% (8)	4.2% (5)	9.3% (4)	5.1% (4)
goldeye	21.5% (2)	64.9% (1)	46.3% (1)	31.9% (1)
northern pike	30.3% (1)	12.5% (2)	18.1% (2)	18.3% (3)
flathead chub	2.1% (7)	0.7% (8)	4.5% (6)	1.3% (7)
longnose sucker	8.9% (5)	2.1% (6)	1.1% (7)	6.7% (5)
white sucker	0.2% (9)	0.2% (9)	0.3% (8)	0.7% (8)
burbot	9.8% (4)	1.7% (4)	0.02% (9)	0.7% (9)
walleye	20.7% (3)	4.5% (4)	14.8% (3)	24.3% (2)

[a] Tripp et al. (1981); [b] McLeod et al. (1985); [c] Tallman et al. (1996); [d] Tallman and Low (unpublished)

Table 5.—Similarity matrix of the relative abundance of fish caught among the four studies in Table 4.

	Tripp et al. (1981)	McLeod et al. (1985)	Tallman et al. (1996a)
McLeod et al. 1985	0.5		
Tallman et al. 1996a	0.64	0.76	
Tallman and Low (unpublished)	0.76	0.67	0.79

Table 6.—Rank correlation among the four studies noted in Table 4.

	Tripp et al. (1981)	McLeod et al. (1985)	Tallman et al. (1996a)
McLeod et al. 1985	0.73		
Tallman et al. 1996a	0.58	0.87	
Tallman and Low (unpublished)	0.67	0.88	0.90

most general kind of data, relative abundance of only the most abundant species, not on CPUE or estimates of population size. Numerical abundance can suffer dramatic declines, even though relative abundance remains the same. We believe that the numerical abudance has remained relatively constant because there was no indication from the local fishermen that fishing (CPUE) had declined from previous decades.

The lack of change in the assemblage is probably because fishing and other human activities, such as remote upstream hydroelectric developments have not had a significant impact on the life cycle of various species in the river. The Slave River environment has remained reasonably constant without significant degradation of the habitat.

All indications are that the Slave River is still a relatively pristine system with its fish assemblage intact. Goldeye relative abundance has increased, possibly due to a reduced harvest. While apparent abundances may fluctuate over time due to sampling or natural differences among years in abundances and in timing of life histories among species, the general composition of the Slave River fauna has remained stable, at least over the past 22 years.

References

Battle, H. I., and W. M. Sprules. 1960. A description of the semi-bouyant eggs and early developmental stages of goldeye, *Hiodon alosoides* (Rafinesque). Journal of the Fisheries Research Board of Canada 17(2):245–265.

Baxter, C. V., C. A. Frissell, and F. R. Hauer. 1999. Geomorphology, logging roads, and the distribution of bull trout spawning in a forested river basin: implications for management and conservation. Transactions of the American Fisheries Society 128:854–867.

Bodden, K. 1980. The economic use by native people of the resources of the Slave River Delta. M. A. thesis. University of Alberta, Edmonton.

Bonetto, A. A. 1986. Fish of the Parana system. Pages 573–588 *in* B. R. Davies, and K. F. Walker, editors. The ecology of river systems. Dr. W. Junk Publishers, Dordrecht, Netherlands.

Brunskill, G. J. 1986. Environmental features of the Mackenzie system. Pages 435–471*in* B. R. Davies, and K. F. Walker, editors. The ecology of river systems. Dr. W. Junk Publishers, Dordrecht, Netherlands.

Davies, K. F. 1981. Slave River Delta hydrology. Interim Report. Prepared for the Mackenzie River Basin Study by Water Survey of Canada, Calgary, Alberta.

English, M. C. 1979. Some aspects of the ecology and environment of the Slave River Delta, NWT and some implications of upstream impoundment. M. Sc. thesis. University of Alberta, Edmonton.

English, M. C., R. B. Hill, M. A. Stone and R. Ormson. 1996. Geomorphological and botanical change on the outer Slave River delta, NWT, before and after impoundment of the Peace River. Hydrological Processes 11:1707–1724.

Falk, M. R., and L. W. Dalkhe. 1975. Creel data and biological data for streams along the south shore of Great Slave Lake, 1971–74. Canadian Fisheries Marine Services Data Report CEN/D-75–8.

Geen, G. H., T. G. Northcote, G. F. Hartman, and C. C. Lindsey. 1966. Life histories of two species of catastomid fishes in Sixteenmile Lake, British Columbia, with particular reference to inlet spawning. Journal of the Fisheries Research Board of Canada 23:1761–1788.

Greenwood, H., D. Rosen, S. Weitzman, and G. Myers. 1966. Phyletic studies of teleostean fishes, with a provisional classification of living forms. Bulletin American Museum Natural History 131:339–455.

Harris, R. H. D. 1962. Growth and reproduction of the longnose sucker, *Catostomus catostomus* (Foerster). Great Slave Lake. Journal of the Fisheries Research Board of Canada Bulletin 9:113–126.

Hartman, G. F., and J. C. Scrivener. 1990. Impacts of forestry practices on a coastal stream ecosystem, Carnation Creek, British Columbia. Canadian Bulletin of Fisheries and Aquatic Sciences 223.

Howland, K. L., R. F. Tallman, and W. M. Tonn. 2000. Migration patterns of freshwater and anadromous inconnu, *Stenodus leucichthys*, in the Mackenzie River system. Transactions of the American Fisheries Society 129:41–59.

Hughes, R. M, J. N. Rinne, and B. Calamusso. 2005. Historical changes in fish assemblages of large rivers of the Americas: a synthesis. Pages 603–612 *in* J. N. Rinne, R. M. Hughes, and B. Calamusso, editors. Historical changes in large river fish assemblages of the Americas. American Fisheries Society, Symposium 45, Bethesda, Maryland.

Jalkotsky, M. 1976. Summary of data and personal comments on the Slave River Project, 1976. Government. N. W. T., Fish and Wildlife Service, Preliminary manuscript. Yellowknife, Northwest Territories.

Kennedy, W. A., and W. M. Sprules. 1967. Goldeye in Canada. Fisheries Research Board of Canada Bulletin 161.

Koelz, W. 1929. Coregonid fishes of the Great Lakes. Bulletin of the Bureau of Fisheries 43:297–643.

Lange, M., and R. F. Tallman. 2002. Life history variation in two populations of migratory lake whitefish (*Coregonus clupeaformis*) in the western Northwest Territories, Canada. Advances in Limnology 57:507–515.

Lawler, G. H. 1965. Fluctuations in the success of year-classes of whitefish populations with special reference to Lake Erie. Journal of the Fisheries Research Board of Canada Bulletin 22:1197–1227.

Legendre, P., and L. Legendre. 1998. Numerical ecology. 2nd edition. Elsevier, Amsterdam.

Lindsey, C. C., and J. D. McPhail. 1986. Zoogeography of fishes of the Yukon and Mackenzie basins. Pages 639–674 *in* C. H. Hocutt, and E. O. Wiley, editors. The zoogeography of North American freshwater fishes. Wiley, New York.

Little, A. S. 1997. Food and habitat use within the fish assemblages of the lower Slave River, Northwest Territories. M.Sc. Dissertation. University of Alberta, Edmonton.

Little, A. S., W. M. Tonn, R. F. Tallman, and J. D. Reist. 1998. Seasonal variation in diet and trophic relationships within the fish communities of the lower Slave River, Northwest Territories, Canada. Environmental Biology of Fishes 53:429–445.

Lowe-McConnell, R. H. 1986. Fish of the Amazon system. Pages 339–352 *in* B. R. Davies, and K. F. Walker. The ecology of river systems. Dr. W. Junk Publishers, Dordrecht, Netherlands.

MacDonald, D. D., and S. L. Smith. 1993. An approach to monitoring ambient environmental quality in the Slave River basin, Northwest Territories: toward a consensus. Department of Indian Affairs, Yellowknife, Northwest Territories.

McCart, P. J. 1986. Fish and fisheries of the Mackenzie system. Pages 493–516 *in* B. R. Davies, and K. F. Walker, editors. The ecology of river systems. Dr. W Junk Publishers, Dordrecht, Netherlands.

McLeod, C., G. Ash, D. Fernet, J. O'Neil, T. Clayton, T. Dickson, L. Hildebrand, R. Nelson, S. Matkowski, C. Pattenden, D. Chiperzak, R. McConnell, B. Wareham, and C. Bjornson. 1985. Fall fish spawning habitat survey, 1978–1985. RL&L/EMA Slave River Joint Venture, Edmonton, Alberta.

McLeod, C. L., and J. P. O'Neill. 1983. Major range extensions of anadromous salmonids and first record of Chinook salmon in the Mackenzie River drainage. Canadian Journal of Zoology 61:2138–2184.

McPhail, J. D., and C. C. Lindsey. 1970. Freshwater fishes of northwestern Canada and Alaska. Fisheries Research Board of Canada Bulletin 173:1–380.

Nelson, J. S., and M. J. Paetz. 1972. Fishes of the northeastern Wood Buffalo National Park region, Alberta and Northwest Territories. Canadian Field-Naturalist 86:133–144.

NRBS (Northern River Basin Study) 1996a. The Northern River Basin Study – an overview. Government of Alberta, Synthesis Report No. 20, Edmonton, Alberta, Canada.

NRBS (Northern River Basin Study) 1996b. The Northern River Basin Study – report to the ministers. Government of Alberta, Edmonton, Alberta.

Olund, L. J., F. B. Cross. 1961. Geographic variation in the North American cyprinid fish, *Hybopsis gracilis.* University Kansas Publications of the Museum of Natural History 13:323–348.

Pankhurst, N. W., N. E. Stacey, and G. Van der Kraak. 1986. Reproductive development and plasma levels of reproductive hormones of goldeye, *Hiodon alosoides* (Rafinesque), taken from the North Saskatchewan River during open-water season. Canadian Journal of Zoology 64:2843–2849.

Pflieger, W. L., and T. B. Grace. 1987. Changes in the fish fauna of the lower Missouri River, 1940–1983. Pages 166–177 *in* W. J. Matthews, and D. C. Heins, editors. Community and evolutionary ecology of North American stream fishes. University of Oklahoma Press, Norman.

Pielou, E. C. 1991. After the ice age. University of Chicago Press, Chicago.

Rawson, D. S. 1947. Great Slave Lake. Northwest Canadian fisheries surveys in 1944–45 (various authors). Fisheries Research Board of Canada Bulletin 72:45–68.

Ricker, W. E. 1975. Computation and interpretation of biological statistics of fish populations. Bulletin of the Fisheries Research Board of Canada 191.

Sandheinrich, M. B. and G. J. Atchison. 1986. Fish associated with dikes, revetments, and abandoned channels in the middle Missouri River. Proceedings of the Iowa Academy of Science 93(4):188–191.

Scott, W. B., and E. J. Crossman. 1973. Freshwater fishes of Canada. Fisheries Research Board of Canada Bulletin 184.

Snedocor, G. W., and W. G. Cochran. 1980. Statistical methods. Iowa State University Press. Ames.

Stein, J. N, C. S. Jessop, T. R Porter, and K. T. J. Chang-kue. 1973. Fish resources of the Mackenzie River Valley. Interim Report II. Fisheries and Marine Service of Environment Canada, Winnipeg, Manitoba.

St. Louis, V. L., C. A. Kelly, E. Duchemin, J. W. M. Rudd, and D. M. Rosenberg. 2000. Reservoir surfaces as sources of greenhouse gases to the atmosphere: a global estimate. BioScience 50(9):766–775.

Tallman, R., W. Tonn, and A. Little. 1996a. Diet, food web and structure of the fish community, lower Slave River, June to December, 1994 and May to August, 1995. Government of Alberta, Northern River Basins Study, Project Report No. 119, Edmonton, Alberta.

Tallman, R. F., K.L. Howland, and S. Stephenson. 2005. Stability, change and species composition of fish assemblages in the lower Mackenzie River: a pristine large river. Pages 13–21 *in* J. N. Rinne, R. M. Hughes, and B. Calamusso, editors. Historical changes in large river fish assemblages of the Americas. American Fisheries Society, Symposium 45, Bethesda, Maryland.

Tallman, R. F., W. M. Tonn, and K. L. Howland. 1996b. Migration of inconnu (*Stenodus leucichthys*) and burbot (*Lota lota*), Slave River and Great Slave Lake, June 1994 to July, 1995. Government of Alberta, Northern River Basins Study, Project Report No. 117, Edmonton, Alberta.

Tallman, R. F., W. M. Tonn, and K. L. Howland. 1996c. Life history variation of inconnu (*Stenodus leucichthys*) and burbot (*Lota lota*), lower Slave River, June to December, 1994. Government of Alberta, Northern River Basins Study, Project Report No. 118, Edmonton, Alberta.

Tallman, R. F., W. M. Tonn, K. L. Howland, and A. Little. 1996d. Synthesis of fish distribution, movements, critical habitat and food web for the lower Slave River north of the 60th parallel: a food chain perspective. Government of Alberta, Northern River Basins Study, Synthesis Report No. 13, Edmonton, Alberta.

Tripp, D. B., P. J. McCart, R. D. Saunders, and G. W. Hughes. 1981. Fisheries studies in the Slave River delta, NWT - Final Report. Prepared for Mackenzie River Basin Study. Aquatic Environments Limited, Calgary Alberta.

Van Tongeren, O. F. R. 1987. Cluster analysis. Pages 174–212 *in* R. H. G. Jongman, C. J. F. ter Braak, and O. F. R. Van Tongeren, editors. Data analysis in community and landscape ecology. Pudoc, Wageningen, Netherlands.

Vanderburgh, S., and D. G. Smith. 1988. Slave River delta: geomorphology, sedimentology, and Holocene reconstruction. Canadian Journal Earth Sciences 25:1990–2004.

American Fisheries Society Symposium 45:41–59, 2005

Historical and Current Perspectives on Fish Assemblages of the Snake River, Idaho and Wyoming

Terry R. Maret* and Christopher A. Mebane
U.S. Geological Survey, 230 Collins Road, Boise, Idaho 83702, USA

Abstract.—The Snake River is the tenth longest river in the United States, extending 1,667 km from its origin in Yellowstone National Park in western Wyoming to its union with the Columbia River at Pasco, Washington. Historically, the main-stem Snake River upstream from the Hells Canyon Complex supported at least 26 native fish species, including anadromous stocks of Chinook salmon *Oncorhynchus tshawytscha*, steelhead *O. mykiss*, Pacific lamprey *Lampetra tridentata*, and white sturgeon *Acipenser transmontanus*. Of these anadromous species, only the white sturgeon remains in the Snake River between the Hells Canyon Complex and Shoshone Falls. Today, much of the Snake River has been transformed into a river with numerous impoundments and flow diversions, increased pollutant loads, and elevated water temperatures. Current (1993–2002) fish assemblage collections from 15 sites along the Snake River and Henrys Fork contained 35 fish species, including 16 alien species. Many of these alien species such as catfish (Ictaluridae), carp (Cyprinidae), and sunfish (Centrarchidae) are adapted for warmwater impounded habitats. Currently, the Snake River supports 19 native species. An index of biotic integrity (IBI), developed to evaluate large rivers in the Northwest, was used to evaluate recent (1993–2002) fish collections from the Snake River and Henrys Fork in southern Idaho and western Wyoming. Index of biotic integrity site scores and component metrics revealed a decline in biotic integrity from upstream to downstream in both the Snake River and Henrys Fork. Two distinct groups of sites were evident that correspond to a range of IBI scores—an upper Snake River and Henrys Fork group with relatively high biotic integrity (mean IBI scores of 46–84) and a lower Snake River group with low biotic integrity (mean IBI scores of 10–29). Sites located in the lower Snake River exhibited fish assemblages that reflect poor-quality habitat where coldwater and sensitive species are rare or absent, and where tolerant, less desirable species predominate. Increases in percentages of agricultural land, total number of diversions, and number of constructed channels were strongly associated with these decreasing IBI scores.

Introduction

Large rivers are the least understood and studied of any inland water bodies and are perhaps the resource most degraded by human activity. Large-river investigations have been hampered by sampling difficulties and the lack of operational theoretical models (Reash 1999). Most river investigations have been over short distances and have not provided a largescale view of longitudinal changes in fish assemblages and their habitats (Fausch et al. 2002). The development of North America's rivers has supported many important human uses including navigation, flood control, hydropower, irrigation, waste disposal, and recreation. However, these changes have not come without a cost, specifically, the decline of native aquatic biodiversity including native

* Corresponding author: trmaret@usgs.gov

fish. It has been estimated that more than one-third of the North American native fish species and subspecies (364) are endangered, threatened, or of special concern (Williams et al. 1989). Extinctions have been documented for 3 genera, 27 species, and 13 subspecies of fish in North America during the past 100 years (Miller et al. 1989). In addition, at least 214 stocks of anadromous salmonids are on the decline or at risk of extinction in the Pacific Northwest and California (Nehlsen et al. 1991). Fragmentation of river systems by dams has affected about 77% of the total water discharge in the northern third of the world (Dynesius and Nilsson 1994). There are at least 100 high dams on streams and rivers within the Columbia River basin, which have been especially damaging to the native fish population of the Pacific Northwest (Li et al. 1987). It has been estimated that dams presently block about 56% of the Snake River basin that historically produced anadromous fish (Chandler and Radko 2001).

Because of the arid climate in southern Idaho and western Wyoming, the Snake River has long been valued as a resource for irrigation and hydropower. In fact, Idaho's total offstream water use (16.9 million acre-ft) ranks fourth in the United States behind California, Texas, and Illinois (Solley et al. 1998). About 86% of this water use in Idaho is for irrigation, primarily from surface water sources. Large storage reservoirs have been constructed in the upper portions of the Snake River for water use through the irrigation period. In addition, many smaller impoundments have been constructed along the mainstem Snake River, primarily for hydropower generation. As a result of these uses, many reaches of the Snake River and its tributaries have been transformed into numerous impoundments characterized by extensive flow diversions, increased pollutant loads, and elevated water temperatures (Maret 1997; Clark et al. 1998).

The interbasin transfer of water is a common irrigation practice in southern Idaho. This practice can have ecological consequences such as changes in streamflow, introduction of alien species, and alteration of habitat (Meador 1992). Land use is as important in shaping fish assemblages as are basin size, discharge, elevation, or ecoregions (Maret 1997). The percentage of agricultural land in the upper Snake River basin has been associated with negative biological, chemical, and physical habitat changes (Maret 1997; Mebane et al. 2003).

The Snake River often is referred to as a working river because of its highly regulated nature (Clark et al. 1998). Blocked access, fragmentation, and alterations of habitat caused by different land uses have resulted in a fish assemblage dramatically different from the historical assemblage. Many native species have been extirpated, and alien species, many of which are tolerant to pollutants and habitat modification, have been introduced (Maret 1995).

As a result of the federal Clean Water Act's objective to restore and maintain the physical, chemical, and biological integrity of the nation's water, there has been a growing focus on the development of biocriteria in state water quality standards. Increasingly, biological monitoring programs and biocriteria development have expanded to include large rivers. In the United States, the index of biotic integrity (IBI) is used by the U.S. Environmental Protection Agency (USEPA) and many state agencies to assess fish assemblage structure because this index serves as an indicator of the history and current health or condition of a stream system (Karr 1991). The Idaho Department of Environmental Quality has recently published monitoring protocols and an IBI using aquatic organisms and habitat measures to evaluate large rivers in Idaho (Grafe 2002; Mebane et al. 2003). Zaroban et al. (1999) classified Northwest fish species according to various attributes to facilitate evaluation of surface-water resource conditions.

The objectives of this paper are to summarize the historical (pre-Euro-American settlement) and recent (1993–present) fish assemblages of the mainstem Snake River and a major tributary, the Henrys Fork, in southern Idaho and western Wyoming. Longitudinal changes in the fish assemblages along these large rivers are described. Various measures of the fish assemblages are related to land-use and water-use indicators that are quantified at the basin scale using a geographic information system (GIS). The hypothesis for this analysis is that changes in the native fish assemblage of the Snake River are primarily the result of human activities.

Study Area

The Snake River is the tenth longest river in the United States, extending 1,667 km from its origin in Yellowstone National Park in western Wyoming to its union with the Columbia River at Pasco, Washington (Palmer 1991). The specific study area (Figure 1) addressed in this paper comprises 12 sites on the main-stem Snake River and 3 sites on the Henrys Fork, a major tributary to the Snake River. Snake River sites extend from about 164 km upstream from the Hells Canyon Complex (HCC, consisting of Brownlee, Oxbow, and Hells Canyon dams) at Nyssa, Oregon (site 15 at river km 621), upstream to Flagg Ranch, Wyoming (site 1 at river km 1,632; Figure 1). The Snake River at the most downstream site drains more than 171,000 km^2 of Wyoming, Idaho, Oregon, and Nevada. Sites on the Snake River and Henrys Fork represent 4th-order to 7th-order streams (Strahler 1957; Table 1). Mean channel widths at all sample sites range from about 50–200 m.

The basin area that contains the study sites consists of about 100,000 km of perennial streams. Mean annual discharge at Nyssa (site 15) for the period 1975–2000 was 416 m^3/s (Brennan et al. 2001). Elevation above sea level ranges from 2,073 m at Flagg Ranch to 661 m at Nyssa. Climate in most of the study area is semiarid, and average annual precipitation ranges from 25 to 50 cm. Precipitation occurs primarily as snow, and peak flows generally result from spring snowmelt. In many years, groundwater discharges to the Snake River between Milner Dam (river km 1,028) and King Hill, Idaho (river km 880), provide more than 50% of the streamflow measured in the Snake River at King Hill (Clark et al. 1998).

Commercial fish farms were established along the abundant spring sources in the Snake River Canyon near Hagerman, Idaho (near river km 920), beginning in 1909 (Fiege 1999). Today, there are at least 80 private or state-owned aquaculture facilities between the cities of Twin Falls and Hagerman,

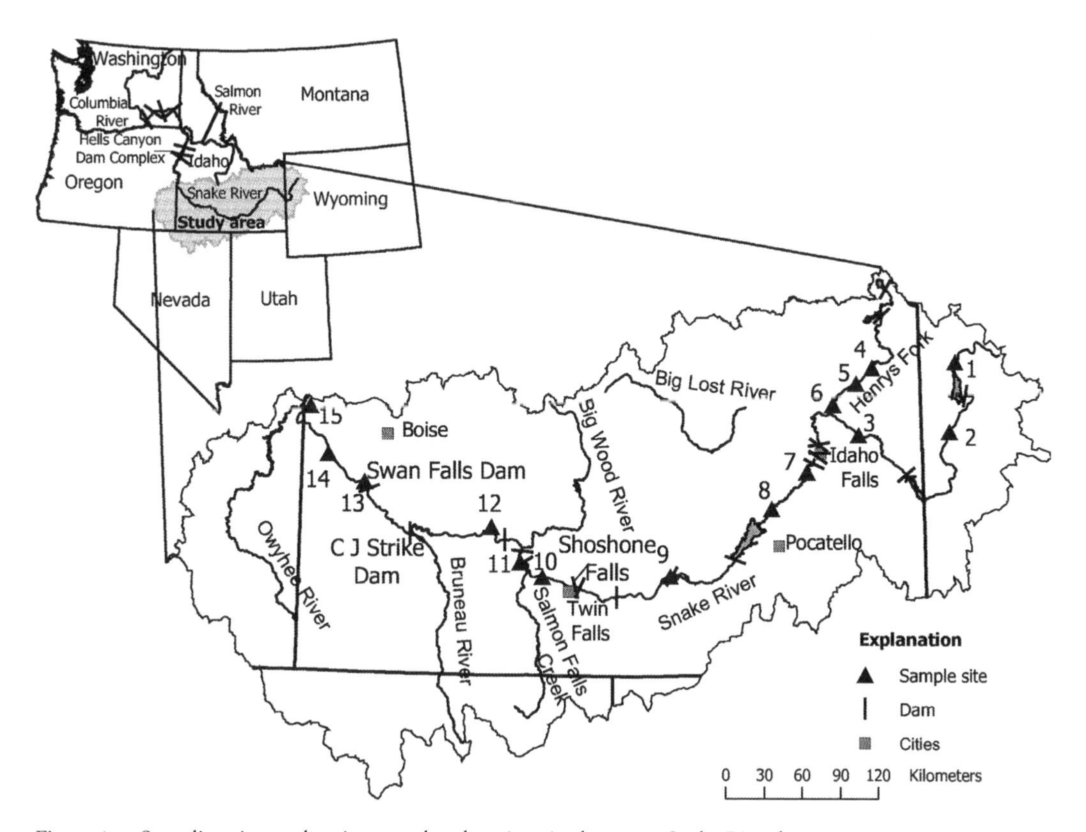

Figure 1.—Sampling sites and main-stem dam locations in the upper Snake River basin.

Table 1.—Basin and landscape characteristics, collection methods, and collection dates for all study sites. Percent land use, number of diversions, and constructed channels are cumulative totals upstream from each study site.

Site number and name[a]	Collection method[b]	Basin area (km²)	River km	Elevation (m)	Stream order	Percent land use: Urban	Agricultural	Range	Forest	Collection date(s)	Number of diversions in basin	Constructed[c] channels (km)
1 Snake River at Flagg Ranch, WY	NAWQA	1,324	1,632	2,073	4	0.0	0.0	8.6	82.8	1993,1994,1995, 2002	1	52
2 Snake River at Moose, WY	NAWQA	4,326	1,553	1,960	4	0.1	0.8	16.6	66.5	1995	115	203
3 Snake River near Heise, ID	NAWQA/ EMAP	14,841	1,374	1,528	6	0.3	5.9	25.3	60.5	1999, 2002	2,783	587
4 Henrys Fork near Ashton, ID	NAWQA	2,832	1,411	1,551	6	0.3	6.3	12.3	76.7	1999	1,630	277
5 Henrys Fork near St. Anthony, ID	NAWQA	5,122	1,392	1,509	6	0.2	12.1	22.2	61.5	1999	2,678	509
6 Henrys Fork near Rexburg, ID	NAWQA	8,337	1,353	1,465	6	0.5	25.8	19.3	49.6	1993,1996,1999	5,809	1,068
7 Snake River near Shelley, ID	EMAP	27,947	1,268	1,402	6	0.6	18.4	26.1	48.5	2002	11,417	2,811
8 Snake River near Blackfoot, ID	NAWQA	31,555	1,207	1,341	7	0.6	19.1	28.7	44.9	1993,1996, 2001	13,345	3,514
9 Snake River near Minidoka, ID	NAWQA	48,830	1,084	1,259	7	0.7	23.0	37.8	32.8	1993, 2002	20,053	4,571
10 Snake River near Buhl, ID	NAWQA	76,104	961	900	7	0.7	21.3	45.8	25.4	1993,1996,1997,1999, 2000	40,098	7,562
11 Snake River near Hagerman, ID	EMAP	82,938	920	880	7	0.6	20.8	48.3	23.9	2002	43,095	8,200
12 Snake River at King Hill, ID	NAWQA	92,941	880	760	7	0.6	20.2	50.2	22.6	1993,1994,1995,1996,1997, 1998,1999,2000,2001,2002	55,227	9,166
13 Snake River near Murphy, ID	NAWQA	129,052	727	692	7	0.7	18.0	56.0	19.8	2000, 2002	61,689	9,862
14 Snake River near Marsing, ID	EMAP	111,235	682	680	7	0.7	18.0	56.3	19.5	2002[d]	63,260	10,109
15 Snake River near Nyssa, OR	NAWQA	171,363	621	661	7	0.7	14.6	59.2	17.2	1997,2000	76,026	13,211

[a] See Figure 1 for locations.
[b] U.S. Geological Survey - National Water-Quality Assessment (NAWQA) Program (Meador et al. 1993); U.S. Environmental Protection Agency, Environmental Monitoring and Assessment Program (EMAP), (Peck et al. 2002).
[c] includes artificial paths constructed by a geographic information system to account for lakes and resevoirs.
[d] Represents two collections on same reach to evaluate intra-annual variability.

Idaho. These include both coldwater facilities, which raise resident rainbow trout *Oncorhynchus mykiss* and steelhead (anadromous rainbow trout), Chinook salmon *O. tshawytscha*, and white sturgeon *Acipenser transmontanus*, and warmwater facilities, which raise channel catfish *Ictalurus punctatus*, tilapia (*Tilapia* spp.), and common carp *Cyprinus carpio*. These facilities along the Snake River supply approximately 80% of the trout consumed in restaurants in the United States (USEPA 2002).

Irrigated agriculture in the study area began in the 1860s when farmers diverted water to their fields to raise crops for sale in nearby mining camps (Fiege 1999). Settlements were built primarily along the main-stem Snake River and adjacent tributaries to access water for domestic and agricultural purposes. Today, land use in the study area comprises about 59% rangeland, 17% forest land, and 15% agricultural land (Table 1). Livestock grazing is common throughout the basin. Almost 70% of the study area is public land.

Migrating fish face both natural and human obstacles along the Snake River. Shoshone Falls, near Twin Falls, Idaho, is higher than Niagara Falls (more than 65 m high) and is a natural barrier to upstream movement of fish (Figure 2). There are currently 25 large dams on the Snake River and on the Henrys Fork; 18 of them are upstream from the HCC (Figure 1). The first recorded dam on the Snake River was constructed in 1903 at Swan Falls, located near site 13 (Chapman 2001). This was the first dam in the study area that effectively blocked Chinook salmon migrations in the Snake River, despite the presence of a fish ladder. The final hydropower dams that resulted in the termination of anadromous fish stocks in the Snake River were Bliss Dam (1949), C.J. Strike Dam (1952) and ultimately, Brownlee Dam (1958), part of the HCC. All these dams formed impassable barriers to migrating salmon, steelhead, white sturgeon, and Pacific lamprey *Lampetra tridentata* because they were constructed without any fish passage facilities (USEPA 2002).

Approximately 76,000 points of diversions (PODs) and more than 13,000 km of channels have been constructed to irrigate agricultural land in the study area (Table 1). These dams, diversions, and resulting irrigation practices have changed the river conditions by inundating free-flowing habitats, hindering fish migration and spawning, and raising water temperatures to the detriment of native coldwater fish (USEPA 2002). At Milner Dam, nearly all streamflow has been diverted for irrigation during times of drought (Clark et al. 1998). This complete diversion of water is most visible at Shoshone Falls, where the falls can be reduced to a mere trickle during the irrigation season (Figure 2B). Low streamflows, combined with instream reservoirs, have transformed some reaches of the Snake River from a high-gradient, coldwater river to a slow-moving, warmwater river supporting primarily alien fish species (Clark et al. 1998). Aquatic plants have reached nuisance levels (>200 g/m^2, dry weight) in some reaches of the river as a result

A

B

Figure 2.—(A) Photo of Shoshone Falls near Twin Falls, Idaho, summer 1996 during high stream discharge (about 566 m^3/s). (B) Photo of Shoshone Falls near Twin Falls, Idaho, summer 1994 during low flow (<3 m^3/s). The photos show the effects of diversions on the Snake River when the water supply is limited.

of nutrient and fine-grained sediment inputs from irrigation return flows and aquaculture discharges (USEPA 2002).

Historical Fishery

An historical perspective is important to the understanding of river systems and fish assemblages. Fish assemblages in the Snake River have been investigated since the late 1800s. Gilbert and Evermann (1895) and Evermann (1896) described the salmon (*Oncorhynchus* spp.) distributions in the Snake River and major tributaries before hydropower development. Fishery studies on the Snake River have consisted primarily of agency reports to assess sportfish populations. Maret (1995) summarized all available literature on fish in the upper Snake River basin (upstream from King Hill). Idaho Power Company has compiled anecdotal historical data on anadromous fish species occurrences and distribution downstream from Shoshone Falls as part of relicensing requirements of their hydropower facilities (Chandler and Radko 2001). Other relicensing studies have described fish assemblages associated with the Snake River and its impoundments in the middle Snake River between Shoshone Falls and Swan Falls Dam and between Swan Falls Dam and Brownlee Reservoir (Lukas et al. 1995; Brink et al. 1997). Most recently (1993–2002), U.S. Geological Survey (USGS) fishery assessments of the Snake River and Henrys Fork using National Water-Quality Assessment (NAWQA) Program and USEPA Environmental Monitoring Program (EMAP) fish collection protocols have provided data to evaluate the status of fish assemblages (Maret 1997; Maret and Ott 2003).

The native fish fauna of the Snake River, such as anadromous salmonids that are known to migrate great distances between saltwater and freshwater, probably were adapted to invade and exploit new or recently disturbed environments. Volcanic eruptions, landslides, tectonic action, glacial movements, and massive floods were all common geologic or hydrologic features of the Snake River basin (Chapman 2001). The catastrophic Bonneville flood, which released more than 400,000 m^3/s of water from Lake Bonneville, dramatically changed the Snake River plain about 12,000–15,000 years ago (Malde 1968). Much of the river morphology that is visible today was the result of this Pleistocene flood.

Historically, the mainstem Snake River within the study area supported at least 26 native species (Table 2). Of these, 14 were found only downstream from Shoshone Falls: white sturgeon, Pacific lamprey, Chinook salmon, rainbow trout, bull trout *Salvelinus confluentus*, largescale sucker *Catostomus macrocheilus*, bridgelip sucker *C. columbianus*, chiselmouth *Acrocheilus alutaceus*, leopard dace *Rhinichthys falcatus*, northern pikeminnow *Pytchocheilus oregonensis*, peamouth *Mylocheilus caurinus*, Shoshone sculpin *Cottus greenei*, shorthead sculpin *C. confusus*, and torrent sculpin *C. rhotheus*. Hubbs and Miller (1948) attributed the lower number of Columbia basin native fish species upstream from the falls to some cataclysmic event such as lava flows or glacial advances. The bull trout has been found in the Lost River drainages above Shoshone Falls as a result of headwater transfer (Behnke 1992), but has not been found in the main-stem Snake River upstream from Shoshone Falls. Bull trout are found in tributaries of the Snake River downstream from Shoshone Falls and likely used the main-stem Snake River as a migratory corridor. In contrast, cutthroat trout *Oncorhynchus clarkii* are present in the main-stem Snake River and tributaries upstream from Shoshone Falls, where they are believed to have been present historically (Thurow et al. 1988; Maret et al. 1997). It is believed that cutthroat trout were replaced naturally by resident rainbow trout and steelhead when these species came into contact with each other downstream from Shoshone Falls (Behnke 1992).

Prior to hydropower development, numerous native, anadromous species inhabited the entire Snake River downstream from Shoshone Falls. In fact, the Snake River was the most important drainage in the Columbia basin for the production of anadromous fish (Chapman 2001). Evermann (1896) summarized numerous accounts of spawning salmon and steelhead and the harvesting of salmon and steelhead by angling or seining on the Snake River in the summer and fall of 1894 between Huntington, Oregon (river km 528), and Shoshone Falls. He also made an interesting comment while observing the difficult time salmon had

Table 2. —Fish species collected by the U.S. Geological Survey from selected sites along the Snake River and Henrys Fork (river kilometer 621 to 1,632) in Idaho and Wyoming, 1993–2002. Also listed are fish species historically found in the Snake River, some of which may no longer be present (Simpson and Wallace 1982; Maret 1995; USEPA 2002; C. Randolph, Idaho Power Company, personal communication). Assignment of geographic origin, tolerance to pollution, and temperature preference of fish was taken from Zaroban et al. (1999); E, eliminated from the study area; SC, Species of Special Concern.

Family / Common name	Species	Origin[a]	Tolerance to pollution[b]	Temperature preference	Sites of occurrence: USGS 1993–2002 (Figure 1 and Table 1)
Acipenseridae					
white sturgeon[c,d] (SC)	*Acipenser transmontanus*	N	I	cold	10,11,12,13,14,15
Catostomidae					
bluehead sucker	*Catostomus discobolus*	N	I	cool	1,7
bridgelip sucker[d]	*C. columbianus*	N	T	cool	12,13,14,15
largescale sucker[d]	*C. macrocheilus*	N	T	cool	10,11,12,13,14,15
mountain sucker	*C. platyrhynchus*	N	I	cool	1,2,15
Snake River sucker (E)	*Chasmistes muriei*	N	unknown	unknown	
Utah sucker	*Catostomus ardens*	N	T	cool	1,2,3,5,6,7,8,9
Centrarchidae					
black crappie	*Pomoxis nigromaculatus*	A	T	warm	12,14
bluegill	*Lepomis macrochirus*	A	T	warm	15
largemouth bass	*Micropterus salmoides*	A	T	warm	12
pumpkinseed	*L. gibbosus*	A	T	cool	15
smallmouth bass	*M. dolomieu*	A	I	cool	9,12,13,14,15
warmouth[c]	*L. gulosus*	A	T	warm	
white crappie	*Pomoxis annularis*	A	T	warm	12
Cichlidae					
tilapia	*Tilapia* sp.	A	T	warm	1
Cottidae					
mottled sculpin	*Cottus bairdii*	N	I	cold	1,2,3,4,5,6,7,8,9,10,12,15
Paiute sculpin	*C. beldingii*	N	I	cold	1,2,4,5,6,7
Shoshone sculpin[d] (SC)	*C. greenei*	N	S	cold	
shorthead sculpin[d]	*C. confusus*	N	S	cold	
torrent sculpin[d]	*C. rhotheus*	N	I	cold	
Cyprinidae					
common carp	*Cyprinus carpio*	A	T	warm	7,8,9,10,11,12,13,14,15
chiselmouth[d]	*Acrocheilus alutaceus*	N	I	cool	10,12,13,14,15
fathead minnow	*Pimephales promelas*	A	T	warm	6,7,15
leopard dace[d]	*Rhinichthys falcatus*	N	I	cool	12
longnose dace	*R. cataractae*	N	I	cool	1,2,3,4,5,7,8

Table 2.—continued

Family Common name	Species	Origin[a]	Tolerance to pollution[b]	Sites of occurrence Temperature preference	USGS 1993–2002 (Fgure 1 and Table 1)
northern pikeminnow[d]	*Ptychocheilus oregonensis*	N	T	cool	10,11,12,13,14,15
peamouth[d]	*Mylocheilus caurinus*	N	I	cool	10,12,13,14,15
redside shiner	*Richardsonius balteatus*	N	I	cool	1,4,5,6,7,8,9,10,11,12,15
speckled dace	*Rhinichthys osculus*	N	I	cool	1,2,3,4,5,6,7,8,9,10,12
tui chub	*Gila bicolor*	A	T	cool	15
Utah chub	*G. atraria*	N	T	cool	2,6,9,10,11,12
Ictaluridae					
black bullhead	*Ameiurus melas*	A	T	warm	
brown bullhead	*A. nebulosus*	A	T	warm	12
channel catfish	*Ictalurus punctatus*	A	T	warm	13,14,15
flathead catfish	*Pylodictis olivaris*	A	T	warm	15
yellow bullhead[c]	*A. natalis*	A	T	warm	
Percidae					
yellow perch	*Perca flavescens*	A	I	cool	8,9,12,14
walleye	*Sander vitreus*	A	I	cool	
Petromyzontidae					
Pacific lamprey[d] (E)	*Lampetra tridentata*	N	I	cool	
Salmonidae					
brook trout	*Salvelinus fontinalis*	A	I	cold	1
brown trout	*Salmo trutta*	A	I	cold	1,3,4,5,6,7,8
bull trout[de] (SC)	*Salvelinus confluentus*	N	S	cold	
Chinook salmon[de] (E)	*Oncorhynchus tshawytscha*	N	S	cold	
cutthroat trout[f] (SC)	*O. clarkii*	N	S	cold	1,2,3,6,7,
mountain whitefish	*Prosopium williamsoni*	N	I	cold	1,2,3,4,5,6,7,8,9,10,11,12,13,15
rainbow trout[d]	*O. mykiss*	N	S	cold	1,3,4,5,6,7,8,9,10,11,12

[a] N = native, A = alien.

[b] I = intermediate species, S = sensitive species, T = tolerant species.

[c] Based on collections by Idaho Power Company (J. Chandler, Idaho Power Company, personal communication, 2003).

[d] Native in the Snake River downstream from Shoshone Falls; rainbow trout includes steelhead and resident redband trout.

[e] Federally listed as threatened.

[f] Cutthroat trout × rainbow trout hybrids collected at sites 3, 4, and 5; considered an alien species upstream from Shoshone Falls.

in ascending lower Salmon Falls, which is about 78 km downstream from Shoshone Falls: "A little blasting at these falls would make it very much easier for the salmon to ascend. The expense would not exceed US$100 to $300 and I believe it would result in a considerable increase in the salmon supply of the Snake River."

Anadromous species included Chinook salmon, steelhead, Pacific lamprey, and white sturgeon. White sturgeon exceeding 250 kg were harvested from the Snake River for market at nearby mining towns (Figure 3).

Methods

Current Fish Collections

Fish sampling was conducted at 12 main-stem Snake River and 3 Henrys Fork sites between 1993 and 2002 (Table 1), following either the USGS-NAWQA (Meador et al. 1993) or USEPA-EMAP (Peck et al. 2002) protocols. Fish collections using

Figure 3.—A 286-kg white sturgeon harvested from the Snake River downstream from Shoshone Falls in 1908 (from the Idaho Historical Society).

USEPA-EMAP protocols were conducted only in 2002. All collections and identifications were made by USGS personnel. The number of collections at each site ranged from 1 to 10 for the study period. Boat electrofishing was done near shore for a distance of about 500–1,000 m, which included repeating geomorphic channel units (i.e., riffles, pools, and runs; Meador et al. 1993), or 40–100 times the channel width (Peck et al. 2002). The boat was equipped with a Smith-Root model VI-A direct current pulsator and a 5,000-W, 240-V generator with bow-mounted electrodes. Both collection methods attempted to capture all fish species from available habitats at each sample site. The USGS-NAWQA collections also were supplemented with backpack electrofishing (Smith-Root model 12) in shallow riffle habitats to collect small benthic fish that are often missed or underrepresented with boat electrofishing. All collections were made during summer–fall (July–October) base flow conditions. Duplicate fish collections were conducted on the Snake River at Marsing, Idaho (site 14), to evaluate intra-annual variability in 2002. Annual fish collections (n = 10) were conducted on the Snake River at King Hill (site 12) for use in evaluating annual variability and long-term trends. All individuals collected were identified to species, counted, measured for total length, examined for external anomalies (deformities, eroded fins, lesions, and tumors), and returned to the stream. When necessary, fish were anesthetized with a dilute solution of clove oil and ethanol. Specimens of selected species were retained for reference and verification of field identifications. Age-class determinations for salmonids and cottids were based on length–frequency distributions and descriptions by Wydoski and Whitney (1979) and Scott and Crossman (1973). A voucher collection of these samples is located in the Orma J. Smith Museum of Natural History, Albertson College of Idaho, Caldwell. Rainbow trout were considered alien upstream from Shoshone Falls, which historically limited their upstream distribution in the Columbia and Snake River system (Behnke 1992). Where distinguishable, hatchery fish were excluded from IBI calculations (Mebane et al. 2003). Wild rainbow trout and redband trout were both considered rainbow trout in this study because of difficulty in field identifications.

River Basin Characteristics

Basin characteristics (drainage area, stream order, stream kilometers, land use, PODs, and constructed channels) were determined by using a GIS. Percent land use, PODs, and constructed channels within each basin were determined to represent surrogates of human stressors. Several sources were used to construct the geographic data layers. Basin boundaries were constructed using digital raster graphics and hydrologic unit boundary data layers (USGS 1994). River km of natural and constructed stream channels were determined from the USGS National Hydrography Dataset (http://nhd.usgs.gov/data.html). Percentage of land use (agricultural, forested, range, and urban) was determined for each basin. Land use was modified from 1:250,000-scale digital data (USGS 1986) consisting of a classification scheme at a 16-ha mapping resolution (Anderson et al. 1976). Points of diversions for Idaho and Wyoming were obtained from the Idaho Department of Water Resources (IDWR), Boise (Wyoming State Engineers Office 1999; M. Ciscell, IDWR, personal communication). Oregon and Nevada PODs were not summarized because they either were not upstream from the study sites (i.e., Owyhee River basin of Oregon) or were insignificant in their total contribution to the hydrology of the Snake River.

Data Analysis

Fish assemblages were analyzed using abundances of individuals and species and 10 fish metrics determined by Mebane et al. (2003) to be useful for evaluating large river conditions in the Pacific Northwest (Table 3). These metrics consisted of the number of coldwater native species, number of cottid age-classes, percent sensitive native individuals, percent coldwater individuals, percent tolerant individuals, number of alien species, percent common carp individuals, number of salmonid age-classes, catch per unit effort (fish captured per minute of electrofishing), and percent selected anomalies (deformities, eroded fins, lesions, and tumors). Metrics are standardized by scoring them continuously from 0 to 1, then weighting them as necessary to produce an IBI score ranging from 0 to 100. According to Mebane et al. (2003), streams with IBI scores from 75 to 100 exhibit high biotic integrity with minimal disturbance and possess an abundant and diverse assemblage of native coldwater species; streams with scores from 50 to 74 have somewhat lower quality, alien species are more frequent, and the assemblage is dominated by coolwater, native species; streams with scores of less than 50 have poor biotic integrity, coldwater and sensitive species are rare or absent, and tolerant fish predominate. For a more detailed description of index development, metric response, and application, see Mebane et al. (2003). Geographic origin (native or alien), tolerance to pollution, and temperature preferences were assigned to each species (Table 2) according to Zaroban et al. (1999). Zaroban et al. (1999) assigned general tolerances and temperature preferences on the basis of information gleaned from state fish books and from their experience of how species distributions and abundances changed with increased temperatures, turbidity, sedimentation, and nutrients.

Fish assemblage data from the Salmon River, a minimally disturbed and undammed large river in central Idaho ($n = 8$), were compared with current fish collections from this study. These Salmon River data were collected by the USGS between 2000 and 2002 (Mebane et al. 2003). Only qualitative comparisons between historical and current Snake River fish populations were made because only anecdotal historical information exists from the journals of early explorers and fur trappers (Pratt et al. 2001).

Major faunal shifts in many streams in the western United States are the result of alien fish species. Often, alien fish species are better adapted than native species to thrive in altered habitats (Moyle 1994). The status of fish assemblages is related partially to the extent of habitat disturbance and the occurrence of native versus alien species. The zoogeographic integrity coefficient (ZIC), an index derived from the ratio of the number of native species to the total number of species, was used to evaluate the degree of habitat disturbance, whereby a value of 1 indicates an undisturbed environment and a value of 0 indicates a highly disturbed environment (Elvira 1995). Relations between the IBI and selected basin characteristics were examined using Spearman's rank correlations. A mean IBI value was calculated for sites with multiple collections prior to comparisons with environmental variables. LOWESS

Table 3.—Summary of the 10 fish metrics and final Index of Biotic Integrity (IBI) scores, number of fish collected, total number of species, number of native species, and the Zoogeographic Integrity Coefficient (ZIC) for the Snake River and Henrys Fork in Idaho and western Wyoming, 1993–2002. Values represent a mean for sites where more than one collection was made.

Site number[a]	1	2	3	4	5	6	7	8	9	10	11	12	13	14[b]	15
Number of collections[c]	4	1	2	1	1	3	1	3	2	5	1	10	2	2	2
Metric and final IBI score															
Number of coldwater native species	3	4	4	3	3	3	4	2	1	1	0	2	1	0	1
Number of cottid age classes	4	3	5	4	4	1	5	3	1	2	0	3	0	0	0
Percent sensitive native individuals	7.8	21.4	12.8	0.0	0.0	3.4	0.6	0.0	0.0	0.0	0.0	0.0	0.0	0.0	0.0
Percent coldwater individuals	65.8	58.7	92.1	89.8	77.6	19.5	20.9	31.9	3.0	5.0	1.2	7.9	24.5	0.0	4.4
Percent tolerant individuals	5.6	18.3	1.7	0.0	1.5	59.8	26.0	25.3	54.0	61.4	93.6	61.5	58.5	55.1	57.0
Number of alien species	2	0	3	3	3	2	5	3	3	1	1	3	3	4	5
Percent common carp individuals	0.0	0.0	0.0	0.0	0.0	0.0	0.3	5.0	3.0	12.0	1.5	8.2	4.8	24.5	5.7
Number of salmonid age classes	4	3	5	4	6	2	3	2	1	1	3	0	0	0	0
Catch per unit effort[d]	3.23	1.85	6.01	7.08	10.24	0.73	1.30	1.56	<0.01	0.19	0.07	0.45	1.04	<0.01	0.20
Percent selected anomalies[e]	0.48	0.79	0.00	0.00	0.50	1.15	0.00	2.05	5.00	2.90	0.00	1.93	0.68	0.56	3.15
Final IBI Score	79	84	84	80	78	49	66	46	15	18	29	22	20	10	11
Other metrics															
Number of fish collected	158	126	600	128	201	104	358	148	145	141	581	253	147	616	139
Number of native species	8	9	7	6	7	7	8	5	5	7	6	9	5	4	6
Total number of species	10	9	10	9	10	9	13	8	8	8	7	12	8	8	11
Zoogeographic Integrity Coefficient[f]	0.80	1.00	0.70	0.67	0.70	0.78	0.62	0.67	0.63	0.88	0.86	0.75	0.63	0.50	0.55

[a] See Figure 1.
[b] Represents duplicate samples within the same year.
[c] See Table 1 for collection years.
[d] Number of fish collected per minute electrofishing.
[e] Includes deformities, eroded fins, lesions, and tumors.
[f] Ratio of native to total species collected.

smoothing, a robust, nonparametric description of data patterns (Helsel and Hirsch 1992), was used to evaluate trends between mean IBI site scores and various measures of human disturbances. All statistical and graphical analyses were performed using SYSTAT (Wilkinson 1999).

Results

At least 46 species of fish have been reported in the Snake River and Henrys Fork, either historically or currently (Table 2). On the basis of our collections, the fish assemblage in the Snake River and Henrys Fork currently consists of 35 species. Sixteen of these are alien species, which is about half (46%) of all species collected. Many of these alien species such as catfish, common carp, and sunfish are adapted for warmwater impounded habitats. Currently, the Snake River and Henrys Fork support 19 native species, compared with 26 species that were present in these rivers historically. The Snake River sucker, anadromous Pacific lamprey, Chinook salmon, and steelhead have been extirpated from the study area. Species of special concern that would be expected to occur in the Snake River include white sturgeon, bluehead sucker *Catostomus discobolus*, Shoshone sculpin, bull trout, cutthroat trout, and redband trout (Maret 1995). Bluehead suckers were collected at sites 1 and 7, and cutthroat trout were collected at sites 1, 2, 3, 6, and 7 (Table 2). Cutthroat × rainbow trout hybrids were collected at sites 3, 4, and 5 on the Henrys Fork and Snake River.

A summary of mean IBI site scores and component metrics (Table 3) reveals a decline in biotic integrity from upstream to downstream on both the Snake River (sites 1, 2, 3, 7–15) and Henrys Fork (sites 4–6). Snake River sites downstream from Blackfoot (site 8) all had mean IBI scores less than 50. The Snake River near Marsing (site 14) exhibited the lowest mean IBI score of 10. No coldwater or sensitive native species were present at this site, and the assemblage was composed of about 73% tolerant individuals. Native benthic-dwelling species, such as dace *Rhinichthys* sp. and sculpins *Cottus* sp. that would typically be found in riffle habitats were absent at this site. Common carp, a tolerant alien associated with turbid, warmwater habitats, was found at all Snake River sites downstream from Shelley, Idaho (site 7), to Nyssa, Oregon (site 15). Smallmouth bass *Micropterus dolomieu* were also present at most of these downstream locations (sites 9, 12, 13, 14, and 15). At Marsing (site 14), common carp and smallmouth bass made up about 20% and 47% of all fish collected, respectively (data not shown). The Henrys Fork at Rexburg (site 6) exhibited an impaired fish assemblage with a mean IBI score of 49, compared with upstream site IBI scores of 78–80 (sites 4–5).

The mean number of fish collected at all sample sites ranged from 104 to 616 (Table 3). The total number of species collected ranged from 7 to 13, and the number of native species ranged from 4 to 9 for all sites. The number of native species collected at sites 1–9 upstream from Shoshone Falls was similar to the number collected at sites 10–15 downstream from Shoshone Falls, with means of 6.9 and 6.2, respectively.

Mean ZIC scores did not show as great a percentage decline as did IBI scores from upstream to downstream. The ZIC scores ranged from 0.50 to 1.00 for all sample sites (Table 3). The most downstream Snake River sites at Marsing (site 14) and Nyssa (site 15) were composed of about 50% alien species with mean ZIC scores of 0.50 and 0.55, respectively.

Longitudinal changes in IBI scores for recent (1993–2002) fish collections ($n = 40$) on the Snake River and Henrys Fork show a steady decline from upstream to downstream (Figure 4). The most precipitous decline in IBI scores on the Snake River begins near Blackfoot just upstream from American Falls Reservoir (site 8, near river km 1,200, Table 3). The IBI scores were quite variable at this site, ranging from 32 to 60. On the basis of mean IBI scores, two distinct groups of Snake River sites are apparent—the upper river sites 1, 2, 3, 7, and 8 upstream from American Falls Reservoir, with relatively high biotic integrity (mean scores of 46–84, Table 3) and the lower river sites 9–15 downstream from American Falls Reservoir, with low biotic integrity (mean scores of 10–29, Table 3). The lowest single IBI score (3) for any fish collection was recorded at Snake River near Minidoka (site 9, river km 1,084). A slight improvement in Snake River

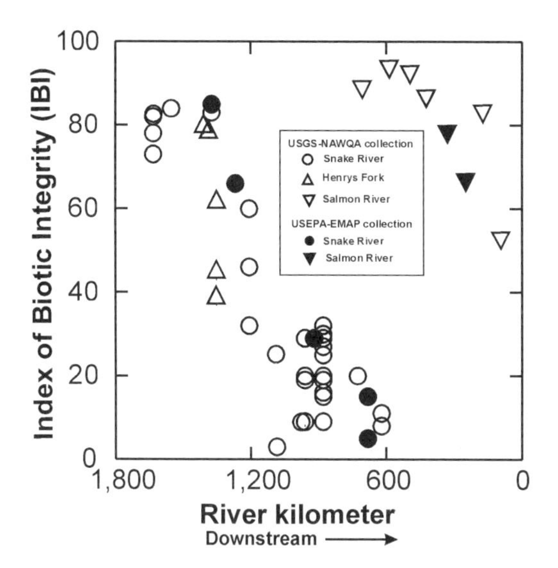

Figure 4.—Longitudinal changes in the index of biotic integrity (IBI) scores and corresponding river kilometers for Snake River and Henrys Fork sites (*n* = 40, lower left), and Salmon River sites (*n* = 8, upper right), Idaho and western Wyoming, 1993–2002. See Table 1 for site names, corresponding river kilometer, and year(s) of collection for Snake River and Henrys Fork. Salmon River data collected by the U.S. Geological Survey in 2000–2002 were taken from Mebane et al. (2003).

IBI scores is apparent near river km 900 (Figure 4), corresponding to collections at Hagerman (site 11) and King Hill (site 12), then scores again decline at the downstream sites at Murphy, Marsing, and Nyssa (river km 727–621). Index of biotic integrity scores for duplicate fish collections at Marsing (site 14) at the beginning and end of the summer were 5 and 15, respectively, and were in close agreement. Annual collections at King Hill (site 12) over the last 10 years indicate little annual or long-term change in the fish assemblage or biotic integrity with IBI scores ranging from 9 to 32 (Figure 4).

Index of Biotic Integrity scores along a longitudinal gradient for the Salmon River reference sites in central Idaho (Figure 4) show some decline from upstream to downstream over about 600 river km; scores ranged from 95 to 51 ($n = 8$). Most of these Salmon River scores ranged from 73 to 85 and are similar to those for upper Snake River sites 1–3 and upper Henrys Fork sites 4–5. The lowest IBI score of 51 for the Salmon River was recorded for the most downstream site (river km 87), Salmon River near White Bird, Idaho. However, this score is still higher than those for all lower Snake River sites 9–15.

Scatterplots (including LOWESS smoothing) of mean IBI scores and three basin-scale measures of human disturbance clearly show a relation between biotic integrity and human disturbance (Figure 5). The decline in IBI scores was more strongly related to the total number of water diversions and km of constructed channels ($r = -0.90$, $p < 0.01$) than to percentage of agricultural land upstream from each site ($r = -0.57$, $p = 0.03$). Henrys Fork near Rexburg, Idaho (site 6), had the highest percentage of agricultural land (about 26%) in its basin, but its mean IBI score was higher than those of many of the Snake River sites with a lower percentage of agricultural land. Mean IBI scores for Snake River sites decline sharply between sites 8 and 9; this decline is evident all the way downstream to site 15 at Nyssa, Oregon (Figure 5). Two distinct groups of sites, as discussed previously, are evident in Figure 5 that correspond to an upper Snake River and Henrys Fork group (sites 1–8) and a lower Snake River group (sites 9–15).

Discussion

Despite the lack of published studies on fish assemblages of the Snake River, we were able to qualitatively describe the historical fish assemblages and assess the current status of the fish assemblages using various measures of biotic integrity. However, these comparisons were hampered by the natural river continuum concept of longitudinal change with river size and gradient (Vannote et al. 1980; Li et al. 1987). In general, fish assemblages change from source to mouth as a result of elevation change and increasing stream size, resulting in a gradient from predominantly coldwater-adapted species upstream to a more diverse, mesothermic assemblage downstream (Li et al. 1987). This natural continuum can mask the impacts of human activities on fish assemblages, impacts which also increase with lower elevations in the arid West. On the basis of the historical and current evidence presented in this study, it is apparent that longitudinal changes in fish as-

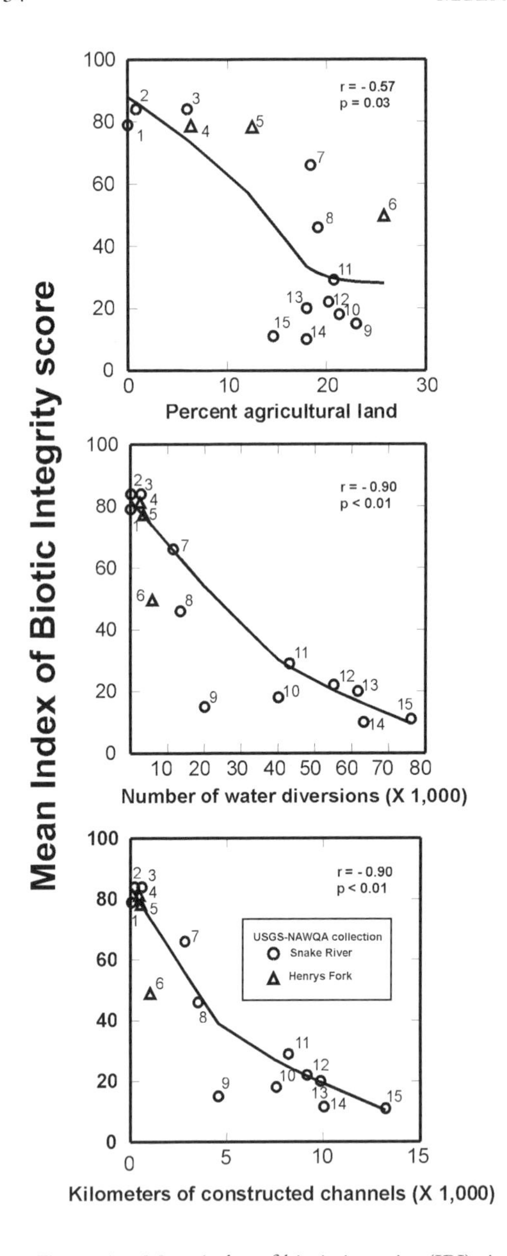

Figure 5.—Mean index of biotic integrity (IBI) site scores in relation to measures of human disturbances that include percent agricultural land, total number of water diversions, and kilometers of channels constructed upstream from each Snake River (n = 12) and Henrys Fork (n = 3) sample site. Trend lines based on LOWESS smoothing techniques (Helsel and Hirsch 1992).

semblages along the Snake River are predominantly the result of human activities and not a continuum of natural river changes.

Two distinct groups of fish assemblages based on IBI scores were identified in the study area that correspond to high (upper river, sites 1–8) and low (lower river, sites 9–15) biotic integrity. The IBI scores at Blackfoot, Idaho (site 8), were highly variable, possibly indicating a transition in the fish assemblage at this location on the Snake River. The upper river sites, with mean IBI scores of 46–84, representing about 400 river km (Flagg Ranch, Wyoming, to Blackfoot, Idaho) generally included two to four coldwater native species. The lower river sites, with mean IBI scores of 10–29 and representing about 460 river km (Minidoka, Idaho, to Nyssa, Oregon) were dominated by tolerant, mesothermic, alien species, and native suckers *Catostomus* sp. Some sites no longer had any coldwater native species. The most common alien species collected at the lower Snake River sites included common carp, smallmouth bass, and channel catfish. Hughes and Gammon (1987) also found that smallmouth bass and common carp were associated with altered habitats in the lower Willamette River, Oregon. The extent to which dams and reservoirs have altered native aquatic assemblages is not widely understood or appreciated by the public because the resulting impoundments often provide popular sportfishing opportunities. Impoundments also serve as recipients for alien introductions and spawning areas for undesirable warmwater species such as the common carp. Reservoirs also can provide habitat for predaceous fish (e.g., smallmouth bass), which greatly increases risk to the native fish fauna (Li et al. 1987).

The mean number of native fish species in the lower Snake River was comparable to the number in the upper Snake River. This comparability would not be a predictable outcome when considering the generalized conceptual model of rivers in the Pacific Northwest unaffected by human activities (Li et al. 1987). This is especially true when considering that, historically, at least 14 native species were present downstream from Shoshone Falls that were not present upstream from the falls. Thus, human activities have decreased the numbers of native species in the lower Snake River to equivalent numbers in the upper Snake River.

The bluehead sucker and cutthroat trout were the only species of special concern collected in this study. The leopard dace, a native cyprinid, was collected at only one location on the Snake River near

King Hill. This species also was recently collected by Maret and Ott (2003) in the lower Salmon River in Idaho. Collection gear limitations were likely the reason no adult white sturgeon were collected during this study, and depressed populations may have been the reason no juvenile white sturgeon were collected. According to recent studies by the Idaho Power Company, this ancient coldwater species is on the decline in the Snake River upstream from the HCC; the only reproduction is occurring between Bliss Dam and C.J. Strike Reservoir on the Snake River (Jager et al. 2000). Following extensive sampling using setlines, angling, and gill nets, Lepla et al. (2001) reported that catch rates and overall numbers of white sturgeon in the Snake River from Swan Falls to the HCC are very low. They attributed this decline in white sturgeon to severe water quality degradation. It has been estimated that about 37% of white sturgeon habitat on the Snake River has been lost as a result of impoundments (Cochnauer 2002).

Our collections of hybrid cutthroat × rainbow trout at some upper sites on the Henrys Fork and Snake River provide evidence that the introduction of alien species upstream from Shoshone Falls is affecting the native cutthroat trout fishery. According to Behnke (1992), rainbow trout introductions pose a serious threat to native cutthroat trout populations.

A comparison of the Salmon River fish assemblage in central Idaho with those in the Snake River and Henrys Fork provided further evidence that human activity has profoundly altered the fish assemblage in the Snake River. This minimally disturbed large river with no impoundments flows for about 700 km between elevations of 300 and 2,500 m. Most IBI scores remained high (>75) along much of this river. In contrast, IBI scores along the Snake River showed a marked decline longitudinally, and a number of sites in the lower river indicated poor-quality fisheries, where coldwater and sensitive species were rare or absent and where tolerant, less desirable species predominated. The lack of native cyprinids and cottids at some of the lower Snake River sites may be associated with the increased number of alien piscivorous smallmouth bass and channel catfish. Alien species introductions are second only to habitat loss as a factor contributing to fish extinctions (Miller et al. 1989).

Historical accounts by Evermann (1896) of spawning Chinook salmon throughout the lower portion of the Snake River support the conceptual model of a native coldwater fish assemblage. Coldwater species typically prefer maximum daily water temperatures less than 22°C (Lyons et al. 1996; Mebane et al. 2003). However, our current fish assemblage collections no longer support this biological template for the lower Snake River. Salmonids are absent or rarely collected in the lower portions of the Snake River. Supporting temperature data provided by Maret et al. (2001) indicate that water temperatures in the lower Snake River (1996–1998) can be excessive for coldwater species. Maximum water temperatures of 23°C or more were recorded at Minidoka, Murphy, and Nyssa. Maximum water temperatures approaching 27°C for this same period were recorded at Murphy and Nyssa. Recent fish kills (summer 2002) composed primarily of mountain whitefish have occurred between these locations on the Snake River as a result of excessively high water temperatures (J. Dillon, Idaho Department of Fish and Game, personal communication). The numerous impoundments on the Snake River may have a cumulative effect of increasing water temperatures in the lower river. Impoundments can increase the surface area exposed to solar radiation, which tends to increase summer water temperatures to levels that may not be favorable to coldwater species normally living in low-elevation streams of the arid West (USEPA 2002). The slight improvement in IBI scores on the Snake River at Hagerman and King Hill may be the result of major spring discharges in the reach, which provide about 150 m^3/s of 14°C groundwater, thus improving the coldwater habitat to support salmonids and cottids (Clark et al. 1998).

Overfishing in the Columbia basin played a role in the decline of Chinook salmon, steelhead, and white sturgeon. Chapman (1986) estimated harvest rates prior to 1958 were about 68% for spring and summer Chinook and 69% for steelhead. These harvest rates would have been prior to construction of most hydropower dams on the Columbia or Snake

Rivers. The first white sturgeon harvest regulations in Idaho were adopted in 1943 when commercial fishing was prohibited (Cochnauer 2002). Catch-and-release regulations for white sturgeon were established on the entire Snake River in 1972.

Seasonal differences cannot be overlooked as a potential contributor to the absence of coldwater species in the lower Snake and Salmon Rivers. For example, the migratory behavior of salmonids is likely an important adaptation to survive periods of warmer conditions. Because fish were collected during the late summer, some coldwater species may have moved into cooler refugia (e.g., tributaries), resulting in lower IBI scores. However, tributary habitats associated with the lower Snake River have been severely fragmented and fluvial stocks may have been lost because of passage barriers and degraded water quality.

Increases in percent agricultural land, total number of diversions, and number of constructed channels were strongly associated with declining IBI scores and lower biotic integrity. Agricultural land use and associated water use in the West are concentrated along rivers and their major tributaries. The majority of the natural flow of the Snake River is withdrawn for irrigation and returned through agricultural drains as wastewater. Hydropower dams and water storage reservoirs also can contribute to the cumulative effects of these water uses on biotic integrity. The absence of a natural flow regime as a result of dams also can have profound effects on fish populations (Poff and Allan 1995; Marchetti and Moyle 2001). The cumulative effects of these activities on the Snake River have only recently been assessed (USEPA 2002). The increasing population in the Snake River basin, with increased demands for energy, irrigation, aquaculture, and dairy feedlots, can place additional burdens on an aquatic resource already substantially changed by human activity.

In conclusion, continued annual monitoring of Snake and Salmon River fish assemblages and associated water quality parameters would improve our understanding of potential limiting factors to fish populations and describe long-term trends in river conditions. Measuring continuous water temperature should be an essential part of this routine monitoring. It was encouraging to see that IBI scores for fish collected by the USGS-NAWQA and USEPA-EMAP methods were similar at sites where both collection methods were used. Additional comparative studies of these large-scale monitoring programs would provide useful information to agencies charged with assessing biotic integrity and water-resource conditions. Much of the Snake River in southern Idaho currently supports a very different fish assemblage than that present historically. The lower Snake River supports fewer native coldwater species, and more than half the assemblage is composed of alien species. The number of alien species and individuals will likely continue to increase in the Snake River as a result of continued alteration of habitats and highly regulated streamflows. We know of no other published studies where entire fish assemblages have been collected and assessed for the Snake River across southern Idaho and western Wyoming. Thus, the data collected in this study should serve as a useful starting point to better understand factors affecting fish populations in the basin and to quantify trends in fish populations of the Snake River and Henrys Fork.

Acknowledgments

Numerous individuals from the U.S. Geological Survey assisted in collecting and processing data during the course of this study: Gregory M. Clark, Ross G. Dickinson, Douglas S. Ott, Dorene E. MacCoy, Terry M. Short, and Kenneth D. Skinner. Colleague reviews by Larry R. Brown, James A. Chandler, Gregory M. Clark, Cyndi S. Grafe, and David A. Peterson improved the quality of the manuscript. Special thanks go to Donald W. Zaroban for assisting with the identification and vouchering of fish specimens. Charles D. Warren and Fred E. Partridge of the Idaho Department of Fish and Game assisted with fish collection on the middle Snake River. Funding was provided by the U.S. Geological Survey, National Water-Quality Assessment Program, and the Idaho Department of Environmental Quality. Any use of trade, product, or firm names in this publication is for descriptive purposes only and does not imply endorsement by the U.S. Government.

References

Anderson, J. R., E. E. Hardy, J. T. Roach, and R. E. Witmer. 1976. A land use and land cover classification system for use with remote sensor data. U. S. Geological Survey Professional Paper 964, Denver.

Behnke, R. J. 1992. Native trout of western North America. American Fisheries Society, Monograph 6, Bethesda, Maryland.

Brennan, T. S., A. M. Campbell, A. K. Lehmann, and I. O'Dell. 2001. Water resources data, Idaho, water year 2000, v. 2, upper Columbia River basin and Snake River basin below King Hill. U.S. Geological Survey Water Data Report ID-00–2, Boise, Idaho.

Brink, S. R., B. Reingold, J. A. Chandler, and C. Randolph. 1997. Fish community and habitat inventory in C. J. Strike Reservoir, Snake River. Idaho Power Company, Technical appendix E. 3.1–A for C. J. Strike Project FERC No. 2055, Boise.

Chandler, J. A. and M. Radko. 2001. Introduction and Overview. Technical appendix E. 3.1–2 *in* J. A. Chandler, editor, Chapter 1. Idaho Power Company, Feasibility of reintroduction of anadromous fish above or within the Hells Canyon Complex, for Hells Canyon Complex Hydroelectric Project FERC No. 1971, Boise.

Chapman, D. 2001. Habitat of the Snake River Plain. Technical appendix E. 3.1–2 *in* J. A. Chandler, editor, Chapter 3. Feasibility of reintroduction of anadromous fish above or within the Hells Canyon Complex, Idaho Power Company for Hells Canyon Complex Hydroelectric Project FERC No. 1971, Boise.

Chapman, D. W. 1986. Salmon and steelhead abundance in the Columbia River in the nineteenth century. Transactions of the American Fisheries Society 115:662–670.

Clark, G. M., T. R. Maret, M. G. Rupert, M. A. Maupin, W. H. Low, and D. S. Ott. 1998. Water quality in the upper Snake River basin, Idaho and Wyoming, 1992–95. U. S. Geological Survey Circular 1160, Boise, Idaho.

Cochnauer, T. G. 2002. Response of white sturgeon population to catch and release regulations—now what? Pages 83–88 *in* W. Van Winkle, editor. Biology, management and protection of North American sturgeon. American Fisheries Society, Symposium 28, Bethesda, Maryland.

Dynesius, M., and C. Nilsson. 1994. Fragmentation and flow regulation of river systems in the northern third of the world. Science 266(4):753–762.

Elvira, B. 1995. Native and exotic freshwater fishes in Spanish river basins. Freshwater Biology 33:103–108.

Evermann, B. W. 1896. A preliminary report upon salmon investigations in Idaho in 1894. Pages 253–275 *in* Bulletin of the United States Fish Commission for 1895. U.S. Government Printing Office, Washington D.C.

Fausch, K. D., C. E. Torgersen, C. V. Baxter, and H. W. Li. 2002. Landscape to riverscape—bridging the gap between research and conservation of stream fishes. Bioscience 52(6):483–498.

Fiege, M. 1999. Irrigated Eden. University of Washington Press, Seattle and London.

Gilbert, C. H., and B. W. Evermann. 1895. A report upon investigations in the Columbia River basin, with descriptions for four species of fishes. Pages 169–179 *in* Bulletin of the United States Fish Commission for 1894. U.S. Government Printing Office, Washington, D.C.

Grafe, C. S., editor. 2002. Idaho rivers ecological assessment framework—an integrated approach. Idaho Department of Environmental Quality, Boise.

Helsel, D. R., and R. M. Hirsch. 1992. Statistical methods in water resources. Elsevier Publishing Company, Inc., New York.

Hubbs, C. L., and R. R. Miller. 1948. The zoological evidence, correlation between fish distribution and hydrographic history in the desert basins of Western United States, Chapter II. The Great Basin, with emphasis on glacial and postglacial times. Salt Lake City, Bulletin of the University of Utah 38 (20), Biological Series 10(7):17–166.

Hughes, R. M., and J. R. Gammon. 1987. Longitudinal changes in fish assemblages and water quality in the Willamette River, Oregon. Transactions of the American Fisheries Society 116(2):196–209.

Jager, H. I., K. Lepla, J. A. Chandler, P. Bates, and W. V. Winkle. 2000. Population viability analysis of white sturgeon and other riverine fishes. Environmental Science and Policy 3:5483–5489.

Karr, J. R. 1991. Biological integrity—a long-neglected aspect of water resource management. Ecological Applications 1:66–84.

Lepla, K., J. A. Chandler, and P. Bates. 2001. Status of Snake River sturgeon associated with the Hells

Canyon Complex. Idaho Power Company, Technical appendices E. 3.1–6 for Hells Canyon Complex Hydroelectric Project, FERC No. 1971, Boise.

Li, H. W., C. B. Schreck, C. E. Bond, and E. Rexstad. 1987. Factors affecting changes in fish assemblages of Pacific Northwest streams. Pages 193–202 *in* W. J. Matthews and D. C. Heins, editors. Community and evolutionary ecology of North American stream fishes. University of Oklahoma Press, Norman.

Lukas, J. A., J. A. Chandler, and C. Randolph. 1995. An evaluation of the fish community and habitat availability for rainbow trout in the Thousand Springs complex. Idaho Power Company, Technical report, appendix E. 3.1-J to the upper Salmon Falls FERC relicense application, Boise.

Lyons, J., L. Wang, and T. D. Simonson. 1996. Development and validation of an index of biotic integrity for coldwater streams in Wisconsin. North American Journal of Fisheries Management 16:241–256.

Malde, H. 1968. The catastrophic late Pleistocene Bonneville flood in the Snake River plain, Idaho. U.S. Geological Survey, Professional Paper 595, Denver.

Marchetti, M. P., and P. B. Moyle. 2001. Effects of flow regime on fish assemblages in regulated California streams. Ecological Applications 11(2):530–539.

Maret, T. R. 1995. Water-quality assessment of the upper Snake River basin, Idaho and western Wyoming—summary of aquatic biological data for surface water through 1992. U.S. Geological Survey Water-Resources Investigations Report 95–4006, Boise, Idaho.

Maret, T. R. 1997. Characteristics of fish assemblages and related environmental variables for streams of the upper Snake River basin, Idaho and western Wyoming. U.S. Geological Survey Water-Resources Investigations Report 97–4087, Boise, Idaho.

Maret, T. R., D. E MacCoy, K. D. Skinner, S. E. Moore, and I. O'Dell. 2001. Evaluation of macroinvertebrate assemblages in Idaho rivers using multimetric and multivariate techniques, 1996–98. U.S. Geological Survey Water-Resources Investigations Report 01–4145, Boise, Idaho.

Maret, T. R. and D. S. Ott. 2003. Assessment of fish assemblages and minimum sampling effort required to determine biotic integrity of large rivers in southern Idaho, 2002. U.S. Geological Survey Water-Resources Investigations Report 03–4274, Boise, Idaho.

Maret, T. R., C. T. Robinson, and G. W. Minshall. 1997. Fish assemblages and environmental correlates in least disturbed streams of the upper Snake River basin. Transactions of the American Fisheries Society 126:200–216.

Meador, M. R. 1992. Inter-basin water transfer—ecological concerns. Fisheries 17(2):17–22.

Meador, M. R., T. E. Cuffney, and M. E. Gurtz. 1993. Methods for sampling fish communities as part of the National Water-Quality Assessment Program. U.S. Geological Survey Open-File Report 93–104, Denver.

Mebane, C. A., T. R. Maret, and R. M. Hughes. 2003. An index of biological integrity (IBI) for Pacific Northwest rivers. Transactions of the American Fisheries Society 132:239–261.

Miller, R. R., J. D. Williams, and J. E. Williams. 1989. Extinction of North America fishes during the past century. Fisheries 14(6):22–38.

Moyle, P. B. 1994. Biodiversity, biomonitoring, and the structure of stream fish communities. Pages 171–186 *in* S. L. Loeb and A. Spacie, editors. Biological monitoring of aquatic systems. Lewis, Boca Raton, Florida.

Nehlsen, W., J. E. Williams, and J. A. Lichatowich. 1991. Pacific salmon at the crossroads—stocks at risk from California, Oregon, Idaho, and Washington. Fisheries 16(2):4–21.

Palmer, T. 1991. The Snake River—window to the West. Island Press, Washington D.C., and Covello, California.

Peck, D. V., D. K. Averill, J. M. Lazorchak, and D. J. Klemm, editors. 2002. Environmental monitoring and assessment. Surface waters—field operations and methods for non-wadeable rivers and streams. U.S. Environmental Protection Agency, Corvallis, Oregon.

Poff, N. L., and J. D. Allan. 1995. Functional organization of stream fish assemblages in relation to hydrological variability. Ecology 76(2):606–627.

Pratt, K. L., M. Kozel, J. Mauser, L. Mauser, and R. Scarpella. 2001. *In* J. A. Chandler, editor. Annotated bibliographies on the chronology of decline of anadromous fish in the Snake River basin above Hells Canyon Dam. Feasibility of reintroduction of anadromous fish above or within the Hells Canyon Complex. Technical appendices E. 3.1–2 for Hells Canyon Complex Hydroelectric Project FERC No. 1971. Idaho Power Company, Boise, Idaho.

Reash, R. J. 1999. Considerations for characterizing Midwestern large-river habitats. Pages 463–473 *in* T. P. Simon, editor. Assessing the sustainability and biological integrity of water resources using fish communities. CRC Press, Boca Raton, Florida.

Scott, W. B., and E. J. Crossman. 1973. Freshwater fishes of Canada. Fisheries Research Board of Canada, Bulletin 184, Ottawa.

Simpson, J. C. and R. L. Wallace. 1982. Fishes of Idaho. University of Idaho Press, Moscow.

Solley, W. B., R. R. Pierce, and H. A. Perlman. 1998. Estimated use of water in the United States in 1995. U.S. Geological Survey Circular 1200, Denver.

Strahler, A. N. 1957. Quantitative analysis of watershed geomorphology. Transactions of the American Geophysical Union 38:913–920.

Thurow, R. F., C. E. Corsi, and V. K. Moore. 1988. Status, ecology, and management of Yellowstone cutthroat trout in the upper Snake River drainage, Idaho. Pages 25–36 *in* R. E. Chandler, editor. American Fisheries Society, Symposium 4, Bethesda, Maryland.

USEPA (U. S. Environmental Protection Agency). 2002. Ecological risk assessment for the middle Snake River, Idaho. National Center for Environmental Assessment, EPA/600/R–01/017, Washington D.C. Available at: http://www.epa.gov/ncea/. (May 2003)

USGS (U.S. Geological Survey). 1986. Land use and land cover digital data from 1:250,000- and 1:100,000-scale maps, data users guide 4. U.S. Geological Survey, Reston, Virginia.

USGS (U.S. Geological Survey). 1994. Hydrologic unit maps of the conterminous United States. Available at: http://water.usgs.gov/lookup/getspatial?huc250. (May 2003)

Vannote, R. L., G. W. Minshall, K. W. Cummins, J. R. Sedell, and C. E. Cushing. 1980. The river continuum concept. Canadian Journal of Fisheries and Aquatic Sciences 37:130–137.

Wilkinson, L. 1999. SYSTAT for Windows—statistics, version 9.0. SYSTAT, Inc., Evanston, Illinois.

Williams, J. E., J. E. Johnson, D. A. Hendrickson, S. Contreras-Balderas, J. D. Williams, M. Navarro-Mendoza, D. E. McAllister, and J. E. Deacon. 1989. Fishes of North America endangered, threatened, or of special concern. Fisheries 14(6):2–20.

Wydoski, R. S., and R. R. Whitney. 1979. Inland fishes of Washington. University of Washington Press, Seattle.

Wyoming State Engineers Office. 1999. Tabulation of adjudicated surface water rights of the State of Wyoming. State of Wyoming Water Division Number Four, Cheyenne.

Zaroban, D. W., M. P. Mulvey, T. R. Maret, R. M. Hughes, and G. D. Merritt. 1999. Classification of species attributes for Pacific Northwest freshwater fishes. Northwest Science 73(2):81–93.

American Fisheries Society Symposium 45:61–74, 2005

Changes in Fish Assemblage Structure in the Main-Stem Willamette River, Oregon

ROBERT M. HUGHES*, RANDALL C. WILDMAN AND STANLEY V. GREGORY

Department of Fisheries and Wildlife, 104 Nash Hall, Oregon State University, Corvallis, Oregon 97331, USA

Abstract.—The Willamette River is Oregon's largest river, with a basin area of 29,800 km^2 and a mean annual discharge of 680 m^3/s. Beginning in the 1890s, the channel was greatly simplified for navigation. By the 1940s, it was polluted by organic wastes, which resulted in low dissolved oxygen concentrations and floating and benthic sludge deposits that hindered salmon migration and boating. Following basin-wide secondary waste treatment and low-flow augmentation, water quality markedly improved, salmon runs returned, and recreational uses increased. However, water pollution remains a problem as do physical habitat alterations, flow modification, and alien species. Fish assemblages in the main-stem Willamette River were sampled systematically, but with different gear, in the summers of 1945, 1983, and 1999. In the past 53 years, tolerant species occurrences decreased and intolerant species occurrences increased. In the past 20 years, alien fishes have expanded their ranges in the river, and four native fish species have been listed as threatened or endangered. We associate these changes with improved water quality between 1945 and 1983, fish migrations, altered flow regimes and physical habitat structure, and more extensive sampling.

Introduction

Large rivers are some of the most altered aquatic ecosystems on earth. Typically, they have been highly modified by navigation, flood control, hydropower generation, recreation, commercial fishing, irrigated agriculture, and industrial and municipal waste disposal. Regier et al. (1989) described this as a predictable process, beginning with excessive harvest of fish and fur bearers, followed by channel alteration and agriculture, then pollution and urbanization. The effects of these modifications on fish assemblages include extirpation of sensitive native species, extirpation or reduction of anadromous or potamodromous species, declines in unique species and large fish, collapse of fisheries and floodplain species, and dominance by tolerant or alien species (Karr et al. 1985; Backiel and Penczak 1989; Carlson and Muth 1989; Ebel et al. 1989; Lelek 1989; Oberdorff and Hughes 1992). Migratory rheophilic fishes are considered by some as the most endangered guild of large regulated rivers (Schiemer 2000) because of dams and declines in channel structure (gravel bars, islands, side channels, alcoves, sloughs, and backwater ponds). Where water quality remains a serious problem in large rivers, sight-feeding fishes are replaced with species adapted for tactile and olfactory feeding (Gammon and Simon 2000).

Massive flow and channel modifications of rivers began in Western Europe in the mid-19th century, and European rivers may be the most highly controlled in the world. The technology was transferred to and expanded in North America, and peaked in the mid-20th century, especially in the arid and semiarid western United States (Reisner 1986). Large-scale river modifications have since shifted to the developing world. The United States is estimated to have over 80,000 dams greater than

* Corresponding author: hughes.bob@epa.gov

2 m high (Graf 2001), but now China leads both in the number of large dams and in new dam starts (Abramovitz 1996). Brazil has the single largest project (Itaipu) and India is planning to overtake it with the Narmada project. Like the earlier European and U.S. projects, none of these projects is seriously concerned with the loss of fish species or fisheries. Ward and Stanford (1989) described the importance of understanding and managing river ecosystems in four dimensions (longitudinally, laterally, vertically, and temporally); however, water engineers appear unconcerned with disrupting all four. Although most industrial nations have greatly improved the water quality of their rivers through waste treatment, the physical habitat, flow, and connectivity of their rivers continue to decline (Schmutz et al. 2000; Graf 2001). For example, Muhar et al. (2000) estimated that 79% of Austrian rivers were impaired by channelization, impoundment, water diversion, or hydropeaking. Jungwirth et al. (2000) described the great importance of longitudinal, lateral, and vertical connectivity for fishes of cool water braided and meandering rivers. Migratory fishes, requiring multiple habitats for spawning, rearing, and maturation, are ideal indicators of habitat connectivity or fragmentation, and such species are the most affected by impoundments (Agostinho et al. 2000; Pringle et al. 2000; Schmutz et al. 2000).

How might we use fish assemblages to assess the ecological effects of anthropogenic disruptions of rivers? Norris and Hawkins (2000) and Joy and Death (2002) believed assessments of species richness are preferable, using best available sites as reference conditions for generating predictive models. Hughes and Noss (1992), Karr and Chu (2000), Oberdorff et al. (2002), and Schmutz et al. (2000) supported assessments at multiple levels of ecological organization, including individual health, species richness, and guilds. Hughes (1995) and Schmutz et al. (2000) also argued for using multiple lines of evidence (such as historical fish and abiotic data, minimally disturbed regional reference sites, and ecological dose-response models) to establish reference conditions against which to compare current conditions. All three approaches are rigorous and informative. Where reference sites and empirical models are inadequate, multiple reference condition approaches provide general descriptors of ecosystem condition and assemblage characteristics. These multiple approaches seem especially suited for rivers, which have been so extensively disturbed that the best available sites are often fundamentally altered. In this chapter, we use historical information, current least-disturbed sites, knowledge of fish species ecology, and both species richness and guild approaches. Our objective is to document the historical changes in the fish assemblage of the Willamette River and to link those changes with key anthropogenic stressors.

Methods

Study Area

The Willamette River (Figures 1 and 2) is the largest river in Oregon, draining a basin area of 29,800 km^2 and having a mean annual discharge of 680 m^3/s. It has the highest runoff per unit drainage area of all large U.S. rivers (USGS 1949), with typical summer and winter flows of 250 m^3/s and 1,800 m^3/s, respectively. Mean annual precipitation in the basin varies from 100 to 250 cm in the Willamette valley and coast range or Cascades Mountain ecoregions, respectively. Based on patterns in fish assemblages, the Willamette River mainstem can be divided into four sections (Hughes and Gammon 1987). The first 43 km of the Willamette River to Willamette Falls are tidal with mid-channel depths of 12 m. The reach from Willamette Falls to Newberg (river kilometer 43–84) is a large pool with mid-channel depths of 8 m. The middle section up to river kilometer 212 has mid-channel depths of 6 m. Mid-channel depths decline to 2 m in the upper section (river kilometer 212–301). Mean channel widths range from 50 to 200 m in the upper and tidal reaches of the main stem, respectively. Hughes and Gammon (1987) found that dominant substrates were sand and cobble (tidal section), claypan, bedrock and sand (pool section), gravel (middle section), and gravel and cobble (upper section).

At the time of European settlement in the 1840s, the main-stem Willamette River flowed from south to north through a savanna roughly 70 km by 200

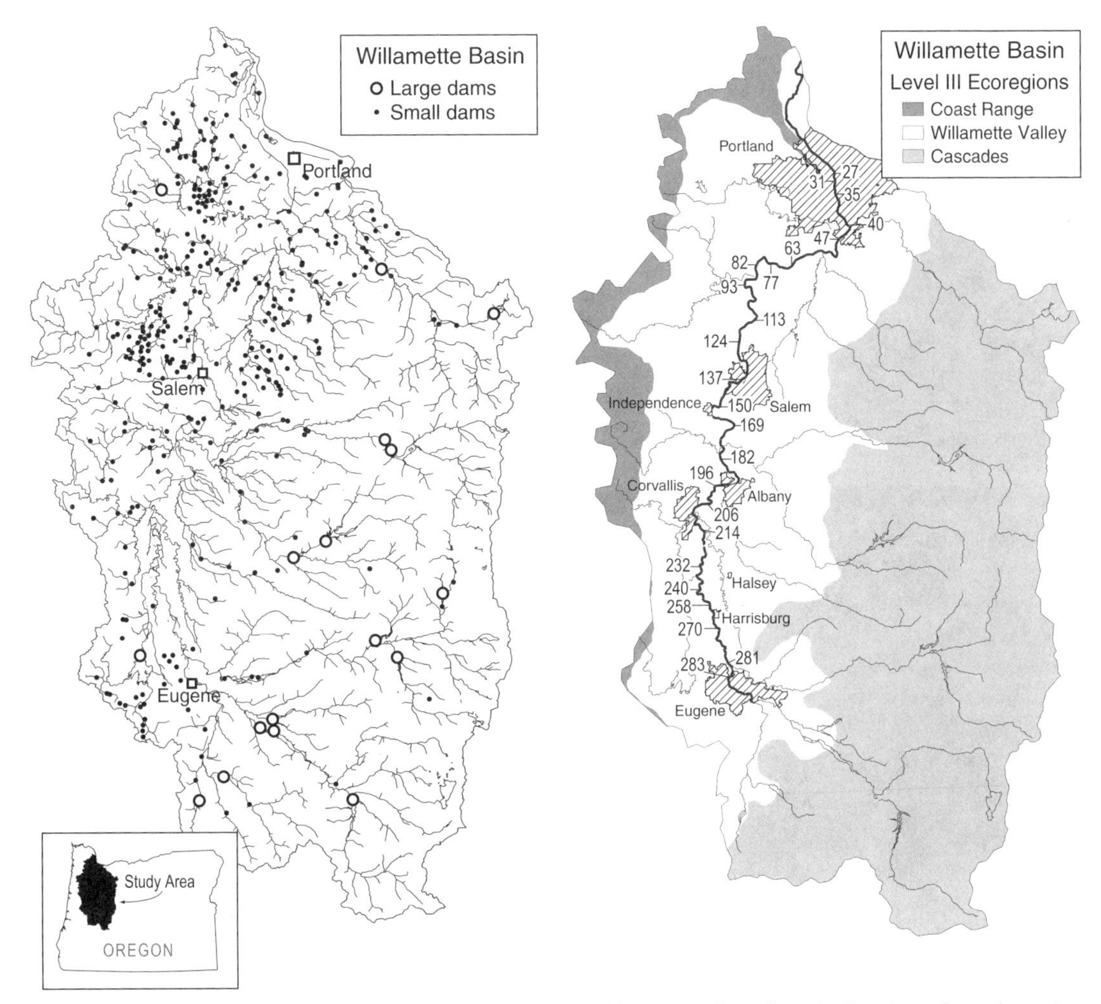

Figure 1.—Locations of small and large dams in the Willamette River basin.

Figure 2.—Sampling site locations along the main-stem Willamette River. Numbers are river kilometer from the mouth.

km, which was maintained by annual fires set by the Calapooia Indians (Johannessen et al. 1971). The river itself was highly braided, meandering, and bordered on both sides by a floodplain forest typically 7 km wide (but ranging from 2 to 9 km). The forest was composed of large hydrophytic trees (white alder, Oregon ash, black cottonwood, bigleaf maple, and several willow species), as well as Douglas-fir, Oregon white oak, and ponderosa pine on progressively drier soils. The Willamette River channel was complex, with multiple heavily shaded channels, sloughs, islands, regularly flooded floodplains, and large numbers of snags and log jams (Sedell and Froggatt 1984; Benner and Sedell 1997; Hulse et al. 2002). During summer low flows, cool water was provided to the river via mountain tributaries and hyporheic inflows from the annually recharged floodplain. Water quality was excellent with little turbidity; and primary production was low with few algae blooms because of extensive wetlands and channel shading (also see Gammon 2004, this volume). During the winter wet season, the floodplain, sloughs, side channels, and snags furnished escape from freshets and served as feeding sites for fishes. This made ideal habitats for a number of riverine and stream fishes, as well as beaver (Scott 1924).

Degradation of the Willamette River followed the general trajectory described by Regier et al.

Figure 3.—Air photos of the middle main-stem Willamette River: a) a complex, remnant, shaded, braided channel with islands and mature Douglas fir in the riparian zone (much of the channel was once more complex than this); b) lower elevation view showing main and side channels, islands, alcoves, and gravel bars (from Steve Cline); c) the active flood plain in a 50-year flood, the active channel is banded by trees in the middle of the photo (from Stan Gregory); d) a simplified, riprapped, single channel with a narrow wooded riparian zone.

(1989). Beavers were decimated by the 1840s, and this was followed by intensive agriculture and channel alterations. To improve navigation, the U.S. Army Corps of Engineers removed over 68,000 snags from the upper main-stem Willamette River between 1870 and 1950, most before 1910. During the same period, side channels, sloughs, and meanders were eliminated by wing dams to form a single major channel, which was hardened with rip-rap at more than half of the meander bends (Figure 3; Benner and Sedell 1997; Hulse et al. 2002). In 100 years, over 75% of channel shoreline along the upper main stem was eliminated (Sedell and Frogatt 1984) and 26% of the channel was rip-rapped on one or both banks;

much of the shoreline along the lower 40 km is bulkhead (Hughes and Gammon 1987). Similar changes occurred along lower reaches of the Willamette's major tributaries. Agriculture replaced hunting and gathering as the dominant land use in the valley, and it was accompanied by growth in canneries, meat packing plants, sawmills, pulp and paper mills, and urbanization. All these industries and cities used the Willamette River for disposal of untreated wastes, so that by 1950, the river had critically low dissolved oxygen levels, extensive *Sphaerotilus natans* beds, and large deposits of floating and benthic sludge. See Alves and Pompeu (2004) for a current example of such conditions. That environment impaired salmon migrations, spawning, and rearing; precluded recreational uses; and eliminated esthetic values. Between the 1940s and 1960s, dams were built on the Willamette's major tributaries to replace the loss of floodplain storage (Figure 1), and Oregon began requiring basin-wide secondary waste treatment.

Water quality in the Willamette River improved markedly after the 1950s. Some small salmon runs have returned, salmon spawning and rearing occur in selected areas of the upper main stem, a resident native trout fishery has been reestablished there as well, boating and swimming are commonly pursued, and the river is considered a positive asset for communities. However, problems remain. The basin houses 70% of Oregon's human population, its three largest cities are located along the Willamette River, and the Portland harbor is designated as a Superfund site. Diffuse pollutants remain untreated, point sources and spills locally alter water quality and fish assemblages, and the river does not meet state temperature and dissolved oxygen criteria. There is also a fish consumption advisory for the Willamette River due to high levels of mercury in northern pikeminnow *Ptychocheilus oregonensis* and bass *Micropterus* (Peterson et al. 2002). As measured at the Albany gauge, channel forming flows at or exceeding bank full changed from the predam average of five per decade to a postdam average of one per decade. The Willamette River's channel, riparian and floodplain integrity remain substantially modified, salmon populations remain diminished, and four fish species or populations are listed as threatened or endangered (Table 1).

Fish Sampling

Sampling sites (identified by river kilometer) were selected to conform with those of Dimick and Merryfield (1945), to bracket major pollution sources and tributaries, and to reduce sampling intervals to approximately 20 km (Figure 2). Fish were sampled in the summer by Dimick and Merryfield (1945), Hughes and Gammon (1987), and by the co-authors in 1999. The initial sampling in 1945 was by beach seine until no new species were encountered at a site. Subsequent sampling in 1987 and 1999 was by boat electrofisher for a distance of 500 m and encompassing areas of slow deep and fast shallow water. A 5-kw generator was used to produce 600–1,000 V DC, a pulse rate of 60 Hz, and 2–3 amps. We fished downstream near one shore and collected fish in 3-mm mesh dip nets. Fish were identified to species and examined for external anomalies, then returned alive to the river. Because of differences in sampling sites, sampling methods, sampling effort, and sampling months, we use no species abundance data in this paper. Instead, we restrict data analyses to native species richness, alien species richness, intolerant species richness, percent alien species, and percent tolerant species (but even these metrics may vary as a result of unequal sampling). Classifications developed by Zaroban et al. (1999) were used to classify species into general tolerance (to increased sediments, nutrients and temperature, and decreased dissolved oxygen) and origin guilds.

Results

There has been no loss of fish species from the Willamette River main stem, but abundances of several species have been reduced, and the fish assemblage has been fundamentally altered by introduced species (Table 1). The Oregon chub *Oregonichthys crameri*, which was once common in floodplain streams and backwaters, is now rare and federally endangered. Several once common salmonid stocks (spring Chinook salmon *Oncorhynchus tshawytscha*, lower Columbia coho salmon *O. kisutch*, and winter steelhead *O. mykiss*) are now much rarer and federally threatened. Twenty-four species, mostly Midwestern pond fishes, have been introduced, and

Table 1.—Native and alien fishes known to occur in or near the main-stem Willamette River (listed alphabetically by family then genus and species).

Scientific, common name	Main-stem occurrence	Location
Acipenser transmontanus, white sturgeon	rare	low-mid
Catostomus macrocheilus, largescale sucker	common	all
C. platyrhynchus, mountain sucker	occasional	mid-upper
Cottus asper, prickly sculpin	common	all
C. beldingii, Paiute sculpin	occasional	mid-upper
C. perplexus, reticulate sculpin	common	all
C. rhotheus, torrent sculpin	common	all
Acrocheilus alutaceus, chiselmouth	common	all
Mylocheilus caurinus, peamouth	common	all
Oregonichthys crameri[a], Oregon chub	rare	floodplain
Ptychocheilus oregonensis, northern pikeminnow	common	all
Rhinichthys cataractae, longnose dace	common	mid-upper
R. falcatus, leopard dace	occasional	mid-upper
R. osculus, speckled dace	common	all
Richardsonius balteatus, redside shiner	common	all
Gasterosteus aculeatus, threespine stickleback	rare	floodplain
Thaleichthys pacificus, eulachon	rare	low
Percopsis transmontana, sand roller	occasional	all
Lampetra ayresii, river lamprey	rare	all
L. richardsoni, western brook lamprey	occasional	all
L. tridentata, Pacific lamprey	common	all
Platichthys stellatus, starry flounder	rare	low
Oncorhynchus clarkii, cutthroat trout	occasional	upper
O. kisutch[a], coho salmon	rare	low
O. mykiss[a], rainbow trout	occasional	all
O. nerka, sockeye salmon	rare	low
O. tshawytscha[a], Chinook salmon	occasional	all
Prosopium williamsoni, mountain whitefish	common	all
Lepomis cyanellus[b], green sunfish	rare	all
L. gibbosus[b], pumpkinseed	rare	all
L. gulosus[b], warmouth	rare	all
L. macrochirus[b], bluegill	rare	all
L. microlophus[b], redear sunfish	rare	all
Micropterus dolomieu[b], smallmouth bass	occasional	all
M. salmoides[b], largemouth bass	occasional	all
Pomoxis annularis[b], white crappie	rare	all
P. nigromaculatus[b], black crappie	rare	all
Alosa sapidissima[b], American shad	rare	low
Carassius auratus[b], goldfish	rare	low
Ctenopharyngodon idella[b], grass carp	rare	low
Cyprinus carpio[b], common carp	occasional	all
Notemigonus crysoleucas[b], golden shiner	rare	low
Tinca tinca[b], tench	rare	low
Fundulus diaphanus[b], banded killifish	rare	low
Ameiurus catus[b], white catfish	rare	low
A. melas[b], black bullhead	rare	low
A. natalis[b], yellow bullhead	rare	low
A. nebulosus[b], brown bullhead	rare	all
Ictalurus punctatus[b], channel catfish	rare	all
Perca flavescens[b], yellow perch	rare	low-mid
Sander vitreus[b], walleye	rare	low-mid
Gambusia affinis[b], western mosquitofish	rare	all

[a] Federally threatened or endangered population or species.
[b] Alien species.

three species now occur occasionally in the main channel (common carp *Cyprinus carpio*, smallmouth bass *Micropterus dolomieu*, and largemouth bass *M. salmoides*).

Although differences in sampling methods and sampling sites limit conclusions, the fish assemblage along the Willamette River apparently changed from upper to lower main stem and between 1945 and 1999. In 1999, total fish species richness ranged from 21 in the upper main stem to 8 in the lower river, and similar declines from upper to lower river were observed in 1945 and 1983. Native fish species richness decreased from the upper main stem to the lower, but increased from 1945 to 1999 throughout the main stem (Figure 4). Similarly, intolerant fish species richness decreased from the upper to the lower main stem, but was usually higher in 1983 and 1999 than in 1945 (Figure 5). Especially notable is the occurrence of intolerant species down river of river kilometer 160 in 1983 and 1999, where none were found in 1945.

As one might expect, percent tolerant fish species, alien fish species richness, and percent alien fish species showed an opposite pattern from intolerant species richness, native species richness, and total species richness. The percent tolerant species increased from upper to lower main stem, but decreased from 1945 to 1983 and 1999 (Figure 6). Alien species richness increased from upper to lower main stem, but first decreased from 1945 to 1983, then increased from both 1945 and 1983 to 1999 (Figure 7). The percent of alien fish

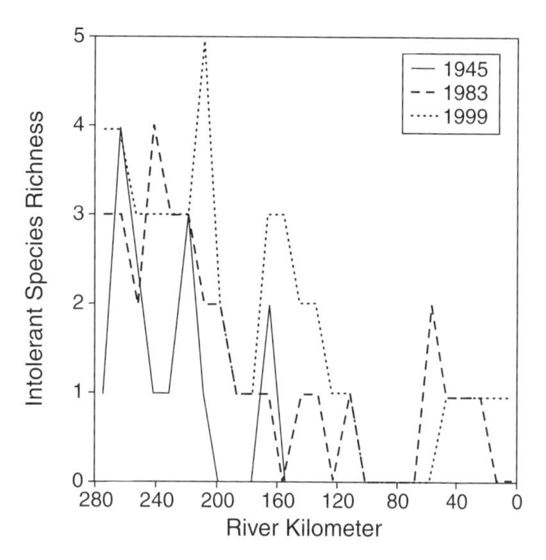

Figure 5.—Richness of intolerant fish species along the main-stem Willamette River, 1945–1999. Note: lines are for clarity and do not indicate values between points.

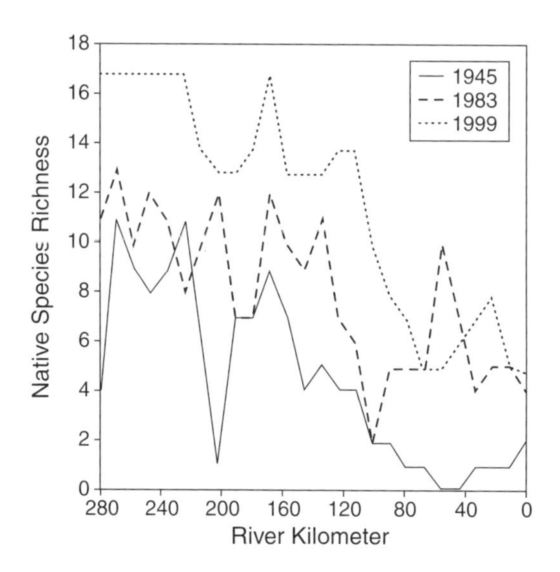

Figure 4.—Native fish species richness along the main-stem Willamette River, 1945–1999. Note: lines are for clarity and do not indicate values between points.

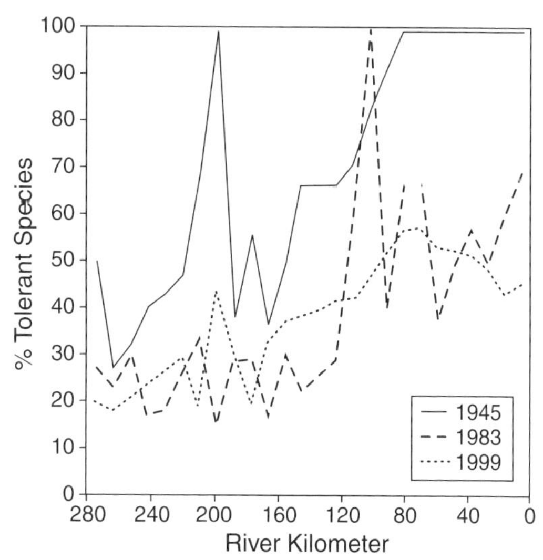

Figure 6.—Percent of tolerant fish species along the main-stem Willamette River, 1945–1999. Note: lines are for clarity and do not indicate values between points.

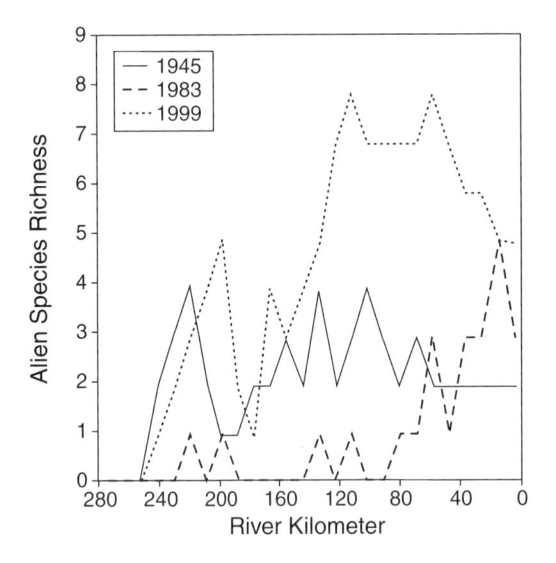

Figure 7.—Richness of alien fish species along the main-stem Willamette River, 1945–1999. Note: lines are for clarity and do not indicate values between points.

species also increased from upper to lower main stem, and decreased from 1945 to 1983, but has increased from 1983 to 1999 (Figure 8). Apparently, alien species are cumulative and largely irreversible.

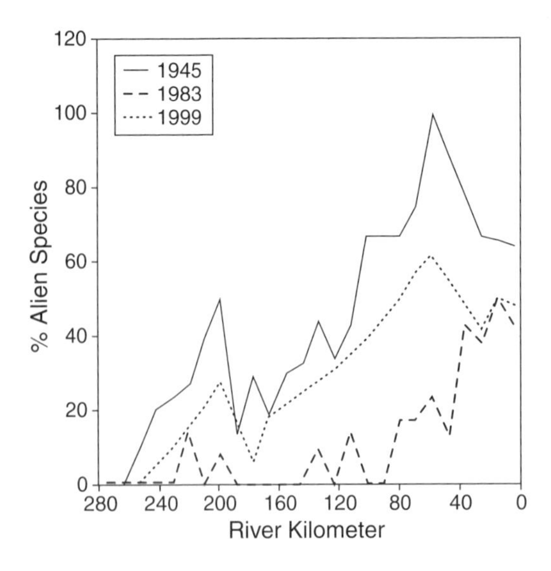

Figure 8.—Percent of alien fish species along the main-stem Willamette River, 1945–1999. Note: lines are for clarity and do not indicate values between points.

Discussion

As suggested by their federal listing, four once common fish species are now only rarely or occasionally encountered. In 6 years of sampling multiple sites on the main-stem Willamette River and its major tributaries, we have encountered Oregon chub only once (unpublished data). Scheerer et al. (1999) are attempting to restore Oregon chub by protecting extant populations and through transfers to habitats lacking alien predators but rich in habitat complexity. We have yet to collect lower Columbia coho salmon in the main stem, both because of its rarity and the timing of its fall migration. Adult and juvenile Chinook salmon and winter steelhead are occasionally encountered, but not consistently. Electrofishing is purposely timed to avoid collecting the listed adult salmonids as a condition of our federal and state fish collecting permits, but we sample on other Oregon rivers in the same season and encounter salmon and steelhead much more often. We, therefore, have no data indicating that the species' federal listings are unwarranted.

In contrast to the infrequent occurrence of listed species, the frequency of common carp and largemouth and smallmouth bass increased in our collections (unpublished data). In particular, the greater occurrence frequency of smallmouth bass in the Willamette River in the past 16 years suggests a potential hazard to native fish. Smallmouth bass is an alien piscivore that was not collected in the middle or upper main-stem Willamette River or its tributaries in 1983. It is now a dominant species in one upper river tributary, common in other tributaries, and occasionally collected in the upper and middle main stem. Smallmouth bass was introduced in the Willamette in the late 1800s. It is unclear why it has not become dominant in the upper and middle reaches of the main-stem Willamette River. Smallmouth bass is a dominant species in at least two other large Oregon rivers (John Day, Umpqua), but these are somewhat warmer than the Willamette River. It is also a dominant species in the lower Snake and lower Verde rivers, which are markedly warmer than the Willamette River (Maret and Mebane 2004; Rinne 2004). Miller et al. (1989) listed alien species as a detrimental factor in 68% of fish extinctions, and Whittier et al. (1997) reported that alien fishes were associated with regional extirpation of native minnows. In the Willamette River or its tributaries, potential prey of smallmouth bass, such as native cyprinids, cottids, and juvenile salmonids, may be increasingly at risk.

We observed marked decreases in total species richness, native species richness, and intolerant species richness from the upper to the lower main-stem Willamette River (Figures 4 and 5). Although Li et al. (1987) presented a species subtraction model with increased river size, this clearly was not the case in the main-stem Willamette, which supports more species and in greater abundance than its major tributaries (Table 2). The decreased species richness in the lower river may have resulted from poorer water quality, natural changes, sampling inefficiencies, or a combination of these factors. Hughes and Gammon (1987) reported marked increases in nutrients and turbidity from upper to lower river, and down river of point sources. Water temperature increased gradually along the main-stem river (Hughes and Gammon 1987), which is associated with natural warming, as well as reduced hyporheic inputs

Table 2.—Native species addition in the Willamette drainage from headwaters to main stem. Dashes represent uncommon presence; X represents core habitat.

Species	Large river	Small river	Large stream	Small stream	Headwater	Comment
tailed frog				————	XXX	cold
Pacific giant salamander				———X	XX	cool
cutthroat trout	————	————	————	——XX	X	cold
shorthead sculpin *Cottus confusus*			—	—XXX		cold
western brook lamprey			——XX	X		cool
mottled sculpin *C. bairdii*			—	—XXX		cool
Oregon chub	-	———	-XXX-	-		cover
coho salmon	-	———X	XX—			cold
Pacific lamprey	————	———X	XX—			cool
bull trout *Salvelinus confluentus*	-	—XXX	————	——XX	X	cold
threespine stickleback	————	————	——XX	XX-		cover
rainbow trout	—	—XXX	—XXX	-		cold
redside shiner	—X—	—X—	—X—	X-		cool
speckled dace	—X—	—X—	—X—	——		cool
reticulate sculpin	—X—	—X—	—X—	—		cool
torrent sculpin	—XX	—X—	X—			cold
Paiute sculpin	—X	XX——				cold
Chinook salmon	—XX	X———	—			cold
mountain sucker	—XX	X—				cool
mountain whitefish	—XXX	X—				cold
northern pikeminnow	—XXX	———	——			cool
largescale sucker	-XXX-	———	—			cool
longnose dace	-XXX-	————				cool
sandroller	XXX—	———	————			cover
chiselmouth	-XXX	———				cool
leopard dace	-XXX-	—				cool
peamouth	-XXX-	—				cool
prickly sculpin	XXX—	-				cool
white sturgeon	XXX—					deep
river lamprey	XX—					cool
eulachon	———					tidal
starry flounder	———					tidal

and lower channel complexity. Both the latter changes result from natural changes, as well as increased channel incision and channelization by humans. Hughes and Gammon (1987) also indicated that the dominant substrate in the main-stem Willamette River changed from gravel in the upper and middle sections to sand or claypan in the lower river. These substrate changes are associated with natural changes in geology and channel slope, as well as historical aggregate mining in the lower river. The sampling gear employed (seines and boat electrofishers) are most effective in shallow water and the river deepens down river. However, in 1983 and 1999, a sampling design was applied that included both shallow and deep areas at all sites except a Portland harbor site. We therefore conclude that the number of native species and intolerant species decreased as a function of natural and anthropogenic changes from upper to lower river.

The number of native species and intolerant species also increased from 1945 to 1983 and 1999 (Figures 4 and 5). We are aware of no natural changes during that period, so the likely explanations are improved water quality, increased sampling effectiveness, and more accurate species identifications. There have been fundamental improvements in water quality as explained above. However, there were also marked differences in sampling gear since 1945. Boat electrofishers may be more effective sampling gear than seines in deep areas of the Willamette River (Simon and Sanders 1998; Gammon 2004), so the gear differences alone could explain some increase in the number of species collected. Nonetheless, electrofishers are unlikely to double the species richness that was observed, and concurrent beach seining in 1999 produced nearly the same species as boat electrofishing. Sampling effort among years increased, which is likely the explanation for the increases in species collected between 1983 and 1999, because we are aware of little improvement in waste treatment or water quality during that period. Cao et al. (2001) and Hughes et al. (2002) demonstrated how increased sampling effort in Oregon rivers increased the number of species observed, especially for rare and patchy species. Hughes and Gammon (1987) indicated that Dimick and Merryfield (1945) misidentified two fish species; however, these occurred only occasionally in the upper river so they would have had a minimal effect. Consequently, we believe that the observed increases in native species richness and intolerant species richness between 1945 and the present largely resulted from improved water quality and possibly more extensive sampling. Thomas et al. (2004) and Yoder et al. (2004) observed similar increases in fish assemblage condition following advanced waste treatment of point source effluents to the Ohio River and Ohio rivers, respectively.

Alien species increased from upper to lower river sites in 1983 and 1999 (Figure 7). In 1945, there was no consistent pattern in alien species richness between upper and lower reaches. As explained above, we do not think that these spatial differences are a result of more effective sampling in the lower river. Instead, we credit the change to natural and anthropogenic changes in the river, and ultimately to the introduction of the species by humans followed by the fishes' successful dispersion. Moyle and Light (1996) offered several empirical rules for successful invasions by alien species. Those pertinent to the main-stem Willamette River are that aliens are more likely to become established when 1) few native species occur, 2) native assemblages are disrupted, 3) ecosystems are permanently altered by humans, or 4) aquatic systems have extremely low variability. All these conditions are met in the Willamette River today.

The decline in alien species from 1945 to 1983, followed by increases in 1999, may have resulted from changes in water quality or sampling effectiveness. Greater sampling effort in 1999 than 1983 could account for differences during that period, but sampling effort was likely somewhat less in 1945 than 1983. Nor do we believe that sampling effort for alien species was more effective than it was for native and intolerant species between 1945 and 1983. We believe that water quality conditions favored alien species in 1945 throughout the main-stem Willamette River. In 1983, these conditions only occasionally prevailed in the Portland harbor and Newberg pool areas. The increased number of alien species in the middle and lower river in 1999 also may have been influenced by the gradual migration of alien pond fishes upstream from lower river and tributary refugia. We, therefore, support water quality and migration as more likely explanations for most of the temporal changes observed in alien species richness. But increased sampling effort in 1999 is also a factor because it produced more rare species, and the alien species are usually rare or occasional (Table 1).

Unlike alien species richness, both percent tolerant species and percent alien species increased between upper river and lower river sites all 3 years (Figures 6 and 8). This may have been simply a result of percent alien species being less influenced than richness by the addition or loss of a few rare alien species. Similarly, the percent tolerant species metric is likely less variable than the number of tolerant species, if some of those tolerant species are rare (which several are). Percentage metrics in general are less affected by slight differences in sampling effort, so the observed increases in percent alien species and percent tolerant species in the lower river are likely most associated with poorer water quality and physical habitat structure. Poorer chemical habitat quality was also associated with increased external anomalies downstream of river kilometer 82 (Hughes and Gammon 1987).

The percent of tolerant species throughout the main-stem Willamette River declined between 1945 and 1983 or 1999, while the percent of alien species declined between 1945 and 1983, then increased between 1983 and 1999 (Figures 6 and 8). We associate the temporal declines in percent tolerant species with improved water quality throughout the river. As explained for alien species richness, we believe that the percent of alien species decreased from 1945 to 1983 because of improved water quality. The increase in percent alien species between 1983 and 1999 likely resulted from successful colonization of new reaches of the river following migrations from tributary and lower river refugia. As suggested by Moyle and Light (1996), we predict that the percent of alien species in Willamette River fish assemblages will continue to increase. This is likely because the Willamette River, especially the lower main stem, supports few native fish species, has a disrupted fish assemblage as indicated by four federally listed species, has a fundamentally altered channel morphology, and no longer experiences major floods and droughts as a result of dams. In addition, because many alien species are rare in the main stem, increased sampling effort between 1983 and 1999 would produce a greater percentage and number of alien species (Cao et al. 2001; Hughes et al. 2002).

In conclusion, the Willamette River has been altered physically, chemically, and biologically. The active channel has been greatly reduced in both the abundance and quality of habitat types because of channel and flow modifications. Water pollution has been greatly reduced by point source treatment, but persistent toxics remain a problem as does diffuse pollution from agriculture and urban runoff. A projected doubling of the human population near the river over the next 50 years does not bode well for water quality improvements. One of the most ominous problems is that of invasive alien species, particularly piscivores such as bass and walleye. We therefore recommend that Willamette River fish assemblages be monitored periodically to assess status and trends, and that the results be reported both to the scientific community and to the public so that both are better informed about the river's health. In addition, we believe that the sites seined by Dimick and Merryfield (1945) should be both seined and electrofished to calibrate data differences resulting from present and historical sampling gear. Also, we propose that a small pilot study be conducted to evaluate sampling methods, sampling effort, and sampling period so that these components of variance can be accurately assessed and reduced in a sampling protocol for long-term trend monitoring. Because alien introductions appear largely irreversible in the Willamette River, such practices by agencies or anglers should cease, and alien predators should be extirpated to the maximum degree possible. This is especially the case where listed fish species may be desirable prey or competitors. Lastly, it would be wise to survey shallow off-main-stem nursery areas, such as alcoves and side channels (Dixon Landers, USEPA, Corvallis, unpublished data) and major tributaries (Stan Gregory, Department of Fisheries and Wildlife, unpublished data) in both summer and winter should the state of Oregon wish to better understand salmonid refugia and invasive alien fishes.

Acknowledgments

This manuscript was partially funded by the USEPA through contract 68-D01-005 with Dynamac Corporation. It benefited from reviews by Larry Brown, Doug Drake, Jim Gammon, John Rinne, and Kirk Schroeder and was subjected to USEPA review and clearance procedures. Final figures were prepared by Suzanne Pierson.

References

Abramovitz, J. N. 1996. Sustaining freshwater ecosystems. Pages 60–77 *in* State of the world 1996. Norton & Company, New York.

Agostinho, A. A., S. M. Thomaz, C. V. Minte-Vera, and K. O. Winemiller. 2000. Biodiversity in the high Parana River floodplain. Pages 89–118 *in* B. Gopal, W. J. Junk, and J. A. Davis, editors. Biodiversity in wetlands: assessment, function and conservation. Backhuys Publishers, Leiden, The Netherlands.

Alves, C. B. M., and P. S. Pompeu. 2005. Historical changes in the Rio das Velhas fish fauna—Brazil. Pages 587–602 *in* J. N. Rinne, R. M. Hughes, and B. Calamusso, editors. Historical changes in large river fish assemblages of the Americas. American Fisheries Society, Symposium 45, Bethesda, Maryland.

Backiel, T., and T. Penczak. 1989. The fish and fisheries in the Vistula River and its tributary, the Pilica River. Pages 488–503 *in* D. P. Dodge, editor. Proceedings of the international large river symposium (LARS). Canadian Special Publication of Fisheries and Aquatic Sciences 106. Department of Fisheries and Oceans, Ottawa.

Benner, P. A., and J. R. Sedell. 1997. Upper Willamette River landscape: a historical perspective. Pages 23–47 *in* A. Laenen and D. A. Dunnette, editors. River quality: dynamics and restoration. CRC Press, Boca Raton, Florida.

Cao, Y., D. P. Larsen, and R. M. Hughes. 2001. Determining sampling sufficiency in fish assemblage surveys—a similarity-based approach. Canadian Journal of Fisheries and Aquatic Sciences 58:1782–1793.

Carlson, C. A., and R. T. Muth. 1989. The Colorado River: lifeline of the American Southwest. Pages 220–239 *in* D. P. Dodge, editor. Proceedings of the international large river symposium (LARS). Department of Fisheries and Oceans, Canadian Special Publication of Fisheries and Aquatic Sciences 106, Ottawa.

Dimick, R. E., and F. Merryfield. 1945. The fishes of the Willamette River system in relation to pollution. Oregon State College, Engineering Experiment Station Bulletin Series 20:7–55, Corvallis, Oregon.

Ebel, W. J., C. D. Becker, J. W. Mullan, and H. L. Raymond. 1989. The Columbia River: toward a holistic understanding. Pages 205–219 *in* D. P. Dodge, editor. Proceedings of the international large river symposium (LARS). Department of Fisheries and Oceans, Canadian Special Publication of Fisheries and Aquatic Sciences 106, Ottawa.

Gammon, J. R. 2005. Wabash River fishes from 1800 to 2000. Pages 365–382 *in* J. N. Rinne, R. M. Hughes, and B. Calamusso, editors. Historical changes in large river fish assemblages of the Americas. American Fisheries Society, Symposium 45, Bethesda, Maryland.

Gammon, J. R., and T. P. Simon. 2000. Variation in a great river index of biotic integrity over a 20-year period. Hydrobiologia 423:291–304.

Graf, W. L. 2001. Damage control: restoring the physical integrity of America's rivers. Annals of the Association of American Geographers 91:1–27.

Hughes, R. M. 1995. Defining acceptable biological status by comparing with reference conditions. Pages 31–47 *in* W. S. Davis, and T. P. Simon, editors. Biological assessment and criteria: tools for water resource planning and decision making. CRC Press, Boca Raton, Florida.

Hughes, R. M., and J. R. Gammon. 1987. Longitudinal changes in fish assemblages and water quality in the Willamette River, Oregon. Transactions of the American Fisheries Society 116:196–209.

Hughes, R. M., P. R. Kaufmann, A. T. Herlihy, S. S. Intelmann, S. C. Corbett, M. C. Arbogast, and R. C. Hjort. 2002. Electrofishing distance needed to estimate fish species richness in raftable Oregon rivers. North American Journal of Fisheries Management 22:1229–1240.

Hughes, R. M., and R. F. Noss. 1992. Biological diversity and biological integrity: current concerns for lakes and streams. Fisheries 17(3):11–19.

Hulse, D., S. Gregory, and J. Baker. 2002. Willamette River basin planning atlas: trajectories of environmental and ecological change. Oregon State University Press, Corvallis, Oregon.

Johannessen, C. L., W. A. Davenport, A. Millet, and S. McWilliams. 1971. The vegetation of the Willamette Valley. Annals of the Association of American Geographers 61:286–302.

Joy, M. K., and R. G. Death. 2002. Predictive modelling of freshwater fish as a biomonitoring tool in New Zealand. Freshwater Biology 47:2261–2275.

Jungwirth, M., S. Muhar, and S. Schmutz. 2000. Fundamentals of fish ecological integrity and their relation to the extended serial discontinuity concept. Hydrobiologia 422:85–97.

Karr, J. R., and E. W. Chu. 2000. Sustaining living rivers. Hydrobiologia 422:1–14.

Karr, J. R., L. A. Toth, and D. R. Dudley. 1985. Fish communities of midwestern rivers: a history of degradation. BioScience 35:90–95.

Lelek, A. 1989. The Rhine River and some of its tributaries under human impact in the last two centuries. Pages 469–487 *in* D. P. Dodge, editor. Proceedings of the international large river symposium (LARS). Department of Fisheries and Oceans, Canadian Special Publication of Fisheries and Aquatic Sciences 106, Ottawa.

Li, H. W., C. B. Schreck, C. E. Bond, and E. Rexstad. 1987. Factors influencing changes in fish assemblages of Pacific Northwest streams. Pages 193–202 *in* W. J. Matthews and D. C. Heins, editors. Community and evolutionary ecology of North American stream fishes. University of Oklahoma Press, Norman.

Maret, T. R., and C. A. Mebane. 2005. Historic and current perspectives on fish assemblages of the Snake River, Idaho and Wyoming. Pages 41–60 *in* J. N. Rinne, R. M. Hughes, and B. Calamusso, editors. Historical changes in large river fish assemblages of the Americas. American Fisheries Society, Symposium 45, Bethesda, Maryland.

Miller, R. R., J. D. Williams, and J. E. Williams. 1989. Extinctions of North American fishes during the past century. Fisheries 14(6):22–38.

Moyle, P. B., and T. Light. 1996. Biological invasions of fresh water: empirical rules and assembly theory. Biological Conservation 78:149–161.

Muhar, S., M. Schwarz, S. Schmutz, and M. Jungwirth. 2000. Identification of rivers with high and good habitat quality: methodological approach and application in Austria. Hydrobiologia 423:343–358.

Norris, R. H., and C. P. Hawkins. 2000. Monitoring river health. Hydrobiologia 435:5–17.

Oberdorff, T., and R. M. Hughes. 1992. Modification of an index of biotic integrity based on fish assemblages to characterize rivers of the Seine basin, France. Hydrobiologia 228:117–130.

Oberdorff, T., D. Pont, B. Hugueny, and J-P., Porcher. 2002. Development and validation of a fish-based index for the assessment of 'river health' in France. Freshwater Ecology 47:1720–1734.

Peterson, S. A., A. T. Herlihy, R. M. Hughes, K. L. Motter, and J. M. Robbins. 2002. Level and extent of mercury contamination in Oregon, USA, lotic fish. Environmental Toxicology and Chemistry 21:2157–2164.

Pringle, C. M., M. C. Freeman, and B. J. Freeman. 2000. Regional effects of hydrologic alterations on riverine macrobiota in the new world: tropical-temperate comparisons. BioScience 50:807–823.

Regier, H. A., R. L. Welcomme, R. J. Steedman, and H. F. Henderson. 1989. Rehabilitation of degraded river ecosystems. Pages 86–97 *in* D. P. Dodge, editor. Proceedings of the international large river symposium (LARS). Department of Fisheries and Oceans, Canadian Special Publication of Fisheries and Aquatic Sciences 106, Ottawa.

Reisner, M. 1986. Cadillac desert: the American west and its disappearing water. Penguin Books. New York.

Rinne, J. N. 2005. Changes in fish assemblages, Verde River, Arizona, 1974–2003. Pages 115–126 *in* J. N. Rinne, R. M. Hughes, and B. Calamusso, editors. Historical changes in large river fish assemblages of the Americas. American Fisheries Society, Symposium 45, Bethesda, Maryland.

Scheerer, P. D., G. D. Apke, and P. J. McDonald. 1999. Oregon chub research: Middle Fork Willamette and Santiam River drainages. Oregon Department of Fish and Wildlife, Contract Number E96970022, Corvallis.

Schiemer, F. 2000. Fish as indicators for the assessment of the ecological integrity of large rivers. Hydrobiologia 423:271–278.

Schmutz, S., M. Kaufmann, B. Vogel, M. Jungwirth, and S. Muhar. 2000. A multi-level concept for fish-based, river-type-specific assessment of ecological integrity. Hydrobiologia 423:279–289.

Scott, L. M., editor. 1924. John Work's journey from Fort Vancouver to Umpqua River and return in 1834. Oregon Historical Quarterly 24:236–268.

Sedell, J. R., and J. L. Froggatt. 1984. Importance of streamside forests to large rivers: the isolation of the Willamette River, Oregon, USA, from its floodplain by snagging and streamside forest removal. Internationale Vereinigung fuer Theoretische und Angewandte Limnologie Verhandlungen 22:1828–1834.

Simon, T. P., and R. E. Sanders. 1998. Applying an index of biotic integrity based on great-river fish communities: considerations in sampling and interpretation. Pages 475–505 *in* T. P. Simon, editor. Assessing the sustainability and biological integrity of water resources using fish communities. CRC Press, Boca Raton, Florida.

Thomas, J. A., E. B. Emery, and F. H. McCormick. 2005. Detection of temporal trends in Ohio River fish assemblages based on lockchamber surveys (1957–2001). Pages 431–449 *in* J. N. Rinne, R. M. Hughes, and B. Calamusso, editors. Historical changes in large river fish assemblages of American rivers. American Fisheries Society, Symposium 45, Bethesda, Maryland.

USGS (U.S. Geological Survey). 1949. Large rivers of the United States. U. S. Geological Survey Circular 44, Reston, Virginia.

Ward, J. V., and J. A. Stanford. 1989. Riverine ecosystems: the influence of man on catchment dynamics and fish ecology. Pages 56–64 *in* D. P. Dodge, editor. Proceedings of the international large river symposium (LARS). Department of Fisheries and Oceans, Canadian Special Publication of Fisheries and Aquatic Sciences 106, Ottawa.

Whittier, T. R., D. B. Halliwell, and S. G. Paulsen. 1997. Cyprinid distributions in northeast USA lakes: evidence of regional-scale minnow biodiversity losses. Canadian Journal of Fisheries and Aquatic Sciences 54:1593–1607.

Yoder, C. O., M. A. Smith, B. J. Alsdorf, D. J. Altfader, C. E. Boucher, R. J. Miltner, D. E. Mishne, E. T. Rankin, R. E. Sanders, and R. F. Thoma. 2005. Changes in fish assemblage status in Ohio's nonwadeable rivers over two decades. Pages 399–429 *in* J. N. Rinne, R. M. Hughes, and B. Calamusso, editors. Historical changes in large river fish assemblages of the Americas. American Fisheries Society, Symposium 45, Bethesda, Maryland.

Zaroban, D. W., M. P. Mulvey, T. R. Maret, R. M. Hughes, and G. D. Merritt. 1999. Classification of species attributes for Pacific Northwest fishes. Northwest Science 73:81–93.

American Fisheries Society Symposium 45:75–98, 2005

Native Fishes of the Sacramento–San Joaquin Drainage, California: A History of Decline

Larry R. Brown*
U.S. Geological Survey, Placer Hall, 6000 J Street, Sacramento, California 95819–6129, USA

Peter B. Moyle
Department of Wildlife, Fish, and Conservation Biology, University of California, 1 Shields Avenue, Davis, California 95616, USA

Abstract.—In this paper, we review information regarding the status of the native fishes of the combined Sacramento River and San Joaquin River drainages (hereinafter the "Sacramento–San Joaquin drainage") and the factors associated with their declines. The Sacramento–San Joaquin drainage is the center of fish evolution in California, giving rise to 17 endemic species of a total native fish fauna of 28 species. Rapid changes in land use and water use beginning with the Gold Rush in the 1850s and continuing to the present have resulted in the extinction, extirpation, and reduction in range and abundance of the native fishes. Multiple factors are associated with the declines of native fishes, including habitat alteration and loss, water storage and diversion, flow alteration, water quality, and invasions of alien species. Although native fishes can be quite tolerant of stressful physical conditions, in some rivers of the drainage the physical habitat has been altered to the extent that it is now more suited for alien species. This interaction of environmental changes and invasions of alien species makes it difficult to predict the benefits of restoration efforts to native fishes. Possible effects of climate change on California's aquatic habitats add additional complexity to restoration of native fishes. Unless protection and restoration of native fishes is explicitly considered in future water management decisions, declines are likely to continue.

Introduction

Loss of aquatic biodiversity is a worldwide problem. Aquatic conservation in regions with Mediterranean climates is particularly challenging. Such regions are generally arid with highly seasonal precipitation and surface water runoff, while human water demand is relatively constant throughout the year (Moyle and Leidy 1992; Moyle 1995). Mismatches between water supply and demand generally result in development of major water storage and diversion systems. In the arid western United States, these engineering solutions have led to some of the most extensive storage and diversion systems in the world (Reisner 1986). In California, the combination of Mediterranean climate, agricultural development, and growing urban demand in out-of-basin areas has resulted in one of the most complex water systems in the world (Mount 1995). The fish fauna of the western United States has also been "engineered" to a significant degree. Eastern settlers unfamiliar with the native fishes introduced familiar sport and food species (Dill and Cordone 1997). The introduction of alien species is often cited as an important factor in the decline of native species (Williams et al. 1989a; Moyle 2002). This combination of water development and ecological engineering has

* Corresponding author: lrbrown@usgs.gov

had profound effects on the fish fauna of California (Moyle and Williams 1989; Moyle 2002).

The combined Sacramento River and San Joaquin River drainages (hereinafter the "Sacramento–San Joaquin drainage") (Figure 1) is the center of fish evolution in California (Moyle 2002). The complex hydrology and geology of the drainage combined with its isolation from other major river systems for the last 10–17 million years (Minckley et al. 1986) has produced 17 species endemic to the drainage (Moyle 2002). The number of endemic forms increases to between 40 and 50 when subspecies and distinct runs of salmon are considered (Moyle 2002). In addition to its importance as a center of endemism, in pre-European times the aquatic resources of the drainage were highly productive. Resident and anadromous fishes, shellfish, reptiles, amphibians, and waterfowl supported what is believed

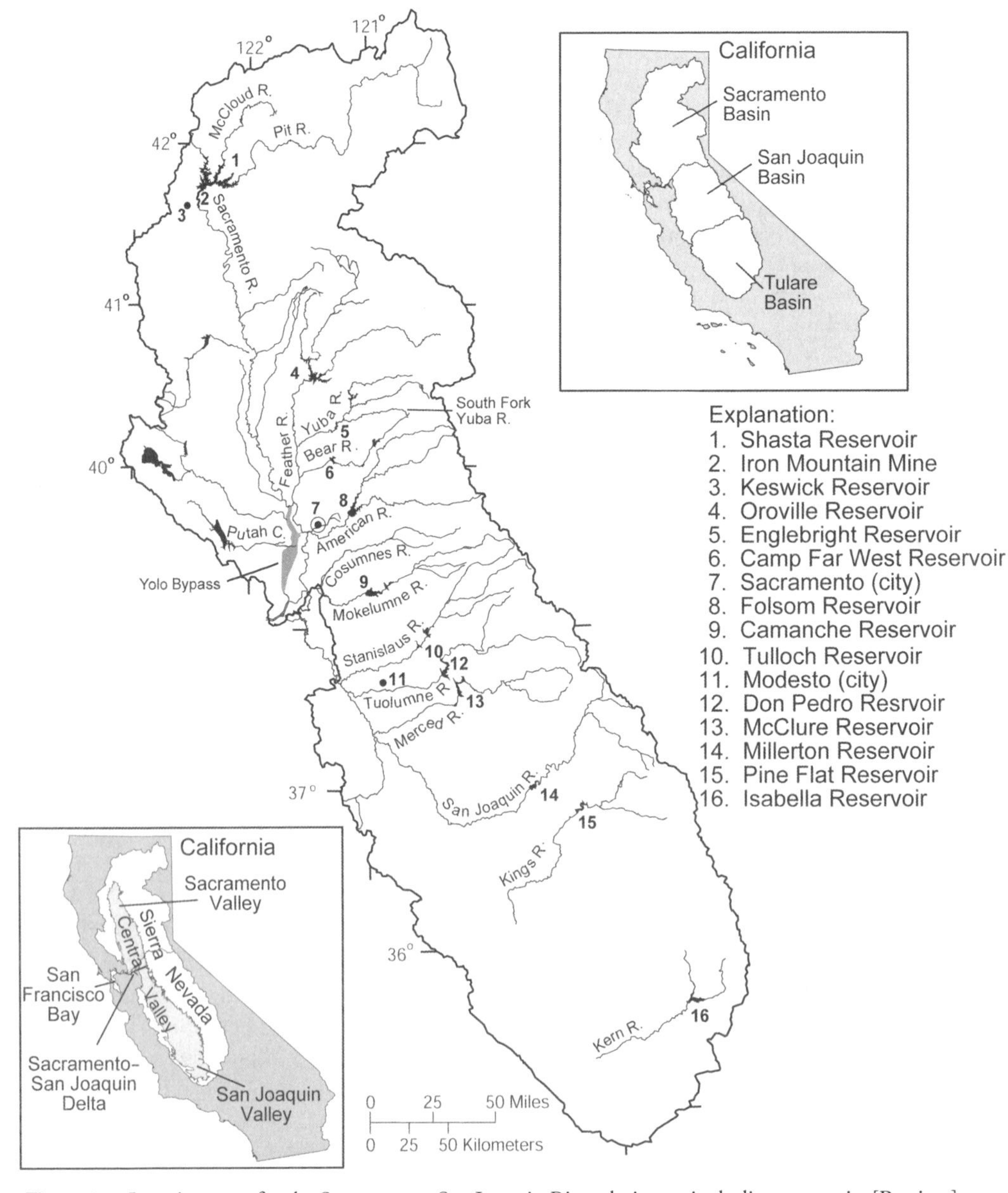

Figure 1.—Location map for the Sacramento–San Joaquin River drainage, including reservoirs [R., river].

to be some of the highest population densities of nonagricultural Native Americans known in North America (Kroeber 1939, 1963; Baumhoff 1963).

Beginning with the Gold Rush in the mid-1800s, land use in the Sacramento–San Joaquin drainage changed rapidly. Mining activities had direct and indirect effects on river systems. Perhaps more important, the rapid influx of people resulted in increased agriculture and urban land uses. All of these activities require water. The population of California continues to grow and is expected to reach 42.4 million people by 2010 (California Institute 1999), resulting in further demands on the water resources of California, including the Sacramento–San Joaquin drainage.

The objective of this article is to review and synthesize existing information on the present status of the fish fauna of the Sacramento–San Joaquin drainage and the human activities associated with changes in the fish fauna since the mid-1800s. Moyle (2002) divided California into six ichthyological provinces, including the Sacramento–San Joaquin Province. Of the seven subprovinces within the Sacramento–San Joaquin Province, this article emphasizes the Central Valley subprovince. Further, this paper emphasizes the nontidal, lower elevation reaches of the larger streams and rivers of the subprovince (Figure 1, but excluding the McCloud River, Pit River, Clear Lake, and the Kern River upstream of Isabella Reservoir). Interested readers are referred to Moyle (2002) and Moyle et al. (1982) for more information on subprovinces and habitats not covered in this paper. Readers interested in more detail on Sacramento–San Joaquin Delta (hereinafter the "Delta") fish issues should consult Bennett and Moyle (1996), Moyle (2002), and Brown (2003a).

Study Area

The Sacramento–San Joaquin drainage is the largest drainage wholly contained in the state of California, with an area of 151,000 km^2. The Sacramento River drainage with 70,000 km^2 is slightly smaller than the San Joaquin River drainage (combined San Joaquin and Tulare basins, Figure 1) with 81,000 km^2. Vertical relief is from sea level in the Sacramento–San Joaquin Delta to over 4,200 m at the top of Mount Whitney, the tallest mountain in the contiguous United States. The mean annual runoff in the Sacramento River drainage is about 27.6 billion m^3 × year^{-1} and in the San Joaquin River drainage is about 11 billion m^3 × year^{-1} (California Department of Water Resources 1993). Precipitation is greatest in the mountains on the eastern border of the drainage, where precipitation falls largely as snow and can exceed 200 cm per year. In the Central Valley (Figure 1), rain is the primary form of precipitation and ranges from about 60 cm per year in the northernmost part of the valley to about 12 cm per year in the southernmost part of the valley. The Sacramento and San Joaquin rivers and their tributaries are generally high gradient with cold water and coarse substrates in the mountains. Runoff is captured by storage reservoirs on all of the larger rivers at the foothill transition between the Sierra Nevada and the Central Valley (Figure 1). Within the Central Valley, rivers become more meandering and progressively warmer with finer substrates until the Sacramento and San Joaquin rivers meet in the tidal Delta, which is largely a freshwater system.

Historically, terrestrial and riparian habitats in the mountains primarily consisted of mixed conifer forest, red fir forest, and lodgepole pine/subalpine forest, and mixed coniferous deciduous forests (Omernik 1987). The foothills and Coast Ranges were dominated by oak woodlands, chaparral, and California steppe (Omernik 1987). The Central Valley consisted of a mosaic of habitat types, including permanent and seasonal tule marsh, riparian forest, valley oak savanna, and native grasslands (San Joaquin Valley Drainage Program 1990). Tulare Lake, the largest (surface area) lake west of the Rocky Mountains, was the dominant feature of the southern San Joaquin Valley. The semiclosed Tulare basin at the southern end of the Central Valley (Figure 1) was dominated by Tulare Lake and two other smaller lakes (not shown). Tulare Lake was connected to the San Joaquin basin by surface water during high flows and probably by groundwater flow through the alluvium forming the basin divide during most other periods. Permanent and seasonal tule marshes were the dominant feature of the Delta and other floodplain wetlands.

Present conditions vary considerably from historic conditions. In general, high mountain areas have not changed much in appearance, in part because much of the land is in national parks, wilderness areas, or national forest. However, water development, logging, mining, grazing, roads, towns, and recreational development have altered the aquatic ecosystems in major ways (Sierra Nevada Ecosystem Project 1996). Such activities continue to expand, especially in the Sierra Nevada foothills. The most obvious manifestations of these changes are large storage reservoirs that have been constructed on most of the rivers.

Land use changes in the Central Valley have been more dramatic, especially in the San Joaquin Valley. Tulare Lake and the smaller lakes have been completely drained and converted to agricultural uses. Less than 10% of the natural habitats remain in the San Joaquin Valley primarily because of conversion to agricultural land uses (San Joaquin Valley Drainage Program 1990). In the Delta, over 95% of the original wetlands have been lost (The Bay Institute 1998). These areas have been leveed and converted to other uses, primarily agriculture.

Fish Fauna and Evidence of Decline

Moyle (2002) lists 28 species as native to the Central Valley subprovince, including the Delta and the drainages around San Francisco Bay (Table 1). Historically, the fishes of the larger rivers and streams were generally organized into three distinct assemblages with somewhat overlapping ranges, depending on the characteristics of particular streams (Figure 2) (Moyle 2002). A fourth assemblage, the California roach assemblage, was characteristic of intermittent streams and is not discussed further in this paper.

Historically, the rainbow trout *Oncorhynchus mykiss* assemblage occurred in steep, cold rivers and streams at higher elevations (roughly above 450 m). The upstream limit of this assemblage was determined by barriers to dispersal. Most of the streams, rivers, and lakes above 1,500 m were fishless and dominated by native amphibians. Riffle sculpin *Cottus gulosus*, Sacramento sucker *Catostomus occidentalis*, and speckled dace *Rhinichthys osculus* are often part of this assemblage. California roach *Hesperoleucus symmetricus* are sometimes included.

The pikeminnow-hardhead-sucker assemblage occurred in the warmer, lower gradient reaches of the rivers and streams as they flow out of the Sierra Nevada, through the foothills, and onto the valley floor. This assemblage occurred from about 30–450 m in the San Joaquin River drainage, where the transition from the mountains to the valley is relatively abrupt. The elevational range was somewhat broader in the Sacramento River drainage. The assemblage was dominated by Sacramento pikeminnow *Ptychocheilus grandis* and Sacramento sucker. Hardhead *Mylopharodon conocephalus* were largely confined to cooler stream reaches with deep, rock-bottomed pools. Tule perch *Hysterocarpus traskii*, speckled dace, California roach, riffle sculpin, and rainbow trout were often found with this assemblage. Chinook salmon *Oncorhynchus tshawytscha*, steelhead (the anadromous form of *O. mykiss*), and lampreys often had major spawning grounds in the regions occupied by this assemblage.

Historically, the deep-bodied fishes assemblage occupied the low elevation rivers (<30 m), oxbows, floodplain lakes, swamps, and sloughs of the Sacramentoan–Joaquin drainage. This assemblage no longer exists for the reasons discussed in the following sections. The assemblage was dominated by Sacramento perch *Archoplites interruptus*, thicktail chub *Gila crassicauda*, tule perch, Sacramento blackfish *Orthodon microlepidotus*, hitch *Lavinia exilicauda*, and splittail *Pogonichthys macrolepidotus*. Large Sacramento pikeminnows and suckers were common and migrated upstream in the spring to spawn. Anadromous fishes also moved through the areas occupied by this assemblage on their way to upstream spawning grounds.

Of the native fishes, one, the thicktail chub, is globally extinct. The thicktail chub was once very abundant in low elevation lakes and rivers. Thicktail chub bones are among the most abundant fish remains in Native American middens (Schulz and Simons 1973; Mills and Mamika 1980).

Three species have been extirpated from the subprovince. The tidewater goby has been extirpated from the tributary streams to San Francisco Bay (Moyle 2002), but the species was probably not present in the upstream, nontidal reaches of the Sacramento and San Joaquin rivers. The bull trout

Table 1.—Native and alien fish species of the Sacramento–San Joaquin River drainage, California.

Common name	Scientific name	Native	Status[a]	Date of introduction[b]	Reason for introduction[c]
lampreys	Petromyzontidae				
Pacific lamprey	*Lampetra tridentata*	yes	WL	NA	NA
river lamprey	*L. ayresii*	yes	WL	NA	NA
western brook lamprey	*L. richardsoni*	yes	WL	NA	NA
Kern brook lamprey	*L. hubbsi*	yes	SC	NA	NA
sturgeons	Acipenseridae				
white sturgeon	*Acipenser transmontanus*	yes	SI	NA	NA
green sturgeon	*A. medirostris*	yes	SC	NA	NA
herrings	Clupeidae				
American shad	*Alosa sapidissima*	no	SI	1871	food
threadfin shad	*Dorosoma petenense*	no	SI	1954	forage
minnows	Cyprinidae				
California roach	*Hesperoleucus symmetricus*[d]	yes	SI[e]	NA	NA
common carp	*Cyprinus carpio*	no	SI	1872	food
fathead minnow	*Pimephales promelas*	no	SI	1953 (?)	forage/bait
golden shiner	*Notemigonus crysoleucas*	no	SI	1891 (?)	forage
goldfish	*Carassius auratus*	no	SI	1860s (?)	ornamental
hardhead	*Mylopharodon conocephalus*	yes	WL	NA	NA
hitch	*Lavinia exilicauda*	yes	WL	NA	NA
Lahontan redside	*Richardsonius egregius*	no[f]	SI	?	bait
Sacramento blackfish	*Orthodon microlepidotus*	yes	SI	NA	NA
Sacramento pikeminnow	*Ptychocheilus grandis*	yes	SI	NA	NA
splittail	*Pogonichthys macrolepidotus*	yes	SC[g]	NA	NA
speckled dace	*Rhinichthys osculus*	yes	SI	NA	NA
thicktail chub	*Gila crassicauda*	yes	extinct	NA	NA
tui chub	*Gila bicolor*	no[f]	SI	?	bait
red shiner	*Cyprinella lutrensis*	no	SI	c. 1950	bait
suckers	Catostomidae				
mountain sucker	*Catostomus platyrhynchus*	no[f]	SI	?	diversion
Sacramento sucker	*C. occidentalis*	yes	SI	NA	NA
catfishes	Ictaluridae				
black bullhead	*Ameiurus melas*	no	SI	1930s	sport/food
blue catfish	*Ictalurus furcatus*	no	SI	1969	sport
brown bullhead	*A. nebulosus*	no	SI	1874	food
channel catfish	*Ictalurus punctatus*	no	SI	1891 (?)	food/sport

Table 1.—continued

Common name	Scientific name	Native	Status[a]	Date of introduction[b]	Reason for introduction[c]
white catfish	*Ameiurus catus*	no	SI	1874	food
pikes	Esocidae				
northern pike	*Esox lucius*	no	SI	c. 1994	sport
smelts	Osmeridae				
delta smelt	*Hypomesus transpacificus*	yes	FT, ST	NA	NA
longfin smelt	*Spirinchus thaleichthys*	yes	SC	NA	NA
wakasagi	*H. nipponensis*	no	SI	1959	forage
salmon and trout	Salmonidae				
brook trout	*Salvelinus fontinalis*	no	SI	1871 or 1872	sport
brown trout	*Salmo trutta*	no	SI	1893	sport
bull trout	*Salvelinus confluentus*	yes	extirp	NA	NA
Chinook salmon	*Oncorhynchus tshawytscha*	yes			
winter run			FE, SE	NA	NA
spring run			FT, ST	NA	NA
fall/late fall run			SC	NA	NA
coho salmon	*O. kisutch*	yes	extirp	NA	NA
cutthroat trout	*O. clarkii*	no	SI	?	sport
kokanee	*O. nerka*	no	SI	1941	sport
lake trout	*Salvelinus namaycush*	no	SI	1889 (?)	sport/food
rainbow trout	*O. mykiss*	yes			
resident			SI	NA	NA
steelhead (anadromous)			FT	NA	NA
killifish	Fundulidae				
rainwater killifish	*Lucania parva*	no	SI	1950s	unintentional
livebearers	Poeciliidae				
western mosquitofish	*Gambusia affinis*	no	SI	1922	insect control
silversides	Atherinidae				
inland silverside	*Menidia beryllina*	no	SI	1967	insect control
sticklebacks	Gasterosteidae				
threespine stickleback	*Gasterosteus aculeatus*	yes	SI	NA	NA
bass and sunfish	Centrarchidae				
black crappie	*Pomoxis nigromaculatus*	no	SI	1891 or 1908	sport/food
bluegill	*Lepomis macrochirus*	no	SI	1908	sport
green sunfish	*L. cyanellus*	no	SI	1891 or 1908	unintentional
largemouth bass	*Micropterus salmoides*	no	SI	1891 or 1895	sport/food
pumpkinseed	*L. gibbosus*	no	SI		

Table 1.—continued

Common name	Scientific name	Native	Status[a]	Date of introduction[b]	Reason for introduction[c]
redear sunfish	*L. microlophus*	no	SI	c. 1950 and 1954	sport
redeye bass	*Micropterus coosae*	no	SI	1962	sport
Sacramento perch	*Archoplites interruptus*	yes	SC	NA	NA
smallmouth bass	*M. dolomieu*	no	SI	1874	sport/food
spotted bass	*M. punctulatus*	no	SI	1936	sport
warmouth	*Lepomis gulosus*	no	SI	1891(?)	sport/food
white crappie	*Pomoxis annularis*	no	SI	1891 or 1908	sport/food
temperate basses	Moronidae				
striped bass	*Morone saxatilis*	no	SI	1879	sport/food
white bass	*M. chrysops*	no	SI	1965	sport
perches	Percidae				
yellow perch	*Perca flavescens*	no	SI	1891	sport/food
bigscale logperch	*Percina macrolepida*	no	SI	1953	unintentional
surfperches	Embiotocidae				
tule perch	*Hysterocarpus traskii*	yes	SI	NA	NA
gobies	Gobiidae				
tidewater goby	*Eucyclogobius newberryi*	yes	extirp[h]	NA	NA
yellowfin goby	*Acanthogobius flavimanus*	no	SI	early 1960s	ballast water
shimofuri goby	*Tridentiger bifasciatus*	no	SI	c. 1980	ballast water
Shokihaze gody	*T. barbatus*	no	SI?	late 1990s	ballast water
sculpins	Cottidae				
Pacific staghorn sculpin	*Leptocottus armatus*	yes	SI	NA	NA
prickly sculpin	*Cottus asper*	yes	SI	NA	NA
riffle sculpin	*C. gulosus*	yes	SI	NA	NA
righteye flounders	Pleuronectidae				
starry flounder	*Platichthys stellatus*	yes	SI	NA	NA

[a] Status: Extinct, globally extinct; Extirp, extirpated from the Sacramento–San Joaquin drainage; FE, federally endangered; SE, state endangered; FT, federally threatened; ST, state threatened; SC, special concern because declining populations could lead to threatened or endangered status; WL, watch list because populations appear to be declining; SI, stable or increasing.

[b] NA, not applicable; ?, date of introduction is unknown or approximate (Dill and Cordone 1997; Moyle 2002).

[c] NA, not applicable. Data from Dill and Cordone (1997) and Moyle (2002).

[d] Moyle (2002) places this species in the genus *Lavinia*.

[e] As the species is presently defined, California roach is secure. The status of genetically and morphologically distinctive subpopulations varies from threatened to stable or increasing (Moyle 2002).

[f] These species are native to California, but not the Sacramento–San Joaquin drainage.

[g] Splittail was removed from the federal list of threatened species in 2003 (U.S. Fish and Wildlife Service 2003).

[h] Tidewater goby has been extirpated from the Sacramento–San Joaquin drainage and is listed as federally endangered in the remainder of California.

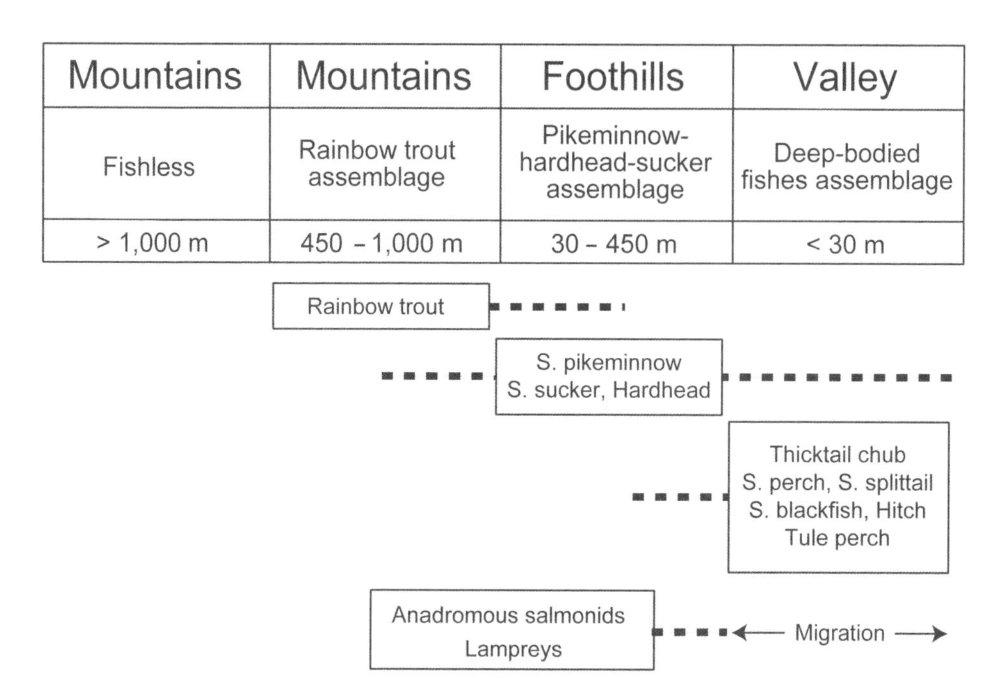

Figure 2.—Distribution of species in the San Joaquin River drainage before European settlement (Moyle 2002). The boxes indicate where the indicated groups of species are most abundant and the dashed lines indicate the approximate range of the species. Only the most common species are included [S., Sacramento].

was native to the McCloud River (a separate subprovince of the Sacramento–San Joaquin Ichthyological Province) but may have occupied similar habitats in the Pit River and upper Sacramento River (Figure 1) (Moyle 2002). Coho salmon was probably never common in the Sacramento–San Joaquin drainage, but there were populations in the tributaries to San Francisco Bay (Leidy 1984) and there were likely small populations in the McCloud River and upper Sacramento River drainages (Moyle 2002).

Another 10 species are considered to be of special concern because of declining populations or have already been listed as threatened or endangered (Table 1) (Moyle 2002). Delta smelt (listed as threatened by state and federal agencies) and longfin smelt (special concern) are primarily estuarine species and rarely enter nontidal freshwaters. The Kern brook lamprey (special concern) is a little known endemic species of nonparasitic lamprey that is mainly of concern because of its limited range (Moyle et al. 1995). Green sturgeon, an anadromous species, is typical of many of the native species because it received little attention until nongovernmental groups petitioned to list the species under the federal Endangered Species Act, a petition that was eventually rejected (NMFS 2003). However, the fact that there were no data available to determine long term population trends and concerns over various threats to the species, especially the Sacramento River population, led the National Marine Fisheries Service (NMFS) to add the species to the list of candidate species.

Similar to thicktail chub, Sacramento perch and splittail were once common in low elevation lakes and rivers of the Central Valley, and the loss and alteration of these habitats presumably led to a reduction in numbers of both species. Sacramento perch is the only native centrarchid west of the Rocky Mountains. Although largely extirpated from its native habitats, it has been widely introduced into ponds and reservoirs in California and elsewhere, where introduced centrarchids are not successful (Moyle 2002). If it were not for these transplanted populations, the species would almost certainly be listed as an endangered species. The status of splittail has been the topic of some controversy (USFWS 2003). The presence of splittail remains in Native American middens near the former large lakes of the Tulare basin (Hartzell 1992; Gobalet and Fenenga 1993) suggests major reductions in population concurrent with losses of habitat. The controversy regarding this species concerns abundance trends in the last several decades. The species was listed as threatened by the U.S. Fish and Wildlife Service (USFWS) in 1999, but was removed from the threatened list in 2003 because of continuing uncertainty over recent trends in abundance and ongoing and future habitat restoration that is expected to improve conditions for the species (USFWS 2003).

The best quantitative evidence of decline for the 10 species of special concern or listed species is for Chinook salmon. Chinook salmon is represented by four different runs in the Sacramento River drainage—winter run, spring run, fall run, and late-fall run. National Marine Fisheries Service includes late-fall run Chinook salmon with fall-run Chinook salmon (Moyle 2002). Only fall-run Chinook salmon occur in the San Joaquin River drainage. Spring-run Chinook salmon, once the dominant run (California Department of Fish and Game 1998; Campbell and Moyle 1991) were extirpated by 1950 (Warner 1991). Sacramento River winter-run Chinook salmon was designated an endangered species by both federal and state agencies (Moyle 2002). Sacramento spring-run Chinook salmon was designated as threatened by both federal and state agencies (Moyle 2002). The fall-run Chinook salmon is a candidate for listing by NMFS. Yoshiyama et al. (1998) compiled annual catch statistics and determined that the Sacramento–San Joaquin in-river fishery alone harvested 4–10 million pounds of Chinook salmon per year for all runs combined (earliest record 1856). On the basis of these and other commercial catch statistics, a conservative estimate for early Central Valley spawning stocks is 1–2 million spawners annually. Recent estimates (1970–2001) of Central Valley spawning escapements range from 94,000 to 566,000 spawners annually (mean = 249,000, SD = 94,000) (Pacific Fishery Management Council 2003). Based on the minimum historical estimate of 1 million spawners annually, the present population corresponds to 9–57% (mean = 25%) of the historic abundance, at best. Also, this estimate is based on natural spawning fish, which includes an unknown, but likely substantial component of hatchery-produced fall-run Chinook salmon. Central Valley steelhead, is listed as federally threatened (Moyle 2002). The principal populations of this anadromous fish are in the Sacramento River drainage, but there is recent evidence of a remnant population in the San Joaquin drainage where the species was once abundant (McEwan 2001). Estimates of historic population size are not possible because historic catch statistics are not available. The population in the 1990s was estimated to be about 10,000 fish for the entire Sacramento–San Joaquin drainage (McEwan 2001). Population estimates at a large diversion dam on the Sacramento River suggest continued declines even in recent times. Annual counts have declined from an average of 11,187 adults for the 10-year period starting in 1967 to 2,202 adults annually in the 1990s (McEwan 2001).

Although extinction, extirpation, and listing of species as threatened, endangered, or special concern are clear indications of a declining fish fauna, there are few quantitative data documenting numeric declines of native species from the mid-1800s to the present, except for the data for Chinook salmon already presented. Presumably the same losses of habitat and other factors (discussed in detail in the next section) leading to extinctions, extirpations, and listings resulted in declines of other native fishes. Further declines have been associated with invasions of alien fishes (see next section for details). Native species have been replaced by alien species in many streams and rivers, particularly in the San Joaquin River drainage (Moyle and Nichols 1973; Saiki 1984; Brown and Moyle 1993; Brown 2000; Saiki et al. 2001a; Moyle et al. 2003). Alien species are not as dominant in the Sacramento River drainage but still occur there and are abundant in some areas (Marchetti and Moyle 2000, 2001; May and Brown 2002).

Invasions of alien fishes have been a major change in California's fish fauna. Moyle (2002) lists a total of 40 alien species as present in the Central Valley subprovince, but new species continue to arrive and the current total is 42 (Table 1). The shokihaze goby, likely a ballast water introduction, has appeared in the Delta and there is a population of northern pike, an illegal introduction, in a reservoir in the headwaters of the Feather River that will spread downstream if not eradicated (Moyle 2002). Dill and Cordone (1997) described the history and motivation for introducing alien species into California in detail. In short, early introductions were usually done for sport or food purposes, and the most recent introductions largely resulted from other human activities, such as ballast water introductions (Table 1).

Causes of Decline and Change

Habitat Alteration and Loss

Some of the first and most dramatic changes in land and water use in California are linked to gold mining. Once easily accessible surface deposits were exhausted, miners turned to large-scale hydraulic mining of ancient alluvial deposits. Hydraulic mining required large quantities of water to supply the water cannons used to expose buried gravels. This water was often supplied via temporary dams and diversions on streams, which often formed barriers to fish passage. The most devastating aspect of this mining was the tremendous quantities of sediment introduced into stream systems. Mount (1995) estimates that 63.7 million m^3 of sediment were washed into the Central Valley from just five heavily mined rivers. The immediate effects of the increased sediment load were to increase fine materials, which smothered spawning gravels. Over the longer term, aggradation of river channels and subsequent incision of rivers isolated many streams from their floodplains. Flooding of surrounding towns and agricultural lands that was due to channel aggradation resulted in leveeing and dredging of many of the most affected river channels. These efforts were successful at moving sediment out of the system (Mount 1995), but further isolated the streams from their floodplains. Indirect effects may also have been important. For example, Herbold and Moyle (1989) suggest that increases in fine sediments may have facilitated invasion of the system by alien striped bass and American shad. These species produce semibouyant eggs that are more likely to survive in systems with high loads of fine sediments than the benthic eggs produced by salmon and many other native California stream fishes (Moyle 2002).

The Gold Rush accelerated the influx of settlers into California (Thomas and Phoenix 1976), stimulating the need for agricultural products. Because of the semiarid Central Valley climate, most early agriculture (early 1800s) occurred near surface water sources (California Department of Water Resources 1982). This demand eventually resulted in construction of thousands of miles of canals used both to drain wetlands, including the large lakes of the Tulare basin, and to supply water to irrigated farmland. Flood control activities allowed agriculture to expand onto the historic floodplains of many rivers, resulting in losses of riparian habitat. Losses of wetland habitat in combination with the introduction of alien predators and competitors were probably key factors in the extinction of thicktail chub and declines of splittail, Sacramento perch, Sacramento blackfish, hitch, and tule perch (Moyle 2002). In addition to the very direct effects of land use change, agricultural activities have a variety of less direct effects on fishes and fish habitat that are covered in subsequent sections.

Flood protection for the Sacramento urban area and other smaller urban areas resulted in channelization and levee construction along many streams and rivers, causing further losses of habitat, particularly for the fishes of the deep-bodied fishes assemblage. Also, several large flood bypasses have been constructed along the Sacramento River. These engineering solutions are successful at providing flood protection but isolate the rivers from their floodplains and associated riparian habitat (Mount 1995). Agriculture and urban development often occur in the isolated floodplains resulting in permanent loss of that habitat. Loss of floodplain habitat that is due to flood control structures and activities has been extensive in the Central Valley. The largest areas of remaining floodplain habitat are largely artificial consisting of the bypasses that are farmed during the summer and may or may not flood during the winter, depending on precipitation and the design of the bypass. For example, the Yolo Bypass only floods when the Sacramento River exceeds 2,000 m^3/s (Sommer et al. 2001a).

Even these artificial floodplains can have important biological functions. Research in the Yolo Bypass has shown that young Chinook salmon grow faster in the warmer food-rich bypass compared with the nearby leveed Sacramento River (Sommer et al. 2001b). The bypass also provides important habitat for splittail in years when it remains flooded for the approximately 30 d needed for spawning and successful recruitment of young of the year (Sommer et al. 1997). The bypass may also serve other less direct functions such as a source of organic matter to the Delta (Sommer et al. 2001a). The importance of bypasses as substitute floodplains provides a strong argument for restoring natural floodplains wherever possible; presumably natural floodplains would have an even higher degree of function. However, some features of floodplain may be less beneficial than anticipated. For example, Feyrer et al. (2004) found that permanent floodplain ponds in the Yolo Bypass primarily supported alien species. Native fishes were less than 1% of fish numbers and less than 3% of fish biomass in the study.

Water Storage and Diversion

Dams constructed for water storage, water diversion, and other purposes are one of the most visible aspects of California's water management system. There are presently over 1,400 reservoirs in California capable of storing almost 60% of the average annual runoff (California Department of Water Resources 1993). This water storage infrastructure supplies water and hydroelectric power for both agricultural and urban needs, and those needs are large. The Central Valley is one of the most productive agricultural areas in the world. In 2000, the agricultural industry of California produced a gross cash income of US$27 billion and supplied more than half the nation's fruits, nuts, and vegetables (California Department of Food and Agriculture 2001). Much of this production occurred in the Central Valley. Most of the agriculture in the Central Valley and elsewhere in California is irrigated. This makes agriculture the largest user of water in the state, taking between 70% and 80% of the stored water (Mount 1995; Moyle 2002). In addition to agricultural demands, urban demands on the water of the Sacramento–San Joaquin drainage are large. Although several urban areas within the Central Valley are growing rapidly, they tend to take relatively little surface water compared with agriculture, at the present time. The bulk of the urban water demand occurs in Southern California and the San Francisco Bay area. Water pumped out of the Delta supplies all or part of the drinking water supply for over 22 million Californians in these two large urban areas (CALFED 2002).

From the perspective of fishes, many of these dams present barriers to movement. Yoshiyama et al. (2001) estimated that only 48% of the over 3,500 km of stream formerly available to anadromous fishes remains accessible. When only Chinook salmon spawning habitat is considered, the loss is much larger with greater than 70% of the habitat now inaccessible. It is likely that the loss for steelhead has been even greater because the species has a greater tendency to spawn in smaller tributary systems and has the ability to ascend higher into many drainages than Chinook salmon (Yoshiyama et al. 2001). The loss for anadromous Pacific lamprey probably parallels that for steelhead. Although large reservoirs on the larger river systems (built in the 1940s or later), such as Shasta Reservoir on the Sacramento River, Oroville Reservoir on the Feather River, and Millerton Reservoir on the San Joaquin River (Figure 1), blocked large areas of habitat, additional reductions in access to the smaller tributary river systems occurred continuously from the Gold Rush era onward (Yoshiyama et al. 2001). For example, access for anadromous fishes to most of the Tuolumne River basin was cut off by the construction in 1893 of a dam just downstream of present day Don Pedro Reservoir (Figure 1) (Brown and Ford 2002).

In addition to blocking access to upstream habitat, dams act as sediment traps, preventing renewal of downstream spawning gravels (Mount 1995). Continued discharge from the dam transports smaller particles, such as gravels, downstream leaving only larger cobbles, boulders, hardpan clay or bedrock. This is mainly perceived as a problem for anadromous fishes, although many other native fishes spawn over gravel riffles (Moyle 2002). Management agencies have attempted to address this problem with gravel augmentation projects and construction of spawning riffles. Such "imported" gravels are often used by Chinook salmon, but not always (Mesick 2001). However, the long-term effectiveness of such programs is questionable. Because rivers continue to move sediment downstream, the positive effects of such projects are transitory (Mount 1995; Kondolf et al. 1996), unless there is a commitment for periodic augmentation over the life of the reservoir.

The impacts of barriers on resident native species are less well known. Sacramento sucker, Sacramento pikeminnow, and Sacramento hitch are all known to make spawning migrations (Moyle 2002); however, it is unclear how important the loss of upstream spawning habitat has been to these species. Perhaps a more important function of barriers and reservoirs has been to fragment the ranges of native stream fishes, particularly the pikeminnow-hardhead-sucker assemblage. Most California stream fishes do not persist in reservoirs for more than 5–10 years because their young do not survive predation by introduced alien fishes such as centrarchid basses and catfishes (Moyle 2002). The reservoirs also provide a source and refuge for alien species that invade upstream reaches of rivers above the reservoirs. For example, electrofishing and snorkeling surveys of the Merced River above McClure Reservoir indicated that alien smallmouth bass dominated the river except for a limited reach near their upstream limit (Brown and Short 1999) (Table 2). Similarly, Gard (1994) found smallmouth bass to be the dominant species in much of the South Fork Yuba River above Englebright Reservoir. This process of fragmentation and isolation by barriers followed by negative interactions with alien species is one of general concern in the Sacramento–San Joaquin drainage (Brown and Moyle 1993).

Entrainment of fishes into water diversion structures has been a major concern in the Sacramento–San Joaquin drainage. In areas accessible to anadromous fishes, Herren and Kawasaki (2001) documented 2,209 diversions in the Delta, 424 diversions on the Sacramento River, and 298 diversions on the San Joaquin River. The large state and federal pumping plants in the Delta are of the most concern because their location is such that they can entrain protected anadromous (Chinook salmon and steelhead) and resident fishes (delta smelt and Sacramento splittail) in large numbers (Arthur et al. 1996; Bennett and Moyle 1996; Brown et al. 1996). Consequently, fish screens and fish salvage facilities were developed at those sites to mitigate their impacts (Brown et al. 1996). Fish screens have also been installed at many of the other larger diversion points in the rivers and in the Delta; however, adequate preproject and postproject assessments are usually not available for evaluating the effectiveness of screening, especially at the population level. Also, fish screens are generally ineffective for larval and small postlarval fishes.

Flow Alteration

Given the number of reservoirs in California and the various purposes they serve—including flood control, water storage, power generation, and water supply—their effects on flow regime vary from stream to stream and from year to year. In general,

Table 2.—Number and percentage of fishes (in parentheses) observed at six sites in the Merced River drainage in 1994 between McClure Reservoir and Yosemite National Park (data from Brown and Short 1999). The elevation of the water surface in McClure Reservoir varies between about 210 and 250 m, depending on water level.

Size	Elevation (m)	Brown trout	Smallmouth bass	Spotted bass	Rainbow trout	Sacramento sucker	Sacramento pikeminnow	Riffle sculpin
Main-stem Merced River								
1	343	0 (0)	104 (79)	0 (0)	0 (0)	10 (8)	15 (11)	0 (0)
2	511	3 (2)	97 (49)	1 (1)	5 (3)	61 (31)	26 (13)	5 (3)
3	556	3 (1)	13 (2)	0 (0)	20 (4)	315 (59)	156 (29)	24 (5)
4	1177	23 (20)	0 (0)	0 (0)	23 (20)	70 (60)	0 (0)	0 (0)
South Fork Merced River								
1	434	0 (0)	76 (99)	0 (0)	0 (0)	0 (0)	1 (1)	0 (0)
2	1268	1 (1)	0 (0)	0 (0)	52 (49)	54 (50)	0 (0)	0 (0)

the larger Sacramento–San Joaquin drainage reservoirs are operated for flood control during the winter, maintaining sufficient storage capacity to moderate peak flows from large storms. In the spring, after the probability of large floods has declined, the reservoirs are allowed to fill, mainly capturing snowmelt runoff. The stored water is then used over the rest of the year. The results of these manipulations are moderated winter peak flows, infrequent inundation of floodplains, and a loss of elevated spring flows (Mount 1995; Bay Institute 1998; Gronberg et al. 1998). In dry years, large reservoirs can store almost an entire year's precipitation (Brown and Ford 2002).

Summer flow effects are different in the San Joaquin and Sacramento River drainages (The Bay Institute 1998). In the San Joaquin River drainage, most of the water is diverted for agriculture at foothill dams, so rivers below the dams are small, containing water from limited flow releases and from return of agricultural drainage water. In the Sacramento River drainage, the larger reservoirs are operated to deliver water at a relatively steady discharge through river channels to the Delta for export to agricultural and urban users via the state and federal pumping plants. In addition, cool water is released from Shasta Reservoir to maintain required water temperatures for maintenance of winter-run Chinook salmon in the Sacramento River. These operations result in much higher summer flows and cooler temperatures in the Sacramento River compared with the San Joaquin River.

Much of the recent research on California fish assemblages has focused on understanding the effects of these changes to the natural flow regime. A number of studies have recognized that high winter–spring discharges favor California native stream fishes over alien species (Baltz and Moyle 1993; Moyle and Light 1996a, 1996b; Brown and Moyle 1997). Brown (2000) noted that flow regime was important in the distribution of native fishes and several groups of alien species in the heavily regulated San Joaquin River drainage. Native species were more common in the lower Stanislaus, Tuolumne, and Merced rivers in a high flow year compared with two low flow years. The importance of flow regime was reinforced by analysis of a 10-year data set from the Tuolumne River, a large tributary to the San Joaquin River (Ford and Brown 2001; Brown and Ford 2002). These analyses indicated that flow regime was a key factor in determining the relative reproductive success of native and alien fishes. High flows favored native species adapted to the cooler water temperatures and more riverine habitat conditions. Alien species did better during low flows when water temperatures were warmer and flow conditions were more conducive to low water velocity nesting species. Marchetti and Moyle (2000, 2001) documented similar relationships in Putah Creek, a small, regulated tributary to the Sacramento River.

May and Brown (2002) observed that, except for agricultural drains, the streams and rivers they sampled in the Sacramento River drainage (not including Putah Creek) generally maintained populations of native fish species, even though alien species were present. They attributed this to minimal flow regulation of the smaller streams sampled and the use of larger rivers for delivery of water to the Delta, rather than diverting water into canal systems. Similarly, Baltz and Moyle (1993) found that the native fish assemblage of an unregulated tributary to the Sacramento River was resistant to invasion by alien species. Overall, the flow alterations in the San Joaquin River drainage appear to create habitat conditions more conducive to alien species and the flow alterations in the Sacramento River drainage are less conducive to alien species, allowing native species to remain common.

Water Quality

Another legacy of the California Gold Rush is contamination of California waters with mercury from mining activities in the Sierra Nevada and from the abandoned mines where the mercury was produced in the coast ranges (Davis et al. 2003). In addition, abandoned hard rock mines for other minerals contribute trace metals and acids. Mercury contamination of fish is a major human and wildlife health concern, especially in the Delta (Davis et al. 2003); however, there is no evidence that mercury accumulation is directly affecting the health of California fish populations. Similarly, much of the concern about trace metals relates to human drinking water quality. A major exception is Iron Mountain Mine, an Environmental Protection Agency Super Fund site, which has been responsible for past fish kills (Finlayson and Wilson 1979). Mitigation work is ongoing, but accumulations of sediments in a debris pond below the mine and in Keswick Reservoir, where the mine discharges to the Sacramento River, are a particular concern. Mount (1995) cites estimates that over 500 kg of copper and 360 kg of zinc are contributed to Keswick Reservoir *per day*. Fish and aquatic insects bioaccumulate metals in the Sacramento River downstream of Keswick Reservoir (Saiki et al. 1995; Cain et al. 2000; Saiki et al. 2001b). Cutthroat trout have been shown to avoid high concentrates of dissolved copper and zinc in laboratory experiments (Woodward et al. 1997), suggesting that these discharges might affect populations of salmonids in the Sacramento River downstream of Keswick Reservoir. A mass movement of these sediment accumulations into the Sacramento River below Keswick Reservoir would likely have severe effects on downstream fish populations including the endangered winter-run Chinook salmon.

Central Valley irrigated agriculture makes extensive use of pesticides, including hundreds of compounds in the thousands of kilograms of active ingredient in any given year (California Department of Pesticide Regulation 2002). Erosion of agricultural soils continues to deliver long-lived organochlorine pesticides and their breakdown products to rivers and streams where they accumulate in biota (Saiki and Schmitt 1986; Brown 1997; MacCoy and Domagalski 1999). Despite this input of toxins, acute mortalities of fishes are rarely documented, with the exception of catastrophic spills (e.g., Cantara spill, Payne and Associates 1998). In many cases, the actual percentage of applied pesticide reaching a surface water body is quite small (Kratzer 1997); however, there is evidence for effects of these chemicals on fishes. Bailey et al. (1994) used bioassays to demonstrate that rice pesticides could account for declines in striped bass populations. Bennett et al. (1995) found histological evidence of liver deformities in striped bass larvae, consistent with pesticide exposure. Chronic effects of pesticides, such as disruption of antipredatory and homing behaviors have been demonstrated (Scholz et al. 2000). Environmental estrogens have been detected in agricultural drains (Johnson et al. 1998) that could affect sexual and functional development of young (Leatherland 1992; Reijnders and Brasseur 1992). Such environmental estrogens might account for feminized male Chinook salmon observed in the Columbia and Sacramento–San Joaquin drainages; however other explanations are possible (Nagler et al. 2001; Williamson and May 2002). Indirect effects on fishes through toxicity to food organisms also seems possible given results of invertebrate bioassays (Foe 1995; Kuivila and Foe 1995).

Selenium in agricultural drainage water in the Central Valley has been a controversial topic for many years. Irrigation of soils derived from marine sediments on the west side of the San Joaquin Valley results in drainage water that is high in concentrations of dissolved salts and trace elements, including selenium. This selenium subsequently entered the food web and caused developmental abnormalities in birds (e.g., Ohlendorf et al. 1988a, 1989; Williams et al. 1989b). Selenium also reached high levels in other biota, and in fishes, reached concentrations known to affect reproduction in some species (Saiki and Lowe 1987; Ohlendorf et al. 1988b; Saiki and May 1988; Saiki et al. 1992a; Saiki and Ogle 1995). Changes in drainage water management have improved the situation to some extent, but selenium is still a management concern, especially in some small tributaries on the west side of the San Joaquin River. Selenium is also a concern in the Delta where the major sources are industrial. Filter-feeding clams accumulate high concentrations of selenium, which are then eaten by some fishes (Linville et al. 2002).

Agricultural drainage water may also contain several seemingly more mundane materials, including dissolved salts and sediment. Saiki et al. (1992b) found that agricultural drainage water from the west side of the San Joaquin Valley could affect survival and growth of juvenile Chinook salmon and striped bass. The effect was attributed to the unusual ionic composition of the drainage water resulting from irrigating soils derived from marine sediments. Although the effects of sediment in the Central Valley have not been well documented, such effects have been recognized as important in many other areas (Waters 1995).

Changes in water quality, as documented above, are probably best viewed as chronic stressors on fish populations rather than causes of acute mortality. Saiki (1984), Brown (2000), and Saiki et al. (2001a) documented the dominance of introduced species over native species in the lower elevation rivers and agricultural drains of the irrigated San Joaquin Valley. May and Brown (2002) found that the principal habitats dominated by alien species in the lower elevation portion of the Sacramento River drainage were agricultural drains. Brown (2000) hypothesized that the harsh and fluctuating environmental conditions associated with agriculturally dominated water bodies favored introduced species from habitats with similar conditions over native species adapted for habitats that no longer exist or are extremely modified. Water clarity and its effects on predation risk and feeding efficiency may also be an important factor in determining the distribution of native and alien species (Brown 2000; Bonner and Wilde 2002).

Alien Invasions

Alien species can be regarded as an irreversible, self-replicating kind of pollution. Alien fishes are most likely to pollute environments already altered by human activity, although they can also invade relatively pristine systems if conditions are favorable (Moyle and Light 1996a, 1996b). In California, introductions of alien fishes have gone hand in hand with major disturbances of the landscape. Among the first successful introductions were American shad (1871), common carp (1872), and striped bass (1879)—species that thrived in the silty habitats created by hydraulic mining. Most of the subsequent 48 introductions were species that thrive in reservoirs, ponds, and stagnant river channels (Moyle 2002). Many of these species, once established, will invade less disturbed habitats and compete with or prey on native fishes. Other species, such as brown trout, brook trout, and redeye bass were introduced as sport fish because they *could* thrive in relatively undisturbed habitats such as mountain or foothill streams. Redeye bass, for example, are now the most abundant fish in long reaches of the Cosumnes River, the last Sacramento–San Joaquin drainage river without a major dam on its main stem (Moyle et al. 2003). In addition, fishes native to various drainages in California have been introduced, often via aqueducts, into drainages to which they are not native. Some of these interbasin transfers have been deliberate, such as the planting of rainbow trout into high mountain lakes and streams. The result of all these invasions, combined with extirpations of native fishes, has been increased homogenization of the fish fauna of California (Marchetti et al. 2001). In some streams, the process of homogenization can be reversed by recreating a natural flow regime that favors native fishes, but even in these cases, the alien fishes are rarely eliminated; they just become uncommon.

Discussion

The decline of the native fish communities of the Sacramento–San Joaquin drainage cannot be attributed to any single change in habitat or water quality condition. Bennett and Moyle (1996) reached a similar conclusion about the fish communities of the San Francisco Estuary. Many species of California native stream fishes are actually quite tolerant of stressful physical conditions (Cech et al. 1990; Brown and Moyle 1993; Moyle 2002). However, invasion of new species in addition to loss and alteration of physical habitat, alteration of physical processes (e.g., flow regime), and changes in water quality have exceeded the ability of the native fishes to adapt. In essence, environmental conditions have been changed to the extent that they are now more suitable for alien species introduced from areas with similar conditions (Moyle and Light 1996a, 1996b; Brown 2000; Marchetti and Moyle 2001).

The changes in the fish assemblages are most obvious in the San Joaquin River drainage, where the physical and water quality changes have been most severe (Saiki 1984; Brown 2000; Saiki et al. 2001a). Of the three native fish assemblages historically present (Figure 2), the deep-bodied fish assemblage is completely gone, and the rainbow trout and pikeminnow-hardhead-sucker assemblages are highly disrupted (Figure 3). Alien species dominate most habitats downstream of the foothill dams. In

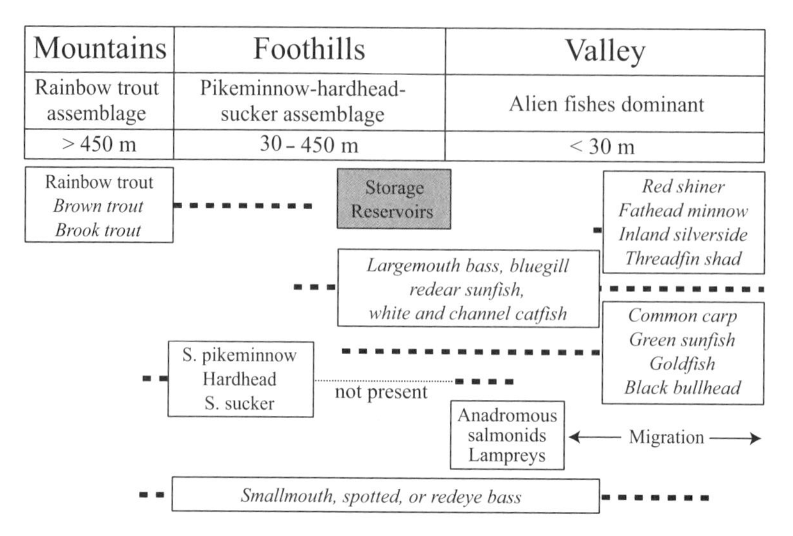

Figure 3.—Distribution of species in the San Joaquin River drainage after European settlement (Brown 2000; Moyle 2002). The ranges of the historical fish assemblages are taken from Figure 2. The boxes indicate where the indicated groups of species are most abundant and the dashed lines indicate the approximate range of the species. Alien species are shown in italics. Only the most common species are included [S., Sacramento]. All the alien species except red shiner, fathead minnow, and inland silverside are commonly found in reservoirs.

the San Joaquin River drainage, the native species that have not disappeared entirely are now largely restricted to the areas just below the reservoirs. These areas resemble their native habitats, with regard to physical habitat and water temperature. The native species still present above the reservoirs are threatened by alien species moving upstream and alien species introduced into higher elevation waters. The situation in the Sacramento River drainage appears to be less severe. Native species still maintain healthy populations in many areas because of different water management practices on dammed streams, the presence of undammed tributaries, and less severe habitat loss and alteration, compared with the San Joaquin River drainage. Low abundances of alien species in many rivers and streams and the dominance of alien fishes in agricultural drains suggests that environmental changes resulting from changing water management operations or invasions of additional alien species could result in rapid changes in fish assemblage composition (May and Brown 2002).

The formal listing of multiple fishes in the drainage as threatened or endangered has stimulated numerous restoration efforts; however, these restoration efforts are generally not focused on native stream fishes. Concerns in the Delta focus on more estuarine species, such as delta smelt and splittail, and migratory anadromous salmonids. In the rivers, efforts focus mostly on threatened anadromous salmonids, although other native fishes are increasingly of concern (e.g., Moyle et al. 1995; Brown and Ford 2002). Management actions that are focused on anadromous salmonids may or may not aid other native fishes. In fact, management actions taken to restore Chinook salmon will not necessarily help restore steelhead in the same river (McEwan 2001). So, how can restoration efforts be directed toward communities rather than species? Moyle (1995, 2002) and Moyle and Yoshiyama (1994) suggest a plan to create a series of aquatic diversity management areas (ADMAs) throughout California. The series of ADMAs would be designed to protect all of California's native aquatic species, not just fishes, and would adopt a watershed approach to management and protection. However, there has been no movement in the management agencies to adopt the ADMA approach.

The largest restoration presently ongoing in the Sacramento–San Joaquin drainage is the Ecosystem Restoration Program (ERP) of the multibillion dollar CALFED Program (CALFED 2002). The purpose of the ERP is to restore ecosystem health of the Delta, its watershed, and northern San Francisco Bay as part of the larger CALFED goals of protecting all beneficial uses of Delta waters, including drinking water quality, and agricultural and urban water supply. The ERP has been conceived as an ecosystem restoration program rather than as a species-specific restoration program. This approach is driven by the need to increase populations of multiple species of threatened and endangered fishes, plants, birds, and mammals. The difficulties of this approach in such an altered system are a major challenge for managers and scientists. For example, one of the initial ERP proposals was to increase populations of native fishes and other plants and animals by restoring tidal wetlands in the Delta. This proposal makes intuitive sense because tidal wetland habitat, once the dominant habitat type in the Delta, is now very limited. However, the benefits of tidal wetland restoration for fishes is complicated by the presence of a variety of alien fishes, other alien animals, and alien plants that interact with each other and native species in unexpected ways (Brown 2003a). There might also be effects on drinking-water quality because of production of organic carbon that can form disinfection byproducts (Brown 2003b) and on accumulation of mercury in food webs (Davis et al. 2003).

Similar complexities can be expected when ecosystem restoration in Central Valley rivers is addressed. For example, Sommer et al. (2001a, 2001b) document the benefits of an engineered seasonal floodplain of the Sacramento River. However, it is unclear if restoring natural floodplain features such as sloughs and oxbow lakes would be beneficial to native fishes. In the Willamette River, Scheerer (2002) found that populations of the endangered Oregon chub *Oregonichthys crameri*, a native floodplain minnow, were negatively affected by increased connectivity among floodplain habitats. Increased connectivity facilitated invasion of Oregon chub habitats by alien species, which apparently suppressed the chub populations. Similarly, Feyrer et al. (2004) found that Yolo Bypass perennial ponds were dominated by alien fishes. It seems likely that such complex interactions in the Sacramento–San Joaquin drainage will require managers and scientists to work closely to ensure that restoration efforts produce the desired outcome.

The need for developing cooperative and innovative approaches to native fish conservation has been made even more urgent by the recent predictions of climate change models. O'Neal (2002) determined that human-caused climate change will result in warming of mean annual temperatures in Central California of at least 3°C, with major consequences to precipitation and water temperatures. Basically, it is expected that there will be less snow and more rain, making run-off even more strongly seasonal than it is today and less predictable from year to year (Aguado et al. 1992; Dettinger and Cayan 1995; O'Neal 2002). In many respects, Sacramento–San Joaquin drainage streams will become more like coastal streams, in which flows rise and fall with rainfall rather than having a long extended period of gradually declining flows in the spring as the result of snowmelt. Because the snowpack in the Sierra Nevada essentially serves as a giant reservoir that slowly sends its water downstream, its depletion will reduce the ability of Californians to store enough water in artificial reservoirs to meet their needs. Increased demand for limited water in turn is likely to greatly decrease the flows of rivers below dams, especially during periods of extended drought. Reduced releases of water that is also warmer (from rain) will cause dramatic changes to the aquatic ecosystems. Without major changes to our water distribution system, we hypothesize some likely consequences to fishes, which in many ways are accelerations of trends caused by other factors:

1. Chinook salmon will likely decline in abundance. The spring-run, which requires coldwater year around, is especially vulnerable because some populations utilize unregulated tributaries. Rainfall runoff is warmer and less sustained than snowmelt runoff, resulting in diminished areas and duration of appropriate coldwater habitat in unregulated streams. The winter and late-fall runs, which also require cold water year around, may be less affected because they utilize the Sacramento River below Shasta Reservoir and reservoir releases are utilized to maintain appropriate temperatures. Declines may occur be cause the coldwater pool available for downstream release is likely to be smaller

and more variable in size and temperature from year to year. Wild fall-run Chinook salmon will presumably persist by shifting the peak of spawning to later in the season because adult upstream migrations will be delayed by low, warm fall flows. There may also be effects on fall-run juveniles if in creased coldwater flows from reservoirs are needed to support juvenile rearing and migration; however, such flows may not be necessary. Connor et al. (2002) found that fall-run Chinook salmon emerged from the gravel sooner, grew faster, and emigrated sooner in response to warmer temperatures.
2. Alien warmwater fishes such as common carp, bluegill, and largemouth bass will be come more abundant as warm, quiet-water habitats become more widespread.
3. Common native fishes, such as Sacra mento sucker and Sacramento pike-minnow, will likely persist. As down stream habitat for these species is lost be cause of reduced flows and warmer temperatures, the ranges of the species will extend into upstream areas of formerly coldwater habitat that previously supported the rainbow trout assemblage.
4. Resident populations of wild rainbow trout will be reduced in range and abundance with decreased flows and increased temperatures of high and mid-elevation streams (O'Neal 2002).
5. Floodplain dependent species, such as splittail, may benefit, assuming flooding during February–April is more frequent and longer in duration.

While many of these effects may be unavoidable, improved management of our water supplies can reduce them. Major changes that will need to be seriously considered include

1. Increase floodplain capacity in the Sacramento–San Joaquin drainage. Increasing the area of floodplain will provide greater flexibility in the management of flood flows and reservoir levels. In addition, flooding floodplains benefits native fishes, particularly Chinook salmon and splittail.
2. Conserve remaining coldwater streams and rivers specifically for fish and other aquatic biota.
3. Develop a system of ADMAs that can, through active management, provide ref uges for native fishes likely to be negatively impacted by climate change and other factors.
4. Improve the efficiency of water use and transfer in order to reduce waste and leave more water instream for fishes.

The native fish assemblages of the Sacramento–San Joaquin drainage have already declined in both species richness and abundance. Failure to consider these assemblages in water management decisions will almost certainly result in further declines of native fishes.

Acknowledgments

We thank Terry Maret, Anne Brasher, Michael Saiki, Robert Calamusso, and other reviewers for the many helpful comments that improved the paper. Special thanks to the many students, friends, and colleagues that have contributed to the information on California native fishes.

References

Aguado, E., D. Cayan, L. Riddle, and M. Roos. 1992. Climatic fluctuations and the timing of west coast streamflow. Journal of Climate 5:1468–1483.

Arthur, J. F., M. D. Ball, and S. Y. Baughman. 1996. Summary of federal and state water project environmental impacts in the San Francisco Bay-Delta Estuary, California. Pages 445–495 *in* J. T. Hollibaugh, editor. San Francisco Bay: the ecosystem, further investigations into the natural history of San Francisco Bay and delta with reference to the influence of man. Pacific Division of the American Association for the Advancement of Science, San Francisco.

Bailey, H. C., C. Alexander, C. Digiorgio, M. Miller, S. I. Doroshov, and D. E. Hinton. 1994. The effect of agricultural discharge on striped bass (*Morone saxatilis*) in California's Sacramento–

San Joaquin drainage. Ecotoxicology 3:123–142.

Baltz, D. M., and P. B. Moyle. 1993. Invasion resistance to introduced species by a native assemblage of California stream fishes. Ecological Applications 3:246–255.

Baumhoff, M. A. 1963. Ecological determinants of aboriginal California populations. University of California Publications in American Archaeology and Ethnology 29:155–236.

Bennett, W. A., and P. B. Moyle. 1996. Where have all the fishes gone? Interactive factors producing fish declines in the Sacramento–San Joaquin estuary. Pages 519–542 *in* J. T. Hollibaugh, editor. San Francisco Bay: the ecosystem, further investigations into the natural history of San Francisco Bay and delta with reference to the influence of man. Pacific Division of the American Association for the Advancement of Science, San Francisco.

Bennett, W. A., D. J. Ostrach, and D. E. Hinton. 1995. Larval striped bass condition in a drought-stricken estuary: evaluating pelagic food limitation. Ecological Applications 5:680–692.

Bonner, T. H., and G. R. Wilde. 2002. Effects of turbidity on prey consumption by prairie stream fishes. Transactions of the American Fisheries Society 131:1203–1208.

Brown, L. R. 1997. Concentrations of chlorinated organic compounds in biota and bed sediment in streams of the San Joaquin Valley, California. Archives of Environmental Contamination and Toxicology 33:357–368.

Brown, L. R. 2000. Fish communities and their associations with environmental variables, lower San Joaquin River drainage, California. Environmental Biology of Fishes 57:251–269.

Brown, L. R. 2003a. Will tidal wetland restoration enhance populations of native fishes? Article 2 *in* L. R. Brown, editor. Issues in San Francisco Estuary tidal wetlands restoration. San Francisco Estuary and Watershed Science 1(1). Available: http://repositories.cdlib.org/jmie/sfews/vol1/iss1/art2 (November 2003).

Brown, L. R. 2003b. Potential effects of organic carbon production on ecosystems and drinking water quality. Article 3 *in* L. R. Brown, editor. Issues in San Francisco Estuary tidal wetlands restoration. San Francisco Estuary and Watershed Science 1(1). Available: http://repositories.cdlib.org/jmie/sfews/vol1/iss1/art3 (November 2003).

Brown, L. R., and T. J. Ford. 2002. Effects of flow on the fish communities of a regulated California river: implications for managing native fishes. River Research and Applications 18:331–342.

Brown, L. R., and P. B. Moyle. 1993. Distribution, ecology, and status of the fishes of the San Joaquin River drainage, California. California Fish and Game 79:96–114.

Brown, L. R., and P. B. Moyle. 1997. Invading species in the Eel River, California: successes, failures, and relationships with resident species. Environmental Biology of Fishes 49:271–291.

Brown, L. R., and T. M. Short. 1999. Biological, habitat, and water quality conditions in the upper Merced River drainage, Yosemite National Park, California, 1993–1996. U.S. Geological Survey, Water Resources Investigations Report 99–4088, Sacramento, California.

Brown, R., S. Greene, P. Coulston, and S. Barrow. 1996. An evaluation of the effectiveness of fish salvage operations at the intake of the California aqueduct, 1979–1993. Pages 497–518 *in* J. T. Hollibaugh, editor. San Francisco Bay: the ecosystem, further investigations into the natural history of San Francisco Bay and delta with reference to the influence of man. Pacific Division of the American Association for the Advancement of Science, San Francisco.

Cain, D. J., J. L. Carter, S. V. Fend, S. N. Luoma, C. N. Alpers, and H. E. Taylor. 2000. Metal exposure in a benthic macroinvertebrate, *Hydropsyche californica*, related to mine drainage in the Sacramento River. Canadian Journal of Fisheries and Aquatic Sciences 57:380–390.

CALFED (CALFED Bay-Delta Program). 2002. CALFED Bay-Delta program annual report 2002. CALFED Bay-Delta Program, Sacramento, California. Available: http://calwater.ca.gov/AboutCalfed/AnnualReport2002.shtml (December 2003).California Institute. 1999. California Institute population data: an online source for information on California and federal policy. Available: http://www.calinst.org/datapages/popproj.html (November 2002).

California Department of Fish and Game. 1998. A status review of the spring-run Chinook salmon (*Oncorhynchus tsawytscha*) in the Sacramento River drainage. California Department of Fish and Game, Candidate Species Report 98–01, Sacramento.

California Department of Food and Agriculture. 2001. Agricultural resource directory 2001. California Department of Food and Agriculture, Sacramento.

California Department of Pesticide Regulation. 2002. Pesticide use information. Available: http://www.cdpr.ca.gov/docs/pur/purmain.htm (November 2002).

California Department of Water Resources. 1982. The hydrologic-economic model of the San Joaquin Valley. California Department of Water Resources, Bulletin 214, Sacramento.

California Department of Water Resources. 1993. California water plan update. California Department of Water Resources, Bulletin 160–93, Volume 2, Sacramento.

Campbell, E. A., and P. B. Moyle. 1991. Historical and recent population sizes of spring-run chinook salmon in California. Pages 155–216 *in* T. Hassler, editor. Proceedings of the 1990 Northeast Pacific Chinook and Coho Salmon Workshop. Humboldt State University, California Cooperative Fishery Research Unit, Arcata, California.

Cech, J. J. Jr., S. J. Mitchell, D. T. Castleberry, and M. McEnroe. 1990. Distribution of California stream fishes: influence of environmental temperature and hypoxia. Environmental Biology of Fishes 29:95–105.

Connor, W. P., H. L. Burge, R. Waitt, and T. C. Bjornn. 2002. Juvenile life history of wild fall chinook salmon in the Clearwater and Snake rivers. North American Journal of Fisheries Management 22:703–712.

Davis, J. A., D. Yee, J. N. Collins, S. Schwarzbach, and S. N. Luoma. 2003. Potential for increased mercury accumulation in the estuary food web. Article 4 *in* L. R. Brown, editor. Issues in San Francisco Estuary tidal wetlands restoration. San Francisco Estuary and Watershed Science 1(1). Available: http://repositories.cdlib.org/jmie/sfews/vol1/iss1/art4 (November 2003).

Dettinger, M. D., and D. R. Cayan. 1995. Large-scale atmospheric forcing of recent trends toward early snowmelt runoff in California. Journal of Climate 8:606–623.

Dill, W. A., and A. J. Cordone. 1997. History and status of introduced fishes in California, 1871–1996. California Department of Fish and Game, Fishery Bulletin 178, Sacramento.

Feyrer, F., T. R. Sommer, S. C. Zeug, G. O'Leary, and W. Harrell. 2004. Fish assemblages of perennial floodplain ponds of the Sacramento River, California (USA), with implications for the conservation of native fishes. Fisheries Management and Ecology 11:335–344.

Finlayson, B., and D. Wilson. 1979. Acid-mine waste–how it affects king salmon in the upper Sacramento River. Outdoor California 40(6):8–12.

Foe, C. 1995. Insecticide concentrations and invertebrate bioassay mortality in agricultural return water from the San Joaquin basin. Central Valley Regional Water Quality Control Board, Sacramento, California.

Ford, T. J., and L. R. Brown. 2001. Distribution and abundance of chinook salmon and resident fishes of the lower Tuolumne River, California. Pages 253–304 *in* R. Brown, editor. Contributions to the biology of Central Valley salmonids. California Department of Fish and Game, Fishery Bulletin 179, Sacramento.

Gard, M. F. 1994. Biotic and abiotic factors affecting native stream fishes in the South Yuba River, Nevada County, California. Doctoral dissertation, University of California, Davis.

Gobalet, K. W., and G. L. Fenenga. 1993. Terminal Pleistocene-Early Holocene fishes from Tulare Lake, San Joaquin Valley, California, with comments on the evolution of Sacramento squawfish (*Ptychocheilus grandis*: Cyprinidae). Paleobios 15(1):1–8.

Gronberg, J. A. M., N. M. Dubrovsky, C. R. Kratzer, J. L. Domagalski, L. R. Brown, and K. R. Burow. 1998. Environmental setting of the San Joaquin–Tulare basins, California. U.S. Geological Survey, Water-Resources Investigations Report 97–4205, Sacramento, California.

Hartzell, L.L. 1992. Hunter-gatherer adaptive strategies and lacustrine environments in the Buena Vista Lake basin, Kern County, California. Doctoral dissertation. University of California, Davis.

Herbold, B., and P. B. Moyle. 1989. The ecology of the Sacramento–San Joaquin Delta: a community profile. United States Fish and Wildlife Service, U.S. Fish and Wildlife Service Report 85, Washington, D.C.

Herren, J. R., and S. S. Kawasaki. 2001. Inventory of water diversions in four geographic areas in California's Central Valley. Pages 343–355 *in* R. Brown, editor. Contributions to the biology of Central Valley salmonids. California Department of Fish and Game, Fishery Bulletin 179, Sacramento.

Johnson, M. L., A. Salveson, L. Holmes, M. S.

Denison, and D. M. Fry. 1998. Environmental estrogens in agricultural drain water from the Central Valley of California. Bulletin of Environmental Contamination and Toxicology 60:609–614.

Kondolf, G. M., J. C. Vick, and T. M. Ramirez. 1996. Salmon spawning habitat rehabilitation on the Merced River, California: an evaluation of project planning and performance. Transactions of the American Fisheries Society 125:899–912.

Kratzer, C. R. 1997. Transport of diazinon in the San Joaquin River basin, California. U.S. Geological Survey Open-File Report 97–411, Sacramento, California.

Kroeber, A. L. 1939. Cultural and natural areas of native North America. University of California Publications in American Archaeology and Ethnology 38:1–242.

Kroeber, A. L. 1963. The nature of land-holding groups in aboriginal California. Pages 81–120 *in* R. F. Heizer, editor. Aboriginal California, three studies in cultural history. Report of the University of California Archaeological Survey, No. 56, Berkeley, California.

Kuivila, K. M., and C. G. Foe. 1995. Concentrations, transport and biological effects of dormant spray pesticides in the San Francisco estuary, California. Environmental Toxicology and Chemistry 14:1141–1150.

Leatherland, J. F. 1992. Endocrine and reproductive function in Great Lakes salmon. Pages 129–146 *in* T. Colborn and C. Clement, editors. Chemically induced alterations in sexual and functional development: the wildlife/human connection. Princeton Scientific Publishing, Princeton, New Jersey.

Leidy, R. A. 1984. Distribution and ecology of stream fishes in the San Francisco Bay drainage. Hilgardia 52:1–175.

Linville, R. G., S. N. Luoma, L. Cutter, and G. A. Cutter. 2002. Increased selenium threat as a result of invasion of the exotic bivalve *Potamocorbula amurensis* into the San Francisco Bay-Delta. Aquatic Toxicology 57:51–64.

MacCoy, D. E., and J. L. Domagalski. 1999. Trace elements and organic compounds in streambed sediment and aquatic biota from the Sacramento River basin, California, October and November 1995. U.S. Geological Survey, Water-Resources Investigations Report 99–4151, Sacramento, California.

Marchetti, M. P., and P. B. Moyle. 2000. Spatial and temporal ecology of native and introduced larval fish in lower Putah Creek (Yolo Co., CA). Environmental Biology of Fishes 58:75–87.

Marchetti, M. P., T. Light, J. Feliciano, T. Armstrong, Z. Hogan, and P. B. Moyle. 2001. Homogenization of California's fish fauna through abiotic change. Pages 269–288 *in* J. L. Lockwood and M. L. McKinney, editors. Biotic homogenization. Kluwer/Academic Press, New York.

Marchetti, M. P., and P. B. Moyle. 2001. Keeping alien fishes at bay: effects of flow regime and habitat structure on fish assemblages in a regulated California stream. Ecological Applications 11:75–87.

May, J. T., and L. R. Brown. 2002. Fish communities of the Sacramento River basin: implications for conservation of native fishes in the Central Valley, California. Environmental Biology of Fishes 63:373–388.

McEwan, D. R. 2001. Central Valley steelhead. Pages 1–44 *in* R. Brown, editor. Contributions to the biology of Central Valley salmonids. California Department of Fish and Game, Fishery Bulletin 179, Sacramento.

Mesick, C. 2001. Studies of spawning habitat for fall-run chinook salmon in the Stanislaus River between Goodwin Dam and Riverbank from 1994 to 1997. Pages 217–252 *in* R. Brown, editor. Contributions to the biology of Central Valley salmonids. California Department of Fish and Game, Fishery Bulletin 179, Sacramento.

Mills, T. J., and K. A. Mamika. 1980. The thicktail chub, *Gila crassicauda*, an extinct California fish. California Department of Fish and Game, Inland Fisheries Endangered Species Program Special Publication 80–2, Sacramento.

Minckley, W. L., D. A. Hendrickson, and C. E. Bond. 1986. Geography of North American freshwater fishes: description and relationships to intercontinental tectonism. Pages 519–614 *in* C. H. Hocutt and E. O. Wiley, editors. The zoogeography of North American freshwater fishes. John Wiley, New York.

Mount, J. F. 1995. California rivers and streams. University of California Press, Berkeley.

Moyle, P. B. 1995. Conservation of native freshwater fishes in the Mediterranean-type climate of California, USA: a review. Biological Conservation 72:271–279.

Moyle, P. B. 2002. Inland fishes of California (2nd edition). University of California Press, Berkeley.

Moyle, P. B., P. K. Crain, K. Whitener, and J. F. Mount.

2003. Alien fishes in natural streams: fish distribution, assemblage structure, and conservation in the Cosumnes River, California, U.S.A. Environmental Biology of Fishes 68:143–162.

Moyle, P. B., and R. L. Leidy. 1992. Loss of biodiversity in aquatic systems: evidence from fish faunas. Pages 127–170 *in* P. L. Fiedler and S. K. Jain, editors. Conservation biology: the theory and practice of nature conservation, preservation, and management. Chapman Hall, New York.

Moyle, P. B., and T. Light. 1996a. Fish invasions in California: do abiotic factors determine success? Ecology 77:1666–1670.

Moyle, P. B., and T. Light. 1996b. Biological invasions of fresh water: empirical rules and assembly theory. Biological Conservation 78:149–162.

Moyle, P. B., and R. D. Nichols. 1973. Ecology of some native and introduced fishes of the Sierra Nevada foothills in central California. Copeia 1973:478–490.

Moyle, P. B., J. J. Smith, R. A. Daniels, T. L. Taylor, D. G. Price, and D. M. Baltz. 1982. Distribution and ecology of stream fishes of the Sacramento–San Joaquin drainage system, California. University of California Publications in Zoology 115, Berkeley.

Moyle, P. B., and J. E. Williams. 1989. Biodiversity loss in the temperate zone: decline of the native fish fauna of California. Conservation Biology 4:275–284.

Moyle, P. B., and R. M. Yoshiyama. 1994. Protection of aquatic biodiversity in California: a five-tiered approach. Fisheries 19(2):6–18.

Moyle, P. B., R. M. Yoshiyama, J. E. Williams, and E. D. Wikramanayake. 1995. Fish species of special concern of California. California Department of Fish and Game, Sacramento.

Nagler, J. J., J. Bouma, G. H. Thorgaard, and D. D. Dauble. 2001. High incidence of a male-specific genetic marker in phenotypic female chinook salmon from the Columbia River. Environmental Health Perspectives 109:67–69.

NMFS (National Marine Fisheries Service). 2003. Endangered and threatened wildlife and plants; 12-month finding on a petition to list North American green sturgeon as a threatened or endangered species. Federal Register 68:19(29 January 2003)4433–4441.

Ohlendorf, H. M., R. L. Hothem, and T. W. Aldrich. 1988b. Bioaccumulation of selenium by snakes and frogs in the San Joaquin Valley, California. Copeia 1988:704–710.

Ohlendorf, H. M., R. L. Hothem, and D. Welsh. 1989. Nest success, cause-specific nest failure, and hatchability of aquatic birds at selenium contaminated Kesterson Reservoir and a reference site. The Condor 91:787–797.

Ohlendorf, H. M., A. W. Kilness, J. L. Simmons, R. K. Stroud, D. J. Hoffman, and J. F. Moore. 1988a. Selenium toxicosis in wild aquatic birds. Journal of Toxicology and Environmental Health 24:67–92.

Omernik, J. M. 1987. Ecoregions of the conterminus United States. Annals of the Association of American Geographers 77:118–125.

O'Neal, K. 2002. Effects of global warming on trout and salmon in U.S. streams. Defenders of Wildlife, Washington, D.C.

Pacific Fishery Management Council. 2003. Review of 2002 ocean salmon fisheries. Pacific Fishery Management Council, Portland, Oregon.

Payne, T. R., and Associates. 1998. Recovery of fish populations in the upper Sacramento River following the Cantara spill of July 1991. 1997 Annual Report. California Department of Fish and Game, Sacramento.

Reijnders, J. H., and S. M. J. M. Brasseur. 1992. Xenobiotic induced hormonal and associated developmental disorders in marine organisms and related effects in humans. Pages 159–174 *in* T. Colborn and C. Clement, editors. Chemically induced alterations in sexual and functional development: the wildlife/human connection. Princeton Scientific Publishing, Princeton, New Jersey.

Reisner, M. 1986. Cadillac Desert. Viking-Penguin, New York.

Saiki, M. K. 1984. Environmental conditions and fish faunas in low elevation rivers on the irrigated San Joaquin Valley floor, California. California Fish and Game 70:145–157.

Saiki, M. K., D. T. Castleberry, T. W. May, B. A. Martin, and F. N. Bullard. 1995. Copper, cadmium, and zinc concentrations in aquatic food chains from the upper Sacramento River (California) and selected tributaries. Archives of Environmental Contamination and Toxicology 29:484–491.

Saiki, M. K., M. R. Jennings, and T. W. May. 1992a. Selenium and other elements in freshwater fishes from the irrigated San Joaquin Valley, California. The Science of the Total Environment 126:109–137.

Saiki, M. K., M. R. Jennings, and R. H. Wiedmeyer. 1992b. Toxicity of agricultural subsurface drainage water from the San Joaquin Valley, California, to juvenile chinook salmon and striped bass. Transactions of the American Fisheries Society 121:78–93.

Saiki, M. K., and T. P. Lowe. 1987. Selenium in aquatic organisms from subsurface agricultural drainage water, San Joaquin Valley, California. Archives of Environmental Contamination and Toxicology 16:657–670.

Saiki, M. K., B. A. Martin, S. E. Schwarzbach, and T. W. May. 2001a. Effects of an agricultural drainwater bypass on fishes inhabiting the Grassland Water District and the lower San Joaquin River, California. North American Journal of Fisheries Management 21:624–635.

Saiki, M. K., B. A. Martin, L. D. Thompson, and D. Welsh. 2001b. Copper, cadmium, and zinc concentrations in juvenile chinook salmon and selected fish-forage organisms (aquatic insects) in the upper Sacramento River, California. Water, Air, and Soil Pollution 132:127–139.

Saiki, M. K., and T. W. May. 1988. Trace element residues in bluegills and common carp from the lower San Joaquin River, California, and its tributaries. The Science of the Total Environment 74:199–217.

Saiki, M. K., and R. S. Ogle. 1995. Evidence of impaired reproduction by western mosquitofish inhabiting seleniferous agricultural drainage water. Transactions of the American Fisheries Society 124:578–587.

Saiki, M. K., and C. J. Schmitt. 1986. Organochlorine chemical residues in bluegills and common carp from the irrigated San Joaquin Valley floor, California. Archives of Environmental Contamination and Toxicology 15:357–366.

San Joaquin Valley Drainage Program. 1990. Fish and wildlife resources and agricultural drainage in the San Joaquin Valley, California. U.S. Department of the Interior, Final Report of San Joaquin Valley Drainage Program, volume 1, Sacramento, California.

Scheerer, P. D. 2002. Implications of floodplain isolation and connectivity on the conservation of an endangered minnow, Oregon chub, in the Willamette River, Oregon. Transactions of the American Fisheries Society 131:1070–1080.

Scholz, N. L., N. K. Truelove, B. L. French, B. A. Borejikian, T. P. Quinn, E. Casillas, and T. K. Collier. 2000. Diazinon disrupts antipredator and homing behaviors in chinook salmon (*Oncorhynchus tshawytscha*). Canadian Journal of Fisheries and Aquatic Sciences 57:1911–1918.

Schulz, P. D., and D. D. Simons. 1973. Fish species diversity in a prehistoric central California Indian midden. California Fish and Game 59:107–113.

Sierra Nevada Ecosystem Project. 1996. Sierra Nevada ecosystem project final report to Congress: status of the Sierra Nevada. University of California, Centers for Water and Wildland Resources, Davis.

Sommer, T., R. Baxter, and B. Herbold. 1997. The resilience of splittail in the Sacramento–San Joaquin Estuary. Transactions of the American Fisheries Society 126:961–976.

Sommer, T., B. Harrell, M. Nobriga, R. Brown, P. Moyle, W. Kimmerer, and L. Schemel. 2001a. California's Yolo Bypass: evidence that flood control can be compatible with fisheries, wetlands, wildlife, and agriculture. Fisheries 26(8):6–16.

Sommer, T., M. L. Nobriga, W. C. Harrell, W. Batham, and W. Kimmerer. 2001b. Floodplain rearing of juvenile chinook salmon: evidence of enhanced growth and survival. Canadian Journal of Fisheries and Aquatic Sciences 58:325–333.

The Bay Institute. 1998. From the Sierra to the sea. The Bay Institute of San Francisco, San Rafael, California.

Thomas, H. E., and D. A. Phoenix. 1976. Summary appraisals of the nation's groundwater resources—California region. U.S. Geological Survey Professional Paper 813-E, Reston, Virginia.

USFWS (U.S. Fish and Wildlife Service). 2003. Endangered and threatened wildlife and plants; notice of remanded determination of status for the Sacramento splittail (*Pogonichthys macrolepidotus*). Federal Register 68:183(22 September 2003)55139–55166.

Warner, G. 1991. Remember the San Joaquin. Pages 61–72 *in* A. Lufkin, editor. California's salmon and steelhead, the struggle to restore an imperiled resource. University of California Press, Berkeley.

Waters, T. F. 1995. Sediment in streams: sources, biological effects, and control. American Fisheries Society, Monograph 7, Bethesda, Maryland.

Williams, J. E., J. E. Johnson, D. A. Hendrickson, S. Contreras-Balderas, J. D. Williams, M. Navarro-Mendoza, D. E. McAllister, and J. E. Deacon. 1989a. Fishes of North America endangered, threatened, or of special concern. Fisheries 14(6):2–20.

Williams, M. L., R. L. Hothem, and H. M. Ohlendorf. 1989b. Recruitment failure in American avocets and black-necked stilts nesting at Kesterson Reservoir, California, 1984–1985. The Condor 91:797–802.

Williamson, K. S., and B. May. 2002. Incidence of phenotypic female chinook salmon positive for the male Y-chromosome-specific marker OtY1 in the Central Valley, California. Journal of Aquatic Animal Health 14:176–183.

Woodward, D. F., J. N. Goldstein, A. M. Farag, and W. G. Brumbaugh. 1997. Cutthroat trout avoidance of metals and conditions characteristic of a mining waste site: Coeur d'Alene River, Idaho. Transactions of the American Fisheries Society 126:699–796.

Yoshiyama, R. M., F. W. Fisher, and P. B. Moyle. 1998. Historical abundance and decline of chinook salmon in the Central Valley region of California. North American Journal of Fisheries Management 18:487–521.

Yoshiyama, R. M., E. R. Gerstung, F. W. Fisher, and P. B. Moyle. 2001. Historical and present distribution of chinook salmon in the Central Valley drainage of California. Pages 71–176 *in* R. Brown, editor. Contributions to the biology of Central Valley salmonids. California Department of Fish and Game, Fishery Bulletin 179, Sacramento.

American Fisheries Society Symposium 45:99–114, 2005

Historical Changes in Fishes of the Virgin–Moapa River System: Continuing Decline of a Unique Native Fauna

Paul B. Holden*
BIO-WEST, Inc., 1063 West 1400 North, Logan, Utah 84321, USA

James E. Deacon
University of Nevada-Las Vegas, 4505 Maryland Parkway, Las Vegas, Nevada 89154, USA

Michael E. Golden
BIO-WEST, Inc., 1063 West 1400 North, Logan, Utah 84321, USA

Abstract.—The Virgin–Moapa River system supports nine native fish species or subspecies, of which five are endemic. Woundfin *Plagopterus argentissimus* and Virgin River chub *Gila seminuda* are endemic to the main-stem Virgin River, whereas cooler and clearer tributaries are home to the Virgin spinedace *Lepidomeda mollispinis*. Moapa dace *Moapa coriacea* and Moapa White River springfish *Crenichthys baileyi moapae* are found in thermal springs that form the Moapa River, and Moapa speckled dace *Rhinichthys osculus moapae* is generally found below the springs in cooler waters. The agricultural heritage of the Virgin–Moapa River system resulted in numerous diversions that increased as municipal demands rose in recent years. In the early 1900s, trout were introduced into some of the cooler tributary streams, adversely affecting Virgin spinedace and other native species. The creation of Lake Mead in 1935 inundated the lower 80 km of the Virgin River and the lower 8 km of the Moapa River. Shortly thereafter, nonnative fishes invaded upstream from Lake Mead, and these species have continued to proliferate. Growing communities continue to compete for Virgin River water. These anthropogenic changes have reduced distribution and abundance of the native Virgin–Moapa River system fish fauna. The woundfin, Virgin River chub, and Moapa dace are listed as endangered, and the Virgin spinedace has been proposed for listing. In this paper we document how the abundance of these species has declined since the Endangered Species Act of 1973. Currently, there is no strong main-stem refugium for the Virgin River native fishes, tributary refugia continue to be shortened, and the Moapa River native fishes continue to be jeopardized. Recovery efforts for the listed and other native fishes, especially in the Virgin River, have monitored the declines, but have not implemented recovery actions effective in reversing them.

Introduction

The Virgin River is tributary to the Colorado River just downstream from the Grand Canyon near the city of Las Vegas, Nevada. The Virgin River basin begins in the mountains and plateaus of southern Utah and drains Zion National Park, along with other portions of the Colorado Plateau, from elevations near 3,050 m before discharging into the Colorado River (Lake Mead) at an elevation of approxi-

* Corresponding author: pholden@bio-west.com

mately 365 m. The total drainage area encompasses about 18,130 km^2. The river starts as two forks; the North Fork (Figure 1) flows from a combination of snowmelt and a large spring, and the East Fork is primarily a snowmelt stream. Several smaller tributaries drain into the Virgin River as the two forks exit canyons and join in a narrow valley just below Zion National Park. From this valley, the river drops into a narrow, steep canyon (Timpoweap Canyon) where it is entrenched for a few kilometers. As it exits the canyon, Pah Tempe Springs, a relatively large, hot spring with high salinity enters the river along with other small, permanent tributary streams. The river remains fairly entrenched as it cuts through Hurricane Fault, then spreads out in a valley near St. George, Utah, where the Santa Clara River enters. Here the Virgin River also exits the Colorado Plateau and enters the northeastern corner of the Mohave Desert (Cross 1975). The Virgin River then enters Arizona, creating the Virgin Gorge, a canyon through the Beaver Dam Mountains. Near the lower end of the Virgin Gorge, a series of springs (Reber Spring and Littlefield Springs) enters the river along with Beaver Dam Wash, which is a large desert wash with a small, permanent flow at its mouth, but dramatically larger flows during spring and summer rainstorms. The Virgin River then flows down Nevada's Virgin Valley where it is joined by the Moapa (or "Muddy") River prior to flowing into the Colorado River. Lake Mead now inundates the lower 80 km of the Virgin River and the lower 8 km of the Moapa River. The main-stem Virgin River from Pah Tempe Springs to the Colorado River is approximately 240 km long. The Moapa River is a short (48-km-long), thermal, spring-fed tributary that is unique because it does not share many flow or water quality attributes with the Virgin River. The Moapa River is the most downstream part of the pluvial White River system that is presently a series of disconnected spring systems in White River Valley, Pahranagat Valley, and Meadow Valley Wash, all of which also contain endemic fishes. In this paper, we discuss only the main-stem Moapa River.

The location and geology surrounding the Virgin River cause its flow to be generated by snow-

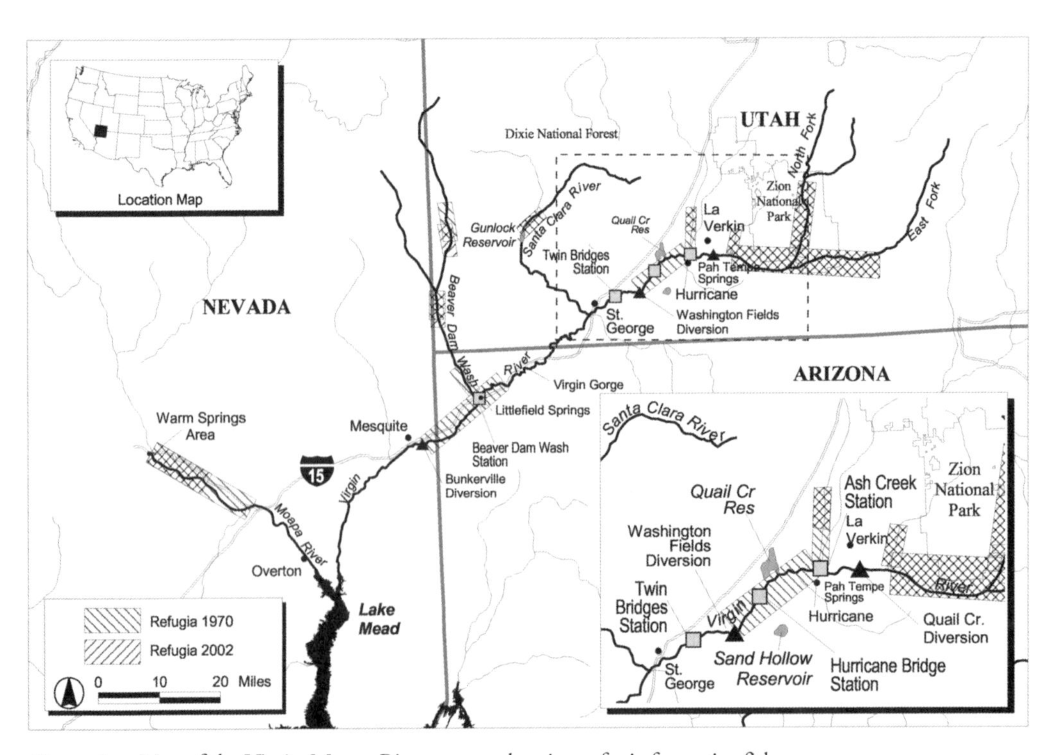

Figure 1.—Map of the Virgin-Moapa River system showing refugia for native fishes.

melt during the spring, thunderstorms during summer and fall, and large springs during base flow conditions. Base flow in the main stem is typically 2.83–5.66 m^3/s, but snowmelt, winter storms, and summer thunderstorms can raise flow to over 566.34 m^3/s in the upper river and over 849.51 m^3/s below Beaver Dam Wash. Spring and late summer flow spikes characterize higher flow years (Figure 2a). Drought years, which have occurred with high frequency over the past 15 to 20 years, are characterized by few, if any, flood spikes and base flows below 2.83 m^3/s. Below Pah Tempe Springs, the Virgin River main stem is highly saline and typically highly turbid, varying from red to brown depending on which portion of the basin is providing the flow. Portions of the Virgin River, primarily in Virgin Gorge, may become dry in mid to late summer. Typical of desert streams, flash floods are common in the gorges and canyons throughout the system. The Zion Narrows, a gorge on the North Fork in Zion National Park, is famous as a dangerous hike during thunderstorm season.

Because of its location on the edge of the Colorado Plateau and within the Mohave Desert, along with the nature of its erosive surroundings, the Virgin River is dominated by sand substrate, which transitions to cobbles in the gorges. Runs dominate the riverine habitat, with interspersed riffles and pools. Therefore, because of location, geology, and Pah Tempe Springs, the Virgin River is characterized by variability in flow, turbidity, and water quality, with conditions that change rapidly.

The Moapa River is formed by a series of thermal springs (collectively called Warm Springs) that are very consistent in both flow (1.1 m^3/s) and temperature (32.0–33.0°C) (Scoppettone et al. 1998). Temperature typically cools below the springs, forming a gradient that is important to the fish fauna. The small upper river drainage basin (about 104 km^2) provides little runoff to the river, but the lower river has a much larger basin (over 20,720 km^2), including Meadow Valley Wash (Figure 1), and flows are flashier (Figure 2b). Variability in flow, turbidity, and water quality, except near Warm Springs, distinguishes the Virgin–Moapa River system from other Colorado River tributaries, and this distinctiveness is exemplified by its native fishes.

The Native Fishes

Main-stem Virgin River fishes include the following: speckled dace *Rhinichthys osculus* and flannelmouth sucker *Catostomus latipinnis*, which are found throughout the Colorado River basin; the desert sucker *Pantosteus* (*Catostomus*) *clarkii* (Note: we agree with Minckley 1973 that *Pantosteus* and *Catostomus* are separate genera), which is found in the Colorado River basin below the Grand Canyon; and the woundfin *Plagopterus argentissimus* and Virgin River chub *Gila seminuda* (Note: DeMarais et al. 1992 have proposed the Virgin River chub as a full species and we follow that proposal). The Virgin River chub is endemic to the Virgin–Moapa River system, and the woundfin was undoubtedly described from the Virgin River even though the type location was erroneously listed as the San Luis Valley of Colorado (Cope and Yarrow 1875). Woundfin were reportedly found in the lower Colorado and Gila rivers (Minckley 1973), but the few records of the species from those other areas all occurred from 1890 to 1895 (Miller 1961), a time when many fish localities were erroneously reported. Since 1895, the woundfin has been found only in the Virgin and Moapa rivers; hence, it may have been endemic to this system. Woundfin is the typical native species that utilizes the sand substrate runs so common in this system. Virgin River chub are primarily found in pool habitats. Speckled dace and desert sucker are typically riffle species, and flannelmouth sucker are found in a variety of habitats. The upstream limit for woundfin and Virgin River chub is Pah Tempe Springs, whereas the other three species penetrate varying distances further up the main stem and tributary streams.

Tributary species include the speckled dace and desert sucker, along with the endemic Virgin spinedace *Lepidomeda mollispinis mollispinis* (Note: We agree with the subspecific taxon for the Virgin spinedace as proposed by Miller and Hubbs 1960), which also occurs in the main stem near the mouths of clear tributary streams it inhabits. Virgin spinedace and woundfin are both members of the tribe Plagopterini, minnows with sharp spines in the dorsal and pelvic fins. Other members of the tribe are found in Arizona's Little Colorado River and parts of the Gila River (Rinn et al. 2005) system. Virgin

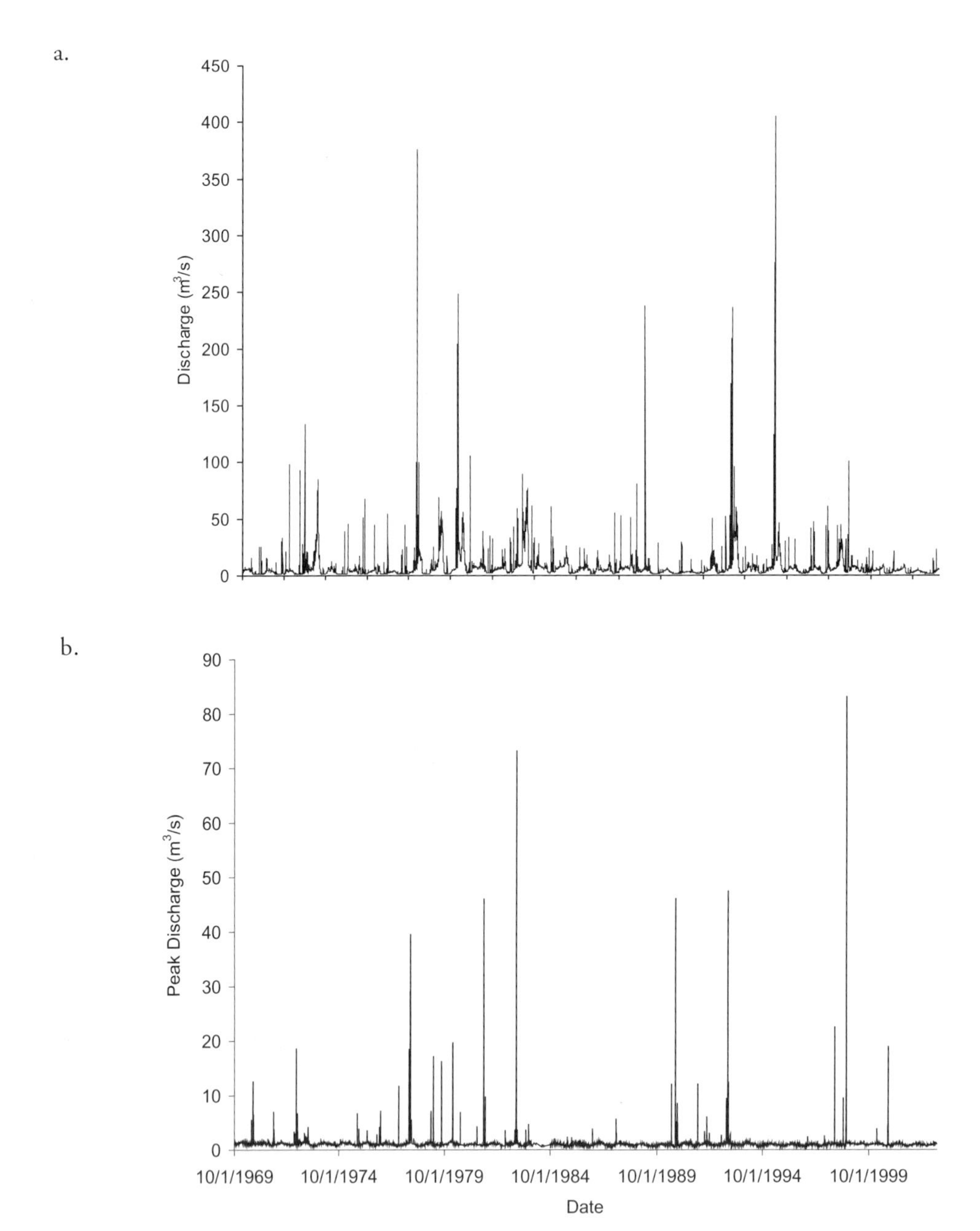

Figure 2. Discharge for (a) the Virgin River at Littlefield, Arizona, and (b) Muddy River at Glendale, Nevada, from 1970 to 2002.

spinedace are most common in the clearer, cooler tributaries and the main stem above Pah Tempe Springs. Virgin spinedace tributary habitat covered approximately 300 km, including the Virgin River above Pah Tempe Springs.

The only trout that may have been in the basin when it was settled in the mid 1800s was the Bonneville cutthroat trout *Oncorhynchus clarkii utah*. It appears that a fairly recent (within the last 2,000 years) stream capture in upper tributaries of the Santa Clara River captured a portion of a small stream (Pinto Creek) that originally flowed into the Bonneville basin and contained Bonneville cutthroat trout (D. Hepworth, Utah Division of Wildlife Resources, personal communication). Behnke (1992) dismissed this probability, but the evidence of a fairly recent stream capture suggests otherwise.

The Moapa River also has a unique native fish fauna. Woundfin, flannelmouth sucker, and most likely the desert sucker occurred in the lower Moapa River, generally below the thermal influence of Warm Springs. The Virgin River chub (Note: DeMarais et al. 1992 combined the Moapa and Virgin River chubs into one species) was found in both the thermal and downstream areas. However, the Warm Springs area provides a unique habitat that is utilized by an endemic goodeid, the Moapa White River springfish *Crenichthys baileyi moapae* (Williams and Wilde 1981), and an endemic minnow, the Moapa dace *Moapa coriacea*. The Moapa speckled dace *Rhinichthys osculus moapae* (Williams 1978), is also endemic and is found below the Warm Springs area. In addition to the fishes, the Moapa River also has endemic snails and aquatic insects (USFWS 1996).

Of the nine native fish species and subspecies in the Virgin–Moapa River system (Table 1), five are endemic (six if the woundfin is considered endemic). This high level of endemism is reflective of the unique aquatic habitats that exist in this system. The objective of this chapter is to summarize what has happened to the Virgin–Moapa River system and its native fish fauna since settlement of the area and what is being done today to recover that fauna.

Methods

This paper is a review of published and unpublished literature describing the major forces that have shaped the Virgin–Moapa River system since settlement in the mid 1800s. Where appropriate, we describe techniques used to collect the data used; however, a detailed description of methods is available in cited literature and reports. Changes in the Virgin River fish fauna from 1976 to the present are fairly well documented because of the Virgin River Fishes Recovery Team's monitoring activities. Starting in 1976, spring (usually the end of April) and fall (usually late September) sampling of main-stem stations consisted of seining 10 distinct habitats, primarily run habitat with sand substrates, with sufficient effort to reduce catch to 10% of the catch of the first seine haul. A standard 4.6 m × 1.2 m × 3.2 mm seine was used. Fishes were identified and counted, native fishes were measured and returned to the river alive, and nonnative fishes were disposed of on the stream bank. Over the years, some variation in this protocol occurred, since a large variety of agencies and personnel were involved. Initially, six stations were sampled, and additional stations have been added recently. Data provided here are from four of the monitoring stations located on the main-stem Virgin River. These four stations have been sampled the most consistently and represent the various reaches of the river. They are as follows: Ash Creek (Figure 1), the Virgin River near the mouth of Ash Creek about 3 km below Pah Tempe Hot Springs; Hurricane Bridge, about 16 km below the Ash Creek station below a highway bridge; Twin Bridges, in the heavily dewatered section below Washington Fields Diversion; and Beaver Dam Wash, the Virgin River near the mouth of Beaver Dam Wash. Data used here are from fall seine collections, when young fish recruitment tends to boost catch and low flows allow efficient sampling.

Results and Discussion

Impacts to the River and Native Fish Abundance

Discussions of changes in native fish abundance in the western United States typically assume that the native fishes were very common and/or abundant prior to anthropomorphic changes to the river of concern. Early surveys that predated major water use were seldom detailed enough to establish fish population levels. The Virgin–Moapa River system is no different. The earliest reports of native fish species are from the 1870s and 1880s, 20 years or so after the area was settled. Cope and Yarrow (1875), reporting on fish collections during military surveys in the early 1870s (Wheeler Survey), noted that Virgin River chub were not scarce, "as several hundreds were observed captured by boys with hook and line" near the present city of St. George, Utah. Woundfin abundance was not as clear in Cope and Yarrow (1875), and the location was misidentified

Table 1.— Fishes present in the Virgin-Moapa River system. X denotes presence in the system. Abundance denotations are as follows: A = abundant, AP = abundant throughout part of the system, C = common, E = extirpated, LA = locally abundant, LC = locally common, R = rare. Endangered Species Act (ESA) status denotations are as follows: E = endangered, CA = conservation agreement.

Species common name	Scientific name	Moapa River	Abundance	Virgin River	Abundance	ESA status
Native						
cutthroat trout[a]	*Oncorhynchus clarkii*			X	LA	
flannelmouth sucker	*Catostomus latipinnis*	X	R	X	C	
desert sucker	*C. clarkii*	X	E	X	C	
speckled dace	*Rhinichthys osculus*			*X*	C	
Moapa speckled dace	*R. osculus moapae*	X	C			
Moapa dace	*Moapa coriacea*	X	R			E
Moapa White River springfish	*Crenichthys baileyi moapae*	X	LC			
Virgin River chub	*Gila seminuda*	X	R	X	LA	E[b]
Virgin River spinedace	*Lepidomeda mollispinis*			X	AP	CA
woundfin	*Plagopterus argentissimus*	X	E	X	R	E
Nonnative						
brown trout	*Salmo trutta*			X	LA	
rainbow trout	*Onchyrhynchus mykiss*	X	E	X	LA	
brook trout	*Salvelinus fontinalis*			X	LA	
bluegill	*Lepomis macrochirus*			X	R	
green sunfish	*L. cyanellus*	X	R	X	R	
largemouth bass	*Micropterus salmoides*	X	R	X	LC	
black bullhead	*Ameiurus melas*	X	R	X	LA	
channel catfish	*Ictalurus punctatus*	X	R	X	LA	
common carp	*Cyprinus carpio*	X	R	X	LA	
red shiner	*Cyprinella lutrensis*	X	R	X	AP	
western mosquitofish	*Gambusia affinis*	X	C	X	AP	
shortfin molly	*Poecilia mexicana*	X	LA			
blue tilapia	*Oreochromis aureus*	X	C	X	LA	
Other historically documented species						
smallmouth bass	*Micropterus dolomieu*			X	R	
yellow bullhead	*Ameiurus natalis*			X	R	
golden shiner	*Notemigonus crysoleucas*	X	R	X	R	
redside shiner	*Richardsonius balteatus*	X	R	X	R	
fathead minnow	*Pimephales promelas*	X	R			
goldfish	*Carassius auratus*			X	R	
spottail shiner	*Notropis hudsonius*			X	R	
longfin dace	*Agosia chrysogaster*			X	R	
mountain sucker	*Catostomus platyrhynchus*			X	R	

[a] See text.
[b] Muddy River *G. seminuda* population is not officially considered endangered.

as they noted, "numerous specimens from the San Luis Valley, Western Colorado."

The Moapa River fish fauna was first described in the 1940s (Hubbs and Miller 1948) and the taxonomy of some of the species has been clarified more recently (Williams 1978; Williams and Wilde 1981). Moapa White River springfish was common to abundant in the warm spring pools forming the headwaters of the Moapa River in the 1940s and it remained so in many of the least disturbed habitats, at least through the 1970s (Deacon and Wilson 1967; Deacon and Bradley 1972; Cross 1976).

Moapa dace was considered "rather common" by Hubbs and Miller (1948) at that time and remained so in a few less-severely altered spring and stream habitats into the 1970s (Cross 1976; Scoppettone 1993). The Virgin River chub, typically found in cooler water downstream from the thermal springs, remained common in a segment of the middle portion of Moapa River through the 1970s (Deacon and Bradley 1972; Cross 1976). Therefore, although hard data are few, the assumption that Virgin–Moapa River system native fishes were likely quite common to abundant in areas that provided good habitat prior to settlement of the region appears to be reasonable.

The Virgin–Moapa River system was used by several groups of indigenous people over the past 2,000 years for many purposes, including watering crops (WCWCD 1997). It is unlikely that this use impaired the native fishes, except in very localized areas. Mormon settlers first arrived in the area in 1854, settling along the Santa Clara River and along the East Fork. As soon as settlers arrived, diversions were built to irrigate crops. Cotton was first raised in 1855, and the area became known as Utah's "Dixie." Within a few years, Mormon settlers expanded into the Littlefield and Mesquite areas (Figure 1). The Moapa River area was settled shortly thereafter. Although no information exists on the impacts to the native fish fauna in the late 1800s, it can be assumed that some portions of tributaries and the main stem directly below diversions probably became sufficiently dewatered that native fish numbers declined, at least locally. However, large areas of the river and tributaries remained relatively intact, and overall abundance of the native fishes probably remained high.

By the early part of the 20th century, several events occurred that we believe had a more dramatic impact on the native fishes. First, diversions became solid structures instead of rocks piled in the stream. The increased efficiency of these diversions resulted in dewatering of larger portions of the Virgin River and its tributaries. The combination of physical diversions and dry stretches of river undoubtedly resulted in fragmentation of the system for native fishes. More agricultural lands were developed, which also increased the amount of water taken from the river. By 1902, more than 20 major irrigation diversions existed on the Virgin–Moapa River system. By 1910, the water in the Virgin–Moapa River system was essentially fully appropriated (USFWS 1995). At that time, it is likely that two major sections of the river—about 70 km of river from the Washington Fields Diversion, located on the Virgin Fault, downstream through the Virgin Gorge, and the lower 120 km of the river below Mesquite, Nevada—were often nearly dewatered in late summer, as was the lower 24 km of the Moapa River (USFWS 1995, 1996; WCWCD 1997). In addition, portions of the Warm Springs area on the Moapa River were enlarged, lined with concrete, and the spring outflows chlorinated to create public swimming pools, likely eliminating some native fishes from those springs (USFWS 1996). By the 1930s, much of the Virgin–Moapa River system was affected by reduced flow and fragmentation caused by irrigation diversions and withdrawals. We believe, based on later fish collections, these impacts undoubtedly reduced populations of native fishes. On a more positive note, the Zion area became a national monument in 1909 and a national park in 1919, thus protecting portions of the North Fork from additional development and water use.

However, Hoover Dam on the Colorado River was closed to form Lake Mead in 1935, which inundated the lower 80 km of the Virgin River and the lower 8 km of the Moapa River. By this time, both areas had already been heavily impaired by water diversion, but inundation by Lake Mead removed any potential for these areas to harbor native stream fishes and disconnected the Virgin and Moapa rivers. On the Santa Clara River, Windsor Dam (1936), Baker Dam (1954), and Gunlock Dam (1970), fragmented and variously altered most of the main stem and dewatered the lower 32 km, impacting Virgin spinedace habitat.

During the late 1930s, Carl Hubbs and R. R. Miller seined fish in the Virgin and Moapa rivers. Their field notes and museum collection records (reported in Cross 1975 and USFWS 1995) provide some of the first real information on native fish occurrence. Seining in the main stem below Mesquite, Nevada, revealed specimens of all the native fishes, but in relatively small numbers (Cross 1975). Sampling efforts in the Moapa River found the Moapa dace and other endemic species for the first time and showed that they were still common in the least-altered portions of the Warm Springs area.

In 1960, the Nevada Division of Wildlife built a small reservoir (Schroeder Reservoir) in the upper portions of Beaver Dam Wash as a sport fish project and, after poisoning the stream above the reservoir, stocked rainbow trout. The stocked trout expansion below the dam and natural recruitment led to predation, which reduced the abundance of Virgin spinedace and other native fish populations in this permanent flowing portion of the wash (Lentsch et al. 1995). Although other nonnative fishes, primarily warmwater species such as channel catfish, black bullhead, largemouth bass, and green sunfish were likely found in farm ponds throughout the basin (Holden et al. 1997) or invaded from Lake Mead (Table 1), the predation of rainbow trout on Virgin spinedace was one of the first documented impacts from a nonnative fish species in the basin. At about this same time (1963), shortfin molly was introduced into the Warm Springs area of the Moapa River, and declines in Moapa dace and Moapa White River springfish were noted shortly thereafter (Deacon and Wilson 1967; Deacon and Bradley 1972; Scoppettone 1993). Therefore, in addition to severe water withdrawal and loss of habitat because of reservoirs, impacts from nonnative fishes were becoming a significant problem by the end of the 1960s.

In the early 1970s, Cross (1975) conducted a thorough seining survey of the Virgin River system (123 collections at 77 sites). For the first time, a comprehensive evaluation of native fish distribution and abundance in the Virgin River was made. Cross (1975) showed two major refuges for the main-stem Virgin River fishes. The refuges were from just below Pah Tempe Springs downstream to the Washington Fields Diversion, a distance of about 25 km, and from the lower part of Virgin Gorge at Reber Spring downstream about 64 km to the Bunkerville Diversion (Figure 1). The amount of main-stem habitat where the native fish fauna remained common was reduced by almost 75% relative to assumed presettlement times. Both of these remaining refuges had large springs and small tributaries that contributed sufficient year-round flow for the native species to maintain relatively strong populations. Other portions of the main stem below each of these areas were still populated by the native fishes but in reduced numbers and with occasional major losses because of river dewatering during dry years (Cross 1975).

Tributary species also saw up to a 60% reduction in habitat by 1970. A major refuge for tributary species remained in the lower East Fork and the North Fork, the latter stream being protected by Zion National Park. In addition, small portions of several other tributaries maintained good native fish populations (Figure 1). However, major losses in habitat had occurred in the Santa Clara River.

Deacon and Bradley (1972) sampled 9 Moapa river sites with seines and electrofishing in 1963–1968 and Cross (1976) electrofished 10 Moapa River sites in 1974–1975. These studies documented declines in abundance of the Moapa River native fishes and showed that all but the upper 24 km of stream were unavailable to them (Figure 1). Loss of flow and water quality issues contributed to this 50% range reduction from the mid-1800s. Within the available 24 km, populations of most native fishes were depressed because of agricultural pollution, water diversion, and, in the warm headwaters, interactions with nonnative western mosquitofish and shortfin molly. Despite these large losses of habitat, none of the native Virgin–Moapa River system fish species disappeared from the system, and in select areas the native fish fauna remained relatively intact and abundant.

The Endangered Species Act was passed in 1973, bringing with it anticipation for recovery of rare native fishes. The woundfin and Moapa dace were both listed as endangered under the first version of the act (The Endangered Species Protection Act of 1967). A woundfin recovery team was established in 1976, and a recovery plan was completed in 1979 (USFWS 1979). A recovery plan for Moapa dace was completed in 1983 (USFWS 1983). The woundfin recovery team initiated annual spring and fall Virgin River monitoring in 1976, and sampling efforts in the Moapa River were increased so that native fish population changes in both streams could be documented. However, neither the Endangered Species Act nor recovery plans were able to keep the Virgin and Moapa rivers free from additional impacts.

Cross (1975) noted that red shiner, a Missouri

River cyprinid that was probably introduced into the lower Colorado River as a baitfish (Miller 1952), was common to abundant in the lower Virgin River below Mesquite, Nevada. Flows were often greatly depleted in this area. Woundfin recovery team data showed that woundfin dominated the river numerically above the lower-most major diversion, the Bunkerville Diversion near Mesquite, Nevada. However, a few red shiner were found in this area and upstream to the Virgin Gorge. The Virgin Gorge was dry throughout most of the summer, which appeared to prevent red shiner upstream movement. In the mid-1980s, a series of events occurred that changed the fish fauna dramatically. Red shiner were found in the Virgin River in Utah in 1984, and they had become abundant around St. George, Utah, by October 1985. At the same time, red shiner abundance exploded in Nevada and Arizona. Concomitant with this increase in red shiner were extensive declines in native species abundance. Woundfin was most impaired by increased red shiner abundance, essentially disappearing from the river in Arizona and Nevada and remaining common only above the Washington Fields Diversion in Utah. Deacon (1988) noted that record high flow during summer 1983 probably produced continuous flow through the Virgin Gorge for the first time since red shiner invaded the river in the early 1960s and was the primary reason for the sudden upstream movement of red shiner.

Reasons why red shiner replaced woundfin are not clear, but several events occurred that may have helped change the fish fauna. In 1985, a relatively large, off-stream reservoir near St. George, Utah (Quail Creek Reservoir), was filled using water diverted from the Virgin River in Timpoweap Canyon at the Quail Creek Diversion. One effect of the operation of this reservoir has been to shift some of the winter flow to summer discharge below St. George, Utah, a consequence of increased irrigation return flow. In addition, flow below St. George has been significantly supplemented throughout the year by growth-related discharge increases from the St. George sewage treatment plant. As a consequence, continuous flow through the Virgin Gorge has been a usual feature of summer discharge since 1985 and has created a sustained period of time for red shiner to move upstream into Utah (Gregory and Deacon 1994). Following the high flows of 1983, a major drought occurred in the basin, with the lowest flows occurring during 1989–1991 (Figure 2a), which may also have produced conditions more favorable to red shiner (Holden et al. 2001). The invasion and proliferation of red shiner in the Utah portion of the Virgin River, and the concomitant decline in woundfin abundance, led the Woundfin Recovery Team to attempt to remove red shiner from the Virgin River in Utah using rotenone in 1988. The first attempt was not detoxified properly, and fishes in the Virgin Gorge and downstream to the mouth of Beaver Dam Wash, possibly as far as Mesquite, Nevada, were killed (DeMarais et al. 1993). Additional treatments were conducted in 1989 and have been conducted nearly annually since 1996 in smaller sections of river, but complete red shiner removal has yet to be sustained in any reach. Additionally, in 1989, a Quail Creek Reservoir dike failed and a flood of more than 1,699.01 m^3/s rushed down the Virgin River. Holden et al. (2001) suggested that a combination of low flow caused by drought, human use of the river, habitat change near the mouth of Beaver Dam Wash, poisoning, reduced upstream woundfin populations, and dramatically increased Utah red shiner populations all played a role in the fish fauna changes seen in the Virgin River. The actual impact of red shiner on woundfin is not known, but it may include predation on larvae, and food and space competition with all sizes of woundfin. In addition, red shiner have a very high reproductive capacity and spawn multiple times within a year, which allows them to increase in number rapidly.

In 1990, the Virgin River chub was listed as endangered because of uncertainties about its abundance, and the Woundfin Recovery Team name was changed to the Virgin River Fishes Recovery Team. In 1994, the Virgin spinedace was proposed for listing as a threatened species, but a conservation plan (Lentsch et al. 1995) eliminated the need for formal listing.

The Moapa River was not exempt from native fish population changes. In 1990, blue tilapia was found in portions of the river just below Warm Springs, and by 1995, Moapa dace and other endemic fishes declined dramatically as the tilapia became more abundant (Scoppettone et al. 1998). The actual impact of tilapia is not well understood. Tilapia herbivory may alter the natural food base for native fishes by reducing algae and other vegetation (USFWS 1996), or tilapia may directly prey on and/or compete for food or space with native species.

Recent Fish Abundance

The fish assemblage at the Beaver Dam Wash station was dominated numerically by woundfin from 1976 through 1984 (Figure 3a), and the other native fish species were common and generally captured each year in Recovery Team monitoring. Red shiner was usually found in relatively low numbers. This portion of the Virgin River has relatively consistent base flows because of Reber and Littlefield springs and the inflow of Beaver Dam Wash, which is reflected in the fish fauna from 1976 to 1984. Even Virgin spinedace was a consistent component of the fish assemblage because of the large population in the cool, clear lower portion of Beaver Dam Wash. However, the substantial increase in red shiner numbers after 1985 changed the composition of the fish assemblage at this site. Virgin spinedace and woundfin essentially disappeared by 1988, and the other native fish species experienced large declines. The other native fishes also declined to where only red shiner and speckled dace were collected in monitoring in 1989. Since 1989, native species, with the exception of woundfin and Virgin spinedace, have returned to this location, but red shiner continue to numerically dominate the fish assemblage. This portion of the Virgin River was disturbed by all the events noted above from 1985 to the early 1990s. In the early 1990s, flannelmouth sucker, desert sucker, speckled dace, and Virgin chub began to again be consistently captured in annual monitoring, but woundfin and Virgin spinedace remained missing from the site and red shiner continued to dominate collections. Hence, the lower refuges for main-stem fishes in the Virgin River lost important components in the late 1980s that were not reestablished. The major reason woundfin have not returned to the Beaver Dam Wash station appears to be intense competition/predation from red shiner and a habitat shift

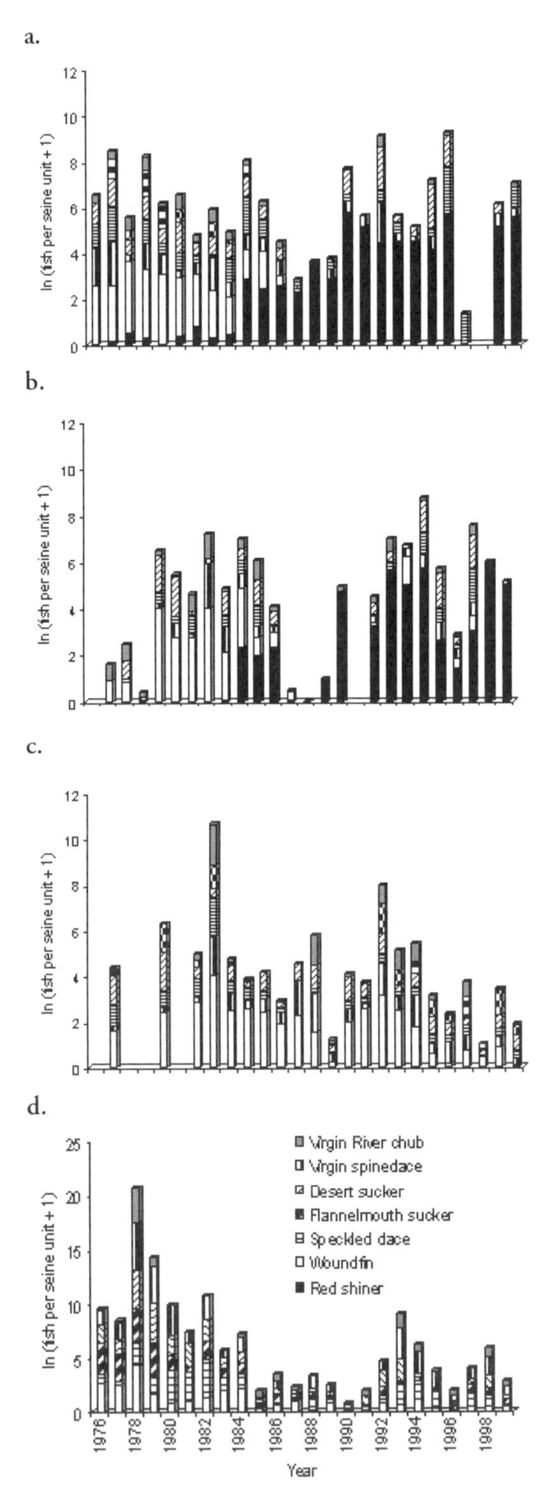

Figure 3.—Fish abundance at the (a) Beaver Dam Wash, (b) Twin Bridges, (c) Hurricane Bridge, and (d) Ash Creek Virgin River Fishes Recovery Team Monitoring Stations from 1976 to 2001.

that replaced sandy runs with more riffles and rocky substrate (Holden et al. 2001). Virgin spinedace disappeared from this station after they disappeared from lower Beaver Dam Wash, likely due to flooding events, confinement of the channel to a ditch through a newly developed golf course, and drainage of ponds in the floodplain to facilitate development. The other native species are less impaired by red shiner but were disturbed in 1988 by the first red shiner poisoning, in 1989 by the Quail Creek Dam flood, and in 1989–1991 by drought; however, their populations have rebuilt since that time.

The Twin Bridges station (Figure 3b) is located in an area of the Virgin River that has been at least partially dewatered by upstream diversions, primarily the Washington Fields Diversion, for many years. Native fish density was somewhat inconsistent from 1976 to 1984, but woundfin tended to be reasonably common and most other main-stem fishes were found here. The fish populations in this area appeared to be maintained by downstream drift of fishes from the refuges above Washington Fields Diversion, along with some reproduction during higher flow years with more consistent summer flow. Red shiner invaded this area in 1984. Similar to the situation at Beaver Dam Wash, as red shiner abundance increased, native fish abundance declined. Almost no fish were found at this station in 1989, which was probably the result of the combined effects of drought-induced low flows, the 1988 poisoning, and the 1989 Quail Creek flood. Apparently, downstream drift of young fish from the area above Washington Fields Diversion resulted in fairly consistent native fish abundance by the mid to late 1990s. Woundfin, which once numerically dominated the catch, persisted in very low abundance, while most of the other native species recovered to near pre-1989 levels. Competition/predation from red shiner, as well as low flows during drought years, are the most likely reasons for woundfin being unable to reestablish at this station (Holden et al. 2001).

Main-stem native fishes, dominated numerically by woundfin, were common at the Hurricane Bridge station in the 1970s and 1980s. Numbers declined noticeably by 1990, recovered in the early 1990s, and then have subsequently declined again in recent years (Figure 3c). Woundfin numbers have been relatively low in fall sampling since 1996 and were very low during fall 2001. Although Virgin spinedace were found at this site during collections in the 1970s and 1980s, their numbers appeared to be related to the presence of a cool, clear tributary (Quail Creek) and springs (Berry Springs) in this portion of the river. In the 1990s, Virgin spinedace became a more abundant component of the main-stem fish population as woundfin declined. The physical barrier provided by Washington Fields Diversion and the chemical barrier provided by frequent poisoning below the diversion during the late 1990s prevented upstream invasion of red shiner into the Hurricane Bridge area. Thus, red shiner interactions did not contribute to native fish declines at Hurricane Bridge. However, this site was affected by the Quail Creek Dam break in 1989, and since 1985, the area has received supplemental flow from Quail Creek Reservoir to maintain 2.44 m^3/s during the summer. We suspect that subtle changes in water quality (turbidity, temperature) caused by the Quail Creek Reservoir Diversion, supplemental cooler inflow from Quail Creek Reservoir, and perhaps drought are responsible for the recent decline in woundfin and increase in Virgin spinedace. Other factors may include declines in woundfin populations upstream from this station and perhaps other unknown changes in this section of the Virgin River (Holden et al. 2001).

Other than water reductions resulting from the Quail Creek Diversion and drought, the Virgin River at the Ash Creek station was not impaired by the major disturbances noted above. Native fish populations at this site were generally strong until 1985 (Figure 3d), at which time they declined. Woundfin and speckled dace abundance showed the greatest declines. Native fish numbers rebounded briefly in the early 1990s and again in 1998, but overall they have remained low since 1985. Woundfin disappeared from fall collections at the station from 1997 to 2000, and very few were collected in 1996 and 2001. The decline that occurred in 1985 was attributed to a strange set of circumstances. For a few months in 1985, Virgin River flow was captured by sink holes that developed in the riverbed near the Quail Creek Diversion. The river flow reappeared as increased, undiluted flow from Pah Tempe Springs, causing high salinity at the Ash Creek station (Deacon 1988; Hardy et al. 1995). Once the sink holes disappeared and the spring flows and downstream salinity returned to normal, woundfin did not recover. Gregory and Deacon (1994) suggested that the continued low woundfin abundance at Ash Creek might be the result of Quail Creek Diversion seriously reducing flows in this river reach. They also suggested that the flows were too low to dilute the highly saline, warmer water from Pah Tempe Springs. In addition, in 1986 a hydroelectric power plant and its discharge point (water diverted at the Quail Creek Diversion) were moved about 1.6 km upstream to within about 0.5 km below Pah Tempe Springs. As a result, the coolwater discharge from the power plant now enters the river in a reach previously unoccupied by native fishes because the high temperature and salinity from the hot springs made this reach undesirable. At its former location, the power plant discharge frequently offered a thermal and salinity refuge for native fishes, but it now is discharged at a point where the river is warmer and the dilution effect of the discharge is diminished. Virgin spinedace, Virgin River chub, and the native suckers are still consistently captured at this station, but speckled dace numbers are now lower.

No refuges for main-stem fishes remain (Figure 1). Only one small portion of the Virgin River main stem continues to consistently have the full component of main-stem native fishes, and that river reach is obviously impaired because woundfin numbers continue to decline. Woundfin is the native species most affected by the various impacts that occurred to the main-stem Virgin River, from red shiner abundance to subtle water quality changes. As noted above, many reasons may exist for woundfin declines at Hurricane Bridge and Ash Creek, and most involve human manipulation of the river system complicated by drought. The other native main-stem species are still found in reasonable densities at most monitoring stations, and Virgin spinedace has actually increased in the Hurricane Bridge reach. Conversely, Virgin spinedace disappeared from the Beaver Dam Wash reach. We suspect the increase in Virgin spinedace at Hurricane Bridge is due to the cooling of the river by the Quail Creek Reservoir inflow, making the main-stem Virgin River temperature in this area more like that of a tributary (Holden et al. 2001).

Virgin River tributaries were not consistently monitored until the last few years. Therefore, it is more difficult to show their long-term trends in distribution and abundance. Valdez et al. (1991) sampled 113 sites from 1986 to 1990 with electrofishing and seines to assess Virgin spinedace distribution and abundance, and threats to the species. Addley and Hardy (1993) surveyed the Virgin River and its tributaries in 1992 for Virgin spinedace using electrofishing or seining at over 500 stations. Both surveys reported that the species was only common in about 35–40% of its original range. Primary centers of abundance all occurred in Utah, including the upper main stem within and just below Zion National Park and small tributaries in the Santa Clara and Beaver Dam Wash drainages. Since 1995, annual fall Virgin spinedace monitoring has been conducted in Nevada, Arizona, and Utah. This monitoring (Fridell and Curtis 2002) has shown that while Virgin spinedace densities have fluctuated in the past few years, overall numbers, primarily in small tributaries, have been down especially during fall 2001 and 2002 (unpublished data, Virgin Spinedace Conservation Committee meetings, January 23 and September 27, 2002). The other tributary species also have been reduced in most areas. The primary reason for the declines is drought and, in some places, irrigation diversion leading to reduced streamflow or dry stream channels. The largest remaining refugium for tributary native fish includes the North Fork, East Fork, and main stem in and below Zion National Park (Figure 1), an area of about 60 km of river (Valdez et al. 1991; Addley and Hardy 1993). These are the largest stream sections that provide habitat for the tributary species and the sections least influenced by irrigation withdrawals; hence they have maintained the best flows during drought conditions.

Inconsistent monitoring of the Moapa River fish fauna precludes graphical depiction of trends. Moapa dace, Moapa speckled dace, and Virgin River chub densities declined dramatically in the river and springs from 1994 to 1997, primarily because of the increase in numbers of blue tilapia and a wildfire in the Warm Springs area (Scoppettone et al. 1998). Removal of a small barrier dam in 1994 allowed blue tilapia to move into the spring pools and threaten Moapa White River springfish populations, which also declined. Surveys in 2000 and 2001 indicated that native fish densities in the Moapa River had stabilized somewhat and that no species had dropped to extremely low levels (J. Heinrich, Nevada Division of Wildlife, personal communication). However, the area where native Moapa River fishes remain common has shrunk since the 1970s and now includes only about 14 km of the upper river (Figure 1) (Scoppettone et al. 1998).

The Future

Recent trends show declines in native fish abundance and distribution in both the Virgin and Moapa rivers, suggesting that the native fish faunas of these systems may be more imperiled now than in the past. Nonnative species invasions, water withdrawal, and other anthropogenic changes in the Virgin–Moapa River system are the primary causative factors, but prolonged drought and a wildfire along the Moapa River (actually in nonnative vegetation) have also contributed to the declines (Scoppettone et al. 1998; Holden et al. 2001). Although introduced parasites have been found on woundfin and Virgin River chub (Heckman et al. 1986, 1987), they have not been considered factors in the declines of these species (USFWS 1995). Rehabilitation efforts continue in both systems. In 2001, the Virgin River Resource Management and Recovery Program (VRRMRP) was established to oversee recovery of Virgin River fishes in Utah, as well as allowing water development to proceed. Rehabilitation activities—such as poisoning red shiner; monitoring woundfin, Virgin River chub, and Virgin spinedace; and evaluation of factors leading to native fish declines—are being funded by this cooperative state/federal program. In Arizona and Nevada, state and federal agencies have initiated the Lower Virgin River Recovery Implementation Program (LVRRIP) to initiate recovery activities in that portion of the system. Not as well funded as the VRRMRP, this program has concentrated on studying the effect of mechanical removal of red shiner on native fish populations, studying red shiner/woundfin interaction, and developing hatchery-reared woundfin stocking protocols. The Moapa River Recovery Team recently initiated barrier construction and poisoning with rotenone to remove blue tilapia from spring pools. Several more years will be required to remove blue tilapia from areas inhabited by the native fishes, if indeed it ever occurs.

Even as these recovery efforts are underway, additional threats hinder the long-term survival of the native fishes. Completed in 2002, Sand Hollow, a new off-stream reservoir, near Hurricane, Utah, is using more water from the Quail Creek Diversion and was built with no environmental review because of agreements involving water rights for Zion National Park. Blue tilapia, which have become very common in Lake Mead, were found in the lower Virgin River in June 2001, and they had moved above the Bunkerville Diversion by late summer. They disappeared over winter 2001–2002, but reappeared in September 2002 after a rain event caused the lower Virgin River to connect to Lake Mead (the lower river had been dry all summer) (unpublished BIO-WEST and Nevada Division of Wildlife data). The LVRRIP is planning barrier construction and conducted a preliminary eradication project in November 2002 to attempt to keep this species from reaching Littlefield Springs, a warm spring that may provide year-round habitat. They again disappeared over winter 2002–2003 and did not move up the river in 2003 due to low flows.

In 2002, red shiner were found above Washington Fields Diversion, the only area below Pah Tempe Springs they had not previously invaded. The source for the red shiner was an offstream pond apparently developed for rearing bait for recreational fishing. Extensive seining during summer and fall 2002 removed over 300 red shiner and reduced red shiner captures to nondetectable levels in over 1,000 seine hauls conducted in October and November 2002 (unpublished Utah Division of Wildlife Resources data). The Virgin River Fishes Recovery Team discussed long-term options for this reach of the Virgin River in January 2003. Poisoning with rotenone is an option, but it will only be conducted if red shiner increase numerically in the reach.

Extensive sampling to mechanically remove red shiner in the area above Washington Fields Diversion revealed several important features of the native fish populations. Adult woundfin in the last remaining wild population probably number less than 1,000 (in a reach where thousands were found when the species was listed as endangered in 1973). Woundfin reproduced in the reach in 2002, but recruitment (i.e., numbers remaining in the fall) was very low (unpublished Utah Division of Wildlife Resources data). Virgin River chub were common in the reach and they had a strong reproduction year, but numbers of young had dropped dramatically by fall 2002 and recruitment was likely very low. The information collected on woundfin and Virgin River chub showed how little was actually known about how spawning, larval production, and young-of-the-year summer abundance relate to recruitment for these species. Studies to understand the factors limiting recruitment are needed in this system, as is a monitoring program that more closely follows reproduction and the fate of young fish. Once limiting factors are understood, rehabilitation projects to correct them need to be designed. In addition, research that clarifies red shiner/woundfin interactions would help future rehabilitation efforts, especially if red shiner cannot be eradicated.

In the Moapa River more consistent monitoring is needed, as are continued efforts to eradicate nonnative fish species such as blue tilapia. Perhaps the major factor needed to ensure future populations of these unique species is increased flow in the rivers and streams that make up the basin. Drought conditions have occurred in the past but fish populations have never been as low as they are today, suggesting that humankind's use of the river may exacerbate drought impacts and prevent population rebounds during average flow years.

Continued growth of the St. George, Utah and Mesquite, Nevada areas will continue to put additional pressures on basin water supplies availability. Rehabilitation of the Virgin–Moapa River system fish fauna is far from ensured at this time, and reversing downward trends is proving difficult. The fish fauna of these systems has survived long-term habitat loss and deterioration, and may be able to survive, at least in small, remnant areas, for some years to come. So far, formal recovery efforts have done little except document the decline of the native fish fauna. More research on limiting factors and the undertaking of effective rehabilitation efforts, as well as more flow in the rivers, will be needed to ensure the long-term survival of the Virgin–Moapa River system native fish fauna.

Acknowledgments

Portions of this study were funded by the Southern Nevada Water Authority. We appreciate the reviews and comments by Gary Scoppettone, Jack Williams, Gerald Smith, and Robert Hughes as they helped the accuracy and clarity of the document immensely. We appreciate the use of the Virgin River Fishes Database, developed by Thomas Hardy of Utah State University, and the use of unpublished data from Jim Heinrich of the Nevada Division of Wildlife and Rick Fridell of the Utah Division of Wildlife Resources. The Virgin River Fishes Recovery Team has been responsible for collection of monitoring data since 1976. Without their efforts, the data presented here would not have been collected.

References

Addley, R. C., and T. B. Hardy. 1993. The current distribution and status of spinedace in the Virgin River basin. Washington County Water Conservancy District, St. George, Utah.

Behnke, R. J. 1992. Native trout of western North America. American Fisheries Society, Monograph 6, Bethesda, Maryland.

Cope, E. D., and H. C. Yarrow. 1875. Report upon the collections of fishes made in portions of Nevada, Utah, California, Colorado, New Mexico, and Arizona 1871, 1872, 1873, and 1874. Report of the Geographic and Geologic Explorations and Surveys West of the 100th Meridian (Wheeler Survey) 5:635–703.

Cross, J. N. 1975. Ecological distribution of the fishes of the Virgin River. Master's thesis. University of Nevada, Las Vegas.

Cross, J. N. 1976. Status of the native fish fauna of the Moapa River (Clark County, Nevada). Transactions of the American Fisheries Society 105:503–508.

Deacon, J. E. 1988. The endangered woundfin and

water management in the Virgin River, Utah, Arizona, and Nevada. Fisheries 13(1):18–24.

Deacon, J. E., and W. G. Bradley. 1972. Ecological distribution of fishes of Moapa (Muddy) River in Clark County, Nevada. Transactions of the American Fisheries Society 101:408–419.

Deacon, J. E., and B. L. Wilson. 1967. Daily activity cycles of *Crenichthys baileyi*, a fish endemic to Nevada. Southwestern Naturalist 12:31–44.

DeMarais, B. D., T. E. Dowling, M. E. Douglas, W. L. Minckley, and P. C. Marsh. 1992. Origin of *Gila seminuda* (Teleostei: Cyprinidae) through introgressive hybridization: implications for evolution and conservation. Evolution 89:2747–2751.

DeMarais, B. D., T. E. Dowling, and W. L. Minckley. 1993. Post-perturbation genetic changes in populations of endangered Virgin River chubs. Conservation Biology 7:334–341.

Fridell, R. A., and R. J. Curtis. 2002. Virgin spinedace (*Lepidomeda mollispinis mollispinis*) population monitoring summary 1994–2001. Utah Division of Wildlife Resources, Publication 02–01, Salt Lake City.

Gregory, S. C., and J. E. Deacon. 1994. Human induced changes to native fishes in the Virgin River Drainage. Pages 435–444 *in* R. A. Marston, and V. R. Hasfurther, editors. Effects of human-induced changes on hydrologic systems. Proceedings Annual Summer Symposium of the American Water Resources Association. Jackson Hole, Wyoming.

Hardy, T. B., R. C. Addley, K. Tarbet, and K. Panja. 1995. Evaluating alternative flow strategies in the Virgin River. Utah State University, Department of Civil and Environmental Engineering, Logan.

Heckman, R. A., J. E. Deacon, and P. D. Greger. 1986. Parasites of the woundfin minnow, *Plagopterus argentissimus*, and other endemic fishes from the Virgin River, Utah. Great Basin Naturalist 46:662–676.

Heckman, R. A., P. D. Greger, and J. E. Deacon. 1987. New host records for the Asian tapeworm, *Bothriocephalus acheilognathi*, in endangered fish species from the Virgin River, Utah, Nevada, and Arizona. Journal Parasitology 73(1):226–227.

Holden, P. B., M. E. Golden, and S. J. Zucker. 2001. An evaluation of changes in woundfin populations in the Virgin River, Utah, Arizona, and Nevada, 1976–1999. BIO-WEST Report PR-735–01. Bureau of Reclamation, Boulder City, Nevada.

Holden, P. B., S. J. Zucker, P. D. Abate, and R. A. Valdez. 1997. Assessment of the effects of fish stocking in the state of Utah: past, present, and future. BIO/WEST Report PR-565–1. Utah Division of Wildlife Resources, Salt Lake City.

Hubbs, C. L., and R. R. Miller. 1948. Two relict genera of cyprinoid fishes from Nevada. Occasional Papers of the University of Michigan Museum of Zoology 507:1–30.

Lentsch, L. D., M. J. Perkins, H. Maddux. 1995. Virgin spinedace conservation agreement and strategy. Utah Division of Wildlife Resources, Publication Number 95–13, Salt Lake City.

Miller, R. R. 1952. Bait fishes of the lower Colorado River from Lake Mead, Nevada, to Yuma, Arizona, with a key for their identification. California Fish and Game 38(1):7–42.

Miller, R. R. 1961. Man and the changing fish fauna of the American Southwest. Papers of the Michigan Academy of Science, Arts, and Letters 46:365–404.

Miller, R. R., and C. L. Hubbs. 1960. The spiny-rayed cyprinid fishes (Plagopterini) of the Colorado River system. Miscellaneous Publications of the Museum of Zoology, University of Michigan 115:1–39.

Minckley, W. L. 1973. Fishes of Arizona. Arizona Game and Fish Department, Phoenix.

Rinne, J. N., J. Simms, and H. Blasius. 2005. Changes in hydrology and fish fauna in the Gila river, Arizona–New Mexico: epitaph for a native fish fauna? Pages 127–138 *in* J. N. Rinne, R. M. Hughes, and B. Calamusso, editors. Historical changes in large river fish assemblages of the Americas. American Fisheries Society, Symposium 45, Bethesda, Maryland.

Scoppettone, G. G. 1993. Interactions between native and nonnative fishes of the upper Muddy River, Nevada. Transactions of the American Fisheries Society 122:599–608.

Scoppettone, G. G., P. H. Rissler, M. B. Nielsen, and J. E. Harvey. 1998. The status of *Moapa coriacea* and *Gila seminuda* and status information on other fishes of the Muddy River, Clark County, Nevada. Southwestern Naturalist 43:115–122.

USFWS (U.S. Fish and Wildlife Service). 1979. Woundfin Recovery Plan. U.S. Fish and Wildlife Service, Albuquerque, New Mexico.

USFWS (U.S. Fish and Wildlife Service). 1983.

Moapa Dace Recovery Plan. U.S. Fish and Wildlife Service, Portland, Oregon.
USFWS (U.S. Fish and Wildlife Service). 1995. Recovery plan for the Virgin River fishes. U.S. Fish and Wildlife Service, Denver.
USFWS (U.S. Fish and Wildlife Service). 1996. Recovery plan for the rare aquatic species of the Muddy River ecosystem. U.S. Fish and Wildlife Service, Portland, Oregon.
Valdez, R. A., W. J. Masslich, R. Radant, and D. Knight. 1991. Status of the Virgin Spinedace (*Lepidomeda mollispinis mollispinis*) in the Virgin River drainage. BIO/WEST Report PR-197–1. Utah Division of Wildlife Resources, Salt Lake City.
WCWCD (Washington County Water Conservancy District). 1997. Virgin River Management Plan (VRMP), St. George, Utah.
Williams, J. E. 1978. Taxonomic status of *Rhinichthys osculus* (Cyprinidae) in the Moapa River, Nevada. Southwestern Naturalist 23:511–518.
Williams, J. E., and G. R. Wilde. 1981. Taxonomic status and morphology of isolated populations of the White River springfish, *Crenichthys baileyi* (Cyprinodontidae). Southwestern Naturalist 25:485–503.

American Fisheries Society Symposium 45:115–126, 2005

Changes in Fish Assemblages, Verde River, Arizona, 1974–2003

JOHN N. RINNE*
USDA Forest Service, Rocky Mountain Research Station, 2500 South Pineknoll Drive, Flagstaff, Arizona 86001, USA

Abstract.—Fish assemblages in the Verde River, Arizona have changed markedly over the last quarter century. Nonnative fishes increase from headwaters toward the mouth and individual native species decrease. River hydrograph and the introduction of nonnative species appear to be the major factors determining fish assemblages, although information is lacking on water quality and other land management impacts. During floods, native species dominated fish assemblages. By contrast, during droughts and sustained base flows, nonnative fishes increased. The threatened spikedace *Meda fulgida* has been collected only in the uppermost reach of this desert river and, even here, has been absent since 1997. Five other native species also have become less abundant or rare. Continued monitoring of fish assemblages, comparison with another large southwestern river, the Gila, and aggressive management are critical to sustain the native fish component of this river.

Introduction

The Verde River flows over 300 km from its headwaters in Big Chino Wash on the Prescott National Forest to its confluence with the Salt River near Phoenix (Figure 1). Linear position in the watershed, changing hydrology, presence of stream gauges, and conceivable human-induced impacts on the river were used in combination as reach designation criteria. The 60 km of river corridor above Sycamore Creek (Reach 1) is relatively undisturbed by humans and contains no flow-altering dams or major diversions. Reach 2 from Sycamore Creek downstream through the Camp Verde area to Beasley Flats supports municipalities, mining, groundwater pumping, diversions, recreational activity, and livestock grazing (Rinne et al. 2005). Reach 3 extends from Beasley Flats to Horseshoe Reservoir. Below this point to the confluence of the Verde with the Salt River, two major mainstream dams, Horseshoe and Bartlett, markedly change the natural hydrograph of the Verde. Reach 4 extends from below the lowermost reservoir, Bartlett, to the Salt River. The river contains one of the few remaining native fish assemblages in Arizona; however, nonnative species of fishes, introduced primarily for sportfishing, also exist throughout the river.

The primary objectives of this paper are to 1) delineate changes in fishes throughout the Verde River from 1974 to 2002, 2) define changes in key species from 1994 to 2002 in Reach 1, 3) document impacts of historic stocking (based on museum collections) of nonnative species to fish assemblage structure, 4) suggest factors that appear paramount cumulatively in changing fish assemblage structure or that need further evaluation, and 5) recommend management alternatives that may help sustain the threatened spikedace *Meda fulgida* and other native fishes in the Verde River.

Methods

The fish data I used were obtained from U.S. Forest Service surveys over the past 10 years, and existing Arizona Game and Fish Department surveys, data-

* E-mail: jrinne@fs.fed.us

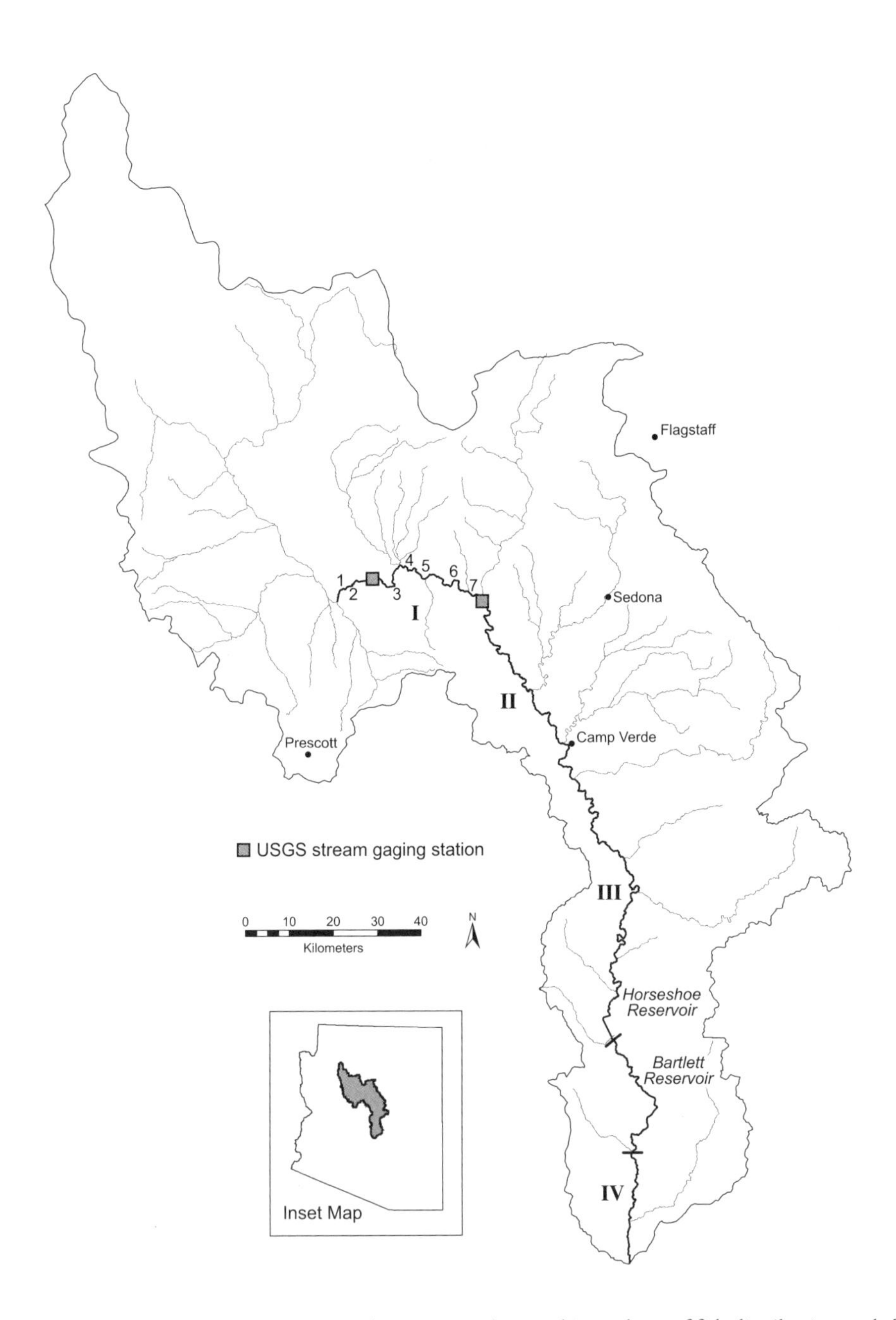

Figure 1.—The Verde River, Arizona, indicating the major reaches used in analyses of fish distribution and abundance.

bases, and administrative reports dating back to 1974 (Table 1). Electrofishing with backpack units and from canoes, seining and trammel netting were used to capture fishes. All collecting techniques with the exception of electrofishing from canoes were used in Reach 1 in my study. Sampling of sites (200–500 m) containing a diversity of habitats were described by Rinne and Stefferud (1996). Using multiple gear and sampling maximum habitat diversity increases the probability of capture of a greater variety of

Table 1.—Native and nonnative fish species of the Verde River. Asterisks following nonnative species indicate those most commonly captured during 1994–2003.

Native fishes	
extirpated	
Gila trout	*Oncorhnychus gilae*
Colorado pikeminnow	*Ptychocheilus lucius*
razorback sucker	*Xyrauchen texanus*
loach minnow	*Rhinichthys cobitis*
flannelmouth sucker	*Catostomus latipinnis*
Gila chub	*Gila intermedia*
extant in 1994	
longfin dace	*Agosia chrysogaster*
speckled dace	*R. osculus*
spikedace	*Meda fulgida*
desert sucker	*C. clarkii*
Sonora sucker	*C. insignis*
roundtail chub	*Gila robusta*
Nonnative fishes	
rainbow trout	*O. mykiss*
brown trout	*Salmo trutta*
brook trout	*Salvelinus fontinalis*
goldfish	*Carassius auratus*
common carp	*Cyprinis carpio*
threadfin shad	*Dorosoma petenense*
fathead minnow*	*Pimephales promelas*
red shiner*	*Cyprinella lutrensis*
golden shiner	*Notemigonus crysoleucas*
tilapia	*Tilapia spp*
northern pike	*Esox lucius*
smallmouth bass*	*Micropterus dolomieu*
largemouth bass	*M. salmoides*
striped bass	*Morone saxitalis*
white crappie	*Pomoxis annularis*
black crappie	*P. nigromaculatus*
green sunfish*	*Lepomis cyanellus*
bluegill	*L. macrochirus*
western mosquitofish*	*Gambusia affinis*
channel catfish	*Ictalurus punctatus*
flathead catfish	*Pylodictis olivaris*
yellow bullhead*	*Ameiurus natalis*

fishes. Notwithstanding, as is often the case in multiple objectives and multi-agency long-term sampling, different units of effort, gear types, and varied sampling in time and space were utilized. Accordingly, because estimates of individual fish numbers and percentages of total catch are not directly comparable through time and space, the results only represent general trends.

Fish data were stratified and categorized for analyses into the four major reaches. Multiple fish data sets (i.e., Arizona Game and Fish, or Game and Fish and Forest Service) within a sample year were combined. Data on fish assemblages are presented as percentages of individuals of native and nonnative species captured within the four major reaches and as changes of individual species numbers in Reach 1. Over 150,000 individuals were collected throughout the entire 300 km of the Verde River over a quarter of a century. Stocking records for sport fishes were obtained from Pringle (1996). Data from the Arizona State University Fish Data Base and museum depositions of nonnative fishes supplemented Forest Service and Game and Fish collections.

Results

Spatial Changes in Fish Assemblages

Among reaches.—The fish assemblage composition varied by reach (Figure 2) and within each reach over time (Figure 3).

Considering all years of sampling, native species dominated estimated total fish assemblages only in Reach 1 and the percent of natives decreased downstream (Figure 2). On average, native species dropped to less than 30% of the fish assemblage in Reaches 2 and 3, before plummeting to about 10% in Reach 4. Years of exceptions for Reaches 2 and 3 occurred in 1994–1995 and 1995 (respectively) when natives comprised 60–80% of the total fish assemblage (Figures 3a, 3b, 3c).

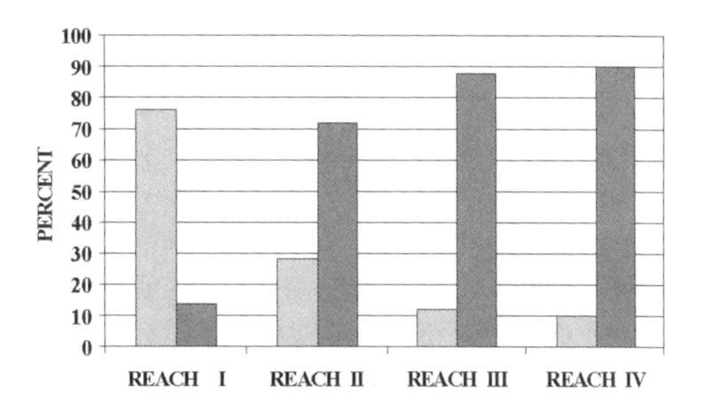

Figure 2.—Proportions of total fish assemblages in the four major river reaches of the Verde River, Arizona, 1974–2003. Native species are indicated by the light bars and nonnatives by the dark bars.

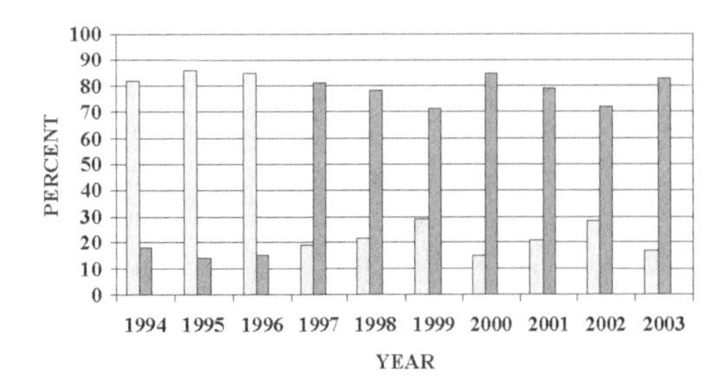

Figure 3.—a) Proportions of total fish assemblages in Reach 1 of the Verde River, 1985–2003. Native species are depicted by light bars and nonnatives by dark bars. These same designations are used in Figures 3b–d.

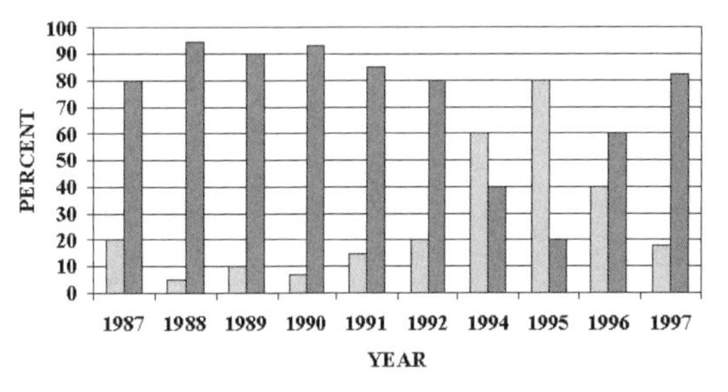

b) Proportional fish assemblages in Reach 2 of the Verde River, 1987–1997.

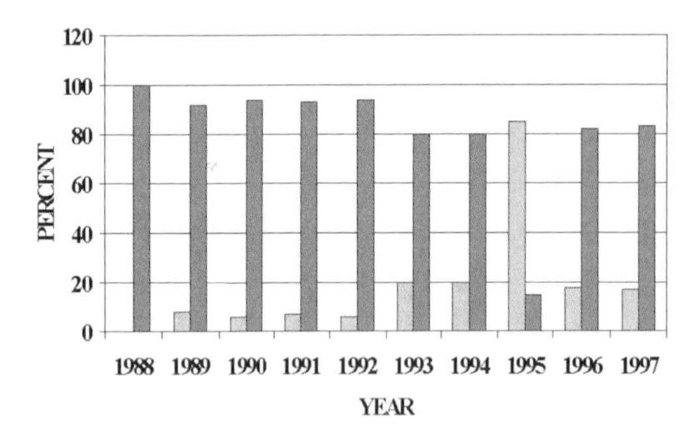

c) Proportional fish assemblages in Reach 3 of the Verde River, 1988–1997.

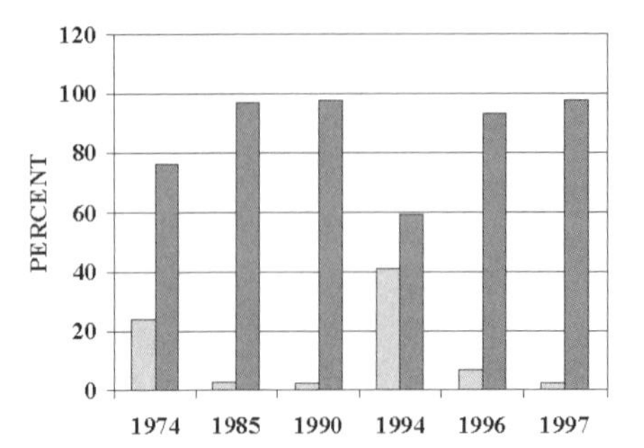

d) Proportional fish assemblages in Reach 4 of the Verde River, 1974–1997.

Within reaches.—Native species were more abundant than nonnative species in 9 of the 19 years for which samples were available for Reach 1 (Figure 3b). Between 1985 and 1987, native species comprised 70–85% of the fishes captured following a series of floods in the early 1980s. In 1988, the native fish assemblage dropped dramatically to less than 20% and remained at 34–53% until 1993 following a 75-year, late winter flood (Figure 4). Native species dominated (60–80%) the Reach I fish assemblage from 1993 to 1996. But, from 1997 to present, native species have comprised only 18–29% of the Reach I assemblage. In Reach 2, natives only exceeded nonnatives in 2 (1994–1995) of 10 sample years (Figure 3b). Native species dominated fish assemblages only in 1995 in Reach 3 and in other years comprised less than 20% of the fishes captured in this reach (Figure 3c). In Reach 4, nonnative fishes have dominated fish collections since 1974 (Figure 3d). However, in 1994, following the 1993 flood (Figure 4), native species comprised 40% of the estimated fish assemblage. Native species again remained a very small component of the fish assemblage by 1996.

Spatial Changes in Native Fish Species

The native Sonora sucker and desert sucker persisted in all sample reaches over the two decades of sampling, although they were collected only half the time in Reach 4 (Table 2). Roundtail chub was taken consistently in Reaches 1 and 2, but during only half the sample years in Reach 3 and was absent in all 6 years of sampling in Reach 4. Longfin dace occurred in all reaches and almost always in Reaches 1 and 2, but was present in only a third to a fifth of sample years in Reaches 3 and 4. Speckled dace often occurred in Reach 1, but was absent from Reaches 3 and 4 during all years of sampling and occurred only in 1 of 10 years in Reach 2. The threatened spikedace was collected only in Reach 1 and has not been collected there since 1997 (Rinne 1999b).

Total numbers of fishes captured in spring 1995–2003 in Reach 1 were markedly less than those collected in 1994 (Figure 5). Individual native species declined in a similar manner—most dramatically between 1994 and 1995 (Figure 6). Of the six native species, the three larger-sized species (desert sucker, Sonora sucker, and roundtail chub) also were

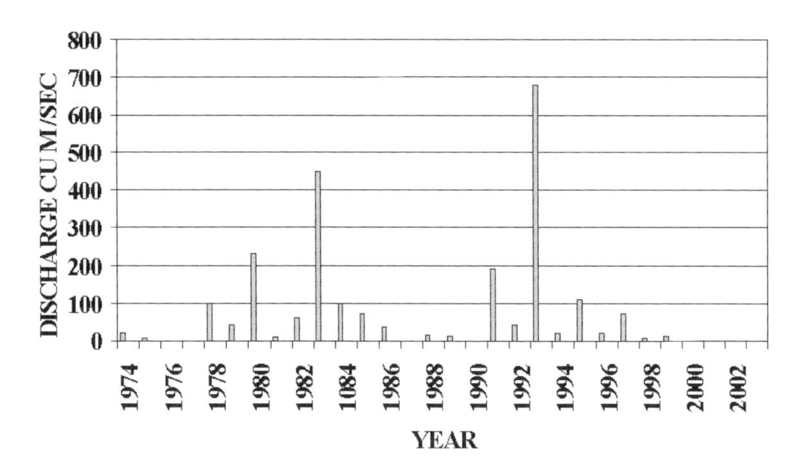

Figure 4.—a) Peak discharges (m³/s) in Reach 1, the Paulden gauge (see Figure 1), Verde River, 1974–2002.

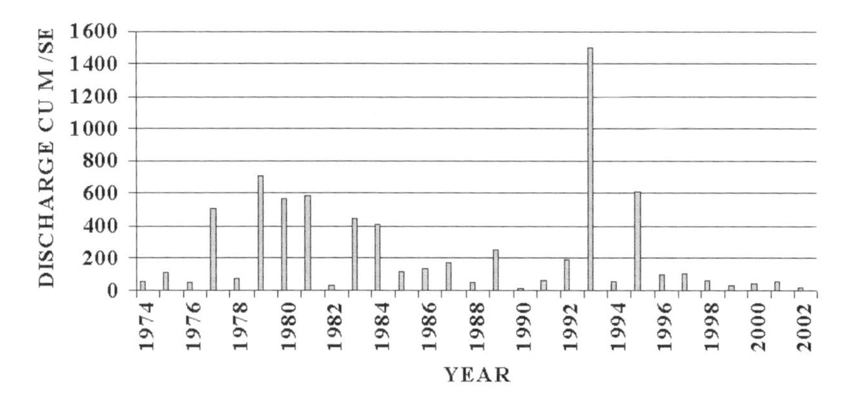

b) Peak discharges (m3/s) in Reach 2, the Clarkdale gauge (see Figure 1), 1974–2002.

reduced markedly from initial estimated numbers in 1994. By 1997, all three species were only 12% of their estimated numbers in 1994. Similarly, the three smaller-sized (<75 mm as adults) species (speckled dace, longfin dace, and spikedace) were rare to absent by 1997 and have remained so to the present.

In comparison, two of the more common non-native, predatory sport species, smallmouth bass and green sunfish have steadily increased in numbers to near 200 individuals at the seven sites (Figure 7). Green sunfish was reduced in 2002, likely because of increased predation by smallmouth bass. Red shiner fluctuated widely in numbers over the 10-year period (Figure 7).

Table 2.—Occurrence of native fishes in annual samples within respective reaches in the Verde River, 1974 to 1997.

	Reaches			
Species	1	2	3	4
Sonora sucker	18/18	10/10	10/10	3/6
desert sucker	18/18	9/10	5/10	3/6
roundtail chub	18/18	10/10	5/10	0/6
longfin dace	17/18	7/10	2/10	2/6
speckled dace	14/18	1/10	0/10	0/6
spikedace	10/18	0/10	0/10	0/6

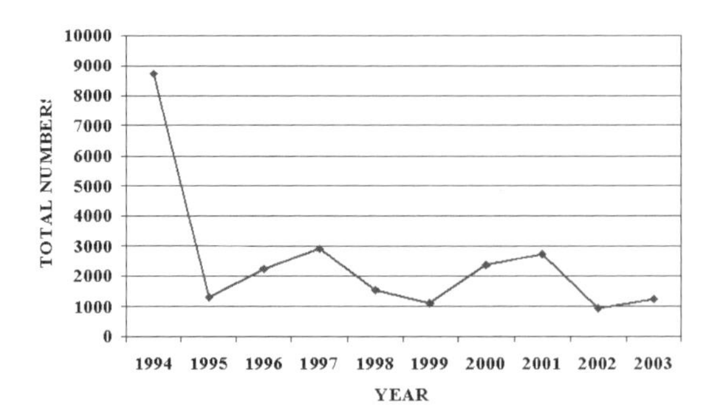

Figure 5.—Changes in total numbers of fishes collected at seven permanent monitoring sites in Reach 1, upper Verde River, 1994–2003.

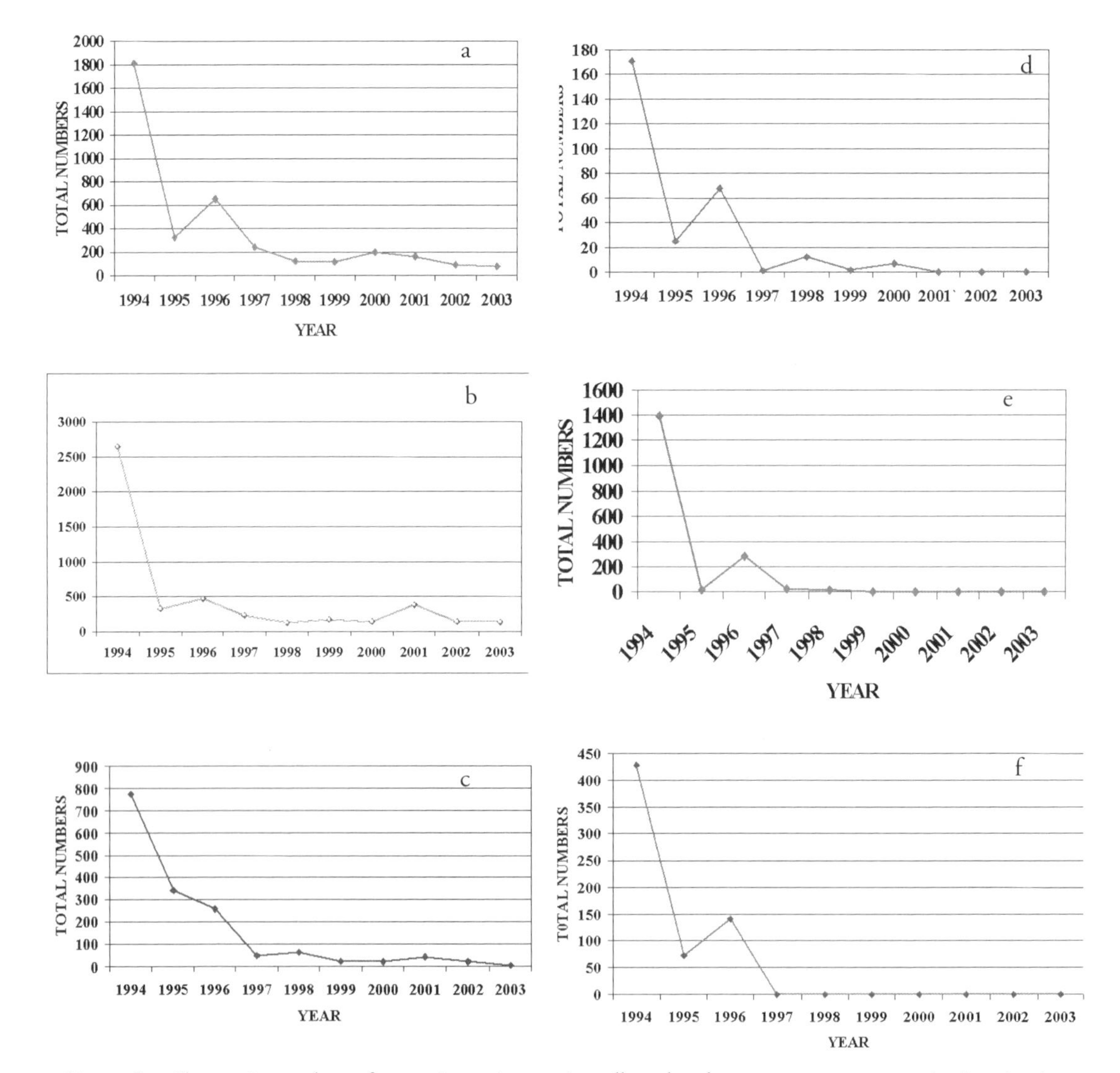

Figure 6.—Changes in numbers of respective native species collected at the seven permanent monitoring sites in Reach 1, Verde River, 1994–2003: Sonora sucker (a), desert sucker (b), roundtail chub (c), speckled dace (d), longfin dace (e), and spikedace (f).

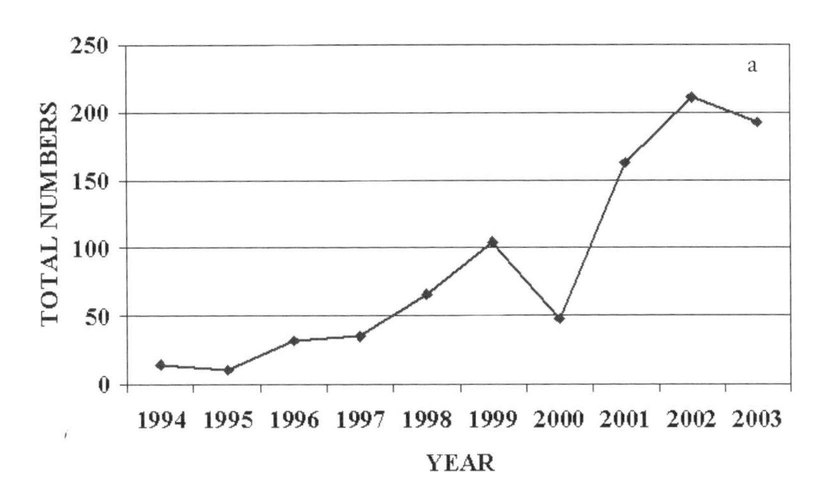

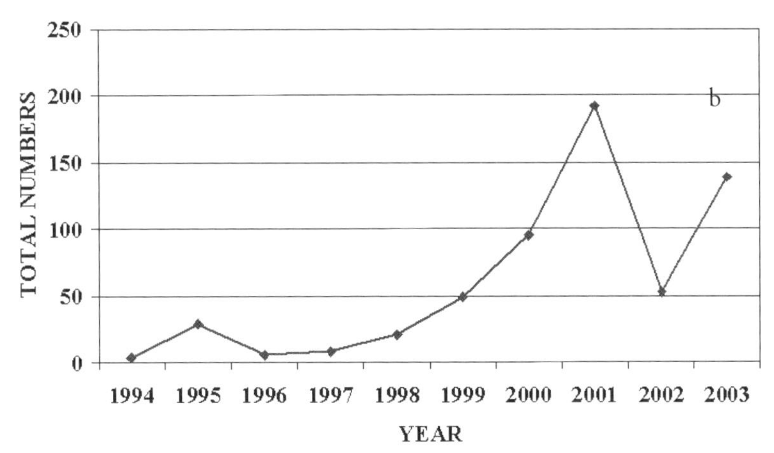

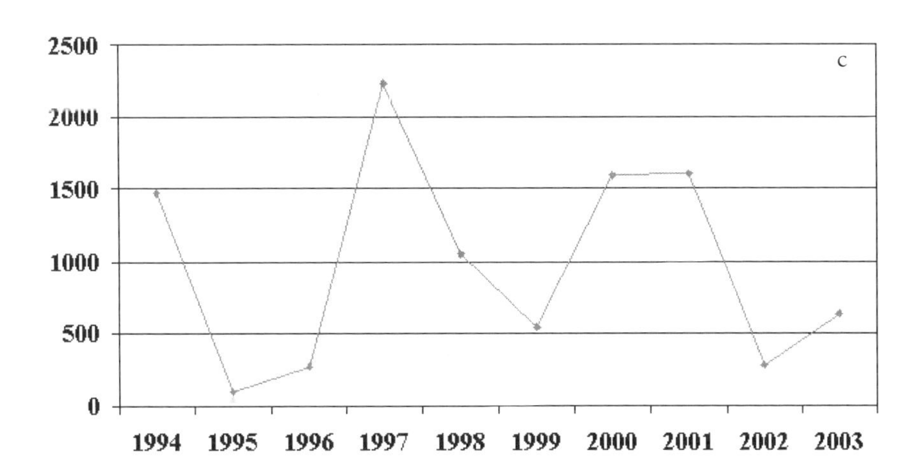

Figure 7.—Changes in numbers of smallmouth bass (a) and green sunfish (b) red shiner (c) in Reach 1, upper Verde River, 1994–2003.

Discussion

Summary of Fish Assemblages

Based on a substantial database, there is a distinct gradient in the Verde River from upstream to downstream in the relative proportions of native and nonnative fish species between 1974 and 1997. In Reach 1, native species dominated the fish assemblage in 1985–1987 and 1993–1996. In contrast, nonnative fishes dominated the fish assemblage increasingly from Reach 2 to 4. The two native suckers and longfin dace occurred in all reaches, and the roundtail chub in all but Reach 4. Speckled dace was common in Reach 1 and rare or absent in the remaining reaches. Spikedace occurred only in Reach 1. By contrast, smallmouth bass and green sunfish have increased steadily in estimated numbers and red shiner is commonly the most abundant species.

Fish Assemblage Stressors

River hydrograph.—Based on peak flows, the Verde River becomes increasingly and cumulatively altered in a downstream direction (Rinne et al. 2005). Fish assemblages also change dramatically as one proceeds downstream (Figure 2). Native fishes respond negatively to these hydrological alterations and nonnative species increase in abundance. Floods can have a marked influence on relative abundance of native and nonnative fishes (Minckley and Meffe 1987; Stefferud and Rinne 1995; Rinne and Stefferud 1997; Rinne 2002; Rinne et al. 2004). Significant floods occurred in the mid-1960s, early 1970s and 1980s, and mid-1990s (Stefferud and Rinne 1995; Figure 4). Native fishes dominated fish assemblages in Reach 1 for 3 years immediately after the 1980s' and 1990s' floods. In 1994–1995, parallel increases in peak flows and relative percent of natives occurred in all reaches, even in the nonnative fish-dominated Reaches 2–4.

In addition to the successive, downstream, cumulative hydrologic changes, periodic floods alter fish assemblages in the Verde. That is, native species comprise an increasingly smaller component of the total fish assemblage as one proceeds downriver with accompanying alteration of quantity of water in time and space. The Arizona Game and Fish Department has documented a proportional increase in numbers of the three larger-sized natives (two suckers and roundtail chub) in Reach 4 in 1998–2002. In part, such increases are likely a response to elevated flows recorded at the Clarkdale gauge between 1996 and 1998 (Figure 4), and, in part, to gear selectivity (i.e., canoe electrofishing), which underestimates smaller nonnative species such as red shiner and western mosquitofish. To reemphasize, nonnative species dominate fish assemblages during stable hydrographs and are reduced in numbers and relative proportions to natives, in response to significant floods.

Groundwater mining or pumping is greatest in the Middle Verde Valley (Reach 2; Rinne et al., 1998; and Rinne et al. 2004), and nonnative species are favored in this reach. Further, hundreds of diversions or stock tanks on the watershed (Sponholtz et al. 1997) are assumed to cumulatively stabilize the Verde hydrograph . More specific information on potential livestock grazing impacts to the river corridor, its riparian area, and ultimately the fish assemblage are needed (Medina and Rinne 1999; Rinne 1999a; 1999b).

Floodplain agriculture, diversions and wells, predominate in Reach 2, and must be evaluated relative to fish assemblages. Probably the most significant lack of information is that of water quality. Suspended sediment and nutrient content in the respective river reaches need evaluation. Numerous diversions in Reach 2 not only reduce base flow, but can increase suspended sediment upon return flow to the mainstream river. Reduced base flow in the main-stem Verde combined with increased suspended sediment, warmer water, and increased nutrients from irrigation return flow may limit native fish species. In contrast, more tolerant, nonnative species such as channel catfish and common carp are less affected by these changes. Accordingly, native species numbers decrease as nonnative species numbers increase commencing in Reach 2.

Human populations.—Historic (pre-1500) human populations along the middle Verde corridor approximated five individuals per square mile (Peter Pilas, U.S. Forest Service, Coconino National

Forest, Flagstaff, personal communication). Currently, human population densities are 20 times historic numbers (Rinne et al. 1998). Human population growth increased 135% in the Verde watershed between 1980 and 1994 and is expected to increase by that same percentage over the next four decades. More significantly, almost half of this population increase occurred in Reach 2.

Changes in native fish populations in the Verde River appear to mirror human-induced changes in river systems on a regional basis. Prior to 1950, the Verde watershed was agrarian and human population was small. Riparian areas were affected mostly by livestock grazing. Stocking of nonnative species began in the 1930s and escalated with completion of dams on the Verde. After 1950, transportation systems and affordable hydropower stimulated urbanization and human population growth. The growth in human population brought about an increased demand for sportfish stocking. The construction of reservoirs followed by introductions of nonnative, predatory sportfish are associated with decline in native species (Rinne 1991, 1994, 1995, 2003a, 2003b; Rinne and Minckley 1991).

Fishery management practices.—Stocking of sport fishes commenced about six decades ago in Bartlett Reservoir on the mainstream Verde, and in tributary streams, lakes, and reservoirs. Over a dozen nonnative, primarily sport fishes were stocked in the Verde basin. During that time, more than a dozen species and 15 million individual, nonnative sport fishes were stocked (Rinne et al. 1998). Predatory species such as catfishes, bass, and sunfish comprised a majority of introduced species. Museum collections of nonnative fishes parallel the stocking history in the Verde basin. Before 1950, all five records of nonnative fishes were from two tributaries of the Verde: Oak and Wet Beaver creeks. Between 1950 and 1964, collection records of nonnative species doubled and 6 of the 11 records were from the main-stem Verde. The occurrence of nonnative species increased four-fold between 1965 and 1979. Again, 59% were from tributary streams. Records of nonnatives between 1980 and 1995 were similar to those of the previous 15-year period, but likely reflect a lack of museum deposition versus actual numbers of nonnative species (Peter Unmack, Arizona State University, personal communication).

Over the past six decades, over 850 stocking events occurred in reservoirs and almost 4,500 in streams, mostly in tributaries to the Verde (Pringle 1996; Rinne et al. 1998). Both coldwater salmonids and warmwater ictalurids and centrarchids were stocked. Usually, trout were stocked in tributaries. In Bartlett Reservoir, almost three million channel catfish, largemouth bass, and sunfish species were stocked in aggregate over a 40-year period.

Conservation and Management Recommendations

The spikedace.—This threatened species was most limited in temporal-spatial distribution, occurring only in Reach 1. Although adapted to the vagaries of southwestern stream dynamics, over the past 6 years, spikedace populations appeared to be dramatically affected by natural variations in the Verde River hydrograph (i.e., alternating floods and drought) (Stefferud and Rinne 1995; Rinne and Stefferud 1997; Rinne 1999b). In addition, nonnative fishes such as introduced red shiner (Rinne 1991a; Douglas et al. 1994) and larger predatory species such as smallmouth bass, green sunfish, and yellow bullhead may have negatively affected spikedace (Rinne 1991a, 2001; Stefferud and Rinne 1995).

Although historically present in the Verde River, spikedace was not rediscovered in samples until the early 1970s from the mouth of Sycamore Creek (Anonymous 1974). Large numbers of spikedace were collected by the U.S. Bureau of Reclamation in the mid 1980s in the upper Verde while examining instream flow needs of native fishes (U.S. Fish and Wildlife Service 1989). Spikedace were common in samples and comprised up to 11% of the total native fish assemblage.

Results of more intensive study of fish populations at seven sites in Reach 1 following a 75-year flood in the winter, 1992–1993 (Stefferud and Rinne 1995) are striking. Large numbers of spikedace were collected in 1994. A 10-year flood occurred in March 1995, and spikedace populations pulsed once more in 1996 (Rinne and Stefferud 1997; Rinne 1999b), only to decline by spring 1997 (Rinne et al. 1998). Additional sampling along the entire course of the upper Verde in 1997 and 1998 resulted in less than a dozen spikedace being captured suggesting the population had declined markedly. In the spring of 1999, sampling in the most upstream reaches of the Verde (Bear Siding upstream) again indicated no spikedace (unpublished data). Based on these data, the species presently is very rare in the upper Verde.

There is a 80–90% chance that the species is extirpated from the upper Verde River (Rinne 1999b). Even if a residual population exists, and even if spawning does occur annually, the probability of survival of young-of-year spikedace is low given the excessive numbers of nonnative, predatory species present in the river (Stefferud and Rinne 1995; Rinne 2001; Table 1). I recommend suppression of nonnative fish populations in the most upstream reaches of the upper Verde. Such action may help sustain the spikedace locally until the next flood event. At best, mechanical, partial suppression of nonnative fishes is conceivably cosmetic, unless instituted immediately following a significant flood (Minckley and Meffe 1987; Rinne and Stefferud 1997).

Isolation of the spikedace population of the upper Verde River from those of the Gila River and Aravaipa Creek (Rinne 1991a) date at least to completion of dams (i.e., Horseshoe and Bartlett) on the Verde in the 1930s. Because of this physical and genetic isolation, introduction of spikedace from other river systems is not a viable management alternative at this time. Based on results of analyses of spikedace populations in the Gila River system, it is reasonable to conclude that the population of spikedace in the upper Verde is genetically distinct (Tibbets and Dowling 1996). Accordingly, no translocation of spikedace should occur unless there is unequivocable evidence that the spikedace has been extirpated from the upper Verde River drainage.

Additional management strategies are ongoing. The first is intensification of survey and monitoring activity for spikedace. Second, effort is needed to locate and evaluate possible refuge streams in the Verde basin. Last, culture techniques must be developed immediately for the species. These last two items should be in place preceeding the next flood event, presumably, when spikedace will reappear in samples and become abundant enough to facilitate transfer to a hatchery for propagation and/or stocking into identified refuge streams or facilities.

All native species.—The opportunity exists to examine population dynamics of not only the threatened spikedace, but several native species (i.e., longfin dace, desert sucker, Sonora sucker, roundtail chub, and loach minnow) relative to hydrograph and in the absence of or reduction in nonnative fishes. A study on the upper Verde is currently examining the response of native fishes relative to mechanical reduction or suppression of nonnative fishes (Rinne 2001).

Future research.—The rarity of spikedace in the upper Verde River, its abundance in the upper Gila River, the abundance of nonnative fishes in the former river and scarcity in the latter, and the short-lived nature of the spikedace are all reasons for a coordinated, regional management-research approach to sustain spikedace in southwestern streams and rivers (Rinne 1991b). At present, there is a window of opportunity to examine flooding and fish assemblage structure and succession (Rinne 1996) from two extant scenarios: 1) abundance of natives and especially spikedace combined with few nonnatives (the Gila River and Aravaipa Creek), and, 2) abundance of nonnatives, few natives, and an apparent lack of spikedace (the Verde River) (Rinne et al. 1998; 1999b; in press). I recommend that the study effort be a coordinated, regional approach, not that of a single river system or jurisdiction (i.e., forest by forest). Federal and state agencies must collaborate on understanding the interrelationships of flooding and native–nonnative fish assemblage structure and succession in the Southwest.

Acknowledgments

I thank the Arizona Game and Fish Department for data and records of past (pre-1994) collections and the many volunteers, students, and state and federal staff who assisted in collection of fishes.

References

Anonymous. 1974. Cyprinidae, *Meda fulgida*, Girard, spikedace. Arizona Game and Fish Department, Phoenix.

Douglas, M. E., P. C. Marsh, and W. L. Minckley. 1994. Indigenous fishes of western North America and the hypothesis of competitive displacement: *Meda fulgida* (Cyprinidae) as a case study. Copeia 1994(1):9–19.

Medina, A. L., and J. N. Rinne. 1999. Ungulate/fishery interactions in southwestern riparian ecosystems: pretensions and realities. Proceedings of the North American Wildlife and Natural Resources Conference 62:307–322.

Minckley, W. L. and G. K. Meffe. 1987. Differential selection by flood in stream-fish communities of the arid American Southwest. Pages 93–104 *in* W. J. Matthews and D. E. Heins, editors. Evolutionary and community ecology of North American stream fishes. University of Oklahoma Press, Norman.

Pringle, T. 1996. Historical stocking summary, 1920 to 1995. Arizona Game and Fish Department, Phoenix.

Rinne, J. N. 1991a. Physical habitat use by spikedace, *Meda fulgida* in southwestern desert streams with probable habitat competition by red shiner, *Notropis lutrensis* (Pisces: Cyprinidae). The Southwestern Naturalist 36:7–13.

Rinne, J. N. 1991b. An approach to management and conservation of a declining regional fish fauna: southwestern USA. Pages 56–60 *in* B. Matamurya, Y. Bobeck, W. Ono, L. Regelin, R. Bartos, and P. R. Rattcliffe, editors. Proceedings of a Symposium on Wildlife Conservation, 5th International Congress on Zoology. Yokohama, Japan.

Rinne, J. N. 1994. Declining southwestern Aquatic habitats and fishes: are they sustainable? Pages 256–265 *in* W. W. Covington and L. F. DeBano, technical coordinators. U.S.D.A., Forest Service, Sustainability symposium, General Technical Report RM-247, Fort Collins, Colorado.

Rinne, J. N. 1995. Interactions of predation and hydrology on native southwestern fishes: Little Colorado spinedace in Nutrioso Creek, Arizona. Hydrology and Water Resources in Arizona and the Southwest 22–25:33–38.

Rinne, J. N. 1996. The effects of introduced fishes on native fishes: Arizona, southwestern United States. Pages 149–159 *in* D. P. Philipp, editor. Protection of aquatic diversity. Proceedings of the World Fisheries Conference, Theme 3. Oxford & IBH Publishing Co. Pvt. Ltd., New Delhi, India.

Rinne, J. N. 1999a. Fish and grazing relationships: the facts and some pleas. Fisheries 24(8):12–21.

Rinne, J. N. 1999b. The status of spikedace, *Meda fulgida*, in the Verde River, 1999. Implications for research and management. Hydrology and Water Resources in the Southwest 29:57–64.

Rinne, J. N. 2001. Nonnative, predatory fish removal and native fish response: Verde River, Arizona, 1999–2000. Hydrology and Water Resources of the Southwest 31: 29-36.

Rinne, J. N. 2002. Hydrology, geomorphology and management: implications for sustainability of native southwestern fishes. Hydrology and Water Resources of the Southwest 32:45–50.

Rinne, J. N. 2003a. Native and introduced fishes: their status, threats and conservation. Pages 194–213 *in* P. F. Ffoliott, M. B. Baker, L. F. Debano, and D. G. Neary, editors. Riparian areas in the southwestern United States: hydrology, ecology, and management. Lewis Publishers, Boca Raton, Florida.

Rinne, J. N. 2003b. Fish habitats: conservation and management implications. Pages 278–297 *in* P. F. Ffoliott, M. B. Baker, L. F. Debano, and D. G. Neary, editors. Riparian areas in the southwestern United States: hydrology, ecology, and management. Lewis Publishers, Boca Raton, Florida.

Rinne, J. N., P. Boucher, D. Miller, R. Pope, A. Telles, B. Deason, and W. Merhage. 2005. Comparative fish community structure in two southwestern desert rivers. Pages 64–70 *in* M. J. Brouder, C. L. Springer, and S. C. Leon, editors. Proceedings of two symposia. Restoring native fish to the lower Colorado river: interactions of native and non-native fishes. July 13–14, 1999, Las Vegas, Nevada. Restoring natural function within a modified riverine environment: the lower Colorado river. July 8–9, 1998. U.S. Fish and Wildlife Service, Southwest Region, Albuquerque, New Mexico.

Rinne, J. N., and W. L. Minckley. 1991. Native fishes of arid lands: a dwindling natural resource of the desert Southwest. U.S.D.A. Forest Service, Rocky Mountain Forest and Range Experiment Station, General Technical Report RM-206:1–45, Fort Collins, Colorado.

Rinne, J. N., L. Riley, B. Bettaso, K. Young, and R. Sorensen. 2004. Managing southwestern native and nonnative fishes: can we mix oil and water and expect a favorable solution. Pages 445–466 *in* J. Nickum, M. Nickum, P. Muzik, and D. MacKinley, editors. Propagated fish in resource management. American Fisheries Society, Symposium 44, Bethesda, Maryland.

Rinne, J. N., and J. A. Stefferud. 1996. Relationships of native fishes and aquatic macrohabitats in the Verde River, Arizona. Hydrology and Water Resources in Arizona and the Southwest 26:13–22.

Rinne, J. N., and J. A. Stefferud. 1997. Factors contributing to collapse yet maintenance of a native fish community in the desert Southwest (USA). Pages 157–162 *in* D. A Hancock, D. C. Smith, A. Grant, and J. P. Beaumer, editors. Developing and sustaining world fisheries resources: the state of science and management. Second World Fish Congress, Brisbane, Australia.

Rinne, J. N., J. A. Stefferud, A. Clark, and P. Sponholtz. 1998. Fish community structure in the Verde River, Arizona, 1975–1997. Hydrology and Water Resources in Arizona and the Southwest 28:75–80.

Sponholtz, P. D. Redondo, B.P. Deason, L. Sychowski, and J. N. Rinne. 1997. The influence of stock tanks on native fishes: Verde River, Prescott National Forest. Pages 156–179 *in* J. M. Feller and D. S. Strouse, editors. Proceedings of a symposium on environmental, economic, and legal issues related to rangeland water developments. Arizona State University, Center for Study of Law, Tempe.

Stefferud, J. A., and J. N. Rinne. 1995. Preliminary observations on the sustainability of fishes in a desert river: the roles of streamflow and introduced fishes. Hydrology and Water Resources in Arizona and the Southwest 22–25:26–32.

Tibbets, C. A., and T. E. Dowling. 1996. Effects of intrinsic and extrinsic factors on population fragmentation in three species of North American minnows (Teleostei:Cyprinidae). Evolution 50(3):1280–1292.

U.S. Fish and Wildlife Service. 1989. Instream flow requirements of fishes of theVerde River system. U.S. Fish and Wildlife Service, Phoenix, Arizona.

American Fisheries Society Symposium 45:127–137, 2005

Changes in Hydrology and Fish Fauna in the Gila River, Arizona–New Mexico: Epitaph for a Native Fish Fauna?

John N. Rinne*
USDA, Forest Service, Rocky Mountain Research Station, Flagstaff, Arizona 86001, USA

Jeffrey R. Simms
USDI, Bureau of Land Management, Tucson, Arizona, USA

Heidi Blasius
USDI, Bureau of Land Management, Safford, Arizona, USA

Abstract.—The Gila River originates in southwestern New Mexico and courses its way for over 700 km to the west before emptying into the main-stem Colorado River near Yuma, Arizona. Historically, this river was a major watercourse across the Sonora Desert of Arizona. At present, main-stem dams and numerous diversions have markedly altered the historic hydrology of the river. Seventeen native species once occupied the main stem of this large southwest desert river. More than twice that number (40) of nonnative fish species have been introduced into the waters of the Gila over the past century. Currently, less than half of the native fauna is present in the main stem and then primarily in the upper three reaches of the river. The majority of the species (70%) are federally listed as threatened, endangered, or sensitive. The combination of hydrological alteration and accompanying introductions of nonnative, principally sport fishes has basically extirpated the native fauna in all but the uppermost reaches of the Gila River main stem.

Introduction

Fish faunas and assemblages in southwestern desert rivers have declined dramatically from historic occurrences (Mueller and Marsh 2002; Calamusso et al. 2005; Mueller et al. 2005; Rinne 2005). These and others (Miller 1961; Minckley 1973; Rinne and Minckley 1991; Rinne 1994, 1996, 2003a) have documented the general loss and modification of aquatic habitat and parallel declines in native fish faunas. In addition, there are many individual surveys and reports in the gray literature. Although such reports provide specific and detailed accounts of declines for species and fish assemblages, they are not readily available. The Gila River is another southwestern desert river that experienced dramatic loss of its fish fauna and for which much of the information resides in agency reports and the gray literature. Accordingly, a primary objective of this chapter is to summarize the historic and current status of the hydrology and fish fauna in this river.

Geography, Topography, and Climate

The Gila River originates in southwestern New Mexico, primarily within the first wilderness area established in the United States, the Aldo Leopold Wilderness (Figure 1). It then courses its way for over 700 km due west across Arizona to its confluence

* Corresponding author: jrinne@fs.fed.us

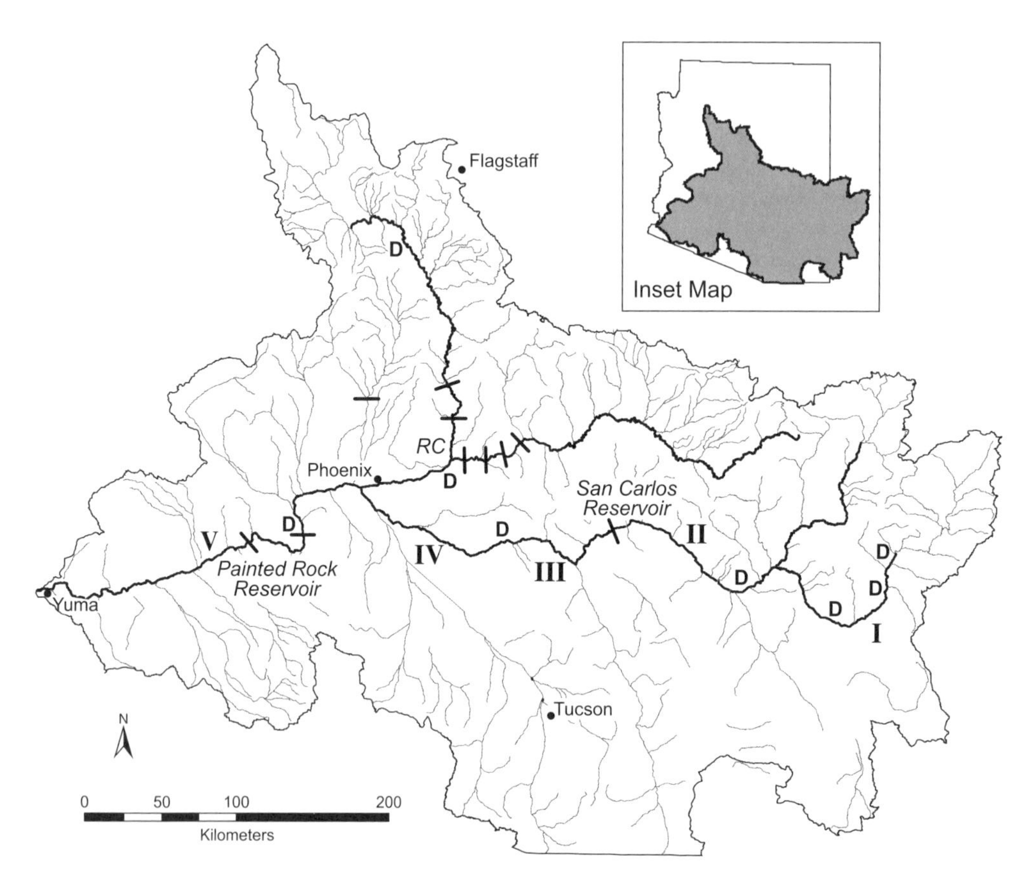

Figure 1.—Gila River basin indicating the five major reaches (I–V), main-stem dams (solid bars), and major diversions (D). San Carlos and Painted Rock reservoirs and a reservoir complex (RC) of main-stem dams on the Salt and Verde are designated.

with the Colorado River near Yuma, Arizona. The Gila River drainage comprises 212,380 km^2 (82,000 mi^2) and is situated in the Basin and Range, Colorado Plateau, and Transition physiographic provinces of the United States. It drains over half of Arizona and nearly 8,000 km^2 in southwestern New Mexico (Figure 1). It heads in the Datil–Mogollon section of the Transition zone in the Gila Wilderness area at elevations of greater than 3,500 m and terminates near sea level in the Basin and Range Province. Primary land uses or land covers for the basin are forested landscapes above 1,500 m and grasslands near and below 1,500 m that historically and presently are used for grazing. Two major urban and irrigated areas occur in the Gila–Cliff Valley and Virden in southwestern New Mexico and the Duncan and Safford reaches in eastern Arizona. The extensive and ever-increasing Phoenix metropolitan area is positioned on the Salt River immediately above the confluence of this major tributary with the Gila. The climate is highly variable, ranging from hot desert to cool-cold montane areas. About half the precipitation falls in summer monsoon events and the other as winter snowfall at upper elevations (>3,000 m) and rain at both lower desert (<1,000 m) and intermediate elevations (1,000–2,000 m). Annual precipitation ranges from over 100 cm in the mountains to near 20 cm in the low desert regions.

Along its course, the river drops in elevation from over 3,000 m to near sea level and traverses the extreme xeric conditions of the upper and lower Sonoran Life Zones (Lowe 1964; Brown and Lowe 1980). Its major tributaries from the south are the Santa Cruz, San Pedro, and Sam Simon rivers. From the north, the San Francisco joins the Gila near the Arizona–New Mexico border and the Salt and

Hassayampa rivers join west of the Phoenix, metropolitan area.

Historic Hydrology

Historically, the Gila River was characterized by surface flow from its headwaters to its confluence with the Colorado (Corle 1951). Beaver were abundant and the river was repeatedly dammed through their activities (Emory 1848). Trappers pursued these resident engineers around the turn of the 20th century (Pattie 1905). The river reach (Reach 4) between Phoenix and Tucson was often over a kilometer wide, characterized by a mosaic of aquatic habitats (Rea 1983), and teeming with fishes. Cottonwood-willow galleries were present but more patchy than at present because of beavers.

Miller (1961) first discussed the overall changes in native fish faunas in the American Southwest. In general, he suggested aquatic habitats in the Southwest have been transformed from "clear and dependable" to "intermittent" and "carrying heavy silt loads." The mainstream Gila was described as a "large, essentially permanent stream of clear to sea-green water" with a "well-defined channel flanked by numerous cottonwoods and set off by a dense growth of willows and cane." Extensive lagoons and marshes abounded in waterfowl, beaver, and fish. In 1846, Emory (1848) described the river just west of present day Gila Bend as "wide, rich, and thickly overgrown with willow and a tall aromatic weed and alive with white brant, geese, and ducks, with many signs of deer and beaver." By 1920, this same area was described by Ross (1923) as "desolate wastes of sand and silt."

Current Hydrology

The Gila River has been drastically altered throughout its course (Figure 1). Within a few kilometers of its issuance from the Aldo Leopold Wilderness area in southwestern New Mexico, flow is diverted into Fort West Ditch, a major canal that has supported agriculture in the Gila–Cliff Valley for a century. Within this broad, alluvial valley, water is again diverted by the Phelps Dodge Diversion into a forebay and pumped to Bill Evans Lake, a storage basin that provides water for mining operations southwest of Silver City, New Mexico. Surface waters persist downstream another 60 km except locally in droughts. The Gila River is diverted once again by Sunset Diversion near the Arizona–New Mexico border. Diverted water then supplies extensive agricultural activity near Virden, New Mexico and Duncan, Arizona. Below Duncan, surface water again appears in the vicinity of Apache Creek and is further supplemented by flow from the San Francisco River near Clifton, Arizona. Surface water persists downstream to the tail of the Gila Box Riparian National Conservation Area just east of Safford, Arizona where the Brown and San Jose diversions continue the process of dewatering the river. Downstream in the Safford Valley, another seven diversions capture the remaining surface flows. Only seasonal, variable surface flows persist in the Gila in the Safford Valley to San Carlos reservoir as a result of irrigation return flows and infrequent input from local washes.

One major dam, Coolidge (1928), impounding San Carlos Reservoir, is present in the upper course of the river (Reach 3, Figure 1) which diverts water into the Florence/Casa Grande Canals. Below San Carlos Reservoir, surface flows in the main-stem Gila are present primarily to supply irrigation waters to the Ashurst–Hayden diversion dam east of Florence at the terminus of Reach 3.

The river is normally dry from Florence downstream to its confluence with the Salt River (Reach 4). In the reaches below the confluence of the Gila and Salt, the river is a mosaic of diversion dams, sewage recharge waters, a flood control dam, and Colorado River water input. At best, physical, hydrological, and climatic factors render the reaches below Phoenix harsh and inhospitable for native fishes. Here, sewage recharge waters produce only limited and poor quality water in the vicinity of the Gila–Salt confluence. Next, Gillespie Dam diverts artificial flows into the Gila Bend Canal. In the upper portion of this reach (Reach 5), water is used for agriculture. Downstream, a few kilometers just west of Gila Bend, a flood control structure, Painted Rock Reservoir, impounds floods from local tributary washes and the Hassayampa and Salt rivers. The last

100 km of river is markedly altered by the Wellton–Mohawk Irrigation system and extensive agricultural development supported by the Wellton–Mohawk and Dome canals.

Historic and Present Fish Fauna of the Basin and Main Stem

Gila River Basin

Twenty fish species were native to the Gila River basin (Table 1). Three quarters of the fauna is comprised of cypriniform species (Minckley 1973; Sublette et al. 1990; Rinne and Minckley 1991). Two native salmonids, two cyprinidontids, and one poeciliid comprise the remainder of the fauna. The Gila trout persists naturally only in less than a half dozen headwater streams in New Mexico (Rinne 1988). The large Colorado pikeminnow and razorback sucker once ran in spawning schools up the Gila and Salt rivers. The pikeminnow was referred to by old timers as the "Colorado salmon." This large, migratory predator was recorded as far upstream as Ft. Thomas in the Safford Valley (Reach 2). Around the turn of the century, it swam with most of the native Gila River fishes in the Salt River, Tempe, Arizona (Minckley and Deacon 1968). The last pikeminnow collected in the Gila River basin was taken in 1950 in the Salt River Canyon, near the U.S. 60 highway crossing (Minckley 1973). Razorback sucker once were pitchforked from canals in the Phoenix Valley because their sheer abundance precluded flow (Minckley 1973). This sucker has been the object of extensive reintroductions (Johnson 1985), but presently, self-sustaining populations are absent in the basin (Minckley 1983).

Presently nonnative fish species dominate the Gila River main stem (Table 2) except for Reach 1 (Rinne et al., 2004). In Reach 2, nonnative species of fishes comprise the majority of fishes captured recently (Minckley 1973; Blasius and Simms 2001). The threatened spikedace and loach minnow are absent from samples. In Reach 3, nonnative fishes are most common with only the desert sucker, Sonora sucker, and longfin dace extant. Fish are generally lacking from Reach 4 because of a complete lack of surface flow. Maps in Minckley (1973) indicate western

Table 1.—Historic fish fauna of the Gila River basin, Arizona and New Mexico, showing current presence in the respective reaches of the main-stem river (see Figure 1), current presence (P) or absence (A) in the Gila basin, and listing status as endangered (E), threatened (T), state listed (S), and extirpated (X).

Common name	Scientific name	Family	P / A	Reaches	Status
Gila trout	*Oncorhynchus gilae*	Salmonidae	P	1	E
Apache trout	*O. gilae apache*		P		T
Monkey Spring chub	*Gila* sp.	Cyprinidae	A		X
roundtail chub	*G. robusta*		P	1	S
headwater chub	*G. nigra*		P		S
Gila chub	*G. intermedia*		P		E
bonytail chub	*G. elegans*		A		E
Colorado pikeminnow	*Ptychocheilus lucius*		A		E
longfin dace	*Agosia chrysogaster*		P	1–3	
spikedace	*Meda fulgida*		P	1	T
speckled dace	*Rhinichthys osculus*		P	1	
loach minnow	*R. cobitis*		A		E
woundfin	*Plagopterus argentissimus*		A	1	
razorback sucker	*Xyrauchen texanus*	Catostomidae	A		E
desert sucker	*Catostomus clarkii*		P	1–3	
Sonora sucker	*C. insignis*		P	1–3	
flannelmouth sucker	*C. latipinnis*			2–4	
Santa Cruz pupfish	*Cyprinodon arcuatus*	Cyprinodontidae	A		X
desert pupfish	*C. macularius*		P		E
Gila topminnow	*Poecilliopsis occidentalis*	Poeciliidae	P		E

Minckley and Clarkson 1979; Minckley 1985

Table 2.—Current distribution of the most common, nonnative, fishes in the Gila River main stem based on recent collections and the literature (Minckley 1973; Sublette et al. 1990).

	Reach of Gila River				
Species	1	2	3	4	5
red shiner *Cyprinella lutrensis*	x	x	x	x	x
fathead minnow *Pimephales promelas*	x	x	x	x	x
common carp *Cyprinus carpio*	x	x	x	x	x
golden shiner *Notemigonus crysoleucas*	x	x	x		x
western mosquitofish *Gambusia affinis*	x	x	x	x	x
channel catfish *Ictalurus punctatus*	x	x	x	x	x
flathead catfish *Pylodictis olivaris*	x	x	x		x
yellow bullhead *Ameiurus natalis*	x	x	x	x	x
black bullhead *A. melas*	x	x	x	x	x
largemouth bass *Micropterus salmoides*	x	x	x	x	x
smallmouth bass *M. dolomieu*	x	x	x		
green sunfish *Lepomis cyanellus*	x	x	x	x	x
bluegill *L. macrochirus*	x	x	x	x	x
black crappie *Pomoxis nigromaculatus*		x			x
sailfin molly *Poecilia latipinna*				x	x
Mozambique tilapia *Oreochromis mossambicus*		x			x
brown trout *Salmo trutta*	x				
rainbow trout *Oncorhynchus mykiss*	x				
brook trout *Salvelinus fontinalis*	x				

mosquitofish as the most commonly-occurring nonnative species in this reach with common carp and catfishes *Ictalurus* and *Ameiurus* also present. In Reaches 4 and 5, no native fishes are present and the fauna is primarily comprised of sunfishes *Centrarchidae,* catfishes *Ictalurus, Amerius, Pylodictus,* and bait fishes (Table 2).

The diminutive and presently endangered Gila topminnow was once the most common fish in the lower Colorado basin (Hubbs and Miller 1941). The species is now absent in the main-stem Gila River and exists only as isolated stream and spring populations (Minckley 1973; Meffe et al., 1983; Table 1). The Monkey springs pupfish *Cyprinodon arcuatus* and chub were lost to predation by largemouth bass and presently exist only in museum jars (Minckley et al. 2002). The desert pupfish is absent in the basin with exception of a few isolated spring and rearing facilities. Loach minnow and spikedace populations remain in the headwaters (Reach 1) and a small number of tributaries: Eagle Creek (tributary to Reach 2), San Francisco River (tributary to Reach 2), and Aravaipa Creek (tributary to Reach 3). Spikedace disappeared from samples in just 5 years in the upper Verde River (Rinne 1999a; Rinne 2005).

Gila River Main Stem

Of the 20 species native to the basin, five, Santa Cruz pupfish, Monkey Springs chub, headwater chub, Apache trout, and Gila trout never inhabited the main stem of the Gila River. Of the 15 remaining species, fewer than half, desert and Sonora sucker, longfin dace, speckled dace, spikedace, loach minnow, and roundtail chub currently occupy the main stem of the Gila (Table 1). Of these seven, only three do not have federal or state listing status. In summary, 12 of the 15 native fishes (80%) that once commonly inhabited the Gila River main stem are federally listed or proposed for listing. Rinne (2005) has demonstrated the almost complete loss of longfin and speckled dace in the Verde River—very likely attributable to altered hydrology and predation by nonnative species (Rinne 2001a).

Primary Factors Altering the Fish Fauna

Dams

The common denominator of change altering the Gila River fish fauna has been human population increase. However, development and settlement of the arid American Southwest was set into motion by the 1902 Reclamation Act. The first dam constructed and completed under this Act was Roosevelt Dam on the Salt River in 1911 (Figure 1). Control of the Salt River and its periodic flooding and damage to irrigational infrastructure and the ability to supply electrical power to the Phoenix Valley set the stage for rapid development and population growth in this region. Three additional dams on the Salt and two on the Verde (Figure 1, "RC") in the next quarter century further augmented this growth and development in central Arizona. Diversions and canal systems stimulated further agriculture. Where dams primarily support irrigation, the sudden discontinuance of flow can result in fish kills as fish are stranded and crowded in the remaining surface water (Hunt et al. 1992). Both dams constructed for reservoirs and diversions for irrigation have affected the potamodromous "big river" fishes such as the Colorado pikeminnow and razorback sucker, essentially cutting them off from spawning and rearing areas essential to their long-term survival (Tyus and Karp 1989; Tyus 1990). The marked increase in artificial impoundments stimulated importation of nonnative fishes and facilitated their establishment and spread (Dill 1944; Rinne et al., 2004).

Diversions

Diversions influencing the native Gila River fish fauna begin in southwestern New Mexico. Historically, and as observed recently (2001), downstream of Fort West Ditch, the river channel becomes spatially intermittent or dry. Diversions at Safford, Florence, and Gillespie on the main-stem Gila and at Granite Reef on the Salt alter the hydrologic character of the Gila. During severe droughts, and in some cases during average years, these diversions limit the extent and quality of aquatic habitat in the natural channel while providing habitat in canals and siphons that ultimately become biological sinks (Marsh and Minckley 1982).

Agriculture

Agriculture parallels irrigational diversion and extends from the headwaters in the Gila–Cliff Valley through the Virden, Duncan, and Safford reaches to the terminus of the Gila in the Wellton–Mohawk Valley. Although intermittent along the course of the Gila, irrigated agriculture has very likely altered native fishes through seasonal loss of stream flow and input of agri-chemicals, nutrients, and sediments via irrigation return flows from fields. However, the effects of these probable stressors on native Gila River fishes have not been specifically documented or studied. Brown and Moyle (2005) also document the decline of fishes in range and abundance resulting from water storage, diversion, flow alteration and water quality, all related to agriculture development in the Sacramento–San Joaquin basin, California. Similarly, Maret and Mebane (2005) report that 86% of the water in the Snake River is used for irrigation with river fragmentation caused by damming and irrigation having negative effects on native fishes.

Watershed Use and Influence

Grazing, logging, mining, and associated activities are continually cited as having an impact on native fish species (Platts 1991; Fleischner 1994). Mine stilling basin spills have been periodically linked to fish kills, but no information is available on the chronic effects of mining contaminants on fishes in the Gila basin. While clear links between grazing and habitat quality for an array of salmonids have been established, no studies are available on the impacts of logging of the extensive Ponderosa pine forest in the headwaters of the Gila basin. Grazing has been and continues to be indicted as a primary cause of loss of native fishes (U.S.D.I. Fish and Wildlife Service 2000, 2002). While clear links between grazing and habitat quality for an array of salmonids have been established, information based on well-designed, specifically linked studies is lacking for species endemic to warmwater habitats in the Southwest United States, including the Gila River (Rinne 1998, 1999b, 2000; Medina et al., 2005). Increased sediment as a primary by-product of livestock grazing and its input into streams has continually been cited as having a negative impact on native fishes. Again, no specific studies are available and Rinne (2001b) could not document a definite impact of fine sediment on the threatened spikedace and loach minnow in the Gila River main stem, New Mexico.

The effects of sedimentation on pool development and quality are perhaps more evident for a few species known to have a preference for these habitats (e.g., Gila chub and roundtail chub). But, as mentioned above, the information used to make the connection between livestock and fish populations is largely anecdotal, based on long-term, often qualitative observations and fails to examine additional, cumulative intrinsic and extrinsic factors. An exception is that Brunson et al. (2001) showed that watershed restoration reduced sediment and initiated pool development that resulted in the reoccupation and population expansion from an adjacent tributary to Hot Springs Creek, Arizona.

Nonnative Fishes

The presence and marked increase in reservoirs facilitated introduction of nonnative fishes, primarily for sportfishing (Rinne 2003a; Rinne et al., 2004). Introduced species, numbering a hundred in the 20th century, quickly and effectively occupied river and stream habitats not dried by diversions and damming (Rinne 1994, 1996). Currently in the Gila main stem 19 nonnative fish species now dominate fish assemblages (Tables 1, 2). Most of these nonnative sport fishes are large predators, new to the native Gila basin fish fauna. In place of the pikeminnow and the roundtail chub, species of catfishes and black basses *Micropterus* are presently the top predators in the Gila River system. The impact of these species on natives is also discussed for the Verde River (Rinne 2005). Indeed, predation (Rinne 2001a) may be the one factor of those discussed here that will render the native fish fauna of the Gila River a remnant of the past.

Sustainability of the Gila River Native Fish Fauna

Conserving and sustaining the native fish fauna of the Gila River basin appears bleak at present. Two species are already extinct. Only 60% of the historic native species are present in the basin. About 70% of all native fishes in the basin are extinct or listed and several are presently candidates for listing. Others, such as Gila trout, spikedace, and loach minnow, persist only in the extreme headwaters of the basin (Table 1). Habitats for large and formerly migratory river species such as Colorado pikeminnow, razorback sucker, and bonytail chub are absent or, where present, are too fragmented to sustain the life history requirements of these species. These species are presently sustained in refuges and have little to no probability of repatriation to the main-stem river. About 50% of native fish species and their dominance in the total fish assemblage occurs only in Reach 1 (Tables 1, 3). Species such as Sonora and desert sucker, and speckled and longfin dace are most common in Reach 1, yet are absent below Reach 3 of the river (Minckley 1973). Although these species appear resilient to the changes listed above, data from the upper Verde (Rinne 2005) suggests cautious optimism especially for the sustainability of the two small-sized, short-lived daces in main-stem river habitats. Even in the upper Gila basin in New Mexico, the speckled dace is rare in main-stem river habitat (Rinne et al. 2004). Minckley (1973) also suggested that despite being present below 1,500 m in large rivers, speckled dace currently occurs in upper elevation (2,000–3,000 m) high-gradient tributary creeks.

If many of the native species are sustainable, they very likely persist only in isolated, natural (i.e., tributary streams) aquatic habitats or in artificial refuges. Reduced riverine habitats and increased reservoir habitat are primary causes of the decline of the native Gila River fish fauna. Superimposed upon this disturbance is the extensive and sustained introduction of nonnative species of fishes (Table 2). Established nonnative species now number about the same as the historic number of native species and they are distributed throughout the basin in most reaches. Past reintroductions of razorback sucker into apparently suitable habitat in the Gila River have met with failure largely or entirely because of predation (Marsh and Brooks 1989).

The major, negative physical, hydrological, and biotic impacts present in the Gila River basin may be irreversible. Accordingly, the sustainability of native fish species (Rinne 1994) lies in part in aggressively managing natural flows, geomorphology, and extant fish assemblages (Rinne 2002). Extant native species in properly functioning riverine habitats that presently occur in the headwaters of the river in New Mexico are a prime example of where to begin (Rinne et al. 2004). Hatchery rearing and holding (Rinne et al. 1986) will always be an essential component of sustaining native Gila River fishes. In addition, refuges in tributary streams and artificial habitats devoid of nonnative species of fishes are the only secure habitats for the historic native fish fauna. In many cases, this will require beheading some stream segments with fish barriers followed by periodic fish removals (Rinne and Turner 1991) when barriers are breached by the public or barrier failure.

The primary questions that need answering by fish management agencies in the Southwest (Rinne 2003b) are 1) Where are we headed with the current sportfish paradigm?, 2) What are the alternatives for continuing native and nonnative sportfish management? (Rinne et al. 2004), 3) Do we proceed to manage rare, declining native species from a biocentric or an anthropocentric perspective?, and 4) How do respective land and riparian uses affect native fish habitats and sustainability? Unless these questions are asked, answered, and acted upon in a multi-agency approach, we may only be documenting the epitaph for native fishes in the Gila and the Southwest in general.

Table 3.—Estimated percent composition of native to nonnative fish assemblage in the five reaches of the Gila River, 2003. Data from Minckley (1973) and Rinne et al. (in press) are used for estimating percentages.

Reach	Native species	Nonnative species
1	95	5
2	10	90
3	10	90
4	0	100
5	0	100

References

Blasius, H. B., and J. Simms. 2001. Gila Box Riparian National Conservation Survey. Region V Fisheries Program Technical Report. Arizona Game and Fish Department, Phoenix.

Brown D. E., and C. H. Lowe. 1980. Biotic communities of the Southwest. Rocky Mountain Forest and Range Experiment Station, USDA Forest Service General Technical Report RM-178, Fort Collins, Colorado.

Brown, L. R., and P. B. Moyle. 2005. Native fish communities in the Sacramento-San Joaquin drainage, California: a history of decline. Pages 75–98 *in* J. N. Rinne, R. M. Hughes and B. Calamusso, editors. Historical changes in large river fish assemblages of the Americas. American Fisheries Society, Symposium 45, Bethesda, Maryland.

Brunson, E., D. D. Gori, and D. Baker. 2001. Watershed improvement to restore riparian and aquatic habitat on the Muleshoe Ranch Cooperative Management Area. The Nature Conservancy of Arizona, Arizona Water Protection Fund Project Number 97–035, Tucson.

Calamusso, B., J. N. Rinne, and R. J. Edwards. 2005. Historic changes in Rio Grande fish fauna: status, threats, and management of native species. Pages 205–223 *in* J. N. Rinne, R. M. Hughes, and B. Calamusso, editors. Historical changes in large river fish assemblages of the Americas. American Fisheries Society, Symposium 45, Bethesda, Maryland.

Corle, E. 1951. The Gila: river of the Southwest. University of Nebraska Press, Lincoln.

Dill, W. A. 1944. The fishery of the lower Colorado River. California Fish and Game 30:109–211.

Emory, W. H. 1848. Notes of a military reconnaissance from Fort Leavenworth, in Missouri, to San Diego, in California, including parts of the Arkansas, del Norte, and Gila rivers. By Lieutenant Col. W. H. Emory, made in 1846–1847, with the advanced guard of the "Army of the West." Wendell and van Benthuysen Print, Washington D.C.

Fleischner, T. L. 1994. Ecological cost of livestock grazing in western North America. Conservation Biology 8:629–644.

Hubbs, C. L., and R. R. Miller. 1941. Studies of the fishes of the order Cyprinodontes. XVII. Genera and species of the Colorado River system. Occasional Papers Museum of Zoology, University of Michigan 433:1–9.

Hunt, W. G., D. E. Driscoll, E. W. Bianchi, and R. E. Jackman. 1992. Ecology of bald eagles in Arizona. Part B. Field studies. Report to the U.S. Bureau of Reclamation. Biosystems Analysis, Inc., Contract 6-CS-30–04470, Santa Cruz, California.

Johnson, J. E. 1985. Reintroducing the natives: razorback sucker. Proceedings of the Desert Fishes Council 13:73–79.

Lowe, C. H. 1964. Vertebrates of Arizona. University of Arizona Press, Tucson.

Maret, T. R., and C. A. Mebane. 2005. Historical and current perspectives on fish assemblages of the Snake River, Idaho and Wyoming. Pages 41–60 *in* J. N. Rinne, R. M. Hughes, and B. Calamusso, editors. Historical changes in large river fish assemblages of the Americas. American Fisheries Society, Symposium 45, Bethesda, Maryland.

Marsh, P. C., and J. L. Brooks. 1989. Predation by Ictalurid catfishes as a deterrent to reestablishment of introduced razorback suckers. The Southwestern Naturalist 14:188–195

Marsh, P. C., and W. L. Minckley. 1982. Fishes of the Phoenix metropolitan area in central Arizona. North American Journal of Fisheries Management 2:395–404.

Medina, A. L., J. N. Rinne, and P. Roni. 2005. Riparian restoration through grazing management: considerations for monitoring project effectiveness. Pages 97–126 *in* P. Roni, editor. Monitoring stream and watershed restoration. American Fisheries Society, Bethesda, Maryland.

Meffe, G. K., Hendrickson D. A., Minckley W. L., Rinne J. N. 1983. Factors resulting in decline of the endangered Sonoran topminnow (Atherinformes:Poeciliidae) in the United States. Biological Conservation 25(2):135–139.

Miller, R. R. 1961. Man and the changing fish fauna of the American Southwest. Michigan Academy of Science, Arts, and Letters 46:365–404.

Minckley, W. L. 1973. Fishes of Arizona. Arizona Game and Fish Department, Phoenix.

Minckley, W. L. 1983. Status of the razorback sucker, *Xyrauchen texanus*, (Abbott) in the lower Colorado River basin. The Southwestern Naturalist 28:65–187.

Minckley, W. L. 1985. Native fishes and natural aquatic habitats of U.S. Fish and Wildlife Service Region II, west Interagency Personnel Act Agreement. Arizona State University, Tempe.

Minckley, W. L., and R. Clarkson. 1979. Fishes. Pages 510–531 *in* W. L. Minckley and M. R. Summerfeld, editors. Resource inventory of the Gila River Complex, eastern Arizona. Final Report for U.S. Bureau of Land Management, Contract YA-512-CT6-216, Arizona State University, Tempe.

Minckley, W. L., and J. E. Deacon, 1968. Southwestern fishes and the enigma of "endangered species." Science 159:1424–1432.

Minckley, W. L., R. R. Miller, and S. K. Norris. 2002. Three new pupfish species, Cyprinodon (Teleostei; Cyprinodontidae) from Chihuahua, Mexico and Arizona, USA. Copeia 2002:687–705.

Mueller, G. A., and P. C. Marsh. 2002. Lost, a desert river and its native fishes: a historical perspective of the lower Colorado River. U.S. Government Printing Office, Information Technical Report USGS/BRD/ITR—2002—0010, Denver, Colorado.

Mueller, G. A., P. C. Marsh, and W. L. Minckley. 2005. A legacy of change: the lower Colorado River, Arizona-California-Nevada, USA and Sonora-Baja Calfornia Norte, Mexico. Pages 139–156 *in* J. N. Rinne, R. M. Hughes, and B. Calamusso, editors. Historical changes in large river fish assemblages of the Americas. American Fisheries Society, Symposium 45, Bethesda, Maryland.

Pattie, J. O. 1905. The personal narrative of J. O. Pattie, of Kentucky. Early western travels, edited by R. G. Thwaites. Arthur H. Clark Co., Cleveland, Ohio.

Platts, W. S., 1991. Livestock grazing. Pages 389–423 *in* W. R. Meehan, editor. Influences of forest and rangeland management on salmonid fishes and their habitats. American Fisheries Society, Special Publication 19, Bethesda, Maryland.

Rea, A. M., 1983. Once a river: bird life and habitat changes on the middle Gila. University of Arizona Press, Tucson.

Rinne, J. N. 1988. Native southwestern (USA) trouts: status, taxonomy, ecology, and conservation. Polish Archives of Hydrobiology 35:305–320.

Rinne, J. N. 1994. Declining southwestern aquatic habitats and fishes: are they sustainable? Pages 256–265 *in* W. W. Covington and L. F. DeBano, technical coordinators. Sustainable ecological systems: implementing an ecological approach to land management. USDA, Forest Service, General Technical Report RM–247, Flagstaff, Arizona.

Rinne, J. N. 1996. The effects of introduced fishes on native fishes: Arizona, southwestern United States. Pages 149–159 *in* D. P. Philipps, editor. Protection of aquatic diversity. Proceedings of the World Fisheries Conference, Theme 3. Oxford and IBH Publishing Company Private Limited, New Delhi.

Rinne, J. N. 1998. Grazing and fishes in the Southwest: confounding factors for research. Pages 75–84 *in* D. F. Potts, editor. Proceedings of the specialty conference: rangeland management and water resources. American Water Resources Association, Middleburg, Virginia.

Rinne, J. N. 1999a. The status of spikedace, *Meda fulgida*, in the Verde River, 1999, Implications for research and management. Hydrology and Water Resources in the Southwest 29:57–64.

Rinne, J. N. 1999b. Fish and grazing relationships: the facts and some pleas. Fisheries 24(8):12–21.

Rinne, J. N. 2000. Fish and grazing relationships in southwestern United States. Pages 329–371 *in* R. Jamison, C. Raish, and D. Finch, editors. Ecological and socioeconomic aspects of livestock management in the Southwest. Elsevier, Amsterdam.

Rinne, J. N. 2001a. Nonnative, predatory fish removal and native fish response: Verde River, Arizona, 1999–2000. Hydrology and Water Resources of the Southwest 31:29–36.

Rinne, J. N. 2001b. Relationship of fine sediment and two native southwestern fish species. Hydrology and Water Resources in the Southwest 31:67–70.

Rinne, J. N. 2002. Hydrology, geomorphology and management: 2002. Implications for sustainability of native southwestern fishes. Hydrology and Water Resources of Arizona and the Southwest 32:45–50.

Rinne, J. N. 2003a. Native and introduced fishes: their status, threat, and conservation. Pages 277–297 *in* P. F. Ffolliott, M. B. Baker, L. F. Debano, and D. G. Neary, editors. Ecology, hydrology and management of riparian areas in the southwestern United States. Lewis Publishers, Boca Raton, Florida.

Rinne, J. N. 2003b. Fish habitats: conservation and management implications. Pages 194–213 *in* P. F. Ffolliott, M. B. Baker, L. F. Debano, and D. G. Neary, editors. Ecology, hydrology and management of riparian areas in the southwestern United States. Lewis Publishers, Boca Raton, Florida.

Rinne, J. N. 2005. Changes in fish assemblages, Verde River, Arizona, 1974–2002. Pages 115–126 *in* J. N. Rinne, R. M. Hughes, and B. Calamusso, editors. Historical changes in large river fish assemblages of the Americas. American Fisheries Society, Symposium 45, Bethesda, Maryland.

Rinne J. N., P. Boucher, D. Miller, A. Telles, J. Montzingo, R. Pope, B. Deason, C. Gatton, and B. Merhage. 2005. Comparative fish community structure in two southwestern desert rivers. Pages 64–70 *in* M. J. Brouder, C. L. Springer, and S. C. Leon, editors. Proceedings of two symposia. Restoring native fish to the lower Colorado River: interactions of native and non-native fishes. July 13–14, 1999, Las Vegas, Nevada. Restoring natural function within a modified riverine environment: the lower Colorado River. July 8–9, 1998. U.S. Fish and Wildlife

Service, Southwest Region, Albuquerque, New Mexico.

Rinne, J. N., J. E. Johnson, B. L. Jensen, A. W. Ruger, and R. Sorenson. 1986. The role of hatcheries in the management and recovery of threatened and endangered species. Pages 271–285 *in* R. H. Stroud, editor. Fish culture in fisheries management. American Fisheries Society, Bethesda, Maryland.

Rinne, J. N., and W. L. Minckley. 1991. Native fishes in arid lands: dwindling resources of the desert Southwest. U.S.D.A., Forest Service, General Technical Report RM–206:1–45, Fort Collins, Colorado.

Rinne, J. N., L. Riley, B. Bettaso, K. Young, and R. Sorensen. 2004. Managing southwestern native and nonnative fishes: can we mix oil and water and expect a favorable solution. Pages 445–466 *in* J. Nickum, M. Nickum, P. Muzik, and D. MacKinley, editors. Propagated fish in resource management. American Fisheries Society, Symposium 44, Bethesda, Maryland.

Rinne, J. N., and P. R. Turner. 1991. Reclamation and alteration as management techniques, and a review of methodology in stream renovation. Pages 219–246 *in* W. L. Minckley and J. E. Deacon, editors. Battle against extinction: native fish management in the American West. University of Arizona Press, Tucson, Arizona.

Ross, C. P. 1923. The lower Gila region. U. S. Geological Survey-Supply paper 498, Reston, Virginia.

Sublette, J. E., J. D. Hatch, and M. Sublette. 1990. The fishes of New Mexico. University of New Mexico Press, Albuquerque.

Tyus, H. M. 1990. Potadromy and reproduction of Colorado squawfish, *Ptychocheilus lucius*. Transactions of the American Fisheries Society 119:1035–1047.

Tyus, H. M., and C. A. Karp. 1989. Habitat use and stream flow needs of rare and endangered fishes, Yampa River. U.S. Fish and Wildlife Service, Biological Reports 89(14):1–27.

U.S.D.I. Fish and Wildlife Service. 2000. Endangered and threatened wildlife and plants; final designation of critical habitat for the spikedace and loach minnow; final rule. Federal Register 65:80(25 April 2000):24328–24372.

U.S.D.I. Fish and Wildlife Service. 2002. Endangered and threatened wildlife and plants: listing the Gila chub as endangered with critical habitat: proposed rule. Federal Register 67:154(9 August 2002):51948–51985.

American Fisheries Society Symposium 45:139–156, 2005

A Legacy of Change: The Lower Colorado River, Arizona–California–Nevada, USA, and Sonora–Baja California Norte, Mexico

GORDON A. MUELLER*

U.S. Geological Survey, Post Office Box 25007
D-8220 Denver, Colorado 80225–0007, USA

PAUL C. MARSH AND W. L. MINCKLEY (DECEASED)

Arizona State University, Department of Biology,
Box 871501, Tempe, Arizona 85287–1501, USA

Abstract.—The lower Colorado is among the most regulated rivers in the world. It ranks as the fifth largest river in volume in the coterminous United States, but its flow is fully allocated and no longer reaches the sea. Lower basin reservoirs flood nearly one third of the river channel and store 2 years of annual flow. Diverted water irrigates 1.5 million ha of cropland and provides water for industry and domestic use by 22 million people in the southwestern United States and northern Mexico. The native fish community of the lower Colorado River was among the most unique in the world, and the main stem was home to nine freshwater species, all of which were endemic to the basin. Today, five are extirpated, seven are federally endangered, and three are being reintroduced through stocking. Decline of the native fauna is attributed to predation by nonnative fishes and physical habitat degradation. Nearly 80 alien species have been introduced, and more than 20 now are common. These nonnative species thrived in modified habitats, where they largely eliminated the native kinds. As a result, the lower Colorado River has the dubious distinction of being among the few major rivers of the world with an entirely introduced fish fauna.

Introduction

The native fish assemblage of the lower Colorado River was one of the most unique in the world. It numbered only nine freshwater species, but all were endemic to the basin. Their adaptations reflected the harsh environment in which they evolved. The lower river was prone to snowmelt fed spring floods, prolonged droughts, extreme water temperatures, turbidity, and sediment loads, and salinity in places that few other species could tolerate. Today, the river runs clear, cool, and deep, conditions foreign to historical settings.

* Corresponding author: Gordon_A_Mueller@USGS.GOV

The lower Colorado River represents the most regulated river of its size in the Western Hemisphere, and possibly the world (Fradkin 1984). Its waters irrigate 1.5 million ha of cropland and provide water to more than 22 million people in the United States and northern Mexico. Nearly 40% of the channel found between Grand Canyon and the international boundary is permanently flooded by storage reservoirs, while long reaches of the river and its major tributaries have been desiccated by upstream diversions.

European settlement of the basin has been devastating to the native ichthyofauna. Seven of the nine native species are federally endangered, and today's fish community more closely resembles that found in the upper Mississippi and Missouri river drainages. The lower Colorado River has the dubious distinction of being among the few major rivers of the world with an entirely introduced fish fauna. The following section describes the historical river, its fish assemblage, the decline of that assemblage, and the fish assemblage that exists today.

Physical Description

The Colorado River of western North America drains portions of seven states in the United States and two in northern Mexico. The upper basin is world renown for its spectacular canyons and mountains, while the lower basin was recognized for its floodplain, immense delta, and surrounding desert. Politically, the upper and lower Colorado River basins are divided at Lee's Ferry, a historical river crossing near the Utah–Arizona border and just downstream from Glen Canyon Dam (Lake Powell). Our description focuses on the lower river and its floodplain, which extends from Grand Canyon downstream almost 700 km to the Gulf of California (Figure 1).

The lower river is bounded by the Mohave and Sonoran deserts, a desolate region that delayed exploration and settlement of the area. Earliest explorers entered the region through the river corridor, where water was plentiful and riparian vegetation offered protection from the desert sun. Francisco De Ulloa, a Spaniard, discovered the Colorado River in 1539 and named it "Rio Colorado" meaning "reddish river." Its distinctive red color came from thick red sediment derived from extensive sandstone layers eroded by its waters. In 1540, the Spanish returned and sailed more than 450 km inland to the present site of Blythe, California (James 1906).

James Ohio Pattie was among the first Europeans to record his downstream journey into the delta (James 1906). Pattie led a group of trappers by canoe down the Gila, which drains much of southern Arizona, and then on to Colorado River in 1827. He described the river as being 60–90 m wide bordered by a lush floodplain, forested with willow and cottonwood that contained thousands of beaver. Kolb (1927), who floated from Grand Canyon to the Gulf of California in 1912–1913 also described the lush willow and cottonwood galleries. Aldo Leopold and his brother canoed through the delta in 1922, and his description of the delta's lush "green lagoons" was immortalized in *A Sand County Almanac* (Leopold 1949).

The river meandered more than 200 km through the delta's 50-km- wide corridor. It has been speculated that the delta contained nearly 1 million ha of tidal mashes, sloughs, old river channels, oxbows, and shallow desert playas (Luecke 1999). The delta region also contained the Salton and Pattie basins, or sinks, which would periodically fill from major floods to create temporary lakes that often exceeded 1,500 km^2 in area. These interim lakes typically supported both freshwater and marine fish assemblages that persisted for months, years, and even decades. As evaporation increased their salinity, they eventually would dry, leaving windrows of dead fish that could extend for 20 km or more (Blake 1857; MacDougal 1917).

The river corridor north of the international border was constricted by a series of short canyons. Going upstream, they increased in frequency while the valleys they separated decreased in size. The river corridor contained three major valleys where its width often extended 5–10 km. Canebrake Canyon throttled the river just north of Yuma, Arizona, from where the Great Colorado Valley extended upstream 170 km to Monument Canyon. Chemehuevi Valley extended another 20 km to Mohave Canyon, and then Mohave Valleys extended 50 km to Pyramid Canyon. Further upstream, the channel narrowed as it flowed from Painted, Black, Boulder, Virgin, and Iceberg canyons and finally from Grand Canyon.

Flow Characteristics

The Colorado River ranks fifth in volume (2.3×10^{10} m^3) among rivers in the lower 48 states (McDonald and Loeltz 1976). The river was renowned for its extremes of flow and sediment (Figure 2). The majority of water originates as spring runoff that begins in late May and peaks in late June or early July. Spring flow at Yuma averaged more than 2,100 m^3 per second (CMS) over the period 1904–1935 (USGS 1978), prior to construction

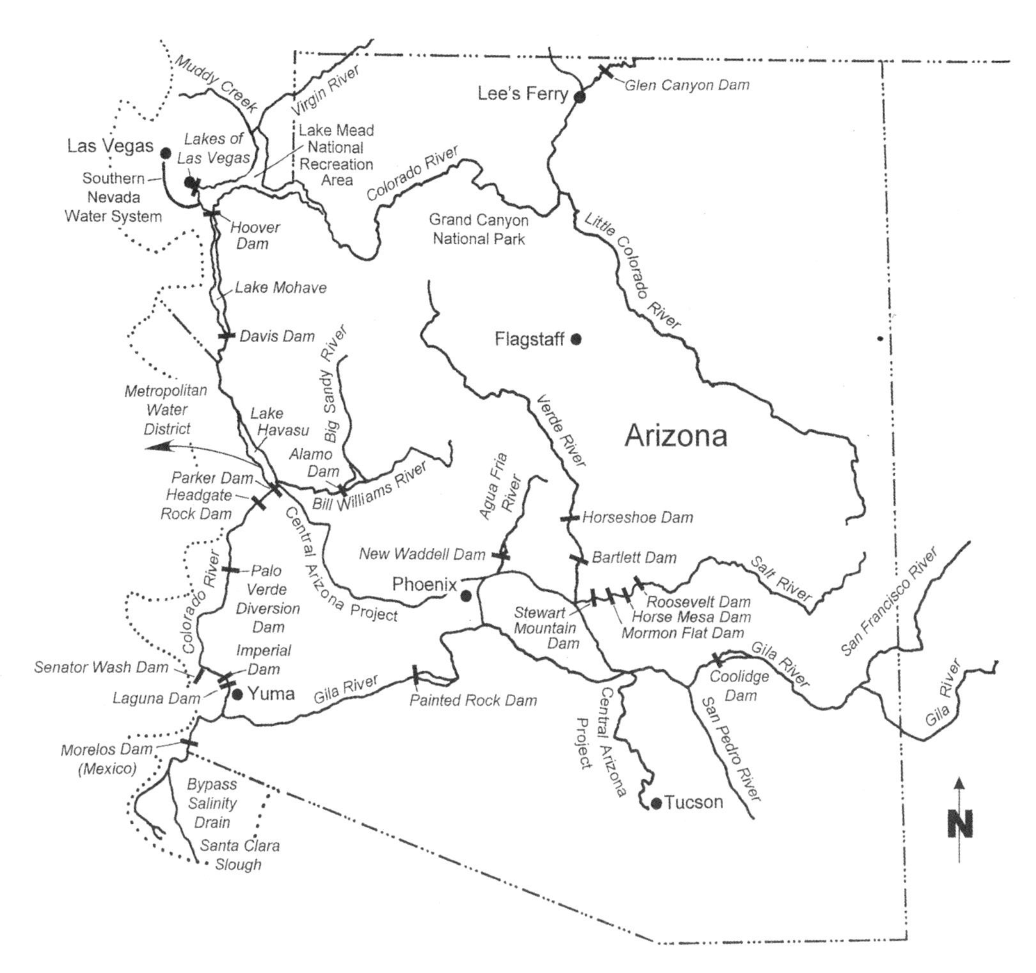

Figure 1.—The lower Colorado River system, showing place names referred to in text.

of Boulder (now Hoover) Dam. Wheeler (1876) reported evidence of a flood event during the early 1870s that reached nearly 11,300 CMS.

Summer brought extreme heat, lower flow, and unpredictable rains of the monsoon season. Summer storms were often intense and flash flooding was common, but there also were periods when rainfall was less generous. The river typically reached low flow in midwinter, long after the monsoon season. Surface flows dwindled or even disappeared entirely in broader alluvial expanses of the river (Kolb 1927). Discharge at Yuma dropped to 15 CMS in August 1934 (USGS 1978), and in many areas low flow was spread across a channel more than a kilometer wide.

Downstream of Grand Canyon, the Colorado River has three major tributaries, the Virgin, Bill Williams, and Gila rivers. All were subject to flash flooding, and their lower reaches would typically dry during summer (Evermann and Rutter 1895; Grinnell 1914). Flows, especially in the Gila River, were altered as early as 1880 by the cattle industry. Miller (1961) suggested that extensive grazing destroyed high desert grasslands and damaged the soil's ability to retain moisture. Bank erosion and head cutting increased in smaller tributaries and once lush cienegas were trampled and drained, all factors led to reductions in groundwater and perennial streams.

The Native Fish Assemblage

The river's estuary was occupied by more than 75 marine species, but only 9 freshwater species were found in the main-stem river in 1880 (Table 1).

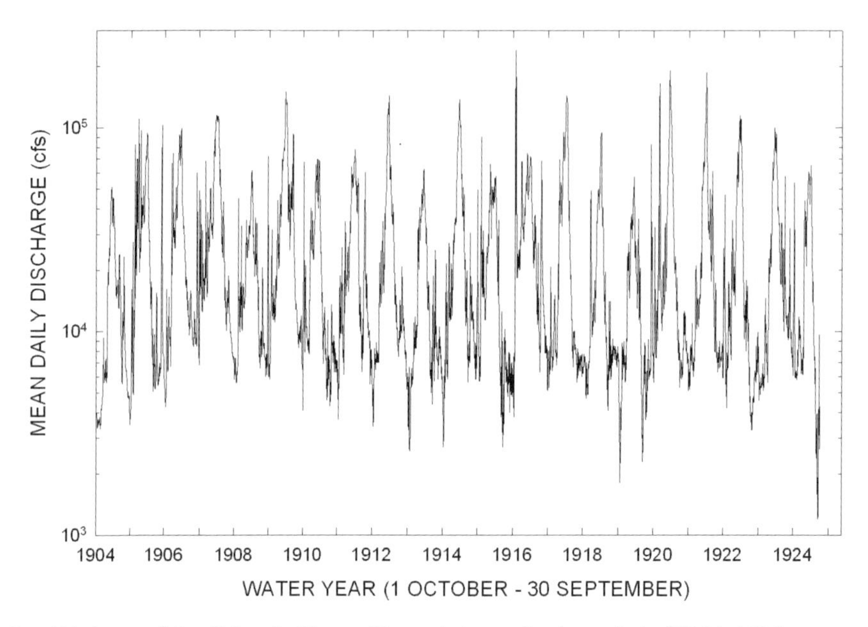

Figure 2.—Discharge of the Colorado River at Yuma, Arizona, for the period of 1904–1925.

These latter represented four families (four cyprinids, two catostomids, one cyprinodontid, and two poeciliids) of freshwater fishes. Two marine species (a mugilid and an elopid) also invaded substantial distances (300 km) inland (Minckley 1973; Mueller and Marsh 2002). Native fishes were unique and often considered bizarre by early visitors. Colorado pikeminnow *Ptychocheilus lucius* reached lengths of 2 m, razorback sucker *Xyrauchen texanus* and humpback chub *Gila cypha* both have prominent dorsal humps or keels, bonytail *G. elegans* is sleek and has a long, pencil-thin caudal peduncle, and desert pupfish *Cyprinodon macularius* and Sonoran topminnow *Poeciliopsis occidentalis occidentalis* have bright, distinctive colors and body shapes.

Abbott (1861), Lockington (1881), and Jordon (1891) were among the first scientists to collect and describe these fish. Gilbert and Scofield (1898) reported that Colorado pikeminnow and razorback suckers were abundant in the lower Colorado and Gila rivers and were "extremely abundant at Yuma and at all points below as far as the Horseshoe Bend, and in Hardee's Colorado [delta area]." Early reports supported later findings that fish were more common off-channel where water was less turbid, food was more abundant, and overall conditions were more favorable (Grinnell 1914; Tyus and Karp 1990). Evermann and Rutter (1895) reported "The lower Colorado is one of the muddiest rivers in America and is unfit for any but mud-loving species."

The intact native fish assemblage was lost before it could be fully studied in its historical habitat. Some ichthyologists believed that native assemblages emanated from the vast nursery habitats found in the delta. Aerial photographs taken there in 1938 showed numerous side channels and deep oxbows throughout the floodplain (Mueller and Marsh 2002). Floods would swell the river, reconnecting the river with its floodplain. Juveniles and reproductive adults migrated upstream to spawn and repopulate areas while larval fishes drifted downstream to encounter critical nursery and rearing areas (Hubbs and Miller 1953; Tyus et al. 1981).

Native assemblages declined dramatically just after the turn of the century, and by 1930, they were considered rare in the lower reaches of the river (Dill 1944). Adult bonytail and razorback sucker remained common further upstream in Mohave Valley (Moffett 1942; Wagner 1952), and Jonez and Sumner (1954)

Table 1.—The native fishes of the lower Colorado River main stem[a].

Common name	Scientific name	Historical abundance	Current status
Freshwater			
Minnows (Cyprinidae)			
bonytail	*Gila elegans*	abundant	endangered-stocked
roundtail chub	*G. robusta*	rare	state listed-extirpated
humpback chub[b]	*G. cypha*	rare (canyons)	endangered-extirpated
Colorado pikeminnow	*Ptychocheilus lucius*	common	endangered-extirpated
woundfin	*Plagopterus argentissimus*	common	endangered-extirpated
Suckers (Catostomidae)			
razorback sucker	*Xyrauchen texanus*	abundant	endangered-stocked
flannelmouth sucker	*Catostomus latipinnis*	rare	uncommon-stocked
pupfish (Cyprinodontidae)			
desert pupfish	*Cyprinodon macularius*	abundant (delta)	endangered-isolated in Mexico and California
live-bearers (Poeciliidae)			
Sonoran topminnow	*Poeciliopsis occidentalis*	common	endangered-extirpated
Gila topminnow	*P. occidentalis occidentalis*	abundant	endangered-extirpated
Marine			
Mullets (Mugilidae)			
striped mullet	*Mugil cephalus*	abundant (delta)	common-present
Tenpounders (Elopidae)			
machete	*Elops affinis*	common (delta)	rare-extirpated

[a] A total of 50 families of native and introduced fish is recorded for both fresh and brackish water of the lower Colorado River, its delta, and the estuary at the Sea of Cortez; there are 125 total species (87 native and 38 nonnative) (Minckley 2002). Shown here are the species that occupied freshwater of the lower river.
[b] A single record from an archaeological site is known for humpback chub below Grand Canyon (Miller 1955).

reported their young were still found in the Davis Dam vicinity in 1950. This was the last recruitment reported for the main-stem species in the flowing portion of the lower river. Bonytail and razorback sucker experienced brief but substantial resurgence during the initial filling of Lake Mohave (1954) and earlier reservoirs (Minckley 1973, 1983). Lake Mohave became the last bastion for these species in the latter half of the 1900s (Minckley 1983; Marsh and Minckley 1992), but even these populations are nearing extirpation (Marsh and Pacey, in press).

Disappearance of their young left many unanswered questions regarding their recruitment strategy. One fact was obvious—native fishes of the Colorado River evolved in a severe, but predictively variable, environment. River flows and habitat conditions were unpredictable and young were susceptible to stranding, predation, and other impacts (Johnson et al. 1993). Main-stem species (i.e., razorback sucker, bonytail, humpback chub, and Colorado pikeminnow) possessed adaptations that enabled them to survive prolonged periods of physical-chemical adversity and still be able to repopulate when conditions allowed.

The longevity of the four main-stem species could exceed 30 years (e.g., McCarthy and Minckley 1987), which provided these fish ample opportunities to reproduce. Growing evidence suggests annual recruitment was seldom great. Annual recruitment could be much lower (<10%) compared to short-lived species, and populations could suffer consecutive years of recruitment failure and still recover (Minckley et al. 1991; Douglas and Marsh 1996; Mueller and Wydoski 2004).

Natives were remarkably adapted to harsh physical conditions. Bonytail and razorback sucker were the most common main-stem species. They could endure high temperatures but were generalists when it came to spawning—both can reproduce in either flowing or standing water (Minckley et al. 1991). This suggests they maintained a presence in both riverine and oxbow assemblages, which may have been essential during prolonged periods of drought or high flow.

Some species have simply survived by their ability to tolerate harsh habitats that are avoided by most alien species. Remaining populations of humpback chub and desert pupfish have retreated into habitats that are too warm, saline, or shallow for most other species (Barlow 1958; Kaeding and Zimmerman 1983; Douglas and Marsh 1996).

Research has examined environmental attributes of Colorado River flood ecology (Schmidt and Graf 1990; Stanford 1994; Stanford et al. 1996); however, the biological aspects of drought ecology have been generally ignored. Flooding may have been a prerequisite for successful spawning, but dispersal and habitat carrying capacity were dictated by low flow. Populations would shrink with river flow as fish were stranded, lost to predators, or forced to seek refuge in deeper or more secure water sources (e.g., scoured pools and springs) (Minckley 1991). Drought devastated populations, but it also reduced the number and distribution of natural predators. Inundation of nursery habitats that contained few, if any predators, undoubtedly improved the chances of survival when spawning conditions improved. Longevity combined with durability allowed their kind to cope with the lower Colorado River.

Decline of Native Fish

Factors that displaced the natives continue to be debated, due in part to the speed and complexity of change that occurred in the basin. Native assemblages had been severely altered before Hoover Dam was built and river flow was regulated (Dill 1944). Isolation and eventual dewatering of the delta eliminated most of the lower basin's wetlands, fish nurseries, and rearing habitat. Most importantly, the delta had served as the basin's ultimate sanctuary, and when it was lost, natives had little chance to recover. Fish also were lost to pollution, commercial harvest, stranding, and migration barriers. Factors more subtle, but no less critical, were the long-term implications of alien introductions and hydrologic alteration. Technologic humans had largely destroyed in a century what took evolution millennia to create. The following section describes various mortality factors that help explain the displacement and loss of native fishes.

Pollution.—Sykes (1937) reported a huge fish kill on the lower Colorado River in 1894 that resulted in dead fish covering the sandbars, "three to every square yard." He estimated the carcasses weighed in the thousands of tons. He later discovered that Wolfley Dam, located on the Gila River near Gila Bend, Arizona, had failed during heavy rains, and it was believed the tainted waters caused the kill. Dill (1944) reported that local residents suggested fish kills were not uncommon. Another massive kill occurred during summer 1939 as Lake Havasu was filling, and it was suspected that suspended sediment and low dissolved oxygen caused the kill.

Native Fishery.—Natives were commercially harvested for meat, fertilizer, and hog feed until 1911 (Miller 1961). Local fishermen supplied Colorado pikeminnow meat to construction crews building the Salt River Project near Phoenix, and pikeminnow for a time were also commercially canned near Yuma (Collier 1999). Dill (1944) reported that bonytail were abundant and easily caught by anglers. Odens (1989) described a story told by George Utley who harvested 147 razorback suckers in an hour in 1909 using a hand axe and 0.22-caliber rifle. The development of large reservoir populations led to commercial harvests in the 1940s. Razorback suckers were commercially taken from Lake Mead as late as 1947 (Mueller and Marsh 2002), and one angler took 5,000 kg from Lake Roosevelt in 1949 (Hubbs and Miller 1953).

Stranding.—Stranding was a natural phenomenon that was intensified by agricultural withdrawal. Diversions to California's Imperial Valley during the 1930s drought totally desiccated a 100-km reach that normally flowed through the delta and permanently destroyed its fish assemblage (Sykes 1937). Off-channel diversions also stranded tens of thousands of bonytail, razorback sucker, and Colorado pikeminnow in canals and on cropland in Imperial

Valley (Miller 1961). Further upstream at Blythe, farmers complained of the stench of bonytail and other dead fish that lay decaying on their fields (Mueller and Marsh 2002).

Migration Barriers.—Construction of Laguna Diversion Dam in 1909 created the first of a series of physical barriers across the Colorado River. Laguna Dam ended steamboat navigation on the lower river, but more importantly it prevented upstream access for fish found in the delta. Spring floods continued to disperse larval fish to delta nurseries, but adults could not return upstream to spawn or repopulate areas. Upstream migrants concentrated below Laguna Dam, where locals harvested thousands of Colorado pikeminnow, striped mullet, and other natives for food, fertilizer, and hog feed (Grinnell 1914; Miller 1961).

Boulder (now Hoover) Dam was constructed in 1935 and provided the flood protection that accelerated construction of smaller structures downstream. Imperial (1938), Parker (1938), Headgate Rock (1944), Davis (1953), Morales (1950), and Palo Verde (1957) dams further blocked fish migrations. Today, the longest unobstructed reach of river is only 140 km.

AlienSpecies and Flow Alteration.—There is little doubt that competition and predation by alien fishes was a problem from the onset (Dill 1944). Natives had disappeared from much of the lower Colorado River by 1930, but they were still common and reproducing in Mohave Valley until 1950 (Jonez and Sumner 1954). Mohave Valley had no significant diversions, and it appears the main-stem natives could better tolerate river conditions (Dill 1944). This changed with closure of Hoover Dam. The integration of storage into the system reduced the frequency and duration of floods and drought-induced low flows. Lentic habitat represented by large reservoirs and perennially cold tailraces created by hypolimnetic discharges below high dams were heretofore unknown along the lower river. These, and stabilized discharge disproportionately benefited aliens, which soon crowded the system and accelerated the displacement of natives (Moyle et al. 1986; Fuller et al. 1999). Tyus and Saunders (2000) suggested that the Colorado River basin provides the most complete record for alien fish problems of any other large river system in the world . . ..

Stocking began in 1881 with common carp *Cyprinus carpio*, bullhead *Ameiurus* spp., and channel catfish *Ictalurus punctatus*. All three were abundant throughout the lower river by 1910 (Grinnell 1914), while natives became increasingly rare. The larger, long-lived native species maintained adult populations further upstream, but their young were rapidly lost to predators and recruitment collapsed. Predation was devastating to smaller, shorter-lived fishes such as woundfin *Plagopterus argentissimus*, desert pupfish, and Sonoran topminnow, which soon were eliminated from the main stem. Miller (1961) reported Sonoran (Gila) topminnow populations were totally displaced shortly after the introduction of western mosquitofish *Gambusia affinis*. Colorado pikeminnow had all but disappeared by the 1930s downstream of Blythe (Dill 1944).

Bonytail, Colorado pikeminnow, and razorback sucker persisted further upstream in Mohave Valley (Moffett 1942; Wagner 1952; Jonez and Sumner 1954); however, impacts of aliens combined with physical and biological changes brought upon by channelization and reservoir storage sealed their fate (Dill 1944; Beland 1953). Dill (1944) described the process:

> Extremely low water would raise the river temperature and strand fish. Such processes have, of course, been going on for many years in the unstable Colorado, and it seems probable that the native fish populations have undergone alternate periods of rise and fall. But each period of destruction was followed by a period during which the population could rehabilitate itself. Before the dams were built, the native fishes were at the mercy of an adverse physical environment, but the deleterious effect of predaceous alien fishes must have been slight. That is, the population of the latter fishes was small before the creation of Boulder Dam, and floods and droughts must have worked just as severe a hardship and probably more on them. Because of the unfavorable water conditions around the early thirties, it seems possible that the popula-

> tion of native fishes sank to one of its low points, and the coincidental advent of clear water following Boulder Dam brought about a heavy production of bass and other alien fishes which preyed upon the already reduced natives. Competition, as well as direct predation may have played a large part in this supposed destruction.

Remarkably, bonytail and razorback sucker experienced brief rebounds when Lake Roosevelt (1911), Lake Mead (1936), Lake Mohave (1953), and Senator Wash Reservoir (1966) were initially filled (Moffett 1943; Miller 1961; Minckley 1983). Minckley et al. (1991) suggested reservoir inundation at first mimicked the physical and biological conditions brought upon by spring runoff. Large expanses of newly flooded habitat were created, which contained few large predators. Initial survival and recruitment were high, and razorback sucker populations in some cases numbered in the hundreds of thousands (Minckley et al. 2004). Recreational species were stocked into the new reservoirs and their populations expanded, and within a few years predators had virtually eliminated native recruitment (Wallis 1951; McCarthy and Minckley 1987). This phenomena also was observed, but to a lesser extent, in canal systems that were seasonally drained. Larval fish, including razorback suckers, were the first to invade these systems when irrigation resumed in early spring. The relative absence of predators allowed some survival and 19 juvenile suckers were salvaged between 1962 and 1988 (Marsh and Minckley 1989). No recruitment could be detected elsewhere in the lower river during the next five decades as these relic populations slowly declined and aging adults died of old age and other ailments (Holden et al. 2001; Marsh and Pacey, in press).

Alien assemblages also expanded in the remaining portions of the river. Prior to dredging, Mohave Valley contained 1,740 surface ha of off-channel wetlands that received an annual angler pressure of 740 fisherman/d per kilometer (Beland 1953). Beginning in 1947, the river was straightened and channellized by dredging and construction of riprap levees, which helped speed water deliveries and reduce evaporative losses (Oliver 1965). As a result, the river became physically separated from its floodplain and the channellized stream was characterized by a greater volume of permanently deep habitat where alien species thrived (Figure 3). The flood and drought cycle that Dill (1944) suggested favoring natives over alien predators was eliminated. Natives simply have been unable to cope with the more aggressive alien species under altered flow regimes (Marsh and Brooks 1989; Tyus and Saunders 2000; Marsh and Pacey, in press).

Recreational Fishery.—Large storage reservoirs and hypolimnetic tail waters created by high dams provided ideal conditions to establish recreational fisheries. Largemouth bass *Micropterus salmoides*, bluegill *Lepomis macrochirus*, white *Pomoxis annularis* and black crappie *P. nigromaculatus*, rainbow trout *Oncorhynchus mykiss*, and brown trout *Salmo trutta* were stocked up- and downstream of Hoover Dam as Lake Mead filled in 1935. Parker and Imperial dams were finished by 1938, providing additional reservoir space. That same year, tens of thousands of largemouth bass, bluegill, and crappie were stocked in most of the river's larger sloughs, backwaters, and oxbows (Dill 1944). The lower basin's largemouth bass fishery was nationally renowned by 1940, as was the trout fishery managed in the tailwater below Hoover Dam (Moffett 1943; Dill 1944). Over a period of only two decades, 36% of the historical river channel between Grand Canyon and the international border was permanently flooded by manmade reservoirs.

The economic prosperity that followed World War II increased the demand for recreational fishing and boating. Hatcheries were built to provide tens of thousands of fish for lower basin waters. Aggressive management programs introduced existing and new fish species to optimize littoral, pelagic, and tailwater production. Native fishes often were considered expendable and were intentionally destroyed in favor of recreational species (Kimsey 1958; Holden 1991). Threadfin shad *Dorosoma petenense* were introduced in 1953 followed by striped bass *Morone saxatilis*, sockeye salmon *O. nerka*, coho salmon *O. kisutch,* and different genetic strains of black bass (Allan and Roden 1978). Nearly 80 alien fish species, representing 17 families, have been introduced through stocking, bait bucket introduction, or by aquarium owners (Minckley et al. 1991). Many now are established, and more than 22 have

Figure 3.—Comparison photographs of the Colorado River near Needles, California, taken prior to canalization in 1938 (left), and in 2000 (right).

become relatively common. Most recreational species are piscivorous and suspected of feeding on natives (Tyus and Saunders 2000; Table 2).

Accidental introductions still occur. For example, gizzard shad *Dorosoma cepedianum* were recently discovered in Lake Powell and are expected to spread through the basin (Mueller 2004). The fishery continues to evolve, in response to new introductions and the unpredictable and often unanticipated shifts of existing populations (Tyus and Sanders 2000). Today, alien fishes comprise 99.9+% of the lower main stem's fish assemblage (Minckley 1979).

Recreational fishing is a popular pastime enjoyed by an estimated 47 million people in this country. Resource management programs typically are funded by revenues generated from fishing license sales and from a federal sport fishery restoration program that collects taxes on sporting equipment. The majority (95%) of this funding is specifically targeted toward sport fish management, with little being diverted for application to native fishes. Nationwide funding for programs that specifically benefit native fishes pales in comparison to sport fishery dollars.

Sport fisheries have become economically important, providing revenue for management agencies and local businesses. Angler use on lakes Mead and Mohave reaches nearly half a million angler-days per year with a harvest exceeding 1.5 million fish (Nevada Department of Wildlife [NDOW], unpublished data), and anglers contribute between

Table 2.—Chronology of the more common or rapidly expanding alien fishes found in the lower Colorado River main stem.

Family	Species	Common name	Year
Clupeidae			
	Dorosoma petenense	threadfin shad	1953
	D. cepedianum[a]	gizzard shad	2000
Cyprinidae			
	Cyprinus carpio	common carp	1881
	Carassius auratus	goldfish	1944
	Pimephales promelas	fathead minnow	1950s
	Cyprinella lutrensis	red shiner	1953
Poeciliidae			
	Gambusia affinis	western mosquitofish	1922
Centrarchidae			
	Pomoxis annularis	white crappie	1934
	P. nigromaculatus	black crappie	1935
	Micropterus salmoides	largemouth bass	1935
	Lepomis macrochirus	bluegill	1937
	L. cyanellus	green sunfish	1937
	Micropterus dolomieu	smallmouth bass	1940s
	Lepomis microlophus	redear sunfish	1951
Cichlidae			
	Heros severus	banded cichlid	1950s
	Cichlasoma nigrofasciatum	convict cichlid	1960s
	Oreochromis aureus	blue tilapia	1960s
	O. mossambicus	Mozambique tilapia	1960s
	Tilapia zilli	redbelly tilapia	1960s
Salmonidae			
	Oncorhynchus mykiss	rainbow trout	pre-1900
	Salmo trutta	brown trout	1924
Ictaluridae			
	Ictalurus punctatus	channel catfish	1892
	Ameiurus natalis	yellow bullhead	1899
	A. melas	black bullhead	1904
	Pylodictis olivaris	flathead catfish	1962
Percichthyidae			
	Morone saxatilis	striped bass	1959

[a] Recently captured in Lake Powell, Utah-Arizona (Mueller and Brooks, in press).

US$16 and $56 million dollars to the local economy (Martin et al. 1980; NDOW unpublished data). A recent (2000) habitat enhancement program on Lake Havasu is being credited with an annual increase of $18 million dollars in local angling revenue (Anderson 2001). Recreational fishing clearly generates more interest and revenue than native species. Unfortunately, despite recognition of many native species as imperiled by state and federal entities, legislation or regulation by themselves seldom protect these fishes, nor do they provide revenues necessary to manage these assemblages.

Current Status of Natives

The future is grim for native fish in the lower Colorado River. Remnant native assemblages continue to decline, except for small refuge populations (Williams 1991). Seven of the nine native fishes are federally listed as endangered. The remaining two are currently being reexamined for possible protection. Wild fish populations are virtually gone from the lower Colorado River main stem. Six species, desert pupfish, Sonoran topminnow, woundfin, Colorado pikeminnow, roundtail chub, and humpback chub are gone from the lower river (humpback chub persists precariously in Grand Canyon). The remaining three, razorback sucker, flannelmouth sucker, and bonytail, are present largely due to recent stocking efforts, although nonrecruiting wild stocks of bonytail and razorback sucker continue to hang on in larger reservoirs. With aggressive management, some members of the native fish community likely can be restored to the lower Colorado River, but without such management, all nine natives soon will be extirpated. The following summarizes their current status.

Bonytail (Federally Endangered)

Bonytail was common throughout the basin and one of the easiest and most entertaining fish to catch (Dill 1944), but range and abundance declined precipitously after major dams were constructed and the species was federally listed as endangered in 1980. Wild broodstock was taken from Lake Mohave in the later 1970s and early 1980s, and hatchery-produced fish have been stocked since then as part of a largely unsuccessful attempt to restore a population (USFWS 1990). The species escaped extirpation, but there are serious genetic concerns because the hatchery stock is based upon production of at most three females (Hedrick et al. 2000). Bonytail are currently being stocked into lakes Mohave and Havasu and portions of the upper basin in efforts to reestablish populations. Although some fish survive, threats have not been mitigated and the factors that led to their ancestors' demise still exist. Early life stages of the native fishes continue to fall prey to nonnative predators, but thrive when stocked by themselves (Mueller and Marsh 2002).

Humpback Chub (Federally Endangered)

Humpback chub was rare downstream of Grand Canyon. Its remains were found at a single archaeological site in Black Canyon, now inundated by Lake Mohave, but the fish was presumed gone from the area by the early 20th century (Miller 1955). The species was federally listed as endangered in 1967. Today, only a few relict populations are found upstream of Lake Mead, the largest (2,000–5,000 fish) is found in the Colorado and Little Colorado rivers. Recent studies suggest that population is declining to dangerously low levels due to habitat reduction and modification, including predation (Marsh and Douglas 1997; Van Haverbeke and Coggins 2003).

Roundtail Chub.—The roundtail chub remains the most common of the three chubs, but it never was common in the main-stem river. It inhabits higher gradient tributaries where populations continue to decline (Brouder et al. 2000). Roundtail chub is extirpated from the lower Colorado mainstream, but small populations can still be found in portions of the Gila, Salt, and Verde rivers in Arizona, and in upper basin streams. The species is currently being reconsidered for federal listing.

Colorado Pikeminnow (Federally Endangered)

Colorado pikeminnow was abundant in the lower Colorado and Gila, Salt, San Pedro, and Verde rivers and were commercially harvested until 1911 (Miller 1961). The last Colorado pikeminnow taken from the main stem downstream of Grand Canyon was captured from Lake Mohave in 1962 (Mueller and Marsh 2002). The species was federally protected as endangered in 1976. Today, the species is only found in portions of the upper basin where they number approximately 5,000 fish (Osmundson and Burnham 1998). There have been discussions of reintroducing the fish into the lower Colorado River main stem, but that habitat is designated only for experimental-nonessential fish (USFWS 1987a) which has caused the plan to stall.

Woundfin (Federally Endangered)

Woundfin was once widely distributed in the lower Colorado main stem and major tributaries from the Virgin through the Salt and Gila rivers, but appears to have disappeared from much of its range by the early 20th century (Minckley 1973). The fish was federally listed as endangered in 1967. Today, it is only found in a small part of the Virgin River, and that population is threatened by continued water development, urbanization, and alien species, especially red shiner *Cyprinella lutrensis* (Deacon 1988).

Flannelmouth Sucker

This species was common throughout the Colorado River system, but in the lower basin was extirpated except from the Virgin River and Grand Canyon reach of the main stem. In 1976, Arizona biologists collected 611 flannelmouth suckers from the Paria River in Grand Canyon and released them in the lower river near Bullhead City. The fish has successfully recolonized a 20-mi reach downstream of Davis Dam and currently numbers between 3,500 and 6,800 fish (Mueller and Wydoski 2004). This population represents the only successful reintroduction of a native fish in the Colorado River main stem. Previously, there were only five confirmed records of the fish being taken from the Colorado River main stem (Minckley 1973). This species is typically found in higher elevation tributary streams. Today, the tailwaters found downstream of main-stem dams are more indicative of habitat conditions they prefer in higher gradient streams. Unlike the razorback sucker, flannelmouth young apparently have been able to avoid predators and survive, but the mechanisms involved have yet to be identified and studied.

Razorback Sucker (Federally Endangered)

This species once was widely distributed and abundant in larger streams throughout the Colorado River basin. Today, fewer than 3,000 wild fish remain basin-wide, and most are found in Lake Mohave and Lake Mead (Modde et al. 1996; Marsh and Pacey, in press).

An extensive stocking program was carried out in central Arizona during the 1980s. The objective was to preclude federal listing by reestablishing the species in the Gila River basin. The program failed because stressors were not addressed and nonnative predators decimated mostly small stocked fish (Marsh and Brooks 1989; Minckley et al. 1991); the razorback sucker finally was listed as endangered in 1991 (USFWS 1991 Federal Register of final rule). Fewer than 200 of the original 12 million-plus stocked fish were ever recaptured. A much attenuated stocking program continues today, but despite utilizing larger fish, a persistent population has not been established. Stressors still have not been addressed.

Continuing decline of Lake Mohave bonytail and razorback sucker populations caused concerned biologists to form the Native Fish Work Group in 1989 (Mueller 1995). The group has captured hundreds of thousands of wild sucker larvae since 1991 and transplanted them to predator-free habitats where they grow to a size large enough (30 cm) to avoid predation. Large bonytail and razorback suckers also are being stocked in Lake Havasu, and razorback sucker is being stocked into sites along the river downstream from Parker Dam. More than 85,000 juvenile razorback suckers have been stocked in those two reservoirs, and preliminary data suggest stocking survival may be less than 10%. Although some adults survive, their offspring (if any) will face the same predation that eliminated wild populations in the first place. Razorback suckers survive and complete their life cycle when nonnative predators are absent. Minckley et al. (2003) contended that maintaining isolated, native-only fish assemblages might be the only practical method of restoring the species to a seminatural state.

Desert Pupfish (Federally Endangered)

Desert pupfish has disappeared from most of its historic range (Minckley 1973; Miller and Fuiman 1987). Its decline is attributed to habitat loss and from direct competition and predation from aliens, such as mosquitofish and cichlids (Schoenherr 1981). It was listed as endangered in 1986. Some relict populations still exist in California and Mexico, mostly due to their ability to survive in the most inhospitable habitats. Desert pupfish can still be found in shallow agricultural drains near the Salton Sea, in San Felipe Creek, La Cienega de Santa Clara, Baja Mexico, and possibly other shallow wetlands in what remains of the Colorado River delta. The U.S. Fish and Wildlife Service maintains desert pupfish refuges at Cibola and Imperial National Wildlife Refuges on the lower Colorado River.

Sonoran Topminnow (Federally Endangered)

Hubbs and Miller (1941) reported the topminnow was one of the most common fish in the lower Colorado River drainage, but the fish rapidly disappeared when mosquitofish was introduced in the early 1920s for mosquito abatement. Mosquitofish fed on topminnow young and their populations disappeared within a couple of years. Gila topminnow (a subspecies *Poeciliopsis occidentalis occidentalis*) was federally listed as endangered in 1967. Fewer than a dozen natural populations exist in the United States today, all in Arizona, and a number of refuge populations have been established in small, isolated habitats (Meffe et al. 1983).

The Future of Native Fish

The fate of native fishes is grim if current downward trends continue. Wild populations are either extirpated or have suffered dramatic reductions in range and number (Minckley and Deacon 1991). Small, short-lived species (woundfin, desert pupfish, and Sonoran topminnow) were the most vulnerable and remain only as small, isolated wild or refuge populations (Williams 1991). Larger, long-lived species persisted for decades; however, most of the old adults have died off and wild populations soon will vanish from the river. Recovery of federally listed main-stem natives (bonytail, humpback chub, Colorado pikeminnow, and razorback sucker) is being led by Fish and Wildlife Service, Region 6 (USFWS 1987b). Activities focus almost exclusively on the upper basin, where impacts are viewed as less severe. During the development of the recovery program, Behnke (1982) warned that

> . . . preservation of endangered fish is a long-term proposition, essentially forever, while political and legal concerns for the fishes may be short-term.

Nevertheless, a consortium of management agencies, water developers, and environmental groups felt recovery was possible within 15 years. The program started in 1988 and was scheduled for completion in 2003. Wydoski and Hamill (1991) reported

> The goal of the recovery program is to recover, delist, and manage the three endangered fishes [bonytail, humpback chub, Colorado pikeminnow] and to free the razorback sucker of a need for protection under the ESA [Endangered Species Act] by the year 2002.

Endangered species recovery was to be achieved in conjunction with further water development.

Proponents have characterized the upper Colorado River Basin Recovery Implementation Program (RIP; USFWS 1987b) as a model program that prevented conflict and developed consensus regarding endangered species objectives (Brower et al. 2001). However, it also has drawn criticism for its lack of biological accomplishments. Targeted species have realized little benefit from more than 15 years and many tens of millions of dollars of recovery effort. Colorado pikeminnow appear slightly more numerous; however, humpback chub has seriously declined, wild bonytail are believed functionally extinct, and razorback sucker was federally listed as endangered 3 years after the program was initiated.

In contrast, the program advanced completion of an unprecedented 200 water development projects without any environmental challenge. Brower et al. (2001) suggested the continued decline of native fish populations could be attributed to a discrepancy in the voting power of participants and that the fishes' fate has become secondary to the recovery process itself. Despite the lack of compelling evidence that recovery, water development, and recreational fisheries are compatible, the program has been extended and funded for another 10 years, which seems to validate Brower's suspicions.

A similar consensus building program is being developed in the lower basin. The lower Colorado River Multi-Species Conservation Program was formed in 1995. Its steering committee is comprised of representatives from 35 federal, state, and public groups who are attempting to develop a long-term mitigation plan to offset any incidental take of listed threatened and endangered species that may occur as a result of their actions and programs (SAIC/Jones and Stokes 2002). However, after 7 years of debate, a conservation plan has yet to be finalized and approved. Whether this approach will actually benefit these species is yet to be determined.

We applaud efforts to save these fish, but it cannot be assumed that recovery is possible (Minckley 1983, 1991; Minckley and Deacon 1991). The philosophy of a quick fix has not proven itself. Predation remains a major, critical problem that has not been confronted or resolved. Minckley and Deacon (1991) stated, "Native fishes of the American West will not remain on earth without active management . . ." The standard approach of saturating habitats with small-hatchery fish failed and has been replaced by stocking fewer, but larger fish. Efforts are focusing on reestablishing spawning populations, but unfortunately, their offspring face even more predators than their ancestors and also have to cope with altered flow regimes and less water. Recruitment failure still remains the primary concern. Several decades of scientific research have provided compelling evidence that native fish recovery is not possible in habitats dominated by alien fishes.

Recovery continues to elude most species as they spiral toward extinction. Captive broodstock is being held, but hatcheries simply do not have the space or resources necessary to protect the genetic diversity needed to adequately represent wild populations. A more concerted integration of conservation management is needed (Minckley and Deacon 1991). Large scale refuge populations may be the only avenue left for larger species, and they may provide conditions similar to historic "oxbow assemblages" that have long since disappeared (Minckley et al. 2004). This is a common management approach used for smaller, short-lived species (e.g., desert pupfish and Sonoran topminnow) (Williams 1991). Space is critically needed to save what genetic diversity still exists and provide research opportunities that no longer exist in the wild.

Steps are being taken to establish refuge populations for the bonytail and razorback sucker in small (2–50 ha) ponds where it is hoped they will establish self-sustaining populations. After four decades of research and debate, the initial step toward the construction of these facilities is slowly making headway. Unfortunately, few management and regulatory agencies embrace the fact that these species will only survive by concerted management through perpetuity.

References

Abbott, C. C. 1861. Descriptions of four new species of North American Cyprinidae. Proceedings of the Philadelphia Academy of Natural Sciences 12(1860):473–474.

Allan, R. C., and D. L. Roden. 1978. Fish of Lake Mead and Lake Mohave. Nevada Department of Wildlife, Biological Bulletin No. 7, Reno.

Anderson, B. E. 2001. The socio-economic impacts of the Lake Havasu fisheries improvement program. Submitted to the Bureau of Land Management, Lake Havasu City, Arizona.

Barlow, G. W. 1958. High salinity mortality of desert pupfish, *Cyprinodon macularius.* Copeia 1958:231–232.

Behnke, R. J. 1982. Realities and illusions of endangered species preservation. Pages 94–95 *in* W. H. Miller, H. M. Tyus, and C. A. Carlson, editors. Fishes of the upper Colorado River system: present and future. American Fisheries Society, Bethesda, Maryland.

Beland, R. D. 1953. The effect of channelization on the fishery of the lower Colorado River. California Fish and Game 39(1):137–140.

Blake, W. P. 1857. Geological report. Pages 1–310 *in* Routes to California, to connect with the routes near the thirty-fifth and thirty-second parallel, exploited by Lieut. R. S. Williamson, Corps Topographical Engineers, in 1853. Explorations and surveys to ascertain the most practicable and economical route for a railroad from the Mississippi River to the Pacific Ocean, 5 (pt. 2), 1856.

Brouder, M. J., D. D. Rogers, and L. D. Avenetti. 2000. Life history and ecology of the roundtail chub *Gila robusta*, from two streams in the Verde River basin. Arizona Game and Fish Department, Technical Guidance Bulletin No. 3, Federal Aid in Sportfish Restoration Project F-14-R, Phoenix.

Brower, A., C. Reedy, and J. Yelin-Kefer. 2001. Consensus versus conservation in the upper Colorado River basin recovery implementation program. Conservation Biology 15:1001–1007.

Collier, M. 1999. Water, earth, and sky, the Colorado River basin. University of Utah Press, Salt Lake City.

Deacon, J. E. 1988. The endangered woundfin and water management in the Virgin River, Utah, Arizona, and Nevada. Fisheries 13(1):18–24.

Dill, W. A. 1944. The fishery of the lower Colorado River. California Fish and Game 1944:109–211.

Douglas, M. E., and P. C. Marsh. 1996. Population estimates/population movements of *Gila cypha*, an endangered cyprinid fish in the Grand Canyon region of Arizona. Copeia 1996(1):15–28.

Evermann, B. W., and C. Rutter. 1895. The fishes of the Colorado basin. U.S. Fish Commission, Washington D.C.

Fradkin, P. L. 1984. A river no more—the Colorado River and the West. University of Arizona Press, Tucson.

Fuller, P. L., L. G. Nico, and J. D. Williams. 1999. Nonindigenous fishes, introduced into inland waters of the United States. American Fisheries Society, Special Publication 27, Bethesda, Maryland.

Gilbert, C. H., and N. B. Scofield. 1898. Notes on the collection of fishes from the Colorado basin in Arizona. Proceedings of the U.S. National Museum 20:487–499.

Grinnell, J. 1914. An account of the mammals and birds of the lower Colorado Valley, with special reference to the distributional problems present. University of California Publications in Zoology 12:51–294.

Hedrick, P. W., T. E. Dowling, W. L. Minckley, C. A. Tibbets, B. D. DeMarais, and P. C. Marsh. 2000. Establishing a captive broodstock for the endangered bonytail chub (*Gila elegans*). Journal of Heredity 91:35–39.

Holden, P. B. 1991. Impacts of Green River poisoning on management of native fishes. Pages 43–54 *in* W. L. Minckley and J. E. Deacon, editors. Battle against extinction, native fish management in the American West. The University of Arizona Press, Tucson.

Holden, P. B., P. D. Abate, and T. L. Welker. 2001. Razorback sucker studies on Lake Mead, Nevada, 2000–2001 annual report. Report PR-578–5, submitted to Nevada Department of Resources, by Bio/West, Inc. Logan, Utah.

Hubbs, C. L., and R. R. Miller. 1941. Studies of the fishes of the order Cyprinodontes. XVII: General and species of the Colorado River system. Occasional Papers of the Museum of Zoology, University of Michigan 443:1–9.

Hubbs, C. L., and R. R. Miller. 1953. Hybridization in nature between the fish genus *Catostomus* and *Xyrauchen*. Papers of the Michigan Academy of Science, Arts, and Letters 38:207–233.

James, G. W. 1906. The wonders of the Colorado Desert. Volumes I and II. Little, Brown, and Company, Boston.

Jonez, A., and R. C. Sumner. 1954. Lake Mead and Mohave investigations. Nevada Fish and Game Commission, Reno.

Johnson, J. E., M. G. Pardew, and M. M. Lyttle. 1993. Predator recognition and avoidance by larval razorback sucker and northern hog sucker. Transactions of the American Fisheries Society 122:1139–1145.

Jordon, D. S. 1891. Report of explorations in Utah and Colorado during the summer of 1889, with an account of the fishes found in each of the river basins examined. Bulletin of the U.S. Fish Commissioner 9:1–40.

Kaeding, L. R., and M. A. Zimmerman. 1983. Life history and ecology of the humpback chub in the little Colorado and Colorado rivers of the Grand Canyon. Transactions of the American Fisheries Society 112:577–594.

Kimsey, J. B. 1958. Fishery problems in impounded waters of California and the lower Colorado River. Transactions of the American Fisheries Society 87:319–333.

Kolb, E. L. 1927. Through the Grand Canyon from Wyoming to Mexico. The Macmillan Company, New York.

Leopold, A. 1949. A Sand County Almanac. Oxford University Press, New York.

Lockington, W. N. 1881. Description of a new species of Catostomus (*Catostomus cypho*) from the Colorado River. Proceedings of the Philadelphia Academy of Natural Sciences 32:237–240.

Luecke, D. F. 1999. A delta once more: restoring riparian and wetland habitat in the Colorado River delta. Environmental Defense Fund, Washington, D.C.

MacDougal, D. T. 1917. The desert basins of the Colorado delta. Bulletin of the American Geographic Society 39(12):705–729.

Marsh, P. C., and J. L. Brooks. 1989. Predation by ictalurid catfishes as a deterrent to reestablishment of introduced razorback sucker. The Southwestern Naturalist 34:188–195.

Marsh, P. C., and M. E. Douglas. 1997. Predation by introduced fishes on endangered humpback chub and other native species in the Little Colorado River. Transactions of the American Fisheries Society 126:343–346.

Marsh, P. C., and W. L. Minckley. 1989. Observations on recruitment and ecology of razorback sucker: lower Colorado River, Arizona–California–Nevada. Great Basin Naturalist 49:71–78.

Marsh, P. C., and W. L. Minckley. 1992. Status of bonytail (*Gila elegans*) and razorback sucker (*Xyrauchen texanus*) in Lake Mohave, Arizona–Nevada. Proceedings of the Desert Fishes Council 23(1991):18–23.

Marsh, P. C., C. A. Pacey, and B. R. Kesner. 2003. Decline of wild razorback sucker in Lake Mohave, Colorado River, Arizona and Nevada. Transactions of the American Fisheries Society 132:1251–1256.

Martin, W. C., R. H. Bollman, and R. Gum. 1980. The economic value of the Lake Mead fishery with special attention to the largemouth bass fishery. Submitted to Water and Power Resources Service, Contract No. 14–06-3000–2719, Boulder City, Nevada.

McCarthy, M. S., and W. L. Minckley. 1987. Age estimation for razorback sucker (Pisces: Catostomidae) from Lake Mohave, Arizona–Nevada. Journal of the Arizona-Nevada Academy of Sciences 21:87–97.

McDonald, C. C., and O. J. Loeltz. 1976. Water resources of the lower Colorado River-Salton Sea area as of 1971, summary report. Government Printing Office, Geological Survey Professional Paper 486-A, Washington D.C.

Meffe, G. K., D. A. Hendrickson, W. L. Minckley, and J. N. Rinne. 1983. Factors resulting in decline of the endangered Sonoran topminnow *Poeciliopsis occidentalis* (Atheriniformes: Poeciliidae) in the United States. Biological Conservation 25:135–159.

Miller, R. R. 1955. Fish remains from archaeological sites in the lower Colorado River basin, Arizona. Papers of the Michigan Academy of Science, Arts, and Letters 40:125–136.

Miller, R. R. 1961. Man and the changing fish fauna of the American Southwest. Papers of the Michigan Academy of Science, Arts, and Letters 46:365–404.

Miller, R. R., and L. A. Fuiman. 1987. Description and conservation status of *Cyprinodon macularius eremus*, a new subspecies of pupfish from Organ Pipe Cactus National Monument, Arizona. Copeia 1987:593–609.

Minckley, W. L. 1973. Fishes of Arizona. Arizona Game and Fish Department, Phoenix.

Minckley, W. L. 1979. Aquatic habitats and fishes of the lower Colorado River, southwestern United States. Arizona State University, Final report for U.S. Bureau of Reclamation Contract 14–06-300–2529, Tempe.

Minckley, W. L. 1983. Status of the razorback sucker, *Xyrauchen texanus* (Abbott), in the lower Colorado River basin. The Southwestern Naturalist 28:165–187.

Minckley, W. L. 1991. Native fishes of the Grand Canyon regions: an obituary? Proceedings of a symposium of Colorado River ecology and dam management, Santa Fe, New Mexico. National Academic Press, Washington D.C.

Minckley, W. L. 2002. Fishes of the lowermost Colorado River, its delta, and estuary. A commentary on biotic change. Pages 63–78 *in* L. Lozano-Vilano, editor. Libro jubilar en honor al Dr. Salvador Contreras Balderas. Universidad Autonoma de Nuevo Leon, Monterrey, Mexico.

Minckley, W. L., and J. E. Deacon. 1991. Battle against extinction: native fish management in the American West. The University of Arizona Press, Tucson.

Minckley, W. L., P. C. Marsh, J. E. Brooks, J. E. Johnson, and B. L. Jensen. 1991. Management toward recovery of the razorback sucker. Pages 303–357 *in* W. L. Minckley and J. E. Deacon, editors. Battle against extinction, native fish management in the American West. The University of Arizona Press, Tucson.

Minckley, W. L., P. C. Marsh, J. E. Deacon, T. E. Dowling, P. W. Hedrick, W. J. Matthews, and G. Mueller. 2004. A conservation plan for the lower Colorado River native fishes. BioScience 53:219–234.

Modde, T., K. P. Burnham, and E. J. Wick. 1996. Population status of the razorback sucker in the middle Green River (U.S.A.). Conservation Biology 10:110–119.

Moffett, J. W. 1942. A fishery survey of the Colorado River below Boulder Dam. California Fish and Game 28:76–86.

Moffett, J. W. 1943. A preliminary report on the fishery of Lake Mead. Transactions of the Eighth North American Wildlife Conference 1943:179–186.

Moyle, P. B., H. W. Li, and B. A. Barton. 1986. The Frankenstein effect: impact of introduced fishes on native fishes in North America. Pages 415–426 *in* R. H. Stroud, editor. Fish culture in fisheries management. American Fisheries Society, Bethesda, Maryland.

Mueller, G. 1995. A program for maintaining the

razorback sucker in Lake Mohave. Pages 127–135 *in* H. L. Schramm, Jr., and R. G. Piper, editors. Uses and effects of cultured fishes in aquatic ecosystems. American Fisheries Society, Symposium 15, Bethesda, Maryland.

Mueller, G. A. 2004. Collection of an adult gizzard shad (*Dorosoma cepedianum*) from the San Juan River, Utah. Western North American Naturalist 64(1):135–136.

Mueller, G. A., and P. C. Marsh. 2002. Lost, a desert river and its native fish: a historical perspective of the lower Colorado River. U.S. Government Printing Office, Information and Technology Report USGS/BRD/INR 2002–010, Denver, Colorado.

Mueller, G. A., and R. Wydoski. 2004. The successful reintroduction of the flannelmouth sucker in the lower Colorado River, Nevada–Arizona–California. North American Journal of Fishery Management 24:41–46.

Odens, P. R. 1989. Southwest corner. The Plain Speaker, Jucumba, California.

Oliver, P. A. 1965. Dredging the braided Colorado: the reclamation era. Bureau of Reclamation (February):4–15.

Osmundson, D. B., and K. P. Burnham. 1998. Status and trends of the endangered Colorado pikeminnow in the upper Colorado River. Transactions of the American Fisheries Society 127:957–970.

SAIC (Science Applications International Corporation)/ Jones and Stokes. 2002. Second Administrative Draft Conservation Plan – Lower Colorado River Multi-Species Conservation Program. (JS 00–450.) January 25. Sacramento, California. Prepared for the LCR MSCP Steering Committee, Santa Barbara, California.

Schmidt, J. C., and J. B. Graf. 1990. Aggradation and degradation of alluvial sand deposits, 1965 to 1986, Colorado River, Grand Canyon national park, Arizona. U.S. Geological Survey Professional Paper 1493, Washington, D.C.

Schoenherr, A. A. 1981. The role of competition in the replacement of native fishes by introduced species. Pages 173–203 *in* R. J. Naiman and D. L. Soltz, editors. Fishes in North American deserts. Wiley, New York.

Stanford, J. A. 1994. Instream flows to assist the recovery of endangered fishes of the upper Colorado River basin. National Biological Survey, Biological Report 24, Washington, D.C.

Stanford, J. A., J. V. Ward, W. J. Liss, C. A. Frissell, R. N. Williams, J. A. Lichatowich, and C. C. Coutant. 1996. A general protocol for restoration of regulated rivers. Regulated Rivers: Research and Management 12:391–413.

Sykes, G. 1937. The Colorado delta. Carnegie Institution of Washington and the American Geographical Society of New York, American Geographical Society Special Publication No. 19, New York.

Tyus, H. M., and C. A. Karp. 1990. Spawning and movements of razorback sucker, *Xyrauchen texanus*, in the Green River basin of Colorado and Utah. The Southwestern Naturalist 35:427–433.

Tyus, H. M., and J. F. Saunders, III. 2000. Nonnative fish control and endangered fish recovery. Fisheries 25(9):17–24.

Tyus, H. M., E. J. Wick, and D. L. Skates. 1981. A spawning migration of Colorado pikeminnow (*Ptychocheilus lucius*) in the Yampa and Green rivers, Colorado and Utah, 1981. Proceedings of the Desert Fishes Council 13:102–108.

USFWS (U.S. Fish and Wildlife Service). 1987a. Endangered and threatened wildlife and plants; proposed determination of experimental population status for an introduced population of Colorado pikeminnow. Federal Register 52()32143–32145.

USFWS (U.S. Fish and Wildlife Service). 1987b. Final recovery implementation program for endangered fish species in the upper Colorado River basin. U.S. Fish and Wildlife Service, Denver.

USFWS (U.S. Fish and Wildlife Service). 1990. Bonytail chub recovery plan. U.S. Fish and Wildlife Service, Denver.

USGS (U.S. Geological Survey). 1978. Water resources data for Arizona water year 1977. U.S. Geological Survey water data report AZ-77–1.

Van Haverbeke, D. R., and L. G. Coggins, Jr. 2003. Stocking assessment and fisheries monitoring activities in the Little Colorado River within Grand Canyon during 2001. Final report submitted to the Grand Canyon Monitoring and Research Center. U.S. Fish and Wildlife Service, Document No. USFWS-AZFRO-FL-02–002, Arizona Fisheries Resource Office, Flagstaff.

Wagner, R. A. 1952. Arizona survey of the wildlife and fishery resources of the lower Colorado River. Arizona Game and Fish Department,

Federal Aid to Fisheries Restoration, Project Completion Report 60-R, Phoenix.

Wallis, O. L. 1951. The status of the fish fauna of the Lake Mead national recreational area, Arizona–Nevada. Transactions of the American Fisheries Society 80:84–92.

Wheeler, G. M. 1876. Annual report on the geographical surveys west of the one-hundredth meridian, in California, Nevada, Utah, Colorado, Wyoming, New Mexico, Arizona, and Montana: Appendix JJ, Annual Report of the Chief of Engineers for 1876. Government Printing Office, Washington, D.C.

Williams, J. E. 1991. Preserves and refuges for native Western fishes: history and management. pages 171–189 in W. L. Minckley and J. E. Deacon, editors. Battle against extinction, native fish management in the American West. The University of Arizona Press, Tucson and London.

Wydoski, R. S., and J. Hamill. 1991. Evolution of a cooperative recovery program for endangered fishes in the upper Colorado River basin. Pages 123–141 *in* W. L. Minckley and J. E. Deacon, editors. Battle against extinction, native fish management in the American West. The University of Arizona Press, Tucson.

American Fisheries Society Symposium 45:157–204, 2005

Ecology and Conservation of Native Fishes in the Upper Colorado River Basin

RICHARD A. VALDEZ*
SWCA, Inc., 172 W. 1275 S., Logan, Utah 84321, USA

ROBERT T. MUTH
Upper Colorado River Endangered Fish Recovery Program, 44 Union Boulevard, Suite 120, Lakewood, Colorado 80228, USA

Abstract.—The upper Colorado River basin supports a native ichthyofauna of 14 species or subspecies that have been impacted by poor land-use practices, altered flows, physical habitat fragmentation, competition and predation from nonnative fish species, and degraded water quality. Five taxa are federally endangered, including the large-river species, Colorado pikeminnow *Ptychocheilus lucius*, humpback chub *Gila cypha*, bonytail *G. elegans*, razorback sucker *Xyrauchen texanus*, and a warm-stream subspecies, Kendall Warm Springs dace *Rhinichthys osculus thermalis*. Two recovery programs, formed through cooperative agreements among federal, state, tribal, and private agencies and stakeholders, coordinate activities in the upper basin that have helped to resolve water resource issues, implement management actions to minimize or remove threats, and conserve endangered species. A cooperative biological management program among state and federal agencies works to protect the Kendall Warm Springs dace. Conservation agreements have also been established for the other native fish species. Continued public and institutional support for these programs is vital to species recovery and to the balance between long-term species conservation and human demands on the Colorado River system.

Introduction

The upper Colorado River basin lies within the states of Colorado, Wyoming, Utah, and New Mexico (Figure 1). Upper basin drainage area is about 289,540 km^2, or less than half the total area of the Colorado River system; average annual historic upper basin discharge is about 93% of average total basin volume (i.e., 12.93 million acre-feet of 13.90 million acre-feet). The upper basin includes the upper Colorado River, Green River, and San Juan River subbasins. Evidence suggests that the Colorado River in the upper basin has been in its present course for more than 5 million years and flowed into one or more closed basins near the upper end of present-day Grand Canyon (Minckley et al. 1986). About 5 million years ago (i.e., late Miocene/early Pliocene), the river began carving its way through the Colorado Plateau forming Grand Canyon, and joined with a lower, more dispersed drainage within the last 2–3 million years (McKee et al. 1967; Luchitta 1990). The ancestral upper Colorado River consisted primarily of a single large river and tributaries in contrast to more dispersed smaller tributaries in the lower basin. Fish species that evolved in the upper basin were mostly large riverine forms and those that evolved in the lower basin were small stream forms. Connection between these two ancestral basins, marked by the river cutting through Grand Canyon, allowed for inter-basin movement of the larger, more mobile species, particularly from the upper basin. These

* Corresponding author: valdezra@aol.com

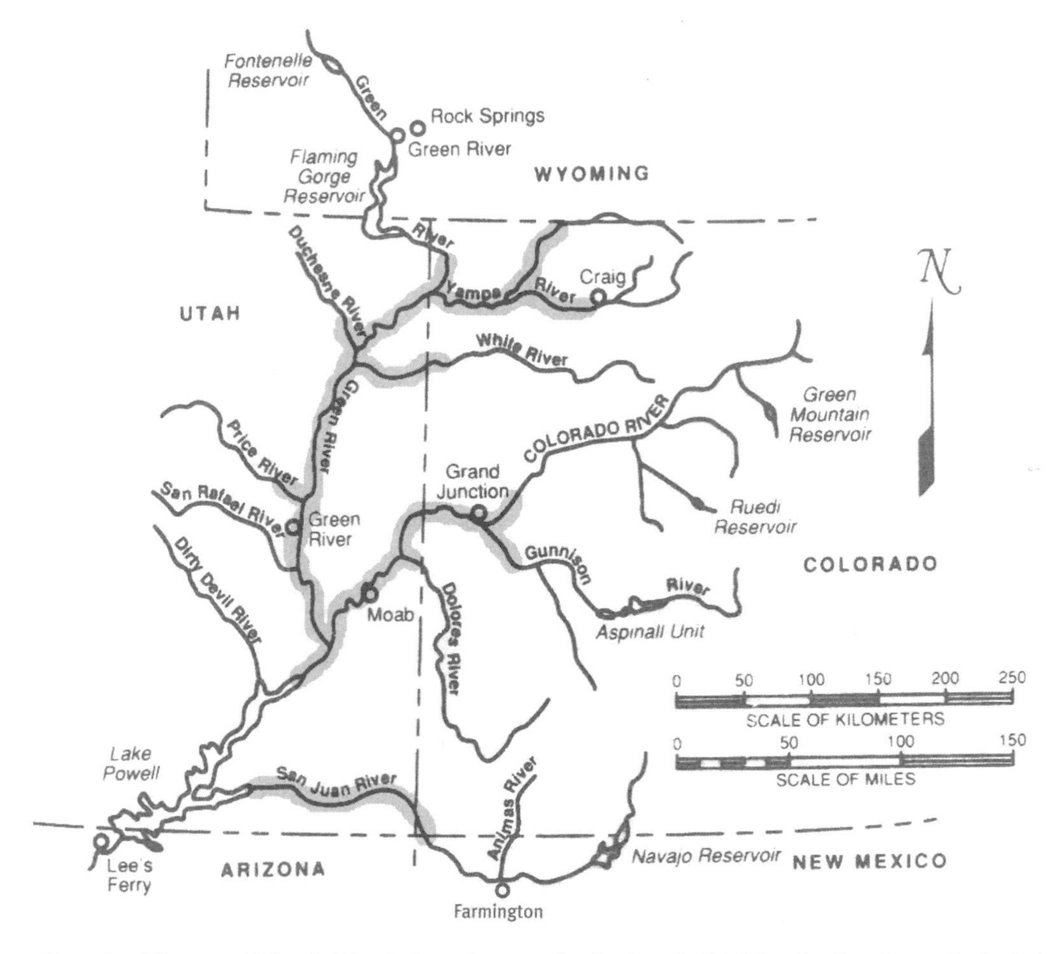

Figure 1.—The upper Colorado River basin and present distribution of wild Colorado pikeminnow *Ptychocheilus lucius* (shaded).

progenitors of modern-day forms further evolved as the Colorado River became a single basin. Many small forms found in the lower basin are unique and have been unable to disperse upstream into the upper basin.

Today, the upper basin originates at elevations of 3,000–4,000 m in high mountain meadows, and the Colorado River flows through a series of mid-elevation sandstone canyons (Figure 2) and intervening deep canyons with isolated upthrusts of hard Precambrian schist and gneiss. The characteristic geomorphic features of the upper basin provide diverse and unique habitats to which the native fishes have adapted over several million years. A long period of geologic isolation, steep stream gradient, high levels of water turbidity and conductance, and extreme seasonal variation in water temperatures and flows have led to unique morphologic and physiologic adaptations. These specialized adaptations, together with low levels of competition and predation, have rendered native fish species highly susceptible to ecological changes from human activities including (a) flow regulation and diversion, (b) physical habitat destruction, alteration, and fragmentation, (c) introduction of nonnative fishes, and (d) degraded water quality (Miller 1961; Carlson and Muth 1989).

Current Status and Ecology

Fourteen species or subspecies of native fishes inhabit the upper basin of which eight (57%) are

Figure 2.—The upper Colorado River downstream of Moab, Utah, December 1981. Photo by R. A. Valdez.

endemic; eight are primarily large-river warmwater inhabitants, five are coolwater or coldwater tributary inhabitants, and one is found in a warm stream (Tyus et al. 1982; Muth et al. 2000; McAda 2003; Table 1). Many native fishes in the upper basin have declined in range and abundance since the early 1900s (Carlson and Muth 1989). Five are federally endangered: Colorado pikeminnow, humpback chub, bonytail, razorback sucker, and Kendall Warm Springs dace. Concerns over declines in native fish populations in the mid-1900s prompted studies to assess the status and life history of these little-known southwestern fishes (e.g., Miller 1955, 1959, 1961; Vanicek 1967; Holden 1973), and passage of the Endangered Species Act of 1973, as amended (ESA; 16 U.S.C. 1531 et seq.), led to initiation of more comprehensive studies in the late 1970s (e.g., U.S. Fish and Wildlife Service 1982a).

Colorado Pikeminnow

Colorado pikeminnow is the largest minnow in North America, with estimated maximum size of 1.8 m total length (TL) and 36 kg (Miller 1961); largest confirmed weights are 12.2 and 15.5 kg from the lower basin (Wallis 1951), and about 12 kg from the upper basin (Figure 3). Maximum age,

Table 1.—Common and scientific names, legal status, current distribution, and relative abundance of native fish of the upper Colorado River basin.

Species	Status[a]	Distribution and abundance
Cyprinidae (minnows)		
Colorado pikeminnow *Ptychocheilus lucius* Girard, 1856	EN En-NM Th-CO Sp-UT	Found as two populations: Green River and upper Colorado River subbasins. Wild fish incidental in San Juan River.
humpback chub *Gila cypha* Miller, 1946	EN Th-CO Sp-UT	Found as five populations: Black Rocks, Westwater, Cataract, Desolation/Gray, Yampa canyons
bonytail *G. elegans* Baird and Girard, 1853	EN Ex-NM En-CO Sp-UT	Wild fish incidental; fewer than 15 specimens from Black Rocks, Cataract and Desolation/Gray canyons since 1980
roundtail chub *G. robusta* Baird and Girard, 1853	En-NM Co-CO, NM, UT, WY	Common to locally abundant in mid-elevation rivers
speckled dace *Rhinichthys osculus* Girard, 1856		Common to abundant throughout
Kendall Warm Springs dace *R. o. thermalis* (Hubbs and Kuhne, 1937)	EN En-WY	Found as single population only in Kendall Warm Springs Creek, Wyoming
Catostomidae (suckers)		
razorback sucker *Xyrauchen texanus* (Abbott, 1860)	EN En-CO Sp-UT	Wild fish incidental to rare in upper Colorado, Gunnison, Green, Yampa, Duchesne, San Juan rivers
flannelmouth sucker *Catostomus latipinnis* Baird and Girard, 1853	Co-CO, NM, UT WY	Common to abundant in mid- and low elevation rivers
bluehead sucker *C. discobolus* Cope 1871	Co-CO NM, UT WY	Common to abundant in rocky riffles of mid-elevation rivers
mountain sucker *C. platyrhynchus* (Cope, 1874)	Sp-CO	Common in high and mid-elevation streams
Cottidae (sculpins)		
mottled sculpin *Cottus bairdii* Girard, 1850		Common to abundant in mid- and high elevation streams and rivers
Paiute sculpin (*Cottus beldingii*) Eigenmann, 1891		Uncommon in mid- and high elevation streams and rivers
Salmonidae (trout and salmon)		
mountain whitefish *Prosopium williamsoni* (Girard, 1856)	Sf-CO, UT, WY	Common to abundant in high elevation streams

Table 1.—Continued.

Species	Status[a]	Distribution and abundance
Colorado River cutthroat trout *Oncorhynchus clarkii pleuriticus*	Co-CO, UT, WY Sf-CO	Small local populations in high elevation streams

[a] EN = federally endangered; status by indicated state: En = endangered, Ex = extirpated, Th = threatened, Sp = species of special concern or sensitive species, Co = conservation species; Sf = sport fish.

determined from scale annuli, is up to 18 years (Vanicek and Kramer 1969; Seethaler 1978; Musker 1981; Hawkins 1992). However, Osmundson et al. (1997) cautioned that scale-based estimations are probably unreliable for Colorado pikeminnow beyond about age 10, and concluded that growth-rate data indicated that large fish (e.g., more than 900 mm TL) averaged 47–55 years old with a minimum of 34 years. Growth and size in the upper basin appear limited by colder temperatures and a shorter growing season (Kaeding and Osmundson 1989). The species was listed as endangered in 1967 (32 FR 4001), and protected by the ESA in 1973 (39 FR 1175), with critical habitat designated in 1994 (59 FR 13374). A recovery plan was approved in 1978 (U.S. Fish and Wildlife Service 1978), revised in 1991 (U.S. Fish and Wildlife Service 1991), and amended and supplemented with recovery goals in 2002 (U.S. Fish and Wildlife Service 2002a). As for humpback chub, bonytail, and razorback sucker, these recovery goals provide demographic criteria and management actions to minimize or remove threats.

Colorado pikeminnow is endemic to the Colorado River system, and was once widespread and abundant in warm main-stem rivers and tributaries (Kirsch 1889; Evermann and Rutter 1895; Jordan and Evermann 1896; Banks 1964; Vanicek 1967; Holden and Stalnaker 1975a; Holden and Wick 1982; Tyus 1991; Quartarone 1995). In the upper basin, the species was first reported in 1825 by Colonel William H. Ashley (Morgan 1964; Seethaler

Figure 3.—Adult Colorado pikeminnow captured in the Redlands fish passage on the Gunnison River in 2002, approximate weight 12 kg. Photo courtesy of Bob Burdick, U.S. Fish and Wildlife Service.

1978). In the upper Colorado River subbasin, it was historically found as far upstream as Rifle, Colorado, on the upper Colorado River (Beckman 1963); Delta, Colorado, on the Gunnison River (Burdick 1995); and Paradox Valley on the Dolores River (Lynch et al. 1950). In the Green River subbasin, it was reported as far upstream as Green River, Wyoming, on the Green River (Ellis 1914; Baxter and Simon 1970); Craig, Colorado, on the Yampa River; Rangely, Colorado, on the White River; and in the lower Price and Duchesne rivers (Tyus and Haines 1991; Cavalli 1999). In the San Juan River subbasin, Colorado pikeminnow were historically found upstream to Farmington, New Mexico, and the lower Animas River (Holden 1999).

Wild populations of Colorado pikeminnow are presently found only in the upper basin in about 25% of historic range basin-wide (Figure 1). Adults occur in the Green River from Lodore Canyon to the confluence of the Colorado River (Tyus 1991; Bestgen and Crist 2000); Yampa River downstream of Craig, Colorado (Tyus and Haines 1991); Little Snake River into Wyoming (Marsh et al. 1991; Wick et al. 1991); White River downstream of Taylor Draw Dam , Colorado (Tyus and Haines 1991); lower 143 km of the Price River (Cavalli 1999); lower Duchesne River; upper Colorado River from Palisade, Colorado, to Lake Powell (Valdez et al. 1982b; Osmundson et al. 1997, 1998); lower 54 km of the Gunnison River (Valdez et al. 1982a; Burdick 1995); lower 2 km of the Dolores River (Valdez et al. 1992); and 241 km of the San Juan River from Shiprock, New Mexico, to the Lake Powell inflow (Jordan 1891; Koster 1960; Propst 1999; Holden 1999).

Colorado pikeminnow is adapted to warm rivers and requires uninterrupted passage and a hydrologic cycle characterized by large spring peaks of snowmelt runoff and lower, relatively stable base flows. Adults are potadromous and may move up to 950 km to and from spawning sites in summer (Tyus and McAda 1984; Tyus 1990; Irving and Modde 2000). Juveniles and adults use deep, low-velocity eddies, pools, and runs, but move into flooded habitats and bottomlands during spring runoff (Tyus and McAda 1984; Valdez and Masslich 1989; Tyus 1990, 1991; Osmundson et al. 1995). Average fecundity is about 66,000–77,000 eggs/female (Hamman 1986), and females broadcast adhesive eggs over cobble bars during June–August at water temperatures of 16°C or higher (Vanicek and Kramer 1969; Hamman 1981). The eggs incubate for 90–121 h at 20–24°C (Hamman 1981; Marsh 1985), and larvae drift up to 200 km to nursery backwaters where survival is critical to recruitment (Holden 1977; Tyus and Karp 1989; Haines and Tyus 1990; Tyus 1991; Tyus and Haines 1991; Bestgen et al. 1997, 1998; Converse et al. 1999). Young Colorado pikeminnow consume zooplankton and midge (chironomid) larvae (Vanicek 1967; Jacobi and Jacobi 1982; Muth and Snyder 1995), and piscivorous adults eat soft-rayed native and nonnative fish (Osmundson 1999), as well as a variety of insects and animals, including Mormon crickets *Anabrus migratorius* (Tyus and Minckley 1988), mice, birds, and rabbits (Beckman 1963).

Humpback Chub

Humpback chub is an endemic cyprinid of the Colorado River system with an evolutionary history of about 3 million years (Miller 1946, 1955; Minckley et al. 1986). Maximum size is 480 mm TL and 1,165 g (Valdez and Ryel 1997), and maximum age is over 20 years (Hendrickson 1993). The body is deep and laterally-compressed, tapering abruptly to a narrow caudal peduncle with a deeply forked tail fin and large falcate paired fins. Head length divided by caudal peduncle depth is usually less than 5.0, compared to greater than or equal to 5.0 for bonytail, and greater than or equal to 3.0 for roundtail chub (Minckley 1973). Introgressive hybridization may be part of their evolutionary history (Dowling and DeMarais 1993), and high phenotypic plasticity exists with morphologic intergrades in all sympatric populations of humpback chub, bonytail, and roundtail chub (Holden 1968; Holden and Stalnaker 1970; Smith et al. 1979; Valdez and Clemmer 1982; Kaeding et al. 1990; Wick et al. 1991; McElroy and Douglas 1995; Douglas et al. 1989, 1998). Humpback chub was listed as endangered in 1967 (32 FR 4001) and protected by the ESA in 1973 (39 FR 1175), with critical habitat designated in 1994 (59 FR 13374). A recovery plan was approved in 1979 (U.S. Fish and Wildlife Service 1979), revised in 1990 (U.S. Fish and Wildlife Service 1990a), and amended and supplemented with recovery goals in 2002 (U.S. Fish and Wildlife Service 2002b).

Historic abundance of humpback chub is unknown, and historic distribution is surmised from various reports and collections, which indicate the species presently occupies about 68% of its historic habitat of about 756 km of river. The species exists primarily in relatively inaccessible canyons of the Colorado River basin and was rare in early collections (Tyus 1998). Common use of the name "bonytail" for all six Colorado River species or subspecies of the genus *Gila* confounded an accurate early assessment of distribution and abundance (Holden and Stalnaker 1975a, 1975b; Valdez and Clemmer 1982). Also, human alterations throughout the basin prior to faunal surveys may have depleted or eliminated the species from some river reaches before its occurrence was documented.

Five populations of humpback chub are currently known in the upper basin (i.e., Black Rocks, Westwater Canyon, Cataract Canyon, Desolation/Gray Canyons, Yampa Canyon; Figure 4), and one from Grand Canyon in the lower basin (U.S. Fish and Wildlife Service 2002b). Small numbers have also been reported from Moab Canyon (Taba et al. 1965; Valdez and Clemmer 1982), Debeque Canyon (Valdez and Clemmer 1982), Cross Mountain Canyon (Wick et al. 1981), Whirlpool and Split Mountain canyons (Holden and Stalmaker 1975a), Little Snake River (Wick et al. 1991), and White River (Lanigan and Berry 1979; U.S. Fish and Wildlife Service 1982a). Based on historic collections, populations have been extirpated from Hideout Canyon in Flaming Gorge (Bosley 1960; Gaufin et al. 1960; McDonald and Dotson 1960; Smith 1960), and

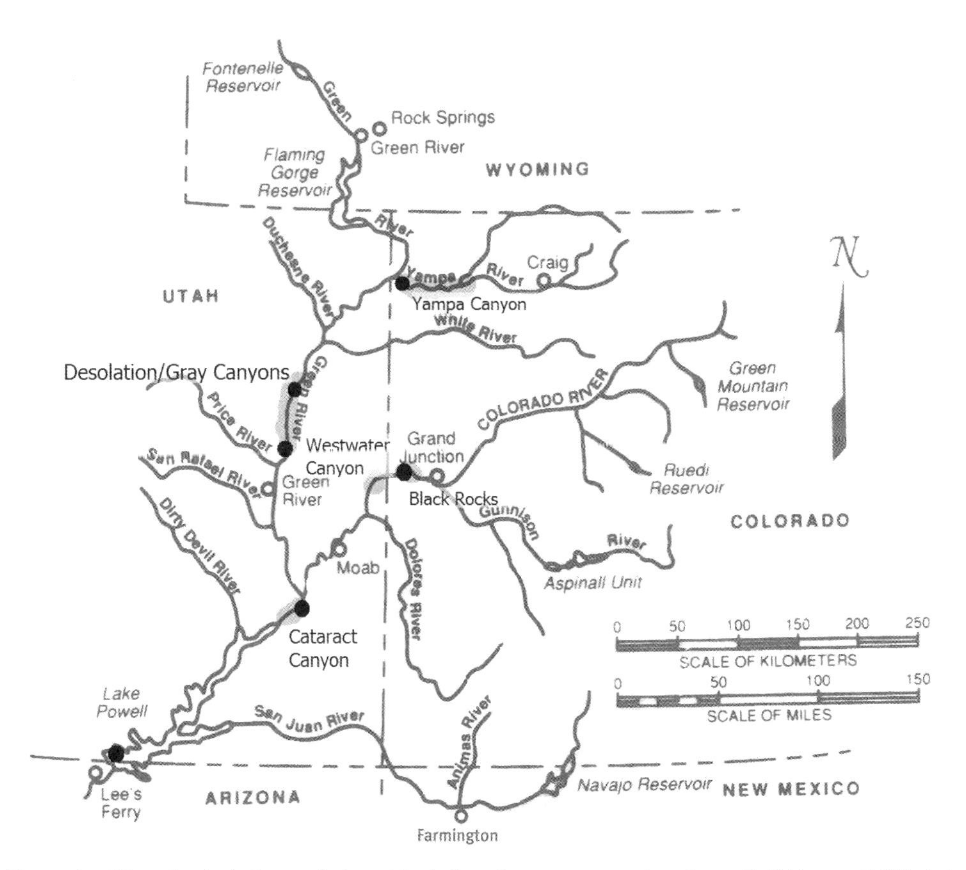

Figure 4.—Humpback chub populations (shaded) and recent capture locations of wild bonytail (filled circles) in the upper Colorado River basin.

Narrow and lower Cataract canyons (Holden and Stalnaker 1970, 1975a; Valdez 1990).

Humpback chub evolved in seasonally warm and turbid water and is highly adapted to extreme hydrologic conditions (Valdez and Carothers 1998). Although not a strong swimmer (Bulkley et al. 1982), the species is extraordinarily specialized for life in torrential water of canyon-bound reaches with a fusiform body, expansive fins, enlarged stabilizing nuchal hump, coarse skin, deeply embedded scales, and small eyes. Adults and juveniles in the upper basin occupy deep eddies and pools along rocky shores, and young use sheltered shorelines and low-velocity habitats (Valdez and Clemmer 1982; Karp and Tyus 1990a; Valdez et al. 1990; Chart and Lentsch 2000). Humpback chub move substantially less than other Colorado River native fishes and exhibit strong fidelity for restricted reaches of river, generally less than 2 km (Valdez and Clemmer 1982; Kaeding et al. 1990; Valdez and Ryel 1997).

Humpback chub mature in 2–3 years at 200 mm TL, and average fecundity is about 2,500 eggs per female (Hamman 1982). Spawning occurs during spring runoff at water temperatures of 16–22°C on large cobble bars or shorelines (Valdez and Clemmer 1982; Kaeding et al. 1986; Tyus and Karp 1989; Karp and Tyus 1990a; Valdez and Williams 1993). Larvae hatch at 18–21°C in about 6 d (Muth 1990) but do not drift extensively and use shorelines close to natal areas. Humpback chub feed opportunistically on drifting food entrained in recirculating eddies, as well as invertebrates or detritus on the river bottom, including planktonic crustacea, larvae of blackfly (simulid) and midges, filamentous green algae (primarily *Cladophora glomerata*), aquatic invertebrates, terrestrial invertebrates, and occasionally other fish and reptiles (Minckley 1973; Kaeding and Zimmerman 1983; Kubly 1990; Valdez and Ryel 1997). They become engorged by feeding on emergences of aquatic insects (e.g., mayfly hatches), grasshopper infestations, or migrations of Mormon crickets (Tyus and Minckley 1988). Parasites include the external parasitic copepod *Lernaea cyprinacea* and Asian tapeworm *Bothriocephalus acheilognathi* (Brouder and Hoffnagle 1997; Clarkson et al. 1997).

Bonytail

Bonytail attain a maximum size of 550 mm TL and 1,100 g (Vanicek 1967). Originally collected and described from the Zuni River, New Mexico (Sitgreaves 1853; Girard 1856), the species is commonly called "bonytail chub," a name that has also been applied to humpback chub and roundtail chub and led to taxonomic confusion. Bonytail are streamlined with a small head, slender body, and thin caudal peduncle. Maximum age of fish from the Green River was 7 years (Vanicek 1967), and bonytail from Lake Mohave in the lower basin were 32–49 years (Ulmer 1983; Rinne et al. 1986). Bonytail was listed as federally endangered in 1980 (45 FR 27710) with critical habitat designated in 1994 (59 FR 13374). A recovery plan was approved in 1984 (U.S. Fish and Wildlife Service 1984), revised in 1990 (U.S. Fish and Wildlife Service 1990b), and amended and supplemented with recovery goals in 2002 (U.S. Fish and Wildlife Service 2002c).

Bonytail was once reported from various regions of the Colorado River system (Cope and Yarrow 1875; Jordan 1891; Jordan and Evermann 1896; Gilbert and Scofield 1898; Kirsch 1889; Chamberlain 1904). The species experienced an apparently dramatic but poorly documented decline starting in about 1950 that was attributed to construction of main-stem dams, introduction of nonnative fishes, poor land-use practices, and degraded water quality (Miller 1961). Its population trajectory over the past century is unclear because of a lack of quantitative, historic basin-wide fishery investigations. Interchangeable nomenclature between "bonytail" and sympatric *Gila* species (Valdez and Clemmer 1982; Quartarone 1995) has also led to confusion in species status. Ellis (1914) synonymized bonytail with roundtail chub "...since intermediate forms and those agreeing with the descriptions of both species were taken from the same station in the Grand [Colorado] River at Grand Junction."

The first record of bonytail from the upper basin was "One specimen taken in the Gunnison at Delta; five in the Green River..." (Jordan 1891; Bookstein et al. 1985). Gaufin et al. (1960) and Smith (1960) reported bonytail from Hideout Canyon before it was inundated by Flaming Gorge Reservoir, but numbers and sizes are unknown because they were grouped with humpback chub and roundtail chub. "Bonytail chub," roundtail chub, and humpback chub were reported in the Green River from the mouth of the Black's Fork River downstream through Flaming Gorge (Bosley 1960), and composed 7.3% of all fish from Green River, Wyoming, to the Utah–Colorado state line (McDonald and Dotson 1960). Individuals collected from the base of Flaming Gorge Dam and from Little Hole (10 km below the dam) in 1962 are held at the University of Michigan (R. Miller, University of Michigan, personal communication; Bookstein et al. 1985). Bonytail outnumbered roundtail chub in the Green River for the 1959, 1960, and 1961 year classes with 67 bonytail more than 200 mm TL collected during 1964–1966 (Vanicek and Kramer 1969). Holden and Stalnaker (1975b) reported 36 bonytail from the lower Yampa River and middle and lower Green River during 1967–1973. Bonytail declined dramatically in the Green River through the 1960s. Reasons for the decline are unknown, but were likely related to the closure of Flaming Gorge Dam in 1964. Before filling Flaming Gorge Reservoir, about 725 km of the Green River and its tributaries were treated with rotenone to poison nonnative carp, catfish, shiners, and perch in advance of stocking the reservoir with rainbow trout *Oncorhynchus mykiss* and kokanee *O. nerka* (Holden 1991). Fish surveys after the closure of Flaming Gorge Dam revealed that the rotenone had not killed all of the fish in the treatment area and did not eliminate the native forms, including bonytail, roundtail chub, humpback chub, Colorado pikeminnow, and razorback sucker (Banks 1964). These surveys concluded that subsequent reductions in native fish populations occurred primarily as a result of reservoir flooding of habitat, and changes in river flows and water temperatures from dam operations.

By the late 1970s, few bonytail were reported from the upper basin (Figure 4), including two from the Green River below Jensen, Utah (Joseph et al. 1977). Bonytail were seen in Lake Powell soon after Glen Canyon Dam was closed in 1962 (K. Miller, Utah Division of Wildlife Resources, personal communication), and two (330 and 380 mm TL) were caught by anglers near Wahweap Bay on September 4, 1977 (Gustafeson et al. 1985) and in 1985 (R. Radant, Utah Division of Wildlife Resources, personal communication); the latter fish was identified by Dr. Mark Rosenfeld (University of Utah, personal communication), and a taxidermy mount is on display at the university's natural history museum. One adult bonytail was captured in the lower Yampa River in 1979 (Holden and Crist 1981), and one adult was caught and released at Coal Creek Rapid in the Green River in Gray Canyon in 1981 (Tyus et al. 1982). Kaeding et al. (1986) captured and released one adult bonytail (458 mm TL) in the Colorado River at Black Rocks on July 17, 1984. Two adult bonytail were captured, photographed, and released in Desolation/Gray Canyons in 1985 (Moretti et al. 1989), and four adults and one juvenile were reported from Cataract Canyon in 1985–1988 (Valdez 1990; Valdez and Williams 1993).

Preferred habitat of bonytail is undetermined, but large fins and a streamlined body suggest adaptation to torrential flows (Beckman 1963). Of 11 wild adults captured in the upper basin since 1977, nine were in deep, swift, rocky canyons (i.e., Yampa Canyon, Black Rocks, Cataract Canyon, and Coal Creek Rapid), and two were in Lake Powell. Vanicek (1967) reported that bonytail were generally found with roundtail chub in pools and eddies in the absence of, but adjacent to, strong current and at varying depths over silt and boulder substrates. Natural reproduction of bonytail was last documented in the Green River for the year classes 1959, 1960, and 1961 (Vanicek and Kramer 1969). Ripe spawning adults, 5–7 years of age, were captured from mid-June to early July at a water temperature of 18°C (Vanicek 1967). Average fecundity is about 25,090 eggs per female, and incubation was shortest (99–174 h) and egg survival, hatching success, and larval survival were highest at 20–21°C (Hamman 1985).

Razorback Sucker

Razorback sucker is a robust fish with maximum size of 1 m TL and 5–6 kg (Minckley 1973; Minckley et al. 1991); maximum age is up to 44 years

(McCarthy and Minckley 1987). Adults are slightly compressed laterally with a bony, predorsal keel behind the occiput. The keel is formed by the growth and fusion of interneural bones. Size and shape of these bones are diagnostic characteristics for young of the species (Snyder and Muth 2004). Scales are well developed with 68–87 in the lateral line. Razorback sucker hybridizes with native flannelmouth sucker and bluehead sucker (Hubbs and Miller 1953; Suttkus et al. 1976; Maddux et al. 1987; Douglas and Marsh 1998), as well as with nonnative white sucker *C. commersonii* (McAda and Wydoski 1980; Buth et al. 1987). Razorback sucker was listed as federally endangered in 1991 (56 FR 54957) with critical habitat designated in 1994 (59 FR 13374). A recovery plan was completed in 1998 (U.S. Fish and Wildlife Service 1998a), and amended and supplemented with recovery goals in 2002 (U.S. Fish and Wildlife Service 2002d).

Razorback sucker was historically common to abundant in most warm regions of the Colorado River system during the 19th and 20th centuries (Jordan and Evermann 1896; Minckley et al. 1991). In the upper basin, the species was common in the Green and upper Colorado rivers and in some warm tributaries, including the White, Duchesne, Little Snake, Yampa, and Gunnison rivers (Burdick 1995; Holden 1999), and possibly as far up the San Juan River as the Animas River (Jordan 1891; Minckley et al. 1991; Holden 2000). Razorback sucker declined throughout the 20th century, and the species now exists naturally in only a few locations. Natural reproduction has occurred with little recruitment over the past 40–50 years, and wild populations are composed primarily of old, senescent adults (Bestgen 1990; Bestgen et al. 2002). Reproduction has been documented in the Green River with collection of larvae (Muth et al. 1998; Chart et al. 1999). Small numbers of juveniles and young adults provide evidence of some recruitment attributed to unusually high spring flows during 1983–1986 that provided critical floodplain nurseries (Modde et al. 1996).

Numbers of wild razorback sucker captured in the upper basin have decreased dramatically since 1974. Razorback sucker are found in small numbers in the middle Green River and in lower reaches of the Yampa, Duchesne, White, and San Rafael rivers (Tyus 1987; Figure 5). The middle Green River population was estimated at 1,000 wild adults in 1985 (Lanigan and Tyus 1989) and 524 in 1995 (Modde et al. 1996), and data from 1998–1999 suggest that about 100 wild adults remained at that time (Bestgen et al. 2002). The wild population is considered extirpated from the Gunnison River, where only two fish were reported in 1976 (Burdick and Bonar 1997). Only 52 individuals, all old adults, were captured during a 3-year study (1979–1981) in 465 km of the upper Colorado River from Hite, Utah, to Rifle, Colorado (Valdez et al. 1982b); and only 12 were captured in the Grand Valley, Colorado, during 1984–1989 (Osmundson and Kaeding 1989). Wild razorback sucker in the San Juan River are limited to two fish captured in 1976 in a riverside pond near Bluff, Utah, and one fish captured in the river in 1988, also near Bluff (Ryden 2000). Large numbers were anecdotally reported in a drained pond near Bluff in 1976, but identification was not verified. Wild razorback sucker were not found in a 7-year study of the San Juan River (1991–1997; Holden 1999).

Adult razorback sucker in the upper basin occupy low-velocity pools, runs, and slackwaters in alluvial reaches and less frequently in canyon-bound areas (Tyus 1987; Lanigan and Tyus 1989; Tyus and Karp 1990; Bestgen 1990). Adults summer in deep eddies, slow runs, and backwaters with silt or sand substrate, depths of 0.6–3.4 m, and velocities of 0.3–0.4 m/s (Valdez et al. 1982b; Tyus 1987; Tyus et al. 1987; Osmundson and Kaeding 1989; Tyus and Karp 1990; Osmundson et al. 1995). During winter, adults use depths of 0.6–2.16 m in slow runs, pools, eddies, and slack water (Osmundson and Kaeding 1989; Valdez and Masslich 1989). Razorback sucker is a moderately migratory, potadromous species that moves to and from spawning and overwintering areas (Tyus and Karp 1990; Modde and Wick 1997; Modde and Irving 1998). Greatest movements are associated with spawning in spring and may account for historic reports of large concentrations of adults (Jordan 1891; Hubbs and Miller 1953; Sigler and Miller 1963; McAda and Wydoski 1980). Adults are relatively sedentary outside of

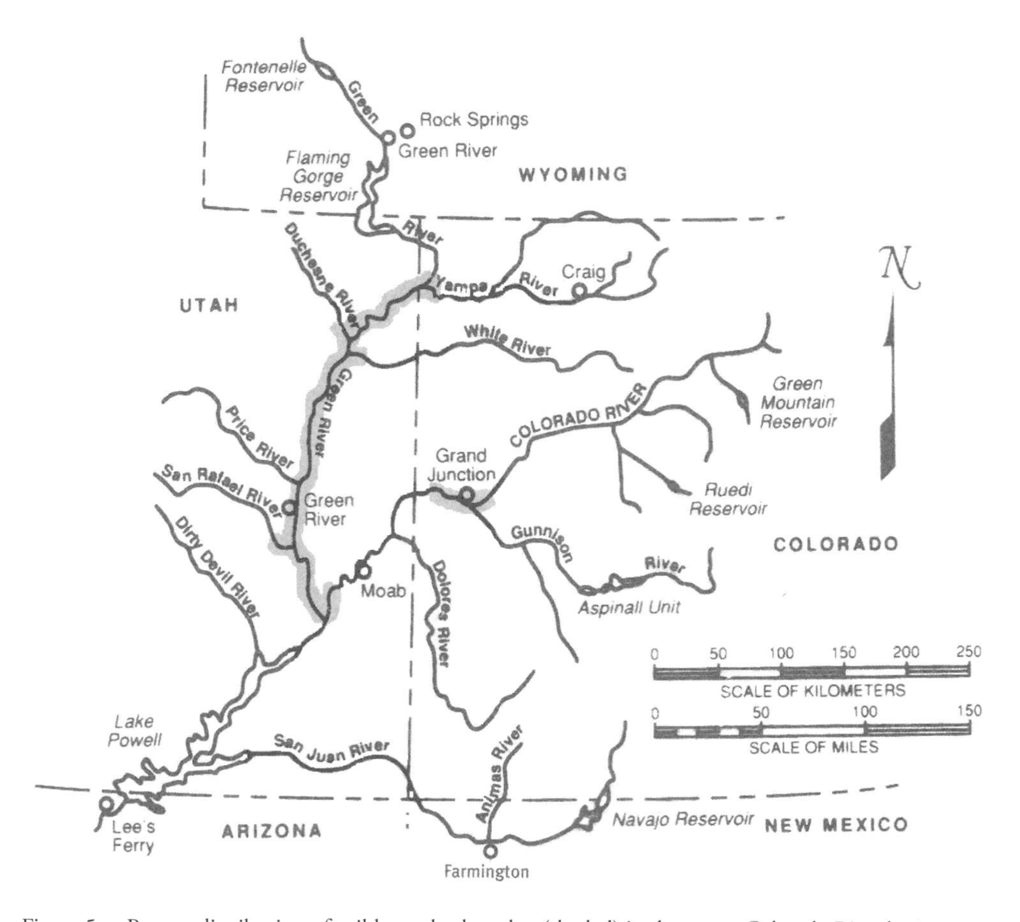

Figure 5.—Present distribution of wild razorback sucker (shaded) in the upper Colorado River basin.

spawning periods (Tyus 1987; Valdez and Masslich 1989; Tyus and Karp 1990). In upper basin riverine environments, spawning is during spring runoff from mid-April to June. Average fecundity is about 46,740 eggs per female and maximum is 103,000 (Inslee 1982; McAda and Wydoski 1980). Adults stage in floodplains, gravel pits, large backwaters, and impounded tributaries near spawning sites (Holden and Crist 1981; Valdez and Wick 1983; Tyus 1987; Osmundson and Kaeding 1989; Tyus and Karp 1990; Modde and Wick 1997; Modde and Irving 1998; Osmundson et al. 1995), and spawn over large mid-channel cobble bars at an average water temperature of about 15°C in velocities less than 1.0 m/s and depths of less than 1.0 m (McAda and Wydoski 1980; Tyus 1987; Tyus and Karp 1990; Bestgen 1990; Wick 1997). Incubation at 18–20°C is 6–7 d (Snyder and Muth 2004), and larvae drift with river currents into food-rich floodplains, where densities of benthos and zooplankton may be 157 times greater than in the main channel (Mabey and Shiozawa 1993). Razorback sucker are omnivorous and consume principally insects, zooplankton, phytoplankton, algae, and detritus (Bestgen 1990). Larvae develop a terminal mouth at 10–11 mm TL and feed on planktonic cladocerans, rotifers, algae and midge larvae that decrease in importance as the mouth migrates to a subterminal position for benthic feeding (Marsh and Langhorst 1988; Muth et al. 1998). Pathogens include bacteria *Erysipelothrix rhysiopathiae,* protozoa (*Myxobolus* sp., can cause blindness; Minckley 1983) and the parasitic copepod *Lernaea cyprinacea* (Flagg 1982).

Roundtail Chub

Roundtail chub have a cylindrical body that is slightly compressed laterally and is silvery-green with a reddish nuptual tint. The back and belly/breast regions are moderately to heavily scaled and a nuchal hump is absent or poorly developed. Maximum size is 500 mm TL; there are 75–85 scales on the lateral line that are small, thin, and slightly imbricated. Roundtail chub does not receive federal protection under the ESA, but is one of three species (also flannelmouth sucker and bluehead sucker) included in a conservation agreement among six western states (Colorado Fish and Wildlife Council 2004).

Roundtail chub was historically distributed in upper reaches of main rivers and throughout most tributaries up to about 2,300 m elevation. In the upper basin, roundtail chub was widely distributed and abundant until the early 1960s, when construction of main-stem dams fragmented and inundated habitats, and altered flow regimes.

Roundtail chub in the upper basin remains as 15 populations in about 55% of its historic habitat (Bezzerides and Bestgen 2002), and is found in the Green, Colorado, and San Juan River drainages (Lentsch et al. 1998; Propst 1999; Bestgen and Crist 2000). It is abundant in the upper Colorado River from Rifle, Colorado, downstream to Moab, Utah, and in the Gunnison and Dolores rivers in Colorado (Valdez 1990; Valdez and Williams 1993). It is rare in Lake Powell and occurs in small numbers in the San Juan River, Utah, and as enclaves in the Animas, La Plata and Mancos rivers. In the Green River, the species occurs in low to moderate numbers. Roundtail chub was once common in the Price, Duchesne, San Rafael, Dirty Devil, and Fremont rivers (McAda et al. 1980), but water depletion, nonnative fish species, and degraded water quality have nearly eliminated it from these tributaries.

Roundtail chub use rocky shorelines and substrate and are rare in sand-bed reaches, and young may use backwaters, if available. Adults occur at depths of up to 20 m, and typically suspend themselves in low-velocity regions of large eddies adjacent to shear zones. Adults are generally sedentary, except for spawning-related movements (Kaeding et al. 1990), although some may make extensive movements at night (Beyers et al. 2001). Maximum reported movement is 80 km in one year (Holden and Crist 1981). Spawning in the upper basin is in May and June, shortly after peak runoff. Adhesive eggs are broadcast over cobble and gravel at 14–22°C (Sigler and Miller 1963; Vanicek and Kramer 1969; Holden 1973; Kaeding et al. 1990; Karp and Tyus 1990b; Brouder et al. 2000), and hatch as 6–7 mm TL larvae in about 5 d (Muth 1990). Roundtail chub typically mature in 3–5 years at 150–300 mm TL, and average life span is 8–10 years (Sigler and Sigler 1996; Brouder et al. 2000). Fecundity of young adults is 1,000–4,300 eggs per female (100–260 mm TL; Neve 1976) and about 14,160–45,120 eggs for fish 4–7 years old (Bestgen 1985). Newly emerged larvae feed on diatoms and filamentous algae (Neve 1976), and juveniles eat mainly immature midges and mayflies (Vanicek and Kramer 1969), but may also consume algae, caddisflies, and ostracods (Bestgen 1985). Juveniles and adults consume aquatic insects, crustaceans, fish, plant matter, snails, ants, beetles, crickets, grasshoppers, and lizards (Koster 1957; McDonald and Dotson 1960; Vanicek and Kramer 1969; Schreiber and Minckley 1981; Tyus and Minckley 1988; Karp and Tyus 1990b).

Flannelmouth Sucker

Flannelmouth sucker is a large, streamlined fish that commonly reaches lengths of 435–500 mm TL with a maximum of 765 mm TL and 3.5 kg in favorable habitat. Flannelmouth sucker mature at age 4–6 and may live 15 or more years (McAda and Wydoski 1980; Douglas and Marsh 1998). The caudal peduncle is thick and robust, and the mouth of adults is subterminal, enlarged, and covered by large fleshy papillae. There are 90–116 small, embedded scales along the lateral line. Flannelmouth sucker hybridizes with sympatric native bluehead sucker and razorback sucker (Hubbs et al. 1942; Hubbs and Miller 1953; Holden 1973; Holden and Stalnaker 1975a; Minckley 1973; McAda and Wydoski 1980; Holden and Crist 1981; Buth et al. 1987), and nonnative white sucker. Flannelmouth sucker does not currently receive federal protection under the ESA, but is one of three species included in a conservation agreement among six western states (Colorado Fish and Wildlife Council 2004).

In the upper basin, flannelmouth sucker was

historically found in the upper Colorado, Green, and San Juan river drainages (Bezzerides and Bestgen 2002). It currently occupies about 50% of historic upper basin habitat where it persists as eight populations in the upper Colorado, Green, and San Juan rivers and their tributaries, including the Escalante, Fremont, San Rafael, Price, Duchesne, White, Yampa, Little Snake, Animas, and La Plata rivers (Bezzerides and Bestgen 2002). It is found in the upper Colorado River from Glenwood Springs, Colorado, to Lake Powell; the Green River from Daniel, Wyoming, to the Colorado River confluence (except Flaming Gorge and Fontenelle reservoirs); and the San Juan River from Bloomfield, New Mexico, to Lake Powell. Temperature range of flannelmouth sucker is 10–27°C (Sublette et al. 1990).

Flannelmouth sucker typically occupy pools, eddies, and deep runs and may congregate to feed at the base of cobble riffles (McAda et al. 1980; Valdez et al. 1982a; Sublette et al. 1990). Larvae and young use low-velocity habitats along shallow shorelines and backwaters, eddies, and side channels over silt substrates (Banks 1964; Minckley 1973; McAda 1977; Snyder and Muth 2004; Childs et al. 1998; Gido and Propst 1999). Juveniles use deep shorelines and shallow gravel/cobble riffles, and adults are common over rocky substrates (Holden and Stalnaker 1975a; McAda et al. 1980). Flannelmouth sucker were found in newly formed impoundments, such as Flaming Gorge Reservoir, Lake Powell, and Kenny Reservoir, but declined dramatically from predation, and reproductive and recruitment failure (Baxter and Stone 1995; Binns 1967; Minckley 1973; McAda 1977; Chart 1987). It is a moderately migratory, potadromous species with long-distance movements usually associated with spawning (Vanicek 1967; Holden 1973; Holden and Crist 1981; Chart and Bergersen 1992; Douglas and Marsh 1998; Beyers et al. 2001). Movements of up to 62 km were reported between the Price River and the Green River in 1997 and 1998 (Cavalli 1999). Fecundity is 4,000–33,000 eggs per female (McAda 1977; McAda and Wydoski 1980), and spawning in the upper basin is in May and June at water temperatures of 16–18.5°C (Holden 1973; Minckley 1973; Snyder and Muth 2004). Adhesive eggs are deposited over sand and gravel bars, they incubate 6–7 d at 15.5–17.8°C, and larvae hatch at 10–11 mm TL (Snyder and Muth 2004). Larvae emerge from cobble substrates and are transported downstream by river currents to low-velocity, sheltered, shoreline nursery habitats (Childs et al. 1998) with greatest drift densities at night and along shorelines (Valdez et al. 1985). Flannelmouth sucker are omnivorous and consume seeds, plant debris, algae, aquatic invertebrates, phytoplankton, and organic detritus (Minckley 1973; Grabowski and Hiebert 1989; Muth and Snyder 1995).

Bluehead Sucker

Bluehead sucker is a small to medium size fish (300–450 mm TL) with a short, broad, bluish head. The sucker mouth is subterminal with strong jaws and cartilaginous scraping ridges, more pronounced on the maxillary. The body is elongate and tapers to a caudal peduncle of varying thickness, generally more robust and deep in fish from tributaries and low-velocity regions than the more slender, streamlined form from mainstream habitats. Bluehead sucker hybridizes with native flannelmouth sucker, razorback sucker, and mountain sucker, and nonnative white sucker (Hubbs et al. 1942; Smith 1966; Holden and Stalnaker 1975a, 1975b; Holden and Crist 1981). The species does not currently receive federal protection under the ESA, but is one of three species included in a conservation agreement among six western states (Colorado Fish and Wildlife Council 2004).

In the upper basin, bluehead sucker was found up to about 2,300 m elevation in the upper Colorado, Green, and San Juan river drainages (Smith 1966; Sublette et al. 1990; Baxter and Stone 1995). It currently occurs as 10 populations in approximately 45% of historic upper basin habitat (Bezzerides and Bestgen 2002). Reservoir inundation and cold releases from dams account for most losses of abundance and distribution. Some populations have been fragmented or isolated by dams and reservoirs, including the upper Green River above Flaming Gorge Reservoir, Gunnison River above the Aspinall Unit, and White River above Taylor Draw Dam (Martinez et al. 1994).

Bluehead sucker feed in cobble/gravel riffles

and rest in pools, eddies, and deep runs (Beyers et al. 2001). Larvae and juveniles use shallow, low-velocity shorelines and backwaters (Haines and Tyus 1990; Robinson et al. 1998). Adults prefer large cool streams of 20°C or less and occupy areas with moderate to fast current and rocky substrates (Banks 1964; Vanicek 1967; Holden and Stalnaker 1975b; McAda et al. 1980; Tyus et al. 1982; Sublette et al. 1990). Adults move little and tend to remain in home ranges (Vanicek 1967; Holden and Crist 1981; Cavalli 1999; Beyers et al. 2001). Spawning in the upper basin occurs in spring and early summer at water temperatures of 15.6–24.6°C (Holden 1973; Sigler and Sigler 1996). Females broadcast adhesive eggs over mid-channel cobble/gravel bars that incubate 7–8 d at 15.6–17.7°C, and emerging larvae 9–11 mm TL are transported downstream by river currents (Valdez et al. 1985). Fecundity is 5,000–20,000 eggs per female (Smith 1966; McAda and Wydoski 1980). Larvae and young feed primarily on diptera larvae, diatoms, and zooplankton (Maddux et al. 1987; Grabowski and Hiebert 1989; Muth and Snyder 1995; Childs et al. 1998). Juveniles and adults use the cartilagenous ridge inside each lip to scrape algae, organic and inorganic debris, and small aquatic insects from rocks and boulders (Simon 1935; Banks 1964; Vanicek 1967; Maddux and Kepner 1988; Muth and Snyder 1995).

Mountain Sucker

Mountain sucker reach a maximum size of 305 mm TL (Sigler and Sigler 1996). The body is elongate with a moderately thickened caudal peduncle, and the mouth is subterminal and sucker-like with distinct lateral notches. Like the bluehead sucker, it has cartilagenous scraping ridges inside each lip. Mountain sucker is a species of special concern in Colorado and has no direct federal protection.

Mountain sucker was once common in high elevation streams of the Colorado River system. In the upper basin, it is found in many tributaries at elevations of 1,220–3,050 m (Tyus et al. 1982). Its popularity as a forage and baitfish expanded its range, and by 1940, mountain sucker was in many reservoirs and tributaries (Sigler and Miller 1963). During 1950–1970, mountain sucker expanded its range to drainages of the upper Green River subbasin, including Ashley Creek, White River, and Price River. Adults inhabit cool, clear tributaries where they are tolerant to occasional turbidity and are common over gravel, rubble, or boulder substrates. Mountain sucker does not occur naturally in lakes but survives in lakes if introduced. Adults in winter and early spring are found adjacent to pools in velocities of 0.5 m/s and depths of 1.5 m (Sigler and Sigler 1987). Fecundity is 2,100–4,000 eggs per female, and spawning is in late spring and early summer when water temperatures are 17–19°C (Hauser 1969). Males mature at ages 2–4 and 127–145 mm TL; and females mature at ages 3–5 and 122–130 mm TL. The diet of young and adults is mostly algae and diatoms scraped from rock surfaces; and diptera, higher plants, animals, and debris (Sigler and Sigler 1996).

Speckled Dace

Speckled dace is a small fish with an elongate body, rounded fins, and a strong, thick caudal peduncle. The mouth is subterminal, often with a small barbel at the end of each maxillary. Adults rarely exceed 75 mm TL. The species has no special state or federal protection. Its status in the upper basin is not well known, but genetic clades unique to specific drainages may be threatened by habitat alterations, stream impoundment and dewatering, degraded water quality, and nonnative fishes (Oakey et al. 2004).

Speckled dace is widespread in the upper basin, except where habitat has been eliminated or degraded. The species has been extirpated locally by predation and competition from nonnative species, although its tolerance to low flows, high temperatures, and low oxygen enables it to persist where other species might perish. It is found in a variety of habitats and tolerates a wide range of water conditions, including cool mountain streams, medium and large rivers, small impoundments, small isolated desert springs, and intermittent streams.

Speckled dace mature at 2 years of age, and fecundity is 174–514 eggs per female (Sigler and Sigler 1987). Spawning in the upper basin occurs in summer with peak activity in June and July at 18.3°C. It often spawns in sequence with two distinct high water events, spring runoff and late summer rain spates. Fertilized eggs are deposited over a large spawning bed and hatch in 6 d at 11–19°C. Larvae remain in the gravel an additional 7–8 d before emerging to congregate in warm shallows. Young fish feed on midwater zooplankton and algae, whereas juveniles and adults are bottom dwellers and feed on benthic insects or plant material (Sigler and Sigler 1996).

Kendall Warm Springs Dace

Kendall Warm Springs dace has a similar shape, size, and appearance to speckled dace (Baxter and Simon 1970; U.S. Fish and Wildlife Service 1982b). Kendall Warm Springs dace was listed as federally endangered in 1970 (35 FR 16047–16048) and received protection under the ESA in 1973. Critical habitat was designated in 1975 (40 FR 21499), and a recovery plan was approved in 1982 (U.S. Fish and Wildlife Service 1982b).

The species is endemic to one small stream, Kendall Warm Springs Creek, Wyoming, which flows into the Green River near Pinedale, Wyoming (U.S. Fish and Wildlife Service 1982b). The stream flows about 300 m before dropping over a travertine embankment into the Green River that naturally provides isolation and protection from predator invasion. The species is associated with numerous seeps and springs and the outflow stream along the north face of a small limestone ridge. Little information is available on the species' life history, and the population is believed to be stable with several thousand individuals. Kendall Warm Springs dace occur in pools and eddies (Binns 1978). Instream plants provide important escape cover, protection from the main current, and nursery areas for larvae (Gryska et al. 1998). Average stream width is 1.8 m, depth is less than 0.31 m, gradient is about 4%, and average streamflow is about 0.2 m^3/s. Kendall Warm Springs dace are absent at the spring source and increase in numbers downstream with increased dissolved oxygen (DO) and decreased carbon dioxide. Kendall Warm Springs dace are nonmigratory and remain within the 300-m length of stream, although individuals usually form small schools that may reflect behavior or space limitations (U.S. Fish and Wildlife Service 1982b). The temperature of the springs is a constant 29.5°C, and spawning may occur year around. Individuals raised in a laboratory matured at 2 years of age, and the number of eggs was generally several hundred per female (Baxter and Simon 1970). Kendall Warm Springs dace are omnivorous but prefer vegetable matter and insects (U.S. Fish and Wildlife Service 1982b).

Mottled Sculpin and Paiute Sculpin

Mottled sculpin and Paiute sculpin are small, stout fish with a large dorso-ventrally compressed head and slender tapering body (Sigler and Sigler 1996). They are usually 100–150 mm TL, their heads are large and proportionate to their bodies, and they have large mouths and eyes. Mottled sculpin lack scales and have small prickles that give the body a rough texture, whereas Paiute sculpin have a smooth body free of prickles. Paiute sculpin have a single, large preopercular spine that is upturned, flattened, and sharp, instead of the two short blunt preopercular spines of mottled sculpin. These species have no special state or federal protection and do not appear to be in decline in the upper basin, although their status is generally unknown.

Mottled sculpin inhabit cool tributaries of the upper basin at elevations of 1,500–3,000 m (Tyus et al. 1982). The species is rare in warm reaches but increasingly abundant in higher elevation streams with rocky substrates. Paiute sculpin inhabit cool streams and deep cold lakes. The species prefers moderate stream gradients, and is not found in headwaters with extreme gradient or in warm areas with low gradient. Males and females of both species mature at 2 years of age. Both species spawn at about 12°C in May and June (Ebert and Summerfelt 1969) in specific habitat along rocky lakeshores and on stream riffles composed of gravel and cobble 20–30 cm diameter. Females typically attach clusters of 100–200 eggs to the underside of stones where they are kept clean and aerated by the male who generally remains until the fry disperse. Eggs incubate 3–4 weeks, and larvae 10 mm TL remain in gravel interstices for about 2 weeks until the yolk sac is absorbed. Nocturnal drift of young occurs in streams and may be followed by a second drift period in response to population density (Sigler and Miller 1963). Mottled sculpin and Paiute sculpin are bottom dwellers and feed mostly at night on snails, amphipods, oligochaetes, insect larvae, and planktonic crustaceans, incidentally consuming detritus, and filamentous green algae (Sigler and Sigler 1996).

Mountain Whitefish

Mountain whitefish are typically 150–450 mm TL and 500–1,300 g (Sigler and Sigler 1996). The species is trout-like in appearance, but has a much smaller head, larger adipose fin, large scales, and no teeth. The snout is pointed and extends past the mouth. Mountain whitefish is a game fish with bag and possession limits in most upper basin states. It is found in cold mountain streams and is common to abundant above 2,100 m elevation. Mountain whitefish inhabits swift streams and cold, deep lakes. Newly hatched fry use shallow water along shorelines, at stream edges, or in protected backwaters, but they move into deeper water as they grow. Mountain whitefish prefers temperatures of 14–16°C, but tolerates temperatures far above and below this range, which gives it greater survival abilities than trout or salmon. It is also able to thrive in water with lower DO than most trout species. Mountain whitefish usually mature in 3–4 years and live up to 17–18 years. Spawning occurs at night between October and December over gravel or rocks in streams or in shallow lake shores at water temperatures of 5–6°C. Fecundity is about 1,500–7,000 eggs per female. Eggs are broadcast over cobble substrate and hatch in early spring after about 5 months of incubation. Mountain whitefish usually eat aquatic insect larvae, small molluscs, eggs, and sometimes fish.

Colorado River Cutthroat Trout

Colorado River cutthroat trout are currently a conservation species in Utah, Colorado, and Wyoming. A conservation agreement and strategy was developed to provide a collaborative strategy for conservation and to allow more flexibility in management (CRCT Task Force 2001). The Colorado River cutthroat trout is classified as a sensitive species by Regions 2 and 4 of the U.S. Forest Service (USFS) and by the Bureau of Land Management (BLM).

Colorado River cutthroat trout historically occupied portions of the Colorado River system in Wyoming, Colorado, Utah, Arizona, and New Mexico (Behnke 1992), including portions of larger streams, such as the Green, Yampa, White, Colorado, and San Juan rivers, but it was probably absent from the lower reaches of many large rivers because of high summer temperatures (Simon 1935; Behnke 1979). Distribution and abundance of Colorado River cutthroat trout have declined, and the species is limited to small populations in less than 1% of its historic range (Binns 1977; Behnke 1979; Martinez 1988; Young 1995). Like other inland forms, this subspecies evolved in the absence of other trouts. It is highly susceptible to hybridization with rainbow trout and competitive replacement by brown trout *Salmo trutta* and brook trout *Salvelinus fontinalis*. Pure Colorado River cutthroat trout remain as small populations in Colorado, Utah, and Wyoming (Behnke 1992). Many adfluvial stocks have been lost (Young 1995), and some populations have been reestablished. The largest pure population of Colorado River cutthroat trout in Trapper's Lake, Colorado, was recently hybridized by rainbow trout (Behnke 2002). Fortunately, a 1931 shipment of pure Trapper's Lake fish was traced to Williamson Lakes, California, and 300 fish were procured for transport to Bench Lake, Colorado, for development of pure populations (Martinez 1988; Pister 1990). Remaining populations of Colorado River cutthroat trout occur mostly in headwater streams and lakes. Most lotic populations are in isolated headwater streams with average daily flow less than 0.85 m^3/s, stream gradients that usually exceed 4%, and elevations above 2,290 m (Young 1995).

Colorado River cutthroat trout hybridizes with other subspecies of cutthroat trout or with rainbow trout in many areas of its historic range, compromising genetic integrity. The Colorado River cutthroat trout conservation team has developed a database that is updated annually to track genetic information for each population. Seven categories have been identified for determining the genetic status of each population and a determination of appropriate management actions. Populations with genetic purity ratings of B, B+, A- or A (i.e., slightly hybridized to essentially pure) are defined as conservation populations for which management actions are implemented, and in 2003 included totals of 1,648 km (1,024 stream miles) and 455 ha (1,124 acres of lakes). These results show that both pure and essentially pure populations are still present in many waters in Utah, Colorado, and Wyoming.

Spawning occurs in spring in streams and tributary inflows, and eggs hatch in 28–40 d at 8–12°C. Colorado River cutthroat trout feed largely on plankton throughout life, and their growth is slower than that of Yellowstone cutthroat trout *O. clarkii bouvieri*. Maximum growth in Trapper's Lake was in midsummer (Drummond 1966). Colorado River cutthroat trout is susceptible to common salmonid diseases, especially whirling disease *Myxobolus cerebralis* (Nehring 1998). Transmission of diseases to wild populations by hatchery stocks is recognized as the most significant threat, and policies and regulations in Wyoming and Utah address fish health status, disease certification of stocked and imported fish, and stocking protocols. Fish testing positive for whirling disease in Utah and Wyoming hatcheries are not stocked, and in Colorado, a policy clearly designates native cutthroat trout waters and other wild trout habitats that are negative for whirling disease as the most protected category.

Threats to Native Fishes

The following threat descriptions apply primarily to endangered and other native fishes in the main rivers and tributaries of the upper basin. Threats to the Kendall Warm Springs dace and Colorado River cutthroat trout are described under *Current Status and Ecology* and *Species Conservation Programs*.

Flow Regulation and Diversion

Flow regulation in the upper basin began in the mid-1800s as small tributary impoundments and irrigation diversions. Ratification of the Colorado River Compact of 1922 by the seven basin states divided the Colorado River into upper and lower basins and allocated 7.5 million acre-feet of water to each basin, based on estimated annual flow of 16.8 million acre-feet for the period 1896–1921 (Fradkin 1981; Reisner 1986). However, average annual flow of the Colorado River, 1922–1976, was only 13.9 million acre-feet, and compact allocation was based on an overestimate of available water. The Upper Basin Compact of 1948 further apportioned water to each of the upper basin states by percentage of available annual volume (i.e., Colorado, 51.75%; Utah, 23%; Wyoming, 14%; New Mexico, 11.25%). All seven basin states faced the problem of constructing costly storage and delivery systems to realize their full compact allocation.

Completion of Boulder (Hoover) Dam in 1935 marked the beginning of dam construction on the Colorado River. Thirteen main-stem dams regulate flow of the Colorado River and hundreds of smaller dams control virtually every stream in the basin. Major dams in the upper basin include Flaming Gorge on the Green River; Blue Mesa, Morrow Point, and Crystal (Aspinall Unit) on the Gunnison River; Dillon, Green Mountain, Shadow Mountain, and Ruedi on tributaries of the upper Colorado River; McPhee on the Dolores River; Taylor Draw on the White River; and Glen Canyon on the main-stem upper Colorado River. The larger dams were completed in the early 1960s under authority of the Colorado River Storage Project Act of 1956. Glen Canyon Dam is the largest in the upper basin and is located 25 km upstream of Lees Ferry, the Compact dividing point between upper and lower basins. Annual discharge from the upper basin at Lees Ferry exceeded 18 million acre-feet in 1929, and lowest discharge was only 4.4 million acre-feet in 1934.

The historic upper basin was characterized by dramatic annual and seasonal flow variation (i.e., exceptionally high flows in spring and early summer, and lower flows in late summer through winter). Flows typically began rising in March with low elevation snowmelt, were highest in late May and early June with snowmelt runoff, receded in late June and July, and were relatively low and stable from August through March, except for flow spikes from periodic storms. Year-to-year flow variation depends on mountain snowpack. Historic average annual flow at Lees Ferry, 1922–1962, was highly variable from about

150 m³/s to nearly 800 m³/s. Closure of Glen Canyon Dam in spring 1963 interrupted river flows, and dam releases no longer reflect natural flows.

Mean daily flow for the Green River at Green River, Utah (1923 and 1997) reflects changes to seasonal flow patterns as a result of human activities in the upper basin (Figure 6). Average peak flows of the Green River in June have decreased by 31% (631 to 433 m³/s) and base flows in January have increased by 81% (48 to 87 m³/s). Average peak flows of the upper Colorado River near Cisco, Utah, in June have decreased by 30% (759 to 530 m³/s) and average base flows in January have increased by 40% (72 to 101 m³/s). These changes in stream hydrology have affected various parts of the river ecosystem on which the native fauna and flora depend.

Physical Habitat Destruction, Alteration, and Fragmentation

Dams, diversions, and local channelizations account for most aquatic habitat destruction or modification in the upper basin. Nine dams impounded major rivers or tributaries of the upper basin starting in 1962 and inundate a total of about 940 km of riverine habitat. These dams, years of completion, and length of inundation are: Glen Canyon Dam (1963, 325 km of the upper Colorado River), Flaming Gorge Dam (1964, 265 km of the Green River), Navajo Dam (1962, 120 km of the San Juan River), Aspinall Unit (i.e., Blue Mesa, Crystal, and Morrow Point dams; 1965, 60 km of the Gunnison River), Taylor Draw Dam (1984, 60 km of the White River), McPhee Dam (1984, 80 km of the Dolores River), and Fontenelle Dam (1964, 30 km of the Green River). These dams have also disrupted the river continuum and converted about 300 km of seasonally warmed river reaches into cold, isothermal tailwaters. Dams have fragmented habitats, blocked passage of migrating fish, reduced high channel reshaping flows, and increased base flows. Local channelizations for highway construction, flood control, and community development have straightened the river channel, further reducing habitat diversity.

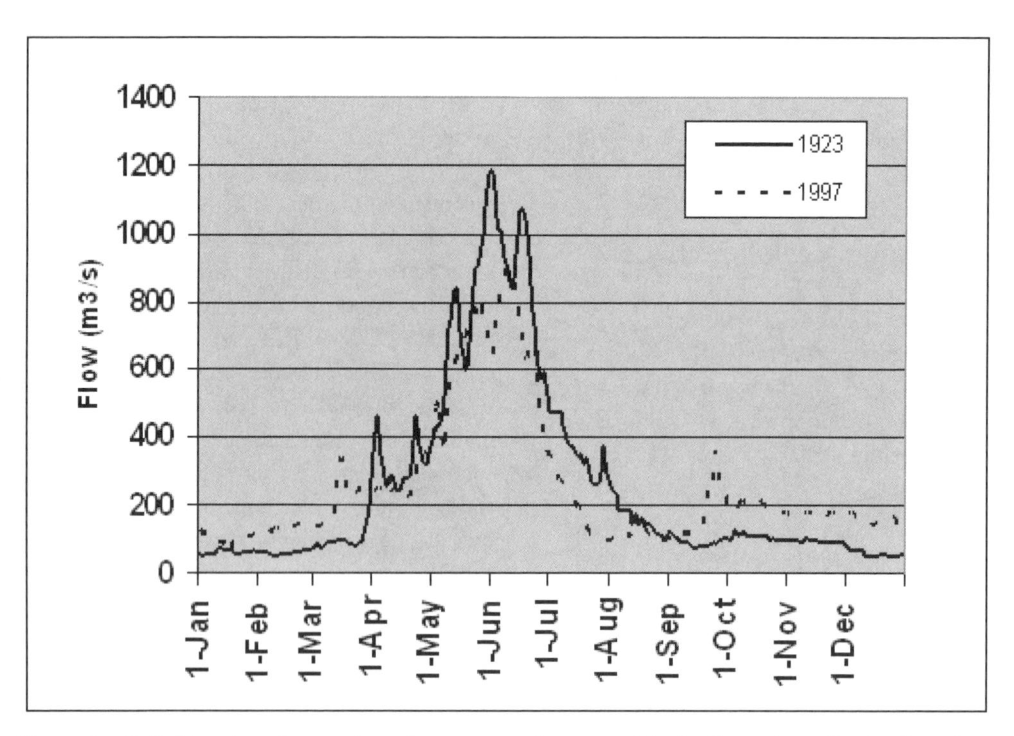

Figure 6.—Daily historic and recent flows for representative years of similar flow volume for the Green River at Green River, Utah. U.S. Geological Survey stream gage data.

Nonnative Fish

Nonnative fishes currently dominate the ichthyofauna of the Colorado River system, and certain species have been implicated in reductions of native fish populations (Carlson and Muth 1989). At least 67 species of nonnative fish have been introduced into the Colorado River system during the last 100 years (Tyus et al. 1982; Carlson and Muth 1989; Minckley 1991; Minckley and Deacon 1991; Lentsch et al. 1996; Pacey and Marsh 1998). About 50 are found in the upper basin (Table 2; e.g., Tyus et al. 1982; Lentsch et al. 1996). Many of these species were intentionally introduced as game or forage fish and others inadvertently gained access with game fish stockings and baitfish releases. Channel catfish were introduced into the upper basin in 1892 (Tyus and Nikirk 1990) and are now widespread and common to abundant with documented predation on native species (Tyus et al. 1982; Hawkins and Nesler 1991; Nelson et al. 1995; Lentsch et al. 1996; Tyus and Saunders 1996; Brooks et al. 2000; Chart and Lentsch 2000). Northern pike escaped from Elkhead Reservoir into the Yampa River in the early 1980s and have expanded into the middle Green River (Wick et al. 1985; Tyus and Karp 1989; Tyus and Beard 1990; Hawkins and Nesler 1991; Nesler 1995). Smallmouth bass also escaped from upstream reservoirs and riverside ponds, and have increased in distribution in the Yampa, Green, and upper Colorado rivers.

Negative effects of nonnative fishes are principally predation, competition, antagonistic behavior, and vectors of parasites and diseases (Karp and Tyus 1990b; Hawkins and Nesler 1991; Tyus 1991; Muth and Nesler 1993; Muth and Snyder 1995; Lentsch et al. 1996; Tyus and Saunders 1996; Bestgen 1997; Bestgen et al. 1997; Holden 1999; McAda and Ryel 1999; Valdez et al. 1999). Red shiner, common carp, fathead minnow, channel catfish, northern pike, green sunfish, black bullhead, and largemouth bass were of greatest concern because of suspected or documented negative interactions with native fish (Osmundson 1987; Hawkins and Nesler 1991; Ruppert et al. 1993; Lentsch et al. 1996). Recent increases in distribution and abundance of smallmouth bass and northern pike have raised concern over the impact of these predators on native fish communities and launched more aggressive management efforts. White sucker have also increased in numbers in the upper basin and incidence of hybridization with native suckers may be increasing.

Degraded Water Quality

Water quality in the upper basin has been substantially altered since the late 1800s. Land-use practices have increased sedimentation; agricultural returns have increased concentrations of pesticides and herbicides and other elements through leaching; water diversion has reduced the dilution capacity of the river; and dams have trapped elements and nutrients in reservoirs and changed element concentrations and water temperatures below outlets. The greatest changes in water quality have occurred in tailwaters below main-stem dams, where there is a measurable reduction in seasonal variability of streamflow and temperature, increase in daily stream fluctuation, reduction in sediment load, and increased nutrient and ionic concentrations (Carlson and Muth 1989; Muth et al. 2000). For example, before Flaming Gorge Dam on the Green River, river temperature ranged from summer highs of 28°C to persistent ice cover in winter (Vanicek et al. 1970). After the dam, hypolimnetic releases have ranged from 4–13°C (Muth et al. 2000).

High spring snowmelt flows and spurious, intense late-summer rainstorms within a sparsely vegetated and arid basin have historically produced high sediment loads and low water clarity. Historically, suspended sediment was highest during three distinct periods. Spring runoff produced a consistent period of moderate sediment from late February through June; summer (July–September) rainstorms produced short, spurious, and sometimes high sediment loads; and midwinter rainstorms or intermittent snowmelt produced minor peaks in suspended sediment. Before Glen Canyon Dam, average sediment load at Lees Ferry (i.e., total upper basin load) was about 140 million tons per year (range, 50–500 million tons); average postdam load is about 15 million tons per year, or a reduction of 89% (Cole and Kubly 1976). Sediments once carried by the Colorado River are now deposited in Lake Powell, and in 1986 ranged in depth

Table 2.—Common and scientific names and relative abundance of nonnative fishes in the upper Colorado River basin.

Common name	Scientific name	Relative abundance
Catostomidae (suckers)		
Utah sucker	*Catostomus ardens* Jordan & Gilbert, 1881	Common in some reservoirs; incidental to rare in some river reaches
longnose sucker	*C. catostomus* (Forster, 1773)	Incidental to rare in some reservoirs; locally common in some river reaches
white sucker	*C. commersonii* (Lacepède, 1803)	Common in some reservoirs; locally common in some river reaches
Centrarchidae (sunfishes)		
green sunfish	*Lepomis cyanellus* Rafinesque, 1819	Locally common
bluegill	*L. macrochirus* Rafinesque, 1819	Common in some reservoirs; incidental to rare in some river reaches
smallmouth bass	*Micropterus dolomieu* Lacepède, 1802	Abundant in some reservoirs; common to abundant in some river reaches
largemouth bass	*M. salmoides* (Lacepède, 1802)	Abundant in some reservoirs; incidental to rare in some river reaches
white crappie	*Pomoxis annularis* Rafinesque, 1818	Incidental
black crappie	*P. nigromaculatus* (Lesueur, 1829)	Abundant in some reservoirs; incidental to rare in some river reaches
Clupeidae (herrings)		
gizzard shad	*Dorosoma cepedianum* (Lesueur, 1818)	Found in Lake Powell
threadfin shad	*D. petenense* (Günther, 1867)	Found in Lake Powell and tributary inflows
Cyprinidae (minnows)		
goldfish	*Carassius auratus* (Linnaeus, 1758)	Incidental
grass carp	*Ctenopharyngodon idella* (Valenciennes, 1844)	Incidental
red shiner	*Cyprinella lutrensis* (Baird & Girard, 1853)	Widespread, common to abundant
common carp	*Cyprinus carpio* Linnaeus, 1758	Widespread, common to abundant
Utah chub	*Gila atraria* (Girard, 1856)	Abundant in Flaming Gorge Reservoir; incidental to rare in some river reaches
brassy minnow	*Hybognathus hankinsoni* Hubbs, 1929	Incidental
plains minnow	*H. placitus* Girard, 1856	Incidental
golden shiner	*Notemigonus crysoleucas* (Mitchill, 1814)	Incidental
sand shiner	*Notropis stramineus* (Cope, 1865)	Common to abundant
fathead minnow	*Pimephales promelas* Rafinesque, 1820	Widespread, common to abundant
bullhead minnow	*P. vigilax* (Baird & Girard, 1853)	Incidental
longnose dace	*Rhinichthys cataractae* (Valenciennes, 1842)	Incidental
redside shiner	*Richardsonius balteatus* (Richardson, 1836)	Rare to common in upper reaches of some rivers

Table 2.—Continued.

Common name	Scientific name	Relative abundance
creek chub	*Semotilus atromaculatus* (Mitchill, 1818)	Incidental to rare
leatherside chub	*Snyderichthys copei* (Jordan & Gilbert, 1881)	Incidental
Cyprinodontidae (pupfishes)		
plains topminnow	*Fundulus sciadicus* Cope, 1865	Incidental
plains killifish	*F. zebrinus* Jordan & Gilbert, 1883	Incidental
rainwater killifish	*Lucania parva* (Baird & Girard, 1855)	Incidental
Esocidae (pikes)		
northern pike	*Esox lucius* Linnaeus, 1758	Common in the Yampa River, rare in the middle Green River
Gadidae (cods)		
burbot	*Lota lota* (Linnaeus, 1758)	Incidental in Flaming Gorge Reservoir and river upstream
Gasterosteidae (sticklebacks)		
brook stickleback	*Culaea inconstans* (Kirkland 1840)	Locally incidental
Ictaluridae (catfishes)		
black bullhead	*Ameiurus melas* (Rafinesque, 1820)	Incidental to locally common
yellow bullhead	*A. natalis* (Lesueur, 1819)	Found in Lake Powell
brown bullhead	*A. nebulosus* (Lesueur, 1819)	Incidental
channel catfish	*Ictalurus punctatus* (Rafinesque, 1818)	Widespread, common to abundant
Moronidae (temperate basses)		
white bass	*Morone chrysops* (Rafinesque, 1820)	Incidental to rare
striped bass	*M. saxatilis* (Walbaum, 1792)	Found in Lake Powell and tributary inflows
Percidae (perches)		
Iowa darter	*Etheostoma exile* (Girard, 1859)	Incidental
johnny darter	*E. nigrum* Rafinesque, 1820	Incidental
yellow perch	*Perca flavescens* (Mitchill, 1814)	Common in some reservoirs; incidental to rare in some river reaches
walleye	*Sander vitreus* (Mitchill, 1818)	Common in some reservoirs; incidental in some river reaches
Poeciliidae (livebearers)		
western mosquitofish	*Gambusia affinis* (Baird & Girard, 1853)	Incidental to locally common
Salmonidae (trout and salmon)		
Yellowstone cutthroat trout	*Oncorhynchus clarkii bouvieri*	Incidental to rare in cool streams
greenback cutthroat trout	*O. c. stomias*	Incidental to rare in cool streams
coho salmon	*O. kisutch* (Walbaum, 1792)	Found in cold reservoirs
rainbow trout	*O. mykiss* (Walbaum, 1792)	Rare to common in cool river reaches
kokanee	*O. nerka* (Walbaum, 1792)	Common in cold reservoirs
brown trout	*Salmo trutta* Linnaeus, 1758	Rare to common in cool river reaches
brook trout	*Salvelinus fontinalis* (Mitchell, 1814)	Rare to common in cool streams
lake trout	*S. namaycush* (Walbaum, 1792)	Found in cold reservoirs

from 11 m near the base of Glen Canyon Dam to 55.5 m near the mouth of Dark Canyon, about 290 km upstream of the dam (Ferrari 1988).

Other water-quality parameters have been variously affected by human activities in the upper basin. High concentrations of radionuclides (i.e., uranium and uranium daughter products) were reported from uranium mill wastes that spilled and killed fish in the San Miguel and Dolores rivers in the 1960s (Sigler et al. 1966), and the Atlas Mills tailings pile on the banks of the Colorado River near Moab, Utah, releases ammonia and other toxins. Heavy metals, such as mercury, lead, zinc, iron, copper, and cadmium were released in high concentrations by extensive mining activities in the San Miguel River and Red Creek, tributaries of the Dolores River in Colorado. High concentrations of polyaromatic hydrocarbons (PAHs) are associated with oil and gas extraction in the San Juan River subbasin (Holden 1999), and high concentrations of selenium are in the San Juan, Green, and upper Colorado rivers in drainages, seeps, and floodplains associated with Mancos Shale formations. Selenium is hypothesized as an inhibitor of reproduction and studies suggest deleterious effects on razorback sucker and possibly Colorado pikeminnow (Hamilton and Wiedmeyer 1990; Stephens et al. 1992; Hamilton and Waddell 1994; Hamilton et al. 1996; Stephens and Waddell 1998; Osmundson et al. 2000).

Species Conservation Programs

Four species conservation programs currently coordinate activities in the upper basin to protect and conserve five federally endangered fishes and one unlisted subspecies. Two of these are recovery programs that encompass much of the upper basin. The Upper Colorado River Endangered Fish Recovery Program (UCRRP) and the San Juan River Basin Recovery Implementation Program (SJRIP) were formed in 1988 and 1992, respectively, under cooperative agreements to resolve water resource issues in concert with conservation of endangered species. The UCRRP is working to recover the Colorado pikeminnow, humpback chub, bonytail, and razorback sucker. The SJRIP focuses on recovery of the Colorado pikeminnow and razorback sucker. Public laws 106-392 and 107-375 provide authorities for capital construction projects and ongoing operations and maintenance funding for both recovery programs. Costs of these programs are shared and total agency contributions for the UCRRP from 1989 through 2005 were about $150 million, and about $27 million for the SJRIP from 1992 through 2005. Kendall Warm Springs dace is managed under an endangered species management program. Conservation agreements have been developed among state and federal agencies for unlisted species. One agreement and strategy includes Colorado River cutthroat trout, and another agreement includes roundtail chub, flannel-mouth sucker, and bluehead sucker, with strategies to be developed by the states.

Upper Colorado River Endangered Fish Recovery Program (UCRRP)

The UCRRP was established under a cooperative agreement in 1988 as a coordinated effort of state and federal agencies, water users, energy distributors, and environmental groups to recover four species of endangered fish in the upper basin while water development proceeds in compliance with federal and state laws (U.S. Department of the Interior 1987; Wydoski and Hamill 1991; Evans 1993). Activities and progress of the UCRRP are intended to serve as the reasonable and prudent alternative to avoid the likelihood of jeopardizing the continued existence of the endangered fishes and destruction or adverse modification of critical habitat. The UCRRP is coordinated by the U.S. Fish and Wildlife Service (USFWS) and functions under the principles of adaptive management. A Recovery Implementation Program Recovery Action Plan (RIPRAP) provides an operational plan to implement the UCRRP, including development of the work plan and future budget needs. The RIPRAP includes the following six program elements.

Habitat management.—Identification, acquisition, and legal protection of instream flows are key elements to secure, protect, and manage habitat for self-sustaining populations of endangered and other native fishes (Tyus 1992; Stanford 1994). The first step in instream-flow protection is identification of flows necessary for species' life histories. Necessary flow regimes seek to mimic natural flow patterns including high spring runoff to reshape the habitat and lower stable flows the remainder of the year. Flow recommendations are developed, implemented, and evaluated through scientific investigations (often include test releases from dams) and adaptive management. Flow recommendations have been developed for most reaches in the upper basin, including the Green River (Muth et al. 2000), Yampa River (Modde and Smith 1995; Modde et al. 1999; Roehm 2004), Duchesne River (Modde and Keleher 2003), White River (Irving et al. 2003), and upper Colorado and Gunnison rivers (Osmundson et al. 1995; McAda 2003).

Water for endangered fishes in the upper basin is being managed through a variety of means, including water leases and contracts, coordinated releases from upstream reservoirs, improvements to irrigation systems, and reoperation of federal facilities (e.g., Flaming Gorge Dam, Aspinall Unit). Legal protection of flows is consistent with state and federal laws related to the Colorado River system (referred to as "Law of the River"), including the ESA, state water laws, and interstate compacts.

In the Green River subbasin, the USFWS entered into a cooperative agreement in January 2005 with the Colorado River Water Conservation District (District) and the states of Colorado and Wyoming to implement the *Management Plan for Endangered Fishes in the Yampa River Basin* (Roehm 2004). The plan will help ensure that current and future water needs are met for people and endangered fishes in the Yampa River basin in northwest Colorado. The UCRRP is funding 5,000 acre-feet (permanent water for endangered fish) of a 12,000 acre-foot enlargement of Elkhead Reservoir in northwest Colorado to make water available to augment late-summer flows in the Yampa River (Modde and Smith 1995; Modde et al. 1999). The District is funding the remaining 7,000 acre-feet, which will help meet future human demands in the Yampa River basin. Construction was initiated in 2005 with completion scheduled for 2007. Local irrigation companies and state and federal agencies formed a work group to implement flow recommendations for the Duchesne River in northeast Utah (Modde and Keleher 2003), and an updated biological opinion was completed in 2005. A final environmental impact statement (EIS) and biological opinion on the operation of Utah's Flaming Gorge Dam on the Green River to meet flow and temperature recommendations for the endangered fishes (Muth et al. 2000) are slated for completion in 2005. Also, the state of Utah has implemented a policy that prioritizes water rights appropriations for endangered fish during certain seasons in the middle Green River (Utah Division of Water Rights 1994).

In the upper Colorado River subbasin, coordinated reservoir operations allow upstream reservoir owners and operators to voluntarily bypass inflows, without affecting project yield, to enhance spring peaks in habitat occupied by endangered fish. The Bureau of Reclamation (BOR) is near completion of the Grand Valley Water Management Project. This project improves water delivery efficiency and plays a major role in managing water resources to meet human and endangered fish needs recommended for the Colorado River (Osmundson et al. 1995; McAda 2003). The Grand Valley Project canal system in western Colorado was retrofitted with internal canal flow control structures and automation, which reduced irrigation diversions by 16% or 45,000 acre-feet in 2002, 12% or 33,000 acre-feet in 2003, and 10% or 29,000 acre-feet in 2004, while meeting all irrigation demands. With completion of the Highline Lake pump station in 2005 and full automation of the seven canal checks, additional water will be saved each year. The state of Colorado has secured two instream flow rights for endangered fish, and continues to evaluate future filing options. An EIS is being developed for releases from the Aspinall Unit to meet flow recommendations for the Gunnison River (McAda 2003). Flow management alone is not sufficient to ensure self-sustaining populations of the endangered fishes and combined flow and non-flow actions are necessary.

Habitat development.—Human activities in the upper basin since the mid-1850s have modified, destroyed, or fragmented historic riverine fish habitat. Dam construction and reservoir inundation account for the majority of habitat loss. Strategies to improve fish habitat include providing fish passage to restore access to historic habitat and complete life histories, screening canals and water outtakes to minimize loss of fish from the main stem, acquiring and restoring floodplain habitats, and remediating contaminants. Historic river habitat is being made accessible by building fish passages around dams and diversions, enabling endangered and other native fish to migrate up and downstream. A 107-m selective fish passage was built at the Redlands Diversion Dam on the lower Gunnison River in 1996, giving endangered and native fishes access to 92 km of historic habitat. The fish passage is operated annually by the USFWS, and as of 2004, 67 Colorado pikeminnow, 9 razorback sucker, 1 bonytail, and more than 62,000 other native fish had passed through the facility, and thousands of nonnative fish had been selectively removed.

Fish passage is also being reinstated for the first time in nearly a century to 90 km of historic habitat in the upper Colorado River above Grand Junction, Colorado, with modification of three diversion dams. The Grand Valley Irrigation Company (GVIC) Diversion Dam was modified for fish passage in January 1998. In 2005, a 4-m wide notch was cut in the concrete crest of the Grand Valley Project Diversion Dam to facilitate construction of a 113-m long selective fish passage. The UCRRP funded the $4.5 million construction project, which is a cooperative effort of the Grand Valley Water Users Association, UCRRP, BOR, and USFWS. Construction of fish passage at the intervening Price-Stubb Diversion Dam is scheduled for the near future.

Canals and water outtakes are also being screened to minimize entrainment and loss of fish from the river. A fish screen was installed at the head of the GVIC canal in 2002 to prevent fish entrainment and similar screens are being constructed in 2005 at the Grand Valley Project and Redlands canals. Design of a fish screen for the Tusher Wash diversion canal on the Green River in eastern Utah is completed and construction is scheduled for the near future.

Floodplains are being made accessible to all life stages of endangered fish by breaching or removing natural or man-made levees on property leased by, or under agreement with, the UCRRP. From 1992 through 2002, the UCRRP inventoried floodplains in the upper basin (Valdez and Nelson 2004, 2005) and acquired 13 private property sites totaling 440 ha (Nelson and Soker 2002). An additional easement on 184 ha (134 ha of floodplain) was acquired in 2003 for Thunder Ranch on the Green River, the first major floodplain downstream of the known spawning bar of razorback sucker. Four floodplain sites on the upper Colorado River have been restored, perpetual easements have been acquired on four other properties (32 ha), and two properties have been acquired in fee (69 ha). Perpetual easements have also been acquired on three properties (80 ha) on the Gunnison River.

Floodplains acquired by the UCRRP have been evaluated to ensure suitable water quality for the endangered fish. High levels of selenium have been found in some floodplains and remediation has been implemented to reduce these levels. A joint effort of the UDWR, BOR, and USFWS at the Stewart Lake Waterfowl Management Area near Jensen, Utah, involves inlet and outlet channels with water control gates, drainage tiles, and water management, all designed to reduce concentrations of selenium and detrimental effects on fish and wildlife. Studies are ongoing to evaluate fish use, growth, and survival in these floodplains.

Nonnative species and sportfishing.—Negative interactions with certain warmwater nonnative fish species have contributed to declines in endangered and other native fish populations. For several years, the UCRRP has worked cooperatively with state and federal partners to identify management actions to minimize the threat of nonnative fish to survival of endangered fish. In spring 2004, UCRRP partners adopted a policy to identify and implement nonnative fish management actions needed to recover the endangered fishes. The policy was a landmark event demonstrating that these diverse organizations recognize that management of nonnative fish is essential to achieve and sustain recovery of the endangered fishes. The policy also recognizes the dual responsibilities of state and federal fish and wildlife agencies to conserve listed and other native fish species while providing recreational sportfishing opportunities.

The UCRRP has implemented several actions to reduce threats from nonnative fishes, including mechanical removal, screening off-river impoundments to prevent escapement of fish to the river, chemical removal of nonnative fish in small off-river impoundments, implementation of nonnative fish stocking procedures, and changes in state bag and possession limits. Scientific evidence demonstrates that northern pike, smallmouth bass, and channel catfish are nonnative fish species that pose significant threats to survival of endangered fish because they prey upon them and compete for food and space. In 2004, the UCRRP revised its nonnative fish management program using what was learned in 2002 and 2003. Biologists from the states of Colorado and Utah, USFWS, and Colorado State University conducted work in 772 km of the Colorado, Green, and Yampa rivers in Colorado and Utah to reduce the abundance of northern pike and smallmouth bass. Efforts to manage channel catfish continued in Yampa Canyon, where effective removal has been demonstrated, but were postponed in other river reaches until methods to improve sampling efficiency are developed. Management of northern pike in the Yampa and Green rivers showed signs of success during 1999–2002. Biologists reported a 60–68% within-year decrease in abundance of northern pike in the targeted river sections, and have implemented studies to determine if these reductions will persist, or if northern pike populations will rebound as fish are replaced through natural production or movement into the targeted river sections from upstream areas. Where feasible, nonnative fish are relocated to area ponds to provide sportfishing opportunities.

Efforts to manage smallmouth bass have had mixed results. Depending on the section of river and methods being employed, within-year reductions in numbers of smallmouth bass in 2004 ranged from 8% to 69%. Biologists use different sampling methods to increase capture efficiency and improve overall catch rates. These changes include the use of new sampling gear to collect fish more effectively in shallow-water habitats and during times of low river flows, extending the sampling period into the fall when smallmouth bass are more vulnerable to capture, and expanding management efforts to include smaller smallmouth bass.

The UCRRP funded placement of a barrier net at Highline Lake State Park in western Colorado in 1999. The net was designed to control escapement of nonnative fish into critical habitat in the upper Colorado River and to ensure that sportfishing opportunities continue at this popular reservoir. A nonnative fish control structure was also installed at Bottle Hollow Reservoir near Roosevelt, Utah, to prevent escapement of fish into the middle Green River and allow the Ute Indian Tribe to place sportfish in its Elders Pond. Fish screens will be installed on outlets of the enlarged Elkhead Reservoir to prevent escapement to the Yampa River. Chemical reclamation has been used to reduce sources of nonnative fish to riverine habitats. Altogether, 104 ponds were surveyed and 19 were chemically treated to remove nonnative fish along the upper Colorado and Gunnison rivers in 1998–1999.

Current stocking of nonnative fish species (mostly sportfish) in the upper basin is generally confined to areas where there is little potential conflict with endangered fish. In 1996, federal and state wildlife agencies in Colorado, Utah, and Wyoming finalized an agreement on stocking nonnative sportfish (U.S. Fish and Wildlife Service 1996). This agreement prohibits stocking of nonnative fish within the 100-year floodplain in designated critical habitat, but does not affect trout stocked in dam tailwaters where native fish are uncommon. Another aspect of nonnative fish management is removal of bag and possession limits for nonnative fish in designated critical habitat. The state of Colorado has removed bag and possession limits on all nonnative, warmwater sportfish in critical habitat. Colorado also has closed river reaches to angling where, and when, angling mortality to native fish is determined to be significant.

Endangered fish propagation and stocking.—The endangered fish propagation program in the upper basin has evolved over the past decade based on initial needs for research and later needs for population augmentation. Initially, hatcheries were designed to maintain razorback sucker in refuges and begin development of broodstocks for the upper Colorado River and Green River subbasins. In the mid 1990s, hatcheries produced various sizes of razorback sucker and bonytail for experimental stocking to evaluate size at stocking, cohort survival, and time of stocking. In the late 1990s, the program increased production of large numbers of small razorback sucker, bonytail, and Colorado pikeminnow for individual state stocking plans. These plans were revised and hatcheries were asked to produce larger fish for greater survival in the wild. Smaller fish were marked with coded wire tags and larger fish (greater than 150 mm TL) were marked with PIT tags. Since 2003, the hatcheries have been producing fish to meet the current integrated stocking plan (Nesler et al. 2003) and to maintain broodstocks.

A genetics management plan (Czapla 1999) provides guidance for culture, propagation, and stocking, and annual operations plans identify numbers of fish to be stocked. The UCRRP funds operations of four hatchery facilities in Colorado and Utah that culture and raise endangered fish:

1. The state of Colorado's J.W. Mumma Native Aquatic Species Restoration Facility (Alamosa, Colorado) raises bonytail andColorado pikeminnow.
2. The state of Utah's Wahweap Fish Hatch ery (Big Water, Utah) raises bonytail.
3. The Ouray National Fish Hatchery (Ouray, Utah) raises razorback sucker.
4. The Recovery Program's Grand Valley Endangered Fish Facility (Grand Junction, Colorado) raises razorback sucker.

Guidance for stocking endangered fish is provided by an integrated stocking plan (Nesler et al. 2003) for the upper Colorado River basin designed primarily to expedite reestablishment of razorback sucker and bonytail populations and to reestablish Colorado pikeminnow in presently restricted or inaccessible reaches of historic habitat. This stocking plan integrates the separate state stocking plans and ensures consistency throughout the upper Colorado River basin for stocking endangered fish and evaluating success. Stocking priorities are:

- Razorback sucker are to be reestablished in the upper Colorado River and Green River subbasins (i.e., one population in the Colorado and Gunnison rivers, one in the middle Green River, and a redundant population in the lower Green River). Target size of stocked razorback sucker is 300 mm TL (i.e., age 2+), and stocking is scheduled for fall with 9,930 fish stocked per population annually for 6 years.
- Bonytail populations are to be reestablished in alluvial reaches of the upper Colorado River and Green River subbasins (i.e., one population in the Colorado River, one in the middle Green River, and a redundant population in the lower Green River). Target size of stocked bonytail is 200 mm TL (i.e., age 2+), and stocking is scheduled primarily in fall with 5,330 fish stocked per population annually for 6 years. By accepting the State of Colorado's stocking plan, the UCRRP has deemed that stocking bonytail in the proximity of existing humpback chub populations is an "acceptable" risk regarding the potential of hybridization between the species.
- Colorado pikeminnow are stocked in restricted or inaccessible reaches of historic habitat in the Colorado River above the Grand Valley Water Project Diversion Dam and in the Gunnison River above the Redlands Diversion Dam. Target size of stocked Colorado pikeminnow is 150 mm TL (i.e., age 3+), and stocking is scheduled primarily in fall with 1,125 fish stocked per reach annually for 8 years. This effort will be reevaluated if stocked Colorado pikeminnow are not retained within the stocking reaches.
- Humpback chub is not anticipated to be stocked. However, augmentation of existing small populations may become necessary. Relocation of young from nearby populations or stocking to expand populations of humpback chub into the Yampa, Lodore, Whirlpool, and Split Mountain complex may be desirable in the future to meet recovery needs.

Stocking of hatchery fish is an important element of recovery program activities, and the most suitable stocking strategies and growth and survival of stocked fish continue to be evaluated. Razorback sucker were first stocked in the Gunnison River near Delta, Colorado, in 1995, and 5 stocked fish used the Redlands fish passage in 2001 and one in 2002. Stocking plans were revised to stock fewer but larger fish in 2001, and larval razorback sucker were discovered in the Gunnison River in 2002 and 2003, indicating that stocked fish had successfully reproduced. Ripe stocked adults have been found on the spawning bar in the Green River near Jensen, Utah, and the presence of razorback sucker larvae suggests successful reproduction by these hatchery fish. Bonytail were first reintroduced in the Green River in 2000 and 2001, in Lodore Canyon in 2000, and in the upper Colorado River in 1996. These fish are being recaptured indicating survival, but reproduction has not been confirmed. During September–November, 2003, 16 stocked bonytail were recaptured in large recirculating eddies and talus shores in Cataract Canyon after about 1 year in the wild, providing evidence of survival by stocked fish (Utah Division of Wildlife Resources 2004). Total numbers of PIT-tagged razorback sucker, bonytail, and Colorado pike-minnow stocked in the upper basin from 1995 through 2004 are 89,730; 44,472; and 4,772, respectively.

Research, monitoring, and data management.—Ongoing research, reliable population monitoring, and assimilation and management of data are vital to UCRRP success. Results of research and monitoring are used to measure progress toward achieving recovery criteria for self-sustaining populations. Considerable research has been conducted in the upper basin and an electronic database is maintained for all data and reports. Results of studies are also available in open literature for a better understanding of life history requirements and conservation strategies. The UCRRP has an ongoing monitoring program to assess population status and trends in response to management actions, including flow protection, habitat restoration, nonnative fish management, and stocking of hatchery fish. Monitoring includes annual sampling of age-0 Colorado pikeminnow and regular mark-recapture population estimates for all populations of Colorado pikeminnow and humpback chub. Survival of hatchery-reared razorback sucker and bonytail released into the upper basin is evaluated to identify and implement stocking strategies that yield maximum growth and survival.

Information, education, and public involvement.—An effective public relations program ensures public awareness, understanding, involvement, and support of UCRRP activities. The UCRRP works with local communities to establish interpretive exhibits and participate in public events that offer opportunities to observe and learn about the endangered fishes. It also provides information at major water user conferences in Colorado, Utah, and Wyoming. The UCRRP holds public meetings and produces a wide range of educational materials, including newsletters, fact sheets, interpretive exhibits, and a web site. The UCRRP issues an annual publication, *Swimming Upstream*, that provides updates of recovery activities.

San Juan River Basin Recovery Implementation Program (SJRIP)

The SJRIP was established under a cooperative agreement in 1992 to conserve populations of Colorado pikeminnow and razorback sucker in the San Juan River subbasin (U.S. Department of the Interior 1995). The SJRIP is coordinated by the USFWS. Program elements, actions, and accomplishments, as defined in the Long Range Plan, are intended to assist species recovery and provide reasonable and prudent alternatives that avoid the likelihood of jeopardy and/or destruction or adverse modification of critical habitat. The SJRIP goals are:

- To conserve populations of Colorado pikeminnow and razorback sucker in the San Juan River subbasin consistent with recovery goals established under the ESA.
- To proceed with water development in the San Juan River subbasin in compliance with federal and state laws, interstate compacts, Supreme Court decrees, and federal trust responsibilities to the Southern Ute, Ute Mountain Ute, Jicarilla, and Navajo tribes.

The following are the main program elements of the SJRIP:

Protection of genetic integrity and management and augmentation of populations.—This element involves completing genetics management and augmentation plans, establishing refuges with wild broodstock, and augmenting wild populations of endangered fish species. Most fish for the SJRIP are produced at the Dexter National Fish Hatchery and Technology Center. A genetics management plan (Crist and Ryden 2003) provides guidance for maintaining genetic integrity of hatchery broodstock and fish stocked into the wild, and a razorback sucker augmentation plan (U.S. Fish and

Wildlife Service 1997) provides guidance on best stocking strategies. In 1999, about 1,000 Colorado pikeminnow were captured from age-0 fish stocked in 1996 and 1997 (Archer et al. 2000). To date, about 10,850 subadult and adult razorback sucker have been stocked in the San Juan River. Larval razorback sucker, which have been found in the river for the last 7 years, indicate that previously stocked fish are surviving and spawning at separate locations (Ryden 2000). Since 2002, over 668,000 juvenile Colorado pike-minnow have been stocked in the San Juan River, and about 300,000 are scheduled to be stocked in fall 2005. Survival of stocked fish provides encouraging prospects for establishing populations of razorback sucker and Colorado pikeminnow in the San Juan River.

Protection, management, and augmentation of habitat.—This element involves identifying important reaches of the San Juan River for different life stages of the endangered fishes by mapping current conditions, determining relationships between flow and habitat, and determining flow needs. Flow recommendations for the San Juan River have been developed (Holden 1999) and provide flow criteria for flushing of sediments and channel reshaping, adult and juvenile habitat, and nursery habitat for Colorado pikeminnow and razorback sucker. A final EIS and biological opinion on operation of New Mexico's Navajo Dam and Reservoir to implement the San Juan River flow recommendations for endangered fish are slated for completion in 2005. The proposed preferred alternative in the EIS will fully meet the flow recommendations. The biological opinion will address the issue of "ongoing effects" of reservoir operations.

Another important component of habitat augmentation is providing fish passage around migration barriers (Masslich and Holden 1996). The Cudei Diversion has been removed, and the Hogback Diversion was modified with a rock channel to provide non-selective fish passage. The Public Service Company of New Mexico (PNM) Weir was fitted with a 122-m selective fish passage, and similar structures are being considered for the Arizona Public Service Company (APS) Weir and the Fruitland Diversion. These modifications will allow for range expansion of Colorado pikeminnow, razorback sucker, and other native fishes. In 2004, 5 razorback sucker and 4 Colorado pikeminnow used the fish passage at the PNM Weir. Some of the Colorado pikeminnow that used the ladder in 2003 were collected more than once indicating that there was downstream movement over the PNM Weir by fish released upstream. All razorback sucker collected in the fish passage in 2004 were captured for the first time.

Entrainment of fish in diversion canals is also considered a threat to native fish. In 1996 and 1997, age-0 Colorado pikeminnow stocked near the Hogback Diversion, about 4.8 km below Shiprock, New Mexico, became entrained in the diversion canal (Trammell and Archer 2000). The effect of this entrainment to the overall population is unclear and construction of a fish screen in the canal is planned.

Water quality protection and enhancement.—This element involves monitoring water quality conditions, evaluating historic information, identifying types and sources of contaminants, investigating changes in water chemistry, and pursuing actions to diminish or eliminate water quality problems that limit recovery. A review of water quality and contaminants, and a detailed study of selenium and selected constituents in water, sediment, soil, and biota in irrigation drainages of the San Juan River subbasin were conducted in 1991–1995. Water quality monitoring is conducted at 5-year intervals.

Interactions between native and nonnative fish species.—This element involves determining the distribution and abundance of nonnative fish species, identifying and characterizing habitats used by nonnative fish, discontinuing stocking of nonnative fish species in areas where endangered fish occur, and control of nonnative fishes through removal. Although 19 of 26 species in the San Juan River are nonnative (Ryden 2000), native fish comprised 75% of all fish collected in primary channels from 1991 to 1997. The most abundant nonnative species were channel catfish (13%), common carp (9%), and red shiner (2%; Ryden 2000). Small numbers of walleye and striped bass were collected in 1995–1997 after gaining access from Lake Powell following inundation of a waterfall barrier in 1995.

Efforts to control nonnative fishes have been

underway in the San Juan River since 1998 and are showing signs of success. Some species, such as channel catfish, striped bass, walleye, and common carp are being removed by raft-mounted electrofishing, whereas control of other species, such as red shiner, is being attempted through restoration of natural flow regimes and river habitat. The SJRIP continues to work with the Navajo Nation and the state of New Mexico to translocate channel catfish from the river to area lakes to enhance recreational fishing opportunities. Totals of 12,660 channel catfish and 10,016 common carp were removed during 1995–1997 (Brooks et al. 2000; Ryden 2000), and over 9,000 channel catfish were translocated during 1998–2004. A shift toward smaller channel catfish was noted by 1997 and was attributed to more efficient capture of large fish by electrofishing (Propst and Hobbes 2000). Results indicate that those efforts have successfully reduced river-wide abundance of channel catfish to the lowest level ever observed, changing the size structure of the channel catfish population to one now dominated by juvenile fish, thereby lessening the potential for channel catfish reproduction and predation on large native fish.

Another nonnative fish control strategy implemented in the San Juan River in 1992 was release of high flows from Navajo Dam to reduce numbers of red shiner, channel catfish, and other nonnative fish species. Nonnative fish species are ill-adapted to flooding characteristics of southwestern streams where native species evolved (Meffe and Minckley 1987; Minckley and Meffe 1987). Declines in red shiner, sand shiner, and fathead minnow were reported in the upper basin following high flows of 1983–1985 (McAda and Kaeding 1989; Osmundson and Kaeding 1989; Valdez 1990; Muth and Nesler 1993; Lentsch et al. 1996; McAda and Ryel 1999). Propst and Hobbes (2000) noted reduced red shiner numbers in secondary channels of the San Juan River in years when a summer flood event occurred.

Monitoring and data management.—Monitoring is necessary to evaluate status and trends of endangered fishes as well as other native and nonnative fish species to assure the SJRIP's overall success in achieving recovery. A long-term monitoring plan provides for native fish assemblage monitoring, including larvae, young, and adults; physical-feature monitoring related to key habitat maintenance; and continued evaluation of flow recommendations. The monitoring plan defines baseline monitoring approaches for fish and habitat, especially those related to flow recommendations. The SJRIP has developed an electronic database of all data and reports.

Kendall Warm Springs Dace Biological Management Program

The USFWS and USFS are responsible for management and conservation of the endangered Kendall Warm Springs dace. The USFWS is responsible for species conservation under the ESA, and the USFS is responsible for integrating management, protection, and conservation of federally listed species into the Forest Planning Process (36 CFR 219.19 and 219.20). Management practices are prohibited that may cause detrimental changes in water temperature or composition, water course blockage, or sediment deposits within 30 m of perennial streams, lakes, or other water bodies (36 CFR 219.27(e)). The Kendall Warm Springs dace recovery plan (U.S. Fish and Wildlife Service 1982b) contains the following recovery efforts and recovery objectives:

1. Maintain the existing population and habitat by monitoring population levels and maintaining biological and physical integrity of stream habitat;
2. Determine the taxonomic status of the Kendall Warm Springs dace; and
3. Complete additional research needs.

In the past, Kendall Warm Springs was subject to human activities within the Bridger-Teton National Forest. Cattle grazed and trampled plant life in and around the spring area. Passage was blocked by rock dams built to create small pools for bathing and washing clothes, and soaps and detergents in the water harmed aquatic organisms. A road built across the spring in 1934 includes a 7.5-m section of culverts that may have prevented the Kendall Warm Springs dace from moving upstream and isolated the upper half of the population. The species was used as bait by anglers for many years and "take" was not regulated because of inadequate laws (Baxter and Simon 1970; U.S. Fish and Wildlife Service 1982b). Recently, these activities have been regulated and new provisions implemented to protect the species. The Wyoming Game and Fish Department stopped issuing permits to seine dace for bait in the 1960s. The USFS has identified 158 acres as the Kendall Warm Springs Biological Management Unit. This area was fenced to prevent cattle access and the springs are closed to wading, bathing, and the use of soap or detergents. Vehicle access has also been blocked along the stream (U.S. Fish and Wildlife Service 1982b).

Colorado River Cutthroat Trout Conservation Agreement and Strategy

A conservation agreement (Agreement) was developed for the Colorado River cutthroat trout in 1999 by Colorado, Wyoming, and Utah; the USFS; BLM; and USFWS (CRCT Task Force 2001). This Agreement was developed to expedite implementation of conservation measures for the Colorado River cutthroat trout in Colorado, Utah, and Wyoming as a collaborative and cooperative effort. Funding for the Agreement is provided by a variety of sources, including federal, state, and local. Threats that warrant listing of Colorado River cutthroat trout as a state special status species could lead to listing under the ESA, and will be eliminated or reduced through implementation of this Agreement and a related conservation strategy (Strategy). The goals of this Agreement are to:

1. Rehabilitate Colorado River cutthroat trout throughout its historic range by establishing two self-sustaining meta populations, each consisting of five separate, viable but interconnected subpopulations, in each Geographic Management Unit (GMU) within the historic range. The short-term goal is to establish one metapopulation in each GMU;
2. Maintain areas that currently support abundant Colorado River cutthroat trout and manage other areas for increased abundance;
3. Maintain the genetic diversity of the species; and
4. Increase the distribution of Colorado River cutthroat trout, where ecologically, sociologically, and economically feasible.

Objectives of the Agreement are:

1. Maintain and restore 383 conservation populations in 2,822 km (1,754 stream miles) and 18 populations in 264 ha (652 lake acres) in 15 GMUs within historic range; and
2. Eliminate or reduce threats to Colorado River cutthroat trout and its habitat to the greatest extent possible.

The Agreement is administered by a Coordination Team, that consists of one designated representative from each signatory and may include technical and legal advisors and others as deemed necessary by the signatories. A total of 10 years is anticipated for completion of all actions described in the Strategy. Conservation actions are scheduled and reviewed annually by the signatory agencies based on recommendations from the Coordination Team.

Aquatic biologists have initially selected a total of 126 streams and lakes in Colorado, 52 in Utah, and 223 in Wyoming for protection, restoration, or conservation planning. A total of 26 strategies within the conservation strategy address threats identified under each of the five listing factors from Section 4 of the ESA. Stream habitat protection and enhancement by the USFS and BLM have greatly increased opportunities to recover the Colorado River cutthroat trout in many historic streams. This has led to increased stream surveys and genetic testing to better define genetic purity of existing stocks.

The long-term objectives set in 1998 (i.e., 2,822 stream km of Colorado River cutthroat trout conservation populations; 523 Colorado, 864 Utah, and 1,437 Wyoming) were exceeded in 2003 in Colorado, but not in Wyoming. Utah exceeded objectives for all GMUs except one. The number of Colorado River cutthroat trout conservation populations increased by 49% from 843 stream km (161 waters) and 243 lake ha (12 waters) in 1998 to 1,648 km and 455 ha in 2003. These increases were due primarily to restoration efforts (160 km) and genetic results identifying pure populations.

Conservation populations continue to be found mainly in short headwater stream sections.

Rangewide Conservation Agreement for Roundtail Chub, Flannelmouth Sucker, and Bluehead Sucker

Six basin states signed a conservation agreement in 2004 (Colorado Fish and Wildlife Council 2004) to expedite implementation of conservation measures for roundtail chub, flannelmouth sucker, and bluehead sucker, and to ensure persistence of these species throughout their ranges. Signatories include Arizona, Colorado, Nevada, New Mexico, Utah, and Wyoming. Each state will develop conservation and management strategies for any or all three species that occur naturally within its authority. Each signatory agrees to (a) develop and finalize a conservation and management strategy, (b) establish and/or maintain populations of all three species to ensure persistence, (c) establish and/or maintain viable metapopulations, and (d) identify, significantly reduce, and eliminate threats. It is believed that conservation actions to protect and enhance these species and their habitats will contribute to conservation of other native fish species with similar distributions.

Conservation and Recovery Prospects

Colorado Pikeminnow

According to the 2002 recovery goals (U.S. Fish and Wildlife Service 2002a), recommended criteria for recovery and long-term conservation of bonytail are:

1. Finalization and implementation of site-specific management tasks to minimize or remove threats to attain necessary levels of protection.
2. Maintenance of an upper basin metapopulation with two genetically and demographically viable, self-sustaining populations, one each in the Green River subbasin and upper Colorado River subbasin, as well as the San Juan River subbasin, if target numbers are not met in the upper Colorado River subbasin. Each population is maintained such that:
   - trends in annual adult (age 7+) point estimates do not decline significantly,
   - mean estimated recruitment of age-5 and age-6 naturally produced fish equals or exceeds average annual adult mortality, and
   - each annual point estimate for the Green River subbasin exceeds 2,600 adults.

Habitat of Colorado pikeminnow in the lower basin is too fragmented and modified to allow completion of life history needs, and recovery of this species in the upper basin is believed to provide long-term species viability.

Colorado pikeminnow persist as self-sustaining populations in the Green River and upper Colorado River subbasins, and concerted efforts are underway to restore the species in the San Juan River subbasin. Preliminary numbers of adults in 819 km of the Green River subbasin range from about 2,300 in 2003 to 3,100 in 2001; and from about 450 in 1992 to 780 in 2003 in 282 km of the upper Colorado River subbasin. Numbers of young and juveniles vary within and between years. Populations in the Green River and upper Colorado River subbasins increased following a series of wet years during 1983–1986, and a population viability analysis declared that the species was viable for 200 years (Gilpin 1993). Similar pulses in recruitment were seen following subsequent high water years (e.g., 1993). These population in creases appear linked to high water years, and are attributed to high channel reshaping flows that diversified habitat, cleansed the substrate of sediment, infused large amounts of food into the river, and diminished nonnative fish populations (Osmundson and Burnham 1998). Reoperation of Flaming Gorge Dam at about the same time provided less fluctuating flows in summer that stabilized nursery backwaters and increased survival of young. An apparent recent decline in Colorado pikeminnow remains unexplained but may be attributed to periods of low flows since the late 1980s and drought conditions since 2000. This drought has allowed increases in nonnative predatory fish, and populations of most native fish have declined (Anderson

2004). Population estimates are available for the upper Colorado River subbasin for 1992–1994, 1998–2000, and 2003–2004; and for the Green River subbasin for 2000–2003. This history of estimates is insufficient to determine if recent declines are attributable to normal population cycles or environmental threats.

The life history of Colorado pikeminnow is reasonably understood and environmental stressors that affect populations continue to be investigated. The upper basin recovery programs have implemented aggressive management actions to address threats, but response by endangered species may not be direct or immediate because of the complexity of environmental correlates that affect population dynamics. Application and evaluation of management actions are on parallel courses such that when threats are removed or minimized, populations are expected to increase.

Humpback Chub

According to the 2002 recovery goals (U.S. Fish and Wildlife Service 2002b), recovery and long-term conservation of humpback chub depend on:

1. Finalization and implementation of site-specific management tasks to minimize or remove threats to attain necessary levels of protection.
2. Maintenance of six genetically and demographically viable, self-sustaining populations, including five in the upper Colorado River basin and one in the lower basin. Each population is maintained such that:
   - trends in annual adult (age 4+) point estimates do not decline significantly,
   - mean estimated recruitment of age-3 naturally produced fish equals or exceeds average annual adult mortality, and
   - three genetically and demographically viable, self-sustaining core populations are maintained, such that each annual point estimate exceeds 2,100 adults.

Concurrent estimates are not available for all five populations, but preliminary numbers of adults during 1998–2003 are Black Rocks (478–921); Westwater Canyon (2,201–4,744); Desolation/Gray Canyons (948–2,193); Yampa Canyon (391); and Cataract Canyon (150). Population estimates for humpback chub tend to be less precise than for Colorado pikeminnow because of the logistical difficulty of sampling whitewater canyons inhabited by this species. As with Colorado pikeminnow, numbers of adult humpback chub apparently declined recently concurrent with extended periods of low flow and increases in nonnative predatory fish. However, linkages between year-class strength and river flow are less clearly defined as for Colorado pikeminnow. A primary threat to humpback chub in the upper basin is predation by channel catfish and smallmouth bass in Desolation/Gray and Yampa canyons. Efforts to mechanically remove nonnative fish from these population centers are ongoing and are being evaluated.

Bonytail

Bonytail is the most imperiled fish species of the Colorado River system. Wild populations are biologically extinct in the upper basin and only a few wild fish remain in the lower basin. According to the 2002 recovery goals (U.S. Fish and Wildlife Service 2002c), recovery and long-term conservation of bonytail depend on:

1. Finalization and implementation of site-specific management tasks to minimize or remove threats to attain necessary levels of protection.
2. Establishment and maintenance of four genetically and demographically viable, self-sustaining populations, two in the lower Colorado River basin and two in the upper basin; one each in the Green River subbasin and upper Colorado River subbasin. Each population is maintained such that:
   - trends in annual adult (age 4+) point estimates do not decline significantly,
   - mean estimated recruitment of age-3 naturally produced fish equals or exceeds average annual adult mortality,
   - annual point estimates for each of the four populations exceeds 4,400 adults, and

- a genetic refuge is maintained in the lower basin.

Use of hatchery fish is vital to recovery of bonytail. A broodstock has been developed from a small number of wild fish that is believed to represent the wild genome for establishment of new populations (Minckley et al. 1989), and fish are being successfully cultured in hatcheries (Hamman 1985). Initial releases of large fish into the wild were unsuccessful (Chart and Cranney 1991), but smaller fish released more recently are being recaptured after 1–2 years in the river, indicating good growth and survival in the wild. Specific life history aspects of bonytail are unknown, such as habitat requirements. It is hypothesized that bonytail use inundated floodplains as nursery areas and that restoration of these habitats will benefit most native species.

Razorback Sucker

According to the 2002 recovery goals (U.S. Fish and Wildlife Service 2002d), recommended criteria for recovery and long-term conservation of razorback sucker are:

1. Finalization and implementation of site-specific management tasks to minimize or remove threats to attain necessary levels of protection.
2. Establishment and maintenance of four genetically and demographically viable, self-sustaining populations, two in the lower Colorado River basin and two in the upper basin; one each in the Green River subbasin and upper Colorado River subbasin, as well as the San Juan River subbasin, if target numbers are not met in the upper Colorado River subbasin. Each population is main tained such that:
   - trends in annual adult (age 4+) point estimates do not decline significantly,
   - mean estimated recruitment of age-3 naturally produced fish equals or exceeds average annual adult mortality,
   - annual point estimates for each of the four populations exceed 5,800 adults, and
   - a genetic refuge is maintained in Lake Mohave.

The population of the middle Green River in 1999 was fewer than 100 wild adults, and numbers of wild fish throughout the upper basin are few and scattered. Large numbers of hatchery fish have been released in the upper basin since 1995 to augment wild populations and some of these fish have been recaptured as ripe adults on an established spawning bar in the middle Green River. Increased collection of larvae in the Green River and first collections of larvae in the Gunnison River indicate successful reproduction by the stocked fish.

The principal reason for decline of razorback sucker in the upper basin is believed to be reduced availability of floodplains that are used by all life stages and serve as nurseries for larvae emerging from mid-channel cobble bars during spring runoff. Levees have been breached to allow the river to connect and inundate these floodplains during critical larval stages and some recruitment is evident. Floodplain management plans for the Green River subbasin (Valdez and Nelson 2004) and the upper Colorado River subbasin (Valdez and Nelson 2005) provide guidance and strategies for maximizing available floodplain habitat. Implementation of flow recommendations is necessary to floodplain restoration so that releases from Flaming Gorge Dam can coincide with Yampa River flows to maximize flooding over breached levees. The current strategy allows floodplains to become inundated seasonally for 2–3 flood cycles to allow for entrainment of larvae and 1–2 years of growth that minimizes the threat of predation by large main-stem predators. Tests with hatchery-reared fish stocked in floodplains show good growth and moderate survival (Birchell and Christopherson 2004). The alternative strategy of repatriation by isolating floodplains and removing or poisoning nonnative fish before release of hatchery razorback sucker (Minckley et al. 2003) is not currently being used in the upper basin.

Kendall Warm Springs Dace

The only population of Kendall Warm Springs dace is extremely localized and prospects for recovery and long-term conservation of this species have been greatly improved with establishment of the Kendall Warm Springs Biological Management Unit. The unit is surrounded with fencing that protects the springs, stream, and riparian area from grazing and human activity. Signs describe this unique species and the importance of protecting the area, and there is general public support for this conservation action. The population is self-sustaining and viable, and sampling and analytical techniques continue to be refined for a better understanding of population dynamics (Gryska 1997).

Other Species

Colorado River cutthroat trout.—Considerable progress has been made toward the long-term goals and objectives of the conservation agreement and strategy for Colorado River cutthroat trout in Colorado, Utah, and Wyoming. Known stream populations in 1998 were about 30% of the long-range objectives for stream miles. All three states increased occupied stream miles by at least 29% and the overall increase was 95% between 1998 and 2003. Conservation populations (less than 10% introgression with nonnative trout) increased much more in lakes and reservoirs than anticipated. Eighty-seven percent of conservation populations occupied streams less than 11 km long, and 96% were in streams 16 km or less; some populations occupied streams up to 34 km in length. Seventy-one percent of known conservation populations are core populations with less than 1% introgression. Efforts are underway to establish two metapopulations consisting of five interconnected populations in each GMU, but these have been difficult to establish because of the simple and limited structure of the drainages. Each state has, or is working to, establish a brood population for each GMU, and distribution of Colorado River cutthroat trout has increased by reduced stockings of nonnative trout, removal of nonnative fishes from occupied waters, and use of brood or donor populations to expand and increase numbers within historic range.

The combined efforts of the conservation agreement and strategy signatories have greatly expanded the number of populations and occupied stream miles of Colorado River cutthroat trout since the agreement was formalized in 1999. Many long-range objectives have been met and all signatories continue to work toward achieving remaining objectives and ensuring long-term species conservation. State administrators continue to support and fund these conservation efforts and promote the success of the program. Because Colorado River cutthroat trout is not listed under the ESA, local governments and private landowners continue to support these conservation efforts and have become important partners in this effort.

Roundtail chub, flannelmouth sucker, and bluehead sucker.—Six western states have signed a conservation agreement to expedite and implement conservation measures for these three species throughout their respective ranges as a collaborative and cooperative effort among resource agencies. Conservation strategies are being developed by each state and an assessment of conservation prospects is not possible at this time. Studies to assess status and trends of these three species in the upper basin are few and localized, and these strategies will help to synthesize information about the species and develop and implement appropriate conservation measures.

Discussion

Species conservation takes time and money, especially in highly altered aquatic systems replete with complex institutional and legal constraints and difficult biotic logistics, such as the Colorado River system. Human changes to the system over the last 150 years have led to the endangerment of some native fishes, and it will take substantial effort to restore habitat components necessary for their recovery and long-term conservation. It is a foregone conclusion that species recovery on the scale and complexity of the Colorado River system will require ongoing public involvement and commitment with reliable support and funding. It is also evident that large-scale ecosystem restoration is not achievable for the Colorado River system, given the long history of complex habitat changes and ongoing human demands, and the most prudent approach to long-term species conservation is wise management of available resources through involvement by all parties with vested interests. Species recovery programs in the upper basin have adopted a multi-stakeholder approach in which federal and state agencies work cooperatively and collaboratively with public and private interests. These stakeholders have realized that a balanced approach is necessary to conserve imperiled fish species while providing for human needs. The most effective stakeholder union is represented by federal and state agencies, local governments, and various land, water, electrical power, wildlife, and environmental interests that can substantially benefit species by providing vital habitat elements through management of resources under their respective authorities.

No single restoration or rehabilitation strategy will simultaneously improve the status of every riverine resource and hence, managers are faced with an intractable dilemma that requires wise choices and ongoing management decisions (Schmidt et al. 1998). An important component of recovery programs in the upper basin has been implementation of the general principles of adaptive management (Walters 1986), whereby stakeholders learn by doing and refine decisions and directions according to the outcome of prior management actions. Recovery programs have succeeded in bringing stakeholders together, uniting conservation efforts, and striking necessary balances between species conservation and human needs, and the ongoing success of these programs is testimony to their effectiveness (Poff et al. 2003). This paradigm of natural resource management in balance with human needs is vital in today's society.

Stakeholder involvement is vital even prior to federal listing of species. Conservation agreements and strategies are being developed and implemented for unlisted species to expedite actions that remove or minimize threats. Federal and state agencies, as well as stakeholders have discovered that species can sometimes be more effectively managed and conserved before they are federally listed. Conservation agreements and strategies require continued and long-term stakeholder involvement and commitment, especially by state wildlife agencies that have the vested authority to manage those species within their jurisdictional boundaries. The Colorado River cutthroat trout conservation agreement and strategy is an example of a multi-stakeholder program working to improve the status of a species and preclude the need for federal listing. Of particular importance and key to the success of these conservation agreements will be a demonstrated and ongoing commitment by the states to assure the public that these species will continue to be protected and conserved in the future. The rangewide conservation agreement and individual state strategies for roundtail chub, flannelmouth sucker, and bluehead sucker are expected to have similar success, given the involvement by many of the same stakeholders and individuals responsible for the success of the Colorado River cutthroat trout conservation agreement and strategy.

Recovery programs are not without difficulties and they will continue to receive a great deal of attention and scrutiny from stakeholders, the American public, and the U.S. Congress that currently helps to fund them. Public skepticism and mistrust for these programs has turned to increasing support with a better understanding of achievements in species conservation, and an increasing recognition that public involvement is vital to species conservation. Critics view participation by water user groups and public utilities in these programs as compromising to the principles of species conservation, since these groups sponsor activities that may have contributed to species decline (Brower et al. 2001). However, it is this realization that has prompted various stakeholders to form these recovery programs and work jointly toward species conservation. Given the complex human interests and demands on the Colorado River system, this balanced approach to species conservation and meeting human needs is vital and increases the scale and magnitude of available management options.

References

Anderson, R. 2004. Riverine fish-flow investigations. Job Progress Report, Federal Aid Project F-288-R7. Colorado Division of Wildlife, Denver.

Archer, E., T. A. Crowl, and M. Trammel. 2000. Abundance of age-0 native fish species and nursery habitat quality and availability in the San Juan River New Mexico, Colorado, and Utah. Final Report to San Juan River Basin Recovery Implementation Program. U.S. Fish and Wildlife Service, Albuquerque, New Mexico.

Banks, J. L. 1964. Fish species distribution in Dinosaur National Monument during 1961 and 1962. Master's thesis. Colorado State University, Fort Collins.

Baxter, G. T., and J. R. Simon. 1970. Wyoming fishes. Wyoming Game and Fish Department, Bulletin 4, Cheyenne.

Baxter, G. T., and M. D. Stone. 1995. Fishes of Wyoming. Wyoming Game and Fish Department, Cheyenne.

Beckman, W. C. 1963. Guide to the fishes of Colorado. Colorado State Museum, Boulder.

Behnke, R. J. 1979. Monograph of the native trout of the genus *Salmo* of Western North America. U.S. Forest Service, U.S. Fish and Wildlife Service, U.S. Bureau of Land Management, Washington, D.C.

Behnke, R. J. 1992. Native trout of Western North America. American Fisheries Society Monograph 6, Bethesda, Maryland.

Behnke, R. J. 2002. Trout and salmon of North America. The Free Press, New York.

Bestgen, K. R. 1985. Distribution, biology and status of the roundtail chub, *Gila robusta*, in the Gila River basin, New Mexico. Master's thesis. Colorado State University, Fort Collins.

Bestgen, K. R. 1990. Status review of the razorback sucker, *Xyrauchen texanus.* Final Report of Colorado State University Larval Fish Laboratory to U.S. Bureau of Reclamation, Salt Lake City, Utah.

Bestgen, K. R. 1997. Interacting effects of physical and biological factors on recruitment of age-0 Colorado squawfish. Doctoral dissertation. Colorado State University, Fort Collins.

Bestgen, K. R., D. W. Beyers, G. B. Haines, and J. A. Rice. 1997. Recruitment models for Colorado squawfish: tools for evaluating relative importance of natural and managed processes. Final Report of Colorado State University Larval fish Laboratory to U.S. National Park Service Cooperative Parks Unit and U.S. Geological Survey Midcontinent Ecological Science Center, Fort Collins, Colorado.

Bestgen, K., and L. W. Crist. 2000. Response of the Green River fish community to construction and re-regulation of Flaming Gorge Dam, 1962–1996. Final Report of Colorado State University Larval Fish Laboratory to Upper Colorado River Endangered Fish Recovery Program, Denver.

Bestgen, K. R., G. B. Haines, R. Brunson, T. Chart, M. Trammell, R. T. Muth, G. Birchell, K. Christopherson, and J. M. Bundy. 2002. Status of wild razorback sucker in the Green River basin, Utah and Colorado, determined from basinwide monitoring and other sampling programs. Final Report of Colorado State University Larval Fish Laboratory to Upper Colorado River Endangered Fish Recovery Program, Denver.

Bestgen, K. R., R. T. Muth, and M. A. Trammell. 1998. Downstream transport of Colorado squawfish larvae in the Green River drainage: temporal and spatial variation in abundance and relationships with juvenile recruitment. Final Report of Colorado State University Larval Fish Laboratory to Upper Colorado River Endangered Fish Recovery Program, Denver.

Beyers, D. W., C. Sodergren, J. M. Bundy, and K. R. Bestgen. 2001. Habitat use of bluehead sucker, flannelmouth sucker, and roundtail chub in the Colorado River. Department of Fishery and Wildlife Biology, Larval Fish Laboratory, Contribution 121, Colorado State University, Fort Collins.

Bezzerides, N., and K. Bestgen. 2002. Status review of roundtail chub *Gila robusta*, flannel-mouth sucker *Catostomus latipinnis*, and bluehead sucker *Catostomus discobolus* in the Colorado River basin. Colorado State University Larval Fish Lab, Contribution 118, Fort Collins.

Binns, N. A. 1967. Effects of rotenone on the fauna of the Green River, Wyoming. Wyoming Game and Fish Commission. Fisheries Technical Bulletin 1:1–114.

Binns, N. A. 1977. Present status of indigenous populations of cutthroat trout, *Salmo clarki*, in southwest Wyoming. Wyoming Game and Fish Department, Fisheries Technical Bulletin No. 2, Cheyenne.

Binns, N. A. 1978. Habitat structure of Kendall Warm Springs, with reference to the Endangered Kendall Warm Springs dace (*Rhinichthys osculus thermalis*, Hubbs and Kuhne). Wyoming Game and Fish Department, Fisheries Technical Bulletin No. 3., Cheyenne.

Birchell, G. J. and K. Christopherson. 2004. Survival, growth and recruitment of larval and juvenile razorback suckers (*Xyrauchen texanus*) introduced into floodplain depressions of the Green River, Utah. Final Report to Upper Colorado River Endangered Fish Recovery Program, Denver.

Bookstein, F. L., B. Chernoff, R. L. Elder, J. M. Humpries, Jr., G. R. Smith, and R. E. Strauss. 1985. Morphometrics in evolutionary biology: the geometry of size and shape change, with examples from fishes. The Academy of Natural Sciences of Philadelphia, Special Publication 15, Philadelphia.

Bosley, C. E. 1960. Pre-impoundment study of the Flaming Gorge Reservoir. Wyoming Game and

Fish Commission, Fisheries Technical Report 9:1–81, Cheyenne.

Brooks, J. E., M. J. Buntjer, and J. R. Smith. 2000. Non-native species interactions: management implications to aid in recovery of the Colorado pikeminnow and razorback sucker in the San Juan River. U.S. Fish and Wildlife Service, Albuquerque, New Mexico.

Brouder, M. J., and T. L. Hoffnagle. 1997. Distribution and prevalence of the Asian tapeworm, *Bothriocephalus acheilognathi*, in the Colorado River and tributaries, Grand Canyon, Arizona, including two new host records Journal of Helminthological Society of Washington 64:219–226.

Brouder, M. J., D. D. Rogers, and L. D. Avenetti. 2000. Life history and ecology of the roundtail chub *Gila robusta*, from two streams in the Verde River basin. Arizona Game and Fish Department Research Branch, Technical Bulletin 3, Phoenix.

Brower, A., C. Reedy, and J. Yelin-Kefer. 2001. Consensus versus conservation in the upper basin recovery implementation program. Conservation Biology 15:1001–1007.

Bulkley, R. V., C. R. Berry, R. Pimentel, and T. Black. 1982. Tolerance and preferences of Colorado River endangered fishes to selected habitat parameters. Pages 185–241 *in* Colorado River Fishery Project, final report, Part 3: Contracted studies. U.S. Fish and Wildlife Service and Bureau of Reclamation, Salt Lake City, Utah.

Burdick, B. D. 1995. Ichthyofaunal studies of the Gunnison River, Colorado, 1992–1994. Final Report of U.S. Fish and Wildlife Service, Grand Junction, Colorado, to Upper Colorado River Endangered Fish Recovery Program, Denver.

Burdick, B. D., and R. B. Bonar. 1997. Experimental stocking of adult razorback sucker in the upper Colorado and Gunnison rivers. U.S. Fish and Wildlife Service, Grand Junction, Colorado.

Buth, D. G., R. W. Murphy, and L. Ulmer. 1987. Population differentiation and introgressive hybridization of the flannelmouth sucker and of hatchery and native stocks of the razorback sucker. Transactions of the American Fisheries Society 116:103–110.

Carlson, C. A., and R. T. Muth. 1989. The Colorado River: lifeline of the American Southwest. Canadian Special Publication of Fisheries and Aquatic Sciences 106:220–239.

Cavalli, P. A. 1999. Fish community investigations in the lower Price River, 1996–1997. Final Report of Utah Division of Wildlife to Upper Colorado River Endangered Fish Recovery Program, Denver.

Chamberlain, F. M. 1904. Notes on fishes collected in Arizona, 1904. Unpublished manuscript, U.S. National Museum, Washington, D.C.

Chart, T. E. 1987. The initial effect of impoundment on the fish community of the White River, Colorado. Master's thesis. Colorado State University, Fort Collins.

Chart, T. E., and E. P. Bergersen. 1992. Impact of mainstream impoundment on the distribution and movements of the resident flannelmouth sucker (Catostomidae: *Catostomus latipinnis*) population in the White River, Colorado. Southwestern Naturalist 37:9–15.

Chart, T. E., and J. S. Cranney. 1991. Radiotelemetered monitoring of stocked bonytail chubs (*Gila elegans*) in the Green River, Utah, 1988–1989. Final Report. Utah Division of Wildlife Resources, Salt Lake City.

Chart, T. E., and L. Lentsch. 2000. Reproduction and recruitment of *Gila* spp. and Colorado pikeminnow (*Ptychocheilus lucius*) in the middle Green River; 1992–1996. Final Report of Utah Division of Wildlife Resources to Upper Colorado River Endangered Fish Recovery Program, Denver.

Chart, T. E., D. P. Svenndson, and L. D. Lentsch. 1999. Investigation of potential razorback sucker (*Xyrauchen texanus*) and Colorado pikeminnow (*Ptychocheilus lucius*) spawning in the lower Green River, 1994 and 1995. Final Report of Utah Division of Wildlife Resources to Upper Colorado River Endangered Fish Recovery Program, Denver.

Childs, M. R., R. W. Clarkson, and A. T. Robinson. 1998. Resource use by larval and early juvenile native fishes in the Little Colorado River, Grand Canyon, Arizona. Transactions of the American Fisheries Society 127:620–629.

Clarkson, R. W., A. T. Robinson, and T. L. Hoffnagle. 1997. Asian tapeworm, *Bothriocephalus acheilognathi*, in native fishes from the Little Colorado River, Grand Canyon. Arizona Great Basin Naturalist 57:66–69.

Cole, G. A., and D. M. Kubly. 1976. Limnologic studies on the Colorado River from Lees Ferry to Diamond Creek. Colorado River Research Program, Arizona Game and Fish Department, Technical Report No. 6., Phoenix.

Colorado Fish and Wildlife Council. 2004.

Rangewide conservation agreement for roundtail chub, flannelmouth sucker, and bluehead sucker. Prepared by Utah Department of Natural Resources, Salt Lake City.

Converse, Y. K., L. D. Lentsch, and R. A. Valdez. 1999. Evaluation of size-dependent overwinter growth and mortality of age-0 Colorado pikeminnow. Final Report of Utah Division of Wildlife Resources to Upper Colorado River Endangered Fish Recovery Program, Denver.

Cope, E. D., and H. C. Yarrow. 1875. Reports upon the collections of fishes made in portions of Nevada, Utah, California, Colorado, New Mexico, and Arizona during the years 1871, 1872, 1873, and 1874. Report of Geographical and Geological Explorations West of the 100th Meridian (Wheeler Survey) 5:635–703.

CRCT Task Force. 2001. Conservation agreement and strategy for Colorado River cutthroat trout (*Oncorhynchus clarki pleuriticus*) in the states of Colorado, Utah, and Wyoming. Colorado Division of Wildlife, Fort Collins.

Crist, L. W., and D. W. Ryden. 2003. Genetics management plan for the endangered fishes of the San Juan River. Prepared for the San Juan River Recovery Implementation Program. U.S. Fish and Wildlife Service, Albuquerque, New Mexico.

Czapla, T. E. 1999. Genetics management plan. Upper Colorado River Endangered Fish Recovery Program, Denver.

Douglas, M. E., and P. C. Marsh. 1998. Population and survival estimates of *Catostomus latipinnis* in northern Grand Canyon, with distribution and abundance of hybrids with *Xyrauchen texanus.* Copeia 1998:915–925.

Douglas, M. E., W. L. Minckley, and H. M. Tyus. 1989. Quantitative characters, identification of Colorado River chubs (Cyprinidae: genus *Gila*) and "the art of seeing well." Copeia 1993:334–343.

Douglas, M. E., W. L. Minckley, and H. M. Tyus. 1998. Multivariate discrimination of Colorado Plateau *Gila* spp.: the art of seeing well revisited. Transactions of the American Fisheries Society 127:163–173.

Dowling, T. E., and B. D. DeMarais. 1993. Evolutionary significance of introgressive hybridization in cyprinid fishes. Nature 362:444–446.

Drummond, R. A. 1966. Reproduction and harvest of cutthroat trout at Trappers Lake, Colorado. Colorado Department of Game, Fish and Parks, Fisheries Research Division Special Report 10, Denver.

Ebert, V. W., and R. C. Summerfelt. 1969. Contributions to the life history of the Paiute sculpin, *Cottus beldingii* Eigenmann and Eigenmann, in Lake Tahoe, California. California Fish and Game 55(2):100–120.

Ellis, M. M. 1914. Fishes of Colorado. University of Colorado Studies 11:1–136.

Evans, P. 1993. A "recovery" partnership for the upper Colorado River to meet ESA Section 7 needs. Natural Resources and Environment 71:24–25.

Evermann, B. W., and C. Rutter. 1895. The fishes of the Colorado basin. U.S. Fish Commission Bulletin 14:473–486.

Ferrari, R. L. 1988. 1986 Lake Powell survey. Report No. REC-ERC-86–6, Salt Lake City, Utah.

Flagg, R. 1982. Disease survey of the Colorado River fishes. Pages 177–184 *in* Colorado River Fishery Project, Final Report, Part 3: Contracted Studies. U.S. Fish and Wildlife Service, Salt Lake City, Utah.

Fradkin, P. L. 1981. A river no more: the Colorado River and the West. University of Arizona Press, Tucson.

Gaufin, A. R., G. R. Smith, and P. Dotson. 1960. Aquatic survey of the Green River and tributaries within the Flaming Gorge Reservoir basin, Appendix A. Pages 139–162 *in* A.M. Woodbury, editor. Ecological studies of the flora and fauna of Flaming Gorge Reservoir basin, Utah and Wyoming. University of Utah Anthropological Papers 48.

Gido, K. B., and D. L. Propst. 1999. Habitat use and association of native and nonnative fishes in the San Juan River, New Mexico and Utah. Copeia 1999(2):321–332.

Gilbert, C. H., and N. B. Scofield. 1898. Notes on a collection of fishes from the Colorado basin in Arizona. Proceedings of the U.S. National Museum 20:487–499.

Gilpin, M. 1993. A population viability analysis of the Colorado squawfish in the upper basin. Department of Biology, University of California at San Diego, La Jolla.

Girard, C. 1856. Researches upon the cyprinoid fishes inhabiting the fresh waters of the United States of America, west of the Mississippi Valley, from specimens in the museum of the Smithsonian Institution. Academy of Natural Science of Philadelphia Proceedings 8:165–213.

Grabowski, S. J., and S. D. Hiebert. 1989. Some

aspects of trophic interactions in selected backwaters and the main channel of the Green River, Utah, 1987–1988. Final Report of U.S. Bureau of Reclamation Research and Laboratory Services Division, Denver, Colorado, to U.S. Bureau of Reclamation, Salt Lake City, Utah.

Gryska, A. D., W. A. Hubert, and K. G. Gerow. 1997. Use of power analysis in developing monitoring protocols for the endangered Kendall Warm Springs dace. North American Journal of Fisheries Management 17:1005–1009.

Gryska, A. D., W. A. Hubert, and K. G. Gerow. 1998. Abundance and lengths of Kendall Warm Springs dace captured from different habitats in a specially designed trap. Transactions of the American Fisheries Society 127:309–315.

Gustafeson, A. W., B. L. Bonebrake, S. J. Scott, and J. E. Johnson. 1985. Lake Powell fisheries investigations, five-year completion report and annual performance report. Utah Division of Wildlife Resources, Publication No. 85–12, Salt Lake City.

Haines, G. B., and H. M. Tyus. 1990. Fish associations and environmental variables in age-0 Colorado squawfish habitats, Green River, Utah. Journal of Freshwater Ecology 5:427–435.

Hamilton, S. J., K. J. Buhl, F. A. Bullard, and S. F. McDonald. 1996. Evaluation of toxicity to larval razorback sucker of selenium-laden food organisms from Ouray NWR on the Green River, Utah. Final Report to Upper Colorado River Endangered Fish Recovery Program, Denver.

Hamilton, S. J., and B. Waddell. 1994. Selenium in eggs and milt of razorback sucker (*Xyrauchen texanus*) in the middle Green River, Utah. Archives of Environmental Contamination and Toxicology 27:195–201.

Hamilton, S. J., and R. H. Wiedmeyer. 1990. Bioaccumulation of a mixture of boron, molybdenum, and selenium in Chinook salmon. Transactions of the American Fisheries Society 119:500–510.

Hamman, R. L. 1981. Spawning and culture of Colorado squawfish in raceways. Progressive Fish-Culturist 43:173–177.

Hamman, R. L. 1982. Spawning and culture of humpback chub. Progressive Fish-Culturist 44:213–216.

Hamman, R. L. 1985. Induced spawning of hatchery-reared bonytail. Progressive Fish-Culturist 47:239–241.

Hamman, R. L. 1986. Induced spawning of hatchery-reared Colorado squawfish. Progressive Fish-Culturist 48:72–74.

Hauser, W. J. 1969. Life history of the mountain sucker, *Catostomus platyrhynchus*, in Montana Transactions of the American Fisheries Society 98:209–224.

Hawkins, J. A. 1992. Age and growth of Colorado squawfish from the upper basin, 1978–1990. Master's thesis. Colorado State University, Fort Collins.

Hawkins, J. A., and T. P. Nesler. 1991. Nonnative fishes of the upper basin: an issue paper. Final Report of Colorado State University Larval Fish Laboratory to Upper Colorado River Endangered Fish Recovery Program, Denver.

Hendrickson, D. A. 1993. Progress report on study of the utility of data obtainable from otoliths to management of humpback chub (*Gila cypha*) in the Grand Canyon. Non-Game and Endangered Wildlife Program, Arizona Game and Fish Department, Phoenix.

Holden, P. B. 1968. Systematic studies of the genus *Gila* (Cyprinidae) of the Colorado River basin. Master's thesis. Utah State University, Logan.

Holden, P. B. 1973. Distribution, abundance and life history of the fishes of the upper basin. Doctoral dissertation. Utah State University, Logan.

Holden, P. B. 1977. Habitat requirements of juvenile Colorado River squawfish. Western Energy and Land Use Team, U.S. Fish and Wildlife Service, Fort Collins, Colorado.

Holden, P. B. 1991. Ghosts of the Green River: impacts of Green River poisoning on management of native fishes. Pages 43–54 *in* W.L. Minckley and J.E. Deacon, editors. Battle against extinction: native fish management in the American Southwest. University of Arizona Press, Tucson.

Holden, P. B., editor. 1999. Flow recommendations for the San Juan River. San Juan River basin Recovery Implementation Program, U.S. Fish and Wildlife Service, Albuquerque, New Mexico.

Holden, P. B. 2000. Program evaluation report for the 7-year research period (1991–1997). San Juan River Basin Recovery Implementation Program, U.S. Fish and Wildlife Service, Albuquerque, New Mexico.

Holden, P. B., and L. W. Crist. 1981. Documentation of changes in the macroinvertebrate and fish populations in the Green River due to inlet modifications of Flaming Gorge Dam. Final Report PR-16–5 of Bio/West, Inc., Logan, Utah, to U.S. Fish and Wildlife Service, Salt Lake City, Utah.

Holden, P. B., and C. B. Stalnaker. 1970. Systematic studies of the cyprinid genus *Gila* in the upper basin. Copeia 1970:409–420.

Holden, P. B., and C. B. Stalnaker. 1975a. Distribution and abundance of mainstream fishes of the middle and upper basins, 1967–1973. Transactions of the American Fisheries Society 104:217–231.

Holden, P. B., and C. B. Stalnaker. 1975b. Distribution of fishes in the Dolores and Yampa river systems of the upper Colorado basin. Southwestern Naturalist 19:403–412.

Holden, P. B., and E. J. Wick. 1982. Life history and prospects for recovery of Colorado squawfish. Pages 98–108 *in* W.H. Miller, H.M. Tyus, and C.A. Carlson editors. Fishes of the upper Colorado River system: present and future. Western Division, American Fisheries Society, Bethesda, Maryland.

Hubbs, C. L., L. C. Hubbs, and R. C. Johnson. 1942. Hybridization in nature between species of catostomid fishes. Contributions of the University of Michigan Laboratory of Vertebrate Biology 22:1–76.

Hubbs, C. L., and E. R. Kuhne. 1937. A new fish of the genus *Apocope* from a Wyoming warm spring. Museum of Zoology, Occasional Papers, University of Michigan 343:1–21.

Hubbs, C. L., and R. R. Miller. 1953. Hybridization in nature between the fish genera *Catostomus* and *Xyrauchen*. Papers of the Michigan Academy of Arts, Science and Letters 38:207–233.

Inslee, T. D. 1982. Spawning and hatching of the razorback sucker (*Xyrauchen texanus*). Proceedings of the Annual Conference of the Western Association of Fish and Wildlife Commissioners 62:431–432.

Irving, D., B. Haines, and T. Modde. 2003. White River base flow study, Colorado and Utah, 1995–1996. Final Report of U.S. Fish and Wildlife Service, Vernal, Utah, to Upper Colorado Endangered Fish Recovery Program, Denver.

Irving, D., and T. Modde. 2000. Home-range fidelity and use of historical habitat by adult Colorado squawfish (*Ptychocheilus lucius*) in the White River, Colorado and Utah. Western North American Naturalist 60:16–25.

Jacobi, G. Z., and M. S. Jacobi. 1982. Fish stomach content analysis. Pages 285–324 *in* Colorado River Fishery Project, Final Report, Part 3: Contracted Studies. U.S. Fish and Wildlife Service, Salt Lake City, Utah.

Jordan, D. S. 1891. Report of explorations in Utah and Colorado during the summer of 1889, with an account of fishes found in each of the river basins examined. Bulletin of the U.S. Fish Commission 9:1–40.

Jordan, D. S., and B. W. Evermann. 1896. The fishes of North and Middle America. Bulletin of the U.S. National Museum 47:1–1240.

Joseph, J. W., J. A. Sinning, R. J. Behnke, and P. B. Holden. 1977. An evaluation of the status, life history, and habitat requirements of endangered and threatened fishes of the upper Colorado River system. U.S. Fish and Wildlife Service Report FWS/OBS/77/ 62:1–168.

Kaeding, L. R., B. D. Burdick, P. A. Schrader, and C. W. McAda. 1990. Temporal and spatial relations between the spawning of humpback chub and roundtail chub in the upper Colorado River. Transactions of the American Fisheries Society 119:135–144.

Kaeding, L. R., B. D. Burdick, P. A. Schrader, and W. R. Noonan. 1986. Recent capture of a bonytail (*Gila elegans*) and observations of this nearly extinct cyprinid from the Colorado River. Copeia 4:1021–1023.

Kaeding, L. R., and D. B. Osmundson. 1989. Interaction of slow growth and increased early-life mortality: an hypothesis on the decline of Colorado squawfish in the upstream regions of its historic range. Environmental Biology of Fishes 22:287–298.

Kaeding, L. R., and M. A. Zimmerman. 1983. Life history and ecology of the humpback chub in the Little Colorado and Colorado Rivers of the Grand Canyon. Transactions of the American Fisheries Society 112:577–594.

Karp, C. A., and H. M. Tyus. 1990a. Humpback chub (*Gila cypha*) in the Yampa and Green Rivers, Dinosaur National Monument, with observations on roundtail chub (*G. robusta*) and other sympatric fishes. Great Basin Naturalist 50:257–264.

Karp, C. A., and H. M. Tyus. 1990b. Behavioral interactions between young Colorado squawfish and six fish species. Copeia 1990:25–34.

Kirsch, P. H. 1889. Notes on a collection of fishes obtained in the Gila River at Fort Thomas, Arizona, Lt. W.L. Carpenter, U.S. Army. Proceedings of the U.S. National Museum 11:555–558.

Koster, W. J. 1957. Fishes of New Mexico. University of New Mexico Press, Albuquerque.

Koster, W. J. 1960. *Ptychocheilus lucius* (Cyprinidae)

in the San Juan River, New Mexico. Southwestern Naturalist 5:174–175.

Kubly, D. M. 1990. The endangered humpback chub (*Gila cypha*) in Arizona: a review of past studies and suggestions for future research. Arizona Game and Fish Department, Phoenix.

Lanigan, S. H., and C. R. J. Berry. 1979. Distribution and abundance of endemic fishes in the White River in Utah. Utah Cooperative Fishery Research Unit, Utah State University, Logan.

Lanigan, S. H., and H. M. Tyus. 1989. Population size and status of razorback sucker in the Green River basin, Utah and Colorado. North American Journal of Fisheries Management 9:68–73.

Lentsch, L. D., R. T. Muth, P. D. Thompson, B. G. Hoskins, and T. A. Crowl. 1996. Options for selective control of nonnative fishes in the upper basin. Final Report of Utah Division of Wildlife Resources to Upper Colorado River Endangered Fish Recovery Program, Denver.

Lentsch, L. D., C. A. Toline, T. A. Crowl, and Y. Converse. 1998. Endangered fish interim management objectives for the Upper Basin Recovery and Implementation Program. Final Report of Utah Division of Wildlife Resources to Upper Colorado River Endangered Fish Recovery Program, Denver.

Luchitta, I. 1990. History of the Grand Canyon and of the Colorado River in Arizona. Pages 311–332 *in* S. S. Beus and M. Morales, editors. Grand Canyon geology. Oxford University Press, New York.

Lynch, T. M., S. Bessire, and J. Gray. 1950. Elementary survey of Dolores River, from Utah line to Paradox Valley, Colorado. Colorado Game and Fish Department, Denver.

Mabey, L. W., and D. K. Shiozawa. 1993. Planktonic and benthic microcrustaceans from floodplain and river habitats of the Ouray Refuge on the Green River, Utah. Department of Zoology, Brigham Young University, Provo, Utah.

Maddux, H. R., and W. G. Kepner. 1988. Spawning of bluehead sucker in Kanab Creek, Arizona (Pisces: Catostomidae). Southwestern Naturalist 33:364–365.

Maddux, H. R., D. M. Kubly, J. C. deVos, W. R. Persons, R. Staedicke, and R. L. Wright. 1987. Evaluation of varied flow regimes on aquatic resources of Glen and Grand Canyon. Final Report of Arizona Game and Fish Department to U.S. Bureau of Reclamation, Glen Canyon Environmental Studies, Salt Lake City, Utah.

Marsh, P. C. 1985. Effect of incubation temperature on survival of embryos of native Colorado River fishes. Southwestern Naturalist 30:129–140.

Marsh, P. C., M. E. Douglas, W.L. Minckley, and R. J. Timmons. 1991. Rediscovery of Colorado squawfish, *Ptychocheilus lucius* (Cyprinidae), in Wyoming. Copeia 1991:1091–1092.

Marsh, P. C., and D. R. Langhorst. 1988. Feeding and fate of wild larval razorback sucker. Environmental Biology of Fishes 21:59–67.

Martinez, A. M. 1988. Identification and status of Colorado River cutthroat trout in Colorado. Pages 81–89 *in* R. E. Gresswell, editor. Status and management of interior stocks of cutthroat trout. American Fisheries Society, Symposium 4, Bethesda, Maryland.

Martinez, P. J., T. E. Chart, M. A. Trammel, J. G. Wullschleger, and E. P. Bergersen. 1994. Fish species composition before and after construction of a main-stem reservoir on the White River, Colorado. Environmental Biology of Fishes 40:227–239.

Masslich, W., and P. B. Holden. 1996. Expanding distribution of Colorado squawfish in the San Juan River: a discussion paper. San Juan River Basin Recovery Implementation Program, U.S. Fish and Wildlife Service, Albuquerque, New Mexico.

McAda, C. W. 1977. Aspects of the life history of three catostomids native to the upper basin. Master's thesis. Utah State University, Logan.

McAda, C. W. 2003. Flow recommendations to benefit endangered fishes in the Colorado and Gunnison rivers. Final Report of U.S. Fish and Wildlife Service, Grand Junction, Colorado to Upper Colorado River Endangered Fish Recovery Program, Denver.

McAda, C. W., C. R. Berry, Jr., and C. E. Phillips. 1980. Distribution of fishes in the San Rafael River system of the upper basin. Southwestern Naturalist 25:41–50.

McAda, C. W., and L. R. Kaeding. 1989. Relations between the habitat use of age-0 Colorado squawfish and those of other sympatric fishes in the upper basin. Final Report. U.S. Fish and Wildlife Service, Colorado River Fishery Project, Grand Junction, Colorado.

McAda, C. W., and R. J. Ryel. 1999. Distribution, relative abundance, and environmental correlates for age-0 Colorado pikeminnow and sympatric fishes in the Colorado River. Final Report to Upper Colorado River Endangered Fish Recovery Program, Denver.

McAda, C. W., and R. S. Wydoski. 1980. The razorback sucker, *Xyrauchen texanus*, in the upper basin, 1974–76. U.S. Fish and Wildlife Service, Technical Papers 99, Denver.

McCarthy, M. S., and W. L. Minckley. 1987. Age estimation for razorback sucker (Pisces: Catostomidae) from Lake Mohave, Arizona and Nevada. Journal of the Arizona-Nevada Academy of Science 21:87–97.

McDonald, D. B., and P. A. Dotson. 1960. Pre-impoundment investigation of the Green River and Colorado River developments. *In* Federal aid in fish restoration investigations of specific problems in Utah's fishery. State of Utah, Department of Fish and Game, Federal Aid Project No. F-4-R-6, Departmental Information Bulletin No. 60–3, Salt Lake City.

McElroy, D. M., and Douglas, M. E. 1995. Patterns of morphological variation among endangered populations of *Gila robusta* and *Gila cypha* (Teleostei: Cyprinidae) in the upper basin. Copeia 3:636–649.

McKee, E. D., R. F. Wilson, W. J. Breed, and C. S. Breed. 1967. Evolution of the Colorado River in Arizona. Museum of Northern Arizona, Museum of Northern Arizona Bulletin 44, Flagstaff.

Meffe, G. K., and W. L. Minckley. 1987. Persistence and stability of fish and invertebrate assemblages in a repeatedly disturbed Sonoran Desert stream. American Midland Naturalist 117:117–191.

Miller, R. R. 1946. *Gila cypha*, a remarkable new species of cyprinid fish from the Colorado River in Grand Canyon, Arizona. Journal of the Washington Academy of Sciences 36:409–415.

Miller, R. R. 1955. Fish remains from archaeological sites in the lower Colorado River basin, Arizona. Papers of the Michigan Academy of Science, Arts, and Letters 40:125–136.

Miller, R. R. 1959. Origin and affinities of the freshwater fish fauna of western North America. Pages 187–222 *in* C.L. Hubbs, editor. Zoogeography. American Association for the Advancement of Science, Publication 51 (1958), Washington, D.C.

Miller, R. R. 1961. Man and the changing fish fauna of the American Southwest. Papers of the Academy of Sciences, Arts, and Letters 46:365–404.

Minckley, W. L. 1973. Fishes of Arizona. Arizona Game and Fish Department, Phoenix.

Minckley, W. L. 1983. Status of the razorback sucker, *Xyrauchen texanus* (Abbott), in the lower Colorado River basin. Southwestern Naturalist 28:165–187.

Minckley, W. L. 1991. Native fishes of the Grand Canyon region: an obituary? Pages 124–177 *in* National Research Council Committee editors. Colorado River ecology and dam management. National Academy Press, Washington, D.C.

Minckley, W. L., D. G. Buth, and R. L. Mayden. 1989. Origin of broodstock and allozyme variation in hatchery-reared bonytail, and endangered North American Cyprinid fish. Transactions of the American Fisheries Society 118:131–137.

Minckley, W. L., and J. E. Deacon. 1991. Battle against extinction: native fish management in the American West. The University of Arizona Press, Tucson.

Minckley, W. L., D. A. Hendrickson, and C. E. Bond. 1986. Geography of western North American freshwater fishes: description and relationships to intracontinental tectonism. Pages 519–613 *in* C.H. Hocutt and E.O. Wiley, editors. The zoogeography of North American freshwater fishes. Wiley-Interscience, New York.

Minckley, W. L., P. C. Marsh, J. E. Brooks, J. E. Johnson, and B. L. Jensen. 1991. Management toward recovery of the razorback sucker. Pages 303–357 *in* W. L. Minckley and J. E. Deacon, editors. Battle against extinction: native fish management in the American West. University of Arizona Press, Tucson.

Minckley, W. L., P. C. Marsh, J. E. Deacon, T. E. Dowling, P. W. Hedrick, W. J. Matthews, and G. Mueller. 2003. A conservation plan for native fishes of the lower Colorado River. BioScience 53:219–234.

Minckley, W. L., and G. K. Meffe. 1987. Differential selection by flooding in stream-fish communities of the arid American Southwest. Pages 93–104 *in* W. J. Matthews and D. C. Heines, editors. Community and evolutionary ecology of North American stream fishes. University of Oklahoma Press, Norman.

Modde, T., K. P. Burnham, and E. J. Wick. 1996. Population status of the razorback sucker in the middle Green River. Conservation Biology 10:110–119.

Modde, T., and D. B. Irving. 1998. Use of multiple spawning sites and seasonal movement by razorback sucker in the middle Green River, Utah. North American Journal of Fisheries Management 18:318–326.

Modde, T., and C. Keleher. 2003. Flow recommendations for the Duchesne River with a synopsis of information regarding endangered fishes. Final Report of U.S. Fish and Wildlife Service Vernal, Utah, to Upper Colorado River Endangered Fish Recovery Program, Denver.

Modde, T., W. J. Miller, and R. Anderson. 1999. Determination of habitat availability, habitat use, and flow needs of endangered fishes in the Yampa River between August and October. Final Report of U.S. Fish and Wildlife Service Vernal, Utah, to Upper Colorado River Endangered Fish Recovery Program, Denver.

Modde, T., and G. Smith. 1995. Flow recommendations for endangered fishes in the Yampa River. Final Report of U.S. Fish and Wildlife Service to Upper Colorado River Endangered Fish Recovery Program, Denver.

Modde, T., and E. J. Wick. 1997. Investigations of razorback sucker distribution movements and habitats used during spring in the Green River, Utah. Final Report of U.S. Fish and Wildlife Service, Vernal, Utah, to Upper Colorado River Endangered Fish Recovery Program, Denver.

Moretti, M., S. Cranney, and B. Roberts. 1989. Distribution and abundance of *Gila* spp. in Desolation and Gray canyons of the Green River during 1985 and 1986. Final Report of Utah Division of Wildlife Resources, Salt Lake City, Utah, to Upper Colorado River Endangered Fish Recovery Program, Denver.

Morgan, D. L., editor. 1964. The west of William H. Ashley. The Old West Publishing Company, Denver.

Musker, B. 1981. Results of a fish aging study for the U.S. Fish and Wildlife Service. U.S. Fish and Wildlife Service, Salt Lake City, Utah.

Muth, R. T. 1990. Ontogeny and taxonomy of humpback chub, bonytail, and roundtail chub larvae and early juveniles. Doctoral dissertation. Colorado State University, Fort Collins.

Muth, R. T., L. W. Crist, K. E. LaGory, J. W. Hayse, K. R. Bestgen, T. P. Ryan, J. K. Lyons, R. A. Valdez. 2000. Flow and temperature recommendations for endangered fishes in the Green River downstream of Flaming Gorge Dam. Final Report to Upper Colorado River Endangered Fish Recovery Program, Denver.

Muth, R. T., G. B. Haines, S. M. Meismer, E. J. Wick, T. E. Chart, D. E. Snyder, and J. M. Bundy. 1998. Reproduction and early life history of razorback sucker in the Green River, Utah and Colorado, 1992–1996. Final Report of Colorado State University Larval Fish Laboratory to Upper Colorado River Endangered Fish Recovery Program, Denver.

Muth, R. T., and T. P. Nesler. 1993. Associations among flow and temperature regimes and spawning periods and abundance of young of selected fishes, lower Yampa River, Colorado, 1980–1984. Final Report of Colorado State University Larval Fish Laboratory to Upper Colorado River Endangered Fish Recovery Program, Denver.

Muth, R. T., and D. E. Snyder. 1995. Diets of young Colorado squawfish and other small fish in backwaters of the Green River, Colorado and Utah. Great Basin Naturalist 55:95–104.

Nehring, B. 1998. Stream fisheries investigations. Special regulations evaluations. Colorado Division of Wildlife, Job 2 Final Report, Federal Aid Project F-237, Colorado Division of Wildlife, Denver.

Nelson, P., C. McAda, and D. Wydoski. 1995. The potential for nonnative fishes to occupy and/or benefit from enhanced or restored floodplain habitat and adversely impact the razorback sucker: an issue paper. U.S. Fish and Wildlife Service, Denver.

Nelson, P., and D. Soker. 2002. Habitat restoration program: a synthesis of current information with recommendations for program revisions. Upper Colorado River Endangered Fish Recovery Program, Denver, Colorado.

Nesler, T. P. 1995. Interactions between endangered fish and introduced gamefishes in the Yampa River, Colorado 1987–1991. Colorado Division of Wildlife, Fort Collins.

Nesler, T. P., K. Christopherson, J. M. Hudson, C. W. McAda, F. Pfeifer, and T. E. Czapla. 2003. An integrated stocking plan for razorback sucker, bonytail, and Colorado pikeminnow for the Upper Colorado River Endangered Fish Recovery Program. Upper Colorado River Endangered Fish Recovery Program, Denver.

Neve, L. L. 1976. The life history of the roundtail chub, *Gila robusta grahami*, at Fossil Creek, Arizona. Master's thesis. Northern Arizona University, Flagstaff.

Oakey, D. D., M. F. Douglas, and M. R. Douglas. 2004. Small fish in a large landscape: diversification of *Rhinichthys osculus* (Cyprinidae) in western North America. Copeia 2004:207–221.

Osmundson, B. C., T. W. May, and D. B. Osmundson. 2000. Selenium concentrations in the Colorado pikeminnow (*Ptychocheilus lucius*): relationship with flows in the upper Colorado River. Archives of Environmental Contamination and Toxicology 38:479–485.

Osmundson, D. B. 1987. Growth and survival of Colorado squawfish (*Ptychocheilus lucius*) stocked in riverside ponds, with reference to largemouth bass (*Micropterus salmoides*) predation. Master's thesis. Utah State University, Logan.

Osmundson, D. B. 1999. Longitudinal variation in fish community structure and water temperature in the upper Colorado River. Final Report of U.S. Fish and Wildlife Service, Grand Junction, Colorado, to Upper Colorado River Endangered Fish Recovery Program, Denver.

Osmundson, D. B., and K. P. Burnham. 1998. Status and trends of the endangered Colorado squawfish in the upper Colorado River. Transactions of the American Fisheries Society 127:957–970.

Osmundson, D. B., and L. R. Kaeding. 1989. Studies of Colorado squawfish and razorback sucker use of the "15-mile reach" of the upper Colorado River as part of the conservation measures for the Green Mountain and Ruedi Reservoir water sales. Final Report. U.S. Fish and Wildlife Service Colorado River Fishery Project, Grand Junction, Colorado.

Osmundson, D. B., P. Nelson, K. Fenton, and D. W. Ryden. 1995. Relationships between flow and rare fish habitat in the "15-Mile Reach" of the upper Colorado River. Final Report. U.S. Fish and Wildlife Service, Grand Junction, Colorado.

Osmundson, D. B., R. J. Ryel, and T. E. Mourning. 1997. Growth and survival of Colorado squawfish in the upper Colorado River. Transactions of the American Fisheries Society 126:687–698.

Osmundson, D. B., R. J. Ryel, M.E. Tucker, B.D. Burdick, W.R. Elmblad, and T. E. Chart. 1998. Dispersal patterns of subadult and adult Colorado squawfish in the upper Colorado River. Transactions of the American Fisheries Society 127:943–956.

Pacey, C. A., and P. C. Marsh. 1998. Resource use by native and non-native fishes of the lower Colorado River: literature review, summary, and assessment of relative roles of biotic and abiotic factors in management of an imperiled indigenous icthyofauna. Final Report of Arizona State University, to U.S. Bureau of Reclamation, Boulder City, Nevada.

Pister, P. 1990. Pure Colorado trout saved by California. Outdoor California 51:12–15.

Poff, N. L., J. D. Allan, M. A. Palmer, D. D. Hart, B. D. Richter, A. H. Arthington, K. H. Rogers, J. L. Meyer, and J. A. Stanford. 2003. River flows and water wars: emerging science for environmental decision making. Frontiers in Ecology 1:298–306.

Propst, D. L. 1999. Threatened and endangered fishes of New Mexico. New Mexico Department of Game and Fish. Technical Report No. 1, Santa Fe.

Propst, D. L., and A. L. Hobbes. 2000. Seasonal abundance, distribution, and population size-structure of fishes in San Juan River secondary channels 1991–1997. Conservation Services Division, New Mexico Department of Game and Fish, Santa Fe.

Quartarone, F. 1995. Historical accounts of upper basin endangered fish. Information and Education Committee, Upper Colorado River Endangered Fish Recovery Program, Denver.

Reisner, M. 1986. Cadillac desert. Viking Penquin, Inc., New York.

Rinne, J. N., J. E. Johnson, B. L. Jensen, A. W. Ruger, and R. Sorenson. 1986. The role of hatcheries in the management and recovery of threatened and endangered fishes. Pages 271–285 *in* R.H. Stroud, editor. Fish culture in fisheries management. American Fisheries Society, Bethesda, Maryland.

Robinson, A. T., R. W. Clarkson, and R. E. Forrest. 1998. Dispersal of larval fishes in a regulated river tributary. Transactions of the American Fisheries Society 127:722–786.

Roehm, G. W. 2004. Management plan for endangered fishes in the Yampa River basin and environmental assessment. U.S. Fish and Wildlife Service, Denver.

Ruppert, J. B., R. T. Muth, and T. P. Nesler. 1993. Predation on fish larvae by adult red shiner, Yampa and Green rivers, Colorado. Southwestern Naturalist 38:397–399.

Ryden, D. W. 2000. Monitoring of experimentally stocked razorback sucker in the San Juan River: March 1994–October 1997. U.S. Fish and Wildlife Service, Grand Junction, Colorado.

Schmidt, J. C., R. H. Webb, R. A. Valdez, G. R. Marzolf, and L. E. Stevens. 1998. Science and values in river restoration in the Grand Canyon. BioScience 48:735–747.

Schreiber, D. C., and W. L. Minckley. 1981. Feeding interrelations of native fishes in a Sonoran Desert stream. Great Basin Naturalist 41:409–426.

Seethaler, K. 1978. Life history and ecology of the Colorado squawfish (*Ptychocheilus lucius*) in the upper basin. Master's thesis. Utah State University, Logan.

Sigler, W. F., W. T. Helm, J. W. Angelovic, D. W. Linn, and S. S. Martin. 1966. The effect of uranium mill wastes on stream biota. Utah Agriculture Experimental Station, Bulletin 462. Utah State University, Logan.

Sigler, W. F., and R. R. Miller. 1963. Fishes of Utah. Utah Department of Fish and Game, Salt Lake City.

Sigler, W. F., and J. W. Sigler. 1987. Fishes of the Great Basin. University of Nevada Press, Reno.

Sigler, W. F., and J. W. Sigler. 1996. Fishes of Utah; a natural history. University of Utah Press, Salt Lake City.

Simon, J. R. 1935. A survey of the waters of the Wyoming National Forest. U.S. Bureau of Fisheries, Department of Commerce. Washington, D.C.

Sitgreaves, L. 1853. Report of an expedition down the Zuni and Colorado Rivers. 32nd Congress, 2nd Session, Executive No. 59, Washington, D.C.

Smith, G. R. 1960. Annotated list of fishes of the Flaming Gorge Reservoir basin, 1959. Pages 163–168 *in* A.M. Woodbury, editor. Ecological studies of the flora and fauna of Flaming Gorge Reservoir basin, Utah and Wyoming. University of Utah, Anthropological Paper 48, Salt Lake City.

Smith, G. R. 1966. Distribution and evolution of the North American Catostomid fishes of the Subgenus *Pantosteus*, Genus *Catostomus*. Miscellaneous Publications of the Museum of Zoology, University of Michigan, No. 129. Museum of Zoology, University of Michigan, Ann Arbor.

Smith, G. R., R. R. Miller, and W. D. Sable. 1979. Species relationships among fishes of the genus *Gila* in the upper Colorado drainage. Pages 613–623 *in* R.M. Linn, editor. Proceedings of the First Conference on Scientific Research in the National Parks. Transactions and Proceedings Series No. 5. U.S. Department of the Interior, National Park Service, Washington, D.C.

Snyder, D. E., and R. T. Muth. 2004. Catostomid fish larvae and early juveniles of the upper Colorado River basin—morphological descriptions, comparisons, and computer-interactive key. Contribution 139 of the Larval Fish Laboratory, Colorado State University, Technical Publication No. 42, Colorado Division of Wildlife, Fort Collins.

Stanford, J. A. 1994. Instream flows to assist the recovery of endangered fishes of the upper basin. U.S. Department of Interior. National Biological Survey, Biological Report 24, Washington, D.C.

Stephens, D. W., and B. Waddell. 1998. Selenium sources and effects on biota in the Green River basin of Wyoming, Colorado, and Utah. Pages 183–203 *in* W.J. Frankenberg and R.A. Engberg, editors. Environmental chemistry of selenium. Marcel Dekker, New York.

Stephens, D. W., B. Waddell, L. A. Peltz, and J. B. Miller. 1992. Detailed study of selenium and selectee elements in water, bottom sediment, and biota associated with irrigation drainage in the middle Green River basin, Utah, 1988–90. U.S. Geological Survey, Water-Resources Investigation Report 92–4084, Salt Lake City, Utah.

Sublette, J. E., M. D. Hatch, and M. Sublette. 1990. The fishes of New Mexico. University of New Mexico Press, Albuquerque.

Suttkus, R. D., G. H. Clemmer, C. Jones, and C. Shoop. 1976. Survey of the fishes, mammals and herpetofauna of the Colorado River in Grand Canyon. Grand Canyon National Park, Colorado River Research Series Contribution 34, Grand Canyon, Arizona.

Taba, S. S., J. R. Murphy, and H. H. Frost. 1965. Notes on the fishes of the Colorado River near Moab, Utah. Utah Academy Proceedings 42:280–283.

Trammell, M. A., and E. Archer. 2000. Chapter 4: evaluation of reintroduction of young of year Colorado pikeminnow in the San Juan River 1996–1998. Pages 4–1 to 4–33 *in* E. Archer, T.A. Crowl, and M. Trammell, editors. Age-0 native species abundances and nursery habitat quality and availability in the San Juan River, New Mexico, Colorado, and Utah. Utah Division of Wildlife Resources, Salt Lake City.

Tyus, H. M. 1987. Distribution, reproduction, and habitat use of the razorback sucker in the Green River, Utah, 1979–1986. Transactions of the American Fisheries Society 116:111–116.

Tyus, H. M. 1990. Potamodromy and reproduction of Colorado squawfish in the Green River basin, Colorado and Utah. Transactions of the American Fisheries Society 119:1035–1047.

Tyus, H. M. 1991. Ecology and management of Colorado squawfish. Pages 379–402 *in* W. L. Minckley and J. E. Deacon, editors. Battle against extinction: native fish management in the American west. The University of Arizona Press, Tucson.

Tyus, H. M. 1992. An instream flow philosophy for recovering endangered Colorado River fishes. Rivers 3:27–36.

Tyus, H. M. 1998. Early records of the endangered fish *Gila cypha* Miller from the Yampa River of the Colorado with notes on its decline. Copeia 1998:190–193.

Tyus, H. M., and J. Beard. 1990. *Esox lucius* (Esocidae) and *Stizostedion vitreum* (Percidae) in the Green River basin, Colorado and Utah. Great Basin Naturalist 50:33–39.

Tyus, H. M., B. D. Burdick, R. A. Valdez, C. M. Haynes, T. A. Lytle, and C. R. Berry. 1982. Fishes of the upper basin: distribution, abundance, and status. Pages 12–70 *in* W. H. Miller, H. M. Tyus, and C. A. Carlson, editors. Fishes of the upper Colorado River system: present and future. Western Division, American Fisheries Society, Bethesda, Maryland.

Tyus, H. M., and G. B. Haines. 1991. Distribution, habitat use, and growth of age-0 Colorado squawfish in the Green River basin, Colorado and Utah. Transactions of the American Fisheries Society 120:79–89.

Tyus, H. M., R. L. Jones, and L. A. Trinca. 1987. Green River rare and endangered fish studies, 1982–1985. Final Report. U.S. Fish and Wildlife Service, Vernal, Utah.

Tyus, H. M., and C. A. Karp. 1989. Habitat use and streamflow needs of rare and endangered fishes, Yampa River, Colorado and Utah. U.S. Fish and Wildlife Service Biological Report 89:1–27.

Tyus, H. M., and C. A. Karp. 1990. Spawning and movements of razorback sucker, *Xyrauchen texanus*, in the Green River basin of Colorado and Utah. Southwestern Naturalist 35:427–433.

Tyus, H. M., and C. W. McAda. 1984. Migration, movements, and habitat preferences of Colorado squawfish, *Ptychocheilus lucius*, in the Green, White, and Yampa rivers, Colorado and Utah. Southwestern Naturalist 29:289–299.

Tyus, H. M., and W. L. Minckley. 1988. Migrating Mormon crickets, *Anabrus simplex* (Orthoptera: Tettigoniidae), as food for stream fishes. Great Basin Naturalist 48:25–30.

Tyus, H. M., and N. J. Nikirk. 1990. Abundance, growth, and diet of channel catfish, *Ictalurus punctatus*, in the Green and Yampa rivers, Colorado and Utah. Southwestern Naturalist 35:188–198.

Tyus, H. M., and J. F. Saunders. 1996. Nonnative fishes in the upper basin and a strategic plan for their control. Final Report of University of Colorado Center for Limnology to Upper Colorado River Endangered Fish Recovery Program, Denver.

Ulmer, L., editor. 1983. Endangered Colorado River fishes newsletter. California Department of Fish and Game, Blythe.

U.S. Department of the Interior. 1987. Recovery implementation program for endangered fish species in the upper basin. U.S. Fish and Wildlife Service, Denver.

U.S. Department of the Interior. 1995. San Juan River basin recovery implementation program. U.S. Fish and Wildlife Service, Albuquerque, New Mexico.

U.S. Fish and Wildlife Service. 1978. Colorado squawfish recovery plan. U.S. Fish and Wildlife Service, Denver.

U.S. Fish and Wildlife Service. 1979. Humpback chub recovery plan. U.S. Fish and Wildlife Service, Denver.

U.S. Fish and Wildlife Service. 1982a. Colorado River Fishery Project, Final Report. U.S. Fish and Wildlife Service, Denver.

U.S. Fish and Wildlife Service. 1982b. Kendall Warm Springs dace recovery plan. U.S. Fish and Wildlife Service, Denver.

U.S. Fish and Wildlife Service. 1984. Bonytail chub recovery plan. U.S. Fish and Wildlife Service, Denver.

U.S. Fish and Wildlife Service. 1990a. Humpback chub recovery plan. U.S. Fish and Wildlife Service, Denver.

U.S. Fish and Wildlife Service. 1990b. Bonytail chub recovery plan. U.S. Fish and Wildlife Service, Denver.

U.S. Fish and Wildlife Service. 1991. Colorado squawfish recovery plan. U.S. Fish and Wildlife Service, Denver.

U.S. Fish and Wildlife Service. 1996. Procedures for stocking nonnative fish species in the upper basin. Upper Colorado River Endangered Fish Recovery Program, Denver.

U.S. Fish and Wildlife Service. 1997. Razorback sucker augmentation plan. U.S. Fish and Wildlife Service, Albuquerque, New Mexico.

U.S. Fish and Wildlife Service. 1998a. Razorback sucker recovery plan. U.S. Fish and Wildlife Service, Denver.

U.S. Fish and Wildlife Service. 2002a. Colorado pikeminnow (*Ptychocheilus lucius*) recovery goals: amendment and supplement to the Colorado squawfish recovery plan. U.S. Fish and Wildlife Service, Denver.

U.S. Fish and Wildlife Service. 2002b. Humpback chub (*Gila cypha*) recovery goals: amendment and supplement to the humpback chub recovery plan. U.S. Fish and Wildlife Service, Denver.

U.S. Fish and Wildlife Service. 2002c. Bonytail (*Gila elegans*) recovery goals: amendment and supplement to the bonytail chub recovery plan. U.S. Fish and Wildlife Service, Denver.

U.S. Fish and Wildlife Service. 2002d. Razorback sucker (*Xyrauchen texanus*) recovery goals: amendment and supplement to the razorback sucker recovery plan. U.S. Fish and Wildlife Service, Denver.

Utah Division of Water Rights. 1994. Policy regarding applications to appropriate water and change applications which divert water from the Green River between Flaming Gorge Dam, downstream to the Duchesne River. Policy adopted on November 30, 1994, State Water Engineer, Robert L. Morgan, Salt Lake City.

Utah Division of Wildlife Resources. 2004. Population estimates of humpback chub in Cataract Canyon. Final Report to Upper Colorado River Endangered Fish Recovery Program, Denver.

Valdez, R. A. 1990. The endangered fish of Cataract Canyon. Final Report. U.S. Bureau of Reclamation, Salt Lake City, Utah.

Valdez, R. A., and S. W. Carothers. 1998. The aquatic ecosystem of the Colorado River in Grand Canyon. Final Report of SWCA, Inc. to U.S. Bureau of Reclamation, Salt Lake City, Utah.

Valdez, R. A., J. G. Carter, and R. J. Ryel. 1985. Drift of larval fishes in the upper Colorado River. Proceedings of the Western Association of Fish and Wildlife Agencies 171–185.

Valdez, R. A., and G. C. Clemmer. 1982. Life history and prospects for recovery of the humpback chub and bonytail chub. Pages 109–119 *in* W. H. Miller, H. M. Tyus, and C. A. Carlson, editors. Fishes of the upper Colorado River system: present and future. Western Division, American Fisheries Society, Bethesda, Maryland.

Valdez, R. A., B. R. Cowdell, and L. D. Lentsch. 1999. Overwinter survival of age-0 Colorado pikeminnow in the Green River, Utah, 1987–1995. Final Report to Upper Colorado River Endangered Fish Recovery Program, Denver.

Valdez, R. A., P. B. Holden, and T. B. Hardy. 1990. Habitat suitability index curves for humpback chub of the upper basin. Rivers 1:31–42.

Valdez, R. A., P. Mangan, M. McInerny, R. B. Smith. 1982a. Fishery investigations of the Gunnison and Dolores rivers. Pages 321–365 *in* Colorado River Fishery Project, Final Report, Part 2: Field Investigations. U.S. Fish and Wildlife Service, Salt Lake City, Utah.

Valdez, R. A., P. Mangan, R. Smith, B. Nilson. 1982b. Upper Colorado River investigations (Rifle, Colorado to Lake Powell, Utah). Pages 100–279 *in* Colorado River Fishery Project, Final Report, Part 2: Field Investigations. U.S. Fish and Wildlife Service, Salt Lake City, Utah.

Valdez, R. A., and W. J. Masslich. 1989. Winter habitat study of endangered fish - Green River: winter movement and habitat of adult Colorado squawfish and razorback suckers. U.S. Bureau of Reclamation, Salt Lake City, Utah.

Valdez, R. A., W. J. Masslich, and A. Wasowicz. 1992. Dolores River native fish habitat suitability study. Final Report, Utah Division of Wildlife Resources, Salt Lake City.

Valdez, R. A., and P. Nelson. 2004. Green River subbasin floodplain management plan. Upper Colorado River Endangered Fish Recovery Program, Denver.

Valdez, R. A., and P. Nelson. 2005. Upper Colorado River subbasin floodplain management plan. Upper Colorado River Endangered Fish Recovery Program, Denver.

Valdez, R. A., and R. J. Ryel. 1997. Life history and ecology of the humpback chub in the Colorado River in Grand Canyon, Arizona. Pages 3–31 *in* C. van Riper, III and E. T. Deshler, editors. Proceedings of the Third Biennial Conference of Research on the Colorado Plateau. National Park Service Transactions and Proceedings Series 97/12, Denver.

Valdez, R. A., and E. J. Wick. 1983. Natural vs. manmade backwaters as native fish habitat. Pages 519–536 *in* V. D. Adams and V. A. Lamarra, editors. Aquatic resources management of the Colorado River ecosystem. Ann Arbor Science Publications, Ann Arbor, Michigan.

Valdez, R. A., and R. D. Williams. 1993. Ichthyofauna of the Colorado and Green rivers in Canyonlands National Park, Utah. Proceedings of the First Biennial Conference on Research in Colorado Plateau National Parks National Park Service, Transactions and Proceedings Series 1993:2–22.

Vanicek, C. D. 1967. Ecological studies of native Green River fishes below Flaming Gorge Dam, 1964–1966. Doctoral dissertation. Utah State University, Logan.

Vanicek, C. D., and R. Kramer. 1969. Life history of the Colorado squawfish, *Ptychocheilus lucius*, and the Colorado chub, *Gila robusta*, in the Green River in Dinosaur National Monument 1964–1966. Transactions of the American Fisheries Society 98:193–208.

Vanicek, C. D., R. H. Kramer, and D. R. Franklin. 1970. Distribution of Green River fishes in Utah and Colorado following closure of Flaming Gorge Dam. Southwestern Naturalist 14:297–315.

Wallis, O. L. 1951. The status of the fish fauna of the Lake Mead National Recreation Area, Arizona-Nevada. Transactions of the American Fisheries Society 80:84–92.

Walters, C. J. 1986. Adaptive management of renewable resources. MacMillan Publishing, New York.

Wick, E. J. 1997. Physical processes and habitat critical to the endangered razorback sucker on the Green River, Utah. Doctoral dissertation. Colorado State University, Fort Collins.

Wick, E. J, J. A. Hawkins, and C. A. Carlson. 1985. Colorado squawfish and humpback chub population and habitat monitoring, 1983–1984. Colorado Division of Wildlife, Endangered Wildlife Investigations Final Report SE 3–7, Denver.

Wick, E. J., J. A. Hawkins, and T. P. Nesler. 1991. Occurrence of two endangered fishes in the Little Snake River, Colorado. Southwestern Naturalist 36:251–254.

Wick, E. J., T. A. Lytle, and C. M. Haynes. 1981. Colorado squawfish and humpback chub population and habitat monitoring, 1979–1980. Endangered Wildlife Investigations, SE-3–3, Colorado Division of Wildlife, Denver.

Wydoski, R. S., and J. Hamill. 1991. Evolution of a cooperative recovery program for endangered fishes in the upper basin. Pages 123–139 *in* W. L. Minckley and J. E. Deacon, editors. Battle against extinction: native fish management in the American West. University of Arizona Press, Tucson.

Young, M. K., editor. 1995. Conservation assessment for inland cutthroat trout. General Technical Report RM-256, U.S. Forest Service, Fort Collins, Colorado.

American Fisheries Society Symposium 45:205–223, 2005

Historic Changes in the Rio Grande Fish Fauna: Status, Threats, and Management of Native Species

BOB CALAMUSSO*

USDA, Tonto National Forest, 2324 East McDowell Road Phoenix, Arizona 85025, USA

JOHN N. RINNE

USDA, Rocky Mountain Research Station, 2500 South Pineknoll Drive, Flagstaff, Arizona 86001, USA

ROBERT J. EDWARDS

The University of Texas-Pan American, Department of Biology, Edinburg, Texas 78541, USA

Abstract.—The Rio Grande is the fourth longest river in North America and the 22nd longest in the world. It begins as a cold headwater stream in Colorado, flows through New Mexico and Texas, where it becomes warm and turbid and finally empties into the Gulf of Mexico. The diversity of native fishes is high in the Rio Grande ranging from freshwater salmonids in its upper reaches to coastal forms in the lower reaches. Historically, about 40 primary freshwater species inhabited the waters of the Rio Grande. Like many rivers throughout North America, the native fish fauna of this river has been irrevocably altered. Species once present are now extinct, others are threatened or endangered, and the majority of the remaining native fishes are declining in both range and numbers. Today, 17 of the 40 primary native freshwater fishes have been either extirpated in part or throughout the Rio Grande drainage. This chapter examines the river, its fauna, and its current plight.

Introduction

The Rio Grande (Río Bravo del Norte as it is called in Mexico) is the fourth longest river in North America, traveling 2,830 km from its headwaters in the San Juan Mountains of southern Colorado, bisecting the state of New Mexico, and forming the shared border between Texas and Mexico (Figure 1). The Rio Grande begins as a cold, clear stream in the southern Colorado Rocky Mountains and descends to warm, estuarine waters near the Gulf of Mexico. During its course, the river travels through a wide variety of habitats ranging from mountain fir and pine forests to chaparral, desert scrub, and lowland brush country. The watershed encompasses about 870,000 km^2. A large proportion of the river's basin is arid or semiarid, with a number of endorheic (inwardly draining) basins such that only about half of the total area, or about 456,000 km^2, actually contributes to the river's overall flow. Its primary tributaries are found in the montane reaches occurring in Colorado and New Mexico. Tributaries downstream are less frequent, more widely spaced, intermittent, and today much more erratic (Smith and Miller 1986). The river is often dry in the reach from El Paso, Texas to the mouth of the Rio Conchos, Mexico. It is not until the Rio Conchos and Pecos River enter the Rio Grande in Texas that the river (Reach 3) receives substantial surface and subsurface recharge to sustain perennial flow. During its course, the river passes through several major impoundments, numerous main-stem irrigation diversion

* Corresponding author: Rcalamusso@fs.fed.us

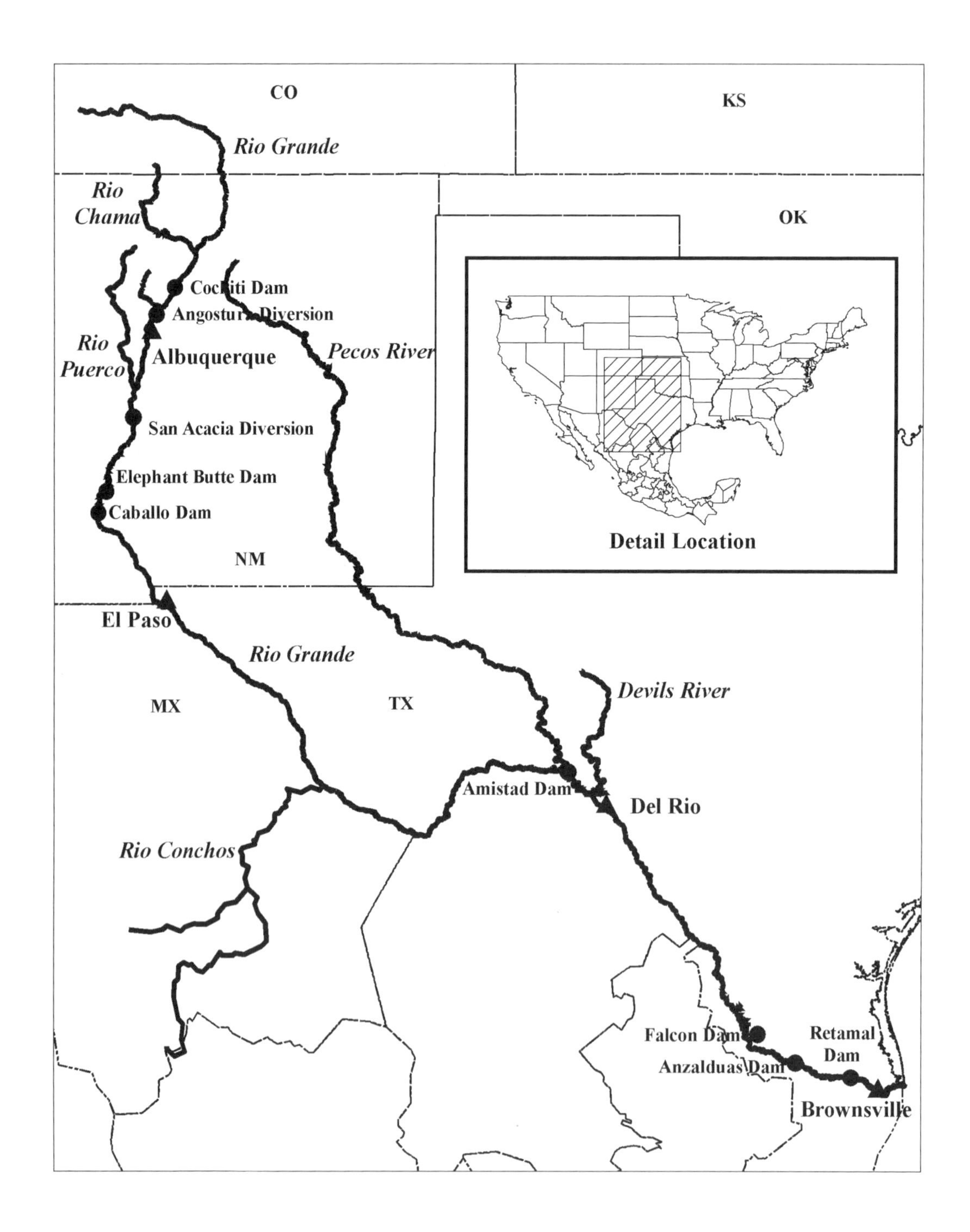

Figure 1.—Rio Grande drainage. Upper Rio Grande Reach—headwaters to the confluence with the Rio Chama; middle Rio Grande Reach—confluence of the Rio Grande and Rio Chama to Caballo Dam; lower Rio Grande Reach Caballo Dam to Gulf of Mexico.

dams, and sustains massive groundwater pumping of its aquifer, especially near major metropolitan areas. In 1993, the Rio Grande was classified as one of the most endangered or imperiled rivers in North America (American Rivers 1993).

In this chapter, we discuss changes from the historic to the current fish faunas, status, and threats to species and suggest conservation strategies to sustain native species. We divide the river into three reaches: the upper Rio Grande (Reach 1) extends from southern Colorado to its confluence with the Rio Chama in northern New Mexico (Figure 1). The middle Rio Grande (Reach 2) flows from the Rio Chama to Caballo Dam in southern New Mexico. The lower Rio Grande (Reach 3) runs from Caballo Reservoir to the Gulf of Mexico.

Study Area

Upper Rio Grande

The upper Rio Grande is a perennial clear and cold river with moderate to high gradients. It begins in a glacial valley near Spring Creek Pass (elevation 3,374 m) in the San Juan Mountains and flows southward through the San Luis Valley for about 112 km (Figure 1). As the river enters New Mexico, it descends into a deeply incised (305 m depth) and narrow canyon (mean floodplain width = 15 m) known as the Rio Grande Gorge (Sublette et al. 1990). Stream gradient in this reach is more than 2%. The river exits the canyon at Embudo, New Mexico and flows through a lower gradient (0.3%) river valley joining the Rio Chama, which enters from the northwest. Precipitation in the upper Rio Grande ranges from 5.0 to 127 cm/year. Most of this comes as snowfall, and snowmelt from winter storms provides the majority of water for surface flows and recharge of groundwater resources. The upper and middle Rio Grande are typical of river basins in semiarid environments with high rates of evaporation. Approximately 98% of the precipitation received evaporates from the Rio Grande watershed from Colorado to west Texas (Scurlock 1998). In general, however, Reach 1 gains water, whereas the downstream reaches lose water.

Of the three reaches, the upper reach has been least disturbed by humans. Localized diversion of water for irrigation began with the entry of the Spanish around 1540. It was not until a large contingent of Anglo-American settlers moved into the region (ca. 1878) that large-scale development of diversion structures and conveyance canals began (Scurlock 1986, 1998). Pine *Pinus* sp., spruce *Abies* sp., aspen/cottonwood *Populus* sp., and willow *Salix* sp. riparian areas dominate in this reach. Introduction of nonnative plants (salt cedar *Tamarisk* sp., Russian olive *Elaeagnus angustifolia*) along with disruption of natural flow patterns have altered the native willow/cottonwood riparian communities in this reach.

Middle Rio Grande

Valley gradient of the middle reach is about 0.3%. Mean precipitation in this reach is about 29 cm/year, but can be less than 15 cm/year during drought years (Scurlock 1998). Most precipitation in the middle Rio Grande comes from summer monsoon rains that provide the majority of the water for surface flows and recharge of groundwater resources. Peak flows from snowmelt in this reach occur in April–May and a second peak occurs during the July and August monsoon. Similar to the upper reach, evaporation is high in the middle reach; however, this reach exhibits a net loss of water.

Prior to Spanish settlement, Pueblo dwelling Indians used limited irrigation systems. Spanish settlement of the region in 1540 initiated the *acequia* (irrigation) network to supply water for domestic and agricultural uses. Beginning in 1846, large numbers of Anglo-American settlers moved into the valley contributing to the improvement, development, and expansion of water diversions and conveyance structures. Diversion of surface flow and alteration of streams and rivers coincided with agricultural development and was the beginning of successive modifications to historic stream courses and flows that continue at present.

Three modern major mainstream Rio Grande reservoirs, Cochiti, Elephant Butte, and Caballo, completed in 1975, 1916, and 1938, respectively, and three major diversions, (Angostura, Isleta, San Acacia) are present in the middle reach. Water is diverted into 1,280 km of levees, drains, and canals between Espanola and the Bosque de Apache National Wildlife Refuge near Socorro. Two of these diversions, Isleta and San Acacia, have the capability under low-flow conditions to eliminate all surface flow from a 177-km reach between Isleta and Elephant Butte Reservoir.

The major tributaries in the middle reach, the Rio Puerco and Jemez River, both enter the Rio Grande from the west, the former upstream of Bernalillo and the later south of Bernardo, New Mexico. Riparian vegetation has declined dramatically with changes in flow (Howe and Knopf 1991), and nonnative tamarisk and Russian olive dominate the riparian vegetation.

Lower Rio Grande

This reach encompasses the Rio Grande from the tailwaters of Caballo Reservoir to the Gulf of Mexico. Four major dams are found in the lower reach: two major diversions and four large-scale dams are found in the lower reach. Leasburg and American Diversions are found in New Mexico and west Texas, respectively and Amistad, Falcon, Anzalduas, and Retamal dams are located in Texas in the lower reaches of Reach 3 (Figure 1). From the New Mexico/west Texas border to Falcon Dam, the river flows through the Chihuahuan Desert and is characterized by low flows, few tributaries, and high evaporation rates (>98%). Mean precipitation is approximately 20.0 cm per year with a range of 8.0 to 30.5 cm. Historically, the river became dry naturally near El Paso during drought (Scurlock 1998). During the drought of 2003, the river was dry from Caballo Dam to the Rio Conchos because of retention in Elephant Butte and Caballo reservoirs. Settlement of this area began in the 1500s with large concentrations of humans and livestock in the El Paso/Juarez area. Currently, the population of these two cities combined is about 1.6 million.

Amistad, Falcon, Anzalduas, and Retamal dams are used for flood control and supply irrigation water for agriculture in the United States and Mexico. Amistad Dam, the largest of the impoundments with a storage capacity of 502,000,000 m^3 of water is located about 1.6 km below the confluence of the Devils River.

Characteristics of the lower reach of the Rio Grande change from Falcon Dam (storage capacity = 325,000,000 m^3) downstream to its mouth at an area locally known as Boca Chica. Two principal tributaries enter the Rio Grande in this region, Río Alamo and Río San Juan. Droughts occur periodically, but for the first time in recorded history, flow ceased below Brownsville during the severe drought of the 1950s (Treviño-Robinson 1955, 1959). This occurred again in 2001. In February 2001, flow declined to such an extent that natural sediment deposits formed a 100-m-wide sandbar which persisted until October 2002.

The delta is surrounded by semiarid brushland and exhibits irregular rainfall patterns with average annual precipitation ranging from 38 to 75 cm (Crosswhite 1980, Jahrsdoefer and Leslie 1988) often coming in the form of torrential rains from tropical storms. The Rio Grande is bounded by a dense riparian forest with trees as high as 21 m in this area (Thornton 1977). Less than 5% of this community has been left, due to the massive clearing for agriculture.

Flow in the lower reach has diminished over time (Edwards and Contreras-Balderas 1991), as more water has been removed from the river for irrigation and municipal use. Water once removed, does not reenter the river, instead entering one of the Laguna Madres of Texas or Mexico or evaporating from irrigated fields. During severe droughts, the amount taken from the river increased dramatically (Contreras-Balderas 1972, 1975; Schmandt et al. 2000).

Historic Fish Fauna

Upper Rio Grande

One of the first surveys of the upper reaches of the Rio Grande was conducted in 1782 by Reverend Fray Juan Agustin de Morti (Langlois et al. 1994). About 100 years later, Cope and Yarrow (1875) collected fishes in the Rio Grande of Colorado reporting the capture of dace, minnows, suckers, and trout. Jordan (1891) made the first systematic survey of the Rio Grande drainage in Colorado in 1889. Subsequently, Ellis (1914) conducted surveys in the area from 1912 to 1913. At the time of Spanish exploration, nine native fishes were found in the upper Rio Grande. This assemblage, however, contained a wide variety of types such as native cutthroat trout *Oncorhynchus clarkii virginalis*, longnose gar *Lepisosteus osseus*, shovelnose sturgeon *Scaphirhynchus platorynchus*, and American eel *Anguilla rostrata* (Table 1; Zuckerman 1983).

Middle Rio Grande

Historic collections of fishes by Cope and Yarrow (1875) and Koster (1930–1950) documented a relatively diverse and largely endemic middle Rio Grande fish fauna ranging from large river species such as freshwater drum to small endemic cyprinids (Table 1) . The middle reach exhibited a more robust assemblage than the upper Rio Grande. As many as 21 native fish species were found in this reach when Spanish settlements were first appearing along the banks of the Rio Grande (Koster 1957; Sublette et al. 1990; Rinne and Platania 1995). By the early 1960s, nine species of native fish were extirpated and two were extinct in this reach (Table 1).

Lower Rio Grande

Of the three reaches, the lower Rio Grande contains the greatest species richness. About 37 fish species were present at the time of Spanish exploration. Of the 37 species, 7 were restricted to the upper portion of the reach in what is now New Mexico and west Texas, 14 were found in the lower portion of the reach (below modern-day Falcon Dam), and about 16 species were found from the upper portion of the reach to the Gulf of Mexico (Table 1).

Caballo Dam to Falcon Dam.—The Pope expedition (1853–1855) surveyed along the 32nd parallel at about the present day New Mexico–Texas boundary and Koster collected in the mid-20th century (Sublette et al. 1990). Historically, 23 native fish species occurred in the reach between present-day Caballo Reservoir in New Mexico and Falcon Dam in west Texas (Table 1). Seven of these (roundnose minnow, Rio Grande chub, bluntnose shiner, Rio Grande sucker, blue sucker, smallmouth buffalo, and flathead catfish) were not found in the lower portion of the reach (below Falcon Dam). Species diversity was high with large river fishes such as freshwater drum, shovelnose sturgeon, and catadromous American eel, and small cyprinids being common. Rio Grande chub and Rio Grande sucker, species associated with the upper reaches of the Rio Grande, were also found in this portion of the lower Rio Grande reach.

Falcon Dam to Gulf of Mexico.—The first records of fishes in the lower Rio Grande were from collections taken during the 1850s near Brownsville, Texas by Clark and others for the U.S. and Mexican Boundary Survey and reported by Baird and Girard (1853, 1854a, 1854b), Girard (1856, 1859a, 1859b) and Evermann and Kendall (1894). The original fish assemblage in the lower Rio Grande included a relatively large number of minnows, catfish, and sunfish along with more subtropical species such as Mexican tetra, Amazon molly, and the Rio Grande cichlid. There were about 30 primary native freshwater fish inhabiting this portion of the lower Rio Grande when these surveys were conducted (Table 1). Of the 30 native fishes present, 13 (alligator gar, spotted gar, threadfin shad, Texas shiner, ghost shiner, bullhead minnow, channel catfish, Amazon molly, green sunfish, largemouth bass, slough darter, bigscale logperch, and Rio Grande cichlid) occurred primarily in the reach between Falcon Dam and the Gulf of Mexico. Nearly 100 years later, Treviño-Robinson (1955) sampled throughout the study area immediately prior to the establishment of Falcon Reservoir. Rodríguez-Olmos (1976) resurveyed the area in 1975 and Edwards and Contreras-Balderas (1991) made a larger series of collections throughout the study reach in 1981 to 1989. Besides these general surveys, a number of other investigations on selected species or parts of the river have also been conducted (e.g., Hubbs 1951, 1964; Schlicht 1959; Contreras-Balderas 1972, 1975; Kornfield and Koehn 1975; Contreras-Balderas et al. 1976; Chernoff et al. 1982; Echelle and Mosier 1982; Pezold and Edwards 1983; White et al. 1983; Andreason 1985; Edwards et al. 1986; Wood 1986; Ciomperlik 1989; Anderson et al. 1995; Estrada 1999).

Table 1.—Historic freshwater fish assemblage and current status by reach in the Rio Grande.

Taxa		Upper	Middle	Lower
shovelnose sturgeon	*Scaphirhynchus platorynchus*	Extr	Extr	
alligator gar	*Atractosteus spatula*			C[b]
longnose gar	*Lepisosteus osseus*	Extr	Extr	C[c, d]
spotted gar	*L. oculatus*			C[b]
American eel	*Anguilla rostrata*	Extr	Extr	R[c]
gizzard shad	*Dorosoma cepedianum*		C	C[c]
threadfin shad	*D. petenense*			C[b]
Rio Grande cutthroat trout	*Oncorhynchus clarkii virginalis*	Extr	Extr	
Mexican tetra	*Astyanax mexicanus*			C[c,d]
red shiner	*Cyprinella lutrensis*	C	C	C[c]
roundnose minnow	*Dionda episcopa*			Extr[a]
Rio Grande chub	*Gila pandora*	R En	R En	Extr[a]
Rio Grande silvery minnow	*Hybognathus amarus*		R En	Extr En[c]
speckled chub	*Macrhybopsis aestivalis*		Extr	C[c, d]
Rio Grande shiner	*Notropis jemezanus*		Extr En	Extr En[c]
Texas shiner	*N. amabilis*			Extr[b]
Tamaulipas shiner	*N. braytoni*			C[b]
ghost shiner	*N. buchanani*			C[b]
phantom shiner	*N. orca*		Ext En	Ext En[c]
bluntnose shiner	*N. simus*		Ext	Ext[a]
fathead minnow	*Pimephales promelas*	C	C	C[c]
bullhead minnow	*P. vigilax*			C[b]
flathead chub	*Platygobio gracilis*		C	
longnose dace	*Rhinichthys cataractae*	C	C	
river carpsucker	*Carpiodes carpio*		C	C[c]
Rio Grande sucker	*Catostomus plebeius*	R	R	Extr[a]
blue sucker	*Cycleptus elongatus*		Extr	Extr[a]
smallmouth buffalo	*Ictiobus bubalus*		C	C[a]
gray redhorse	*Moxostoma congestum*			C[c, d]
blue catfish	*Ictalurus furcatus*		Extr	C[c, d]
channel catfish	*I. punctatus*			C[b]
flathead catfish	*Pylodictis olivaris*			C[a]
western mosquitofish	*Gambusia affinis*			C[c]
Amazon molly	*Poecilia formosa*			C[b]
green sunfish	*Lepomis cyanellus*			C[b]
bluegill	*L. macrochirus*			C[c]
largemouth bass	*Micropterus salmoides*			C[b]
slough darter	*Etheostoma gracile*			C[b]
bigscale logperch	*Percina macrolepida*			C[b]
freshwater drum	*Aplodinotus grunniens*		Extr	C[c, d]
Rio Grande cichlid	*Cichlasoma cyanoguttatum*			C[b]

Current status: Extr = extirpated, Ext = extinct, C = common, R = rare, En = Eedemic, [a] occupies lower Rio Grande from Caballo Reservoir to Falcon Dam, [b] occupies lower Rio Grande below Caballo Dam to Gulf of Mexico, [c] occupies entire lower Rio Grande from Caballo Dam to Gulf of Mexico, [d] extirpated or rare in New Mexico portion of lower Rio Grande; common below Falcon Dam, Texas.

Current Fish Assemblages

Upper Rio Grande

Two thirds of the native fishes have been extirpated from the main-stem upper Rio Grande. Currently, 39 fish species, 34 alien (87%) and 5 native (13%), occur in the upper reach of the Rio Grande main stem (Table 2) (Zuckerman 1983). Shovelnose sturgeon, American eel, and longnose gar, all once thought to be common, have been extirpated in this reach and become rare or extirpated in the middle reach of the Rio Grande. Rio Grande cutthroat trout, Rio Grande sucker, a species listed as endangered by the state of Colorado, and Rio Grande chub have been functionally eliminated in the Rio Grande main stem; however, viable populations exist in isolated reaches of middle to upper elevation headwater streams (Rinne 1995a 1995b,1995c; Calamusso 1996; Calamusso and Rinne 1995, 1999, 2004; Calamusso et al. 2002). Only three out of the nine native species (33%; fathead minnow, red shiner, and longnose dace) are viable in the main stem of the upper Rio Grande. Of the alien fishes present, 17 (50%) are game fishes, with the others ranging from North American cyprinids to fishes from Africa and South America. Many of the latter two groups may not occur in any great abundance or have disappeared after initial introduction.

Middle Rio Grande

Currently, 32 fish species are found in the middle reach. Alien species, which account for the majority of the fish fauna (*n* = 21, 64%), have been introduced for sport, bait, or accidentally (Table 2). Of the 21 original native fish species, 9 (43%) have been extirpated, 2 (10%) are extinct, 3 (14%) are rare, and 7 (33%) remain common (Table 1) (Sublette et al. 1991; Propst et al. 1987; Platania 1991a, 1991b, 1993; Rinne and Platania 1995). The last collections of two mainstream cyprinids, speckled chub, and Rio Grande bluntnose shiner were in 1964, just downstream of the present location of Cochiti Dam (Bestgen and Platania 1988, 1989, 1990; Platania and Bestgen 1988). Rio Grande silvery minnow is the only short-lived endemic broadcast spawning cyprinid that survives in the Rio Grande in New Mexico (Bestgen and Platania 1991; Cook et al. 1992). Rio Grande cutthroat trout were extirpated from the Rio Grande main-stem in the beginning of the 20th century. Rio Grande sucker and Rio Grande chub were found to be declining in the 1980s (Calamusso 1996; Propst 1999; Swift-Miler et al. 1999; Calamusso et al. 2002) and today are believed to be extirpated in the main channel.

Among the surviving species, the Rio Grande silvery minnow is federally and state listed as endangered (USDI 1993, 1994; New Mexico Game and Fish Department 1987, 1988) and the Rio Grande shiner is a federal "notice of review" species (USDI 1991). Rio Grande cutthroat trout, the most southerly occurring of the cutthroat trout complex (Behnke 1980), has been eliminated from the main stem and now occupies only about 10% of its original range (Behnke 1992; Rinne 1995c; Stumpff and Cooper 1996; Young 1995; Calamusso and Rinne 2004). Rio Grande sucker and Rio Grande chub may warrant listing since they are believed to be extirpated from the main-stem Rio Grande in this reach (Propst 1999). These fishes persist in the montane tributaries to the middle reach where they are considered vulnerable. Native fishes that are common (*n* = 8 of 21, 38%) include gizzard shad, red shiner, fathead minnow, flathead chub, longnose dace, river carpsucker, smallmouth buffalo, and western mosquitofish. In summary, about 62% of the native fish species of the middle Rio Grande have been eliminated or are considered rare in this reach of river, and no doubt, the decline will continue.

Lower Rio Grande

Currently, 50 primary freshwater fishes occur from the tail waters of Caballo Reservoir to the Gulf of Mexico. Alien species (*n* = 23) account for 46% of the total fauna. Of the 37 original native fishes present, 21 (57%) are still common and retain their historic distribution throughout the reach, 7 (19%) native fish have become rare or extirpated (freshwater drum, longnose gar, shovelnose sturgeon, American eel, Mexican tetra, roundnose minnow, speckled chub, Rio Grande chub, Rio Grande Silvery minnow, Rio Grande shiner, blue sucker, gray redhorse, and blue catfish), 2 (5%) are extinct, (phantom

Table 2.—Current freshwater fish assemblage and status by reach in the Rio Grande.

Taxa		Reach 1	Reach 2	Reach 3
alligator gar	*Atractosteus spatula*			N[b] C
longnose gar	*Lepisosteus osseus*			N[c] C[d]
spotted gar	*L. oculatus*			N[b] C
American eel	*Anguilla rostrata*			N[c] C[d]
gizzard shad	*Dorosoma cepedianum*		N C	N[c] C
threadfin shad	*D. petenense*		A C	N[b] C
cutthroat-rainbow trout hybrid	*Oncorhynchus clarkii x mykiss*	A R		
cutthroat trout	*O. clarkii sbsp.*	A R		
westslope cutthroat trout	*O. clarkii lewisi*	A R		
rainbow trout	*O. mykiss*	A C	A R	A R
sockeye salmon	*O. nerka*	A R		
brown trout	*Salmo trutta*	A C	A R	
brook trout	*Salvelinus fontinalis*	A R	A R	
northern pike	*Esox lucius*	A C		A C
Mexican tetra	*Astyanax mexicanus*			N[c] C[d]
longfin dace	*Agosia chrysogaster*		A R	A R
central stoneroller	*Campostoma anomalum*	A	AC	AC
goldfish	*Carassius auratus*	AC	AC	AC
grass carp	*Ctenopharyngodon idella*	A R		A R
common carp	*Cyprinus carpio*	A C	A C	A C
Rio Grande chub	*Gila pandora*	N R	N R	
Rio Grande silvery minnow	*Hybognathus amarus*		N En R	N[c] En R
speckled chub	*Macrhybopsis aestivalis*			N[c] C[d]
Tamaulipas shiner	*Notropis braytoni*			N[b] C
ghost shiner	*N. buchanani*			N[b] C
red shiner	*Cyprinella lutrensis*	N C	N C	N[c] C
bullhead minnow	*Pimephales vigilax*	A R	A R	N[b] C
fathead minnow	*Pimephales promelas*	N C	N C	N[c] C
flathead chub	*Platygobio gracilis*		N C	
longnose dace	*Rhinichthys cataractae*	N C	N C	A R
tench	*Tinca tinca*	AR	A	
river carpsucker	*Carpiodes carpio*		N C	N[c] C
Rio Grande sucker	*Catostomus plebeius*	N R	N R	
white sucker	*C. commersonii*	A C	A C	A C
blue sucker	*Cycleptus elongatus*			N[c] C[d]
gray redhorse	*Moxostoma congestum*			N[c] C[d]-
smallmouth buffalo	*Ictiobus bubalus*		N C	N C
black bullhead	*Ameiurus melas*	A C	A C	
yellow bullhead	*A. natalis*	A C	A C	A C
blue catfish	*Ictaluarus furcatus*			N[c] C[d]
channel catfish	*I. punctatus*	A C	A C	A C
flathead catfish	*Pylodictis olivaris*			N[a] C
golden shiner	*Notemigonus crysoleucas*			A R
western mosquitofish	*Gambusia affinis*	A C	A C	N[c] C
Amazon molly	*Poecilia formosa*			N[b] C
plains killifish	*Fundulus zebrinus*			A R
sailfin molly	*Poecilia latipinna*	A R		
shortfin molly	*P. mexicana*	A R		
green swordtail	*Xiphophorus helleri*	A R		
southern platyfish	*X. maculatus*	A R		
brook stickleback	*Culea inconstans*	A R		
white bass	*Morone chrysops*		A R	A R
striped bass	*M. saxatilis*		A C	
green sunfish	*Lepomis cyanellus*	A C	A C	A C

Table 2.—Continued.

Taxa		Reach 1	Reach 2	Reach 3
bluegill	*L. macrochirus*	A C	A C	N C
longear sunfish	*L. megalotis*		A C	A C
redbreast sunfish	*L. auritus*			A C
warmouth	*L. gulosus*			A C
redear sunfish	*L. microlophus*			A C
white crappie	*Pomoxis annularis*		A C	A C
black crappie	*P. nigromaculatus*		A C	A C
largemouth bass	*Micropterus salmoides*	A C	A C	A C
yellow perch	*Perca flavescens*	A C	A C	A C
walleye	*Sander vitreus*	A C	A C	A C
mozambique tilapia	*Oreochromis mossambicus*	A R		
blue tilapia	*O. aureus*	A R		
angelfish	*Pterophyllum Sp.*	A R		
black tetra	*Gymnocorymbus ternetzi*	A R		
head and light tetra	*Memigrammus ocellifer*	A R		
neon tetra	*Paracheirodon innesi*	A R		
suckermouth catfish	*Hypostomus plecostomus*	A R		
plated catfish	*Corydoras sp.*	A R		
slough darter	*Etheostoma gracile*			N C
bigscale logperch	*Percina macrolepida*			N[b] C
Rio Grande cichlid	*Cichlasoma cyanoguttatum*			N[b] C
rudd	*Scardinius erythropththalmus*			A C
freshwater drum	*Aplodiontus grunniens*			N[c] C[d]

N = native, A = alien, En = endemic, C = common, R = rare, [a] occupies lower Rio Grande in New Mexico and west Texas, [b] occupies lower Rio Grande below Falcon Dam to Gulf of Mexico,[c] occupies entire lower Rio Grande from Caballo Dam to Gulf of Mexico, [d] extirpated or rare in New Mexico portion of lower Rio Grande; common below Falcon Dam, Texas.

shiner, bluntnose minnow), and 7 (19%) have reduced ranges and numbers.

Caballo Reservoir to Falcon Dam.—Today, only 13 (57%) of the 23 native fishes originally found in this section of the lower Rio Grande still remain. Of these 13, 8 (62%) are considered common and the remaining 6 (46%) are either extirpated or extremely rare (Tables 1 and 2). Currently, 45 primary freshwater fishes (21 native, 23 alien) occur from the tailwaters of Caballo Reservoir to the Gulf of Mexico (Table 2). Similar to the middle Rio Grande, nonnative fishes were introduced for sport, bait, food, or incidentally. The most common remaining native fishes (8, 36%) in this reach are gizzard shad, red shiner, fathead minnow, river carpsucker, smallmouth buffalo, flathead catfish, western mosquitofish, and bluegill.

Fish samples in the Tamaulipan Reach during the 1980s and 1990s still contained two major ecological faunas (Wood 1986; Edwards and Contreras-Balderas 1991; Hubbs et al. 1991; Estrada 1999) a primarily freshwater fish fauna below Falcon Dam downstream to Brownsville and an estuarine and marine influenced fauna below the city of Brownsville to the mouth of the river. Some species have declined or disappeared throughout the lower Rio Grande in the last 150 years (Table 2). These include the speckled chub, Rio Grande silvery minnow, Rio Grande shiner, and phantom shiner. In the lower portion of this reach near the Gulf of Mexico, the predominant fishes include many highly tolerant species such as threadfin shad, red shiner, bullhead minnow, Mexican tetra, western mosquitofish, sailfin and Amazon mollies , bluegill, Rio Grande cichlid, and the alien inland silverside *Menidia beryllina* and blue tilapia.

Below the cities of Brownsville and Matamoros, the fish assemblage changes markedly. Few upriver species persist and estuarine and marine species have replaced most of the primary freshwater fishes. Some of the more common species are bay anchovy *Anchoa mitchilli*, and striped anchovy *A. hepsetus*; scaled sardine *Harengula jaguana*; tidewater silversides *Menidia peninsulae*; white mullet *Mugil curema*; several game species such as common snook *Centropomus undecimalis*, red drum *Sciaenops ocellatus* and southern flounder *Paralichthys lethostigma*; crevalle jack *Caranx hippos*; spotfin mojarra *Eucinostomus argenteus;* and a relatively diverse goby fauna, including darter goby *Ctenogobius boleosoma*, highfin goby *Gobionellus oceanicus*, frillfin goby *Bathygobius soporator*, and lyre goby *Evorthodus lyricus*. When the river ceased flowing in 2001 to 2003, the number of species inhabiting the estuary declined dramatically with inland silversides, sheepshead minnow *Cyprinodon variegatus*, and Gulf killifish *Fundulus grandis* predominating.

Threats to and Impacts on Native Fish Fauna

Upper Rio Grande

Threats to the native fish fauna of the upper Rio Grande include the introduction of alien fishes, overharvest, habitat alteration, and dewatering (Sublette et al. 1990; Calamusso and Rinne 1999; Rinne 2004). The introduction and range expansion of alien fishes is probably the single most devastating factor to the persistence of native fish assemblages in the upper reach today. The replacement of native fishes by alien forms has dramatically altered the native fish assemblage and its ecological dynamics in the main stem and in the upper watershed. Currently, relict populations and assemblages of native fishes are confined to remote middle to high elevation streams, most of which occur on U.S. Forest Service lands and other public properties.

Loss of native species in relation to the presence of alien species occurs through the mechanisms of competition, aggression, hybridization, and direct predation. Introduced piscivorous esocids, salmonids, ictalurids, percichthyids, and centrarchids have a direct negative impact on native fishes through predation. Alien *Oncorhynchus* have eliminated native Rio Grande cutthroat trout through hybridization and competition for food and space throughout the Rio Grande watershed. Today, brown trout and, to a much lesser degree, rainbow trout are the most common salmonids in the main-stem Rio Grande, whereas the native Rio Grande cutthroat trout has been essentially extirpated (Calamusso and Rinne 2004). Alien white sucker, common throughout the Rio Grande main stem and its tributaries, has been indicated in the replacement of Rio Grande sucker throughout its range in Colorado and New Mexico (Calamusso and Rinne 1995, 1999; Calamusso 1996; Calamusso et al. 2002). While the mechanisms of replacement are not well understood, it may be that the white sucker has a competitive advantage due to its greater fecundity and higher probability of recruiting young into the breeding population (Calamusso 1996). Native wild populations of Rio Grande chub are now only found in tributaries (Calamusso and Rinne 1999; Propst 1999). The elimination of this species in the Rio Grande main stem has been attributed to predation by northern pike, brown trout, largemouth bass, sunfishes, and perches. Longnose gar and shovelnose sturgeon succumbed to recreational and commercial fishing at the turn of the century.

With respect to habitat alteration and water diversion, small-scale diversions and reservoirs have contributed to the loss of native fishes. The conversion of a riverine system into a lentic system by reservoirs deprives some native species key habitat components to complete their life cycles, thus creating the potential for decline or extirpation of sensitive species. However, the construction of these small-scale projects have altered only a small portion of the reach. And while water diversion from these small-scale projects have no doubt decreased habitat size, they have not been great enough to eliminate entire populations of native fish from this reach.

Middle Rio Grande

In contrast to the upper Rio Grande, the decline and extirpation of native fishes in the middle Rio Grande has been attributed mainly to the construction of a series of large-scale main-stem dams and diversion structures with their attendant water storage, water use, flow and channel reconfiguration (Rinne and Platania 1995). Alien fishes, no doubt, have contributed to this decline. Considering the former, there are three major impoundments and three major water diversion structures in this reach that have irrevocably altered the native aquatic environment (Figure 1). The impoundments have changed riverine habitats to lentic habitats, thereby eliminating necessary habitats required for the completion of many native fish life cycles. Water temperature, dissolved oxygen (DO), and pH were altered from natural conditions, making the habitat unsuitable for native fishes. Fish trapped in intermittent pools may ultimately succumb due to declining water quality (especially excess temperature and insufficient DO). Concentration and crowding at the base of dams increases the probability of losing major portions of the native fish fauna during natural events such as deoxygenation or human-caused activities such as spills of toxic materials.

Major water diversions and groundwater mining for agriculture and municipalities decrease river volume and desiccate large portions of the river. Typical of many U.S. rivers, discharge is generally highest during spring runoff between March and June. During this season, water supply exceeds demand. Unfortunately, as occurs with many rivers and streams in the Southwest, the period of lowest discharge in the middle Rio Grande occurs from July to October and coincides with peak irrigation demand. During these months, the need for irrigation water combined with variable monsoon precipitation and resulting streamflow may result in complete loss of surface flow and drying of extensive reaches of the main-stem river channel. For example, flow in the Rio Grande downstream of Isleta Diversion Dam is mainly the result of significant summer monsoon convectional storms, supplemented only by unpredictable irrigation return flow. Sustained flow in this reach sometimes returns only following termination of irrigation in November. Further, water diversion reduces both water quantity and quality (increases in temperature, nutrients, and total dissolved solids) in upper elevation tributary streams and rivers at the northern extent of the middle Rio Grande (Rinne 1995c). The sporadic and cyclic desiccation and rewetting of the main-stem Rio Grande channel severely alters habitat availability, reproductive and general life cycles, and population levels of fishes throughout the middle Rio Grande.

While nonnative fishes are considered a secondary factor in the decline of the native fish fauna in this reach, the number of introduced sport fishes in this reach is large (Table 2). Threat from predation in this reach may be greatest during low-flow periods when native fish are often trapped in pools where they may more readily fall prey to game fishes. Additionally, desert fish have a tendency to move upstream during periods of lowflow, thereby concentrating populations below mainstream diversions. Below these areas, there is not only a greater probability of encountering predation, but also increased disease due to stress. Predation on natives by aliens may be minimized in the Rio Grande because of its highly regulated flows. Game species reach their greatest abundance immediately after cessation of spring runoff and decline in abundance throughout the summer during periods of reduced and potential loss of surface flow. As summer progresses, water volumes of intermittent pools decrease, water temperatures increase, and DO levels decrease. Predatory species (particularly centrarchids and percicthyids) seem less tolerant of harsh physical-chemical conditions than native species. Therefore, game fish are not currently considered the major reason for decline of native fish in the middle reach (Rinne et al. 1998; Rinne and Deason 2000).

Habitat degradation and loss due to livestock grazing, although more difficult to quantify, may also have had a large impact on native Rio Grande fishes throughout its course in Colorado, New Mexico, and Texas. Domestic livestock grazing occurred around the middle Rio Grande since the arrival of the Spanish expeditions (Scurlock 1986, 1998). By 1830, wealthy ranchers were annually herding hundreds of thousands of sheep to feed miners in northern New Mexico (Williams 1986). In 1860, 830,000 sheep were being grazed in northern New Mexico; by 1880, that number had increased to about 4 million. This overstocking had profound negative effects on southwestern U.S. watersheds. Additional factors such as severe drought throughout the southwest United States during the late 1800s and early 1900s exacerbated the degraded range conditions (Scurlock 1986).

Currently, livestock grazing is marginally controlled by permit on National Forest lands. However, this land use can negatively affect headwater streams (Behnke and Zarn 1976; Rinne 1985, 1988; Behnke 1992). These impacts include trampling of streambanks and removal of streamside vegetation (Platts 1981, 1982, 1991). Loss of streamside vegetation facilitates the elevation of stream temperatures in the summer (Brown and Krygier 1970) and the development of anchor ice in winter. Collapse of stream banks and reduction in riparian vegetation increase silt loads in tributaries, which adds to the silt load of the main stem.

Irrigation diversions accompanying the immigration of early settlers into northern New Mexico (Scurlock 1986; Sayles and Williams 1986) probably resulted in habitat losses for Rio Grande cutthroat trout, which increased its decline in the Rio Grande basin (Sublette et al. 1990).

Lower Rio Grande

Caballo Reservoir to Falcon Dam.—The reach below Caballo Dam has been greatly altered since Spanish and Anglo settlement. Use of the area by stockmen degraded riparian areas. Drought added to this degradation. The river has repeatedly been straightened, channelized, and diked. Stream flow is highly regulated for agriculture and is dependent on annual snow pack and precipitation. This reach is frequently desiccated due to retention of water in Elephant Butte and Caballo reservoirs, which eliminates the fish fauna. Thus, the dramatic alteration in flows and physical habitat quantity and quality are the major factors in eliminating native fishes. Predation by alien sport fish species adds to this mortality.

Falcon Dam to the Gulf of Mexico.—The modern lower Rio Grande region is far different from what it was over a century ago. The river is regulated by a series of dams throughout the basin and its flow is severely reduced by pumping for irrigation and other human uses (Edwards and Contreras-Balderas 1991). Decreased flows along with organic inputs from incompletely treated wastewater, especially from Mexico, have stimulated dominance by alien aquatic plants such as hydrilla and water hyacinth in some parts of the river, further reducing flows. Flow reductions caused the river to completely cease flowing to the Gulf of Mexico in 2001 and 2002 for a longer period than occurred during the record drought in the 1950s. Much of the riparian habitat has also been extensively modified and replaced with irrigated agricultural lands. Since 1950, agricultural chemicals and petrochemicals have had an increasing impact in the lower Rio Grande region. Presently, a number of chemicals can be found throughout the lower reach in quantities approaching or exceeding guidelines of the U.S. Environmental Protection Agency for the safety of fishes and other aquatic organisms (White et al. 1983; TNRCC 1994). Although changes in faunal abundances due to the presence of these chemicals alone have not been quantified, Edwards and Contreras-Balderas (1991) suspected that chemicals, together with altered flows limited the indigenous fauna. The accumulation of toxic chemicals by some fishes of the lower Rio Grande has been presented by White et al. (1983), Andreason (1985), TNRCC (1994), and Webster et al. (1998).

Management and Conservation to Sustain Rio Grande Native Fishes

Upper Rio Grande

Native fish in this reach are protected by state regulations; however, proactive efforts to repatriate native fishes in the Rio Grande main stem are minimal except for annual stocking of hatchery-raised Rio Grande chub. However, establishment of a self-sustaining population of Rio Grande chub in the main stem still eludes fish managers. Most efforts in this reach are aimed at middle to high elevation tributaries where populations of native fishes and/or suitable habitat remain. New Mexico and Colorado both have an active program protecting and repatriating waters for Rio Grande cutthroat trout, Rio Grande sucker, and Rio Grande chub. Management actions within the historic range of these species consist of constructing dams in streams with suitable habitat, removing alien fishes with piscicides (recently, New Mexico Commission disallowed use of piscides), and stocking with native fishes (Stumpff and Cooper 1996; Calamusso and Rinne 1996, 1999, 2004). Rio Grande cutthroat trout was listed as a Forest Service sensitive species and a "management indicator species" (Stefferud 1988; Calamusso and Rinne 1995, 1999, 2004). The American Fisheries Society listed the subspecies as "protected" (Johnson 1987) and of "special concern" (Williams et al. 1989). In New Mexico, Rio Grande cutthroat trout is managed as a sport species and is therefore subject to State Game Commission sportfishing regulations. The New Mexico Department of Game and Fish has the legislative mandate to "preserve the natural diversity and distribution patterns of the state's native ichthyofauna." Under this directive, the state has the dual objectives of maintaining Rio Grande cutthroat trout while ensuring that its populations are not diminished to the point of special regulations. The state's program is coordinated with the Forest Service's land and resource management plans (U.S. Forest Service 1986, 1987). The national forests, through best management practices, monitor water quality in cutthroat trout streams in order to meet state water quality standards. The New Mexico Department of Game and Fish initiated a broodstock program in 1987 that is currently being refined. Both the New Mexico Department of Game and Fish and the Colorado Division of Wildlife have draft management plans for Rio Grande cutthroat trout (Colorado Division of Wildlife 1992; New Mexico Department of Game and Fish 2002: Stumpff 1992). Based on available information, Rinne (1995c) suggested that the following areas of research should be pursued (not listed in order of importance):

1. Distribution and genetic analyses of populations.
2. Habitat (spawning, rearing, overwintering) evaluation.
3. Effects of alien salmonids.
4. Basic life history (reproduction, age growth, production, parasites and diseases, food) delineation.
5. Response of this subspecies to land management activities (e.g., fire, grazing, roads).
6. Fish–habitat relationships.

Research efforts should be closely meshed and integrated with management plans of the U.S. Forest Service and the management plan for the middle Rio Grande. Generation of information in the above six areas will facilitate management activities to restore this rare native trout to its former range and abundance. Further, researchers should be opportunistic and proactive in synchronizing research efforts with those of the New Mexico Game and Fish Department and the University of New Mexico.

Extensive efforts have been expended and are ongoing to survey streams to locate populations of Rio Grande cutthroat trout and determine their genetic purity (Calamusso and Rinne 1996, 1999). A priority should be to continue these efforts. It is critical to know the size and distribution of the resource across the landscape before it can be either properly managed or effectively conserved. Great efforts should be made to conduct this research in the concept of "ecosystem," or the newly adopted "ecology-based multiple use management" philosophy in Region 3 (U.S. Forest Service 1992). Research needs to be conducted on whether and how alien salmonids (principally brown and rainbow trout) limit Rio Grande cutthroat trout populations. In addition, interactions of the Rio Grande cutthroat trout with the other members of the fish assemblage should be investigated.

Middle Rio Grande

Currently, the only native fish that is actively managed to maintain self-sustaining populations in the middle Rio Grande main stem is the Rio Grande silvery minnow. It is the last remaining endemic middle reach Rio Grande cyprinid. The Rio Grande silvery minnow historically occurred in the Rio Grande from Abiquiu, New Mexico to the Gulf of Mexico. It now occurs only in about 5% of its former range; a reach restricted to the middle Rio Grande between Cochiti Dam and Elephant Butte Reservoir. Long-term studies designed to determine the species' life history attributes, habitat associations, and relative abundances were initiated in 1987. Additional research activities on reproductive biology and the early life history of the Rio Grande silvery minnow were initiated at the University of New Mexico in summer 1994. A recovery plan for this species was developed (USDI 1999) and is currently being revised.

Similar to the upper reach, middle to high elevation tributaries to the middle reach are currently being managed for the conservation and repatriation of native fishes, in particular Rio Grande cutthroat trout, Rio Grande chub, and Rio Grande sucker.

Lower Rio Grande

With the exception of planned reintroduction of Rio Grande silvery minnow, there are no active management activities focused on the conservation of native fishes in this reach.

Epilogue

Throughout the Rio Grande main stem, native fishes have experienced dramatic declines in ranges and numbers that continue today. Regional alteration of the rivers' physical habitat and flow resulting from large-scale dams and diversions, along with intentional and unintentional releases of nonnative fishes, have fundamentally altered the Rio Grande ecosystem. The logistics of applying treatments to large river reaches and similarly large lentic areas is daunting. The probability of successfully removing fishes is low, while the likelihood of recontamination is high. Repatriation efforts for native fishes in the main stem Rio Grande are unwarranted without naturalizing flows and installing fish passage structures on main stem dams. The greatest probability of success for repatriating native fishes lies in the tributaries. There, physical aspects of streams are manageable (i.e. barrier placement) and management of alien fishes feasible (i.e. chemical/mechanical removal).

Because of the regulated and adjudicated nature of the Rio Grande along the international boundary, management options to sustain native fishes are limited. However, increased efforts are needed to assure sufficient quantities of water in the lower river to support native fishes. Additional efforts are also needed to isolate and remediate water quality problems in the lower basin, an especially troublesome problem due to the international nature of the river. Naturalized flow regimes in the lower reach require both interstate and international cooperation and planning efforts to a much greater degree than is now occurring, and current laws are presently insufficient to accomplish these goals.

In retrospect, the issues of native fish conservation in the Rio Grande are similar to those in other large rivers. By their very nature, large rivers are difficult to manage. Given its collective history of physical change and biological contamination, the future for sustainable native fish assemblages in the main-stem Rio Grande is doubtful. Currently, efforts and dollars are best focused on its tributaries.

Acknowledgments

The authors express their gratitude to the reviewers who improved the manuscript through their thorough and thoughtful comments.

References

American Rivers. 1993. The nation's ten most endangered rivers and fifteen most threatened rivers for 1993. American Rivers, Washington, D.C.

Anderson, A. A., C. Hubbs, K. O. Winemiller, and R. J. Edwards. 1995. Texas freshwater fish assemblages following three decades of environmental change. Southwestern Naturalist 40: 314-321.

Andreason, M. K. 1985. Insecticide resistance in mosquitofish of the lower Rio Grande valley of Texas—an ecological hazard? Archives Environmental Contamination Toxicology 14:573-577.

Baird, S. F., and C. Girard. 1853. Descriptions of new species of fishes collected by Mr. John H. Clark, on the United States and Mexican Boundary Survey, under Lt. Col. Jas. D. Graham. Proceedings of the National Academy of Science Philadelphia 6:387-390.

Baird, S. F., and C. Girard. 1854a. Descriptions of new species of fishes collected in Texas, New Mexico, and Sonora, by Mr. John H. Clark, on the United States and Mexican Boundary Survey, and in Texas by Capt. Steward Van Vliet, U.S.A. Proceedings of the National Academy of Science Philadelphia 7:24-29.

Baird, S. F., and C. Girard. 1854b. Notice of a new genus of Cyprinidae. Proceedings of the National Academy of Science Philadelphia 7:158.

Behnke, R.J. 1980. Report on collections of cutthroat trout from north-central New Mexico. New Mexico Department of Game and Fish, Santa Fe.

Behnke, R.J. 1992. Native trout of western North

American. American Fisheries Society Monograph 6, Bethesda, Maryland.

Behnke, R. J. and M. Zarn. 1976. Biology and management of threatened and endangered western trout. U.S. Forest Service, General Technical Report RM-28, Fort Collins, Colorado.

Bestgen, K.R., and S. P. Platania. 1988. The status of bluntnose and phantom shiner in the Rio Grande drainage of New Mexico. New Mexico Department of Game and Fish, Santa Fe.

Bestgen, K. R., and S. P. Platania. 1989. Inventory and microhabitat association of fishes of the Middle Rio Grande, New Mexico: Year I Progress report: survey of the fishes and their habitats in the middle Rio Grande and in the low-flow conveyance canal. New Mexico Department of Game and Fish (Contract 516.6-74-23) and U.S. Bureau of Reclamation (Intergovernmental Agreement 8-AG-53-06920), Santa Fe.

Bestgen, K. R., and S. P. Platania. 1990. Extirpation and notes on the life history of *Notropis simus simus* and *Notropis orca* (Cypriniformes: Cyprinidae) from the Rio Grande, New Mexico. Occasional Papers of the Museum of Southwestern Biology 6:1–8.

Bestgen, K. R. and S. P. Platania. 1991. Status and conservation of the Rio Grande silvery minnow, *Hybognathus amarus*. Southwestern Naturalist 36:225–232.

Brown, G. W., and J. T. Krygier. 1970. Effects of clear-cutting on stream temperature. Water Resources Research 6:1133–1139.

Calamusso, B. 1996. Distribution, abundance, and habitat of Rio Grande sucker (*Catostomus plebeius*) in the Carson and Santa Fe national forests, New Mexico. Master's thesis. New Mexico State University, Las Cruces.

Calamusso, B. and J. N. Rinne. 1996. Distribution of Rio Grande cutthroat trout (*Oncorhynchus clarki virginalis*) and its co-occurrence with Rio Grande sucker (*Catostomus plebeius*), and Rio Grande chub (*Gila pandora*) on the Carson and Santa Fe national forests. Pages 157–167 *in* D. W. Shaw, and D. M. Finch, technical coordinators. Proceedings of a symposium. Desired future conditions for southwestern riparian ecosystems: bringing interests and concerns together. U.S. Forest Service, General Technical Report RM-272, Fort Collins, Colorado.

Calamusso, B. and Rinne J. N. 1999. Native montane fishes of the middle Rio Grande ecosystem: status, threats, and conservation. Pages 231–237 *in* D. M. Finch, J. C. Whitney, J. F. Kelly, and S. R. Loftin. Rio Grande ecosystems: linking land, water, and people. Toward a sustainable future for the Middle Rio Grande basin. Proceedings RMRS-P-7. U.S. Forest Service, Fort Collins, Colorado.

Calamusso, B., and J. N. Rinne. 2004. Distribution and Abundance of the Rio Grande cutthroat trout (*Oncorhynchus clarki virginalis*), relative to an introduced salmonid in northern New Mexico. Pages 31–37 *in* G. J. Scrimgeour, G. Eisler, B. McCulloch, U. Silinis, and M. Monita, editors. Forest Fish Conference II – Ecosystems Stewardship through Collaboration. Proceedings of the Forest –Fish Conference II, April 26–28, 2004, Edmonton, Alberta.

Calamusso, B., J. N. Rinne, and P. R. Turner. 2002. Distribution and abundance of the Rio Grande sucker in the Carson and Santa Fe National Forests, New Mexico. Southwestern Naturalist 47:182–186.

Chernoff, B., R. R. Miller, and C. R. Gilbert. 1982. *Notropis orca* and *Notropis simus*, cyprinid fishes from the American Southwest, with description of a new subspecies. Occasional Papers of the Museum of Zoology, University of Michigan 698:1–49.

Ciomperlik, M. A. 1989. Seasonal distribution of the ichthyofauna of Santa Ana National Wildlife Refuge, Alamo, Texas. Master's thesis. Pan American University, Edinburg, Texas.

Colorado Division of Wildlife. 1992. Rio Grande cutthroat trout management plan. Colorado Division of Wildlife, Denver.

Contreras-Balderas, S. 1972. *Agonostomus monticola* (Bancroft):primer registro de la familia Mugilidae en Nuevo León, México. Cuadernos del Instituto Investigaciones Científicas Universidad Nuevo León, México 16:1–5.

Contreras-Balderas, S. 1975. Impacto ambiental de Obras Hidráulicas, Informe Tecnico, Plan Nacional Hidráulico, Secretaría de Recursos Hidráulicos, México.

Contreras-Balderas, S., V. L. Salinas, T. V. Gaytan, and G. R. Olmos. 1976. Peces, piscicultura, presas, polución, planificación pesquera y monitoreo en México, o la danza de las P. Memorias del Simposio sobre Pesquerias en Aguas Continentales 1:315–346.

Cook, J. A., K. R. Bestgen, D. L. Propst, and T. L. Yates. 1992. Allozymic divergence and systematics of the Rio Grande silvery minnow,

Hybognathus amarus (Teleostei: Cyprinidae). Copeia 1992:36–44.

Cope, E. D., and H. C. Yarrow. 1875. Report upon the collections of fishes made in portions of Nevada, Utah, California, Colorado, New Mexico, and Arizona, during 1871, 1872, 1873, and 1874. Report of the Geographic and Geologic Exploration Survey West of the 100th Meridian (Wheeler Survey) 5(6):637–700.

Crosswhite, F. S. 1980. Dry country plants of the south Texas plains. Desert Plants 2:141–179.

Echelle, A. A., and D. T. Mosier. 1982. *Menidia clarkhubbsi*, n. sp. (Pisces: Atherinidae), an all female species. Copeia 1982:533–540.

Edwards, R. J. and S. Contreras-Balderas. 1991. Historical changes in the ichthyofauna of the lower Rio Grande (Rio Bravo del Norte), Texas and Mexico. Southwestern Naturalist 36:201–212.

Edwards, R. J., T. S. Sturdivant, and C. S. Linskey. 1986. A report of *Awaous tajasica* from the lower Rio Grande, Texas and Mexico. Texas Journal of Science 38:191–192.

Ellis, M.M. 1914. Fishes of Colorado. University of Colorado Studies 112(1):5–135.

Estrada, M. 1999. The ecology and life history of the Mexican tetra, *Astyanax mexicanus,* (Teleostei: Characidae) in the lower Rio Grande Valley, Texas. Master's thesis. University of Texas-Pan American. Edinburg, Texas.

Evermann, B. W., and W. C. Kendall. 1894. The fishes of Texas and the Rio Grande basin, considered chiefly with reference to their geographic distribution. Bulletin of the U.S. Fish Commission 1892:57–126.

Girard, C. 1856. Researches upon the cyprinoid fishes inhabiting the fresh waters of the United States of America west of the Mississippi Valley, from specimens in the Museum of the Smithsonian Institution. Proceedings of the Philadelphia National Academy of Science 8:165–218.

Girard, C. 1859a. Ichthyology of the boundary. Report of the United States and Mexican Boundary Survey, made under the direction of the Secretary of Interior, by William H. Emory, Major, First Cavalry, and United States Commissioner. Washington, 1858(3):1–85.

Girard, C. 1859b. Ichthyological notices, Nos. I-LXXVII. Proceedings of the National Academy of Science, Philadelphia 10:1859.

Howe, W. H., and F. L. Knopf. 1991. On the imminent decline of Rio Grande cottonwoods in central New Mexico. Southwestern Naturalist 36:218–224.

Hubbs, C. 1951. Minimum temperature tolerances of fishes of the genera *Signalosa* and *Herichthys* in Texas. Copeia 1951:297.

Hubbs, C. 1964. Interactions between a bisexual fish species and its gynogenetic sexual parasite. Bulletin of the Texas Memorial Museum 8:1–72.

Hubbs, C., R. J. Edwards, and G. P Garret. 1991. An annotated checklist of the freshwater fishes of Texas with keys to the identification of species. The Texas Journal of Science, Supplement, 43(4):1–56.

International Boundary and Water Commission (IBWC). 2002. Impact assessment. USIBWC trench excavation through a sandbar at the mouth of the Rio Grande. International Boundary and Water Commission, El Paso, Texas.

Jacobs, J. L. 1981. Soil survey of Hidalgo County, Texas. Soil Conservation Service, Washington, D.C.

Jahrsdoefer, S. E., and D. M. Leslie, Jr. 1988. Tamaulipan brushland of the lower Rio Grande Valley of south Texas: description, human impacts, and management options. U.S. Department of the Interior, Fish and Wildlife Service, Biological Report 88(30):1–63. Washington, D.C.

Johnson, J. E. 1987. Protected fishes of the United States and Canada. American Fisheries Society, Bethesda, Maryland.

Jordan, D. S. 1891. Report of explorations in Colorado and Utah during the summer of 1889, with an account of the fishes found in each of the river basins examined. Bulletin of the U.S. Fish Commission, 9(1889):1–40.

Kornfield, I. L., and R. K. Koehn. 1975. Genetic variation and speciation in new world cichlids. Evolution 29:427–437.

Koster, W. J. 1957. Guide to the fishes of New Mexico. University of New Mexico Press, Albuquerque.

Langlois, D., J. Alves, and J. Apker. 1994. Rio Grande sucker recovery plan. Colorado Division of Wildlife, Montrose.

New Mexico Department of Game and Fish. 1987. Operation plan: management of New Mexico aquatic wildlife. New Mexico Game and Fish Department, Santa Fe.

New Mexico Department of Game and Fish. 1988. Handbook of species endangered in New Mexico. New Mexico Game and Fish Department, Santa Fe.

New Mexico Department of Game and Fish. 2002. Long range plan for the management of Rio Grande cutthroat in New Mexico. Fisheries Management Division. Santa Fe, New Mexico.

Pezold, F. L., and R. J. Edwards. 1983. Additions to the Texas marine ichthyofauna, with notes on the Rio Grande Estuary. Southwestern Naturalist 28:102–105.

Platania, S. P. 1991a. Fishes of the Rio Chama and upper Rio Grande, New Mexico, with preliminary comments on their longitudinal distribution. Southwestern Naturalist 36:186–193.

Platania, S. P. 1991b. Interim report of the middle Rio Grande Fishes Project: inventory and habitat associations of the fishes of the middle Rio Grande, New Mexico. Survey of the fishes in the upper reach of the middle Rio Grande. New Mexico Department of Game and Fish (contract 516.6-74-23) and U.S. Bureau of Reclamation (cooperative agreement O-FC-40-08870), Santa Fe.

Platania, S. P. 1993. The fishes of the Rio Grande between Velarde and Elephant Butte Reservoir and their habitat associations. New Mexico Department of Game and Fish (contract 516.6-74-23) and U.S. Bureau of Reclamation (cooperative agreement 0-FC-40-08870), Santa Fe.

Platania, S. P., and K. R. Bestgen. 1988. A survey of the fishes in a 8 km reach of the Rio Grande below Cochiti Dam, July, 1988. Report to the U.S. Army Corps of Engineers, Albuquerque, District. New Mexico Department of Game and Fish, Santa Fe.

Platts, W. S. 1981. Effects of sheep grazing on a riparian-stream environment. U.S. Forest Service Research Note INT-307: 1–6. Ogden, Utah.

Platts, W. S. 1982. Livestock and riparian fishery interactions: what are the facts? Transactions of the North American Wildlife and Natural Resources Conference 47:507–515.

Platts, W. S. 1991. Livestock grazing. Pages 389–424 *in* W. R. Meehan, editor. Influences of forest and rangeland management on salmonid fishes and their habitats. American Fisheries Society, Special Publication 19, Bethesda, Maryland.

Propst, D. L. 1999. Threatened and endangered fishes of New Mexico. New Mexico Department of Game and Fish, Technical Report No.1, Santa Fe.

Propst, D. L., G. L. Burton, and B. H. Pridgeon. 1987. Fishes of the Rio Grande between Elephant Butte and Caballo reservoirs, New Mexico. Southwestern Naturalist 32: 408–411.

Rinne, J. N. 1985. Livestock grazing effects on southwestern streams: a complex research problem. Pages 295–299 *in* R. R. Johnson, C. D. Ziebell, D. R. Patton, P. F. Ffolliott, and R. H. Hamre, editors. Riparian ecosystems and their management: reconciling and conflicting uses. U.S. Forest Service, General Technical Report RM-120, Fort Collins, Colorado.

Rinne, J. N. 1988. Grazing effects on stream habitat and fishes: research design considerations. North American Journal Fisheries Management 8:240–247.

Rinne J. N. 1995a. Reproductive biolgy of Rio Grande chub (*Gila pandora*) in a montane stream, New Mexico. Southwestern Naturalist 40:107–110.

Rinne J. N. 1995b. Reproductive biology of the Rio Grande sucker (*Catostomus plebeius*) in a montane stream, New Mexico. Southwestern Naturalist 40:237– 241.

Rinne J. N. 1995c. Rio Grande cutthroat trout. Pages 24–27 *in* M. K. Young, editor. Conservation assessment for inland cutthroat trout. USDA Forest Service, General Technical Report 256 , Fort Collins, Colorado.

Rinne, J. N. 2004. Native fishes: their status, threats, and conservation. Pages 277–297 *in* P. F. Ffolliot, M. B. Baker, L. F. Debano, and D. G. Neary, editors. Hydrology ecology and management of riparian areas in the southwestern United States. Lewis Press, Boca Raton, Florida.

Rinne, J. N., and S. P. Platania. 1995. Fish fauna. Pages 165–175 *in* D. H. Finch, editor. Ecology, diversity, and sustainability of the Middle Rio Grande basin. U.S. Forest Service, General Technical Report RM-GTR-268, Fort Collins, Colorado.

Rodríguez-Olmos, G. R. 1976. Cambios en la composición de especies de peces en comunidades del bajo Río Bravo, México–Estados Unidos. Master's thesis. Universidad Autónoma Nuevo León, Monterrey, México.

Sayles, S., and J. L. Williams. 1986. Land Grants. Pages 105–107 *in* J. L. Williams, editor. New Mexico in maps, 2nd edition. University of New Mexico Press, Albuquerque.

Schlicht, F. G. 1959. First records of the mountain mullet, *Agonostomus monticola* (Bancroft), in Texas. Texas Journal of Science 11:181–182.

Schmandt, J., I Águilar-Barajas, M Mathis, N.

Armstrong, L. Chapa-Alemán, S. Contreras-Balderas, R. Edwards, J. Hazleton, J. Navar-Chaidez, E.Vogel, and G. Ward. 2000. Water and sustainable development in the binational lower Rio Grande/Río Bravo basin. Center for Global Studies, Houston Advanced Research Center, Final Report to EPA/NSF Water and Watersheds grant program (Grant No. R 824799-01-0), Center for Global Studies, Houston Advanced Research Center.

Scurlock, D. 1986. Settlement and missions. Pages 92–94 *in* J. L. Williams, editor. New Mexico in maps, 2nd edition. University of New Mexico Press, Albuquerque.

Scurlock, D. 1998. From Rio to the Sierra: an environmental history of the middle Rio Grande basin. U.S. Forest Service, General Technical Report RMRS-GTR-5, Fort Collins, Colorado.

Smith, M. L., and R. R. Miller. 1986. The evolution of the Rio Grande basin as inferred from its fish fauna. Pages 457–485 *in* C. H. Hocutt, and E. O. Wiley, editors. Zoogeography of North American freshwater fishes. John Wiley and Sons, New York.

Stefferud, J. A. 1988. Rio Grande cutthroat trout management in New Mexico. Pages 90–92 *in* R. E. Gresswell, editor. Status and management of interior stocks of cutthroat trout, American Fisheries Society, Bethesda, Maryland.

Stumpff, W. K. 1992. Stabilization of native trout populations. Federal Aid Report F-22-R-33:1–20, New Mexico Department of Game and Fish, Santa Fe.

Stumpff, W. K., and J. Cooper. 1996. Rio Grande cutthroat trout, *Oncorhynchus clarki virginalis*. Pages 74 – 86 *in* D. E. Duff, editor. Conservation assessment for inland cutthroat trout. U.S. Forest Service, Ogden, Utah.

Sublette, J. E., M. D. Hatch, and M. Sublette. 1990. The fishes of New Mexico. University of New Mexico Press, Albuquerque.

Swift-Miller, S. M., B. M. Johnson, R. T. Muth, and D. Langlois. 1999. Distribution, abundance, and habitat use of Rio Grande sucker (*Catostomus plebeius*) in Hot Creek, Colorado. Southwestern Naturalist 44:148–156.

TNRCC (Texas Natural Resource Conservation Commission). 1994. Regional assessment of water quality in the Rio Grande basin including the Pecos River, the Devils River, the Arroyo Colorado and the lower Laguna Madre. October, 1994, AS-34.

Thornton, O. W., Jr. 1977. The impact of man upon herpetological communities in the lower Rio Grande Valley, Texas. Master's thesis. Texas A&M University, College Station.

Treviño-Robinson, D. 1955. The ichthyofauna of the lower Rio Grande river, from the mouth of the Pecos to the Gulf of Mexico. Masters thesis. University of Texas, Austin.

Treviño-Robinson, D. 1959. The ichthyofauna of the lower Rio Grande, Texas and Mexico. Copeia 1959:255–256.

USDI (U.S. Department of the Interior). 1991. Endangered and threatened wildlife and plants; animal candidate review for listing as endangered or threatened species, proposed rule. Federal Register 58:38(21 November 1991):58804–58836.

USDI (U.S. Department of the Interior). 1993. Endangered and threatened wildlife and plants; Proposed rule to list the Rio Grande silvery minnow as endangered, with critical habitat. Federal Register 58:38(1 March 1993):11821–11828.

USDI (U.S. Department of the Interior). 1994. Endangered and threatened wildlife and plants: Final rule to list the Rio Grande silvery minnow as an endangered species. Federal Register 59:138(20 July 1994):36988–36995.

USDI (U.S. Department of the Interior). 1999. Rio Grande silvery minnow *Hybognathus amarus* recovery plan. U.S. Fish and Wildlife Service, Albuquerque, New Mexico.

U.S. Forest Service. 1986. Carson National Forest Plan. Taos, New Mexico.

U.S. Forest Service. 1987. Santa Fe National Forest Plan. Santa Fe, New Mexico.

U.S. Forest Service. 1992. Ecology-based multiple use management. Albuquerque, New Mexico.

Webster, C. F., T. A. Buchanan, J. Kirkpatrick, and R. M. Miranda. 1998. Polychlorinated biphenyls in Donna Reservoir and contiguous waters. Texas Natural Resource Conservation Commission, Agency Studies AS-161, Austin.

White, D. H., C. A. Mitchell, H. D. Kennedy, A. J. Krynitsky, and M. A. Ribick. 1983. Elevated DDE and toxaphene residues in fishes and birds reflects local contamination in the lower Rio Grande Valley, Texas. Southwestern Naturalist 28:325–333.

Williams, J. E., J. E. Johnson, D. A. Hendrickson, S. Contreras-Balderas, J. D. Williams, M. Navarro-Mendoza, D. E. McAllister, and J. E.

Deacon. 1989. Fishes of North America endangered, threatened, or species of special concern: 1989. Fisheries 14(6):2–21.

Williams, J. L. 1986. Ranching and conflicts. Pages 12–122 *in* J. L. Williams, editor. New Mexico in maps. University of New Mexico Press, Albuquerque.

Wood, M. G. 1986. Life history characteristics of introduced blue tilapia, *Oreochromis aureus* in the lower Rio Grande. Master's thesis. Pan American University, Edinburg, Texas.

Young, M. K. 1995. Conservation assessment for inland cutthroat trout. U.S. Forest Service General Technical Report 256: 1–61, Fort Collins, Colorado.

Zuckerman, L. D. 1983. Rio Grande fishes management: progress report, November 1982 to June 1983. Colorado State University, Fort Collins.

American Fisheries Society Symposium 45:225–237, 2005

Historical Changes in the Index of Biological Integrity for the Lower Río Nazas, Durango, México

Salvador Contreras-Balderas*, María de Lourdes Lozano-Vilano, and María E. García-Ramírez

BIOCONSERVACIÓN, A.C., and Universidad A. De Nuevo León. San Nicolás, N. L., México

Abstract.—The interior Río Nazas basin is located in arid north-central México. It is an interior drainage, subject to dewatering since the early 20th century, and sustains wide fluctuations in runoff. It drains 85,530 km^2 and has a major dam in the middle reaches, producing a highly controlled river, with 100% consumption for agriculture and urban use. Hydrologic gauge reports at Torreón from the Comisión Nacional del Agua indicate a 10-year average runoff of 581.9 million m^3 from 1936 to 1945, and only 66.4 million m^3 in 1972, the last year of recorded runoff. Its 13 known native fish species are of Rio Grande/Rio Bravo origin. Eleven are endemic to the basin complex (only one absent from the study area), seven species have been listed by the Mexican federal government as threatened or endangered, and three are undescribed. The basin has 13 invasive alien species. An index of biological integrity (IBI), based on historical data, was applied to the current fish assemblage at 10 localities in the lower basin, below El Palmito reservoir. The IBI ranged from 50 to 57 at sites in the northern branch, to 39–61 in the southern branches, and to 0–57 from below their junction to the lower reaches, and averaged 37 or very poor. The overall biotic integrity is very low, especially near reservoirs and in the lower reaches of the river, where human activities consume all available water. The main causes of fish loss from this interesting fish fauna are alien invasive species, habitat disruption, pollution, and dewatering.

Introduction

The riverscapes of México need ecological evaluation. Scanty data are available for methods now commonly used, such as the index of biological integrity (IBI; Karr 1981). Because little has been published on the Mexican fish fauna, it is difficult to access information on the biological characters of the fish fauna. Hence, one of us (SCB) converted his data from former studies on freshwater fish assemblages, incorporating the views of Karr (1981).

We reject the idea of delaying ecological evaluations until the necessary bioassays or ecological studies are completed. The concept of integrity held here is as expressed in Webster's dictionary "the quality or state of being complete; unbroken condition; wholeness, entirety, … unimpaired…", and we refer it to the recent evolutionary state, and not only as system health. As a consequence, alien species have a negative effect on integrity (Contreras-Balderas et al. 1976; Moyle et al. 1986; Angermeier 1994; Contreras-Balderas et al. 2000), although their absence in itself does not mean undisturbed conditions (Weaver and Garman 1994). Eradication of species is also regarded as loss of integrity. It is possible that a locality may have many alien fish that are well fed, not diseased, and flourishing, that may be considered healthy. On the other hand, to properly address ecological and evolutionarily integrity, it should contain the natural complement of species, as reflected by the local long standing previous conditions, with few or no human impacts. The change in dominance from na-

* Corresponding author: saconbal@axtel.net

tive to alien species in fish assemblages has been considered a measure of disturbance (Meffe 1991; Hughes and Noss 1992; Weaver and Garman 1994) or low IBIs (Fausch et al. 1990).

Aquatic indexes have shown high ecological sensitivity elsewhere and are capable of early detecting of environmental impacts, both aquatic and terrestrial (Karr 1981; Karr and Chu 1990). Given the high degree of endemism of Mexican fishes (Lyons et al. 1995), it is difficult to compare areas that may be very close in the same basin yet have distinct species. Accordingly, Mexico needs a method that may be applied to areas not having exactly the same fish fauna, but a parallel and compatible one. That is, a candidate index must be capable of basing predictions that are here derived from our other IBI projects and previous work on the bioecology of Mexican fishes. To properly assess assemblage changes, we must consider zoogeographical and other information and conditions before the impact of human activities (Fausch et al. 1984). Having little such information published for Mexico, we resorted to fish collections and field notes.

Objectives

Our main objective was to use a relative measure of the aquatic integrity of the lower Río Nazas (Smith and Miller 1985; Contreras-Balderas and Ramirez-Flores 2000) to evaluate the impact of humans on the Rio Nazas. The index could in turn be used for both planning for restoration and conservation in the region and for monitoring the best known group of bioindicators (fishes) in the area. The best available information was based on experience and unpublished data. Given the minimal availability of adequate ecological data and the urgent need to have evaluations that consider exploitation rates and rising impacts, available data should be integrated in a workable and informative method to the best of our knowledge.

Study Area

The Rio Nazas basin is one of the few rivers in North México (Figure 1). The basin is an interior drainage located in north-central México, in the state of Durango. Its watershed is known as Comarca Lagunera, North of Torreón, Coahuila. The Nazas is an old basin, losing water in geological time, and now sustaining little runoff as a result of overuse of its waters except during the rainy season. The river drains 85,530 km^2 of mostly mountainous terrain, contains little pluvial fans, is highly fertile, and has a mean annual discharge of 1,661 million m^3 that empties into Laguna Mayrán (Tamayo 1962; Tamayo and West 1964). However, an independent examination of recent water runoff records indicates an average of only 233.9 million m^3 of all the years recorded at Torreón (1936–1972; BANDAS 1997). Hydrologic gauge reports at Torreón from the Comisión Nacional del Agua indicate a 10-year average runoff of 581.9 million m^3 from 1936 to 1945, compared to only 66.4 million m^3 for 1961–1972. Since 1971–1972, gauges have recorded no surface flow. Along its course, two major dams have been constructed; Lázaro Cárdenas, in the middle section, and Francisco Zarco in the lower basin near the towns of Lerdo, Gomez Palacio, and Torreón. In addition, the basin has several smaller diversion dams for irrigation. Dams are known to have multiple environmental impacts, including altered water temperatures, oxygen content, turbidity, siltation, pH and timing, and volume of flows. Dams usually stimulate human colonization and pollution, because of poor public utilities (Ackerman et al. 1973; Contreras-Balderas 1976). The region has also experienced extensive agricultural development and a growing human population that depends on the water from the river.

The basin is located between the Rio Grande (Río Bravo, north), the Sierra Madre Occidental (west), Río Santiago (south), and Río Aguanaval (east). The river was sampled for this study from Presa L. Cárdenas down to Torreón, Coahuila (Figure 1). Samples were taken at El Salvador in the Middle Río Nazas proper. In the lowermost tributary, Río Peñón de Covadonga, we collected from La Concha and Peñón Blanco; in the lower basin, we collected from Presa Francisco Zarco, Puentes Cuates, Parque Raymundo, and Torreón. Published records are scarce. Only Parque Raymundo, Durango, and San Pedro, Coahuila, have been studied for a long time. There are no fish records for the lowermost irrigation channels, which are highly modified from the original Laguna Mayrán, an area that was not

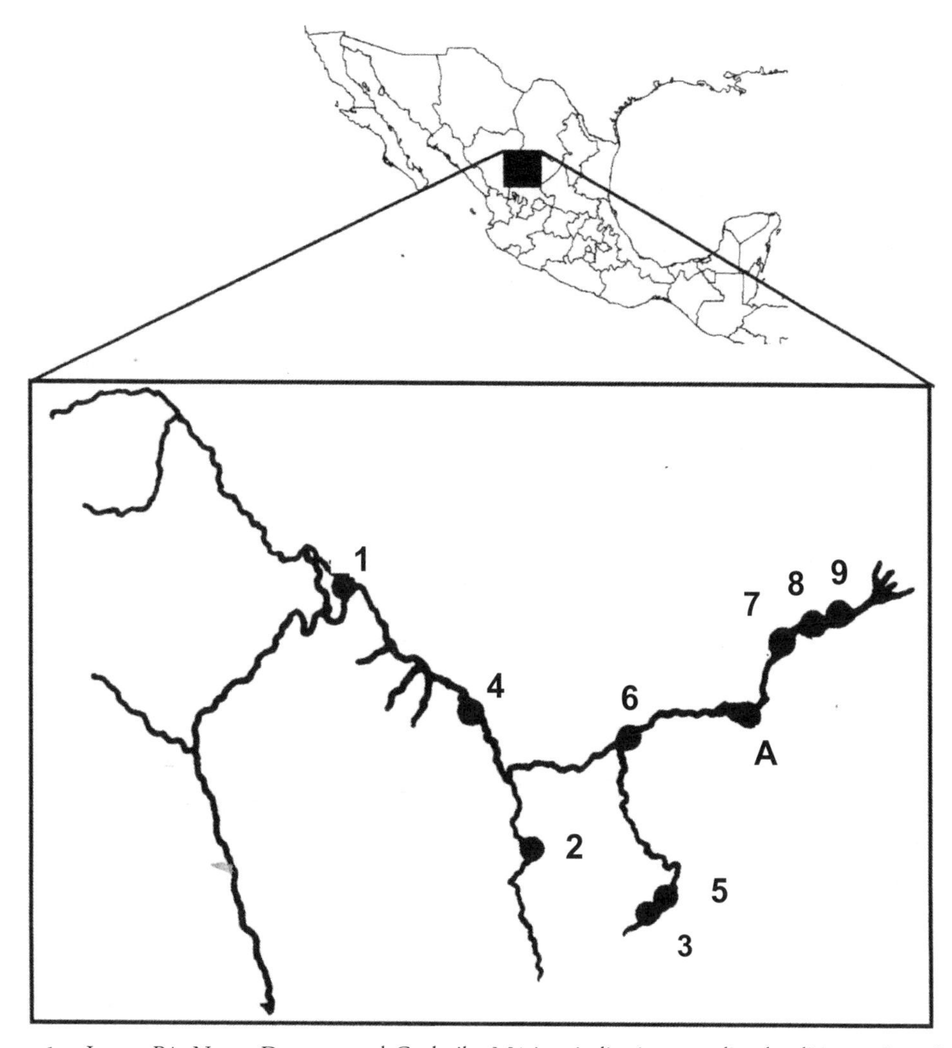

Figure 1.—Lower Río Nazas, Durango and Coahuila, México, indicating sampling localities: 1. Presa Lázaro Cárdenas ; 2. Creek at Primo Verdad; 3. Río del Peñón de Covadonga at La Concha; 4. San Salvador; 5. Río Peñón Blanco; 6. Río Nazas at Nazas; 7. Río Nazas at Puentes Cuates; 8. Río Nazas at Parque Raymundo; 9. Río Nazas at Torreón. A. Reservoir Francisco Zarco, only two fish specimens collected, IBI not calculated. Number sequence from higher headwaters downstream.

sampled in this study because of a lack of or highly polluted water.

Thirteen native fish species are known from the Río Nazas (Table 1). The fauna is derived from Río Bravo forms (Meek 1904), although less complex, consisting of numerous endemics and a number of alien invasive species. Alien fishes and hydraulic works are associated with fewer natives (Contreras-Balderas 1976; Contreras-Balderas et al. 1976) for the fishes at Torreón and Lerdo, and elsewhere (Contreras-Balderas and Escalante-Cavazos 1984; Contreras-Balderas 1999). The local fish assemblages are characterized by species with divergent ecological niches within an ecosystem highly sensitive to changes.

Methods

To assess the condition of the fish fauna with only a scanty database, one of us (SCB) modified the index of biological integrity (Karr 1981, 1997) to a historical version to reflect assemblage changes that we have been monitoring since 1964. We also applied such

Table 1.—Native fish fauna of Rio Nazas, México: listing, endemism, ecological preferences, and status on Norma Oficial Mexicana 059–2002. The indications in the following list are * = basin endemic, + = regional endemic (Tarahumaran), E = endangered, T = threatened, Sp = special protection.

* Sp	*Astyanax* sp.		Omnivorous to carnivorous, mid-water, temperate to tropical.
+ T	*Campostoma ornatum*	Mexican stoneroller	Herbivorous, benthic, usually in cool clean waters.
+ T	*Codoma ornata*	Sardina adornada	Invertivorous, benthic, usually in cool clean waters.
* T	*Gila conspersa*	Nazas chub	Carnivorous, mid-water, preferring cool, clean, rapid waters.
* T	*Notropis nazas*	Nazas shiner	Invertivorous, mid-water, usually in cool, clean waters.
* T	*Cyprinella garmani*	gibbous shiner	Invertivorous, benthic.
* E	*C. alvarezdelvillari*	Tepehuan shiner	Invertivorous, mid-water, thermophilous, withstanding 32°C at La Concha. Possibly extinct.
	Pimephales promelas	fathead minnow	Herbivorous, benthic.
* T	*Pantosteus nebuliferus*	Matalote del Nazas	Microphagous, benthic, usually in cool, clean waters.
* E	*Ictiobus* sp.		Microphagous, benthic, preferring high flow rivers. Occasional or rare.
+ Sp	*Ictalurus pricei*	Yaqui catfish	Carnivorous, benthic, mostly nocturnal, highly tolerant.
* T	*Cyprinodon nazas*	Nazas pupfish	Herbivorous, benthic, standing high salinity or extreme temperatures.
+ T	*Etheostoma pottsi*	Mexican darter	Invertivorous, benthic, in cool, clean, well-oxygenated running waters. A local strain or a different species tolerates 32°C at La Concha. Rare.

an IBI to the lower Rio Grande (Contreras-Balderas et al. 2000). Our version uses 10+ biological parameters known to reflect impacts in the region (Contreras-Balderas 1976; Contreras-Balderas et al. 1976) in order to measure the fish assemblage and habitat changes through time. Our older collections spanned at least 40 years and sometimes included 100 years of literature records. We scaled mean metric scores to 100 and obtained a relative and comparable measure of the integrity of the ecosystem. Species attributes known to reflect changes in habitat in the region are 1) primary (usually preferring cool and clear flowing water), 2) secondary (surviving or even benefiting from warm, stagnant or quiet water, siltation, salinity), 3) food preferences and availability (Contreras-Balderas et al. 1976, 1978). Lack of bioassays on the effect of environmental conditions on Mexican fishes caused us to develop an empirical method for detecting sensitivity of fishes based on our own field data. Such data are often all that are available for evaluations in countries like México.

The IBI of Karr (1981, 1997) consists of a selection of 12 biological parameters deducted from a set of observations on a floral/faunal group, recognized by their values as bioindicators of eco-environmental impacts. The chosen parameters were calibrated with excellent scores. Scores were assigned to the oldest known records (usually 40+ years) at each locality and recent collections were scaled as a percent of the "original" value.

Historical tables were compiled for each locality, with temporary absences regarded as present. Predictions using this modified IBI were 80% precise, considered acceptable for estimating future impacts, ecological planning, and monitoring.

Collecting effort was standardized to 1 h/2 seines (or 2 h/1 seine, 3 × 2 m, 5 mm mesh), except where the size or conditions of the river were poor and strong collecting was not convenient. We did not standardize areas sampled due to the changing nature of desert rivers, with annual droughts and floods, changing the shape and characters of the riverbed. Such changing conditions hinder high precision of ecological data.

In the present study, the following parameters were selected, each expressed in percent values:

1. Original (historical maximum) versus current native species number.
2. Native versus total specimens in each sampling event.

3. Alien versus total specimen number .
4. Primary species number.
5. Secondary species number.
6. Nearctic species number.
7. Neotropicals species number.
8. Microphagous species number.
9. Herbivorous species number.
10. Omnivorous species number.
11. Invertivorous species number.
12. Carnivorous species number.
13. Benthic species number.
14. Midwater species number.
15. Sensitive species number.
16. Tolerant species number.

One of the most valuable parameters is sensitivity. One of us (SCB) developed an empirical method and applied it in the lower Rio Grande (Contreras-Balderas et al. 2000). In this approach, fish species of a given locality disappear in subsets, The first third of species disappearing were termed sensitive, and the lower third was rated tolerant (Table 2).

Calculated percent values obtained for each attribute are added and averaged to provide the IBI for each locality (Table 3). Collection results for each locality are also presented in the tables.

The Nazas fish assemblage is essentially Nearctic and primary, except the Neotropical primary *Astyanax* and the Nearctic secondary Cyprinodon. The fauna is considered strongly threatened due to expansion of the high number of invasive species stocked throughout the basin, dewatering, and pollution.

Recently, a list of the alien fishes in México has been published (Contreras-Balderas 1999) and includes the recommendation of not stocking more alien species (Table 4).

Results

Localities in the main riverbed sustained very high flows due to strong recent rains and the opening of the floodgates at Lázaro Cárdenas dam and a short distance below Zarco dam. Afterwards, flows were greatly reduced and garbage was observed at every locality.

1. Lázaro Cárdenas Reservoir (N 25°36'17"; W 105°01'37").

This dam was completed in 1946 (Figure 2). At the time of our survey, the river was at its lowest recorded level making it possible to drive more than 1 km laterally and 30 m vertically from its

Table 2.—Guilds of native fish species in the Río Nazas basin, Durango, México.

SENS	NATIVES	ORIGIN	ECOL	ZOOGEO	FOOD	HABITAT
1	*Cyprinella alvarezdelvillari*	NAT	PR	NA	IN	MW
1	*Ictiobus cf. bubalus*	NAT	PR	NA	MIC	BEN
1.4	*Codoma ornata*	NAT	PR	NA	IN	MW
1.5	*Campostoma ornatum*	NAT	PR	NA	MIC	BEN
1.6	*Gila conspersa*	NAT	PR	NA	CAR	MW
1.6	*Ictalurus pricei*	NAT	PR	NA	CAR	BEN
1.7	*Pimephales promelas*	NAT	PR	NA	MIC	MW
1.7	*Cyprinodon nazas*	NAT	SE	NT	HER	BEN
2	*Notropis nazas*	NAT	PR	NA	IN	MW
2.2	*Astyanax* sp.	NAT	PR	NT	OM	MW
2.7	*Cyprinella garmani*	NAT	PR	NA	IN	MW
2.8	*Etheostoma pottsi*	NAT	PR	NA	IN	BEN
3	*Pantosteus plebeius*	NAT	PR	NA	MIC	BEN

Note: SENS = sensitivity as calculated from their level of elimination from records, for example 1 = first elimination and most sensitive, NAT = native, ECOL = ecology, PR = primary, SE = secondary, ZOOGEO = zoogeography, NA = nearctic, NT = neotropical, IN = invertivorous, MIC = microphagous, CAR = carnivore, HER = herbivore, OM = omnivore, MW = midwater, BEN = benthic.

Table 3.—Summary of historical index of biotic integrity (IBI) in the lower Río Nazas, Durango, México. Localities: 1. Presa Lázaro Cárdenas ; 2. Creek at Primo Verdad; 3. Río del Peñón de Covadonga at La Concha; 4. San Salvador; 5. Río Peñón Blanco; 6. Río Nazas at Nazas; 7. Río Nazas at Puentes Cuates; 8. Río Nazas at Parque Raymundo (= Lerdo) ; 9. Río Nazas at Torreón. IBI = metric mean per site.

Parameters/stations	1	2	3	4	5	6	7	8	9
% native species	13	40	29	60	63	50	14	17	0
% native individuals	17	100	28	90	90	78	17	4	0
% alien individuals	14	–	0	25	0	0	0	0	0
% primary species	0	40	17	50	71	50	14	17	0
% secondary species	14	–	100	–	0	–	–	–	0
% nearctic species	0	25	17	44	57	42	17	0	0
% neotropical species	0	100	100	100	100	100	100	100	0
% microphagous species	0	0	0	3	100	33	0	0	0
% herbivorous species	0	–	100	–	0	–	–	–	0
% omnivorous species	0	100	100	100	100	100	100	100	0
% invertivorous species	0	0	0	50	67	100	0	0	0
% carnivorous species	50	100	0	50	50	0	0	0	0
% benthic species	20	0	33	33	50	67	100	0	0
% mid-water species	0	67	25	57	75	80	25	33	0
% sensitive species	17	33	20	33	50	25	100	0	0
% tolerant species	0	50	50	75	100	75	25	33	0
IBI	9	50	39	57	61	57	37	22	0
IBI quality	VP	P	VP	P	P	P	VP	VP	VP

Integrity ratings: 95–100 Excellent; 80–94 Good; 65–79 Fair; 50–64 Poor (P); <50 Very poor (VP). 0 = missing in sample; – = not known to be in local assemblage.

Table 4.—Alien fishes of the Río Nazas, Durango/Coahuila, México.

Common name	Scientific name	
Mexican tetra	*Astyanax mexicanus*	Río Bravo species, then southeastward. Not found in this survey, known formerly from the lower Río Nazas region between 1964 and 1994. Origin of introduction not known.
goldfish	*Carassius auratus*	Eurasian species; ornamental, widely introduced, not commonly used as food.
common carp	*Cyprinus carpio*	Eurasian species. Food fish widely stocked through México.
Chihuahua shiner	*Notropis chihuahua*	Endemic species of the Río Conchos, of Río Bravo. Purpose not known.
unknown catfish	*Ictalurus* sp.	Species not identified, similar to *I. punctatus.*
porthole livebearer	*Poeciliopsis gracilis*	Southern México species, around Istmo de Tehuantepec.
bluegill	*Lepomis macrochirus*	North American species, from Río Bravo and northeast. Sports and forage species.
redear sunfish	*L. megalotis*	As former species.
white crappie	*Pomoxis annularis*	Central Texas to northeastern North America. Sports and food species.
largemouth bass	*Micropterus salmoides*	Eastern North American species. Highly regarded sports and food species, good eating also.
tilapia sp.		African and Middle East group of species. Several species cultured for food.
scowling silverside	*Chirostoma aculeatum*	Endemic Mexican species from Chapala/Lerma basins used for food.
slender silverside	*C. attenuatum*	Endemic Mexican species from Chapala/Lerma basins used for food.

Figure 2.—View of Presa Lázaro Cárdenas (= El Palmito) from the dam. Water level is scarcely evident in the far center. The Club de Pesca building at center left, formerly next to the maximum coastline. Just below is the channel leading to the spill. Durango, México. Photo SCB.

maximum historical level to the current shore. Fishing was reported by fishermen as extremely poor, essentially catfish, sunfish, and carp. Water was reasonably clear; however, there were no forage fishes. Our sample was taken from commercial fishermen (Table 5). The IBI here was uninstructive and unrepresentative because of poor collecting and the scarcity of specimens.

Table 5.— Presa Lázaro Cárdenas. Fish faunal history.

	Base*	1964	2002
Natives			
Astyanax cf. Mexicanus	X		
Cyprinodon nazas	X		
Ictiobus sp.	X		
Campostoma ornatum	X	63	
Codoma ornata	X	4	
Gila conspersa	X	40	
Pantosteus nebuliferus	X	20	
Ictalurus pricei	X	?	4
Total native specimens	–	127	4
Aliens			
Carassius auratus	0		1
Lepomis macrochirus	0		14
Lepomis sp.	0		2
Pomoxis annularis	0		1
tilapia sp.	0		1
Total alien specimens			19

? Probably present.
* Probably present before the dam building.

2. Creek in Primo Verdad (N 24°54′10″; W 104°27′09″).

This is a tributary of the middle Río Nazas downstream from El Palmito dam (Figure 3). Flow was highly reduced from that in 1964, changing from clear to muddy. Only two species were collected (Table 6). The IBI changed from 100 in 1964 to 50 in 2002.

3. La Concha (N 24°43′57″; W 104°05′24″).

This site includes a thermal spring on a small tributary that enters a creek and picturesque canyon above the town of Peñón Blanco. Water was abundant in 1988, and subjected to heavy flash floods after torrential rains in 1994 and 1998. The spring and nearby creek have been altered by a park and swimming pools, administered by a group of residents. The spring has at least one endemic species, Tepehuan shiner, which Contreras-Balderas and Lozano-Vilano (1994) listed as endangered (NOM 059 ECOL 2001). Tilapia has been present since the first visit (Table 7). The IBI went from 95 in 1988 to 39 in 2002.

4. Río Nazas at San Salvador (N 25°17′44″; 104°36′58″).

Located downstream from El Palmito dam, this site has a diversion dam for agricultural uses with high

Figure 3.—Arroyo Ramos in the locality Primo Verdad. Formerly the river bed spanned to the shoulder along the sides, with an acceptable flow nearly 0.6 m above actual level. Durango, México. Photo MLLV.

Table 6.—Río Ramos at Primo Verdad, fish faunal history.

Native specimens	1964	2002
Codoma ornata	4	
Campostoma ornatum	63	
Pantosteus nebuliferus	20	
Astyanax cf. mexicanus		14
Gila conspersa	40	28
Native specimens	127	42

flow due to recent rains and the opening of the spillways at El Palmito dam. Water was slightly turbid and with a foul odor.

The bottom changed from sandy to muddy. Faunal changes are presented in Table 8; there was one alien species since the first collecting. The IBI changed from 87 in 1964 to 57 in 2002.

5. Río Peñón Blanco (N 24°47'05"; W 104°02'08").

Located near the town of the same name, this site is a short distance below site 3, at the lower reaches of the canyon, bordered by several cultivated fields. It is highly polluted by organic and agricultural loads. Formerly clean, flowing, and with few weeds, the locality became muddy, green to clear in places, stinky, and with reduced flow. The bottom is muddy, with some small sandy or gravel beds. An alien weed *Elodea* was abundant not found before. The fish fauna reflected similar changes (Table 9), and the IBI changed from 100 in 1968 to 61 in 2002.

Table 7.—Río Peñón Blanco at Balneario La Concha, Durango, México, fish faunal history.

Species	1988	2002
Natives		
Codoma ornata	78	
Cyprinella alvarezdelvillari	478	
Gila conspersa	23	
Pantosteus nebuliferus	1	
Ictalurus pricei	4	
Astyanax cf. mexicanus	193	3
Cyprinodon nazas	89	23
Native specimens	866	26
Aliens		
Tilapia sp.	135	58
Poeciliopsis gracilis		8
Alien specimens	135	66

Table 8.—Río Nazas at San Salvador, Durango, México, fish faunal history.

	1964	1981	2002
Natives			
Codoma ornata	1		
Gila conspersa	31		
Notropis braytoni	6		
Pantosteus nebuliferus	7		
Ictiobus sp.	5	1	
Astyanax cf. mexicanus	10		11
Cyprinella garmani	725	38	239
Notropis nazas	?	?	3
Pimephales promelas	?	?	20
Ictalurus sp.	?	?	5
Native specimens	785	39	278
Aliens			
Chirostoma aculeatum	3		
Tilapia sp. *indet.* sp. scowling silverside	14		
Notropis chihuahua Chihuahua shiner	10	?	7
Lepomis macrochirus	7	28	4
Micropterus salmoides	22	3	11
Pomoxis annularis			5
Lepomis megalotis			2
Ictalurus sp. indetermined			1
Alien specimens	42	45	30

6. Río Nazas at Nazas (25°12'59"; W 104°10'32").

Located downstream of the Ramos and Nazas rivers junction, this site has diversion dams a few kilometers upstream from the town of Nazas. Formerly mostly sandy, it has become mostly muddy and the fish assemblage has been altered (Table 10). The IBI went from 100 in 1968 to 57 in 2002.

7. Río Nazas at Puentes Cuates (N 25°28'02"; W 103°41'34").

This locality had a moderate current, but was strongly polluted, dark colored, and plagued by introduced aquatic weeds like *Elodea*. The bottom was putrid mud, indicating a heavy organic load not present in 1964. The fish fauna also changed (Table 11) with

Table 9.—Río Peñón Blanco at Peñón Blanco, Durango, México, fish faunal history.

Natives	1968	1975	2002
Cyprinella alvarezdelvillari	1		
Ictalurus pricei	5		
Cyprinodon nazas	68	24	
Astyanax cf. mexicanus	136	4	1
Codoma ornata	132	17	222
Gila conspersa	311	234	61
Pantosteus nebuliferus	1	14	13
Etheostoma pottsi	?	1	62
Native specimens	654	294	359
Aliens			
Carassius auratus			1
Poeciliopsis gracilis			19
Lepomis macrochirus			1
Tilapia sp.			1
Notropis chihuahua			18
Alien specimens			40

Table 10.—Río Nazas at Nazas, Durango, México, fish faunal history.

	1968	2002
Natives		
Gila conspersa	66	
Campostoma ornatum	2	
Pantosteus nebuliferus	5	
Ictalurus pricei	1	
Astyanax sp.	2	3
Notropis nazas	213	1
Cyprinella garmani	?	140
Pimephales promelas	?	148
Native specimens	289	292
Aliens		
Cyprinus carpio		1
Notropis chihuahua		57
Lepomis macrochirus		1
Micropterus salmoides		14
Pomoxis annularis		7
Tilapia sp. indetermined		2
Alien specimens		82

Table 11.—Río Nazas at Puentes Cuates, Durango, México, fish faunal history.

	1964	1968a	1968b	2002
Campostoma ornatum	?	?	1	
Pantosteus nebuliferus		6	9	
Gila conspersa		94	93	
Ictalurus pricei	4	?	14	
Notropis nazas	17	544	234	
Cyprinella garmani	136	162	45	
Astyanax cf. mexicanus	59	55	83	24
Native specimens	216	861	479	24
Aliens				
Lepomis macrochirus	2	168	30	5
Micropterus salmoides	1	57	17	40
Tilapia sp. indetermined sp				67
Alien specimens	3	225	47	112

an IBI of 95 in 1964, due to early alien introductions, and 37 in 2002.

8. Río Nazas at Parque Raymundo (N 25°30'42"; W 103°32'13").

This site was at a small city park at the riverside, where the river was very wide and half as shallow as in 1994 (Figure 4). The bottom had changed from sandy to muddy over coarse gravel and rubble, but no aquatic weeds were detected. Marked change was detected over the period of record (Table 12) and the IBI went from 100 in 1903 to 22 in 2002.

Figure 4.—Río Nazas, at Parque Raymundo, Lerdo, Durango, México. Average level in 1964, some 50 cm higher, to just above the car on the center left. Water disappeared from this section of the river in late spring 2004. Photo MLLV.

Table 12.—Río Nazas at Parque Raymundo, Lerdo, Durango, México, fish faunal history.

	Meek 1903	1963	1964	1975	2002
Natives					
Ictalurus pricei	X				
Pantosteus nebuliferus	X				
Gila conspersa	X	1			
Ictiobus sp.	?	?	1		
Cyprinella garmani	X	187	7	19	
Astyanax cf. mexicanus	X	4	34	?	18
Native specimens		192	42	19	18
Aliens					
Lepomis macrochirus		89	16	90	236
Micropterus salmoides		6	3	146	5
Pomoxis annularis					1
Tilapia sp.					211
Chirostoma cf attenuatum					2
Cyprinus carpio					8
Alien specimens		95	19	236	463

9. Río Nazas En Torreón.

This main river site is located between Lerdo and Torreón, where all water is diverted for agricultural uses. The river now has only a few water holes with most of the old river dry year round. In 1968, one of us (SCB) collected near the railroad bridge after a rain, securing an excellent sample (Table 13). The IBI went down from the original 100 in 1903 to 0 in 2002.

Discussion

Biotic integrity of the lower Río Nazas has been much reduced, averaging 31.1% of historical levels (Figure 5). Because of diversions, the river now rarely retains surface flow in reaches near Torreón and San Pedro de las Colonias, although historically surface flows occurred every rainy season. Channel bottoms, formerly sandy or gravelly, are now silted and muddy, products of land and riverbank erosion. Overexploitation of water is observed by lower flows in tributaries, and the absence of flow at downriver localities. Pollution is high, both point and nonpoint (mainly agricultural).

The fish fauna of the lower Río Nazas consists of 13 species, 11 are endemic to the basin or to the combined basins with Río Aguanaval and Laguna de Tlahualilo. Eleven are listed in NOM 059 ECOL 2001. The levels of endemism and risk are high for the basin. At least two other species are probably endemics and are under study at this time.

Current IBI scores, varied from 9 to 57 for the main-stem Río Nazas in its middle reaches, 50 in its tributary Río Ramos, and 39 to 61 in Río Peñón Blanco. All together, the lower Río Nazas yielded integrity values between 57 and 0 as water extraction and pollution increased downstream. At Presa Lazaro Cardenas, conditions did not allow seining,

Table 13.—Río Nazas at Torreón, Coahuila, Limits with Durango, México, fish faunal history.

	MEEK 1903	1968	2002
Natives			
Ictiobus sp.	X	0	0
Notropis nazas	?	118	0
Gila conspersa	X	32	0
Ictalurus pricei	X	2	0
Cyprinella garmani	X	133	0
Pantosteus nebuliferus	X	13	0
Native specimens		152	0
Alien			
Astyanax cf. mexicanus	X	65	0
Alien specimens		65	0

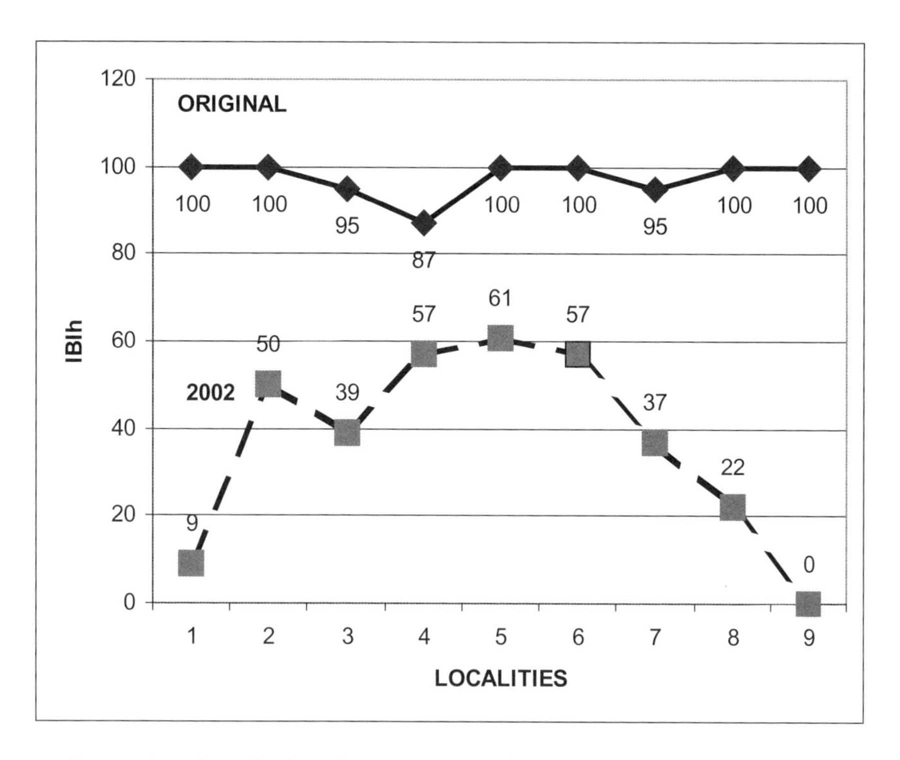

Figure 5.—Changes in index of biological integrity scores along the lower Río Nazas, Durango and Coahuila, México: 1. Presa Lázaro Cárdenas; 2. Creek at Primo Verdad; 3. Río del Peñón de Covadonga at La Concha; 4. San Salvador; 5. Río Peñón Blanco; 6. Río Nazas at Nazas; 7. Río Nazas at Puentes Cuates; 8. Río Nazas at Parque Raymundo; 9. Río Nazas at Torreón. Localities 1, 2, 3/5, represent different heedwaters. 4 is below junction from 1 and 2, 2nd from 6 to 9 are linear main stem.

so the fish sample was taken from fishermen and resulted in extremely low IBI values. At Torreón, water was usually absent. The few pools in this section are occasional and highly polluted.

Besides water quality and quantity problems, the basin experiences severe erosion and consequently turbidity and siltation in the rivers. Silt from Río Nazas is deposited at Presa Lazaro Cardenas, reducing its capacity and probable life span. Turbidity lowers the oxygen content, clogs fish gills, and covers coarse substrates where fish feed and spawn, limiting or preventing reproduction, concealment, and survival of fishes, although they may tolerate short-term turbidity.

Conditions of the lower Río Nazas result from local impacts and from cumulative impacts on the catchment. The basin has exhausted available water, with no water remaining for future growth and development.

Recommendations

Treatment of sewage and industrial waters to ecological protection levels, as defined by the Mexican federal government, should be required. Also, there is a need to control agriculture fertilizers and pesticides to improve water quality. Water-saving irrigation systems should be implemented in the basin to diminish waste and to retain instream flow. Garbage should be collected and not allowed to reach the riverbed. Water levels in the river and reservoirs should be regulated to retain commercial and sport fishing, and other environmental services of the aquatic ecosystem. This aspect is more relevant in reservoirs where discharges to absorb the spring floods are managed in the breeding season of the most important sport and food fishes (e.g., largemouth bass), when they are guarding their nests.

It is of the utmost importance to have a good

erosion control program in the basin, especially in forest and agricultural lands, to prevent turbidity and siltation from stressing resevoir fish assemblages be they native or alien.

References

Ackerman, W. C., G. White, E. B. Worthington, and J. Lorena Ivens, editors. 1973. Man-made lakes: their problems and environmental effects. American Geophysical Union, Geophysical Monograph No. 17, Washington D.C.

Angermeier, P. 1994. Does biodiversity include artificial diversity? Conservation Biology 8(2):600–602.

BANDAS. 1997. Banco nacional de datos de aguas superficiales (National data base of surface waters). Set: 7 CDs. Comisión Nacional del Agua – Instituto Mexicano de Tecnología del Agua. Jiutepec, México.

Contreras-Balderas, S. 1976. Impacto ambiental de obras hidráulicas. Dirección General del Plan Nacional Hidráulico, Secretaría de Recursos Hidráulicos, Informes. (Environmental impact of hydraulic works. General Direction of National Hydraulic Plan, Ministry of Hydraulic Resources, Reports, Mexico City.)

Contreras-Balderas, S. 1978. Speciation aspects and man-made community composition changes in Chihuahuan desert fishes. Pages 405–431 *in* R. H. Wauer, and D. H. Riskind, editors. Biological resources Chihuahuan desert region. U.S. Department of the Interior, National Park Service, Washington, D.C.

Contreras-Balderas, S. 1999. Annotated checklist of introduced invasive fishes in Mexico, with examples of some recent introductions. Pages 31–52 *in* R. Claudi and J.H. Leach, editors, Nonindigenous freshwater organisms: vectors, biology, and impacts. Lewis Publishers, Boca Raton, Florida.

Contreras-Balderas, S., R. J. Edwards, M. L. Lozano Vilano, and M. E. García Ramírez. 2000. Ecology. In J. Schmandt, editor. Water and sustainable development in the lower Rio Grande/Rio Bravo, Texas and Mexico. CD final report, Houston Advanced Research Center and Instituto Tecnológico y de Estudios Superiores de Monterrey: *www.harc.edu/mitchellcenter/*, go to: publications, title, and Ecology chapter.

Contreras-Balderas, S., and M. A. Escalante-Cavazos. 1984. Distribution and known impacts of exotic fishes in Mexico. Pages 102–130 *in* W. R. Courtenay and J. R. Stauffer Jr., editors. Distribution, biology, and management of exotic fishes. Johns Hopkins University Press, Baltimore, Maryland.

Contreras-Balderas, S., V. Landa-Salinas, T. Villegas-Gaytán, and G. Rodríguez-Olmos. 1976. Peces, piscicultura, polución, planificación pesquera y monitoreo en México, o la danza de las. Pages 217–315 *in* Primer Simposio de Planificación Pesquera, México. (Fishes, fish culture, pollution, fisheries planning, and monitoring in Mexico, or the dance of the P's. First Symposium of Fisheries Planning, Tuxtla Gutierres, Chiapas, Mexico.)

Contreras-Balderas, S., and M. L. Lozano-Vilano. 1994. *Cyprinella alvarezdelvillari*, a new cyprinid fish from Río Nazas of México, with a key to the *lepida* clade. Copeia 1994:897–906.

Contreras-Balderas, S., and M. Ramírez-Flores. 2000. Inventario nacional de la ictiofauna dulceacuícola Mexicana. Capítulo V, En: Estado de salud de la acuicultura, Dir. Gral. de Investigación en Acuacultura, Instituto Nacional de la Pesca, Secretaría de Medio Ambiente, Recursos Naturales y Pesca, México. (National Inventory of Mexican freshwater fish fauna. General Direction of Aquaculture, National Institute of Fisheries, Ministry of the Environment, Natural Resources and Fisheries, Mexico City.)

Fausch, K. D., J. R. Karr, and P. R. Yant. 1984. Regional application of an index of biological integrity based on stream fish communities. Transactions of the American Fisheries Society 113:39–55.

Fausch, K. D., J. Lyons, J. R. Karr, and P. Angermeier. 1990. Fish communities as indicators of environmental degradation. Pages 123–144 *in* S. M. Adams, editor. Biological indicators of stress in fish.American Fisheries Society, Symposium 8, Bethesda, Maryland.

Hughes, R. M., and R. F. Noss. 1992. Biological diversity and biological integrity: current concerns for lakes and streams. Fisheries 17(3):11–19.

Karr, J. R. 1981. Assesment of biotic integrity using fish communities. Fisheries 6(6):21–27.

Karr, J. R. 1997. Measuring biological integrity. Pages 483–485 *in* G. K. Meffe and C. R. Carroll, editors. Principles of conservation biology, 2nd Edition. Sinauer, Sunderland, Massachussetts.

Karr, J. R., and E. W. Chu. 1990. Restoring life in running waters. Better biological monitoring. Island Press, Covelo, California.

Lyons, J., S. Navarro-Pérez, Ph. A. Cochran, E. Santana-C., and M. Guzmán-Arroyo. 1995. Index of biotic integrity based on fish assemblages for the conservation of streams and rivers in west-central Mexico. Conservation Biology 9:569–584.

Meek, S. E. 1904. The freshwater fishes of Mexico north of the Isthmus of Tehuantepec. Field Museum of Natural History Publications Zoological Series 5.

Meffe, G. K. 1991. Failed invasions of a southeastern blackwater stream by bluegillls: implications for conservation of native communities. Transactions of the American Fisheries Society 120:333–338.

Moyle, P. B., H. W. Li, and B. A. Barton, 1986. The Frankenstein effect: impact of introduced fishes on native fishes in North America. Pages 415–426 *in* R. H. Stroud, editor. Fish culture in fisheries management. American Fisheries Society, Bethesda, Maryland.

NOM 059 ECOL. 2001. Norma Oficial Mexicana de Especies en Riesgo, Protección ambiental - Especies nativas de México de flora y fauna silvestres - Categorías de riesgo y especificaciones para su inclusión, exclusión o cambio - Lista de especies en riesgo. Diario Oficial de la Federación, Marzo 6, 2002. (Mexican Official Norm of species at risk, environmental protection-native species of wild flora and fauna of Mexico-Risk categories and specifications for inclusion, exclusion, or change-list of species at risk. Federal Official Diary, March 6, 2002, Mexico City.)

Smith, M. L., and R. R. Miller. 1985. The fishes of the Rio Grande basin. Pages 457–485 *in* C. H. Hocutt and E. O. Wiley, editors. Zoogeography of North American freshwater fishes. John Wiley, New York.

Tamayo, J. L. 1962. Geografia General de Mexico, Instituto Mexicano de Investigaciones Economicas, Mexico, 4 vol. and atlas.

Tamayo, J. L., and R. C. West. 1964. Hydrography of middle America. Pages 84–121 *in* R. C. West, editor. Handbook of middle American Indians, volume 1. Natural environment and early cultures. University of Texas Press, Austin.

Weaver, L. A., and G. C. Garman. 1994. Urbanization of a watershed and historical changes in a stream fish assemblage. Transactions of the American Fisheries Society 123:162–172.

American Fisheries Society Symposium 45:239–248, 2005

Historical Changes in Fish Distribution and Abundance in the Platte River in Nebraska

E. J. PETERS*
School of Natural Resource Sciences, University of Nebraska, Lincoln, Nebraska 68583-0814, USA

S. SCHAINOST
Nebraska Game and Parks Commission, Alliance, Nebraska 69301-0725, USA

Abstract.—From its headwaters in the Rocky Mountains, the Platte River drains 230,362 km^2 in Colorado, Wyoming, and Nebraska. The Platte River is formed by the confluence of the North Platte and South Platte near the city of North Platte, Nebraska, and receives additional flow from the Loup and Elkhorn rivers that drain the Sand Hills region of Nebraska. Water diversions for mining and irrigation began in the 1840s in Colorado and Wyoming, and irrigation diversions in Nebraska began in the 1850s. Construction of dams for control of river flows commenced on the North Platte River in Wyoming in 1904. Additional dams and diversions in the North Platte, South Platte, and Platte rivers have extensively modified natural flow patterns and caused interruptions of flows. Pollution, from mining, industrial, municipal, and agricultural sources, and introductions of 24 nonnative species have also taken their toll. Fishes of the basin were little studied before changes in land use, pollution, and introduction of exotic species began. The current fish fauna totals approximately 100 species from 20 families. Native species richness declines westward, but some species find refugia in western headwaters streams. Declines in 26 native species has led to their being listing as species of concern by one or more basin states.

Introduction

The Platte River Basin

The Platte River basin drains an area of 230,362 km^2 in Colorado, Wyoming, and Nebraska. The North Platte River and South Platte River rise in the Rocky Mountains. They flow generally eastward until they merge in western Nebraska near North Platte, where the Platte River continues across the Great Plains until it reaches the Missouri River (Figure 1).

North Platte River.—The North Platte River drains approximately 90,352 km^2 in Colorado, Wyoming, and western Nebraska. Most of the flow in the North Platte River originates as snowmelt runoff, which peaks in May and June. Early efforts at irrigation in Wyoming were noted as early as 1847 (McKinley 1935). Construction of storage reservoirs and diversion canal systems that began in the late 1800s and were completed in the mid 1940s on the North Platte have altered this flow pattern. This has resulted in channel narrowing (Figure 2) and fragmentation of river habitats by dewatering and dams.

South Platte River.—The South Platte River drains approximately 62,888 km^2 from the mountains of central Colorado. Like the North Platte River, the South Platte is primarily a snowmelt river, but some flow results from trans-basin

* Corresponding author: EPETERS2@unl.edu

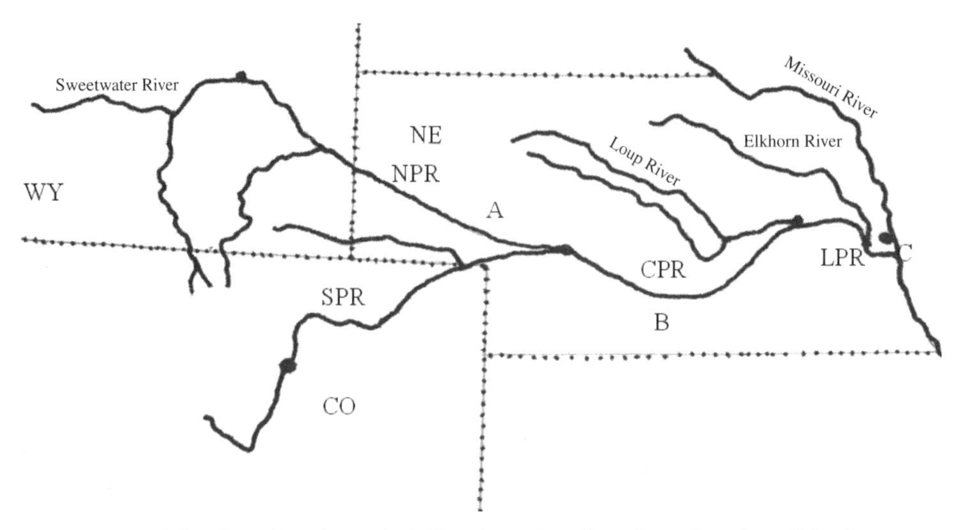

FIGURE 1.—The Platte River basin, including the major tributaries and reaches of the river system (NPR = North Platte River; SPR = South Platte River; CPR = central Platte River; LPR = lower Platte River). Specific localities compared on aerial photos are (A) North Platte River near Keystone, Nebraska (NE), (B) Platte River near Kearney, Nebraska, and (C) Platte River at its confluence with the Missouri River. CO = Colorado; WY = Wyoming.

diversions from the Colorado River drainage. Diversions of water for mining and watering gardens began by the 1840s in Colorado and then expanded to use for raising crops by 1860 (McKinley 1935). Today, virtually all of the flow of the South Platte is diverted in Colorado but restored via sewage treatment plant water return flows downstream from metropolitan areas of the front range.

Platte River.— Two reaches of the Platte River in Nebraska are termed the middle or central Platte and the lower Platte. The middle Platte reach comprises the 340-km section of the river from the North Platte–South Platte confluence downstream to the mouth of the Loup River. Records of irrigation in this reach of the Platte date to the 1850s (Carlson 1963). Today, flow in this reach is highly altered by upstream diversions and irrigation return flows. Encroachment of woody vegetation has changed the river from open shifting sand bars to wooded islands that separate narrow river channels (Figure 3). The lower Platte begins where the Loup River joins its flow with the Platte near the present location of Columbus, Nebraska. This 160-km reach of the river also receives additional water from the Elkhorn River and Salt Creek (Figure 1) before flowing into the Missouri River. The lower Platte reach retains the wide, shifting sand bar habitats and high turbidity characteristic of the pre-European settlement river (Figure 4). This may be due to fluctuating flows from peaking power generation facilities at Columbus. Irrigation and municipal water use in this reach of the Platte River is primarily from wells placed in aquifers fed by percolation from river flows.

Studies of Fishes of the Platte River

Little is known of the fishes from the Platte River basin prior to the time when diversions and damming had already altered flow volumes and patterns. Meek (1895) and Evermann and Cox (1896) are the only 19th century published records we have. Subsequently, during the first half of the 20th century, Ellis (1914), Johnson (1942), and Simon (1946) completed studies of Colorado, Nebraska, and Wyoming fishes, respectively. More recently, studies by Baxter and Stone (1995) and Patton (1997) have documented the status of fish populations in Wyoming. In Colorado, Propst and Carlson (1986), Woodling (1985), and Nesler et al. (1997) have inventoried and evaluated the status of fishes in the South Platte River. In Nebraska, Morris (1960) and Bliss and Schainost

Figure 2.—Aerial photographs showing the North Platte River and the Central Public Power and Irrigation District canal near Keystone, Nebraska, in (A) 1939, with open sand bars in the wide channel, and (B) 1993, with a narrow channel.

Figure 3. —Aerial photographs showing the Platte River near Kearney, Nebraska, in (A) 1939 with open sandbars, and (B) 1993, with tree covered sand islands.

(1973a, 1973b, 1973c, 1973d) inventoried fishes in the North Platte, South Platte, middle Platte, and lower Platte drainages. Lynch and Roh (1996) surveyed the fishes of the North Platte and South Platte rivers in western Nebraska. Chadwick et al. (1997) and Goldowitz (1996) compared fish species composition from their studies to those of Johnson (1942). Peters et al. (1989), Peters and Holland (1994), and Yu and Peters (1997) collected extensively in the lower Platte River for development of habitat suitability index models. Data from unpublished collection permit reports compiled by S. Schainost provided additional distributional information (Schainost and Koneya 1999). Our evaluations about changes in abundance in relation to habitat changes have been limited to Nebraska records, but we gleaned inferences regarding changes in distribution and abundance from literature references to Colorado and Wyoming.

The Platte River fish fauna includes approximately 100 species (Appendix A), of which 26 have been labeled as "species of concern" by at least one of the three basin states. The fishes of the Platte River basin exhibit three general types of distributional patterns. These are widespread species, downstream big river species, and headwaters or upstream species. Imposed upon these general patterns are the effects of anthropogenic habitat alterations that have resulted in the extirpation of several species from all or parts of the basin. On the other hand, introductions of species have spread some distributional boundaries to the west.

Among the big river species, lake sturgeon *Acipenser fulvescens* and pallid sturgeon

FIGURE 4.—Aerial photographs of the Platte River at its confluence with the Missouri River, showing the many channels of the Missouri River in (A) 1939, and the single channel in (B) 1971. Platte River bars and islands are abundant in 1939, but not in 1971.

Scaphirhynchus albus are known only from the lower Platte. Shovelnose sturgeon S. platorynchus was collected in the North Platte drainage as far west as Casper (Evermann and Cox 1896), but today is confined primarily to the lower Platte. Paddlefish *Polyodon spathula*, longnose gar *Lepisosteus osseus*, shortnose gar *L. platostomus*, and goldeye *Hiodon alosoides* are found only in the lower and middle reaches of the Platte, although there are records of goldeye in the North Platte River in Wyoming (Evermann and Cox 1896). In the family Cyprinidae, speckled chub *Macrhybopsis aestivalis*, sturgeon chub *M. geleda*, and silver chub *M. storeriana* are virtually confined to the lower Platte River today, and the sicklefin chub *M. meeki* has been extirpated from the Platte River system. The sturgeon chub's distribution has shrunk most, since it is no longer found in the North Platte or the middle Platte reaches of the basin. Among the suckers, the blue sucker *Cycleptus elongatus*, bigmouth buffalo *Ictiobus niger*, and smallmouth buffalo *I. cyprinellus* are known only from the lower Platte or as far west as the central Platte (bigmouth buffalo). Of the ictalurids, the blue catfish *Ictalurus furcatus* is native only to the lower Platte, but has been stocked more widely in the basin. Flathead catfish *Pylodictis olivaris*, yellow bullhead *Ameiurus natalis*, and tadpole madtom *Noturus gyrinus* reach their upstream limits in the central Platte. Northern pike *Esox lucius* were apparently native to the lower Platte, but they, along with grass pickerel *Esox americanus vermiculatus*, have been introduced farther west into the North Platte drainage. This is probably also the case with several of the centrarchids, like bluegill *Lepomis macrochirus*, largemouth bass *Micropterus salmoides*, and the crappies *Pomoxis* spp. Among the Percidae, sauger *Sander canadensis* has been extirpated from the North Platte and South Platte rivers in Wyoming and Colorado and, today, is uncommon in the lower Platte. By contrast, walleye *S. vitreus*, have been widely introduced and are now an important predator throughout the Platte basin. Finally, freshwater drum *Aplodinotus grunniens* were probably confined to the eastern part of the Platte basin, but today they are found associated with reservoirs in the North Platte and South Platte drainages where they were apparently introduced by accident.

Headwaters or upstream species include a diverse group of fishes, some of which may be relicts of Pleistocene distributions that are now more typical of more northern or northeastern fish assemblages. Other species may have been more widespread in clear, warmwater stream systems of the pre-European Great Plains. Those with more extensive distributions to the north include lake chub *Couesius plumbeus*, common shiner *Luxilus cornutus*, northern redbelly dace *Phoxinus eos*, finescale dace *Phoxinus neogaeus*, hornyhead chub *Nocomis biguttatus*, blacknose shiner *Notropis heterolepis*, blacknose dace *Rhinichthys atratulus*, longnose dace *R. cataractae*, and longnose sucker *Catostomus catostomus*. Two species, mountain sucker *C. platyrhynchus* and greenback cutthroat trout *Oncorhynchus clarki stomias*, have distinctly western affiliations. Both are the farthest eastern representatives of their species range. The clear warmwater species include central stoneroller *Campostoma anomalum*, Topeka shiner *Notropis topeka*, stonecat

Noturus flavus, plains topminnow *Fundulus sciadicus*, plains killifish *Fundulus zebrinus*, brook stickleback *Culea inconstans*, orangespotted sunfish *Lepomis humilus*, Iowa darter *Etheostoma exile*, johnny darter *E. nigrum*, and orangethroat darter *E. spectable*. A common denominator in their habitat requirements seems to be clean gravel substrate, which is in short supply in many plains streams today. All of these species are listed as a species of concern, by one or more of the basin states (Appendix A) because of destruction of their remaining habitat in some part of their range.

Today's widespread species have probably been abundant for a long time. They are opportunists that can adapt to a wide variety of conditions. Red shiner, sand shiner *Notropis stramineus*, fathead minnow, bigmouth shiner, green sunfish, black bullhead, river carpsucker, quillback, and white sucker are widely distributed and noted as tolerant species in many classifications. However, black bullhead is listed as a species of concern by Wyoming because of its limited distribution in streams there.

Collection records in Nebraska prior to 1945 documented 47 species from the Platte River of which one, the common carp, was an introduced species. Comparison of the ranks of frequency of occurrence of these species in collections since 1985 indicated no change in rank for three species, decrease in rank for 29 species, and increase in rank for 15 species (Table 1). We did not attempt to evaluate the influence of using different types of sampling gear on the species collected or their abundance. Three species, red shiner, sand shiner, and speckled chub, did not change rank. The red shiner and sand shiner are arguably the most abundant and widespread of all fish species in the Platte River system. In Nebraska, they were captured in over 60% of the pre-1945 collections and over 70% of the post-1985 collections. Speckled chub are much less abundant but widespread in the Platte River. All three species use shallow water areas (<60 cm) with current velocities up to 60 cm/s, and these habitats are widespread in the Platte River. However, red shiner and sand shiner tend to be advantaged by slightly clearer water (Bonner and Wilde 2002). Of the species changing rank, 14 species declined 10 or less ranks and 12 species increased 10 or less ranks. Included in this near-neutral group are species like plains minnow, river carpsucker, silver chub, and quillback. Plains minnow seems to have declined in abundance in areas where water flows have become more controlled but not in the lower Platte River where water level fluctuations may foster its

Table 1.—Changes in ranks of percent frequency of occurrence for fish species reported in collections from the Platte River in Nebraska before 1945 and after 1985.

Decline more than 10 ranks	Decline 1–10 ranks	No change in rank	Increase 1–10 ranks	Increase more than 10 ranks
Shovelnose sturgeon	Plains minnow	Red shiner	Bigmouth shiner	Shortnose gar
Lake sturgeon	Brassy minnow	Sand shiner	Fathead minnow	Shorthead redhorse
Goldeye	Emerald shiner	Speckled chub	Common carp	Channel catfish
River shiner	Central stoneroller		Silver chub	
Flathead chub	Suckermouth minnow		Creek chub	
Western silvery minnow	Longnose dace		Quillback	
Common shiner	River carpsucker		White sucker	
Sturgeon chub	Longnose sucker		Black bullhead	
Golden shiner	Yellow bullhead		Plains topminnow	
Bigmouth buffalo	Stonecat		Green sunfish	
Tadpole madtom	Plains killifish		Largemouth bass	
Orangespotted sunfish	Orangethroat darter		White crapppie	
Iowa darter	Yellow perch			
Johnny darter	Walleye			
Sauger				

reproduction (Bonner and Wilde 2000). We would expect silver chub to increase in abundance with increases in water clarity since it is a sight-feeding species (Cross and Moss 1987).

The 15 species that declined more than 10 ranks in frequency of occurrence included several species, such as shovelnose sturgeon, goldeye, western silvery minnow, flathead chub, sturgeon chub, and sauger, which are considered typical of large turbid rivers. Shovelnose sturgeon has been extirpated from the North Platte River, is infrequently found in the middle Platte, and is common only in the lower Platte. Goldeye is also only found regularly in the lower Platte and has been extirpated from the North Platte in Wyoming. Sturgeon chub and sauger have also been extirpated from the North Platte River as well as the middle Platte. Populations of flathead chub are declining in the North Platte River, have virtually disappeared from the middle Platte, and are most widespread in the lower Platte. Western silvery minnow was never known from Colorado or Wyoming reaches of the Platte drainage but seems to be holding its own in Nebraska.

The three species that increased more than 10 ranks are channel catfish, shorthead redhorse, and shortnose gar. Channel catfish is widely stocked as a sport fish. Shorthead redhorse populations may be fostered by clear and cool effluents from reservoirs along the river. Shortnose gar collections may be greater because of the use of a wider variety of sampling gear during recent years.

The list of species collected since 1985 includes a growing number of introduced species, some of which have become invasive and may be having an impact on native fish populations. Many of these, like rainbow trout, alewife, brook silverside, and white perch, are not native to any part of the Platte River drainage and were imported to enhance reservoir fisheries. Others, like walleye, largemouth bass, and white crappie, are more widespread today than they were prior to European settlement. The impact of western mosquitofish on plains topminnow populations has been documented by Lynch (1988). Similar studies to evaluate the impacts of species such as bighead carp, brook silverside, white perch, and others on native stream fishes, to our knowledge, have not been conducted.

Acknowledgments

Support to the authors was provided by the Nebraska Game and Parks Commission and the University of Nebraska Agricultural Research Division.

References

Baxter, G. T., and M. D. Stone. 1995. Fishes of Wyoming. Wyoming Game and Fish Department, Cheyenne.

Bliss, Q. P., and S. Schainost. 1973a. South Platte River stream inventory report. Nebraska Game and Parks Commission, Lincoln.

Bliss, Q. P., and S. Schainost. 1973b. North Platte River stream inventory report. Nebraska Game and Parks Commission, Lincoln.

Bliss, Q. P., and S. Schainost. 1973c. Middle Platte River stream inventory report. Nebraska Game and Parks Commission, Lincoln.

Bliss, Q. P., and S. Schainost. 1973d. Lower Platte River stream inventory report. Nebraska Game and Parks Commission, Lincoln.

Bonner, T. H., and G. R. Wilde. 2000. Changes in the fish assemblage of the Canadian River, Texas, associated with reservoir construction. Journal of Freshwater Ecology 15:189–198.

Bonner, T. H., and G. R. Wilde. 2002. Effects of turbidity on prey consumption by prairie stream fishes. Transactions of the American Fisheries Society 131:1203–1208.

Carlson, M. E. 1963. The development of irrigation in Nebraska, 1854–1910: a descriptive survey. Doctoral dissertation. University of Nebraska, Lincoln.

Chadwick, J. W., S. P. Canton, D. J. Conklin, Jr., and P. L. Winkle. 1997. Fish species composition in the central Platte River, Nebraska. Southwestern Naturalist 42(3):279–289.

Cross, F. B., and R. E. Moss. 1987. Historic changes in fish communities and aquatic habitats in plains streams of Kansas. Pages 155–165 *in* W. J. Matthews and D. C. Heins, editors. Community and evolutionary ecology of North American stream fishes. University of Oklahoma Press, Norman.

Ellis, M. M. 1914. Fishes of Colorado. University of Colorado Studies 11:1–136.

Evermann, B. W., and U. O. Cox. 1896. Report on the fishes of the Missouri River basin. Report to the U.S. Fish Commission for 1894 20:325–429.

Goldowitz, B. S. 1996. Qualitative comparison of long term changes in the habitats and fish species composition of the Platte and North Platte rivers in Nebraska. Final Report to the U.S. Fish and Wildlife Service, Platte River whooping crane maintenance trust, Cooperative agreement 14-16-0006-90-917, Grand Island, Nebraska.

Johnson, R. E. 1942. The distribution of Nebraska fishes. Doctoral dissertation. University of Michigan, Ann Arbor.

Lynch, J. D. 1988. Introduction, establishment, and dispersal of western mosquitofish in Nebraska (Actinopterygii: Poeciliidae). Prairie Naturalist 20:203–216.

Lynch, J. D., and B. R. Roh. 1996. An ichthyological survey of the forks of the Platte River in western Nebraska. Transactions of the Nebraska Academy of Sciences 23:65–84.

McKinley, J. L. 1935. The influence of the Platte River upon the history of the valley. Doctoral dissertation. University of Nebraska, Lincoln.

Meek, S. E. 1895. Notes on the fishes of western Iowa and eastern Nebraska. Bulletin of the U.S. Fish Commission 14:133–138.

Morris, L. A. 1960. The distribution of fish in the Platte River, Nebraska. Master's thesis. University of Missouri, Columbia.

Nesler, T. P., R. Van Buren, J. A. Stafford, and M. Jones. 1997. Inventory and status of South Platte native fishes in Colorado. Colorado Division of Wildlife, Final Report, Denver.

Patton, T. M. 1997. Distribution and status of fishes in the Missouri River drainage in Wyoming: implications for selecting conservation areas. Doctoral dissertation. University of Wyoming, Laramie.

Peters, E. J., and R. S. Holland. 1994. Biological and economic analyses of the fish communities in the Platte River: modifications and tests of habitat suitability criteria for fishes of the Platte River. Nebraska Game and Parks Commission, Federal Aid in Fish Restoration, Project F-78-R, Final Report, Lincoln.

Peters, E. J., R. S. Holland, M. A. Callam, and D. B. Bunnell. 1989. Platte River suitability criteria, habitat utilization, preference, and suitability index criteria of fish and aquatic invertebrates in the lower Platte River. Nebraska Game and Parks Commission, Nebraska Technical Series 17, Lincoln.

Propst, D. L., and C. A. Carlson. 1986. The distribution and status of warmwater fishes in the Platte River drainage of Colorado. Southwestern Naturalist 31(2):149–167.

Schainost, S., and M. D. Koneya. 1999. Fishes of the Platte River basin. Nebraska Game and Parks Commission, Lincoln.

Simon, J. R. 1946. Wyoming fishes. Wyoming Game and Fish Department, Bulletin 4, Cheyenne.

Woodling, J. 1985. Colorado's little fish: a guide to the minnows and other lesser known fishes of the State of Colorado. Colorado Division of Wildlife, Denver.

Yu, S. L., and E. J. Peters. 1997. Use of Froude number to determine habitat selection by fish. Rivers 6:10–18.

Appendix A.—Species known to occur in the Platte River as of 1999, with their distribution and abundance by subbasin and status in the basin. LP = lower Platte, MP = middle Platte, SP = South Platte, NP = North Platte; A = abundant, C = common, U = uncommon, R = rare, X = extirpated, N = native, I = introduced, N/I = introduced outside its native range in the basin. An asterisk indicates species is listed as a species of concern by one or more of the basin states.

		Distribution and abundance				
Family	Species	LP	MP	SP	NP	Status
Petromyzontidae	Chestnut lamprey *Ichthyomyzon castaneus*	R				N
Acipenseridae	Lake sturgeon *Acipenser fulvescens*	R				N*
	Pallid sturgeon *Scaphirhynchus albus*	R				N*
	Shovelnose sturgeon *S. platorynchus*	C	R		X	N
Polyondontidae	Paddlefish *Polyodon spathula*	U	R			N
Lepisosteidae	Longnose gar *Lepisosteus osseus*	C	U			N
	Shortnose gar *L. platostomus*	C	U			N
Hiodontidae	Goldeye *Hiodon alosoides*	C	U		X	N
Clupeidae	Alewife *Alosa pseudoharengus*		U	C		I
	Gizzard shad *Dorosoma cepedianum*	A	C	C	C	N/I
Cyprinidae	Central stoneroller *Campostoma anomalum*	U	C	C	C	N
	Goldfish *Carassius auratus*	R	R	R	R	I
	Lake chub *Couesius plumbeus*			E	E	N
	Grass carp *Ctenopharyngeodon idella*	U				I
	Red shiner *Cyprinella lutrensis*	A	A	A	A	N
	Common carp *Cyprinus carpio*	A	A	A	A	I
	Western silvery minnow *Hybognathus argyritus*	C	C		R	N*
	Brassy minnow *H. hankinsoni*	C	C	C	C	N*
	Plains minnow *H. placitus*	C	C	X	R	N*
	Bighead carp *Hypophthalmichthys nobilis*	R				I
	Common shiner *Luxilis cornutus*		R	X	U	N*
	Speckled chub *Macrhybopsis aestivalis*	C	R			N
	Sturgeon chub *M. gelida*	R	X		X	N*
	Sicklefin chub *M. meeki*	X				N*
	Silver chub *M. storeriana*	U	R			N
	Hornyhead chub *Nocomis biguttatus*	X		X	R	N*
	Golden shiner *Notemigonus crysoleucas*	U	U	U	U	N/I
	Emerald shiner *Notropis atherinoides*	C	C	C	C	N/I
	River shiner *N.blennius*	A	C		X	
	Bigmouth shiner *N. dorsalis*	C	C	C	C	N
	Blacknose shiner *N. heterolepis*			X		N*
	Spottail shiner *N. hudsonius*				C	I

Appendix A.—Continued.

Family	Species	Distribution and abundance				Status
		LP	MP	SP	NP	
Cyprinidae	Sand shiner *N. stramineus*	A	A	A	A	N
	Topeka shiner *N. topeka*					N*
	Suckermouth minnow *Phenacobius mirabilis*	U	U	R	R	N*
	Northern redbelly dace *Phoxinus eos*		R	R	R	
	Finescale dace *P. neogaeus*		R		R	
	Fathead minnow *Pimephales promelas*	A	A	A	A	N
	Flathead chub *Platygobio gracilis*	C	R		R	N*
	Blacknose dace *Rhinichthys atratulus*				X	N*
	Longnose dace *R. cataractae*		R	U	C	N*
	Rudd *Scardinius erythrophthalmus*				U	I
	Creek chub *Semotilus atromaculatus*	U	C	A	C	N
Catostomidae	River carpsucker *Carpiodes carpio*	A	C	U	U	N*
	Quillback *C. cyprinus*	C	C	C	U	N
	Longnose sucker *Catostomus catostomus*		R	C	C	N
	White sucker *C. commersonii*	U	C	C	A	N
	Mountain sucker *C. platyrhynchus*				R	N*
	Blue sucker *Cycleptus elongatus*	U				N
	Smallmouth buffalo *Ictiobus bubalus*	R				N
	Bigmouth buffalo *I. cyprinellus*	R	R			N
	Shorthead redhorse *Moxostoma macrolepidotum*	C	C	R	C	N
Ictaluridae	Black bullhead *Ameiurus melas*	A	C	C	C	N
	Yellow bullhead *A. natalis*	U	U			N
	Blue catfish *Ictalurus furcatus*	R				N/I
	Channel catfish *I. punctatus*	A	C	C	C	N/I
	Stonecat *Noturus flavus*	U	U	R	U	N*
	Tadpole madtom *N. gyrinus*		R			N
	Flathead catfish *Pylodictis olivaris*	U	R			N
Esocidae	Northern pike *Esox lucius*	R	R			N/I
	Grass pickerel *E. americanus vermiculatus*	R	R		R	
Salmonidae	Brook trout *Salvelinus fontinalis*			U	U	I
	Greenback cutthroat trout *Oncorhynchus clarkii stomias*			U	U	N/I
	Rainbow trout *O. mykiss*			U	U	I
	Brown trout *Salmo trutta*			U	U	I

Appendix A.—Continued.

Family	Species	Distribution and abundance				Status
		LP	MP	SP	NP	
Fundulidae	Plains topminnow *Fundulus sciadicus*	R	U	U	U	N*
	Plains killifish *F. zebrinus*	U	C	U	R	N
Poeciliidae	Western mosquitofish *Gambusia affinis*	U	C	C	C	I
Atherinidae	Brook silverside *Labidesthes sicculus*	U	U	U		I
Gasterosteidae	Brook stickleback *Culaea inconstans*	R	C	R	C	N
Moronidae	White perch *Morone americana*	U	R			I
	White bass *M. chrysops*	R	R	R	R	N/I
Centrarchidae	Rock bass *Ambloplites rupestris*	R	R		R	I
	Green sunfish *Lepomis cyanellus*	C	C	C	C	N
	Pumpkinseed *L. gibbosus*			R	R	I
	Orangespotted sunfish *L. humilis*	R	U	U	R	N*
	Bluegill *L. macrochirus*	C	C	U	U	N/I
	Smallmouth bass *Micropterus dolomieu*		R	R	U	I
	Largemouth bass *M. salmoides*	C	C	C	U	N/I
	White crappie *Pomoxis annularis*	C	U	U	U	N/I
	Black crappie *P. nigromaculatus*	C	U	U	U	N/I
Percidae	Iowa darter *Etheostoma exile*	R	R	C	C	N*
	Johnny darter *E. nigrum*	U	U	U	C	N*
	Orangethroat darter *E. spectabile*		U		U	N*
	Yellow perch *Perca flavescens*	U	C	C	C	N/I
	Sauger *Sander canadensis*	U	R	X	X	N*
	Walleye *S. vitreus*	C	C	C	C	N/I
Sciaenidae	Freshwater drum *Aplodinotus grunniens*	C	C	R	U	N/I

American Fisheries Society Symposium 45:249–291, 2005

Spatiotemporal Patterns and Changes in Missouri River Fishes

David L. Galat*

U.S. Geological Survey, Cooperative Research Units, 302 ABNR Building, University of Missouri, Columbia, Missouri 65211, USA

Charles R. Berry

U.S. Geological Survey, Cooperative Research Units, Box 2104B, South Dakota State University, Brookings, South Dakota 57007, USA

William M. Gardner

Montana Fish, Wildlife and Parks, Post Office Box 938, Lewistown, Montana 59457, USA

Jeff C. Hendrickson

North Dakota Game and Fish Department, Riverdale, North Dakota 58565, USA

Gerald E. Mestl

Nebraska Game and Parks Commission, 2200 North 33rd Street, Lincoln, Nebraska 68503, USA

Greg J. Power

North Dakota Game and Fish Department, 100 North Bismarck Expressway, Bismarck, North Dakota 58501, USA

Clifton Stone

South Dakota Department of Game, Fish, and Parks, American Creek Fisheries Station, 1125 North Josephine, Chamberlain, South Dakota 57325-1249, USA

Matthew R. Winston

Missouri Department of Conservation, Columbia Fish & Wildlife Research Center, 1110 South College Avenue, Columbia, Missouri 65201, USA

Abstract.—The longest river in North America, the Missouri, trends southeast from Montana across the mid continent of the United States, 3,768 km to its confluence with the Mississippi River near St. Louis, Missouri. Frequent flooding, a shifting, braided channel, and high turbidity characterized the precontrol "Big Muddy." Major alterations occurred over the past century primarily for flood protection, navigation, irrigation, and power production. Today, the middle one-third of its length is impounded into the largest volume reservoir complex in the United States and the lower one-third is channelized, leveed, and its banks stabilized.

* Corresponding author: galatd@missouri.edu

Spatial and temporal patterns of Missouri River fishes are reviewed for the main channel, floodplain, and major reservoirs. Twenty-five families, containing 136 species, compose its ichthyofauna. Seven families represent 76% of total species richness, with Cyprinidae (47 species), Catostomidae (13), Centrarchidae (12), and Salmonidae (10), the five most specious. Native fishes compose 79% of the river's ichthyofauna with representatives of four archaic families extant: Acipenseridae, Polyodontidae, Lepisosteidae, and Hiodontidae. Fifty-four percent of Missouri River fishes are classified as "big river" species, residing primarily in the main channel, and 93% of these are fluvial dependent or fluvial specialists. Significant floodplain use occurs for 60 species. Many of its big river fishes are well adapted for life in turbid, swift waters with unstable sand-silt bottoms.

Populations of 17 species are increasing and 53% of these are introduced, primarily salmonids, forage fishes, and Asian carps. Ninety-six percent of the 24 species whose populations are decreasing are native. Fishes listed as globally critically imperiled and federally endangered (G1) or globally vulnerable (G3) include pallid sturgeon *Scaphirhynchus albus* (G1), lake sturgeon *Acipenser fulvescens*, Alabama shad *Alosa alabamae*, sturgeon chub *Macrhybopsis gelida*, and sicklefin chub *M. meeki* (G3). Eleven fishes are listed by two of more of the seven main-stem states as imperiled; all are big river species.

Richness increases going downriver from 64 species in Montana to 110 species in Missouri with 36% of widely distributed taxa absent below one or more reservoir. Long-term fish collections from several states show declines in sauger *Sander canadensis* throughout the river and decreases in the lower river of several big river fishes (e.g., sturgeons, chubs, *Hybognathus* spp.). Spatiotemporal changes in Missouri River fishes reflect interactions between natural (climate, physiography, hydrology, and zoogeography) and anthropogenic (impoundment, geomorphic, flow, and temperature alterations, and introduced species) factors. Recurrent droughts and floods and persistent stakeholder conflicts over beneficial uses have recently directed national attention to Missouri River issues. Acquisition of floodplain lands and channel and floodplain rehabilitation programs are underway to improve habitat in the lower river. Unfortunately, many are site specific and few have included explicit ecological objectives and performance evaluations. Several proposals for flow normalization are being considered, but remain controversial.

Introduction

Great rivers, defined as having basins greater than 3,200 km^2 (Simon and Emery 1995), are the most profound flowing-water features of the continental landscape. Human societies developed and flourished along the banks of many of the world's great rivers, sustained by their plentiful resources and fertile floodplains. Rivers also provided efficient highways for transport of people and their products, facilitating exchange among diverse cultures. Expansion of human populations and alterations of their environments have affected aquatic resources of many of the world's great rivers, particularly in the north-temperate zone (Dynesius and Nilsson 1994). The Missouri River exemplifies such a system, and its fishes serve as indicators to assess natural and anthropogenic change.

Ideal indicators that provide a defensible ecological assessment for rivers should meet several guidelines: (1) characterize current river health (status), (2) track changes in river health at multiple spatial and temporal scales (trends), (3) identify and respond to major stressors and rehabilitation programs, and (4) interact across ecological, economic, and social realms. Additionally, they should show conceptual relevance, be logistically feasible to implement, differentiate natural from anthropogenic variability, and produce results that are clearly understood and accepted by scientists, policy makers, and the public (Jackson et al. 2000). Fishes meet all of these requirements (Fausch et al. 1990; Simon and Lyons 1995). Fish and fishing are socially relevant. They provide valuable goods and services to the public, and fish have a long tradition as bellwethers of water quality (Bayley and

Li 1996; Schmutz et al. 2000). Fishes are particularly useful ecological indicators for large rivers because their various guilds integrate a wide range of riverine conditions, ranging from properties of bed sediments for egg development at the micro-scale to longitudinal integrity for spawning migrations at the landscape scale (Copp 1989; Scheimer 2000). As migratory organisms, fishes are ideal indicators of longitudinal and lateral connectivity across the riverscape (Jungwirth 1998; Fausch et al. 2002). At the habitat scale, fishes use clearly defined, life stage specific habitats. Their longevity helps fishes "register" environmental alterations across time. Fishes are often top predators in rivers and thereby subsume trophic conditions across the food chain (Schmutz et al. 2000). Additionally, cost-effective and standardized collection methods exist to characterize fishes at multiple spatiotemporal levels (Murphy and Willis 1996).

Our goal is to employ fishes as indicators of spatiotemporal patterns and change for the Missouri River. We first provide a brief background of the system's physiography and review its approximately 200 years of Euro-American occupation. Zoogeography of the basin's fishes is then summarized along with an abbreviated history of fish introductions.

Our primary objectives are to review

1. composition, distribution, relative abundance, general habitat use, conservation status, and relative population status of fishes in the Missouri River channel, its floodplain, and main-stem reservoirs;
2. spatial and temporal patterns in relation to natural (i.e., zoogeographic) and anthropogenic (e.g., impoundment, channelization, and flow regulation) factors; and
3. activities underway to conserve and restore the Missouri River and its native fishes.

Study Area

The 3,768-km-long Missouri River is the longest river in North America from its named origin at the confluence of the Madison, Jefferson, and Gallatin rivers near Three Forks, Montana, to its confluence with the Mississippi River, near St. Louis, Missouri. Its basin is second in size (1,371,017 km^2) in the United States only to the Mississippi River. Elevation of the named river extends over 1,000 m from about 1,226 m above sea level (asl) at Three Forks, Montana, to about 122 m asl at the mouth, and its main channel covers nearly 10° latitude (48.17°N to 38.53°N); (Galat et al. 2005). The southern limit of continental glaciations largely defines the Missouri's modern course. Its channel flows from northwest to southeast across the mid-continent, bisecting or contiguous to seven states (Figure 1).

Metcalf (1966) recognized three major preglacial (Pliocene) components in the evolution of the contemporary Missouri River basin: the southward-flowing Teays-Mississippi, an Arctic ("Hudson Bay") component to the northwest, and a preglacial plains stream system that flowed southward, but independent of the preglacial Mississippi. Metcalf (1966), Pflieger (1971), and Cross et al. (1986) summarize Missouri River drainage history and zoogeography of its fishes in greater detail.

The Missouri River begins in the Northern Rocky Mountain physiographic province, flows through highly erodible soils of the glaciated Great Plains and Central Lowlands provinces and then through the unglaciated, limestone-dolomite Ozark Plateaus. Its main channel traverses 6 of 13 terrestrial ecoregions within the basin, beginning in the North Central Rockies forests, passing through four grassland ecoregions, and then the Central Forest/Grassland Transition Zone (Ricketts et al. 1999). Convergence of four major air masses and its mid-continent location produce extreme seasonal and daily fluctuations in climate (Galat et al. 2005). About 70% of the river's basin lies within the semiarid Great Plains. Thus, the Missouri River is largely a dryland river with about one-half of its basin receiving less than 41 cm/year of precipitation, coming largely as rainfall during the growing season. Precipitation in the remainder of the basin averages more than 80 cm/year in the Rocky Mountains (~11% of basin total area), ranges from about 40–102 cm/year in the Central lowlands (~17%), and exceeds 100 cm/year in the Ozark plateaus (~2%).

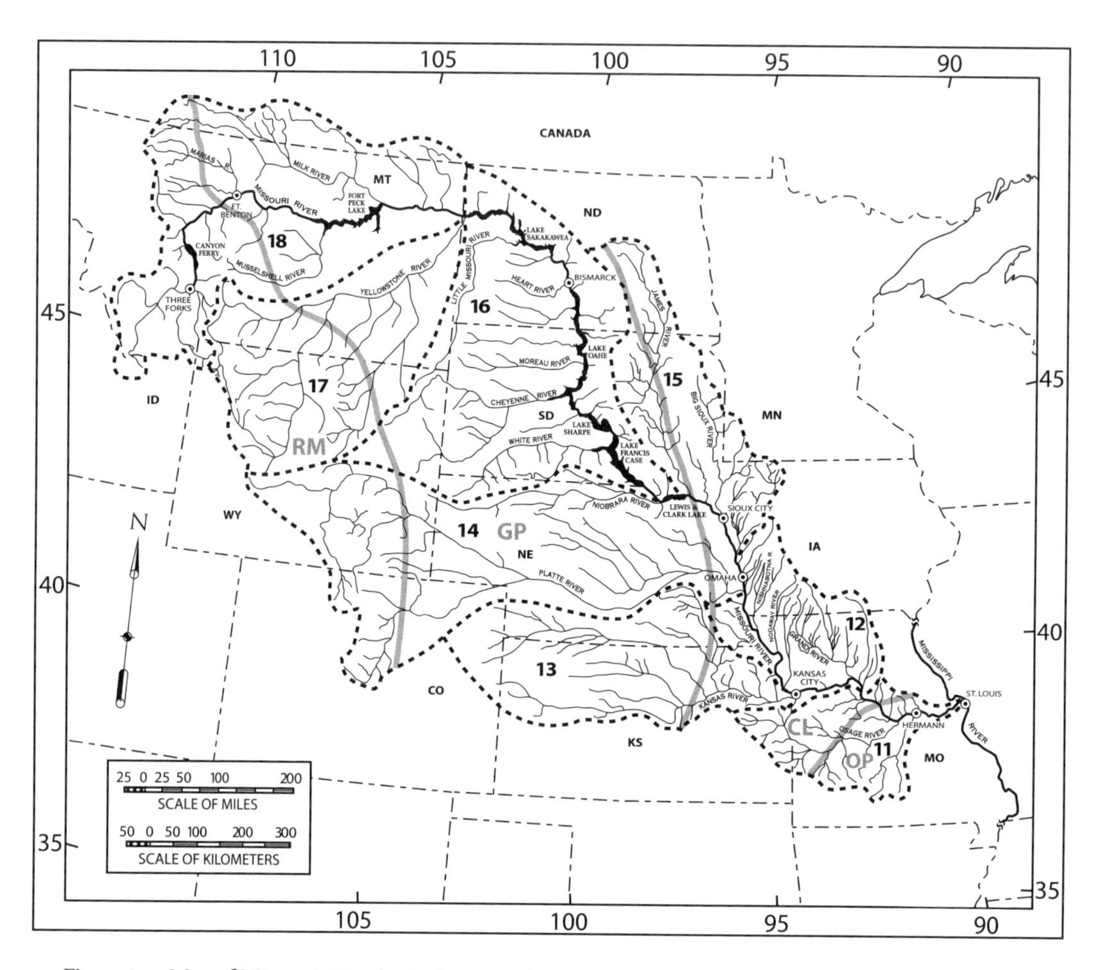

Figure 1.—Map of Missouri River basin showing physiographic provinces and inclusive drainage units (adapted from Cross et al. 1986), selected major tributaries, main-stem dams and reservoirs, and locations mentioned in text. Physiographic provinces: RM = Rocky Mountains, GP = Great Plains, CL = Central Lowlands, OP = Ozark Plateaus. Drainage units: 18, upper Missouri; 17, Yellowstone; 16, Little Missouri–White; 15, James-Sioux; 14, Niobrara-Platte (15 and 14 combined herein); 13, Kansas; 12, Nishnabotna-Chariton; 11, lower Missouri. Drainage units 17 and 13 were not analyzed as neither contains the Missouri River main channel.

Land uses in the floodplain are primarily cropland (37%) or grassland (30%), with about 9% of the area developed (Revenga et al. 1998).

Three freshwater ecoregions (Upper Missouri, Middle Missouri, and Central Prairie; Abell et al. 2000) and 47 tributaries with drainage basins greater than 1,000 km^2 contribute to the Missouri River. Cross et al. (1986) divided the basin into eight drainage units based on tributary groupings (Figure 1), and we organized our treatment of fishes around a subset of these, as well as freshwater ecoregions and physiographic provinces. Largest tributaries to the Missouri River include the Platte, Yellowstone, and Kansas rivers, which are ranked from 13th to 15th in descending order, respectively, of drainage area in the United States (van der Leeden 1990). Selected tributaries are described in Galat et al. (2005).

Spatiotemporal patterns in runoff and hydrology reflect climatic and physiographic diversity of the basin (Galat et al. 2005). The historical annual hydrograph was bimodal with the first peak or "spring rise" in March–April, corresponding to ice-out in the mid and upper basins and prairie snowmelt. The second, larger flow pulse or "June rise" occurred as a result of Rocky Mountain snowmelt and precipitation in the lower basin. Flows declined in July and were generally low until the following spring rise. Runoff to the Missouri River originates primarily at the upper and lower ends of the river in the Rocky Mountains (8.6 cm/year from basin above Fort Benton, Montana), and the Central lowlands and Ozark plateaus (29.9 cm/year between Kansas City and Hermann, Missouri; Galat et al. 2005). More than 1,200 reservoirs have been constructed within the basin, with the six U.S. Army Corps of Engineers (Corps) main-stem reservoirs being the most significant (Table 1). These six reservoirs account for 69% of the basin's total water storage and collectively are the largest volume water storage project in the United States (~90.5 km^3). Their most significant effect on downriver seasonal flow patterns has been to dampen flood pulses, create artificially high summer–autumn flows (Galat and Lipkin 2000), and reduce flow variability (Pegg et al. 2003). Galat and Lipkin (2000) detail pre- and post-impoundment magnitude, timing, frequency, duration, and rate of change of the Missouri River hydrograph at multiple locations. Pegg and Pierce (2002a) classified the river into six hydrologically distinct units based on flow characteristics, and Pegg et al. (2003) statistically showed that flows were most highly altered in the middle portion of the river. Galat et al. (2005) pro-

Table 1.—Features of Missouri River main-stem reservoirs. Canyon Ferry is a Bureau of Reclamation reservoir, whereas the remaining reservoirs were constructed and are maintained by the Corps of Engineers. Location of dam is kilometers upriver from Missouri River mouth. Sources: U.S. Army Corps of Engineers (1994a), Galat and Frazier (1996).

	Reservoir						
Feature	Canyon Ferry	Fort Peck	Sakakawea	Oahe	Sharpe	Francis Case	Lewis & Clark
Name of dam	Canyon Ferry	Fort Peck	Garrison	Oahe	Big Bend	Fort Randall	Gavins Point
Location of dam (km)	3,688	2,850	2,236	1,725	1,589	1,416	1,305
Year dam closed	1953	1937	1953	1958	1963	1952	1955
Total drainage area (10^3 km^2)	41.2	148.9	469.8	630.6	645.8	682.4	723.9
Water surface elevation (m above sea level)	1,157	680.9	560.2	490.1	432.8	411.5	366.1
Surface area (km^2)	142.5	858.0	1,242	1,263	230.7	319.7	101.2
Length (km)	40	216	286	372	129	172	40
Mean width (km)	4.0	3.2	5.3	3.9	1.9	2.1	3.2
Shoreline length[a] (km)	122.3	2,446	2,156	3,620	321.8	868.9	144.8
Gross volume (km^3)	2.5	23.1	29.4	28.5	2.3	6.8	0.61
Mean depth (m)	17.7	24.1	18.9	19.2	9.4	15.2	4.9
Maximum depth (m)	50.0	65.8	53.9	61.9	22.9	41.5	17.1
~Mean annual water level change (m)	3.7	3.0	3.0	1.5	0.6	10.7	1.2
Mean annual summer surface water temperature (°C)		19.2	18.9	22.2		24.4	24.2
Exchange rate (years)	0.38	2.8	1.4	1.6	0.12	0.50	0.04

[a] At top of carryover multiple use level; base of flood control level.

vide more detail on the river's physiography, climate, land use, geomorphology, hydrology, chemistry, and ecology.

The predevelopment Missouri River exhibited a braided, sandbar, and island filled channel that had continually shifting banks due to sediment erosion and deposition. Overbank floods were common, turbidity was high, and enormous quantities of sediment were transported to the Mississippi River. Lewis and Clark (1804–1806) were the first to formally document the Missouri River's ichthyofauna, describing specimens that later would be identified as goldeye *Hiodon alosoides*, blue catfish *Ictalurus furcatus*, channel catfish *I. punctatus*, interior cutthroat trout *Oncorhynchus clarkii lewis*, sauger *Sander canadensis*, and possibly golden shiner *Notemigonus crysoleucas* (Moring 1996). They remarked on a catch of blue catfish, "Two of our men last night caught nine catfish that would together weigh three hundred pounds" (Cutright 1969). The river became the first great highway for settlement and development of the West once steamboat travel began in 1819 (Thorson 1994). Societal interests to improve navigation, irrigate the arid Great Plains, control devastating floods, and generate hydropower began in earnest in the early 1900s and are well described elsewhere (Ferrell 1993, 1996;Thorson 1994; Schneiders 1999).

The contemporary Missouri River can be divided into three riverine zones plus an additional reservoir zone based on anthropogenic influences. The *upper unchannelized zone* extends 739 km from Three Forks to the first of the six major main-stem impoundments, Fort Peck Lake, Montana. Portions of this section are the most free-flowing remnants of the Missouri River and retain many of its historical features. Several small Bureau of Reclamation impoundments occur here (Canyon Ferry being the largest, Table 1), but their impact is small relative to the Corps reservoirs (Scott et al. 1997). The middle section is about 1,850 km long, and we recognize two zones: the six Corps reservoirs (*reservoir zone*) and an "*inter-reservoir*" riverine zone between the impoundments. We will include in the inter-reservoir zone a short, highly flow-altered reach extending 127 km below the lowermost impoundment, Lewis and Clark Lake, to Sioux City, Iowa, where channelization begins (Figure 1). Some authors (e.g., Hesse et al. 1993) consider this flow-regulated but unchannelized reach to be a unique zone and refer to it as the *lower unchannelized zone* or segment. The remaining 1,178 km of river below Sioux City to the Mississippi is referred to as the *channelized zone*. Here the main stem is altered by flow regulation from upriver main-stem reservoirs, numerous sub-basin reservoirs (e.g., on the Platte and Kansas rivers; Galat et al. 2005), channelization, bank stabilization, and floodplain levees.

Most of the Missouri River is a warmwater river with average July–October water temperatures increasing from about 21.5°C to 27°C between the upper unchannelized and channelized zones. However, water temperatures in the river below the stratifying, large volume reservoirs (Table 1) are depressed due to hypolimnetic water releases (Galat et al. 2001).

Collectively, basin and channel-floodplain development have transformed nearly one-third of the Missouri River's lotic habitat into lentic reservoirs. Longitudinal and lateral connectivity has been fragmented, sediment transport and turbidity below reservoirs drastically reduced, and channel geomorphology altered through bed and bank degradation. Sediment aggradation occurs in reservoirs at the mouths of inflowing tributaries (e.g., Niobrara River into Lewis and Clark Lake) and produces shallow deltas at the upstream ends of reservoirs. The complex of delta sandbars and braided, shallow channels encourages riparian flooding and fosters growth of emergent and submergent vegetation. These delta wetland complexes attract waterfowl and influence fish assemblage structure. Summer water temperatures are depressed below large reservoirs, and contaminants, nutrients, and nonnative biota have been introduced. Much of the channelized river's in-channel habitat complexity and floodplain native vegetation are gone and inter-reservoir and channelized river flow patterns are highly altered. These changes in the river's structure, function, and processes and their effects on river health are further described elsewhere (Funk and Robinson 1974; Hesse et al. 1989; Schmulbach et al. 1992; Galat et al. 2001, 2005; National Research Council 2002).

Methods

A list of fishes documented in the Missouri River basin was revised and expanded from Cross et al. (1986). We abbreviated the Cross et al. (1986) list by including only those species reported from the main-stem Missouri River channel, floodplain, and reservoirs (Figure 1), excluding the remainder of the drainage basin (i.e., tributaries and non-main-stem reservoirs). Occurrence, distribution, and relative abundance were summarized spatially for each species within three geographic strata: drainage unit, freshwater ecoregion, and physiographic province. Boundaries and spatial cross-referencing for these geographic categories are summarized in Table 2. Common names of fishes will be used hereafter; scientific names are listed in Appendix A.

Species were grouped into five "sections" of the main stem that were contained within, or contiguous to, the eight Missouri River subbasin drainage units identified by Cross et al. (1986), (Figure 1, units 11–18). We follow the standard convention for the Missouri River of proceeding from upstream to downstream when referencing locations along the river (e.g., drainage units, river kilometer). For example, the White–Little Missouri drainage unit listed in Cross et al. (1986) was changed to Little Missouri–White. Yellowstone (17) and Kansas (13) drainage units were excluded, as they do not contain any of the main-stem Missouri River. We also combined the Sioux–James (15) and Platte Niobrara (14) drainage units from Cross et al. (1986) into a single unit: James–Sioux–Niobrara–Platte (15 + 14), since they overlap and are contiguous to the main channel. The five drainage units were also grouped into approximate freshwater ecoregions (Abell et al. 2000; Table 2). The status of each species within a drainage unit/ecoregion section of the river was classified as N = native; I = introduced, but could be native to another drainage unit/ecoregion section; U = uncertain if native or introduced; D = diadromous, migrates to or from ocean (not reservoir).

Relative abundance for each species included in the Missouri River for the four physiographic provinces through which it flows was summarized as P = prevalent or abundant; M = marginal or uncommon; S = sporadic or rare; X = present, but relative abundance unknown; E = extinct or an introduced species that is no longer stocked or collected. Species for which there is questionable validity of its presence or identification from the Missouri River at a single location were excluded (e.g., Sacramento perch *Archoplites interruptus*). Occurrence, distribution, and relative abundance data were updated from Cross et al. (1986) using more recent publications (e.g., Hesse et al. 1989, 1993; Galat and Frazier 1996; Patrick 1998; National Research Council 2002; Berry et al. 2004; Berry and Young, 2004); books about fishes from Missouri River states (e.g., Bailey and Allum 1962; Holton and Johnson 1996; Pflieger 1997); agency reports (e.g., Riis et al. 1988; Johnson et al. 1992; U.S. Army Corps of Engineers 1994a, 1994b, 2001; Power and Ryckman 1998; Hendrickson and Lee 2000; U.S. Fish and Wildlife Service 2001); data files of Montana Fish, Wildlife, and Parks; North Dakota Game and Fish Department; South Dakota Department of Game, Fish and Parks; Nebraska Game and Parks Commission; Rivers Corporation (Hesse 2001); Missouri Department of Conservation; and personal knowledge of authors and reviewers (see Acknowledgments). We deleted fishes in the Cross et al. (1986) species list that have not been reported from the main-stem Missouri River by any of the above sources since 1970 and are presumed extinct. Other species were deleted from various drainage units (e.g., striped bass *Morone saxatilis* from drainage unit between James and Platte rivers) for the same reasons. Species deleted were historically rare in the Missouri River, introduced, or reported only from tributaries.

Fishes were assigned to one of three general habitat-use guilds developed by Kinsolving and Bain (1993) to assess their relative dependence on riverine or floodplain habitats (see also Travnichek et al. 1995 and Galat and Zweimüller 2001). *Fluvial specialists* are fishes that are almost always found only in streams and rivers or are described as using flowing water habitats throughout life. These species may be occasionally found in a reservoir or lake, but most information on them pertains to lotic systems. *Fluvial dependent* fishes are found in a variety of habitats, but require flowing water at some point in their life cycle. These species may have significant lake, reservoir, or estuary populations, but use tributary streams or rivers for some life history trait, typically reproduc-

Table 2.—Cross references among physiographic provinces (Cross et al. 1986), freshwater ecoregions (Abell et al. 2000, upper Missouri, *middle Missouri*, <u>Central Prairie</u>), and drainage units (Cross et al. 1986) for main-stem Missouri River. Boundaries between physiographic provinces are approximate locations along main stem. Tributaries and main-stem reservoirs (**bold**) are listed in order from upriver to downriver. Drainage units (unit number 18–11 from Cross et al. 1986) apply to antecedent tributaries from unit number in parentheses. Distances (km) begin with 0 at confluence with Mississippi River.

Physiographic province	Approximate boundaries	Inclusive tributary confluences and freshwater ecoregion (normal, *italic* or <u>underlined</u> font)	Drainage unit
Northern Rocky Mountains km 3,734–3,386	Three Forks, MT, to between Ft. Benton and Great Falls, MT (Belt Creek is most downriver tributary in province)	Madison, Jefferson, Gallatin, **Canyon Ferry**, Sun, Smith	Upper Missouri (18)
Great Plains km 3,386–1,283	between Great Falls, MT, and Ft. Benton, MT, (below Belt Ck., includes Highwood Ck.)to and including confluence of James R., SD	Teton, Marias, Judith, Musselshell, **Ft. Peck**, Milk (18), Yellowstone (17) L. Missouri, **Sakakawea**, Knife, Heart, Cannonball, Grand, Moreau, Cheyenne, **Oahe**, Bad, **Sharpe**, White **Francis Case**, (16) *Niobrara*, ***Lewis & Clark*** (14), *James*	Upper Missouri (18) km 3,734-–2,545 Little Missouri–White (16), km 2,545–1,357
Central Lowland km 1,283–274	from below James R., SD, to and including Chariton R., MO, (~Perche Ck, MO)	*Vermillion, B. Sioux, Floyd, L. Sioux*, (15)*Platte*, (14) *Nishnobotna, L. Nemaha, Nodaway, Platte (MO)* (12), *Kansas (*13),<u>Grand</u>, <u>Chariton</u> (12), <u>Lamine</u>	James–Sioux (15)-Niobrara–Platte (14) km 1,357–957 Nishnabotna–Chariton (12) km 957–384
Ozark Plateaus km 274–0	from below Perche Ck., MO, to Mississippi R. confluence.	<u>Osage</u>, <u>Gasconade</u> (11)	Lower Missouri (11), km 384–0

tion. *Macrohabitat generalists* include species that are commonly found in lakes, reservoirs, floodplain water bodies, and flowing waters, but are capable of completing their life cycle in lentic systems. Species were placed into habitat-use guilds based on literature (e.g., Pflieger 1971, 1997; Lee et al. 1980; Theiling et al. 2000; National Research Council 2002; Berry et al. 2004) and the authors' experience.

Habitat distribution of Missouri River fishes was further summarized by placing each species into one or more of five categories: main channel, main channel margin, floodplain, reservoir, and waif. The three channel-use categories include *main channel* = C, including "big river" fishes (defined in Pflieger 1971; Cross et al. 1986; Simon and Emery 1995), "sandbar," and "main channel" fishes (National Research Council 2002); *channel margin* (edge, border) = B, found in main channel, but generally along the edge out of current; and *waif* = W, not a large river fish, although records exist, the species is seldom collected from the main stem and specimens are likely washed out of a tributary or reservoir. *Floodplain* fishes are species found in off-channel habitats, including marshes, backwaters, backups, oxbows, scours, and so forth. Velocity is low or absent in floodplain habitats, and the waterbody may infrequently connect to the main channel. *Reservoir* fishes include those species reported within a reservoir and tailwater areas below dams.

Conservation status of fishes is reported as Global Heritage Status (NatureServe 2003) and listings under the U.S. Endangered Species Act (ESA, as reported in NatureServe 2003). Global Heritage ranks we include are critically imperiled

(G1), imperiled (G2), vulnerable (G3), apparently secure (G4), secure (G5), or numeric range rank (G#G#), where uncertainty exists about the exact status of a species (e.g., G2G3). Similar state rankings (S) are also summarized. Species are listed under the ESA as endangered, threatened, or a candidate for listing.

We report relative temporal trends in population status over the main-stem Missouri River as increasing (+), stable (0), or decreasing (–). Species too uncommon in catches to rank are identified as U, and blanks indicate a species population status was unknown. Population information does not exist for many fishes, particularly small species (e.g., minnows, chubs, shiners) and species difficult to identify (e.g., *Hybognathus* spp.). Population trends are also difficult to assess across the entire Missouri River since a species may be increasing in one section and decreasing in another (see subsequent section on temporal trends for selected states). We report intermediate or differing intra-river trends as pairwise combinations of +, 0, and – to account for these patterns. Criteria for assigning population status were based on literature (e.g., National Research Council 2002) and file sources previously listed for relative abundance, as well as the author's collective professional judgment. Consequently, population trends reported here should be viewed as qualitative and approximate because consistent and river-wide benchmark fish sampling was absent prior to major changes in flow and habitat.

Spatial Changes

We review longitudinal distribution of fishes from research that evaluated distribution, habitat use, and population structure of benthic fishes over 3,217 km of the warmwater, unimpounded Missouri River (Berry et al. 2004; Berry and Young 2004). This study (hereafter referred to as the benthic fishes study) employed five standardized collection gears (drifting trammel net, benthic trawl, bag seine, stationary gill net, and electrofishing) to collect a diversity of fishes using all major main-channel habitat types. Gears were deployed using standardized effort in replicates of six macrohabitats (outside bend, inside bend, channel cross-over, tributary mouth, nonconnected secondary channel, and connected secondary channel) to target both small and large benthic fishes (Sappington et al. 1998). Sampling occurred from about mid-July to early October over 3 years (1996–1998) within macrohabitats from 15 representative segments of the 27 segments identified. Longitudinal distribution information is illustrated for taxa with greater than or equal to 100 individuals collected during the 3-year study or from information reported in Pegg and Pierce (2002b).

Temporal Changes

No standardized, long-term fish-monitoring program currently exists on the Missouri River as it does for the nearby upper Mississippi River (e.g., Theiling et al. 2000). Consequently, our treatment of temporal changes to its ichthyofauna is uneven. Trends for selected species and reservoirs or river reaches within several Missouri River states are summarized from published papers, Federal Aid in Sport Fish Restoration reports, and unpublished data from agency files.

Reservoir fishes.—Comprehensive, multi-year assessments of impoundment and postimpoundment activities on Missouri River fishes were conducted in the Corps reservoirs by the North Central Reservoir Investigations. Technical reports describe long-term changes in fish populations for Lake Oahe (1965–1974, Beckman and Elrod 1971), Lake Francis Case (1954–1975, Walburg 1977), Lewis and Clark Lake (1956 1974, Walburg 1976), and multiple reservoirs (e.g., Benson 1980). We illustrate differences in temporal trends for selected species within the six main-stem reservoirs by summarizing patterns from two morphologically very different impoundments: Lake Sakakawea and Lewis and Clark Lake (Table 1). Lake Sakakawea has a low exchange rate, it having the second largest surface area and largest volume of the six main-stem reservoirs. Lewis and Clark Lake has the smallest area and volume and the highest exchange rate (Table 1). Lake Sakakawea is managed as a cold, cool, and warmwater fishery, whereas Lewis and Clark Lake is managed as a cool and warmwater fishery (U.S. Army Corps of Engineers 1994a; see Wehrly et al.

2003 for a review of thermal terms relevant to stream fish assemblages). Temporal changes in Lewis and Clark Lake fishes were evaluated from the period following dam closure in 1956–2001 (45 years) by summarizing CPUE (mean catch per overnight gill net set) from Walburg (1976) (1965 missing) and annual surveys of Missouri River reservoirs conducted by the South Dakota Department of Game, Fish and Parks for 1983–1987 (Riis et al. 1988), 1988–1991 (Johnson et al. 1992), 1992–1996 (Wickstrom 1997), and 1997–2001 (Wickstrom 2002). Trends for nearly the same species in Lake Sakakawea were adapted from Hendrickson and Power's (1999) synthesis of changes in species abundance between 1956, 3 years after filling began, and 1998 (42 years). Linear regression was used following Hendrickson and Power's (1999) example with year as the independent variable and CPUE as the dependent variable to test for slopes significantly different from 0.0.

Riverine fishes.—Temporal changes in fish abundance and distribution in the main-stem Missouri River are best documented in Montana, South Dakota–Iowa–Nebraska, and Missouri, as these are the states where main-stem reservoirs are few or absent and reasonable historical databases exist.

Montana.—The few remaining largely free-flowing portions of the Missouri River are above Ft. Peck Lake, 240 km of which is a designated National Wild and Scenic River. The segment below Ft. Peck Lake and between the confluence of the Yellowstone River and the North Dakota border is also somewhat natural with minimal geomorphic alteration. Effects of Ft. Peck Lake flow regulation are dampened by discharge from the Yellowstone River, which is about 1.2 times that of the Missouri River above its confluence (U.S. Geological Survey 2002). The fish assemblage contains many coolwater species from about the origin of the named Missouri River at Three Forks to between Great Falls and Fort Benton, the approximate boundary between the Rocky Mountain and Great Plains physiographic provinces. The Missouri is considered a warmwater river below the Rocky Mountain province.

We contrasted catch per unit effort (CPUE) and percent composition of the total catch for selected fishes along a Wild and Scenic River segment above Fort Peck Lake (km 3,363–3,097) during two time intervals (1976–1979 and 1997–2002). Fishes were collected from five locations within this reach during the first period by Berg (1981) using boat electrofishing in autumn. Gardner (1998, 1999, 2001, 2003) repeated this sampling from the same sites using similar techniques about 20 years later. Potential differences between the two time periods were compared among sites for each species using a students paired *t*-test.

South Dakota–Iowa–Nebraska.—Three distinct segments occur in this section of the Missouri River (see Figure 1 in Hesse et al. 1993). There is a 58-km *upper unchannelized* riverine segment that is flow regulated between Ft. Randall Dam (Lake Francis Case) and Lewis and Clark Lake (South Dakota–Nebraska). This segment is a designated National Recreational River and the Niobrara River is the major tributary. A 127-km *lower unchannelized* and flow regulated segment extends from Gavins Point Dam to Sioux City, Iowa (South Dakota–Iowa–Nebraska); 93 km of this section below Lewis and Clark Lake is also a National Recreational River. The James and Big Sioux rivers are major tributaries entering the Missouri River within this reach (Figure 1). The last segment in Nebraska and Iowa is both *channelized* and flow regulated and extends about 288 km between Sioux City, Iowa, and the Missouri border. Hesse (1993a) and Hesse et al. (1993) summarized status of selected fishes from the Missouri River in these reaches. More specific trend information was synthesized in a series of articles on individual species or groups of species, including paddlefish (Hesse and Mestl 1993), burbot (Hesse 1993b), channel catfish (Hesse 1994a), flathead and blue catfish (Hesse 1994b), chubs and minnows (Hesse 1994c), and sauger (Hesse 1994d). We summarize spatial and temporal changes for several of these groups. We specifically analyzed data for slopes different from 0.0 as described above on Hesse's (1994d) data of temporal changes in experimental gill netting CPUE from unchannelized Missouri River segments in South Dakota, Iowa, and Nebraska.

Missouri.—Pflieger and Grace (1987) summarized changes in fishes of the Missouri River, Missouri, at multiple locations (usually 13) over three time periods 1940–1945 (Fisher 1962), 1962–1972, and 1978–1983. Bag seines were the major gear used to collect fishes, primarily from July to October. Gelwicks et al. (1996) and Grady and Milligan (1998) resampled most of the same sites in 1994 and 1997 using similar bag seines. Their primary interest was to update the status of five minnow species of concern: sicklefin chub, sturgeon chub, flathead chub, plains minnow, and western silvery minnow. Gelwicks et al.'s (1996) collections were made in November and Grady and Milligan's (1998) in July and August.

Lack of sampling effort data for early collections prevented statistical evaluation of changes in catch rate over time. However, Grady and Milligan (1998) used logistic regression to examine trends in fish distribution over time and among sites. They tested the probability of presence or absence of the five target species and six additional minnows over time and among sites for all collections.

Commercial harvest.—Trends in commercial catch provide another perspective to illustrate temporal changes in Missouri River fishes. We summarize trends in number of licensed commercial fishers and composition and weight of reported catches for the Missouri River, Missouri (V. Travnichek, Missouri Department of Conservation, personal communication).

Results

Our list of fishes includes 136 species from 25 families that presently occur in the Missouri River main stem, its floodplain, and reservoirs (Appendix A). This constitutes 79% of the 173 species listed by Cross et al. (1986) for the entire Missouri River basin. Seven families are represented by at least 5 species: Cyprinidae (47 species), Catostomidae (13), Centrarchidae (12), Salmonidae (10), Ictaluridae (9), Percidae (7), and Clupeidae (5). These families compose 76% of total species richness. Ten families are represented by only one species: Polyodontidae, Amiidae, Anguillidae, Percopsidae, Osmeridae, Gadidae, Poeciliidae, Gasterosteidae, Sciaenidae, and Cottidae. Native species of four archaic families are extant in the river: Acipenseridae, Polyodontidae, Lepisosteidae, and Hiodontidae.

We added seven species (four nonnative: alewife, cisco, big head carp, and silver carp; three rare natives: spotted gar, channel shiner, and Mississippi silvery minnow) for which we have reliable records for the main stem that were not reported in Cross et al. (1986) for any Missouri River drainage unit (Appendix A). Forty-four species were deleted that Cross et al. (1986) included (Table 3). The majority of these fishes (34 species) are presumed tributary species, and we have no records for them from the main-stem Missouri River, its floodplain, or reservoirs. Coho salmon were stocked into the Missouri River above Lake Sakakawea and main-stem reservoirs in North and South Dakota in the 1970s, but are no longer captured and presumed extirpated.

Native and Introduced Fishes

Native fishes comprise 78% of the Missouri River's main-stem fauna (106 species) in comparison with 80% (138 species) for the basin (Cross et al. 1986). Two diadromous fishes are present below the reservoirs, American eel and Alabama shad. The native status of some fishes to portions of the Missouri River is uncertain for several reasons: (1) the edge of their range occurs along the river, (2) their preEuro-American distribution is unclear, or (3) unrecorded stockings in the 1800s may have transferred sport and forage fishes before the native fauna was well catalogued.

Bailey and Allum (1962) indicated that largemouth bass, bluegill, and pumpkinseed were not native to South Dakota and that the status of black and white crappie was uncertain. Cross et al. (1986) list largemouth bass and both crappies as native to the James and Sioux rivers of South Dakota and bluegill and pumpkinseed as uncertain. We combined the James, Sioux, Niobrara, and Platte drainages where they were contiguous to the Missouri River channel and list these species as native to the main stem if Cross et al. (1986) classified them as native to any of these drainages. We follow Bailey and Allum (1962) and list pumpkinseed as introduced, since Cross et al. (1986) are uncertain for the James and Sioux rivers, but considered it introduced elsewhere in the Missouri River basin.

Uncertainty exists over the native status of walleye within drainages of the middle Missouri River and the main stem (Tyus 2002). North Dakota con-

Table 3.—Fishes reported by Cross et al. (1986) in drainage units of the Missouri River basin that are not considered present in the Missouri River. Reasons for exclusion: T = presumed tributary species; E = introduced, presumed extirpated; U = originally listed as uncertain in Sioux–James drainage unit.

Species name	Common name	Reason
Ichthyomyzon fossor	northern brook lamprey	T
I. gagei	southern brook lamprey	T
Oncorhynchus mykiss aguabonita	golden trout	T
O. kisutch	coho salmon	E
Thymallus arcticus	arctic grayling	E
Umbra limi	central mudminnow	E
Luxilus zonatus	bleeding shiner	T
Nocomis biguttatus	hornyhead chub	T
Notropis anogenus	pugnose shiner	U
N. greenei	wedgespot shiner	T
N. heterodon	blackchin shiner	U
N. heterolepis	blacknose shiner	T
N. nubilus	Ozark minnow	T
N. topeka	Topeka shiner	T
Richardsonius balteatus	redside shiner	T
Minytrema melanops	spotted sucker	T
Moxostoma anisurum	silver redhorse	T
M.carinatum	river redhorse	T
M.duquesnei	black redhorse	T
Ameiurus nebulosus	brown bullhead	E
Typhlichthys subterraneus	southern cavefish	T
Fundulus catenatus	northern studfish	T
F. olivaceus	blackspotted topminnow	T
Poecilia mexicana	shortfin molly	E
Xiphophorus hellerii	green swordtail	E
X. variatus	variable platyfish	E
Ambloplites constellatus	Ozark bass	T
Lepomis gulosus	warmouth	E
Crystallaria asprella	crystal darter	T
Etheostoma blennioides	greenside darter	T
E. caeruleum	rainbow darter	T
E. chlorosoma	bluntnose darter	T
E. flabellare	fantail darter	T
E. gracile	slough darter	T
E. microperca	least darter	T
E. nianguae	Niangua darter	T
E. punctulatum	stippled darter	T
E. spectabile	orangethroat darter	T
E. tetrazonum	Missouri saddled darter	T
E. zonale	banded darter	T
Percina cymatotaenia	bluestripe darter	T
P. evides	gilt darter	T
P. maculata	blackside darter	T
Cottus carolinae	banded sculpin	T

siders it as native to the state (Power and Ryckman 1998), but (Tyus 2002) questions walleye's native status to the Missouri River based on its lacustrine habits, widespread stocking before its native range was fully documented, and misidentifications with sauger of early records. Bailey and Allum (1962) list walleye as native to eastern South Dakota, including the Missouri River. Cross et al. (1986) list it as native to the White–Little Missouri (16) and James–Sioux drainage units (15, exclusive of the Missouri River main stem) and as uncertain within the Niobrara-Platte drainage unit (14, including the Missouri River main stem). However, Bailey (R. M. Bailey, University of Michigan, personal communication) now considers walleye to be introduced to the main- stem Missouri River within the White-Little Missouri drainage and its status as uncertain in the James River. Our designation reflects the walleye's now uncertain status throughout most of the Missouri River (Appendix A). We have also listed yellow perch as uncertain for the same reasons.

The presence of fossil walleye from the Illinoian glacial cycle in Kansas where they are now listed as uncertain (Cross et al. 1986) raises the question of what temporal criteria should be used to designate a species as native or nonnative to a drainage basin?

Northern pike is another problematic species. Cross et al. (1986) list northern pike as native to the James–Sioux (15) and Niobrara–Platte (14) drainages, but as introduced to the Little Missouri–White drainage unit (16). However, North Dakota Game and Fish consider it a native species as records indicate its presence at the time Bismarck was settled (1882–1883, Barrett 1895). Rail service arrived in central North Dakota in the 1860s and U.S. Bureau of Fisheries records beginning in 1880 do not indicate northern pike stockings until 1899. We therefore list its status to the Little Missouri–White drainage unit drainage as native.

General Habitat Use

Fifty-four percent of Missouri River fishes reside primarily in the main channel (73 species) and are thereby categorized as big river fishes. We classified an additional 18 fishes as more common along the main-channel border (e.g., *Esox* spp., *Ameiurus* spp., *Lepomis* spp., *Pomoxis* spp.; Appendix A). Another 38 species are included in our list, but are considered waifs to the main channel from tributaries or reservoirs (Appendix A). Combining channel and channel border fishes yields a total of 91 species classified as main channel Missouri River fishes.

Significant use of floodplain habitats occurs for 60 species, including many that also frequent the channel border. Genera that include numerous floodplain representatives are *Lepisosteus* spp., *Esox* spp., *Carpiodes* spp., *Fundulus* spp., and *Lepomis* spp. (Appendix A). Use of reservoirs occurs for 72 species with 20 of them largely restricted to reservoirs or waifs from reservoirs to the main channel (16 of the 20 species). The majority of these (11 species) are fishes introduced into reservoirs for sport fishing (e.g., 7 species of salmonids) or forage for sport fishes (e.g., rainbow smelt and spottail shiner).

Roughly one-half of Missouri River fishes (68 species) require flowing water for some life-stage activity (i.e., fluvial specialist or fluvial dependent) and 21% (28 species) are categorized as fluvial specialists (Appendix A). Representative fluvial dependent native genera are *Acipenser, Polyodon, Alosa, Hiodon, Hybognathus, Moxostoma*, and *Morone*. Fluvial specialist fishes are predominately from the genera *Scaphirhynchus, Macrhybopsis, Notropis*, and *Noturus*. The 69 species of macrohabitat generalists contain numerous representatives from the families Clupeidae, Esocidae, Cyprinidae, Catostomidae (*Ictiobus*), Ictaluridae, Fundulidae, and Centrarchidae.

As their name implies, macrohabitat generalists show the widest distribution among drainage units, physiographic regions, and river habitats. However, proportion of macrohabitat generalists was higher in the Upper Missouri–Little Missouri–White drainage units (60.8%) than in remaining drainages (range: 49.1–54.8) with a corresponding decline in fluvial specialists. One-half of main channel fishes are macrohabitat generalists (44% of waifs) and 23% of the remaining species are fluvial specialists (23% of waifs). Macrohabitat generalists also dominate floodplains (78%) and reservoirs (65%). All fluvial specialists in the Missouri River are native fishes, whereas 54% of macrohabitat generalists and 46% of fluvial dependent species are nonnatives.

Population and Conservation Status

Populations of 33% of the 95 fishes we were able to classify are considered either stable to increasing or increasing (0, + or +, Appendix A), whereas 45% (43 species) of fishes are either stable to decreasing or decreasing (0, – or –, Appendix A). More specifically, we list populations of 17 species as increasing (+, Appendix A), 13 species as stable to increasing (0, +), 19 species as stable (0), 19 species as stable or decreasing (0, –), 24 species as decreasing (–), and 3 species as increasing in some reaches and decreasing in others (+, –). Forty-one species were too uncommon to rank or their status is unknown (Appendix A). Fifty-three percent of the species whose populations are increasing (+) were introduced into the river or reservoirs, and 18% are not present in the main channel except as waifs. Ninety-six percent of species with decreasing populations (–) are native or diadromous fishes, 72% of these reside predominately in the main channel (C) and 63% are fluvial specialist or fluvial dependent fishes.

Populations of some fishes we identify as declining include species whose conservation need is reported by others. Pallid sturgeon is federally listed as endangered and Alabama shad is a candidate for federal listing. The U.S. Fish and Wildlife Service recently concluded listing of sicklefin and sturgeon chubs as endangered was not warranted, despite a 56% reduction in their distribution along the Missouri River main stem (Department of Interior 2001).

Conservation status is judged globally secure (G5) for 120 native and nonnative Missouri River fishes included in Appendix A (NatureServe 2003). We identified populations of 13 native fishes (12% of native and diadromous species) as less than secure (Appendix A) using NatureServe's global rankings in comparison with our site-specific classification of 23% as declining (–). Pallid sturgeon is ranked as globally critically imperiled (G1) and no Missouri River fishes are classified as globally imperiled (G2). Lake sturgeon, Alabama shad, sturgeon chub, and sicklefin chub are ranked as globally vulnerable (G3). Speckled chub and blue sucker are classified as globally vulnerable to apparently secure (G3G4), and six native, main-channel fishes are apparently globally secure (G4): chestnut lamprey, silver lamprey, shovelnose sturgeon, paddlefish, western silvery minnow, and plains minnow. Highfin carpsucker is considered uncertain between apparently globally secure and secure (G4G5). Eleven fishes are listed by two or more main-stem states as imperiled (S1, S2): lake sturgeon, pallid sturgeon, paddlefish, western silvery minnow, plains minnow, sturgeon chub, sicklefin chub, silver chub, flathead chub, quillback, and highfin carpsucker. All are big river fishes, 45% are fluvial specialists, and another 36% are fluvial dependent species.

Longitudinal Spatial Patterns

There is a general trend of increasing species richness going downriver. Total numbers of native/introduced species by drainage unit section are as follows: Upper Missouri 37/27 (64), Little Missouri–White 53/22 (79), James–Sioux–Niobrara–Platte 73/16 (94), Nishnabotna–Chariton 68/15 (87), and lower Missouri 89/18 (110). Species totals do not always equal the sum of native plus introduced due to status uncertainties and exclusion of diadromous fishes. The proportion of native big river fishes varies little among drainage units, ranging from 48% in the upper Missouri and Little Missouri–White units, where the river is both least altered and the six main-stem reservoirs are located, to 51% in the remaining units. The number of introduced fishes is highest in the Upper Missouri ecoregion (28 species) where the main-stem reservoirs are located. Forty-one and 24% of introduced fishes use reservoirs in the upper Missouri and Little Missouri–White drainage units, respectively. Most introductions are sport fishes from four families: Salmonidae, Esocidae, Centrarchidae, and Percidae, or forage species (e.g., cisco, rainbow smelt, spottail shiner, golden shiner). Fewer introduced fishes occur in the Middle Missouri (19 species) and Central Prairie (18 species) ecoregions, and a much lower percentage of these fishes use reservoirs (9–13%) compared with upper river drainage units. Predominant species include the same sport and forage fishes stocked into upriver reservoirs as well as Asian carps (grass, silver, and bighead). Gavins Point Dam appears to be a barrier to upriver migration of introduced Asian carps at the time of this writing.

Salmonids decrease going downriver as a proportion of total species richness within each drainage unit (unit number): 19% (18), 11% (16), 3% (15 + 14), 0% (12), and 1% (11). Ictalurids make up a slightly higher proportion of species (11.3–12.9%) in the lower two drainage units than the upper three (9.6–9.9%) and clupeids compose between 4.2% and 5.0% of species richness in drainage units below the reservoirs (15–11) compared with 0.0–1.6% in the Little Missouri–White and Upper Missouri units (16 and 18). No clear longitudinal trends occur for species richness among the other major families.

The benthic fishes study collected approximately 114,000 fishes of 106 species from the mainstem Missouri River (Berry et al. 2004). Longitudinal distributions for the 50 taxa with greater than or equal to 100 individuals collected during this study also show an increase in species richness from the upper unchannelized to lower channelized river (Figure 2). Twenty-seven taxa were collected above Ft. Peck Lake, 39 taxa in the inter-reservoir segments below Ft. Peck Lake to Lewis and Clark Lake, and 44 taxa were sampled below Lewis and Clark Lake. Four coolwater fishes, longnose dace, longnose sucker, white sucker, and burbot were only collected above approximately km 1,200. Twenty-one taxa were absent above about km 3,000, and three species (speckled chub, blue catfish, and mosquitofish) were collected only downriver from approximately km 1,100. Thirty-six percent of widely distributed taxa were absent in one or more river segments below a reservoir.

Structure, morphology, functional composition, and life history features of the river's benthic fish assemblage have been evaluated in relation to differences in flow regime among six distinct longitudinal flow units (Pegg and Pierce 2002a, 2002b). Pegg and Pierce (2002b) showed that gizzard shad, emerald shiner, red shiner, river shiner, plains minnow, flathead catfish, and freshwater drum were most associated with the lower channelized river (they defined this as from Kansas City to the mouth), whereas longnose suckers, white suckers, and sicklefin chubs were most associated with the upper unchannelized and inter-reservoir segments. Fish assemblages from the uppermost segments above Fort Peck Lake (km 3,217–3,029) had a Bray-Curtis similarity of less than 45% with fishes from the lower channelized river.

Fish trophic guilds above Ft. Peck Lake are composed largely of invertivores (e.g., flathead chubs and sturgeon chubs) and herbivorous species (e.g., *Hybognathus* spp.). These guilds decline precipitously and omnivores and benthic invertivores increase below Ft. Peck Lake and Lake Sakakawea. Planktivorous fishes, predominantly gizzard shad, are most abundant in the channelized river. Bergsted et al. (2004) further summarize feeding and reproductive guilds of benthic fishes throughout the warmwater Missouri River.

Riverine fishes above Lake Oahe are dominated by species exhibiting a more elongate body shape (e.g., flathead chubs, *Hybognathus* spp., and longnose suckers) than elsewhere, whereas prolonged swimmers like gizzard shad were most abundant in the channelized river (Pegg and Pierce 2002b). Water velocities where fishes are generally found based on the literature also varied among longitudinal flow units. Welker and Scarnecchia (2003) compared catostomid fishes in two upper reaches of the Missouri, North Dakota: a Yellowstone River confluence to Lake Sakakawea (YSS), a moderately altered reach, and the Missouri river between Garrison Dam and Lake Oahe (GOS), a highly altered segment. Differences in sucker species composition, prey density and composition, and sucker feeding ecology were associated with the major anthropogenic disturbances in the GOS reach.

Species associated with either fast (e.g., shovelnose sturgeon, blue sucker, and sicklefin and sturgeon chubs) or moderate (e.g., walleye, sauger, and emerald shiner) water velocities are most abundant above Fort Peck Lake. Fishes frequenting moderate and slow velocities (e.g., bigmouth buffalo, carpsuckers, and freshwater drum) are more prevalent in the channelized river. Fishes routinely collected over sand substrate predominated above Ft. Peck Lake and Lake Sakakawea (e.g., emerald shiners and *Hybognathus* spp.). Species use of gravel (e.g., blue suckers and shorthead redhorse) was greater than 40% below these two reservoirs, but well below 20% elsewhere in the river. Fishes that use gravel (e.g., longnose and white suckers and shovelnose sturgeon), sand (e.g., *Notropis* spp. and river

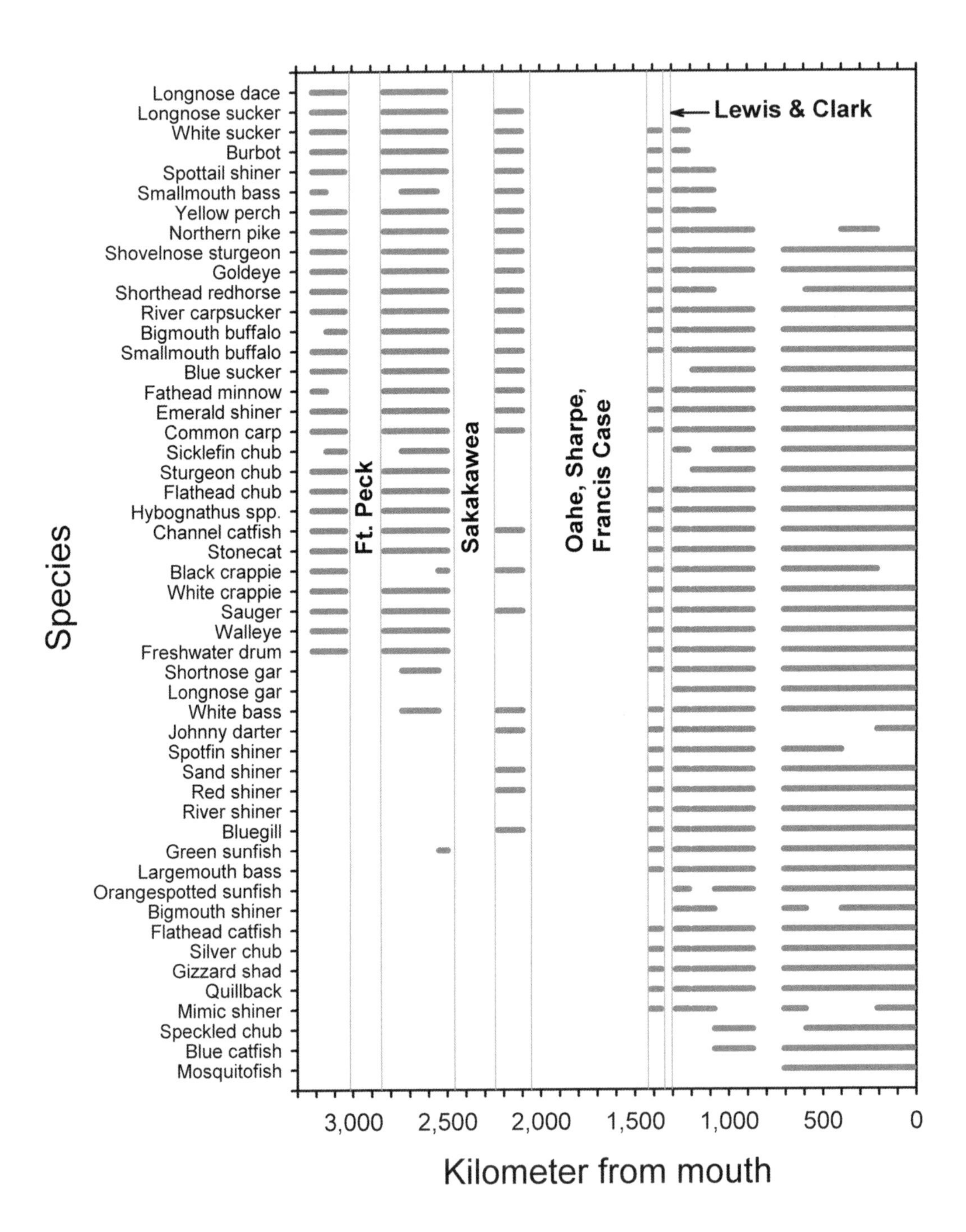

Figure 2.—Longitudinal distribution of major (3,100 collected) Missouri River fishes based on standardized sampling between July and October 1996–1998, from main-channel habitats exclusive of reservoirs. A taxon was presumed present in segments not sampled if it was collected from segments sampled above and below the missing reach. Vertical lines represent reservoir boundaries. *Hybognathus* spp. represents Western silvery, brassy, and plains minnows combined. Based on data from Berry et al. (2004).

carpsucker), and structure for spawning predominate above the reservoirs; gravel spawners dominate in inter-reservoir segments; and a high percentage of general and pelagic spawners (e.g., gizzard shad and freshwater drum) occur in channelized segments. Pegg and Pierce (2002b) suggest that fishes from river segments with a high degree of flow alteration tend to be deeper bodied and not well suited for the more natural flow patterns that still exist in some portions of the river.

Braaten and Guy (2002) also used the benthic fishes' database to examine species-specific life history characteristics for two short-lived (emerald shiner and sicklefin chub) and three long-lived fishes (freshwater drum, river carpsucker, and sauger) across a large portion of the Missouri River's latitudinal (48°03'N to 38°47'N) and thermal gradients. Mean water temperature and number of days in the growing season averaged 1.3 times greater and growing-season degree-days were twice as high in southern than northern latitudes of the main stem. Longevity for all species, except freshwater drum, increased significantly from south to north. Other population variables like length at age, growth coefficients, and growth rates during the first year of life varied differently with latitude among the species studied. They demonstrated that latitudinal variations in thermal regime across the Missouri's broad spatial gradient greatly influence growth and other life history characteristics of fishes studied.

Pegg and Pierce (2001) conducted a similar latitudinal analysis as Braaten and Guy (2002) on channel catfish, emerald shiner, freshwater drum, river carpsucker, and sauger. They observed significant latitudinal trends in growth rate coefficients only for emerald shiners, while results were inconclusive for the other four species. Growth coefficients were higher for age 1+ emerald shiners at higher latitudes. They concluded that natural (e.g., differences in biological communities and variability in individual growth rates) and anthropogenic (e.g., flow alteration, impoundment, and channelization) factors may have been responsible for the few river-wide latitudinal trends observed in growth rates, despite large differences in growing season and cumulative degree-day gradients.

Bergsted et al. (2004) used the benthic fishes data to develop a preliminary index of biotic integrity for the warmwater Missouri River to evaluate changes related to channelization and impoundment. The least-altered zone served as the reference condition with all sites rated excellent to good. The inter-reservoir zone showed the greatest variability in ratings with 60% of sites rated fair to very poor. The regulated-unchannelized zone (Gavins Point Dam to Ponca, Nebraska) had 60% of sites rated "good," and the channelized zone had 89% of its sites rated between good and fair.

Hesse and Mestl (1993) and Hesse (1994a, 1994b, 1994c, and 1994d) reported trends for several Missouri River fishes at a smaller spatial scale within the upper unchannelized, lower unchannelized, and channelized segments previously described. Drift net collections for larval fishes between 1983 and 1991 yielded the highest mean total CPUE (number/1,000 m^3) in the channelized section (539), followed by the lower unchannelized section (316), with much lower CPUE from the upper unchannelized section between Lake Francis Case and Lewis and Clark Lake (49.4, Hesse 1994d). Although reservoir fragmentation was associated with reduced density of larval fishes in the upper unchannelized section of the lower river, the pattern of lowest larval CPUE in the inter-reservoir reach was not necessarily similar for individual species. For example, mean larval paddlefish CPUE was highest in the upper unchannelized reach (0.30), lower in the channelized reach (0.10), and larvae were generally absent from the lower unchannelized river in Nebraska (CPUE = 0.005, Hesse and Mestl 1993). The spatial trend observed for larval sauger was different with highest CPUEs in the lower unchannelized section (2.3), and nearly similar catches from the channelized (1.1) and upper unchannelized segments (0.9, Hesse 1994d).

Hesse (1994b) compared CPUE from electrofishing collections of flathead catfish from the same three river segments in South Dakota, Iowa, and Nebraska. Mean CPUE (number/min) between 1981 and 1991 was 1.1 for the channelized reach, 0.6 for the lower unchannelized reach, 0.0 for the upper unchannelized reach upstream from the Niobrara River, and 0.1 downstream from the Niobrara River.

Standardized seine collections of small fishes from years with comparable data (1986–1993) showed a different pattern with highest mean CPUEs (number/seine haul) in the lower unchannelized section (67.4) and the other two sections about the same (upper unchannelized: 37.7, channelized: 35.0, Hesse 1994c). Sicklefin, sturgeon, flathead, and speckled chubs were absent from small fish collections from the unchannelized Missouri River above and below Lewis and Clark Lake in 1976 and between 1983 and 1993. Average composition of total seine catches was also very low from the lower unchannelized and upper unchannelized segments for silver chubs: 0.03% and 0.14%, respectively, and for plains/silvery minnows: 0.04% and 0.07%. These chubs and minnows were collected from the channelized reach as well, with sicklefin, sturgeon, and flathead chubs composing less than 0.06% of catch. Silver chub (6.6%), plains/western silvery minnow (2.1%), and speckled chub (0.21%) were slightly more abundant in the channelized reach. Despite its being channelized, highest CPUEs or percent composition for total larval and small fishes and flathead catfish were observed in the channelized river section, the reach farthest from the influence of reservoirs.

Temporal Changes

Reservoir fishes.—Mean yearly CPUE for goldeye, common carp, black and white crappie combined, and yellow perch significantly declined in Lake Sakakawea over the 42 years examined, whereas catches of spottail shiner, white bass, and johnny darter increased (Hendrickson and Power 1999). No significant differences in CPUE were observed over time for 10 other taxa, 6 of which we compared with Lewis and Clark Lake (Table 4). Mean yearly CPUE significantly declined over the 45 years examined for six taxa from Lewis and Clark Lake, and increased for four taxa (Table 4). Goldeye and common carp CPUE significantly declined, whereas only walleye CPUE significantly increased in both reservoirs.

Trends in walleye were similar for Lake Sakakawea (Figure 3 in Hendrickson and Power 1999) and Lewis and Clark Lake (Figure 3) with a gradual increase in CPUE occurring from the mid-1950s to mid-1970s and then the increase becoming more variable thereafter. Sauger showed a similar temporal pattern in Lewis and Clark Lake, but CPUE declined in the Missouri River below this reservoir (see *Riverine fishes*).

Table 4.—Regression statistics showing temporal trends in mean yearly catch per unit effort CPUE (number/standard gill-net set) for Missouri River fishes in Lake Sakakawea (1956–1998; various years missing; N = 16 to 22 years; source: Hendrickson and Power [1999]) and Lewis & Clark Lake (1956–2001; 1965, 1973–1983 missing; N = 35 years; see text for sources). Bigmouth and smallmouth buffalos compose "buffalos" and black and white crappies compose "crappies." Gizzard shad are absent from Lake Sakakawea. See text for collection methods. Only slopes significantly different from 0.0 at P less than 0.05 are shown.

	Lake Sakakawea			Lewis & Clark Lake		
Species	r^2	P	slope	r^2	P	slope
gizzard shad				0.16	0.019	+0.277
goldeye	0.22	0.027	–0.031	0.44	<0.001	–0.027
common carp	0.33	0.005	–0.012	0.58	<0.001	–0.412
river carpsucker	0.19	0.090		0.32	<0.001	–0.099
buffalos	0.02	0.635		0.60	<0.001	–0.049
shorthead redhorse	0.04	0.387		0.42	<0.001	–0.017
channel catfish	0.12	0.111		0.28	0.001	–0.061
white bass	0.65	<0.001	+0.001	0.03	0.295	
crappies	0.35	0.004	–0.006	0.02	0.411	
yellow perch	0.34	0.004	–0.027	0.00	0.903	
sauger	0.13	0.096		0.43	<0.001	+0.167
walleye	0.77	<0.001	+0.028	0.38	<0.001	+0.139
freshwater drum	0.05	0.334		0.48	<0.001	+0.211

Catch per unit effort for goldeye, common carp, river carpsucker, and buffalos in Lewis and Clark Lake showed a precipitous decline in the 16 years following impoundment, then became more stable thereafter. This pattern is illustrated by a regression of buffalo CPUE against year for the 1956–1972 period yielding a highly significant decrease in CPUE (Figure 3, compare with buffalos in Table 4). However, CPUE did not change significantly with time thereafter. The decline was more gradual for shorthead redhorse (Figure 3) and channel catfish. We were not able to determine if the pattern was similar in Lake Sakakawea because fish sampling did not begin immediately following impoundment.

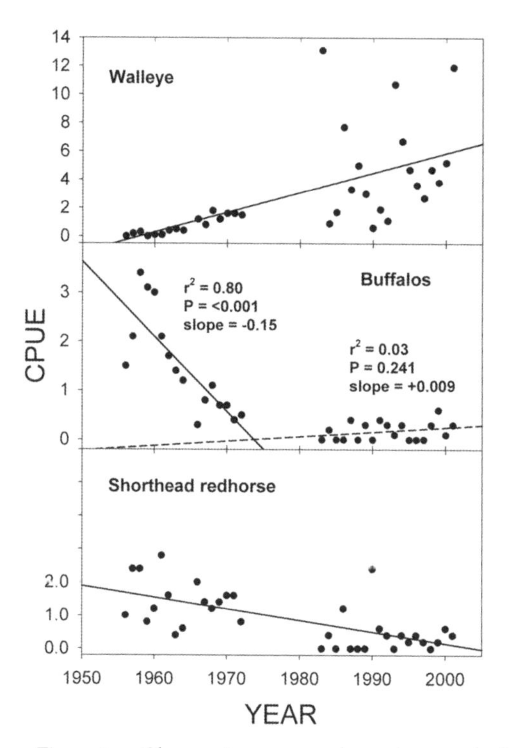

Figure 3.—Changes in mean yearly catch per unit effort (CPUE, number/standard gill-net set) for walleye, bigmouth and smallmouth buffalos combined, and shorthead redhorse from Lewis and Clark Lake, 1956–2001. Year-CPUE regressions for buffalos were divided into two periods, 1956–1972 and 1983–2001, to illustrate their rapid decline following impoundment followed by fairly stable catches. See Table 5 for regression statistics for the three taxa over the entire 1956–2001 interval. See text for data sources.

Riverine fishes. Montana.—Sufficient CPUE and percent composition data were available to contrast temporal changes for nine fish species between the 1976–1979 and 1997–2002 periods at five locations above and below Fort Benton (Table 5). Most species are fluvial dependent, big river fishes and two species (common carp and walleye) were introduced to the Upper Missouri drainage unit. Catch per unit effort over the 20-year interval decreased significantly ($P \leq 0.05$) only for longnose sucker and increased for river carpsucker, shorthead redhorse, walleye, and freshwater drum. No significant change in CPUE between the two periods was observed for goldeye, common carp, white sucker, and sauger. The proportion a species contributed to the total catch decreased significantly for goldeye and longnose sucker, increased for walleye, and freshwater drum, and did not change during the approximately 20 years for common carp, white sucker, and sauger.

Our analysis and that of McMahon and Gardner (2001) show that saugers have declined in the Missouri River main stem above Fort Benton, but not below. McMahon and Gardner (2001) also reported declines of sauger in Ft. Peck Lake, the Missouri River between Ft. Peck Dam and the North Dakota border, the Yellowstone River, and in spawning runs up important tributaries (Marias, Milk, and Tongue rivers). They estimated a 22% range reduction of sauger in the main-stem Missouri River and a 75% reduction in occupancy of tributaries. Sauger populations at the peripheries of their distributions (i.e., tributary streams, upper and lower ends of their distribution) appear to have been most altered. Factors attributed to sauger declines over the past 20 years in Montana include low reservoir levels and river flows, dams and water diversions, hybridization with introduced walleye, interactions with nonnative piscivorus walleye and smallmouth bass, and overfishing.

South Dakota–Iowa–Nebraska.—Declines in CPUE from 1983 to 1991 occurred for 6 of the 22 species Hesse (1993a), Hesse et al. (1993), and Hesse (1994d) reported from the unchannelized river, whereas increasing trends occurred for only three species, and the remainder showed no clear pattern (Table 6). Percent composition of total catch declined over the 9 years for 5 species and increased for 3 species. White sucker, white crappie, and sauger

Table 5.—Change in catch per unit effort (CPUE, number fish electrofished per hour) and percent composition (%, percent composition of species relative to the entire catch for the year) of nine fishes from five locations along the Missouri River, Montana, between two time periods 1976–1979 (76–79) and 1997–2002 (97–02). Data for the 1976–1979 period are from Berg (1981), and data for the 1997–2002 period are from W. Gardner files. T = Trace, replaced with 0.05 for statistics. Comparisons significantly different at *P* less than 0.05 are in **bold**.

		Location and kilometers above St. Louis, Missouri							
Species	Statistic	Morony Dam 3,363	Fort Benton 3,339	Coal Banks 3,266	Judith Landing 3,189	Robinson Bridge 3,097	Mean	Paired *t*-statistic	*P* value
goldeye	CPUE 76–79	22.7	13.2	29.3	13.9	19.5	19.7		
	CPUE 97–02	25.4	27.2	22.9	13.3	16	21.0	−0.35	0.743
	% 76–79	27	20.7	38.2	27.9	59.8	34.7		
	% 97–02	16	17.6	18.1	13.1	20.6	17.1	2.91	**0.044**
common carp	CPUE 76–79	1.5	3.7	6.5	3.2	6.3	4.2		
	CPUE 97–02	9	7	6.8	6.3	3.6	6.5	−1.35	0.247
	% 76–79	1.8	5.8	8.5	6.4	19.3	8.4		
	% 97–02	6.4	7.2	7.3	7.3	5.3	6.7	0.52	0.634
longnose sucker	CPUE 76–79	17.9	11.9	8	4.3	0.1	8.4		
	CPUE 97–02	13.6	4.7	4	2	0.4	4.9	2.84	**0.047**
	% 76–79	21.3	18.7	10.4	8.6	T	11.8		
	% 97–02	9.5	3.3	3.3	1.6	0.5	3.6	3.06	**0.038**
river carpsucker	CPUE 76–79	0.3	0.9	2.2	3	0.9	1.5		
	CPUE 97–02	2.1	2.4	5.8	6.5	7.5	4.9	−3.75	**0.020**
	% 76–79	T	1.4	2.9	6	2.8	2.6		
	% 97–02	1.4	3.1	5.8	7.6	10.9	5.8	−2.46	0.069
shorthead redhorse	CPUE 76–79	7.8	22	21.4	16.5	2.1	14.0		
	CPUE 97–02	53.2	60.7	41.5	24.1	14.9	38.9	−3.39	**0.027**
	% 76–79	8.7	34.5	27.9	33.1	6.4	22.1		
	% 97–02	35.1	42.7	34.5	26.2	19	31.5	−1.75	0.155
white sucker	CPUE 76–79	1.8	1	0.3	0.5	0	0.7		
	CPUE 97–02	4.3	11.6	3.3	0.2	T	3.9	−1.61	0.183
	% 76–79	2.1	1.6	T	1	0	0.9		
	% 97–02	2.7	7	1.7	0.3	T	2.3	−1.31	0.261
sauger	CPUE 76–79	20.1	6.7	3.6	3.6	2.9	7.4		
	CPUE 97–02	1.9	3.6	6	7.7	9.6	5.8	0.36	0.734
	% 76–79	23.9	10.5	4.5	7.2	8.9	11.0		
	% 97–02	1.4	2.1	5.8	9	14.4	6.5	0.88	0.428
walleye	CPUE 76–79	0.3	0.1	T	T	T	0.1		
	CPUE 97–02	2.1	1.8	2.9	3.9	1.7	2.5	−5.50	**0.005**
	% 76–79	T	T	T	T	T	T		
	% 97–02	1.6	1.5	2.8	3.7	2.5	2.4	−5.83	**0.004**
freshwater drum	CPUE 76–79	2.6	0.6	0.2	0.2	0.1	0.7		
	CPUE 97–02	6.1	3.6	5.9	2.7	1.2	3.9	−4.21	**0.014**
	% 76–79	3.1	1	T	T	T	0.8		
	% 97–02	4.4	2.9	5.7	2.8	1.4	3.4	−3.20	**0.033**

declined both in CPUE and percent composition. These data were collected 30 and 28 years, respectively, after Francis Case and Lewis and Clark reservoirs began filling, and it is likely many changes in fish composition and abundance had already taken place. There is evidence supporting this claim for sauger as Hesse (1994d) reported declining electrofishing catch rate in the unchannelized river between 1963 and 1991 (Figure 4). These data show the same precipitous decrease following dam closure (1963–1975) with a less steep decline thereafter (1983–1991), as was observed for several riverine fishes in Lewis and Clark Lake (Figure 3).

Density of larval sauger from the upper unchannelized segment also declined, from 10.6/1,000 m^3 in 1965 to an average of 0.9/1,000 m^3 between 1983 and 1991 (Hesse 1994d). Similarly, a 93% reduction in sauger larvae was observed between 1974 and 1985–1991 from the channelized segment. Sportfishing harvest of sauger from the Gavins Point Dam tailwater also decreased from 49% of total harvest in 1961 to 0.2% in 1992 (Hesse 1994d).

Small fishes were seined from the channelized reach of the Missouri River, Iowa–Nebraska, each year from 1970 to 1975 and again from 1986 to 1993 enabling comparison of these two periods separated by about a decade (Hesse 1994c). There was no significant difference in CPUE for all small fishes combined (t-test assuming unequal variances, $t = 0.74$, $P = 0.50$) between the 1970s (mean ± 1 SE = 46.5 ± 14.4) and the 1980–1990s (35.0 ± 6.1). However, fewer flathead chubs ($t = 2.83$, $P = 0.066$) and plains/western silvery minnows ($t = 3.29$, $P = 0.046$) were collected in the second period. Sicklefin and sturgeon chub numbers were so low by 1970 that they were collected in only 1 of the subsequent 11 years sampled (1988, 0.2 fish per standard haul). Some small fishes apparently did benefit from changes in the river as Hesse (1994c) reports emerald shiners increased from about 17% of the catch in 1971–1975 to 69% of catch in 1989.

Table 6.—Regression statistics showing temporal trends in catch per unit effort (CPUE) and percent composition of fishes from experimental gillnetting in the unchannelized Missouri River, Nebraska, 1983–1991. Only slopes significantly different from 0.0 at $P <$ 0.10 are shown. Source of data: Hesse (1994d).

	CPUE			Percent composition		
Species	r^2	P	slope	r^2	P	slope
shortnose gar	0.14	0.315		0.02	0.702	
gizzard shad	0.39	0.073	−0.187	0.32	0.114	
goldeye	0.06	0.538		0.19	0.237	
northern pike	0.61	0.014	0.117	0.53	0.026	1.232
common carp	0.43	0.056	−0.317	0.34	0.102	
river carpsucker	0.10	0.402		0.31	0.121	
white sucker	0.67	0.007	−0.015	0.54	0.024	−0.052
smallmouth buffalo	0.27	0.151		0.24	0.185	
bigmouth buffalo	0.03	0.681		0.02	0.742	
shorthead redhorse	0.33	0.106		0.34	0.099	2.392
black bullhead	0.01	0.841		0.00	0.948	
channel catfish	0.00	0.908		0.02	0.720	
white bass	0.67	0.007	−0.015	0.63	0.011	−0.065
rock bass	0.19	0.236		0.35	0.093	−0.050
largemouth bass	0.01	0.810		0.03	0.646	
smallmouth bass	0.39	0.070	0.017	0.49	0.036	0.080
black crappie	0.28	0.146		0.28	0.139	
white crappie	0.44	0.052	−0.072	0.44	0.053	−0.318
yellow perch	0.10	0.407		0.09	0.423	
sauger	0.87	0.000	−0.568	0.78	0.002	−2.332
walleye	0.20	0.225		0.06	0.537	
freshwater drum	0.15	0.302		0.15	0.310	

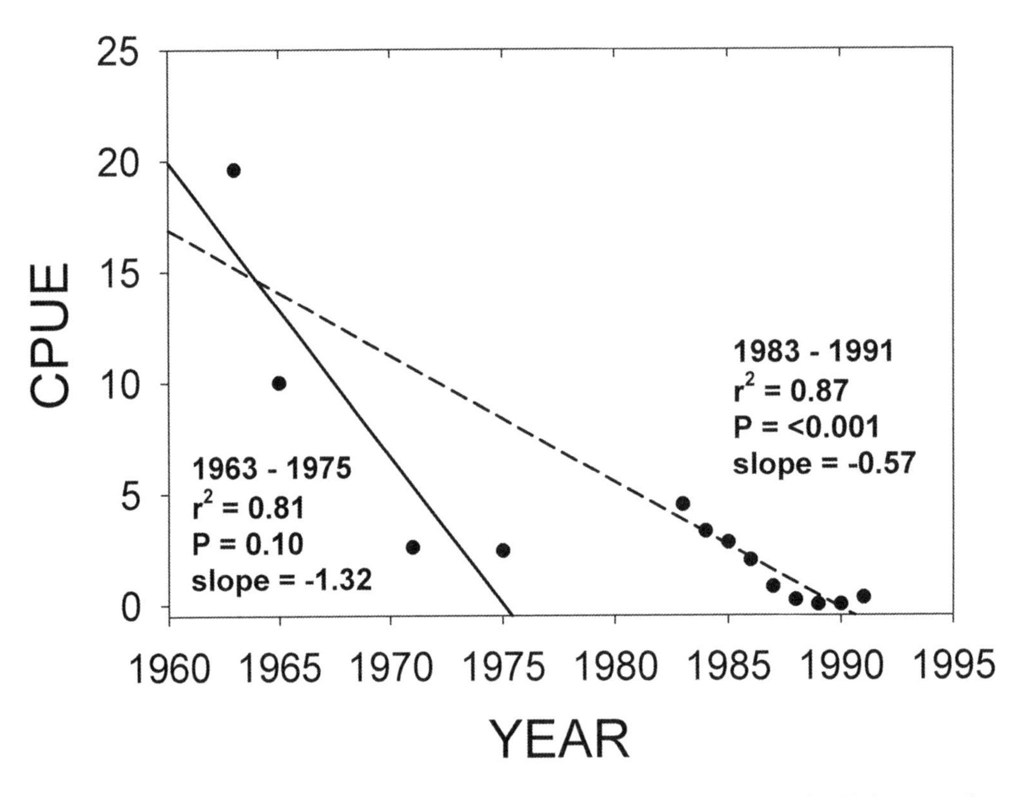

Figure 4.—Changes in mean yearly electrofishing catch per unit effort (CPUE, number/h) for sauger from the channelized Missouri River, Nebraska, downstream from Lewis and Clark Lake, which began filling in 1955. Year-CPUE regressions are calculated separately for two time periods: 1963–1975 and 1983–1991 to illustrate different trends between the two periods (data from Hesse 1994d).

Missouri.—The 890 km of Missouri River passing through Missouri to its confluence with the Mississippi River are both channelized and flow regulated. However, numerous large volume tributaries discharge to the river here because it is within the more mesic Central Lowlands and Interior Highlands physiographic provinces. Approximately one-half of the Missouri River's total discharge to the Mississippi River enters the main stem within the state of Missouri. Consequently, flood pulses have not changed dramatically within this reach since upstream impoundment. However, low flows and other ecologically relevant hydrologic variables have been more altered (Galat and Lipkin 2000). Fishes from the Ozark Highlands and Mississippi River are added to the ichthyofauna in this segment of river accounting for some of the observed increase in species richness.

Pflieger and Grace (1987) summarized changes in fishes in the Missouri River, Missouri, from surveys at approximately 20-year intervals between 1940 and 1983. Gizzard shad substantially increased in numbers, and goldeye, bluegill, channel catfish, white crappie, sauger, and freshwater drum may also have increased. Catches of common carp, river carpsucker, and bigmouth buffalo decreased markedly.

Pallid sturgeon composed 3% of river sturgeons (*Scaphirhynchus*) from 1940s collections, although river sturgeons were rare (Fisher 1962). No pallid sturgeons were collected in the 1960s or 1980s collections (Pflieger and Grace 1987). Carlson et al. (1985) examined 1,806 river sturgeons during 1978–1979 from the lower Missouri and middle Mississippi rivers, and reported only 0.3% were pallid sturgeons and four specimens were pallid–shovelnose hybrids. Pallid sturgeons were also very rare in these areas nearly 20 years later. Grady et al.

(2001) collected 5,197 sturgeons from the lower Missouri and middle Mississippi rivers between 1996 and 2000; 8 (0.15%) were wild origin pallid sturgeons and 22 were pallid–shovelnose hybrids.

Small fishes whose prominence increased included red shiners and several *Notropis* spp.: emerald, river, and sand shiners (Pflieger and Grace 1987). Relative abundance of several chubs (speckled chub, sturgeon chub, sicklefin chub, and silver chub) also was reported to increase slightly in contrast to the trends reported above for Nebraska. Small fishes whose numbers declined greatly included flathead chub and plains minnow.

Pflieger and Grace (1987) designated 26 species as "stragglers" in the Missouri River, Missouri (generally identified as sporadic or rare and waifs on our species list). More of these species occurred during the 1962–1966 and 1978–1983 periods than during the 1945–1949 period. Six species apparently were added to the Missouri River's fish fauna in Missouri after 1945. Skipjack herring is an anadromous big river fish from the Mississippi River that was first recorded from Missouri in approximately 1954, but numbers increased in the 1970s. Its increase in the Missouri River coincided with a decline in the upper Mississippi River following construction of locks and dams and a reduced suspended sediment load (a surrogate for turbidity) in the lower Missouri River following impoundment (Cross 1975). White bass is a nonnative species and was intentionally introduced into Lewis and Clark Lake in 1959. Alien grass carp first appeared in commercial catches in 1971 and striped bass, stocked into Lake of the Ozarks on the Osage River in 1967, first appeared in 1975. Rainbow smelt was introduced into Lake Sakakawea, North Dakota in 1971 and was first collected in Missouri in 1978. Silver carp first appeared in commercial fish catches in 1982.

In general, Pflieger and Grace (1987) observed that fishes that became more abundant were mostly pelagic planktivores and sight-feeding carnivores: skipjack herring, gizzard shad, white bass, bluegill, white crappie, river shiner, and red shiner. Fishes that decreased in abundance included big river species adapted for life in turbid waters with specialized habitat or feeding requirements (e.g., pallid sturgeon and flathead chub) or species more common in backwaters (e.g., western silvery minnow, plains minnow, and river carpsucker).

Grady and Milligan (1998) quantitatively examined changes in the minnow fauna following Pflieger and Grace's (1987) study. Presence of minnows increased from the 1940s to 1990s for five species, decreased for five species, and remained about the same for one species (Table 7). Populations of sturgeon chubs and western silvery minnows continued to decline from the 1980s. These results generally corroborated those of Pflieger and Grace (1987). Grady and Milligan (1998) and Berry et al. (2004) report that recent adoption of boat-mounted benthic trawls in channel habitats has yielded higher catches of many minnows and chubs than did historical shoreline bag seining. This suggests that channel populations of some small riverine cyprinids might not be as low as previously reported.

Commercial fishery.—The number of commercial fishers in the state of Missouri gradually decreased from 1948 (968) to 1963 (350), and then gradually increased to a peak of 1,039 in 1982 (Figure 5). Commercial fishers declined nearly continuously thereafter to 67 in 2001 (Figure 5). Factors contributing to these fluctuations through 1990 include increases in permit fees and health advisories against consumption of Missouri River fishes (Robinson 1992). Closure of the commercial catfish fishery on the Missouri River in 1992 and record flooding during the 1990s contributed to the further decreases in permit sales (V. Travnichek, personal communication).

Prior to 1997, total reported harvest was highest in 1945 at 222 metric tons (mt), and then declined gradually to 35 mt in 1966, paralleling the decline in number of fishers. Methods of estimating annual harvest changed in 1967, providing a more accurate, but higher reported harvest. Total harvest generally increased from the late 1960s until 1990 when it peaked at 432.5 mt (Figure 5). The precipitous decline thereafter is attributed to closure of the commercial catfish fishery.

Eleven groups of fishes comprised 98% of the total catch between 1945 and 2001 (Figure 6). In decreasing order of percent composition, these were

Table 7.—Summary of changes in presence-absence of cyprinid minnows in Missouri River, Missouri, between 1945 and 1997 (Grady and Milligan 1998). Grady and Milligan (1998) used logistic regression to examine if there was a significant positive (+), negative (–), or no (0) relationship between the probability of collecting a species over time (year) and distance (upriver to downriver within Missouri).

Species	Year	Distance	Notes
sicklefin chub	+	+	More abundant below Kansas City than above.
sturgeon chub	0	0	Rare at all times and locations.
plains minnow	–	0	Absent only in 1994 collections, stable otherwise.
flathead chub	–	0	Only one fish collected since 1980s.
western silvery minnow	–	–	Drastic decline in numbers collected over time.
sand shiner	+	+	Uncommon in general.
ghost shiner	–	+	More abundant below Kansas City than above.
river shiner	+	–	Increase in numbers over time.
emerald shiner	+	0	Most abundant minnow collected.
bigmouth shiner	+	–	Slight increase over time.
bluntnose minnow	–	+	More abundant in quiet backwaters.

common carp (39.0%), buffalo spp. (24.2%), flathead catfish (9.7%), channel catfish (6.5%), freshwater drum (4.5%), carpsucker spp. (4.1%), blue catfish (3.4%), grass carp (2.1%), paddlefish (1.9%), other Asian carps (*Hypophthalmichthys* spp., 1.5%), and sturgeon spp. (1.2%). Catches of grass, silver, and bighead carps have been gradually increasing (Figure 6). Shovelnose sturgeons are the only other group where catch has not shown a precipitous decline since about 1990. It declined to less than 1,800 kg during the early 1990s flood years, but gradually increased to more than 5,600 kg in 2001 (Figure 6). Increased commercial interest in sturgeons is a result of high prices paid for

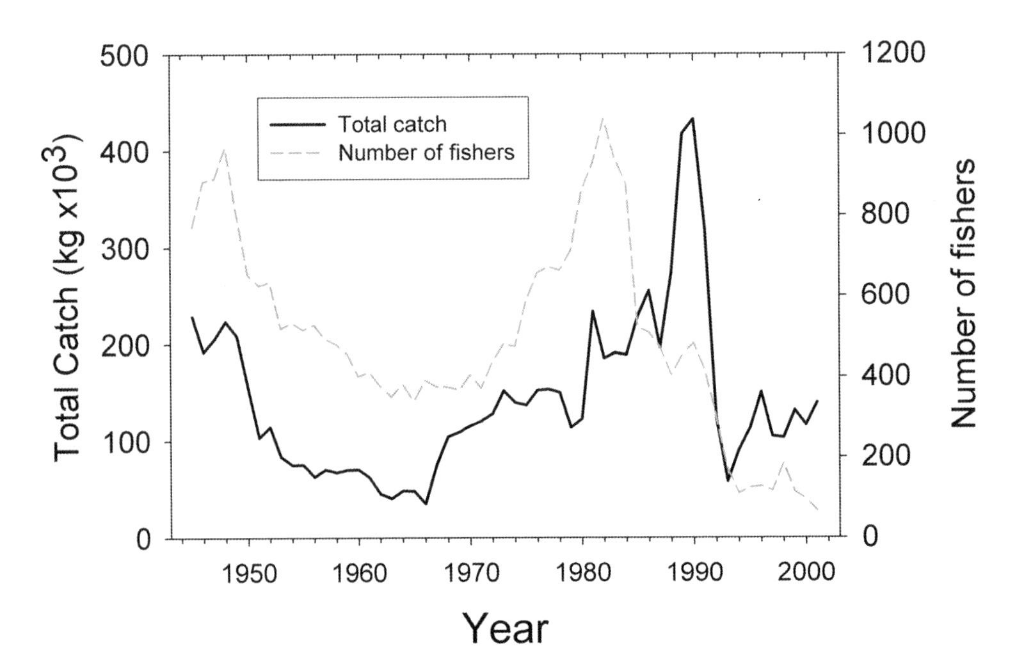

Figure 5.—Changes in number of commercial fishers and their reported total annual harvest of all fish species from Missouri River, Missouri between 1945 and 2001 (data from V. Travnichek, Missouri Department of Conservation).

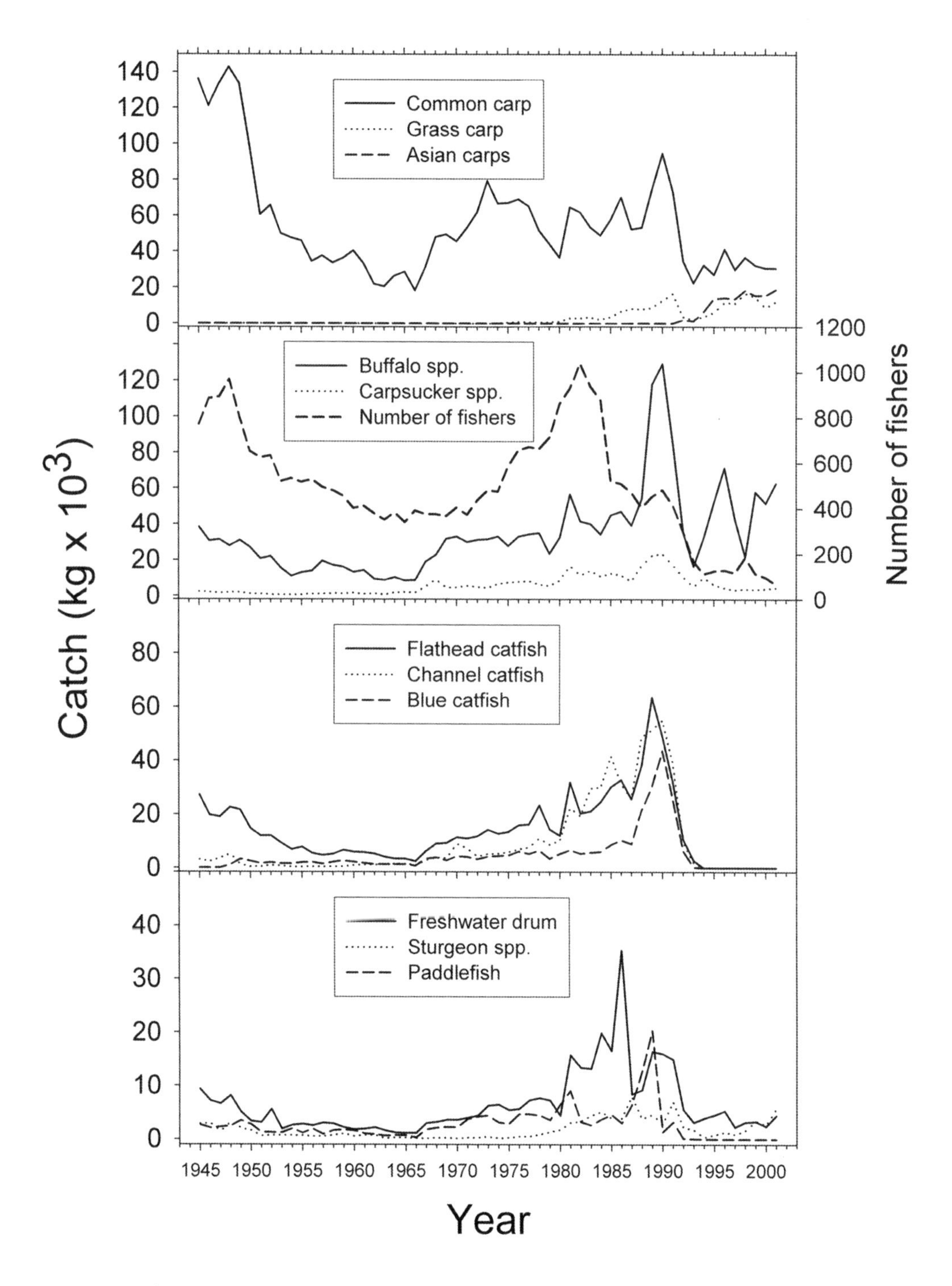

Figure 6.—Changes in reported total annual harvest of 11 fish taxa from Missouri River, Missouri, between 1945 and 2001 (data from V. Travnichek, Missouri Department of Conservation).

caviar and is a cause for concern about over harvest (Quist et al. 2002).

A 61% decline in commercial harvest of channel catfish as a result of overfishing was reported for Nebraska (Zuerlein 1988 in Hesse 1994a) over four, 10-year periods between 1944 and 1983. Closure of the commercial catfish fishery in 1992 resulted in noticeable changes in population structure of Missouri River channel catfish in Nebraska (Mestl 1999) and of channel, flathead, and blue catfishes in the Missouri River recreational fishery in Missouri (Stanovick 1999). Mean length and percentage of quality-sized channel catfish (>410 mm) in Nebraska catches increased significantly after the closure (Mestl 1999). Percentage of ages four to seven catfish made up 9% of standing crop before closure (1974–1975) and 58% after closure (1996–1997). Although annual production decreased from 199 to 83.5 kg/100 kg of standing stock between these periods, most of this production was in fish below harvestable size. Seventy-two percent of production of age 2–3 channel catfish was lost through mortality during 1974–1975, whereas only 35% was lost through mortality in 1996–1997.

Size of channel, flathead, and blue catfish in the Missouri River, Missouri, recreational harvest also increased significantly from before (1991–1992) to after (1995–1996) the commercial ban (Stanovick 1999). Additionally, angler harvest rates and release rates for all three species were higher after the commercial ban. This was most noticeable for flathead catfish where harvest rates more than tripled and release rates more than doubled following the commercial ban.

Discussion

The Missouri River possesses a diverse ichthyofauna. It is distinguished by native species well adapted for life in an environment with continuous high turbidity, a swift current, a scarcity of quiet backwaters, and an unstable sand-silt bottom (Pflieger 1971). The term "big river" fishes in common use today was coined by Pflieger (1971) to describe this distinctive assemblage of fishes inhabiting the Missouri–Mississippi system. Here resides a preponderance of benthic specialists exhibiting a panoply of ecomorphological adaptations, including inferior mouth position, dorsoventral flattening of the head, streamlined or deep humpbacked body shape, sickle-shaped or enlarged pectoral fins, reduced eyes, and an array of well-developed electrosensory and chemosensory organs (e.g., sturgeons, paddlefish, chubs, buffaloes, carpsuckers, blue sucker, catfishes, burbot, and freshwater drum). Environmental factors that molded ecomorphology of Missouri River's fishes are perhaps most similar to those operating in other largely turbid, dryland rivers like the Colorado (Mueller et al. 2005) and Rio Grande (Calamusso et al. 2005). However, unlike the Colorado, Missouri River fishes also evolved within a community rich in native piscine predators and competitors. Indeed, flathead catfish, one of the most damaging, nonnative piscivores to the lower Colorado River basin's native fishes (Minckley et al. 2003) is an archetypical Missouri–Mississippi big river predator.

Spatiotemporal changes across the Missouri's riverscape are a consequence of complex interactions between natural and anthropogenic factors. Predominant landscape factors that shape its present-day fish distributions include basin climate, physiography, zoogeography, and hydrology. Collectively, these constitute the "historical ecology" of a river's fish assemblage (Matthews 1998). Distribution patterns of native fishes along the Missouri River well illustrate the classic phenomenon of "longitudinal zonation" in riverine fish assemblages (Hawkes 1975; Matthews 1998). Additions of more Ozarkian (e.g., black redhorse and golden redhorse) and Mississippian fishes (e.g., Alabama shad, flathead, and blue catfish,) and replacement of coolwater species (e.g., longnose dace and longnose sucker) by warmer-water species (e.g., speckled chub and quillback) in the lower river contribute to the biogeographic longitudinal patterns observed.

These abiotic and biotic determinants constitute a series of filters (sensu Tonn 1990; Poff 1997) upon which human activities have further molded the patterns reported herein. Impoundments have fragmented the main-stem Missouri River and created longitudinal barriers to long-distance migrants like pallid sturgeon. Similarly, several small-bodied chubs are now extirpated from over more than one-half of the river's length. Temperature reductions below hypolimnetic-release dams impounding Ft. Peck Lake and Lake Sakakawea have been implicated in lack of recruitment of pallid sturgeon (Gardner and Stewart 1987; Dryer and Sandvol 1993), chubs (Galat and Clark 2002), and native larval fishes in general (Wolf et al. 1996). Reduction in channel complexity and changes to the natural flow regime have been particularly damaging to its big river fauna as is evidenced by the predominance of imperiled fluvial specialist and fluvial dependant fishes. Lateral fragmentation of the river from its floodplain by levees has been implicated in population declines of fishes that use the floodplain for spawning and/or nursery (Galat et al. 1998). Clear water and dense submergent vegetation associated with delta formation in reservoirs supports a greater density of centrarchids (e.g., largemouth bass and bluegill) and certain cyprinids (e.g., spotfin shiners) than other segments of the river or reservoirs (VanZee 1997; Wickstrom 2001) and may have contributed to range expansion of grass pickerel (VanZee and Scalet 1997).

The most recent anthropogenic impact to the Missouri River may be the population explosion of introduced Asian carps, particularly bighead and silver carp below Lewis and Clark Dam. They are often the most abundant larval fish collected (Galat et al. 2004) and adults also frequently dominate in experimental gill net catches (D. Chapman, U.S. Geological Survey, personal communication). Pflieger (1997) warned that bighead carp may compete for food with native planktivores, including paddlefish and bigmouth buffalo, as well as larvae of most native Missouri River fishes. Research is currently underway to determine if diet overlap exists in the lower Missouri River (D. Chapman, personal communication).

Richness of the Missouri River's native fish fauna remains relatively intact despite these assaults; no native fishes have yet been extirpated. Proportion of native fishes is higher in the Missouri River than the Colorado or Columbia rivers, and it had the lowest proportion of imperiled fishes of eight north-temperate rivers reviewed by Galat and Zweimüller (2001). Nevertheless, the widespread and long history of human intervention has contributed to spatiotemporal declines of about 25% to its ichthyofauna. Our review substantiates consistent population declines throughout much of the main channel Missouri River for the following species that are not federally listed: sicklefin chub, sturgeon chub, plains minnow, western silvery minnow, highfin carpsucker, and sauger. In contrast, over one-half of additions to fish biodiversity of the Missouri River since Cross et al. (1986) have been intentional and unintentional introductions, contributing to a homogenization of its ichthyofauna (sensu Rahel 2002). These changes, in part, prompted the National Research Council (2002) to warn that, "Degradation of the Missouri River ecosystem will continue unless some portion of the hydrologic and geomorphic processes that sustained the preregulation Missouri River and floodplain ecosystem are restored—including flow pulses that emulate the natural hydrograph, and cut-and-fill alluviation associated with river meandering. The ecosystem also faces the prospect of irreversible extinction of species."

Restoration Activities

Several events during the past decade have directed national attention to impairment of the Missouri River. Declines in populations of archetypical Missouri River fishes and birds resulted in listing the least tern *Sterna antillarum* and the before-mentioned pallid sturgeon as endangered and the piping plover *Charadrius melodus* as threatened under the Endangered Species Act. The conservation organization, American Rivers, designated the Missouri River as the nation's most endangered river in 1997 and again in 2001. Basin-wide droughts in the late 1980s and early 2000s and catastrophic flooding in the lower river in the early to mid 1990s have highlighted conflicts over water allocation. Socioeconomic values for the river and floodplain are changing from primarily transportation and agriculture uses, respectively, to an increase in reservoir and river based recreation, and there is a recognized need for more balance among all of the river's designated beneficial uses. The Lewis and Clark "Corps of Discovery" bicentennial in 2004–2006 is anticipated to boost tourism and general public interest in the history and status of the "Big Muddy." Activities along the Missouri River are moving from chronicling offenses to its ecological integrity towards designing and implementing rehabilitation and restoration programs (Galat et al. 1998; U.S. Fish and Wildlife Service 1999; U.S. Army Corps of Engineers 2004) within an adaptive management framework (National Research Council 2002; Prato 2003).

Three recent publications have provided a catalyst for this shift. First, is the U.S. Fish and Wildlife Service's (USFWS) *Biological Opinion on Operation of the Missouri River Main-Stem Reservoir System and Operation and Maintenance of the Bank Stabilization and Navigation Project* (U.S. Fish and Wildlife Service 2001). The USFWS concluded that past and current operating plans and actions have jeopardized the continued existence of federally listed least tern, piping plover, and pallid sturgeon. To avoid jeopardy, the USFWS directed the U.S. Army Corps of Engineers to implement five "reasonable and prudent alternatives" (RPAs): (1) implement flow enhancement below Ft. Peck and Gavins Point dams (i.e., variability, volume, timing, and temperature) to provide hydrologic conditions necessary for species reproduction and recruitment; (2) restore, create, or acquire shallow-water, low-velocity channel and floodplain habitats, and sandbars in the lower river; (3) unbalance storage among the upper three reservoirs to benefit spawning fishes and increase availability of tern and plover sandbar habitat in riverine segments below reservoirs; (4) implement an interagency coordination team and a robust monitoring program to allow adaptive modification and implementation of management actions; and (5) increase pallid sturgeon propagation and augmentation efforts. The philosophy behind these recommendations is that both flow and habitat restorations are required to reduce jeopardy and recover the listed species. Evaluation of these projects is also necessary to assess progress towards these goals, and flexibility in actions is needed to respond to the uncertainty inherent in ecological systems.

Second, the U.S. Army Corps of Engineers' *Master Water Control Manual* directs operation of the main-stem dams on the Missouri River and is undergoing its first major revision since the reservoir system became operational in 1967. Its revision, as described in the *Revised Draft Environmental Impact Statement* (RDEIS, U.S. Army Corps of Engineers 2001), will affect future Missouri River flow management and be an integral part of implementing the USFWS's RPAs. Six alternative operating plans for the reservoirs were evaluated. One is the status quo or Current Water Control Plan. A second alternative contains four features: increased drought conservation measures, changes in Ft. Peck Dam releases to provide warmer water temperatures and a spring flow pulse for pallid sturgeon spawning, unbalancing the upper three reservoirs to benefit recreational fisheries, and adaptive management. The four other alternatives include a range of spring-rise enhancements and decreased summer releases from Gavins Point Dam to more closely approximate historical conditions. Differences among the four alternatives relate largely to magnitude of release modifications and downstream river stage changes (see Jacobson and Heuser [2001] for a comparison of flow alternatives). As of this writing, no final alternative has been selected for implementation.

A third publication that should influence Missouri River management is a National Research Council Report: *The Missouri River Ecosystem: Exploring the Prospects for Recovery* (2002). It recommends four steps to lay the groundwork for Missouri River recovery: (1) legitimize and empower Missouri River managers with the authority and responsibility to actively experiment with river operations to enhance ecological resources; (2) convene a representative stakeholder committee to develop a basin-wide strategy, conduct assessments, review plans, and provide oversight of implementation of adaptive management strategies; (3) develop long-term goals and short-term measurable objectives for adaptive management actions; and (4) work with stakeholders to build commitment to, and acceptance of, changes in current patterns of benefits delivered from the river and reservoir system.

Numerous flow and habitat enhancement projects are ongoing within the Missouri River corridor. Instream flow reservations for fisheries have been secured for the Missouri and Yellowstone rivers and many of their tributaries in the upper Missouri drainage unit. Negotiations are ongoing between Montana Fish Wildlife and Parks and the U.S. Bureau of Reclamation to provide spring flow pulses out of Canyon Ferry Dam. A new program called the Conservation Reserve Enhancement Program (CREP) expands the Conservation Reserve Program by allowing the U.S. Department of Agriculture to work with states and local interests to meet specific conservation objectives. Land along part of the Madison and Missouri rivers to Fort Peck Lake (919 km total) can be enrolled for conservation easements such as riparian protection from poor agriculture practices and for development assistance to farmers so they do not have to subdivide their land to survive. Most land is eligible that falls within 3.2-km along the river corridor.

Habitat acquisition and improvement projects in the lower Missouri River accelerated during the past 20 years and particularly since the floods of the 1990s. Over 20,000 ha of floodplain in Missouri have come under public ownership since the 1993 flood. Most of these lands will be managed using a combination of intensive and passive strategies to enhance river–floodplain connectivity and expand river channel top width (Galat et al. 1998). The newly established Big Muddy National Fish and Wildlife Refuge is targeting a total of 24,300 ha purchased from willing sellers. An additional 48,000 ha are authorized for acquisition and development by the Corps under the Water Resources Development Act of 1999 for the expanded Missouri River Bank Stabilization and Navigation Mitigation Project (U.S. Army Corps of Engineers 2004). Numerous Corps habitat mitigation projects have created or enhanced secondary channels and sandbars along the lower Missouri River, and expansion of this program is in progress. Should planned programs reach their acquisition and development goals, nearly 20% of the former lower Missouri River floodplain may be managed for natural resource benefits. Flood stages will be reduced as the river is allowed to expand laterally. Sandbars and braided channels will once again be common riverine features. Urban areas will be protected from devastating floods and the majority of the floodplain will remain dedicated to agriculture. Populations of native river and floodplain fishes are expected to benefit greatly from these actions.

Unfortunately, Missouri River rehabilitation efforts have seldom included explicit ecologically based objectives and performance measures. They are often site specific and driven by political realities rather than recovery of ecological processes. Equally important, they generally lack adequate pre- and postproject appraisals to evaluate progress towards restoration objectives and their outcomes. The result is that the "learning-by-doing" feedback loop essential to adaptive management is often missing. Other river (e.g., Kondolf 1995) and native fish (e.g., Minckley et al. 2003) restoration programs have experienced mixed success for similar reasons. Management agencies are encouraged to adopt a more holistic perspective for their activities to benefit the biological integrity of the Missouri River hydrosystem, rather than the single species approach emphasized by endangered species recovery plans. The National Research Council (2003) recently recommended that the Corps of Engineers adopt a set of principles and guidelines for successful restoration programs. We urge Missouri River restorationists to consider these in their project planning, execution, and evaluation. A well-designed, performance-based, restoration program should include relevant stakeholders and treat habitat rehabilitation and flow reregulation as an adaptive management experiment. Perhaps then, the public may one day experience a Missouri River more similar to what Lewis and Clark witnessed while enhancing diversity of contemporary socioeconomic benefits the river provides.

Acknowledgments

Information not referenced in this chapter reflects the expertise and opinions of its authors and is not the policy of their agencies. Vince Travnichek provided commercial fisheries data and John Stanovick provided information on responses of catfishes to closure of the commercial fishery in Missouri. Gust Annis made available fish species lists for Missouri. Larry Hesse provided access to the historical database being developed for the Missouri River (South Dakota, Iowa, and Nebraska). Sandy Jo Clark helped research and prepare conservation status and habitat use information for fishes. Donna Farris prepared the Missouri River basin map. We thank the following persons for commenting on drafts of this chapter: Pat Braaten, JoAnn Grady, Chris Guy, Jim Milligan, and Mark Pegg. The final draft was improved by reviews from John Chick, Ed Peters, Hal Schramm, Jr., and Vince Travnichek. Funding for this chapter was provided to the senior author by the U.S. Army Corps of Engineers through the benthic fishes project and the Missouri Department of Conservation through its Cooperative Agreement with the University of Missouri. This chapter is a contribution from the Missouri and South Dakota Cooperative Fish and Wildlife Research Units (U.S. Geological Survey; Missouri Department of Conservation; South Dakota Department of Game, Fish, and Parks; University of Missouri; South Dakota State University; and Wildlife Management Institute cooperating).

References

Abell, R. A., D. M. Olson, E. Dinerstein, P. T. Hurley, J. T. Diggs, W. Eichbaum, S. Walters, W. Wettengel, T. Allnutt, C. J. Loucks, and P. Hedao. 2000. Freshwater ecoregions of North America: a conservation assessment. Island Press, Washington, D.C.

Bailey, R. M., and M. O. Allum. 1962. Fishes of South Dakota. University of Michigan, Museum of Zoology, Miscellaneous Publications No. 119, Ann Arbor.

Barrett, W. W. 1895. First biennial report of the state fish and game commissioner. Letter to North Dakota Governor Eli Shortridge, Bismarck, North Dakota.

Bayley, P. B., and H. W. Li. 1996. Riverine fishes. Pages 92–122 *in* G. Petts, and P. Calow, editors. The rivers handbook. Volume 1. Blackwell Scientific Publications, Oxford, England.

Beckman, L. G., and J. H. Elrod. 1971. Apparent abundance and distribution of young-of-year fishes in Lake Oahe, 1965–69. Pages 333–347 *in* G. E. Hall, editor. Reservoir fisheries and limnology. American Fisheries Society, Bethesda, Maryland.

Benson, N. G. 1980. Effects of post-impoundment shore modifications on fish populations in Missouri reservoirs. U.S. Fish and Wildlife Service, Research Report 80, Washington, D.C.

Berg, R. K. 1981. Fish populations of the wild and scenic Missouri River, Montana. Montana Fish, Wildlife, and Parks, Federal Aid to Fish and Wildlife Restoration Project FW-3-R, Job Ia, Helena.

Bergsted, L. C., R. White, and A. V. Zale. 2004. Development of an index of biotic integrity for measuring biological condition on the Missouri River. Volume 7. Population structure and habitat use of benthic fishes along the Missouri and lower Yellowstone rivers. U.S. Geological Survey, Cooperative Research Units, Montana State University, Bozeman. Available: www.nwo.usace.army.mil/html/pd-e/benthic_fish/benthic_fish.htm. (January 2005).

Berry, C. R., Jr., and B. A. Young. 2002. Introduction to the benthic fishes study. Volume 1. Population structure and habitat use of benthic fishes along the Missouri and lower Yellowstone rivers. U.S. Geological Survey, Cooperative Research Units, South Dakota State University, Brookings. Available: www.nwo.usace.army.mil/html/pd-e/benthic_fish/benthic_fish.htm. (January 2005).

Berry, C. R., Jr., and B. A. Young. 2004. Fishes of the Missouri national recreational river, South Dakota and Nebraska. Journal of Great Plains Research 14:89–114.

Berry, C. R., Jr., M. L. Wildhaber, and D. L. Galat. 2004. Fish distribution and abundance. Volume 3. Population structure and habitat use of benthic fishes along the Missouri and lower Yellowstone rivers. U.S. Geological Survey, Cooperative Research Units, South Dakota State University, Brookings. Available: www.nwo.usace.army.mil/html/pd-e/benthic_fish/benthic_fish.htm. (January 2005).

Braaten, P. J., and C. S. Guy. 2002. Life history attributes of fishes along the latitudinal gradient of the Missouri River. Transactions of the American Fisheries Society 131:931–945.

Calamusso, R., J. R. Rinne, and R. J. Edwards. 2005. Historic changes in Rio Grande fish fauna: status, threats, and management of native species. Pages 205–223 *in* J. N. Rinne, R. M. Hughes, and B. Calamusso, editors. Historical changes in large river fish assemblages of the Americas. American Fisheries Society, Symposium 45, Bethesda, Maryland.

Carlson, D. M., W. L. Pflieger, L. Trail, and P. S. Haverland. 1985. Distribution, biology and hybridization of *Scaphirhynchus albus* and *S.*

platorynchus in the Missouri and Mississippi rivers. Environmental Biology of Fishes 14:51–59.

Copp, G. H. 1989. The habitat diversity and fish reproductive function of floodplain ecosystems. Environmental Biology of Fishes 26:1–26.

Cross, F. B. 1975. Skipjack herring, *Alosa chrysochloris*, in the Missouri River basin . Copeia 1975:382–385.

Cross, F. B., R. L. Mayden, and J. D. Stewart. 1986. Fishes in the western Mississippi basin (Missouri, Arkansas, and Red rivers). Pages 367–412 *in* C. H. Hocutt, and E. O. Wiley, editors. The zoogeography of North American freshwater fishes. Wiley, New York.

Cutright, P. R. 1969. Lewis and Clark: pioneering naturalists. University of Nebraska Press, Lincoln.

Department of Interior. 2001. Endangered and threatened wildlife and plants: 12-month finding for a petition to list the sicklefin chub (*Macrhybopsis meeki*) and sturgeon chub (*Macrhybopsis gelida*) as endangered. Fish and Wildlife Service, 50 CFR Part 17. Federal Register 66:75(18 April 2001):19910–19914.

Dryer, M. P., and A. J. Sandvol. 1993. Recovery plan for the pallid sturgeon (*Scaphirhynchus albus*). U.S. Fish and Wildlife Service, Denver.

Dynesius, M., and C. Nilsson. 1994. Fragmentation and flow regulation of river systems in the northern third of the world. Science 266:753–762.

Fausch, K., J. Lyons, J. Karr, and P. Angermeier. 1990. Fish communities as indicators of environmental degradation. Pages 123–144 *in* S. Adams, editor. Biological indicators of stress in fish. American Fisheries Society, Symposium 8, Bethesda, Maryland.

Fausch, K. D., C. E. Torgersen, C. V. Baxter, and H. W. Li. 2002. Landscapes to riverscapes: bridging the gap between research and conservation of stream fishes. BioScience 52:483–497.

Ferrell, J. 1993. Big dam era. U.S. Army Corps of Engineers, Omaha, Nebraska.

Ferrell, J. 1996. Soundings: 100 years of the Missouri River navigation project. U.S. Army Corps of Engineers, Omaha, Nebraska.

Fisher, H. J. 1962. Some fishes of the lower Missouri River. American Midland Naturalist 106:372–378.

Funk, J. L., and J. W. Robinson. 1974. Changes in the channel of the lower Missouri River and effects on fish and wildlife. Missouri Department of Conservation, Aquatic Series 11, Jefferson City.

Galat, D. L., C. R. Berry, Jr., E. J. Peters, and R. G. White. 2005. Missouri River basin. Pages 427–480 *in* A. C. Benke and C. E. Cushing, editors. Rivers of North America. Elsevier, Amsterdan.

Galat, D. L., and S. J. Clark. 2002. Fish spawning and discharge-temperature coupling along the Missouri River: implications for environmental flows. Sixth Annual Missouri River Natural Resources Conference, South Sioux City, Nebraska.

Galat D. L., and A. G. Frazier, editors. 1996. Overview of river-floodplain ecology in the upper Mississippi River basin. Volume 3 *of* J. A. Kelmelis, editor. Science for floodplain management into the 21st century. U.S. Government Printing Office, Washington, D.C.

Galat, D. L., L. H. Fredrickson, D. D. Humburg, K. J. Bataille, J. R. Bodie, J. Dohrenwend, G. T. Gelwicks, J. E., Havel, D. L. Helmers, J. B. Hooker, J. R. Jones, M. F. Knowlton, J. Kubisiak, J. Mazourek, A. C. McColpin, R. B. Renken, and R. D. Semlitsch. 1998. Flooding to restore connectivity of regulated, large-river wetlands. BioScience 48:721–733.

Galat, D. L., and R. Lipkin. 2000. Restoring ecological integrity of great rivers: historical hydrographs aid in defining reference conditions for the Missouri River. Hydrobiologia 422/423:29–48.

Galat, D. L., J. W. Robinson, and L. W. Hesse. 1996. Restoring aquatic resources to the lower Missouri River: issues and initiatives. Pages 49–71 *in* D. L. Galat, and A. G. Frazier, editors. Overview of river-floodplain ecology in the upper Mississippi River basin. Volume 3 *of* J. A. Kelmelis, editor. Science for floodplain management into the 21st century. U.S. Government Printing Office, Washington, D.C.

Galat, D. L., G. W. Whitledge, and G. T. Gelwicks. 2004. Influence of lateral connectivity on larval fish assemblage structure and habitat use in lower Missouri River floodplain waterbodies. Missouri Cooperative Fish and Wildlife Research Unit, Final Report to Missouri Department of Conservation, Columbia.

Galat, D. L., M. L. Wildhaber, and D. J. Dieterman. 2001. Spatial patterns of physical habitat. Volume 2. Population structure and habitat use of benthic fishes along the Missouri and lower Yellowstone rivers. U.S. Geological Survey, Cooperative Research Units, University of Missouri,

Columbia. Available: www.nwo.usace.army.mil/html/pd-e/benthic_fish/benthic_fish.htm. (January 2005).

Galat, D. L., and I. Zweimüller. 2001. Conserving large-river fishes: is the highway analogy an appropriate paradigm? Journal of the North American Benthological Society 20:266–279.

Gardner, W. M. 1998. Middle Missouri River fisheries evaluations. 1997 annual report. Montana Fish, Wildlife, and Parks, Federal Aid to Fish and Wildlife Restoration Project F-78-R-4, Helena.

Gardner, W. M. 1999. Middle Missouri River fisheries evaluations. 1998 annual report. Montana Fish, Wildlife, and Parks, Federal Aid to Fish and Wildlife Restoration Project F-78-R-5, Helena.

Gardner, W. M. 2001. Middle Missouri River fisheries evaluations. 1999–2000 annual report. Montana Fish, Wildlife, and Parks, Federal Aid to Fish and Wildlife Restoration Project F-78-R-6, F-113-R1, F-113-R2, Helena.

Gardner, W. M. 2003. Middle Missouri River fisheries evaluations. 2001–02 annual report. Montana Fish, Wildlife, and Parks, Federal Aid to Fish and Wildlife Restoration Projects F-113-R3, & F-113-R4, Helena.

Gardner, W. M., and P. A. Stewart. 1987. The fishery of the lower Missouri River, Montana. Montana Fish, Wildlife and Parks, Federal Aid to Fish and Wildlife Restoration Project FW-2-R; Job I-b, Helena.

Gelwicks, G. T., K. Graham, and D. L. Galat. 1996. Status survey for sicklefin chub, sturgeon chub, and flathead chub in the Missouri River, Missouri. Missouri Department of Conservation, Fish and Wildlife Research Center, Final Report, Columbia.

Grady, J., J. Milligan, C. Gemming, D. Heroz, G. Mestl, L. Miller, D. Hernig, K. Hurley, P. Wills, and R. Sheehan. 2001. Pallid and shovelnose sturgeon in the lower Missouri and middle Mississippi rivers. Final Report Prepared for MICRA. U.S. Fish and Wildlife Service, Columbia Fishery Resources Office, Columbia, Missouri.

Grady, J. M., and J. Milligan. 1998. Status of selected cyprinid species at historic lower Missouri River sampling sites. U.S. Fish and Wildlife Service, Columbia Fisheries Resources Office, Columbia, Missouri.

Hawkes, H. A. 1975. River zonation and classification. Pages 312–374 *in* B. A. Whitton, editor. River ecology. University of California Press, Berkeley.

Hendrickson, J. C., and J. Lee. 2000. Aquatic investigations on the Missouri mainstem in North Dakota. North Dakota Game and Fish Department, Division Report 36, Bismarck.

Hendrickson, J. C., and G. J. Power. 1999. Changes in fish species abundance in a Missouri River main stem reservoir during the first 45 years. Journal of Freshwater Ecology 14:407–416.

Hesse, L. W. 1993a. Floral and faunal trends in the middle Missouri River. Pages 73–90 *in* D. L. Galat and A. G. Frazier, editors. Overview of river-floodplain ecology in the upper Mississippi River basin. Volume 3 *in* J. A. Kelmelis, editor. Science for floodplain management into the 21st century. U.S. Government Printing Office, Washington, D.C.

Hesse, L. W. 1993b. The status of Nebraska fishes in the Missouri River. 2. Burbot (Gadidae: *Lota lota*). Transactions of the Nebraska Academy of Sciences 20:67–71.

Hesse, L. W. 1994a. The status of Nebraska fishes in the Missouri River. 3. Channel catfish (Icatluridae: *Ictalurus punctatus*). Transactions of the Nebraska Academy of Sciences 21:73–87.

Hesse, L. W. 1994b. The status of Nebraska fishes in the Missouri River. 4. Flathead catfish, *Pylodictus olivaris*, and blue catfish, *Ictalurus furcatus* (Ictaluridae). Transactions of the Nebraska Academy of Sciences 21:89–98.

Hesse, L. W. 1994c. The status of Nebraska fishes in the Missouri River. 5. Selected chubs and minnows (Cyprinidae): sicklefin chub (*Macrhybopsis meeki*), sturgeon chub (*M. gelida*), silver chub (*M. storeriana*), speckled chub (*M. aestivalis*), flathead chub (*Platygobio gracilis*), plains minnow (*Hybognathus placitus*), and western silvery minnow (*H. argyritis*). Transactions of the Nebraska Academy of Sciences 21:99–108.

Hesse, L. W. 1994d. The status of Nebraska fishes in the Missouri River. 6. Sauger (Percidae: *Stizostedion canadense*). Transactions of the Nebraska Academy of Sciences 21:109–121.

Hesse, L. W. 2001. MRHD, Missouri River historical database for the middle reach. Rivers Corporation, Crofton, Nebraska.

Hesse, L. W., and G. E. Mestl. 1993. The status of Nebraska fishes in the Missouri River. 1. Paddlefish (Polyodontidae: *Polyodon spathula*). Transactions of the Nebraska Academy of Sciences 20:53–65.

Hesse, L. W., G. E. Mestl, and J. W. Robinson. 1993. Status of selected fishes in the Missouri River in Nebraska with recommendations for their recovery. Pages 327–340 *in* L. W. Hesse, C. B.

Stalnaker, N. G. Benson, and J. R. Zuboy, editors. Proceedings of the symposium on restoration planning for the rivers of the Mississippi River ecosystem. National Biological Survey, Biological Report 19, Washington, D.C.

Hesse, L. W., J. C. Schmulbach, K. D. Keenlyne, D. G. Unkenholz, J. W. Robinson, and G. E. Mestl. 1989. Missouri River fishery resources in relation to past, present, and future stress. Canadian Special Publication of Fisheries and Aquatic Sciences 106:352–371.

Holton, G. D., and H. E. Johnson. 1996. A field guide to Montana fishes. Montana Fish, Wildlife and Parks, Helena.

Jackson, L. E., J. C. Kurtz, and W. S. Fisher. 2000. Evaluation guidelines for ecological indicators. U.S. Environmental Protection Agency, Office of Research and Development, EPA/620/R-99/005, Research Triangle Park, North Carolina.

Jacobson, R. B., and J. Heuser. 2001. Visualization of flow alternatives, lower Missouri River. U.S. Geological Survey, Open-file Report OF02–122, Columbia, Missouri.

Johnson, B., D. Fielder, J. Riis, C. Stone, D. Warnick, and G. Wickstrom. 1992. Annual fish population surveys on South Dakota Missouri River reservoirs, 1991. South Dakota Department of Game, Fish and Parks, Dingell-Johnson Project F-21-R-25, Study Number I, Job 2, Pierre.

Jungwirth, M. 1998. River continuum and fish migration - going beyond the longitudinal river corridor in understanding ecological integrity. Pages 19–32 *in* M. Jungwirth, S. Schmutz, and S. Weiss, editors. Fish migration and bypasses. Fishing News Books, Oxford, England.

Kinsolving, A. D., and M. B. Bain. 1993. Fish assemblage recovery along a riverine disturbance gradient. Ecological Applications 3:531–544.

Kondolf, G. M. 1995. Five elements for effective evaluation of stream restoration. Restoration Ecology 3:133–136.

Lee, D. S., C. R. Gilbert, C. H. Hocutt, R. E. Jenkins, D. E. McAllister, and J. R. Stauffer, Jr. 1980. Atlas of North American freshwater fishes. North Carolina State Museum of Natural History, Raleigh.

Matthews, W. J. 1998. Patterns in freshwater fish ecology. Chapman and Hall, New York.

McMahon, T. E., and W. M. Gardner. 2001. Status of sauger in Montana. Intermountain Journal of Sciences 7:1–21.

Mestl, G. 1999. Changes in Missouri River catfish populations after closing commercial fishing. American Fisheries Society, Symposium 24, Bethesda, Maryland.

Metcalf, A. L. 1966. Fishes of the Kansas River system in relation to zoogeography of the Great Plains. University of Kansas Publications, Museum of Natural History 17(3):23–189.

Minckley, W. L., P. C. Marsh, J. E. Deacon, T. E. Dowling, P. W. Hedrick, W. J. Matthews, and G. Mueller. 2003. A conservation plan for native fishes of the lower Colorado River. BioScience 53:219–234.

Moring, J. R. 1996. Fish discoveries by the Lewis and Clark and Red River expeditions. Fisheries 21(7):6–12.

Mueller, G. A., P. C. Marsh, and W. L. Minckley. 2005. A legacy of change: the lower Colorado River, Arizona–California–Nevada, USA and Sonora–Baja California Norte, Mexico. Pages 139–156 *in* J. N. Rinne, R. M. Hughes, and B. Calamusso, editors. Historical changes in large river fish assemblages of the Americas. American Fisheries Society, Symposium 45, Bethesda, Maryland.

Murphy, B. R., and D. W. Willis. 1996. Fisheries techniques. American Fisheries Society, Bethesda, Maryland.

National Research Council. 2002. The Missouri River ecosystem, exploring the prospects for recovery. National Academy Press, Washington, D.C.

National Research Council. 2003. Improving the Corps of Engineers' methods for water resources project planning. National Academy Press, Washington, D.C.

NatureServe. 2003. NatureServe Explorer: An online encyclopedia of life [web application]. Version 1.8. NatureServe, Arlington, Virginia. Available: http://www.natureserve.org/explorer. (February 2003).

Patrick, R. 1998. Rivers of the United States. Volume IV, Part A: the Mississippi River and tributaries north of St. Louis. Wiley, New York.

Pegg, M. A., and C. L. Pierce. 2001. Growth responses of Missouri and lower Yellowstone river fishes to a latitudinal gradient. Journal of Fish Biology 59:1529–1543.

Pegg, M. A., and C. L. Pierce. 2002a. Classification of reaches in the Missouri and lower Yellowstone rivers based on flow characteristics. River Research and Applications 18:31–42.

Pegg, M. A., and C. L. Pierce. 2002b. Fish community structure in the Missouri and lower Yellowstone rivers in relation to flow characteristics. Hydrobiologia 479:155–167.

Pegg, M. A., C. L. Pierce, and A. Roy. 2003. Hydrological alteration along the Missouri River basin: a time series approach. Aquatic Sciences 65:63–72.

Pflieger, W. F. 1971. A distributional study of Missouri fishes. University of Kansas Publications, Museum of Natural History 20(3):225–570.

Pflieger, W. F. 1997. The fishes of Missouri. Missouri Department of Conservation, Jefferson City.

Pflieger, W. L., and T. B. Grace. 1987. Changes in the fish fauna of the lower Missouri River, 1940–1983. Pages 166–177 *in* W. L. Matthews, and D. C. Heins, editors. Community and evolutionary ecology of North American stream fishes. University of Oklahoma Press, Norman.

Poff, N. L. 1997. Landscape filters and species traits: towards a mechanistic understanding and prediction in stream ecology. Journal of the North American Benthological Society 16:391–409.

Power, G. J., and F. Ryckman. 1998. Status of North Dakota's fishes. North Dakota Game and Fish Department, North Dakota Fisheries Investigations, Report Number 27, Bismarck.

Prato, T. 2003. Adaptive management of large rivers with special reference to the Missouri River. Journal of the American Water Resources Association 39:935–946.

Quist, M. C., C. S. Guy, M. A. Pegg, P. J. Braaten, C. L. Pierce, and V. T. Travnichek. 2002. Potential influence of harvest on shovelnose sturgeon populations in the Missouri River system: a case for pro-active management. North American Journal of Fisheries Management 22:537–549.

Rahel, F. J. 2002. Homogenization of freshwater faunas. Annual Review of Ecology and Systematics 33:291–315.

Revenga, C., S. Murray, J. Abramovitz, and A. Hammond. 1998. Watersheds of the world: ecological value and vulnerability. Worldwatch Institute, Washington, D.C.

Ricketts, T. H., E. Dinerstein, D. M. Olson, C. J. Loucks, W. Eichbum, D. DellaSala, K. Kavanagh, P. Hedao, P. T. Hurley, K. M. Carney, R. Abell, and S. Walters. 1999. Terrestrial ecoregions of North America: a conservation assessment. Island Press, Washington, D.C.

Riis, J., D. Fielder, B. Johnson, C. Stone, D. Unkenholz, and G. Wickstrom. 1988. Annual fisheries surveys on the Missouri River reservoirs, 1986–87. South Dakota Department of Game, Fish and Parks, Dingell-Johnson Project F-21-R-21, Study Number I, Job 2, Pierre.

Robinson, J. W. 1992. Missouri's commercial fishery harvest, 1990. Missouri Department of Conservation, Columbia.

Sappington, L., D. Dieterman, and D. Galat. 1998. 1998 Standard operating procedures to evaluate population structure and habitat use of benthic fishes along the Missouri and lower Yellowstone rivers. National Biological Service, Midwest Science Center, Columbia, Missouri. Available: http://infolink.cr.usgs.gov/science/benthicFish. (January 2005).

Scheimer, F. 2000. Fish as indicators for the assessment of the ecological integrity of large rivers. Hydrobiologia 422/423:271–278.

Schmulbach, J. C., L. W. Hesse, and J. E. Bush. 1992. The Missouri River – Great Plains thread of life. Pages 137–158 *in* C. D. Becker, and D. A. Nietzel, editors. Water quality in North American river systems. Battelle Press, Columbus, Ohio.

Schmutz, S. M. Kaufmann, B. Vogel, M. Jungwirth, and S. Muhar. 2000. A multi-level concept for fish-based, river-type-specific assessment of ecological integrity. Hydrobiologia 422/423:279–289.

Schneiders, R. K. 1999. Unruly river; two centuries of change along the Missouri. University of Kansas Press, Lawrence.

Scott, M. L., G. T. Auble, and J. Friedman. 1997. Flood dependency of cottonwood establishment along the Missouri River, Montana, USA. Ecological Applications 7:677–690.

Simon, T. P., and E. B. Emery. 1995. Modification and assessment of an index of biotic integrity to quantify water resource quality in great rivers. Regulated Rivers: Research and Management 11:283–298.

Simon, T. P., and J. Lyons. 1995. Application of the index of biotic integrity to evaluate water resource integrity in freshwater ecosystems. Pages 245–262 *in* W. Davis, and T. Simon, editors. Biological assessment and criteria. Lewis Publishers, Boca Raton, Florida.

Stanovick, J. S. 1999. Recreational harvest rates, release rates, and average length of catfish on the Missouri River, Missouri before and after the commercial catfish ban. American Fisheries Society, Symposium 24, Bethesda, Maryland.

Theiling, C. H., C. Korschgen, H. De Haan, T. Fox, J. Rohweder, and L. Robinson. 2000. Habitat needs assessment for the upper Mississippi River system: technical report. Appendix A. U.S. Geological Survey, Upper Midwest Environmental Science Center, La Crosse, Wisconsin.

Thorson, J. E. 1994. River of promise, river of peril: the politics of managing the Missouri River. University of Kansas Press, Lawrence.

Tonn, W. M. 1990. Climate change and fish communities: a conceptual framework. Transactions of the American Fisheries Society 119:337–352.

Travnichek, V. H., M. B. Bain, and M. J. Maceina. 1995. Recovery of a warmwater fish assemblage after the initiation of a minimum-flow release downstream from a hydroelectric dam. Transactions of the American Fisheries Society 124:836–844.

Tyus, H. M. 2002. Decline of native Missouri River fishes: the introduced fish problem. University of Colorado, Center for Limnology, Cooperative Institute for Research in Environmental Sciences, Boulder.

U.S. Army Corps of Engineers. 1994a. Missouri River master water control manual: review and update study. Volume 7A: environmental studies, reservoir fisheries. U.S. Army Corps of Engineers, Missouri River Division, Omaha, Nebraska.

U.S. Army Corps of Engineers. 1994b. Missouri River master water control manual: review and update study. Volume 7C: environmental studies, riverine fisheries. U.S. Army Corps of Engineers, Missouri River Division, Omaha, Nebraska.

U.S. Army Corps of Engineers. 2001. Missouri River master water control manual: review and update. Revised draft environmental impact statement. Volume 1: main report. U.S. Army Corps of Engineers, Northwest Division, Omaha, Nebraska. Available: www.nwd.usace.army.mil. (January 2005).

U.S. Army Corps of Engineers. 2004. Missouri River bank stabilization and navigation project, fish and wildlife mitigation project, annual implementation report. U.S. Army Corps of Engineers, Omaha District, Omaha, Nebraska, and Kansas City District, Kansas City, Missouri. Available: www.nwd.usace.army.mil. (January 2005).

U.S. Fish and Wildlife Service. 1999. Final environmental impact statement summary for proposed expansion of Big Muddy National Fish & Wildlife Refuge, Missouri. U.S. Fish and Wildlife Service, Puxico, Missouri. Available: http://midwest.fws.gov/planning/bigmuddy/top.htm. (January 2005).

U.S. Fish and Wildlife Service. 2001. Biological opinion on the operation of the Missouri River main stem reservoir system, operation and maintenance of the Missouri River bank stabilization and navigation project, and operation of the Kansas River reservoir system. U.S. Fish and Wildlife Service, Region 6, Denver, Colorado.

U.S. Geological Survey. 2002. Water resources data Montana. Water year 2001. U.S. Geological Survey, Water Resources Division, Helena, Montana.

van der Leeden, F. 1990. The water encyclopedia. Lewis Publishers, Chelsea, Michigan.

VanZee, B. 1997. Assessment of walleye, sauger, and black bass populations in Lewis and Clark Lake, South Dakota. Master's thesis. South Dakota State University, Brookings.

VanZee, B., and C. Scalet. 1997. Range extension of the grass pickerel into South Dakota. The Prairie Naturalist 29:277–278.

Walburg, C. H. 1976. Changes in the fish populations of Lewis and Clark Lake, 1956–74, and their relation to water management and the environment. U.S. Fish and Wildlife Service, Research Report 79, Washington, D.C.

Walburg, C. H. 1977. Lake Francis Case, a Missouri River reservoir: changes in the fish population in 1954–75, and suggestions for management. U.S. Fish and Wildlife Service, Technical Paper 95, Washington, D.C.

Wehrly, K. E., M. J. Wiley, and P. W. Seelbach. 2003. Classifying regional variation in thermal regime based on stream fish community patterns. Transactions of the American Fisheries Society 132:18–38.

Welker, T. L., and D. L. Scarnecchia. 2003. Differences in species composition and feeding ecology of catostomid fishes in two distinct segments of the Missouri River, North Dakota, U.S.A. Environmental Biology of Fishes 68:129–141.

Wickstrom, G. 1997. Annual fish population surveys of Lewis and Clark Lake, 1996. South Dakota Department of Game, Fish and Parks, Dingell-Johnson Project F-21-R-29, Pierre.

Wickstrom, G. 2001. Annual fish population survey of Lewis and Clark Lake and Missouri River creel survey, 2000. South Dakota Department of Game, Fish and Parks, Annual Report No. 01–17, Pierre.

Wickstrom, G. 2002. Annual fish population surveys of Lewis and Clark Lake, 2001. South Dakota Department of Game, Fish and Parks, Dingell-Johnson Project F-21-R-34, Pierre.

Wolf, A. E., D. W. Willis, and C. J. Power. 1996. Larval fish community in the Missouri River below Garrison Dam, North Dakota. Journal of Freshwater Ecology 11:11–19.

Zuerlein, G. 1988. Nebraska commercial fishery statistics – the Missouri River. Nebraska Game and Parks Commission, Final Report, National Marine Fisheries Service Project 2–402-R, Lincoln.

Appendix A.—Distribution of Missouri River fishes by freshwater ecoregion, drainage unit, and physiographic province. Habitat use guild, habitat distribution, global conservation status, and population status are summarized by species. Numbers in parentheses refer to drainage units in Figure 1.

		Freshwater ecoregion/drainage unit																
		Upper Missouri		Middle Missouri		Central Prairie	Physiographic province				Habitat use guild			Habitat distribution			Status	
Species name	Common name	Upper Missouri (18)	Little Missouri-White (16)	James-Sioux-Niobrara- Platte (15+14)	Nishnabotna-Chariton (12)	Lower Missouri (11)	Northern Rocky Mountains	Great Plains	Central Lowland	Ozark Plateaus	Fluvial specialist	Fluvial dependent	Macrohabitat generalist	Main channel	Floodplain	Reservoir	Heritage global	Population
Ichthyomyzon castaneus	chestnut lamprey			N	N	N		S	M	M		X		C			G4	0
I. unicuspis	silver lamprey		N	N				S	S			X		C			G5	
Acipenser fulvescens	lake sturgeon			N	N	N			S	S		X		C	X		G3	–
Scaphirhynchus albus	pallid sturgeon	N	N	N	N	N		S	S	S	X			C		X	G1	–
S. platorynchus	shovelnose sturgeon	N	N	N	N	N		P	P	P	X			C		X	G4	–
Polyodon spathula	paddlefish	N	N	N	N	N		P	P	M		X		C	X	X	G4	–
Lepisosteus oculatus[1]	spotted gar			N	N	N			S	S			X	B	X		G5	U
L. osseus	longnose gar			N	N	N		M	P	P		X		C	X	X	G5	–
L. platostomus	shortnose gar	N	N	N	N	N		S	P	S			X	C	X	X	G5	0,–
Amia calva	bowfin				I	I			S	S			X	B	X		G5	U
Anguilla rostrata	American eel			D	D	D		M	S	P		X		C			G5	U
Alosa alabamae	Alabama shad					D				S		X		C			G3	–
A.chrysochloris	skipjack herring			N	N	N		S	S	S		X		C			G5	–
A. pseudoharengus[a]	alewife			I	I			S					X	C		X	G5	
Dorosoma cepedianum	gizzard shad		N	N	N	N		P	P	S			X	C	X	X	G5	+
D. petenense	threadfin shad					N			M	M			X	C		X	G5	+
Hiodon alosoides	goldeye	N	N	N	N	N		P	P	P		X		C	X	X	G5	0,–
H. tergisus	mooneye				N	N			S	S		X		C			G5	–
Coregonus artedi[a]	cisco, lake herring	I	I					S					X	W		X	G5	0
C. clupeaformis	lake whitefish	I	I					S					X	W		X	G5	0

Appendix A.—Continued.

Species name	Common name	Freshwater ecoregion/drainage unit: Upper Missouri: Upper Missouri (18)	Upper Missouri: Little Missouri-White (16)	~Middle Missouri: James-Sioux-Niobrara Platte (15+14)	~Middle Missouri: Nishnabotna-Chariton (12)	Central Prairie: Lower Missouri (11)	Physiographic province: Northern Rocky Mountains	Great Plains	Central Lowland	Ozark Plateaus	Habitat use guild: Fluvial specialist	Fluvial dependent	Macrohabitat generalist	Habitat distribution: Main channel	Floodplain	Reservoir	Status: Heritage global	Population
Oncorhynchus clarkii	cutthroat trout	N	I				M	M				X		W		X	G4	–
O. mykiss	rainbow trout	I	I	I		I	C	M		S		X		C		X	G5	+
O. nerka	sockeye salmon, kokanee	I					M					X				X	G5	0
O. tshawytscha	Chinook salmon	I	I					S				X		W		X	G5	+
Prosopium williamsoni	mountain whitefish	N					C	S				X		C		X	G5	U
Salmo trutta	brown trout	I	I	I			C	S				X		C		X	G5	0
Salvelinus fontinalis	brook trout	I					S	S				X		W		X	G5	U
S. namaycush	lake trout	I	I					S				X		W		X	G5	U
Percopsis omiscomaycus	trout-perch			N	N	N			S	S			X	W			G5	–
Osmerus mordax	rainbow smelt	I	I	I	I	I		S	S	S		X		C		X	G5	0,+
Esox americanus	grass pickerel			N		N		S	S	S			X	B	X		G5	U
E. lucius[2]	northern pike	I	N	N	I	I	S	P	P	S			X	B	X	X	G5	0,+
E. masquinongy	muskellunge		I	I		I		S		S			X		X	X	G5	U
Campostoma anomalum	central stoneroller		N	N	N	N		M	M	P	X			W		X	G5	+
C. oligolepis	largescale stoneroller					N				P	X			W			G5	
Carassius auratus	goldfish	I		I	I	I	S	S	S	S			X	C	X	X	G5	+
Couesius plumbeus	lake chub	N	N				P	M					X	B			G5	0
Ctenopharyngodon idella	grass carp			I	I	I		S	S	S		X		C	X		G5	+

Appendix A.—Continued.

Species name	Common name	Freshwater ecoregion/drainage unit: Upper Missouri		Middle Missouri		Central Prairie	Physiographic province				Habitat use guild			Habitat distribution			Status	
		Upper Missouri (18)	Little Missouri-White (16)	James-Sioux-Niobrara Platte (15+14)	Nishnabotna-Chariton (12)	Lower Missouri (11)	Northern Rocky Mountains	Great Plains	Central Lowland	Ozark Plateaus	Fluvial specialist	Fluvial dependent	Macrohabitat generalist	Main channel	Floodplain	Reservoir	Heritage global	Population
Cyprinella lutrensis	red shiner		N	N	N	N		P	P	P			X	C	X	X	G5	0,+
C. spiloptera	spotfin shiner			N	N	N		S	S	S	X			C		X	G5	
Cyprinus carpio	common carp	I	I	I	I	I	P	P	P	P			X	C	X	X	G5	0,–
Erimystax x-punctatus	gravel chub					N				P	X			W			G4	–
Gila atraria	Utah chub	I					M						X			X	G5	0
Hybognathus argyritis	western silvery minnow	N	N	N	N	N		P	P	S		X		C	X		G4	0,–
H. hankinsoni	brassy minnow	N	N	N	N		S	S	P				X	C	X	X	G5	0,+
H. nuchalis[a]	Mississippi silvery minnow		N	N		N		S	S	S		X		C		X	G5	U
H. placitus	plains minnow	N	N	N	N	N		S	P	P		X		C			G4	0,–
Hypophthalmichthys molitrix[a]	silver carp				I	I			P	P		X		C	X		G5	+
H. nobilis[a]	bighead carp			I	I	I		P	P	P		X		C	X		G5	+
Luxilus chrysocephalus	striped shiner					N				S	X			W			G5	
L. cornutus	common shiner		N	N	N	N		S	P	S		X		W		X	G5	0
Lythrurus umbratilis	redfin shiner				N	N			P	P	X			W			G5	0
Macrhybopsis aestivalis	speckled chub			N	N	N		P	P	P	X			C			G3G4	+,–
M. gelida	sturgeon chub	N	N	N	N	N		S	S	M	X			C			G3	–
M. meeki	sicklefin chub	N	N	N	N	N		S	S	M	X			C			G3	–
M. storeriana	silver chub			N	N	N		S	P	M			X	C	X		G5	–
Margariscus margarita	pearl dace	N	N					S				X		W			G5	U

Appendix A.—Continued.

Species name	Common name	Freshwater ecoregion/drainage unit: Upper Missouri: Upper Missouri (18)	Little Missouri-White (16)	~Middle Missouri: James-Sioux-Niobrara- Platte (15+14)	Nishnabotna-Chariton (12)	Central Prairie: Lower Missouri (11)	Physiographic province: Northern Rocky Mountains	Great Plains	Central Lowland	Ozark Plateaus	Habitat use guild: Fluvial specialist	Fluvial dependent	Macrohabitat generalist	Habitat distribution: Main channel	Floodplain	Reservoir	Status: Heritage global	Population
Notemigonus crysoleucas	golden shiner	I	U	U	N	N		S	P	S			X	W	X	X	G5	U
Notropis atherinoides	emerald shiner	N	N	N	N	N		P	P	P			X	C	X	X	G5	+
N. blennius	river shiner			N	N	N		M	P	P	X			C			G5	+,–
N. boops	bigeye shiner					N			S	S	X			W			G5	0
N. buchanani	ghost shiner			N	N	N		S	S	S	X			C	X		G5	0,–
N. dorsalis	bigmouth shiner		N	N	N	N		P	P	S	X			C			G5	–
N. hudsonius	spottail shiner	I	I	I		N		P	S	S			X	C	X	X	G5	0,+
N. rubellus	rosyface shiner					N				P		X		W			G5	0,–
N. shumardi	silverband shiner		N	N	N	N		M	P	S	X			C			G5	
N. stramineus	sand shiner	N	N	N	N	N		M	P	M	X			C		X	G5	0,–
N. volucellus	mimic shiner			N	N	N		M	M	M			X	C	X		G5	0
N. wickliffi[a]	channel shiner					N			S	S	X			C			G5	
Phenacobius mirabilis	suckermouth minnow		N	N	N	N	S	P	P	M	X			C			G5	
Phoxinus eos	northern redbelly dace		N	N				S	S				X	W		X	G5	U
P. erythrogaster	southern redbelly dace					N				S		X		W			G5	+
P. neogaeus	finescale dace		N	N				S					X	W		X	G5	
Pimephales notatus	bluntnose minnow		I	N	N	N		S	P	P			X	B	X		G5	0,+
P. promelas	fathead minnow	N	N	N	N	N	M	P	P	S			X	C	X	X	G5	0,–
P. vigilax	bullhead minnow			I		N			M	M			X	C		X	G5	
Platygobio gracilis	flathead chub	N	N	N	N	N	M	P	S	S	X			C			G5	0,–
Rhinichthys atratulus	blacknose dace		N	N	U			S	S			X		W			G5	U
R. cataractae	longnose dace	N	N				P	P					X	C		X	G5	

Appendix A.—Continued.

Species name	Common name	Freshwater ecoregion/drainage unit: Upper Missouri		Middle Missouri		Central Prairie	Physiographic province				Habitat use guild			Habitat distribution			Status	
		Upper Missouri (18)	Little Missouri-White (16)	James-Sioux-Niobrara Platte (15+14)	Nishnabotna-Chariton (12)	Lower Missouri (11)	Northern Rocky Mountains	Great Plains	Central Lowland	Ozark Plateaus	Fluvial specialist	Fluvial dependent	Macrohabitat generalist	Main channel	Floodplain	Reservoir	Heritage global	Population
Semotilus atromaculatus	creek chub	N	N	N	N	N		S	P	P			X	W		X	G5	0,+
Carpiodes carpio	river carpsucker	N	N	N	N	N		P	P	P			X	C	X	X	G5	0,–
C. cyprinus	quillback			N	N	N		P	S	S			X	C	X		G5	–
C. velifer	highfin carpsucker			N	N	N			S	S	X			C	X		G4G5	–
Catostomus catostomus	longnose sucker	N	N				P	P					X	C			G5	0,+
C. commersonii	white sucker	N	N	N	N	N	P	P	P	S		X		C		X	G5	0,+
C. platyrhynchus	mountain sucker	N					M	M			X			C			G5	0
Cycleptus elongatus	blue sucker	N	N	N	N	N		P	P	M	X			C		X	G3G4	0,–
Hypentelium nigricans	northern hog sucker					N				M	X			W			G5	U
Ictiobus bubalus	smallmouth buffalo	N	N	N	N	N		P	P	S			X	C	X	X	G5	–
I. cyprinellus	bigmouth buffalo	N	N	N	N	N		P	P	S			X	C	X	X	G5	–
I. niger	black buffalo			N	N	N			S	M			X	C	X		G5	–
Moxostoma erythrurum	golden redhorse			N	N	N			S	S		X		C			G5	U
M. macrolepidotum	shorthead redhorse	N	N	N	N	N		P	P	S		X		C		X	G5	0
Ameiurus melas	black bullhead	I	N	N	N	N		S	P	S			X	B	X	X	G5	–
A. natalis	yellow bullhead	I	I	N	N	N		M	M	S			X	B	X	X	G5	–
Ictalurus furcatus	blue catfish			N	N	N			P	P	X			C			G5	–
I. punctatus	channel catfish	N	N	N	N	N		P	P	P			X	C	X	X	G5	0,+
Noturus exilis	slender madtom				N	N			M	S	X			W	X		G5	U
N. flavus	stonecat	N	N	N	N	N	M	P	P	S	X			C			G5	0
N. gyrinus	tadpole madtom	N	N	N	N	N		S	P	S			X	B	X		G5	

Appendix A.—Continued.

Species name	Common name	Freshwater ecoregion/drainage unit: Upper Missouri		-Middle Missouri		Central Prairie	Physiographic province				Habitat use guild			Habitat distribution			Status	
		Upper Missouri (18)	Little Missouri-White (16)	James-Sioux-Niobrara- Platte (15+14)	Nishnabotna-Chariton (12)	Lower Missouri (11)	Northern Rocky Mountains	Great Plains	Central Lowland	Ozark Plateaus	Fluvial specialist	Fluvial dependent	Macrohabitat generalist	Main channel	Floodplain	Reservoir	Heritage global	Population
N. nocturnus	freckled madtom					N			P	S	X			W			G5	
Pylodictis olivaris	flathead catfish		N	N	N	N		M	P	P		X		C		X	G5	0,–
Lota lota	burbot	N	N	N	N	N	P	P	S	S			X	C		X	G5	0,–
Fundulus diaphanus	banded killifish		U					S					X	W	X		G5	
F. notatus	blackstripe topminnow					I			M	S			X		X		G5	+
F. sciadicus	plains topminnow		N	N		N		S	S	S			X	W	X		G4	0
F. zebrinus	plains killifish	I	I	N	N	N		S	S	S			X	W	X		G5	0
Gambusia affinis	western mosquitofish				I	I			S	M			X	B	X	X	G5	+
Labidesthes sicculus	brook silverside				N	N			S	S			X	C			G5	0
Menidia beryllina	inland silverside					I			S	S			X	C		X	G5	+
Culaea inconstans	brook stickleback	N	N					M					X	W	X		G5	U
Morone americana	white perch			I	I	I		S	S	S		X		W		X	G5	
M. chrysops	white bass		I	U	N	N		P	P	S		X		C		X	G5	0,+
M. mississippiensis	yellow bass				N	N			S	S		X		C	X		G5	U
M. saxatilis	striped bass				I	I			S	S		X		W	X		G5	+
Ambloplites rupestris	rock bass		I	N		I		S	S	S			X	W		X	G5	0
Lepomis cyanellus	green sunfish	I	U	N	N	N		S	P	P			X	B	X	X	G5	U

Appendix A.—Continued.

		Freshwater ecoregion/drainage unit																
		Upper Missouri		-Middle Missouri		Central Prairie	Physiographic province				Habitat use guild			Habitat distribution			Status	
Species name	Common name	Upper Missouri (18)	Little Missouri-White (16)	James-Sioux-Niobrara- Platte (15+14)	Nishnabotna-Chariton (12)	Lower Missouri (11)	Northern Rocky Mountains	Great Plains	Central Lowland	Ozark Plateaus	Fluvial specialist	Fluvial dependent	Macrohabitat generalist	Main channel	Floodplain	Reservoir	Heritage global	Population
L. gibbosus[b]	pumpkinseed	I	I	I	I		S	S	S				X	W	X	X	G5	
L. humilis	orangespotted sunfish		U	N	N	N		S	P	M			X	B	X	X	G5	U
L. macrochirus	bluegill	I	I	U	U	N		P	P	P			X	B	X	X	G5	0,+
L. megalotis	longear sunfish					N		S	P	S			X	W			G5	0
L. microlophus	redear sunfish			I		I		M	M	S			X	B	X	X	G5	U
Micropterus dolomieu	smallmouth bass	I	I	I		N		P	S	S			X	C		X	G5	+
M. punctulatus	spotted bass			I	I	I		S	S	S			X	C			G5	U
M. salmoides	largemouth bass	I	I	N	N	N		P	P	S			X	B	X	X	G5	–
Pomoxis annularis	white crappie	I	I	N	N	N		P	P	M			X	B	X	X	G5	0,–
P. nigromaculatus	black crappie	I	I	N	I	N		P	S	M			X	B	X	X	G5	0,–
Etheostoma exile	Iowa darter	N	N	N				S	S				X	W	X		G5	
E. nigrum	johnny darter		N	N	N	N		P	S	S			X	B	X	X	G5	+,–
Perca flavescens	yellow perch	I	U	U			M	P	P				X	W	X	X	G5	0,–
Percina caprodes	logperch					N			M	S			X	W			G5	U
P. phoxocephala	slenderhead darter					N			M	S	X			W			G5	0,–
Sander canadensis	sauger	N	N	N	N	N		P	P	M			X	C	X	X	G5	0,–
S. vitreus[b]	walleye	I	U	U	U	U	M	P	P	S			X	C	X	X	G5	+
Aplodinotus grunniens	freshwater drum	N	N	N	N	N		P	P	P			X	C	X	X	G5	0,+
Cottus bairdii	mottled sculpin	N	N			N	P	S		S			X	C			G5	U

[a] species added since Cross et al. (1986); [b] status changed from Cross et al. (1986), see text for explanation. Symbols: N = native, U = uncertain, D = diadromous, I = introduced; P = prevalent/common, M = marginal/uncommon, S = sporadic/rare; C = channel, B = channel border, W = waif, X = present; + = increasing, 0 = stable, – = decreasing, blank = unknown, U = too uncommon to rank. Intermediate or differing intra-river trends reported as pairwise combinations of +, 0, and –. See text for explanation of Heritage Global "G" rankings.

American Fisheries Society Symposium 45:293–321, 2005

Changes in Fish Assemblage Structure of the Red River of the North

LUTHER P. AADLAND*

Minnesota Department of Natural Resources, 1509 First Avenue North, Fergus Falls, Minnesota 56537, USA

TODD M. KOEL

Fisheries and Aquatic Sciences Section, Center for Resources, Post Office Box 168, Yellowstone National Park, Wyoming 82190, USA

WILLIAM G. FRANZIN

Department of Fisheries and Oceans, Freshwater Institute, 501 University Crescent, Winnipeg, Manitoba, R3T 2N6, Canada

KENNETH W. STEWART

Senior Scholar, Department of Zoology, University of Manitoba, Winnipeg, Manitoba, R3T 2N2, Canada

PATRICK NELSON

Department of Zoology, University of Manitoba, Winnipeg, Manitoba, R3T 2N2, Canada

Abstract.—The Red River of the North basin (RRNB) has an area of about 287,000 square kilometers of the upper Midwestern United States and south-central Canada. The river forms the North Dakota–Minnesota boundary and flows into Lake Winnipeg, Manitoba, and then, via the Nelson River, into Hudson Bay. While the Red River main stem remains a sinuous stream similar to early descriptions, the river's watershed has been altered dramatically by intensive agriculture, wetland drainage, channelization of tributary streams, and dam construction. Early land surveys described a landscape largely covered by prairie and wetlands. However, thousands of kilometers of ditches have been excavated to drain wetlands for agriculture in the United States in the late 1800s to the 1920s, and continuing, in Canada, to the present. Over 500 dams have blocked access to critical spawning habitat in the basin starting in the late 1800s. Also, during the mid-1900s, many of the tributaries were channelized, causing the loss of several thousand stream kilometers. While much of RRNB's fish assemblage remains similar to earliest historical records, the loss of the lake sturgeon *Acipenser fulvescens* is a notable change resulting from habitat loss and fragmentation, and overfishing. Additional localized extirpations of channel catfish *Ictalurus punctatus*, several redhorse *Moxostoma* species, sauger *Sander canadensis*, and other migratory fishes have occurred upstream of dams on several tributaries. Presently, efforts are underway to restore migratory pathways through dam removal, conversion of dams to rapids, and construction of nature-like fishways. Concurrently, lake sturgeon is being reintroduced in the hope that restored access to historic spawning areas will allow

* Corresponding author: luther.aadland@dnr.state.mn.us

reestablishment of the species. Proposed construction of new flood control dams may undermine these efforts.

Introduction

The survival of riverine fishes depends on a complex mix of physical and biological factors. Climate, hydrology, geomorphology, water quality, biotic factors, and connectivity interact to determine what species are likely to be present in a river system. Many of these factors have been significantly altered by human activities. Marked changes in population size, extirpation, or extinction of aquatic species often result from these alterations to aquatic and riparian habitat.

Land use changes such as urbanization, agriculture, deforestation, and land drainage can cause substantial increases in surface runoff and peak flow (Striffler 1964; Leopold 1968; Lusby 1970; Verry 1997). Verry (1988) found bank-full flows increased as much as four times when 30% or more of the watershed was drained. Miller (1999) estimated a four-fold increase in bank-full flow rates in present-day south-central Minnesota watersheds compared to pre-European settlement conditions.

Instream habitat is a function of both channel geomorphology (Rosgen 1996) and flow (Bovee 1986). Geomorphology is determined naturally by climate and geology (Leopold et al. 1994), but is altered directly by channelization and indirectly by land-use changes affecting runoff and removal of riparian vegetation. Increases in channel-forming or bank-full flows can cause subsequent increases in channel cross-sectional area (Verry 2000) and decrease sinuosity (Verry and Dolloff 2000). These geomorphic changes can result in reduced habitat diversity and quality, loss of interstitial space due to increased input of fine materials, loss of pool depth due to sedimentation or simplification of channel, loss of cover, and homogenization of habitat.

Changes in flow and flow regime, due to watershed changes, dam regulation, or water extraction, cause shifts in microhabitat types by changing the distribution of depth and velocity (Bovee 1986; Aadland 1993). Horowitz (1978) found increases in species richness as flow variability decreased in Midwestern streams. Stable flow conditions generally favor reproduction and juvenile instream habitat in headwater streams and greater abundance of specialized fishes (Schlosser 1985). Conversely, native fishes of large floodplain rivers depend on seasonal floods that provide lateral connectivity, maintenance of riparian zones, spawning and nursery areas, and temporal cues (Welcomme 1985; Junk et al. 1989; Sparks 1995). Fishes adapted to floodplain habitats may decline in abundance when their use of floodplains is restricted (Ross and Baker 1983; Finger and Stewart 1987)

The ability of fish to use habitat(s) at broader spatial scales depends on the connectivity of the system. Large, low-gradient rivers often lack spawning habitat for certain species; especially riffle spawning fishes. As a result, many species ascend tributaries into headwaters, often migrating hundreds of kilometers to find suitable habitat. Fishes also optimize foraging by shifting habitat selection as they grow (Werner and Hall 1979; Mittlebach 1981). Migrations for spawning or optimal foraging are essential for many riverine fish species, but construction of dams fragment habitat and prevent these migrations from occurring. This may affect fish assemblages through loss of immigrating individuals in tributaries and through reduced reproductive success in the main stem. Because dams typically are built on outcrops of hard materials, they also eliminate critical high gradient habitat by inundation. Reservoir creation also tends to cause increases in lacustrine or undesirable alien species such as common carp *Cyprinus carpio*. Alien species can affect native fishes through displacement, genetic contamination, predation, and competition (Kimsey 1957). Nutrient, sediment, and temperature regimes also are affected by dam construction and can cause a shift towards headwater species in tailwater areas.

The objectives of this chapter are to 1) provide recent and historical data on fish assemblages of the Red River of the North, 2) identify changes in the river and watershed that have affected the system, and 3) identify efforts that could restore extirpated species or assemblages.

Watershed Characteristics

The Red River of the North main stem lies entirely in the bed of former glacial Lake Agassiz. The hydrologic headwaters begin in the glacial moraines of western Minnesota that form the Otter Tail River (Figure 1). The currently recognized headwaters of the Red River are the confluence of the Otter Tail and Bois de Sioux rivers. This was defined largely for political reasons; early developers feared opposition to dam construction on the present Otter Tail River if it carried the name "Red River" since the Red was known as a navigable trade route (Seltz 1999). From 9,900 to 9,200 years ago (YA), the south end of Glacial Lake Agassiz covered what are now the main stems of the Red and Bois de Sioux rivers (Fenton et al 1983; Matsch 1983). The sill between what is now Lake Traverse and Big Stone Lake, near Browns Valley, Minnesota, formed the most recent outlet by which Lake Agassiz spilled into the River Warren. The opening of the Lake Nipigon outlets of Lake Agassiz about 9,000 YA (Clayton 1983; Teller and Thorleifson 1983; Dyke and Prest 1987) caused abandonment of the River Warren outlet.

When the lake level fell below the elevation of the Browns Valley outlet, a northward-draining stream/wetland occupying the exposed lakebed developed. That stream could be viewed as the progenitor of the northward flowing Bois de Sioux River. Lake Agassiz also had an outlet into Lake Superior until about 8,500 YA. About 8,000 YA, Lake Agassiz drained eastward, north of the Great Lakes into Glacial Lake Barlow-Ojibway, which, in turn, drained into the St. Lawrence River (Dyke and Prest 1987). By about 7,700 YA, the Hudson Lobe of the Laurentide ice sheet stagnated and Lake Agassiz began to drain into Hudson Bay via subglacial drainage channels. This quickly caused the Hudson Lobe to founder and break up, resulting in the final drainage of Lake Agassiz about 7,700 YA (Dyke and Prest 1987). As a result of this geologic history, the fish assemblage of the Red River of the North is similar to that of both the upper Mississippi and Great Lakes basins. Much more recent connections have occurred between the upper Mississippi and the Red River of the North. Until a levee was constructed in the 1940s on the continental divide between Lake Traverse (Red River drainage) and Big Stone Lake (Minnesota River drainage), the divide was a marsh that reportedly was navigable during high water.

Average discharge of the Red River of the North near its mouth is about 238 cubic meters per second (cms). The largest tributary is the Assiniboine River, which represents 58% of the watershed but only contributes about 20% of the flow. The second largest tributary is the Red Lake River, which makes up only 5% of the watershed but contributes 14% of the flow. Runoff increases substantially from west to east across the Red River basin ranging from an average of 0.7 cm per year for the Sheyenne River near Cooperstown, North Dakota (USGS 2002a) to 11.2 cm per year for the Clearwater River at Plummer, Minnesota (USGS 2002b). The Red River of the North is very sinuous and low gradient over most of its length, with slopes ranging from 0.04% in the upper reaches to as little as 0.003% near the Canadian border. The river is predominantly a continuous run or pool but there are several riffles in the lower 100 km in Manitoba. All of the Manitoba riffles and rapids are located where the river channel is transected by outcrops of Paleozoic limestone and/or glacial till. Additional riffles occur in the upstream reaches near the confluence of the Bois des Sioux and Otter Tail rivers. Most of the river has a silt/clay/sand bottom except for a few reaches in the far southern and northern ends of the main stem and near the mouths of some tributaries, where it has a cobble/boulder bottom.

Due to its low gradient and silt/clay/sand substrates, most of the Red River main stem lacks suitable spawning habitat for riffle spawning fishes. Higher gradient gravel and boulder bed habitat does occur in the tributaries especially where they pass through the old beach ridges of Glacial Lake Agassiz and glacial moraines. Historically, fish would have had unimpeded access to these tributaries; however, by the late 1800s, dams had been constructed that blocked access to these high gradient reaches. In 1893, Woolman (1895) wrote of the Buffalo River, "The stream is well stocked with fish but is obstructed by several dams which prevent the running

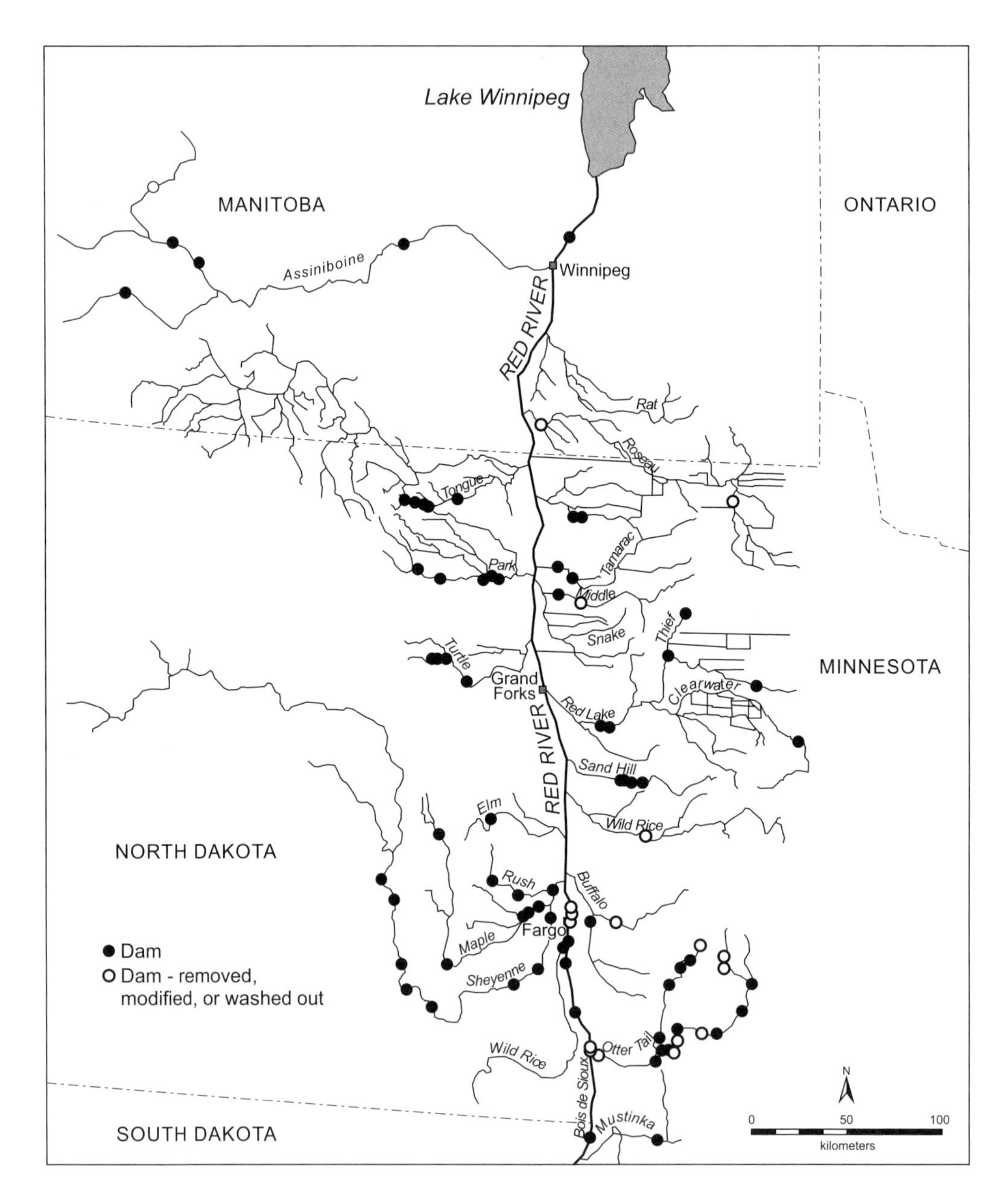

Figure 1.—Red River of the North basin. Only dams providing major barriors to fish migration are indicated.

of the fish" (Woolman, 1895). The Red River main stem presently has 9 dams, and over 500 have been constructed on the tributaries in the United States portion of the basin alone. Many of these have been built in the highest gradient reaches of the streams blocking fish migrations or flooding spawning riffles and rapids.

The most dramatic changes in the Red River of the North basin have been in the river's watershed. This is especially true in the portion of the drainage lying on the bed of former Glacial Lake Agassiz, where former tall grass prairies and wetlands are now intensively farmed and thousands of kilometers are ditched. While the farmland in the ancient lakebed is very fertile, it is also prone to flooding in both Canada and the United States.

Flooding is inherent to the Red River of the North due to its low gradient, but the magnitude and frequency of flooding may have been affected by land use changes. Moore and Larson (1979) found that drainage of wetlands and channelization caused increased peak flow in Minnesota streams. Land use changes over time also have been shown to be a significant predictor of increasing flood magnitude on the Red River of the North (Miller and Frink 1984). Flood magnitude increases in Red River tributaries as the proportion of the watershed that is wetland or lake decreases (Lorenz et al. 1997). Over 95% of the wetlands in the Agassiz Lake Plain ecoregion have been drained (Anderson and Craig 1984). This loss of storage and decreased soil permeability also may reduce base flows by impeding recharge of aquifers (Franke and McClymonds 1972). In addition, water extraction for irrigation, which is in greatest demand during dry conditions, may further reduce base flow.

Dredging and straightening of streams in the United States portion of the Red River basin was a common practice starting in the late 1800s and peaking in the 1950s, when the Army Corps of Engineers channelized a number of the larger tributaries. The same process has taken place in Canada, but it began in the 1930s and is continuing. While channelization projects are less common in recent years, channel maintenance continues on previously dredged and channelized smaller streams to maintain and improve wetland drainage. During recent floods, extensive head cutting degraded tributary streams upstream of previous channelization projects. Thousands of kilometers of tributary streams have been eliminated or degraded by channelization, with further losses resulting from subsequent head cuts. In the United States, most of these straightened reaches have been designated as legal ditches and are routinely dredged and sprayed with herbicides to prevent willow and other vegetation from becoming established. In Canada, changes in the administration of fish habitat protection regulations may lead to the restoration of fish habitat in at least some of the channelized tributaries. However, the competition between advocates of land drainage and those charged with protecting and rehabilitating fish habitat is fierce. Snagging and clearing also have been a common practice in the Red River of the North basin (RRNB), but has become less so recently due to greater restrictions on the practice and recognition of the importance of woody debris for fish habitat.

Methods

The Red River of the North has been sampled with a variety of gear types. Alexander Henry wrote the first accounts of Red River fishes in the years 1799–1806. While not trained as an ichthyologist, Henry took detailed notes, including air temperatures associated with some of his catches, and most of the species to which he refers are easily identifiable. Woolman (1895) conducted the first scientific survey of the Red River in the 1890s for the U.S. Commission of Fish and Fisheries. A comprehensive review of RRNB fishes in the United States was conducted by Koel (1997) and Koel and Peterka (1998). As a part of these analyses, site-specific fish distribution information was compiled for all available surveys (historic and recent). Henry's records were based on collections made from seining, weirs and fish traps, or hook and line, while Woolman (1895) used large seines. More recent sampling has involved DC electrofishing, prepositioned area sampling using alternating current, seining, gill nets, and hoop nets. Although some of these efforts have been quantitative or have been expressed in catch per unit effort, most of the data are not quantitative, tend to be species biased, and do not necessarily measure abundance of a given species. Since sampling methods and efficiency vary, this chapter focuses primarily on species richness measures rather than comparisons of abundance.

Results and Discussion

While much of the present fish assemblage of the Red River of the North basin appears similar to early records, there have been notable changes (Appendix A). The most prominent change in the fish assemblage is the extirpation of lake sturgeon *Acipenser fulvescens*. However, because sampling effort and identification accuracy have increased with time, it is difficult to determine whether additional species may have disappeared from the basin. As a system-level migratory species requiring a large amount of unfragmented riverine habitat, it is probable that the lake sturgeon is a good indicator of changes in the basin that may have adversely affected other migratory riverine species.

Henry referred frequently to lake sturgeon, including the identification of reaches associated with various life stages (Gough 1988). He reported on the catch of 120 fish weighing 27–82 kg from trap nets set on the Pembina River near its confluence with the Red on May 8, 1808. This probably was a spawning migration as we have observed lake sturgeon spawning in early May at the same latitude on the Little Fork River, a Rainy River tributary in northern Minnesota. Juvenile lake sturgeon (referred to as "sturgeon millers") were caught by one of Henry's men in the Red River of the North near the confluence of the Pembina River. Henry also identified the confluence of the Red Lake and Clearwater rivers as rapids, which had "famous sturgeon fishing in the spring season." Henry's Salteaux guide Charlo told him that "great numbers of sturgeon pass the winter" in deep water at the confluence of the Red and Red Lake rivers in Grand Forks, North Dakota. The observation of numerous jumping sturgeon in the Red River is a frequent entry in his journal. During drought conditions on September 8, 1800, he wrote of numerous sturgeon (in the Red River) in shallow water over clay bottom, and that one of his men killed one with an axe.

Figure 2.—Photo of a 184-kg lake sturgeon taken on October 27, 1903 from the Roseau River near Dominion City, Manitoba. The fish was killed with an axe while trapped in a shallow pool. It was estimated to be 150 years old. Photo courtesy of the Queen's Hotel, Dominion City and the Manitoba Museum. George Baraclough photographer.

The largest recorded lake sturgeon (184 kg) was caught in the Roseau River near Dominion City, Manitoba on October 27, 1903 (Figure 2). It was found stranded in a pool in the Roseau River by a farmer, Mr. Sandy Waddell, who roped it by the tail and dragged it from the stream with his horse. It was hung from a tree limb in Dominion City and photographed.

Woolman (1895) makes no mention of lake sturgeon in 1892 surveys of the Red River and its tributaries. While isolated sturgeon catches continued in various large lakes and tributaries within the basin until 1957, it is likely that the species was largely extirpated by the early 1900s (Figure 3). Radio-tagged lake sturgeon in the Rainy River watershed have been observed to move more than 200 km during apparent spawning migrations (Mosindy and Rusak 1991). Lake sturgeon spawn in rapids with boulder or bedrock substrates and turbulent flows (Aadland, unpublished data). These habitats are found primarily in the larger tributaries of the Red River where they pass through glacial moraines and the abandoned beach ridges of glacial Lake Agassiz. Most U.S. tributaries that have suitable spawning habitat had been blocked by dams by the 1870s. In Canada, the St. Andrews Dam (Lockport, est. 1910) has a fishway, but the cells in it are not large enough to pass adult sturgeon. While the gates of this dam are left open in the spring, a concrete sill maintains a barrier during lower spring flow conditions. This is important since blockage at the St. Andrews Dam prevents Lake Winnipeg sturgeon from reaching any of the remaining rapids in the Red River basin. The Red River Floodway Control

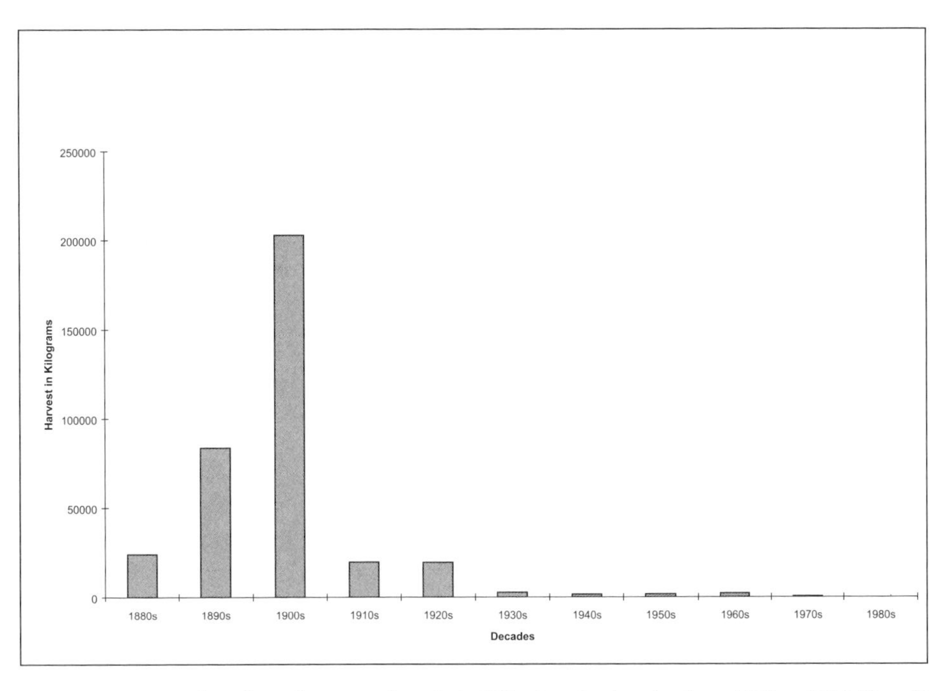

Figure 3.—Mean annual catches of sturgeon from Lake Winnipeg by decades from 1883 to 1990 (from Franzin et al. 2003).

Structure at St. Norbert, Manitoba, also may be a barrier to upstream fish movement during high flows that do not cause a sufficient rise in water level to warrant raising the control gates. In addition, lake sturgeon populations are vulnerable to over harvest. Entries by Henry indicated that the Salteaux constructed trap net barriers on tributaries during spring, and Henry and other early European settlers probably adopted this practice. Minnesota Department of Natural Resources, beginning in 1997, attempted to reintroduce lake sturgeon by stocking subadults, yearlings, and fingerlings in Red River tributaries and large lakes. This occurred concurrent with efforts to remove or modify dams within the basin and restore passage to critical habitat.

Other species mentioned by Henry include "lackaishe" (goldeye *Hiodon alosoides* or mooneye *H. tergius*), "catfish" (channel catfish *Ictalurus punctatus*), "pike" (northern pike *Esox lucius*), "doree (doré)" (walleye *Sander vitreus* or possibly sauger *Sander canadensis*), "piccanan" (shorthead redhorse *Moxostoma macrolepidotum*), "brim" (probably pumpkinseed *Lepomis gibbosus* or bluegill *L. macrochirus* or possibly quillback *Carpiodes cyprinus*), "Pois D'Oile" (unknown species, probably cyprinids) and "male archegan" (malachigan) (freshwater drum *Aplodinotus grunniens*). These are all well represented in recent surveys of the Red River (Appendix A).

Collections by Woolman (1895) included 38 species that generally are abundant in the Red River of the North and its tributaries. The exception is the longnose gar *Lepisosteus osseus*, which he reported collecting in the Otter Tail River near Breckenridge, Minnesota, a species that has not been collected there since. Woolman caught one large individual but noted that locals reported that they were "common in certain deep places in the river." The longnose gar is found in the upper Pomme de Terre watershed close to the Otter Tail River and in the Minnesota River near the continental divide separating it from the Bois de Sioux River. It is unclear why the species would have disappeared from the Otter Tail River since longnose gar populations exist in the adjacent Pomme de Terre, at the same latitude. American eel *Anguilla rostrata* have been reported in the Red River in Fargo, North Dakota by an angler who fished the Red in the 1950s and accurately described the species morphology as distinct from lamprey (John Peterka, North Dakota State University, personal communication). It is probable that individuals crossed the continental divide between Lake Traverse and Bigstone Lake or at other marsh-like locations during high water. While American eel was not a significant member of the fish assemblage, it may have had a periodic presence that would have been eliminated by construction of the Browns Valley Dike in 1941, which separated the two watersheds, and also by construction of dams on the Minnesota River. Despite the presence of this dike, the Little Minnesota River flowed into Lake Traverse as recently as the floods of 1993 and 1997 through culverts designed to provide local drainage. Both the longnose gar and the American eel were collected in the upper Minnesota River by Woolman (1895) and more recently in Minnesota Department of Natural Resources (DNR) surveys (Chris Domeier, Minnesota Department of Natural Resources, personal communication).

Koel (1997) and Koel and Peterka (1988) noted moderate range reductions for several species. For example, the chestnut lamprey has not been collected from the Sheyenne River since Woolman's early collection in 1892 (Woolman 1895). But the species apparently has increased its range southward in the eastern basin. Several chestnut lampreys have been collected in recent years from the Wild Rice River, Minnesota and from the Otter Tail River. It is unclear if the other Petromyzontid, the silver lamprey, has undergone changes in its range. Records of the silver lamprey are exclusively from recent years, suggesting that the species has recently expanded its range in the basin. However, it is possible that the lack of early records of this species is a result of misidentification in the field. The silver lamprey has been collected from the Red, Clearwater, Red Lake, and Buffalo rivers in the United States and from the Rat and Assiniboine rivers and Lake Winnipeg in Manitoba.

The Cyprinidae is the most diverse fish family in the RRNB and several cyprinids have exhibited moderate reductions in range over the past century (Koel 1997; Koel and Peterka 1998). Records from the early 1900s indicate that the brassy minnow once occurred in the Wild Rice River and Shotley Brook, a tributary to upper Red Lake. Later collections have not found this species in these locations. Underhill (1958) described the brassy minnow as common in Minnesota; however, in the Red River basin, brassy minnows have been collected only at 3% of stream sites sampled from 1962 to 1994; typically, collections have consisted of only one or a few individuals (Koel and Peterka 1998). Harbicht et al. (1988) collected the brassy minnow in the Pembina River and other drainages in Manitoba and emphasize its "strangely disjunct (distribution) within the province." Similarly, the pearl dace, while existing in isolated populations in locations such as the headwaters of the South Branch Buffalo, Park, Thief, Roseau, Snake, Middle, and Tamarac rivers (as well as in several escarpment headwater tributaries of the Red and Assiniboine rivers in Manitoba), has evidently been extirpated from the headwaters of the Otter Tail and Wild Rice rivers, as it has not been found in these reaches since 1955 (Bell Museum of Natural History, University of Minnesota, Bloomington, unpublished records). The hornyhead chub once existed in the Sheyenne and Maple rivers and in Daugherty Creek, a tributary to Lake Traverse, but has not been found in recent collections despite its widespread distribution in several eastern basin streams. Recent collections indicate the species remains abundant in the Otter Tail River near Fergus Falls, Minnesota and in the upper reaches of the Clearwater River (Aadland, unpublished data). One record exists from the Red River at Winnipeg (Clarke et al. 1980); however, no recent records exist for hornyhead chub from the main stem of the Red River. The pugnose shiner, only rarely collected in the RRNB, had not appeared in documented collections for more than 20 years and has never been reported from Manitoba waters of the Red River. However, recent collections of two individuals from the Otter Tail River have confirmed its continued existence and rarity there (Aadland, unpublished data 2003). Although the river shiner apparently no longer occurs in the Red Lake or Sandhill rivers, it has increased its range and currently occurs throughout the Sheyenne River in North Dakota (North Dakota Game and Fish Department reported it at one site above Baldhill Dam). The distribution of the blacknose shiner, once widespread in the Sheyenne River, now occurs only in the spring-fed streams at the Mirror Pool Wildlife Management Area.

Additional changes in range and/or apparent abundance were noted for brown bullhead, banded killifish, Iowa darter, and river darter (Koel and Peterka 1998). Overall occurrence of brown bullhead was reduced by 10% from the mid-1950s to the 1900s at sites where it was collected frequently. The banded killifish has not been collected from the Sheyenne River since 1892 (Woolman 1896). The Iowa darter has not been collected from the Park or Tamarac rivers since the early 1900s, and the current distribution of river darter appears restricted to the Thief, Middle, and Roseau rivers in the United States, but the species is still common in Canada.

Species richness is highest in the Otter Tail and Red Lake River drainages (73 and 65 species, respectively) in the eastern basin and in the Sheyenne River drainage (56 species) in the western basin. These streams have a wide range of habitats for stream fishes and have a relatively high, stable hydrological regime.

Many streams in the Red River basin have one or several unique fish species. The Forest River, a relatively small stream with a length of only about 120 km, supports the only documented largescale stoneroller population in the entire Hudson Bay drainage basin, and the only remaining population of hornyhead chub in North Dakota. The hornyhead chub is intolerant of mud or silt substrates and turbid waters (Dalton 1989) that often result from agricultural practices. It does not occur in any of the tributaries of the Red River in Manitoba but does occur in the Brokenhead River that drains directly to Lake Winnipeg and in the Whitemouth River that flows into the Winnipeg River. The upper reaches of the Tongue River, a tributary to the Pembina River, have isolated populations of pearl dace and finescale dace. In the western basin, the northern redbelly dace occurs only in the Rush River and in spring-fed creeks in the Mirror Pool Wildlife Management Area, Sheyenne River. Similarly, pearl dace, northern redbelly dace, and finescale dace occur in western headwaters of the Boyne River, a small tributary that enters the Red River south of Winnipeg. This group of species occurs sporadically in headwater streams with suitable habitat all along the Manitoba Escarpment northwest of the RRBN basin and in tributaries and lakes of the Winnipeg River drainage. The only occurrences of banded killifish in the western basin is in the lower reaches of the Park and Turtle rivers where waters are high in specific conductance (average for the Turtle River is 3,800 mS/cm); the species is known to be salinity tolerant (Houston 1990).

Flow Alterations

Watershed changes such as conversion of grassland to intensive row crop agriculture, wetland drainage, and channelization of tributaries may have affected species richness by increasing hydrologic variability. Climatic and hydrologic records over the past 120 years suggest wet periods in the late 1800s to early 1900s and the late 1900s to present separated by a dry period that included the drought of the 1930s. While the 40-year periods from 1882 to 1921 and 1962 to 2001 had similar winter precipitation averages, peak flows at Grand Forks have averaged 60% higher in the latter time period (Figure 4). Conversely, minimum flows as a proportion of the annual mean are 76% higher in the first time period than in the last at Fargo (Figure 5). Stream flow expressed as a percentage of the mean annual flow has been used as an indicator of flow dependent habitat quality. Species richness in tributaries of the Red River generally increases with the ratio of the lowest monthly average flow to the mean annual flow (Figure 6). Most rivers within the basin have their lowest flows in the winter and their highest flows following spring snowmelt (Harkness et al. 1992; Mitton et al. 2001). Tributaries that have average winter flows near zero still have 20–30 species, probably due to recolonization and because collections are usually taken during suitable flow conditions.

Seasonally variable flows have been shown to disrupt habitat composition and suitability for fishes (Horowitz 1978; Bain et al. 1988) and may be a significant factor contributing to the success of harmful alien species (Koel and Sparks 2002). Data from 1,026 sites on 26 major RRNB tributaries were

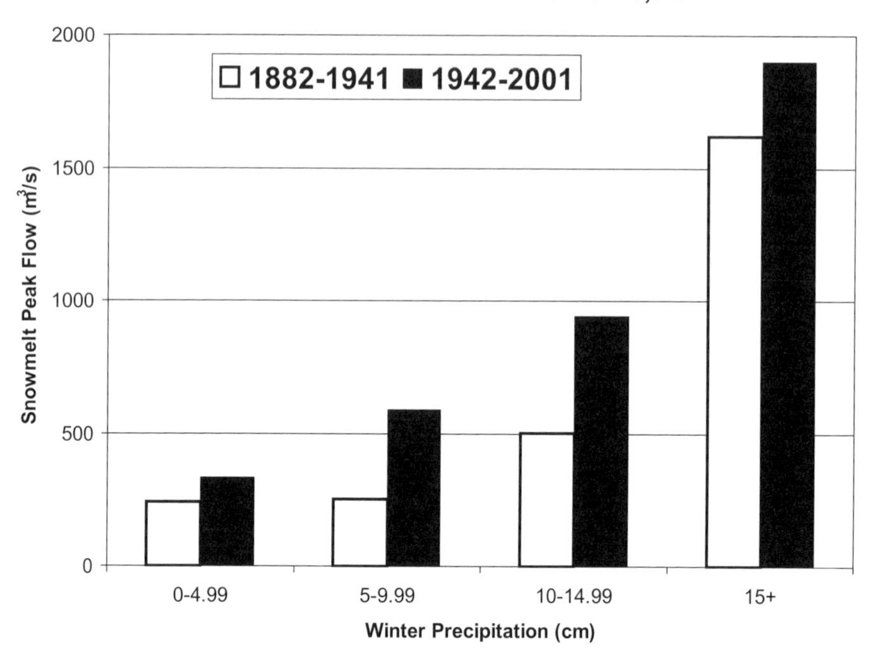

Figure 4.—Peak flows of the Red River of the North in Grand Forks, North Dakota as a function of winter precipitation recorded at Fargo, North Dakota. The Grand Forks gauge is the longest running peak flow gauge (beginning in 1882) on the Red River of the North in the United States. Fargo is the only precipitation station that recorded data continuously from 1882 to present.

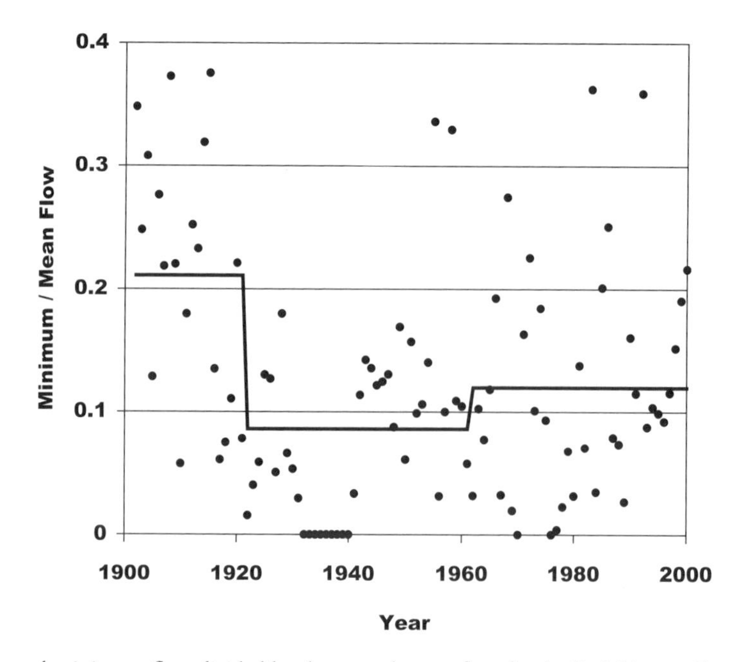

Figure 5.—Annual minimum flow divided by the annual mean flow for the Red River at Fargo, North Dakota with 40-year averages (line). The Fargo gauge is the longest running daily values gauge on the Red River of the North in the United States.

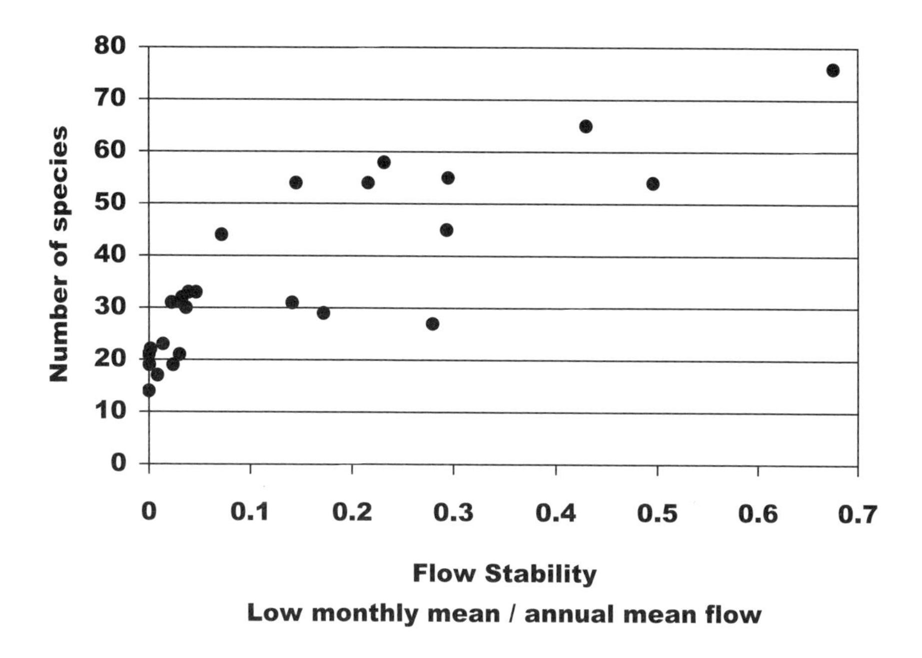

Figure 6.—Fish species richness versus the ratio of the minimum monthly average to the mean annual flow for tributaries of the Red River of the North.

used to relate frequency of occurrence of fish species to several important environmental factors from 1962 to 1994 (Koel and Peterka 2003). Hydrologic variability (coefficient of variation of mean monthly discharge) was the most important factor explaining variation in fish assemblage structure. Examples of species that were most frequent in reaches with high flow variability were black bullhead, fathead minnow, creek chub, brook stickleback, white sucker, and common carp (Figure 7). These species are habitat generalists that are tolerant of rapidly changing flows where habitat conditions are in constant flux. Reaches with these conditions included the Elm, Mustinka, Rabbit, and Park rivers in the Red River Valley (Agassiz Lake Plain) ecoregion (Omernik and Bailey 1997), and the Maple River and Wild Rice River, North Dakota and small rivers on the west side of the river valley in Manitoba in the northern glaciated plains ecoregion. Reaches with low flow variability were considered in a more natural condition and reminiscent of the conditions prior to Euro-American settlement to the RRNB. Examples of these reaches included the Otter Tail River in all ecoregions and the Pelican River in the North Central Hardwoods. Fishes typical of reaches in these and/or other similar streams included species such as weed shiner, logperch, yellow bullhead, and spottail shiner (Figure 7). In the future, as stream flows and overall water quality are altered by a multitude of potential anthropogenic factors at the landscape scale, fish assemblages of the RRNB may shift from abundant habitat specialists (such as exist in many areas of the eastern basin) to those dominated by tolerant habitat generalists. Also, if conditions across the RRNB are allowed to continue to deteriorate, the potential for invasive alien species to become successfully established and displace ecologically important native fishes is high.

Land use generated changes in hydrology also may affect fish habitat by increasing bank erosion and channel enlargement (Leopold 1994). The frequency of bank-full flows of the Red River at Fargo, North Dakota has increased from about 0.75 per year in the early 1900s to about 2.2 events per year in the past 20 years (Figure 8).

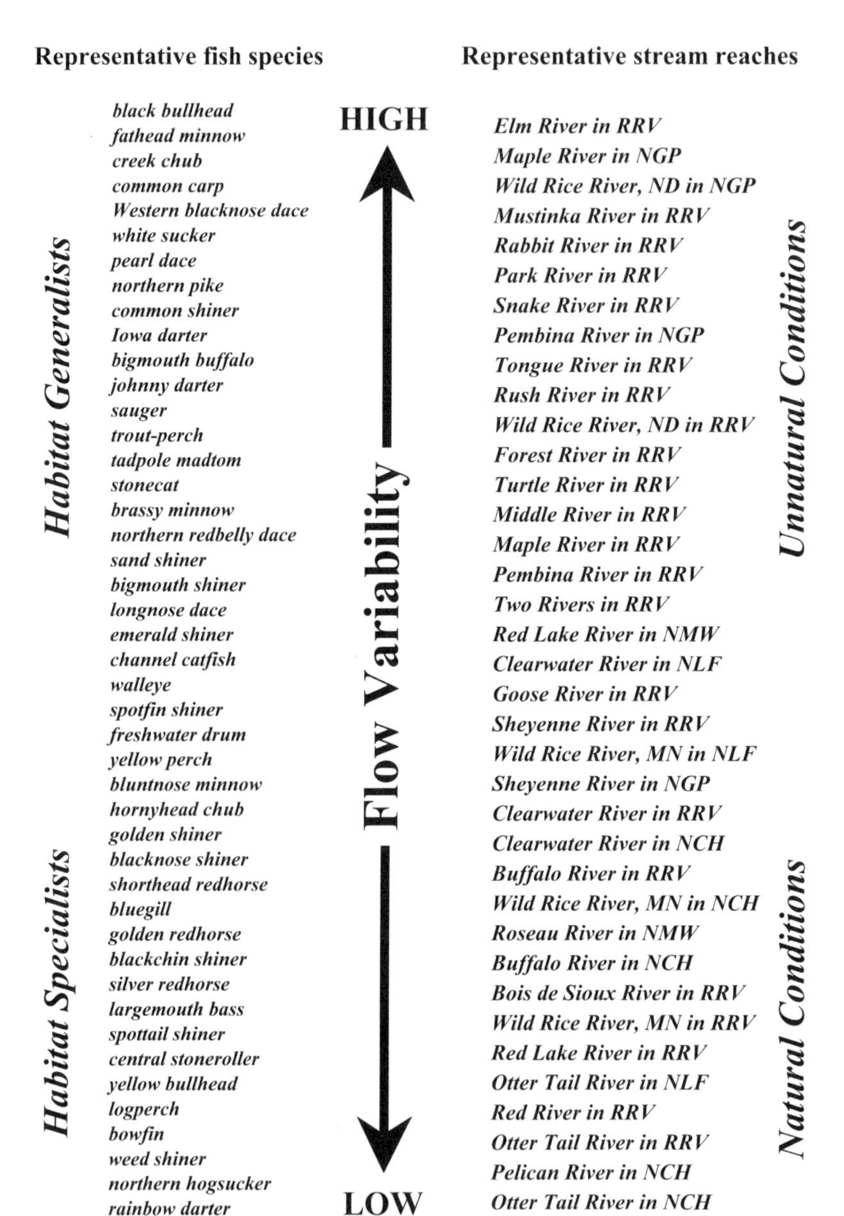

Figure 7.—Representative fish species and stream reaches as they have been described along a flow variability gradient (adapted from Koel and Peterka 2003). Streams include reaches in the North Central Hardwoods (NCH), Northern Glaciated Plains (NGP), Northern Lakes and Forests (NLF), Northern Minnesota Wetlands (NMW), and Red River Valley (RRV) ecoregions (Omernik and Gallant 1988). Although the most important in terms of explaining variation in fish assemblage composition, flow variability is just one of many environmental gradients influencing fish distribution in the Red River of the North basin.

Dams

Abrupt reductions in species richness have been shown at dams in Red River tributaries. Surveys of the Wild Rice River indicated 12 more species downstream of the impassable Heiberg Dam than upstream of it (Table 1; Baily et al. 1996; Barsness 1996; and Huberty et al. 1999). The effect on chan-

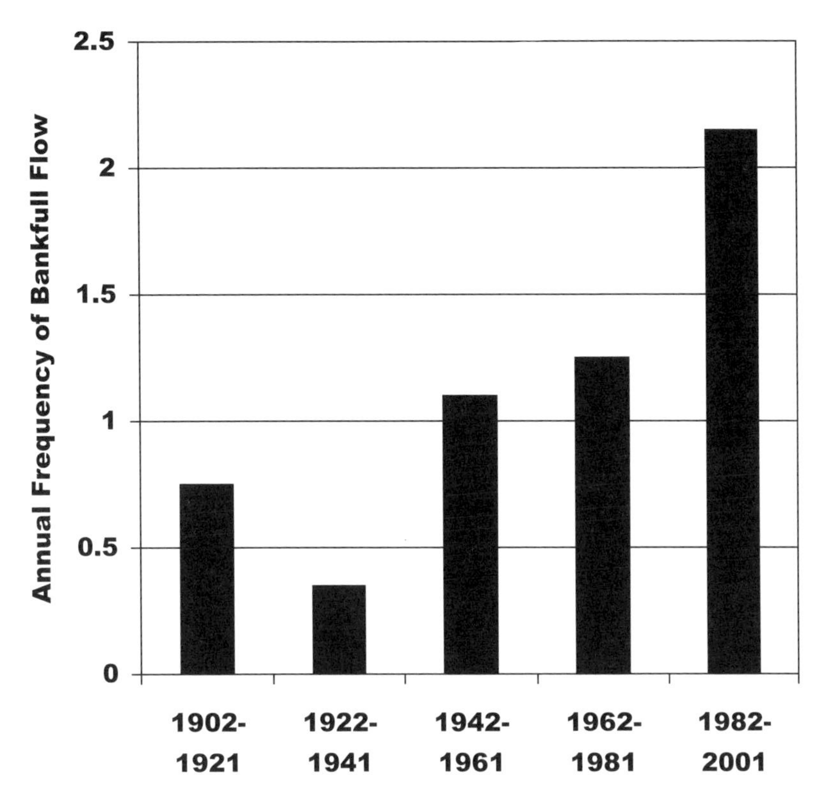

Figure 8.—Changes in recurrence of bank-full flows on the Red River at Fargo, North Dakota.

nel catfish was most notable considering they were one of the most abundant species downstream of the dam, but absent upstream. Following failure of the dam in 1965 at the same site and prior to its reconstruction in 1977, locals claim to have caught channel catfish up to 100 km upstream. The Wild Rice River eroded a new channel around Heiberg dam and diverted into a tributary during summer floods of 2002, and it is no longer a barrier. Large numbers of channel catfish have been caught in trap nets along with walleye, freshwater drum, goldeye, and smallmouth bass (Dave Friedl, Minnesota Department of Natural Resources, personal communication) up to 120 km upstream of the dam following its failure. The Minnesota Department of Natural Resources is currently negotiating with the dam owner and the Federal Emergency Management Agency to prevent reestablishment of this barrier. The Portage la Prairie Dam (built in 1970), 161 km upstream of the confluence with the Red River at Winnipeg on the Assiniboine River, is an absolute barrier to golden redhorse, and bigmouth buffalo migrations. A recent electrofishing survey failed to produce a single observation of either species above the dam, while both species were captured immediately below it. A diversion channel connecting Portage Reservoir to the south end of Lake Manitoba is not passable for upstream migrating fish.

Catches in a trap net set at the upstream end of a fishway through a previously impassable dam on the Otter Tail River included 21 species of upstream migrating fish (Table 2). Since the trap net had a mesh size of 13 mm, the gear was not effective for small-bodied fishes, which account for the bulk of the fish assemblage. Shorthead redhorse comprised the greatest proportion of the catch followed by channel catfish, golden redhorse, and silver redhorse. Sauger, which had not been recorded previously in the Otter Tail River also passed through the fishway.

Channel catfish migrate large distances in the

Table 1.—Fish species found upstream and downstream of Heiberg Dam in the Wild Rice River. Based on Barsness 1994, Baily et. al 1996, and Huberty et al. 1999. Species that were found in 2003 assessments following the dam's failure are indicated by "R."

Species	Upstream of Heiberg Dam	Downstream of Heiberg Dam
Ichthyomyzon castaneus	X	
I. unicuspis		X
Hiodon alosoides	R	X
H. tergisus		X
Cyprinella spiloptera	R	X
Cyprinus carpio	R	X
Luxilus cornutus	X	X
Margariscus margarita	R	X
Nocomis biguttatus	X	X
Notropis atherinoides		X
N. dorsalis		X
N. stramineus	X	X
Phoxinus eos	X	
Pimephales promelas	X	X
Rhinichthys obtusus	X	X
R. cataractae	X	X
Semotilus atromaculatus	X	X
Carpiodes cyprinus		X
Catostomus commersonii	X	X
Moxostoma anisurum	X	X
M. erythrurum	X	X
M. macrolepidotum	R	X
Amieurus melas	X	X
A. nebulosus	X	X
Ictalurus punctatus	R	X
Noturus flavus	X	X
N. gyrinus		X
Esox lucius	X	X
Umbra limi	X	
Percopsis omiscomaycus	X	X
Lota lota		X
Culaea inconstans	X	
Ambloplites rupestris	X	X
Lepomis gibbosus	R	
L. macrochirus		X
Micropterus dolomieu	R	X
Etheostoma exile	X	
E. nigrum	X	X
Percina caprodes	X	X
P. maculata	X	X
Perca flavescens	X	
Sander canadensis	R	X
S. vitreus	R	X
Aplodinotus grunniens	R	X

RRNB, and one individual was caught 500 km from where it was tagged, moving from Fargo to Lake Winnipeg (Hegrenes 1992). Catfish movement on the Red River has been associated with upstream spring migrations to spawning habitat and downstream fall migrations to overwintering habitat (Wendel 1999). Migrations also may result from size-related changes in feeding behavior. Higher gradient tributaries with larger substrates may be favored by juveniles, which are benthic invertivores, while

Table 2.—Upstream migrating fishes caught in a trap net at the upstream end of the Breckenridge fishway on the Otter Tail River in 1998 and 2000. Catches represent 14 net-days from April 7 to June 4, 1998 and 22 net-days from March 23 to June 1, 2000.

Species	Total catch	Percent of total	Date of peak catch	Earliest catch	Latest catch
Hiodon alosoides	2	<1	May 19	May 19	Jun 1
H. tergisus	204	5	May 18	Mar 24	Jun 1
Esox lucius	6	<1	Apr 26	Mar 24	May 3
Cyprinus carpio	5	<1	May 11	Apr 14	May 26
Carpiodes cyprinus	181	4	May 14	Mar 24	Jun 1
Ictiobus cyprinellus	2	<1	May 26	Mar 30	May 26
Catostomus commersonii	75	2	Mar 30	Mar 23	May 25
Moxostoma anisurum	369	9	May 3	Mar 23	May 26
M. erythrurum	435	11	May 3	Mar 23	Jun 4
M. macrolepidotum	1707	43	May 3	Mar 23	May 26
M. valenciennesi	133	3	May 3	Mar 23	Jun 1
Amieurus melas	4	<1	May 11	Apr 23	May 12
A. nebulosus	1	<1	May 3	May 3	May 3
Ictalurus punctatus	679	17	Apr 29	Apr 14	Jun 3
Noturus flavus	4	<1	Apr 15	Apr 14	Apr 15
Ambloplites rupestris	27	1	May 11	Apr 9	Jun 4
Micropterus dolomieu	34	1	Apr 23	Apr 21	Jun 4
Pomoxis nigromaculatus	4	<1	May 25	Apr 26	May 12
Sander canadensis	1	<1	Apr 21	Apr 21	Apr 21
S. vitreus	65	2	Apr 22	Mar 23	May 26
Aplodinotus grunniens	65	2	May 26	Apr 22	Jun 3

larger main-stem reaches may be preferred by piscivorous adults (Dames et al. 1989; Hegrenes 1992). As with most species, microhabitat preferences of channel catfish change with growth; age-0 fish tend to be more riffle oriented, while juveniles and adults favor pools (Aadland 1993). However, opportunistic use of inundated river margins also has been observed for all age-classes of channel catfish, which move off river to feed on abundant terrestrial earthworms (Franzin, unpublished data).

Alien Species

The Red River of the North has been less affected by alien species than many large rivers and the common carp is the most widely established alien. Several species have been stocked outside their North American ranges. White bass and white crappie have been introduced and maintain spotty populations within the river. Common carp were stocked in Minnesota in 1883, but did not appear in the Red River until later, and first records in Manitoba were in 1938 (Eddy and Underhill 1974). Walleye are routinely stocked throughout many of the lakes in the basin. Smallmouth bass, muskellunge, brook trout, northern pike, rainbow trout, black crappie, bluegill, yellow perch, and white bass have been stocked in the Red River basin. Even Chinook salmon *Oncorhynchus tshawytscha* were stocked in the Red River in 1876, though it is unlikely that these fish survived their first summer (Eddy and Underhill 1974).

Lake sturgeon reintroduction efforts by the Minnesota Department of Natural Resources and the White Earth Indian Reservation began in 1997. To date, releases have included 375 subadults, 3,482 yearlings, and 18,000 fingerlings (in 2002) in the Otter Tail, Wild Rice, and Buffalo watersheds. The 20-year reintroduction plan calls for the annual release of 34,000 fingerlings and 600,000 fry in the Roseau, Red Lake, Wild Rice, Buffalo, and Otter Tail watersheds (Table 3; MN DNR 2002).

Prospective Changes

The Red River of the North watershed has been substantially altered physically and biologically. Threats of additional dam construction and even more intensive drainage continue. Several proposals include construction of dams on some of the most pristine river reaches, including a 20-m-high dam on a high gradient gravel and boulder bed reach of the Wild Rice River near Twin Valley, Minnesota. The proposed Garrison Diversion would connect the Missouri River to the Red River through a series of massive ditches, many of which already have been constructed. While some versions of this project involve treating the water, transfer of nonindigenous species and altering the flow regime of the Sheyenne River remain a concern. An outlet to the closed basin of Devils Lake also has been advanced. This project would add substantial flow to the Sheyenne River and has prompted concerns over water quality (Devils Lake has high salinity levels) and increased flooding and erosion.

There are, however, reasons for optimism. Several flood damage reduction projects and proposals have included restoration of the streams, floodplains, and wetlands. One recent project in the Wild Rice Watershed district restored 4.4 km of meanders to ditched portions of a small Wild Rice River tributary in Minnesota. There are additional proposals to restore channelized reaches of several large tributaries, including the Wild Rice River.

The Minnesota Department of Natural Resources, the U.S. Army Corps of Engineers, the U.S. Fish and Wildlife Service, Manitoba Conservation, the North Dakota Game and Fish Commission, and several other agencies, local governments, and organizations are currently working to restore migratory pathways on the Red River and its tributaries. Three dams have been removed, one recently washed out, nine have been converted to rapids (Figure 9), and two nature-like fishways (Figure 10) have been constructed on the Red River and its Minnesota tributaries. These rapids were designed by the senior author (Aadland) to provide fish passage and to mimic the hydraulics and microhabitat of natural rapids where we have observed sturgeon spawning (Figure 11). Similar rapids and a nature-like fishway have been built on the Roseau and Little Saskatchewan rivers in Manitoba (Gaboury et al. 1995). Four of the nine Red River main-stem dams have been converted to rapids. Removal of the Buffalo State Park Dam on the Buffalo River and failure of the Hieberg Dam on the Wild Rice River have restored access to high quality spawning habitat. Conversion of the East Grand Forks Dam on the Red Lake River to rapids and the planned removal of the Crookston Dam on the same river will restore access to the historic sturgeon-spawning habitat described by Alexander Henry. The connection of the Red River to Lake Winnipeg, the world's 11th largest lake, may be essential to the viability of a lake sturgeon population. The barrier created by St. Andrew's Dam at Lockport also should be addressed to provide year-round passage. The success of lake sturgeon recovery and the restoration of historic fish assemblages in the Red River and its tributaries will depend heavily on these efforts.

The fish assemblage of the Red River of the North remains diverse and unique despite habitat fragmentation and major adverse changes in the watershed. Changes to many less conspicuous species are unknown due to lack of reliable data prior to substantial physical changes in the watershed. Since the earliest records focused on species of commercial importance or used gear that was ineffec-

Table 3.—Proposed lake sturgeon stocking locations and operational plans for the 20-year (2002–2022) program from Minnesota Department of Natural Resources (MN DNR 2000).

Stocking locations	Life stage	Number	Frequency	Jurisdiction
Big Detroit Lake	fingerling	4,000	annual	MN DNR
Otter Tail Lake	fingerling	15,000	annual	MN DNR
Round Lake	fingerling	5,000	annual	White Earth
White Earth Lake	fingerling	8,000	annual	White Earth
Otter Tail River	fingerling	1,000	annual	MN DNR
Buffalo River	fingerling	1,000	annual	MN DNR
Roseau River	fry	200,000	annual	MN DNR
Red Lake River	fry	400,000	annual	MN DNR

Figure 9.—A dam on the Red River of the North at Moorhead, Minnesota that has been converted to rapids.

Figure 10.—A nature-like fishway constructed on the Otter Tail River near Breckenridge, Minnesota.

Figure 11.—Deadman's Rapids on the Little Fork River in Northern Minnesota where lake sturgeon spawn.

tive for small, specialized species, it is likely that our reliance on more recent records underestimates the loss and diminished ranges of species in the RRNB. Long-term monitoring of fish distributions will be critical in understanding problems, preventing further species losses, and restoring river functions and diversity.

Acknowledgments

Thanks to the many individuals who assisted in the writing of this paper. Thanks to Arlin Schalekamp, Mike Larson, Dennis Topp, Dave Friedl, Henry Van Offelen, Tom Groshens, Molly McGregor, and Henry Drewes for providing stream survey data. Thanks to Fernand Saurette for his help in translating the fish names in Alexander Henry's Journal into the common names used today for the species. Thanks to Rita and James Mazurski, owners of the Queen's Hotel in Dominion City, Manitoba, Jack Curran, also of Dominion City, Gavin Hanke, Curator of Zoology, Manitoba Museum, Winnipeg, Manitoba, and Cindy Adams of the Kittson County Historical Society, whose combined efforts tracked down the documentation and photographs of the world record lake sturgeon. Thanks to Shelly Buitenwerf for preparing the basin map. Finally, thanks to John Peterka, for reviewing this paper.

References

Aadland, L. P. 1993. Stream habitat types: their fish assemblages and relationship to flow. North American Journal of Fisheries Management 13:790–806.

Anderson, J. P., and W. J. Craig. 1984. Growing energy crops on Minnesota's wetlands: the land use perspective. University of Minnesota Center for Urban and Regional Affairs, Publication CURA 84–3, St. Paul.

Baily, P., L. Hotka, W. Mattison, and M. Feist. 1996. Wild Rice River longitudinal survey preliminary report. Minnesota Pollution Control Agency, St. Paul.

Bain, M. B., J. T. Finn, and H. E. Booke. 1988. Streamflow regulation and fish community structure. Ecology 69:382–392.

Barsness, D. O. 1996. Stream population assessment of the Wild Rice River. Minnesota Department of Natural Resources, Detroit Lakes.

Bovee, K. D. 1986. Development and evaluation of habitat suitability criteria for use in the instream flow incremental methodology. U.S. Fish and Wildlife Service, Instream Flow Information Paper No. 21, Fort Collins, Colorado.

Clarke, R. McV., R. W. Boychuck, and D. A. Hodgins. 1980. Fishes of the Red River at Winnipeg, Manitoba. Department of Fisheries and Oceans, Winnipeg, Manitoba.

Clayton, L. 1983. Chronology of Lake Agassiz drainage to Lake Superior. Pages 291–307 *in* J. T. Teller and L. Clayton, editors. Glacial Lake Agasiz. Geological Association of Canada, Special Paper 23, Winnipeg, Manitoba.

Dalton, K. W. 1989. Status of the hornyhead chub, *Nocomis biguttatus*, in Canada. Canadian Field-Naturalist 103:180–185.

Dames, H. R., T. G. Coon, and J. W. Robinson. 1989. Movements of channel and flathead catfish between the Missouri River and a tributary, Perche Creek. Transactions of the American Fisheries Society 118:670–679.

Dyke, A. S., and V. K. Prest. 1987. Paleogeography of northern North America, 18,000–5,000 years ago. Geological Survey of Canada, Map 1703A, scale 1:12,500,000, Sheets 1–3, Ottawa, Ontario.

Eddy, S., and J. C. Underhill. 1974. Northern fishes. University of Minnesota Press, Minneapolis.

Fenton, M. S. R. Moran, J. T. Teller, and L. Clayton 1983. Quaternary stratigraphy and history in the southern part of the lake Agassiz basin. Pages 49–74 *in* J. T. Teller and Lee Clayton, editors. Glacial Lake Agassiz. Geological Association of Canada, Special Paper 23, Winnipeg, Manitoba.

Finger, T. R., and E. M. Stewart. 1987. Response of fishes to flooding regime in lowland hardwood wetlands. Pages 86–92 *in* W. J. Matthews and D. C. Heins, editors. Community and evolutionary ecology of North American stream fishes. University of Oklahoma Press, Norman.

Franke, O. L., and N. E. McClymonds. 1972. Summary of the hydrologic situation on Long Island, New York, as a guide to water management alternatives. United States Geological Survey, Professional Paper 627-F, Troy, New York.

Franzin, W. G., K. W. Stewart, G. F. Hanke, and L. Heuring. 2003. The fish and fisheries of Lake Winnipeg: the first 100 years. Canadian Technical Report of Fisheries and Aquatic Sciences #2398, Ottawa, Ontario.

Gaboury, M. N., R. W. Newbury, and C. M. Erickson. 1995. Pool and riffle fishways for small dams. Manitoba Natural Resources Fisheries Branch, Winnipeg.

Gough, B. M., editor. 1988. The journal of Alexander Henry the Younger 1799–1814. Volume One. The Champlain Society, Toronto.

Harbicht, S. M., W. G. Franzin, and K. W. Stewart. 1988. New distributional records for the minnows *Hybognathus hankinsoni, Phoxinus eos*, and *P. neogaeus* in Manitoba. Canadian Field-Naturalist 101:171–185.

Harkness, R. E., N. D. Haffield, W. R. Berkas, and S. W. Norbeck. 1992. Water Resources Data North Dakota Water Year 1991. United States Geological Survey, Water-Data Report ND-91-1, Bismarck, North Dakota.

Hegrenes, S. G. 1992. Age, growth, and reproduction of channel catfish in the Red River of the North. Master's thesis. University of North Dakota, Grand Forks.

Horowitz, R. J. 1978. Temporal variability patterns and the distributional patterns of stream fishes. Ecological Monographs 48:307–321.

Houston, J. 1990. Status of the banded killifish, *Fundulus diaphanus*, in Canada. Canadian Field-Naturalist 104:45–52.

Huberty, G., M. Olson, and A. Ostgarden. 1999. Stream survey description summary: Wild Rice River. Minnesota Department of Natural Resources, Detroit Lakes.

Junk, W. J., P. B. Bayley, and R. E. Sparks. 1989. The flood pulse concept in river-floodplain systems. Pages 110–127 *in* D. P. Dodge, editor. Proceedings of the international large river symposium (LARS). Canadian Special Publication of Fisheries and Aquatic Sciences 106. Department of Fisheries and Oceans. Ottawa, Ontario.

Kimsey, J. B. 1957. Fisheries problems in impounded waters of California and the lower Colorado River. Transactions of the American Fisheries Society 87:39–57.

Koel, T. M. 1997. Distribution of fishes in the Red River of the North basin on multivariate environmental gradients. Doctoral dissertation. North Dakota State University, Fargo.

Koel, T. M., and J. J. Peterka. 1998. Stream fishes of the Red River of the North basin, United States: a comprehensive review. Canadian Field-Naturalist 11:631–646.

Koel, T. M., and J. J. Peterka. 2003. Stream fish communities and environmental correlates in the Red River of the North, Minnesota and North Dakota. Environmental Biology of Fishes 67:137–155.

Koel, T. M., and R. E. Sparks. 2002. Historical patterns of river stage and fish communities as criteria for operations of dams on the Illinois River. River Research and Applications 18:3–19.

Leopold, L. B. 1968. Hydrology for urban land planning: a guidebook. United States Geological Survey, Circular 554, Menlo Park, California.

Leopold, L. B. 1994. A view of the river. Harvard University Press, Cambridge, Massachusetts.

Lorenz, D. L., G. H. Carlson, and C. A. Sanocki. 1997. Techniques for estimating peak flow on small streams in Minnesota. U.S. Geological Survey, Water-Resources Investigations Report 97–4249, Mounds View, Minnesota.

Lusby, G. C. 1970. Hydrologic and biotic effects of grazing versus non-grazing near Grand-Junction, Colorado. Journal of Range Management 23:256–260.

Matsch, C. L. 1983. River Warren, the southern outlet of Glacial Lake Agassiz. Pages 231–244 *in* J. T. Teller and Lee Clayton, editors. Glacial Lake Agasiz. Geological Association of Canada, Special Paper 23, Winnipeg, Manitoba.

Miller, J. E., and D. L. Frink. 1984. Changes in flood response of the Red River of the North basin, North Dakota-Minnesota. U.S. Geological Survey Water Supply Paper 2243, Alexandria, Virginia.

Miller, R. C. 1999. Hydrologic effects of wetland drainage and land use change in a tributary watershed of the Minnesota River basin: a modeling approach. Master's thesis. University of Minnesota, St. Paul.

Mittlebach, G. G. 1981. Foraging efficiency and body size: a study of optimal diet and habitat use by bluegills. Ecology 62:1370–1386.

Mitton, G. B., K. G. Guttormson, G. W. Stratton, and E. S.Wakeman. 2001. Water Resources Data Minnesota Water Year 2000. United States Geological Survey, Water-Data Report MN-00-1, Mounds View, Minnesota.

Moore, I. D., and C. L. Larson. 1979. Effects of drainage projects on surface runoff from small depressional wetlands in the North Central Region. University of Minnesota, Water Resources Research Center, Minneapolis.

Mosindy, T., and J. Rusak. 1991. An assessment of lake sturgeon populations in Lake of the Woods and the Rainy River 1987–90. Ontario Ministry of Natural Resources, Lake of the Woods Fisheries Assessment Unit Report 1991:01, Peterborough.

Omernik, J. M., and R. G. Bailey. 1997. Distinguishing between watersheds and ecoregions. Journal of the American Water Resources Association 33:935–949.

Omernik, J. M., and A. L. Gallant. 1988. Ecoregions of the upper midwest states. U.S. Environmental Protection Agency, EPA/600/3-88/037, Corvallis, Oregon.

Rosgen, D. 1996. Applied river morphology. Printed Media Companies, Minneapolis, Minnesota.

Ross, S. T., and J. A. Baker. 1983. The response of fishes to periodic spring floods in a southeastern stream. American Midland Naturalist 109(1):1–14.

Schlosser, I. J. 1985. Flow regime, juvenile abundance, and the assemblage structure of stream fishes. Ecology 66(5):1484–1490.

Seltz, A. 1999. History of the Otter Tail River. Otter Tail Power Company, Fergus Falls, Minnesota.

Sparks, R. E. 1995. Need for ecosystem management of large rivers and their flood plains. BioScience 45:168–182.

Stewart, K. W., and D. A. Watkinson. 2004. The freshwater fishes of Manitoba. University of Manitoba Press, Winnipeg, Manitoba.

Striffler, W. D. 1964. Sediment, streamflow, and land use relationships in northern lower Michigan. U.S. Forest Service, Research Paper LS-16, St. Paul, Minnesota.

Teller, J. T., and L. H. Thorleifson 1983. The Lake Agassiz - Lake Superior connections. Pages 261–290 *in* J. T. Teller and Lee Clayton, editors. Glacial Lake Agasiz. Geological Association of Canada, Special Paper 23, Winnipeg, Manitoba.

Tennant, D. L. 1976. Instream flow regimen for fish, wildlife, recreation, and related environmental resources. Pages 359–373 *in* J. F. Orsborn and C. H. Allman, editors. Instream flow needs, volume II. American Fisheries Society, Bethesda, Maryland.

Underhill, J. C. 1958. The distribution of Minnesota minnows and darters in relation to Pleistocene glaciation. Minnesota Museum of Natural History. Occasional papers Number 7. The University of Minnesota Press, Minneapolis.

USGS (United States Geological Survey). 2002a. Water Resources Data Minnesota Water Year 2002. U.S. Geological Survey, Water-Data Report MN-02-1, Mounds View, Minnesota.

USGS (United States Geological Survey). 2002b. Water Resources Data North Dakota Water Year 2002. U.S. Geological Survey, Water-Data Report ND-02-1, Bismarck, North Dakota.

Van Offelen, H., B. Evarts, M. Johnson, T. Groshens, and G. Berg. 2002. Red River basin stream survey report. Wild Rice Watershed 2000. Minnesota Department of Natural Resources, Detroit Lakes.

Verry, E. S. 1988. The hydrology of wetlands and man's influence on it. Pages 41–61 *in* Symposium on the hydrology of wetlands in temperate and cold regions. Volume 2. Publications of the Academy of Finland 5/1988, Helsinki.

Verry, E. S. 1997. Hydrological processes of natural, northern forested wetlands. Pages 168–188 *in* C. C. Trettin, M. F. Jurgensen, D. F. Grigal, M. R. Gale, and J. K. Jeglum, editors. Northern forested wetlands. Lewis Publishers, New York.

Verry, E. S. 2000. Water flow in soils and streams: sustaining hydrologic function. Pages 99–124 *in* W. S. Verry, J. W. Hornbeck, and C. A. Dollhoff, editors. Riparian management in forests of the continental eastern United States. Lewis Publishers, Boca Raton, Florida.

Verry, E. S., and C. A. Dolloff. 2000. The challenge of managing for healthy riparian areas. Pages 1–22 *in* E. S. Verry, J. W. Hornbeck, and C. A. Dolloff, editors. Riparian management in forests of the continental eastern United States. Lewis Publishers, Boca Raton, Florida.

Welcomme, R. L. 1985. River fisheries. Food and Agriculture Organization, FAO Fisheries Technical Paper Number 262, Rome.

Wendel, J. L. 1999. Habitat use and movements of channel catfish in the Red River of the North. Master's thesis. University of North Dakota, Grand Forks.

Werner, E. E., and D. J. Hall. 1979. Foraging efficiency and habitat switching in competing sunfishes. Ecology 60:256–264.

Woolman, A. J. 1895. A report upon ichthyological investigations in western Minnesota and eastern North Dakota. United States Commission on Fish and Fisheries. Part XIX. Government Printing Office, Washington, D.C.

Appendix A.—Fish species reported in surveys of streams and lakes in the Red River of the North basin during 1962–2000. "E" indicates species extirpated from respective tributaries. Lake sturgeon were extirpated from the entire basin but are being reintroduced into the Assiniboine River, and U.S. tributaries as indicated by Table 4.

	River Drainage																														
Taxon	Red 1	Pembina 2	Tongue 3	Park 4	Forest 5	Turtle 6	Goose 7	Elm 8	Rush 9	Maple 10	Sheyenne 11	Wild Rice, ND 12	Bois de Sioux 13	Mustinka 14	Rabbit 15	Otter Tail 16	Pelican 17	Buffalo 18	Wild Rice, MN 19	Sandhill 20	Red Lake 21	Clearwater 22	Snake 23	Middle 24	Tamarac 25	Two 26	Roseau 27	Assiniboine 28	Souris 29	Little Saskatchewan 30	Qu'appelle 31
Petromyzontidae																															
chestnut lamprey																															
Ichthyomyzon castaneus	X						X									X			X		X	X					X	X			X
silver lamprey																															
I. unicuspis	X																	X			X	X						X			
Acipenseridae																															
lake sturgeon																															
Acipenser fulvescens	E	E		E												E	E		E		E	E					E	E			
Amiidae																															
bowfin																															
Amia calva														X		X	X														
Hiodontidae																															
goldeye																															
Hiodon alosoides	X			X		X					X		X			X			X		X						X	X	X		X
mooneye																															
H. tergisus	X	X									X					X			X		X							X			X
Salmonidae																															
cisco																															
Coregonus artedi[1]																X	X	X	X		X	X						X		X	X
lake whitefish																															
C. clupeaformis[1]	X															X					X							X		X	X
rainbow trout																															
Oncorhynchus mykiss[2]		X	X			X					X							X			X	X					X	X	X	X	
brown trout																															
Salmo trutta[2]		X	X													X	X		X		X	X					X		X		

Appendix A.—Continued.

Taxon	River Drainage																													
	Red 1	Pembina 2	Tongue 3	Park 4	Forest 5	Turtle 6	Goose 7	Elm 8	Rush 9	Maple 10	Sheyenne 11	Wild Rice, ND 12	Bois de Sioux 13	Mustinka 14	Rabbit 15	Otter Tail 16	Pelican 17	Buffalo 18	Wild Rice, MN 19	Sandhill 20	Red Lake 21	Clearwater 22	Snake 23	Middle 24	Tamarac 25	Two 26	Roseau 27	Assiniboine 28	Souris 29	Little Saskatchewan 30	Qu'appelle 31
brook trout																															
Salvelinus fontinalis[2]		X														X	X	X	X		X	X					X		X		
lake trout																															
S. namaycush[2]																														X	
Catostomidae																															
quillback																															
Carpiodes cyprinus	X	X		X	X		X		X	X	X		X		X	X		X	X		X	X				X		X			X
white sucker																															
Catostomus commersonii	X	X	X	X	X	X	X	X	X	X	X	X	X	X	X	X	X	X	X	X	X	X	X	X	X	X	X	X	X	X	X
northern hog sucker																															
Hypentelium nigricans																X															
smallmouth buffalo																															
Ictiobus bubalus																X															
bigmouth buffalo																															
I. cyprinellus	X	X			X					X	X	X	X		X	X	X				X					X		X			X
silver redhorse																															
Moxostoma anisurum	X						X				X	X	X		X	X	X	X	X		X	X				X	X	X			
golden redhorse																															
M. erythrurum	X	X		X			X				X					X	X	X	X	X	X	X		X		X	X	X			
shorthead redhorse																															
M. macrolepidotum	X	X					X			X	X	X	X			X	X	X	X	X	X	X		X		X	X	X	X	X	X
greater redhorse																															
M. valenciennesi	X									X	X					X															
Cyprinidae																															
central stoneroller																															
Campostoma anomalum																X															
largescale stoneroller																															
C. oligolepis					X																										

Appendix A.—Continued.

Taxon	River Drainage																														
	Red 1	Pembina 2	Tongue 3	Park 4	Forest 5	Turtle 6	Goose 7	Elm 8	Rush 9	Maple 10	Sheyenne 11	Wild Rice, ND 12	Bois de Sioux 13	Mustinka 14	Rabbit 15	Otter Tail 16	Pelican 17	Buffalo 18	Wild Rice, MN 19	Sandhill 20	Red Lake 21	Clearwater 22	Snake 23	Middle 24	Tamarac 25	Two 26	Roseau 27	Assiniboine 28	Souris 29	Little Saskatchewan 30	Qu'appelle 31
goldfish																															
Carassius auratus[2]	X	X																										X			
spotfin shiner																															
Cyprinella spiloptera	X			X	X	X	X	X	X	X	X	X	X			X		X	X	X					X		X	X			
common carp																															
Cyprinus carpio[2]	X	X		X	X	X	X	X	X	X	X	X	X	X	X	X	X	X	X		X	X	X	X	X	X	X	X		X	X
brassy minnow																															
Hybognathus hankinsoni		X								X	X					E			X		X		X	X	X		X	X			
common shiner																															
Luxilus cornutus	X	X	X	X	X	X	X	X	X	X	X		X			X	X	X	X	X	X	X	X	X	X	X	X	X	X	X	X
silver chub																															
Macrhybopsis storeriana	X					X					X										X					X	X	X			
pearl dace																															
Margariscus margarita		X	X	X			X									E		X	E		X	X	X	X	X		X	X			
hornyhead chub																															
Nocomis biguttatus	X				X					E	E		E			X	X	X	X		X	X									
golden shiner																															
Notemigonus chrysoleucas	X										X					X	X	X	X		X	X						X			
pugnose shiner																															
Notropis anogenus											E					X	X														
emerald shiner																															
N. atherinoides	X	X							X	X	X		X	X	X	X		X	X		X						X	X			X
river shiner																															
N. blennius	X	X		X	X			X	X	X	X	X								E	E						X	X	X		X
bigmouth shiner																															
N. dorsalis	X	X		X	X	X	X		X	X	X		X			X	X	X	X	X	X	X			X		X	X	X	X	

Appendix A.—Continued.

	River Drainage																													
	Red	Pembina	Tongue	Park	Forest	Turtle	Goose	Elm	Rush	Maple	Sheyenne	Wild Rice, ND	Bois de Sioux	Mustinka	Rabbit	Otter Tail	Pelican	Buffalo	Wild Rice, MN	Sandhill	Red Lake	Clearwater	Snake	Middle	Tamarac	Two	Roseau	Assiniboine	Souris	Little Saskatchewan	Qu'appelle
Taxon	1	2	3	4	5	6	7	8	9	10	11	12	13	14	15	16	17	18	19	20	21	22	23	24	25	26	27	28	29	30	31
blackchin shiner																															
N. heterodon											X					X	X	X	X		X	X					X	X	X		
blacknose shiner																															
N. heterolepis											X					X	X	X	X		X	X					X	X	X		X
spottail shiner																															
N. hudsonius	X										X					X	X	X	X		X	X						X	X	X	X
carmine shiner																															
N. percobromus											X							X	X	X	X	X									
sand shiner																															
N. stramineus	X	X	X	X	X	X	X	X		X	X		X		X	X		X	X	X	X	X				X	X	X	X		X
weed shiner																															
N. texanus																X	X				X										
mimic shiner																															
N. volucellus																X	X	X	X		X	X		X							
northern redbelly dace																															
Phoxinus eos			X	X			X		X		X					X		X	X	X	X	X	X	X	X	X	X	X	X		
finescale dace																															
P. neogaeus			X													X					X		X	X	X		X	X			
bluntnose minnow																															
Pimephales notatus	X				X				X	X	X					X	X		X		X	X									
fathead minnow																															
P. promelas	X	X	X	X	X	X	X	X	X	X	X	X	X	X	X	X	X	X	X	X	X	X	X	X	X	X	X	X	X	X	X
flathead chub																															
Platygobio gracilis	X																											X	X		
western blacknose dace																															
Rhinichthys obtusus		X	X	X	X	X	X			X	X					X	X	X	X	X	X	X	X	X	X	X	X	X	X	X	X

Appendix A.—Continued.

Taxon	River Drainage Red 1	Pembina 2	Tongue 3	Park 4	Forest 5	Turtle 6	Goose 7	Elm 8	Rush 9	Maple 10	Sheyenne 11	Wild Rice, ND 12	Bois de Sioux 13	Mustinka 14	Rabbit 15	Otter Tail 16	Pelican 17	Buffalo 18	Wild Rice, MN 19	Sandhill 20	Red Lake 21	Clearwater 22	Snake 23	Middle 24	Tamarac 25	Two 26	Roseau 27	Assiniboine 28	Souris 29	Little Saskatchewan 30	Qu'appelle 31
longnose dace																															
R. cataractae	X	X	X	X	X	X					X					X	X	X	X	X	X	X					X	X	X	X	X
creek chub																															
Semotilus atromaculatus	X	X	X	X	X	X	X	X	X	X	X					X	X	X	X	X	X	X	X	X	X	X	X	X	X	X	X
Ictaluridae																															
black bullhead																															
Ameiurus melas	X	X	X	X	X	X	X	X	X	X	X	X	X	X	X	X	X	X	X	X	X	X	X	X	X	X	X	X	X		X
yellow bullhead																															
A. natalis	X												X	X		X	X	X	X		X	X					X				
brown bullhead																															
A. nebulosus	X	X		X	X						X		X	X	X	X	X	X	X	X	X	X	X		X	X	X	X	X		
channel catfish																															
Ictalurus punctatus	X	X		X	X	X	X		X	X	X	X	X			X		X	X	X	X	X					X	X		X	X
stonecat																															
Noturus flavus	X	X					X	X			X		X				X	X	X	X	X	X		X	X		X	X	X	X	
tadpole madtom																															
N. gyrinus	X	X			X	X	X			X	X	X			X	X	X	X	X		X	X				X	X	X	X		
Umbridae																															
central mudminnow																															
Umbra limi	X	X	X													X	X	X	X	X	X	X	X	X	X	X	X	X	X		
Esocidae																															
northern pike																															
Esox lucius	X	X	X	X	X	X	X	X	X	X	X	X	X	X	X	X	X	X	X	X	X	X	X	X	X	X	X	X	X	X	X
muskellunge																															
E. masquinongy[2]											X					X	X											X			

Appendix A.—Continued.

Taxon	River Drainage																													
	Red 1	Pembina 2	Tongue 3	Park 4	Forest 5	Turtle 6	Goose 7	Elm 8	Rush 9	Maple 10	Sheyenne 11	Wild Rice, ND 12	Bois de Sioux 13	Mustinka 14	Rabbit 15	Otter Tail 16	Pelican 17	Buffalo 18	Wild Rice, MN 19	Sandhill 20	Red Lake 21	Clearwater 22	Snake 23	Middle 24	Tamarac 25	Two 26	Roseau 27	Assiniboine 28	Souris 29	Little Saskatchewan 30	Qu'appelle 31
Osmeridae																															
rainbow smelt																															
Osmerus mordax	X																														
Cyprinodontidae																															
banded killifish																															
Fundulus diaphanus	X			X		X					E					X	X	X	X			X									
Gadidae																															
burbot																															
Lota lota	X															X		X	X	X	X	X					X	X		X	X
Percopsidae																															
trout-perch																															
Percopsis omiscomaycus	X	X	X	X	X		X	X	X	X	X	X				X		X	X	X	X						X	X	X	X	X
Moronidae																															
white bass																															
Morone chrysops[2]	X										X		X		X	X												X			
Centrarchidae																															
rock bass																															
Ambloplites rupestris	X	X			X	X					X	X			X	X	X	X	X	X	X	X			X	X	X	X	X	X	X
green sunfish																															
Lepomis cyanellus	X									X	X				X	X	X	X	X			X									
pumpkinseed																															
L. gibbosus											X	X	X	X		X	X	X	X		X	X			X	X	X			X	
orangespotted sunfish																															
L. humilis	X										X		X			X															

Appendix A.—Continued.

	River Drainage																														
	Red	Pembina	Tongue	Park	Forest	Turtle	Goose	Elm	Rush	Maple	Sheyenne	Wild Rice, ND	Bois de Sioux	Mustinka	Rabbit	Otter Tail	Pelican	Buffalo	Wild Rice, MN	Sandhill	Red Lake	Clearwater	Snake	Middle	Tamarac	Two	Roseau	Assiniboine	Souris	Little Saskatchewan	Qu'appelle	
Taxon	1	2	3	4	5	6	7	8	9	10	11	12	13	14	15	16	17	18	19	20	21	22	23	24	25	26	27	28	29	30	31
bluegill																															
L. macrochirus	X						X				X		X			X	X	X	X		X	X				X	X				
smallmouth bass																															
Micropterus dolomieu	X	X									X					X	X	X	X		X	X								X	
largemouth bass																															
M. salmoides		X		X							X					X	X	X	X		X	X				X	X			X	
white crappie																															
Pomoxis annularis[2]	X			X						X	X		X	X	X	X															
black crappie																															
P. nigromaculatus	X	X		X	X		X			X	X	X	X	X	X	X	X	X	X		X	X	X			X	X	X			
Percidae																															
rainbow darter																															
Etheostoma caeruleum																X															
Iowa darter																															
E. exile		X				X	X	X	X	X	X	X	X	X	X	X	X	X	X	X	X	X					X	X	X		X
least darter																															
E. microperca																X	X														
johnny darter																															
E. nigrum	X	X	X	X	X	X	X			X	X	X	X	X		X	X	X	X	X	X	X	X	X	X	X	X	X	X	X	X
logperch																															
Percina caprodes	X	X					X									X	X				X	X									
blackside darter																															
P. maculata	X	X	X	X	X	X	X		X	X	X	X	X		X	X		X	X	X	X	X	X	X	X	X	X	X	X	X	X
river darter																															
P. shumardi	E										E							X			X			X			X	X			
yellow perch																															
Perca flavescens	X	X	X	X	X	X	X				X	X	X	X		X	X	X	X		X	X				X	X	X	X	X	X

Appendix A.—Continued.

	River Drainage																													
Taxon	Red 1	Pembina 2	Tongue 3	Park 4	Forest 5	Turtle 6	Goose 7	Elm 8	Rush 9	Maple 10	Sheyenne 11	Wild Rice, ND 12	Bois de Sioux 13	Mustinka 14	Rabbit 15	Otter Tail 16	Pelican 17	Buffalo 18	Wild Rice, MN 19	Sandhill 20	Red Lake 21	Clearwater 22	Snake 23	Middle 24	Tamarac 25	Two 26	Roseau 27	Assiniboine 28	Souris 29	Little Saskatchewan 30	Qu'appelle 31
sauger																															
Sander canadensis	X	X		X	X	X	X		X	X	X								X	X	X					X	X	X			X
walleye																															
S. vitreus	X	X		X	X	X	X		X		X	X	X	X		X	X	X	X	X	X	X				X	X	X	X	X	X
Sciaenidae																															
freshwater drum																															
Aplodinotus grunniens	X				X	X	X		X		X	X	X		X	X	X	X	X	X	X	X	X			X	X	X			
Cottidae																															
slimy sculpin																															
Cottus cognatus																												X			
mottled sculpin																															
Cottus bairdii																X					X	X									
spoonhead sculpin																															
C. ricei																															X
Gasterosteidae																															
brook stickleback																															
Culaea inconstans		X	X	X	X	X	X	X	X	X	X	X	X	X	X	X	X	X	X	X	X	X	X	X	X	X	X	X		X	X
ninespine stickleback																															
Pungitius pungitius																												X			X
River drainage	1	2	3	4	5	6	7	8	9	10	11	12	13	14	15	16	17	18	19	20	21	22	23	24	25	26	27	28	29	30	31
Species richness	57	43	20	31	31	27	33	14	22	30	56	23	32	17	21	73	49	54	58	29	65	55	19	21	22	31	52	60	32	28	35

[1] Species found in headwater lakes.
[2] Alien species.

American Fisheries Society Symposium 45:323–343, 2005

A Comparison of the Pre- and Postimpoundment Fish Assemblage of the Upper Mississippi River (Pools 4–13) with an Emphasis on Centrarchids

JEFFREY A. JANVRIN*

Mississippi River Habitat Specialist, Wisconsin Department of Natural Resources, 3550 Mormon Coulee Road, La Crosse, Wisconsin 54601, USA

Abstract.—An investigation of historical fisheries information for pools 4–13 of the upper Mississippi River (UMR) was conducted to 1) determine the pre-1938 relative abundance and distribution of bluegill *Lepomis macrochirus* and largemouth bass *Micropterus salmoides,* 2) determine the composition and relative abundance of the preimpoundment fish assemblage, and 3) determine if a shift in frequency of occurrence and relative abundance has occurred due to impoundment.

Many of the preimpoundment information sources did not include a detailed description of the fish assemblage, but did yield qualitative statements regarding the preimpoundment abundance of bluegill, largemouth bass, and other species. This qualitative assessment indicated bluegill and largemouth bass were widely distributed and abundant prior to impoundment in pools 4–13 of the UMR.

Preimpoundment (1900–1938) quantitative seining data were obtained from the Wisconsin Department of Natural Resources (WDNR) fisheries database and annual reports of federal fish rescue operations. Postimpoundment (1993–1999) quantitative seining data were obtained from the Environmental Management Program's Long Term Resources Monitoring Program (LTRMP) database, maintained by the U.S. Geological Survey (USGS). Pre- and postimpoundment data were compared for similarity in the rank of 13 groupings of fish based on summaries of catch reported by the federal fish rescue. There was a significant correlation between the preimpoundment datasets but no correlation between the preimpoundment datasets and the USGS postimpoundment dataset. This indicates that the relative abundance has changed for the groups included in this analysis.

A comparison of fish species (n = 75) common to the WDNR and LTRMP database showed significant correlation for ranks of percent frequency of occurrence and relative abundance. In general, what was common and abundant prior to lock and dam construction is common and abundant today. However, the ranks of percent frequency of occurrence and relative abundance have changed for some species.

Introduction

A critical component of successful habitat restoration is obtaining historical information to document changes that have occurred in a project area. This information provides some insight into the physical changes that have occurred and may provide evidence of the physical forces that have altered the ecosystem. For example, beginning in the late 1930s, the upper Mississippi River (UMR) from Minneapolis, Minnesota, to Alton, Illinois, was transformed from a free-flowing large river into a series of impoundments for the purpose of navigation (Merritt

* E-mail: jeff.janvrin@dnr.state.wi.us

1979). The impoundments inundated vast areas of the floodplain and altered the hydrology of the UMR floodplain (USGS 1999; WEST Consultants, Inc. 2000a). Survey maps and aerial photos of UMR floodplain features are available and cover various periods from the late 1800s to the present. The maps and photos provide documentation of changes in land and water ratio, channel location, and so forth. However, documentation of UMR fish and wildlife assemblages prior to the 1930s is limited. Documentation of the animals previously inhabiting an area can be combined with current knowledge of their life requirements to infer the mix of habitats present in the UMR prior to human or naturally induced change. This historical biological information would also be useful for defining "reference" conditions of the UMR ecosystem (Hughes 1995).

The UMR has been the site of several projects designed to reverse human impacts on the environment. These projects have been primarily funded through the Environmental Management Program (EMP) that was authorized by Congress in 1986 (Water Resources Development Act 1986). One component of the EMP provides authority for the Corps of Engineers to construct Habitat Rehabilitation and Enhancement Projects (HREPs) to improve environmental conditions within the floodplains of the UMR and Illinois River. Several of the HREPs have included features for restoration or improvement of habitat for centrarchids (USACE 1997). However, questions have arisen concerning the appropriateness of constructing centrarchid habitat since some believe that centrarchid abundance and amount of available habitat are greater today than prior to locks and dams (i.e., reference condition). The justification for these concerns is primarily based on Fremling and Claflin (1984).

Fremling and Claflin (1984) are often cited when describing changes in fish assemblage due to construction of dams on the UMR. Fremling and Claflin (1984) stated

> The closure of the dams and the resultant inundation of the floodplain created vast new areas of high quality fish habitat. Furthermore, the stabilization of water levels generally improved existing habitat. A few species such as the paddlefish *Polyodon spathula*, which relied upon open river habitat, were adversely affected by the creation of lentic environments. These adverse effects, however, were offset by the beneficial effects observed for such species as largemouth bass *Micropterus salmoides*, crappies *Pomoxis* spp., and sunfishes *Lepomis* spp.

That statement implies there is evidence that lentic fishes, such as centrarchids, benefited by the construction of the locks and dams and that a shift in species dominance and composition occurred. However, Fremling and Claflin (1984) did not cite any references or data comparing pre- and postimpoundment effects on fish communities.

An early reference describing the effects of impoundment on UMR fishery resources summarizes a survey conducted by the Upper Mississippi River Conservation Committee (UMRCC 1946). The UMRCC conducted a survey of commercial fishermen fishing in pools 2–11 in the early 1940s (UMRCC 1946). One-hundred-one commercial fishermen responded to the survey and provided their opinions on the trends in abundance of various species. The results indicated the majority of respondents believed the population of rough fish (common carp *Cyprinus carpio,* buffalo *Ictiobus* spp., and freshwater drum *Aplodinotus grunniens*) had increased since impoundment (UMRCC 1946). The majority of respondents also believed the populations of bullhead *Ameiurus* spp. and "hackleback" (shovelnose) sturgeon *Scaphirhynchus platorynchus* had declined during the same period (UMRCC 1946). However, perceptions regarding any trends in the population of other species were mixed. For example, the perception of some respondents was that bluegill *Lepomis macrochirus* and largemouth bass abundance was declining in some areas and increasing in others (UMRCC 1946). The authors of the report stated

> The marked difference of opinion on some species can possibly be accounted for by varying local conditions and the difference in gear and fishing practices used by various operators. It is felt that when commer-

> cial fisherman are in general agreement on trends that some significance can be attached.

The report did not include quantitative assessments of abundance.

The commercial fishermen's perceptions of largemouth bass appear similar to comments made by Eddy and Surber (1943). Eddy and Surber (1943) may have been referring to a potential decline in largemouth bass on the UMR when they stated, "In Minnesota it is, or was, common in the muddy lakes of the southern and central counties and in sloughs along the Mississippi." They also did not include any quantification or citation for their statement.

The lack of quantifiable comparisons of the distribution and relative abundance of fish species prior to and after construction of the UMR locks and dams has implications for setting goals and objectives for ecosystem restoration. For example, are the HREPs restoring habitats that promote a more "natural," or preimpoundment, fishassemblage , or are they just maintaining the assemblage that developed postimpoundment? Therefore, a literature and data review was undertaken to 1) determine the pre-1938 relative abundance and distribution of bluegill and largemouth bass, 2) determine the composition and relative abundance of the preimpoundment fish assemblage, and 3) determine if a shift in frequency of occurrence and relative abundance has occurred due to impoundment.

Study Area

The focus for this study was aquatic areas of pools 4–13 of the UMR, which extends approximately 442 km (275 mi) from Red Wing, Minnesota, to Clinton, Iowa (Figure 1). Galstoff (1924) described a portion of the study area in pool 9 prior to construction of locks and dams. Galstoff (1924) states

> ...The river winds from one side of the floodplain to the other, numerous islands dividing its channel and forming many sloughs and bays, which often are transformed by the sand bars into pools of stagnant water. A characteristic feature of the Mississippi Flood Plain is the many shallow lakes or pools, which seldom exceed 1 1/2 mi in diameter. When the water is high, the river floods the whole area, covering these lakes and the spaces between, but in summer many of them become almost dry. There are many of these lakes. Martin (1916) counted over 200 of them in an area of about 20 square miles in the Wisconsin section between Lynxville and De Soto, only the lakes that had no connection with the river being counted, the sloughs and bays being excluded. It seems that the number of lakes in the other parts of the river is not less than in the section. Many of them have a rich aquatic vegetation, and as they slowly become filled with detritus they gradually become swamps. All stages of this process can easily be observed in many points of the Mississippi Flood Plain (Figure 2).

This preimpoundment description also applied to much of the river floodplain in the study area prior to locks and dams being placed into operation in 1939 (Galstoff 1924).

Pools 4–13 of the UMR (Figure 1) have changed the river's hydrology and planform. The changes within the floodplain were primarily related to developing and maintaining this section of the UMR for commercial navigation. Actions to improve the UMR for commercial navigation in pools 4–13 from 1824 to 1930 involved clearing of trees, snags, and rocks; construction of wing dams and closing dams; and dredging of sand from the main navigation channel of the river (Merritt 1979; Anfinson 2003). Many of these actions continue today to maintain a navigation channel on the UMR. The historical ecological impacts of these various actions are not well documented, but the physical impacts of the actions can be described. Removal of trees and snags from the main channel resulted in a reduction of woody structure within main channel and main channel border habitats. Wing dams and closing dams were, and still are, designed to increase water velocities in the main navigation channel to promote the transport of accumulated sediments and scour the riverbed to maintain navigation depths (WEST Consultants, Inc. 2000b). A secondary im-

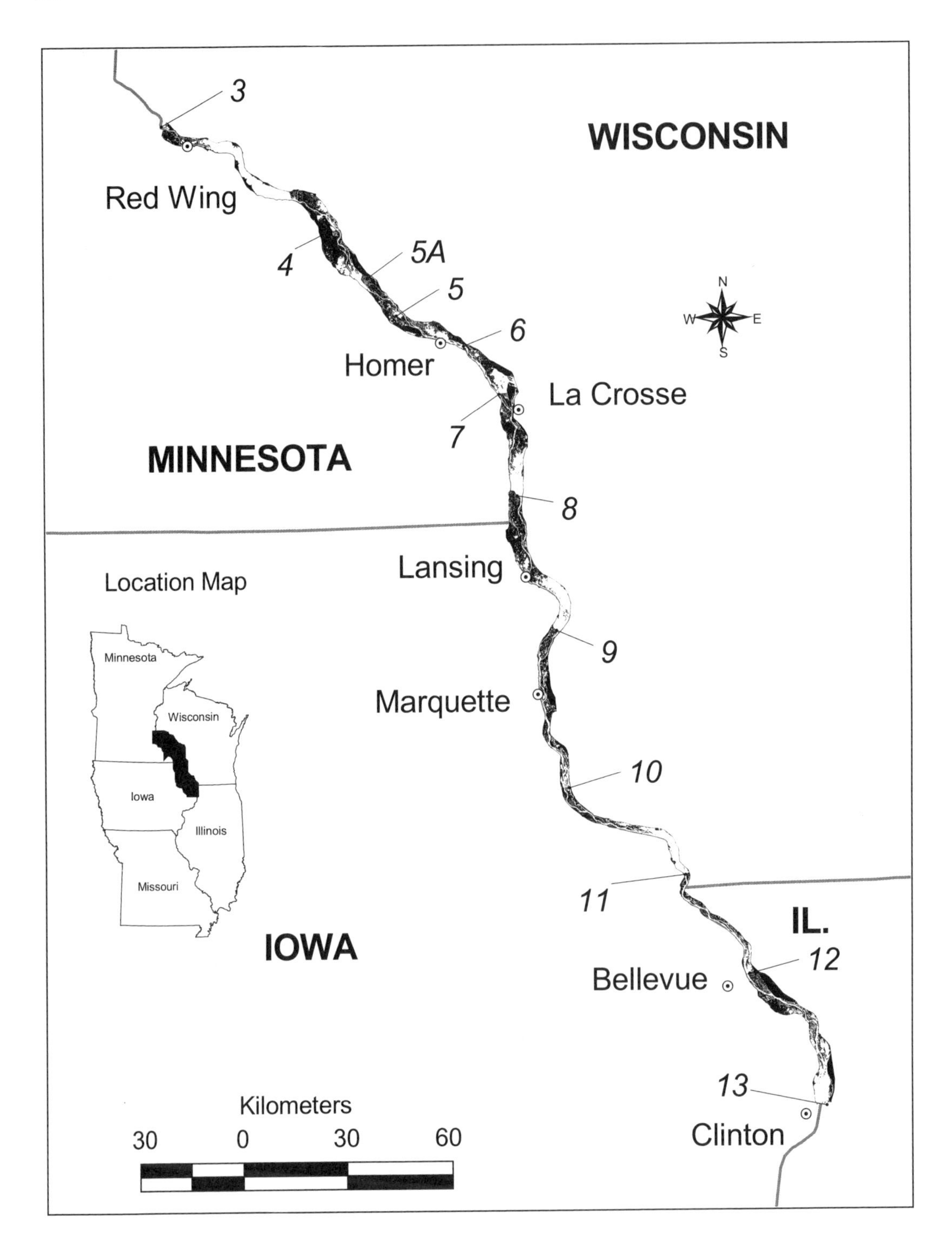

Figure 1.—Geographic extent of the study area. Numbers indicate the locations of locks and dams for this section of the upper Mississippi River (UMR). Pools are named after the lock and dam that creates the impoundment. For example, lock and dam 13 impounds a section of the UMR referred to as pool 13.

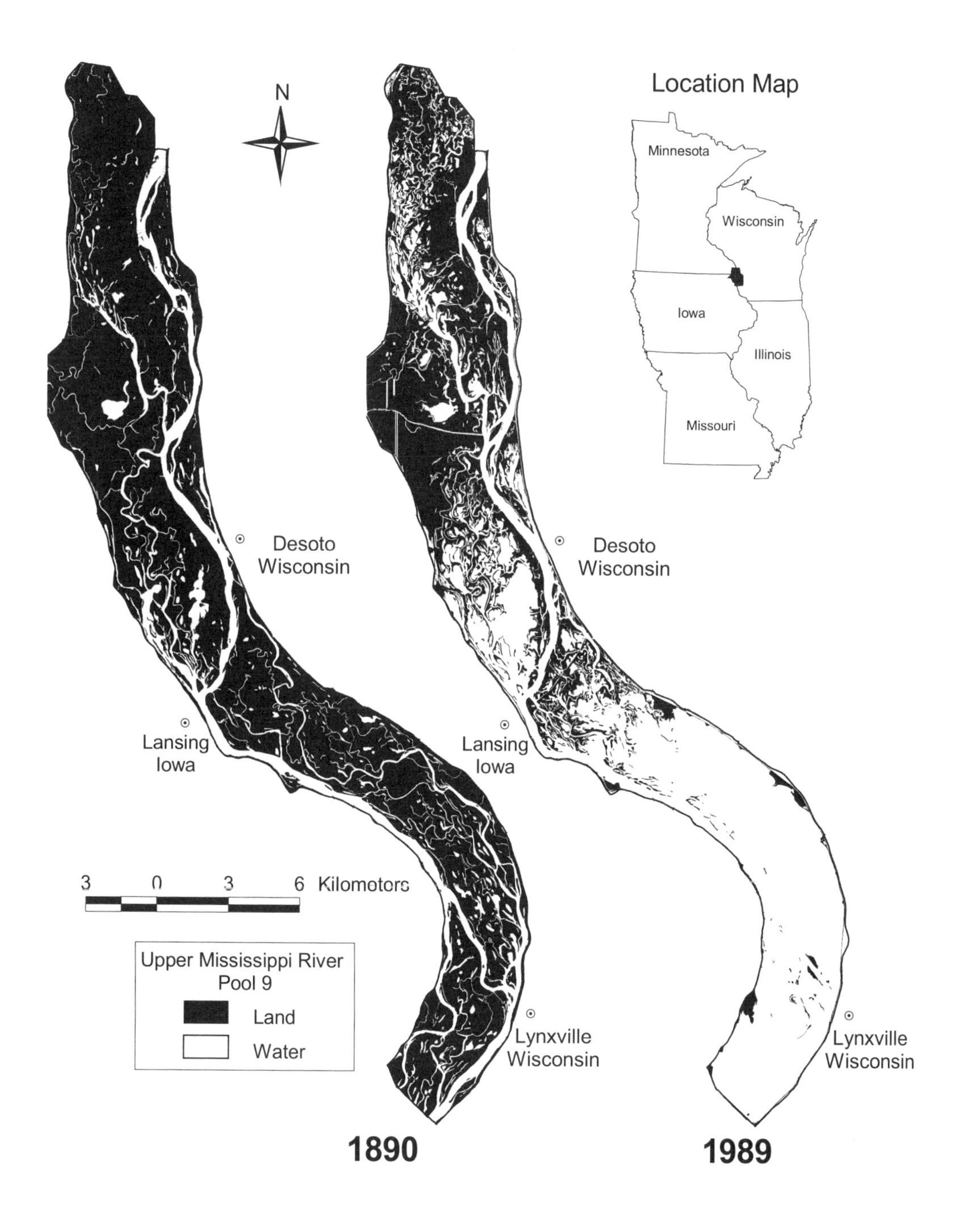

Figure 2. —Pool 9 of the upper Mississippi River (UMR) provides an example of how construction of locks and dams altered the land and water ratio of the impounded sections of the UMR. The locks and dams were completed in 1939, creating 27 impoundments on the UMR to provide a minimum commercial navigation channel depth of 2.74 m.

pact of wing dams is the reduction of velocities in the main channel border. Prior to impoundment, the reduced velocity in the main channel border increased the rate of sedimentation in these areas, transforming water into land at several locations within the river floodplain (Anfinson 2003). Closing dams keep water in the main channel by blocking, or "closing off," flow into side channels. This caused a reduction or elimination of flow into these areas and reduced connectivity of the side channels to the main channel (WEST Consultants, Inc. 2000b). The historical impacts of dredging on the aquatic ecosystem included removal of freshwater mussel *Unionidae* beds and riffle areas and conversion of aquatic areas to land due to disposal of the dredged material (WEST Consultants, Inc. 2000b).

In 1930, Congress authorized the construction of 26 low-head locks and dams on the UMR to provide a 2.74-m (9-ft) deep navigation channel for commercial navigation (Anfinson 2003). All of the locks and dams were completed by 1939, creating a series of pools from Minneapolis, Minnesota, to Alton, Illinois.

Impoundment of the UMR converted approximately 50% of the terrestrial areas in pools 4–13 to aquatic areas. Most of the conversion of land to aquatic areas occurred in the sections of the pools immediately upstream of each lock and dam (Figure 2). Impoundment has resulted in higher and more stable water levels during low discharge periods and increased sediment transport into UMR floodplain backwaters (USGS 1999). The increased sediment rates have contributed to a reduction in habitat quality for fish in many backwater areas of the UMR (USGS 1999). More detailed descriptions of changes due to navigation structures and the status of current conditions have been made in several recent reports (USGS 1999; Theiling et al. 2000; WEST Consultants, Inc. 2000a, 2000b; USACE 2000; Anfinson 2003).

Methods and Results

Qualitative Assessment

A search of published and unpublished preimpoundment fisheries reports was conducted, focused on bluegill and largemouth bass abundance in pools 4–13 of the UMR. The search was by no means exhaustive, yet yielded information to document preimpoundment (prior to 1939) attributes for these species. Many of the sources did not include a detailed description of the fish assemblage, but did yield qualitative statements regarding the composition of the preimpoundment fishery resources. The review of historical literature also involved identification of the species based on terminology common to that period when scientific names were not provided. For example, largemouth bass were often referred to as "black bass" (Cox 1897; U.S. Department of Commerce Bureau of Fisheries 1903–1938; Culler 1920; Surber 1920; Ellis 1931; Barnickol and Starrett 1951). The term "black bass" was sometimes used in its plural form to mean both largemouth and smallmouth bass *Micropterus dolomieu*, but was never found to reference only smallmouth bass except in the form "smallmouth black bass." The following description from the Wisconsin Fish Commission Report (State of Wisconsin 1907) on its fish rescue operations provides a common example of the usage of the term black bass as it refers to largemouth and smallmouth bass and an anecdotal account of species abundance:

> Millions of small black bass of the largemouth variety perish in these shallow ponds every year. Some other kinds are also found in these waters; namely, crappie fry, large dogfish [*Amia calva*], small bullheads, and a few small catfish [*Ictalurus* spp. and *Pylodictis* spp.]. Very few small-mouth bass are found in these ponds. Among the thousands of bass we have taken from these waters, I venture to say we have not caught one-hundred small-mouth bass.

Several authors presented the abundance and distribution of largemouth bass. The earliest reference to largemouth bass found during this study was for the UMR along Minnesota's border (pool 4–upper pool 9) (Figure 1). Tiffany (1876) described largemouth bass as less abundant than smallmouth bass and expressed curiosity over this when they were both present in Illinois and Wisconsin. Cox (1897) reported largemouth bass were common in all parts of Minnesota, much more common than the smallmouth bass. Unfortunately, he did not provide details regarding the distribution of largemouth bass. Cox (1897) stated,

> Specific comments regarding its [largemouth bass] distribution are unnecessary here, since it has been taken in nearly all the regions where the Nat. Hist. Surv. [Natural History Survey of Minnesota] and others have made collections.

Surber (1920) states the following regarding the abundance and distribution of largemouth bass:

> It is a fish of the lakes, sloughs and other sluggish waters, and is perhaps more common in the Mississippi bottom lands than elsewhere in the state (Minnesota)...

Largemouth bass were such an important fisheries resource of the UMR that they were used as one of the primary justifications for establishment of the Upper Mississippi Fish and Wildlife Refuge in 1924. Testimony before both the U.S. House of Representatives and Senate frequently mentioned the need for protection of black bass habitat within the UMR floodplain (U.S. Congress, House, Committee on Agriculture 1924; U.S. Congress, Senate, Committee on Commerce 1924).

Qualitative statements on the abundance and distribution of bluegill within the floodplain of the UMR were similar to those about largemouth bass. Several sources reported them as common and widespread in the UMR (Cox 1897; Surber 1920; Ellis 1931; Greene 1935; Carlander 1954).

Carlander (1954) presented a history of fish rescue on the UMR prior to the construction of the locks and dams. Fish rescue, or salvage of fish from overflowed lands, was routinely carried out by states bordering the UMR and by the U.S. Bureau of Fisheries. These fish were used to stock various water bodies throughout the Midwest (Carlander 1954). Carlander (1954) provided a brief summary of these operations and provided some insight into the relative dominance of some groups of fish prior to the construction of locks and dams.

> In the years for which it was possible to obtain records of the estimated numbers of fish caught [from fish rescue operations of the Bureau of Fisheries], it appears that there were only five species which in any year comprised more than five percent of the numbers of rescued fish. From 14 to 74 percent of the fish were "catfish," including bullheads, "sunfish" comprised 6 to 32 percent of the annual catch and "crappies" varied from 3 to 37 percent. Carp comprised from 0.6 to 39 percent of the catch and "buffalo" from 0.6 to 16 percent.

This statement also provides evidence that centrarchids were numerous on the UMR prior to impoundment.

The qualitative assessment indicated that largemouth bass and bluegill were widely distributed and abundant prior to impoundment of the UMR. This would be expected, for Cavender (1986) and Cross et al. (1986) provided documentation, based on a review of fossil records, that these two species evolved in a large floodplain river system. Cavender (1986) and Cross et al. (1986) noted that the fossilized remains formed in limestone deposited in small lakes of expansive fluvial plains or "lowland habitat." The preimpoundment UMR was described by Galstoff (1924) as having numerous lakes and sloughs, which is similar to the descriptions provided by Cavender (1986) and Cross et al. (1986).

Quantitative Assessment

The preimpoundment quantitative assessment was based on a search for historical data that could be spatially defined, covered a large portion of the UMR, and included information on species abundance. Two preimpoundment databases meeting these criteria were found (Table 1). The first was associated with the U.S. Bureau of Fisheries fish rescue operations in pools 4–13 of the UMR . Fish rescue by the U.S. Bureau of Fisheries was initiated at Quincy, Illinois, in 1889 (Carlander 1954). Rescue operations were carried out by the U.S. Bureau of Fisheries in pools 4–13 of the UMR beginning in 1903 (U.S. Department of Commerce, Bureau of

Table 1.—Comparison of quantitative databases for assessment of pre- and postimpoundment fisheries assemblage on the upper Mississippi River, pools 4–13.

Data source	Years	"Pools" sampled	Number of sampling events	Number of species caught	Total number of fish caught
U.S. Bureau of Fisheries Fish Rescue	1903–1938	4, 5, 5A, 6, 7, 8, 9, 10, 11, 12, 13	Unknown	Thirteen different species groups	1,380,771,496
Wisconsin Department of Natural Resources Statewide Database	1900–1930	4, 5, 5A, 6, 7, 8, 9, 10, 11, 12	81	92	11,462
Environmental Management Program Long Term Resource Monitoring Program	1993–1999	4, 8, 13	945	84	315,025

Fisheries 1903–1938). The main federal rescue stations for pools 4–13 were at Homer, Minnesota, La Crosse and Lynxville, Wisconsin, and North McGregor (Marquette) and Bellevue, Iowa (Figure 1). The federal fish rescue operations were abandoned in 1938 when many of the locks and dams were completed (U.S. Department of Commerce, Bureau of Fisheries 1903–1938). The rescue operations ceased because the impoundments created by the locks and dams permanently flooded many areas that were previously seined by federal and state agencies (Anfinson 1993). The Corps of Engineers "mitigated" for the loss of these fishery resources by constructing several hatcheries along the river (Anfinson 1993).

Culler (1920) summarized the rescue operations. Seine hauls were made in the floodplain of the UMR and included ponds, lakes, and sloughs that were isolated from flowing water or "about to become isolated" (Culler 1920). Culler (1920) emphasized that the areas seined did not include permanently flowing channels since the focus of the seining efforts was to "rescue" fish before the areas dried up or went anoxic. The seines typically used by fish rescue crews were 15.2 and 21.3 m long, 1.8 m deep and made of 6.35-mm mesh (Culler 1920). After a haul was made, the fish were sorted by size and species and placed in 43.8 L (1 and one-fifth bushel) galvanized tubs. The number of fish per tub was calculated based on displacement of water in the tubs. The tubs were "recalibrated" throughout a given year to account for growth of the fish over several months. Some of the fish were placed into milk cans and transported to a holding pond or facility to await loading onto specially built railroad cars or trucks and transport to a variety of stocking locations. The vast majority of the fish were released back to a flowing water section of the river. Gar *Lepisosteus* spp. and bowfin were discarded on shore to reduce their impact on populations of other fish fry (Culler 1920).

Annual reports published by the U.S. Bureau of Fisheries contained summaries of fish caught based on groupings of species (U.S. Department of Commerce, Bureau of Fisheries 1903–1938) (Table 2). The species included in the groupings were confirmed from tables provided in field reports submitted by the U.S. Bureau of Fisheries, La Crosse, Wisconsin, rescue station archived at the U.S. Fish and Wildlife Service, D.C. Booth National Historic Fish Hatchery, Spearfish, South Dakota.

The second set of preimpoundment data were obtained from a database maintained by the Wisconsin Department of Natural Resources (WDNR) (Table 1; Appendix A). Eighty-nine sampling events

Table 2.—Species groups reported by U.S. Bureau of Fisheries fish rescue and common names of species included in each group. These groupings represent those that were commonly reported and do not include special collections that were reported only once during the fish rescue operations. (see Table 3 for scientific names of species).

Fish rescue species groupings	Common name(s) of species included in group
buffalo fish	bigmouth buffalo, smallmouth buffalo, black buffalo
carp	common carp
catfish	flathead catfish, channel catfish, blue catfish, brown bullhead, black bullhead, yellow bullhead, paddlefish
crappie	white crappie, black crappie
freshwater drum	freshwater drum
large mouth black bass	largemouth bass
pike and pickerel	northern pike, walleye, sauger
rock bass	rock bass
smallmouth black bass	smallmouth bass
sunfish	bluegill, green sunfish, pumpkinseed
warmouth bass	warmouth
white bass	white bass
yellow perch	yellow perch

conducted within the floodplain of the UMR, pools 4–12, prior to 1930, were identified. All of the samples included a township, range, and quarter section description to indicate where the sample was taken. Many of the records also included a name of the waterbody sampled. The location names and geographic descriptions in the database indicate that the samples collected represented a variety of areas within the floodplain, including flowing channels and backwater areas. Only collections using seines (n = 81) were used for analysis. These seine samples collected 92 species of fish in the UMR floodplain (Appendix A). The WDNR database did not include specifics on the dimensions of the seines, listing them as either "small-mesh seine" or "combination of many types of seines." The sources of the sampling records were equally obscure. The sources were listed as "Early Wisconsin fish collections (1900–1931) (reported by Greene 1935); Greenbank et al. (1940s) (from the UW-Madison Zoology Museum); UW-Madison's Catalog of Wis. Conserv. Department collections; Dr. Underhill and Univ. Minnesota students; and U.S. Fish and Wildlife Service."

The EMP Long Term Resource Monitoring Program (LTRMP) fisheries database (maintained by the Upper Midwest Environmental Sciences Center of the U.S. Geological Survey) was queried for seining samples in pools 4, 8, and 13 collected from 1993 to 1999 (Table 1; Appendix A). The LTRMP uses stratified random sampling to identify sampling locations, which are weighted by strata (Wilcox 1993; Gutreuter et al. 1995). The floodplain areas sampled by LTRMP represent a variety of aquatic area types present in the UMRS floodplain. The LTRMP standardized sampling protocol used 10.7 m long by 1.8 m high 3-mm-mesh bag seines pulled around a 90-degree arc to the shoreline at stratified random sites (Gutreuter et al. 1995). The LTRMP seining data were used to compare ranks of species abundance and frequency of occurrence of the "current" postimpoundment fish assemblage and any historical database since most of the identified historical records used seining as a collection technique. Seining records from the LTRMP database in pools 4–13, included 84 species of fish caught over 7 years (Appendix A).

The LTRMP and WDNR databases were summarized to match the species groupings of the fish rescue (Table 2) to allow for comparison of the three datasets. Spearman's rho (Conover 1980) was used to test for correlation based on ranks among databases.

The relative abundance of species groups for the preimpoundment databases were significantly correlated when species were grouped into categories used for reporting of the fish rescue operations (Spearman's rho = 0.70, $p < 0.01$) (Tables 2 and 3). The relative abundance of species groups from the

Table 3.—Comparison of number of fish caught for different quantitative databases used for analysis based on groupings reported by the U.S. Bureau of Fisheries' fish rescue operations (U.S. Department of Commerce, Bureau of Fisheries 1903–1938). The relative abundance of fish species groups, based on ranks, was similar for the fish rescue and Wisconsin Department of Natural Resources (DNR) data (Spearman's rho = 0.70, $p < 0.01$). The relative abundance of fish species groups, based on ranks, for the Environmental Management Program Long Term Resources Program (EMP LTRMP) was different from the Fish Rescue and Wisconsin DNR data (Spearman's rho = 0.31, $p > 0.10$, and Spearman's Rho = 0.32, $p > 0.10$, respectively).

U.S. Bureau of Fisheries' fish rescue reported groupings	Fish rescue (1903–1938)		Wisconsin DNR (1900–1930)		EMP LTRMP (1993–1999)	
	Number caught	Rank	Number caught	Rank	Number caught	Rank
warmouth bass	8,385	1	2	1	10	1
rock bass	16,647	2	17	5	250	3
smallmouth black bass	53,187	3	15	4	330	4
white bass	1,144,200	4	49	7	2,565	10
freshwater drum	1,456,933	5	14	3	6,075	12
pike and pickerel	9,822,526	6	138	9	389	5
largemouth black bass	10,017,359	7	124	8	4,524	11
yellow perch	21,829,420	8	227	10	634	7
buffalo fish	74,377,826	9	12	2	496	6
carp	211,826,242	10	20	6	971	8
sunfish	255,327,392	11	370	12	16,051	13
crappie	280,263,998	12	498	13	1,350	9
catfish	439,336,674	13	236	11	242	2

LTRMP database was not correlated with species groups from either the fish rescue database or WDNR database (Spearman's rho = 0.31, $p > 0.10$, and Spearman's rho = 0.32, $p > 0.10$, respectively) (Table 3). This indicates that the relative abundance of some species groups has changed. For example, "catfish" ranked 13th and 11th in the fish rescue and WDNR databases, respectively, but was 2nd in the LTRMP database (a higher rank = greater relative abundance) (Table 3).

A comparison was also made between the ranks of percent frequency of occurrence and relative abundance for species caught in both the WDNR and LTRMP databases ($n = 75$) (Appendix A). There was significant correlation between the databases for ranks of frequency of occurrence and ranks of relative abundance (Spearman's rho = 0.71, $p < 0.01$, and Spearman's rho = 0.63, $p < 0.01$, respectively) (Figure 3). This indicates that, in general, what was common and abundant prior to the locks and dams is also common and abundant today. However, some species (i.e., bullhead, skipjack herring, northern pike, and black crappie) showed a potential decline in ranks of abundance and frequency of occurrence while others (i.e., smallmouth bass, river carpsucker, sauger, and freshwater drum showed an increase) (Appendix A). Some species showed no major change (i.e., longnose gar, golden redhorse, and emerald shiner) based on ranks (Appendix A).

The comparison of ranks between the WDNR and LTRMP databases indicates essentially no change in the frequency of occurrence (rank of 74 versus 71, respectively) or the relative abundance (rank of 70 versus 71, respectively) of bluegill. A comparison for largemouth bass ranks between the WDNR and LTRMP databases also indicates essentially no change in frequency of occurrence (rank of 71.5 versus 69, respectively) but indicates a possible increase in relative abundance (rank of 54 versus 67, respectively).

Discussion

The assessments conducted for the study document that bluegill and largemouth bass were common and abundant prior to construction of the locks and dams and that their ranks of abundance and frequency have not changed. The "beneficial effects"

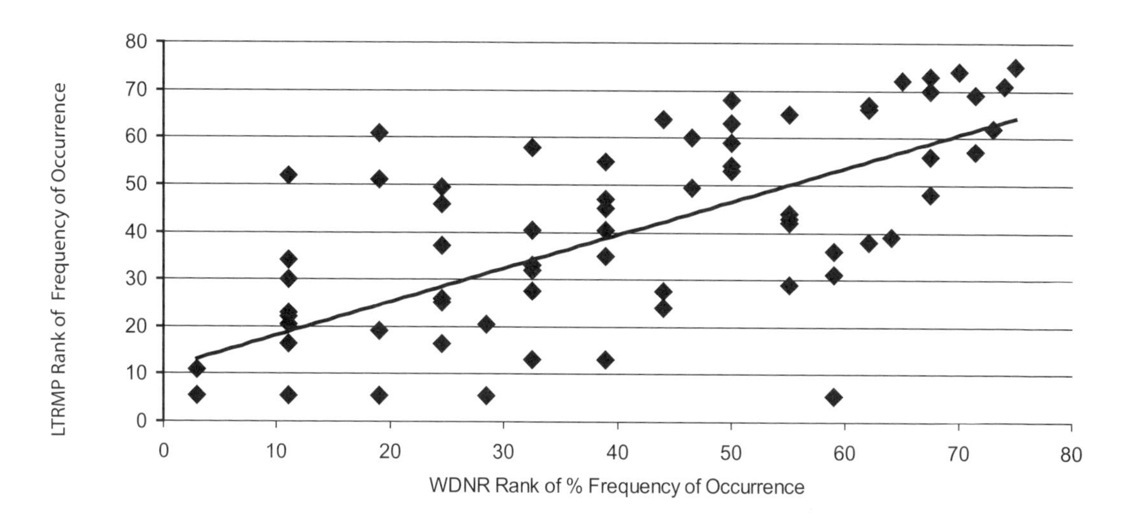

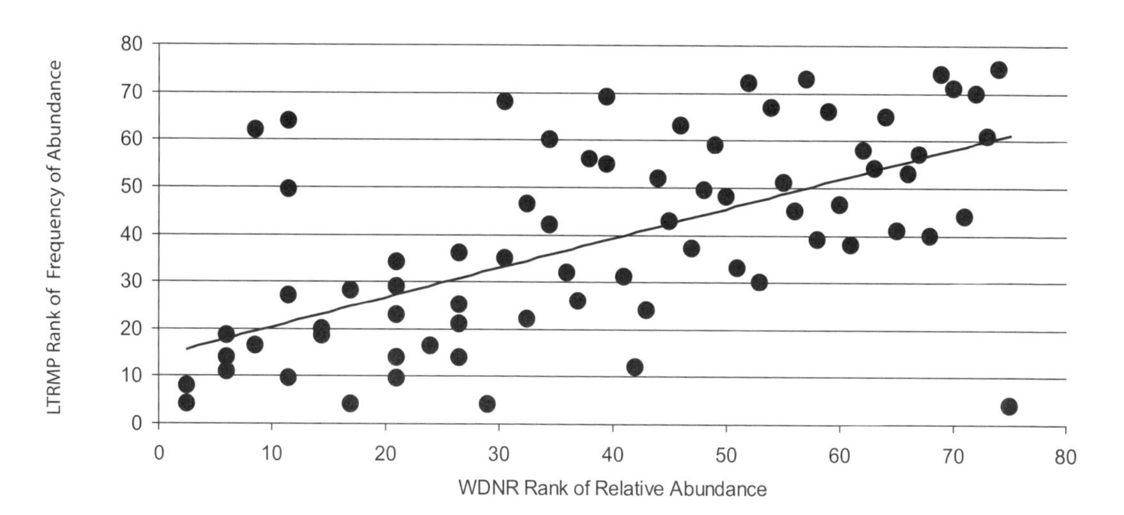

Figure 3. —Comparison of the pre-Wisconsin Department of Natural Resources (WDNR) and post-Long Term Resources Program (LTRMP) impoundment ranks of frequency of occurrence and ranks of relative abundance of species in the fish community.

of the locks and dams for these species, presented by Fremling and Claflin (1984), were most likely short lived increases in overwintering habitat.

Pitlo (1986) and Sheehan et al. (1990) reported overwintering habitat as being critical for centrarchids on the UMRS. Knights et al. (1995) described overwintering habitat as areas having dissolved oxygen levels greater than 3 ppm, water temperatures greater or equal to 1° C, and water velocities less than or equal to 1 cm/s. Water depth is also an important factor, since the northern portion of the UMRS may have greater than 0.6 m of ice in the winter. Therefore, habitat projects designed for meeting overwintering needs of centrarchids also include the need for greater than or equal to 1.2 m of water.

Loss of islands, increased flows to backwaters due to island dissection, and loss of depth from sedimentation have all contributed to a decrease in the quality and number of available overwintering sites for centrarchids (Gutreuter and Theiling 1999). Historical photos show how postimpoundment island erosion was resulting in a loss of overwintering habitat for centrarchids by subjecting more of the floodplain aquatic areas to flow. The loss of islands not only affects aquatic species whose habitat includes low velocity conditions, but island loss also affects the presence and abundance of aquatic vegetation (Fischer and Claflin 1995).

A comparison of pre- and postimpoundment maps and air photos indicates there may be fewer centrarchid overwintering sites today than existed preimpoundment. These images also indicate preimpoundment overwintering habitat may have been relatively evenly distributed throughout lower pool 4–pool 13 at approximately 0.25–2.5-km intervals (Figures 2 and 4). Today, approximately one-third of many pools are wide expanses of open water with few or no islands, resulting in 5–21-km sections of the UMR without any available overwintering habitat due to excessive water velocities (Figures 2 and 4). An investigation of the preimpoundment distribution of overwintering habitat has not been done to date, but such a study would aid the establishment of ecosystem restoration goals and objectives on the UMRS.

The quantitative assessments conducted did not provide any conclusive evidence regarding changes in the fish assemblage as a whole. However, these results should be interpreted with caution considering the limitations of the datasets. Limitations of the WDNR dataset are the unknown configuration of the seines (length and mesh size), lack of details on seining technique, and precise locations of the sample sites. A limitation of the LTRMP database is that only a small proportion of the seining occurred in the sections of the pools most affected by impoundment. Seining in the lower portions of many pools is limited by water depth and/or lack of shoreline. This means that the majority of LTRMP seining occurs in portions of the pool where centrarchid habitat is present and water depths are not as limiting to sampling crews. Limitations of the fish rescue data include lack of information to identify specific seining locations, unknown level of effort, lack of detailed records. Another potential limitation that arose during discussions with river managers and biologists regarding the usefulness of the fish rescue data was the common belief that only isolated backwaters were seined. The WDNR and fish rescue data were significantly correlated, which implies that the fish rescue data may be a representative sample of the habitats in the floodplain, including areas with flow (i.e., channels). This study indicates that additional historical research into the fish rescue may be warranted to see if this database may be suitable for other evaluation purposes.

With the above limitations in mind, the analyses suggest there may have been some changes in the pre- and postimpoundment fish assemblage when analyzed based on species groupings from the fish rescue, and for some individual species. Yellow bullhead and black bullhead appear to have declined in both frequency of occurrence and relative abundance since construction of the locks and dams. The reasons for this decline should be investigated to fully understand the cause and effect relationship. One factor that may be contributing to the reduced abundance of bullhead is a loss in habitat. Becker (1983) describes black bullhead habitat as "quiet backwaters, oxbows, impoundments, ponds, lakes, and low-gradient streams" and stated that yellow bullheads "occurred in quiet, weedy sectors of lakes and reservoirs." Impoundment of the UMR greatly reduced the number of isolated water bodies (Figures 2 and 4). Additional loss of habitat for bullhead may be associated with erosion of remaining islands in several of the backwaters. This process has been called island dissection and has been identified to be occurring in hundreds of locations in pools 4–12 (Theiling et al. 2000). Island dissection allows river flows to enter a greater percentage of aquatic areas in the floodplain, causing further loss in low velocity habitats and aquatic areas with no flow. Bullhead are not often included as indicators of a healthy ecosystem, and indeed are frequently used as indicators of degraded systems. However, bullhead appear to have been more numerous and widespread prior to impoundment and may be two species that should be considered when assessing the ecological health of the UMR, especially related to aquatic areas with low to no flow.

The shifts in rank documented in the quantitative assessment for this study show some similarity to

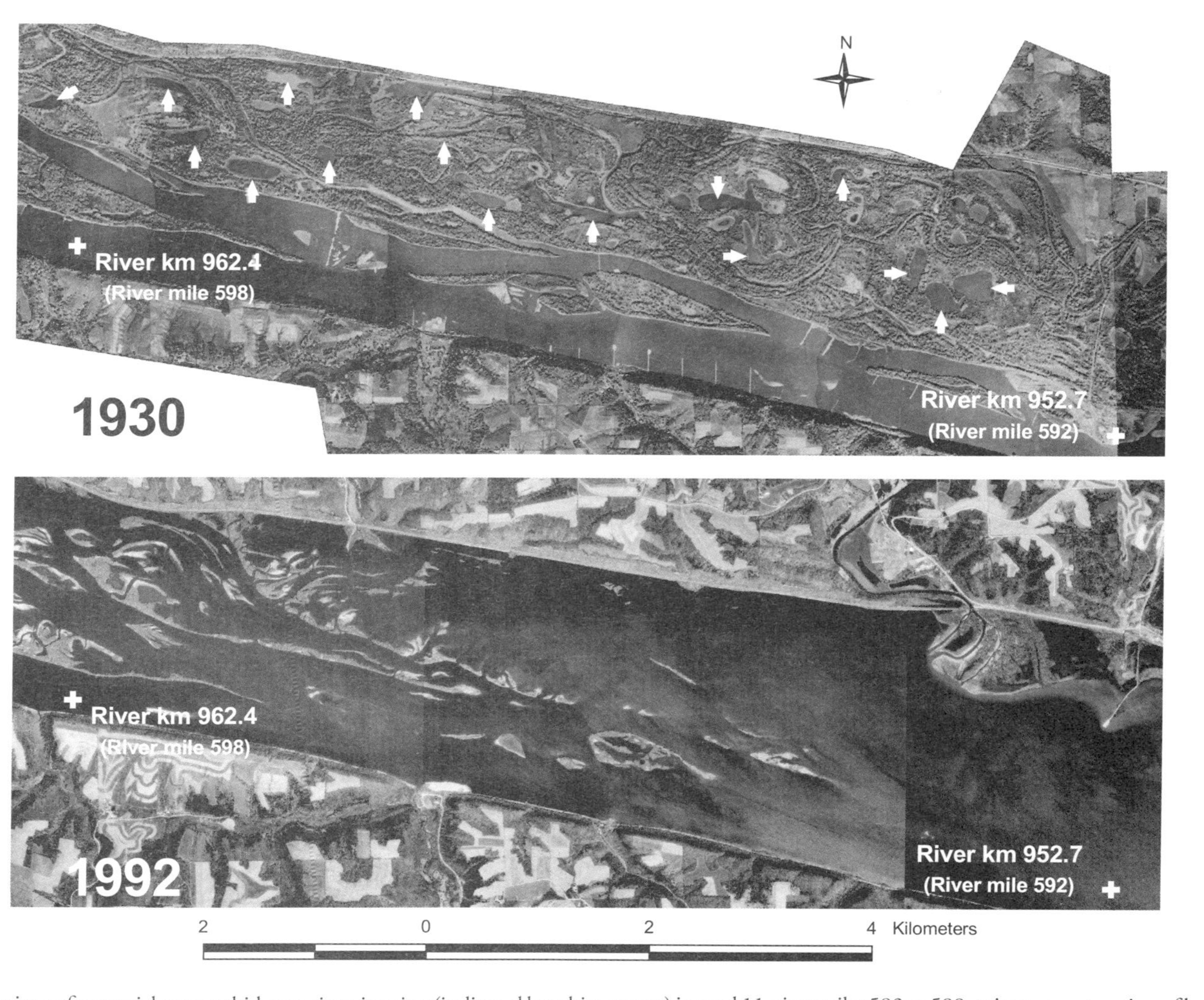

Figure 4.—Locations of potential centrarchid overwintering sites (indicated by white arrows) in pool 11, river miles 593 to 599, prior to construction of lock and dams (1930 photo) and the same section of the upper Mississippi River (UMR) several years postimpoundment. No known centrarchid overwintering sites existed in this area in 1992.

changes in species abundance on the Illinois River. Atwood (1984) duplicated historical hoopnet sampling conducted in 1931–1936, 1942, and 1957–1967. His comparisons between 1978 and 1979 data and historical data

> ...indicated that the hoopnet catches of such species as shortnose gar, goldfish [*Carassius auratus*], black bullhead, bluegill, white crappie and black crappie have declined over the years, while catches of river carpsucker, quillback, smallmouth buffalo, shorthead redhorse, channel catfish, white bass, sauger, and freshwater drum have increased.

Atwood attributed these changes on the Illinois River to increased sedimentation and turbidity.

The search for data in this study presented several challenges, from many promising leads resulting in dead ends to interpretation of the information. For example, a third database identified for potential use in the quantitative assessment provides a case study for a common problem with many databases, the proper archiving of biological data. Cox (1897) often referenced data collected on the UMR by the Minnesota Zoological Survey, which at that time was cited as the Natural History Survey. Fisheries sampling was conducted on the UMR by the Minnesota Zoological Survey along the Minnesota border from about 1895 to 1907 and was supervised primarily by Cox. Records and photographs indicate the sampling gear consisted of hoop nets, seines, and gill nets. In 1899, the Minnesota Zoological Survey had a boat specifically built for sampling on the UMR. The following year, 1900, the University of Minnesota established several field stations along the UMR for the purposes of conducting fish and mussel surveys. Cox left the University of Minnesota for Indiana State University in 1905. His letter of resignation states, "I shall take all my private materials with me and expect to find time next year to work up the systemic part of the Fishes of Minnesota."

Several months of inquiries to archives at both universities revealed that he had a very successful career at the University of Indiana. However, Cox never did publish his report and no copies of his data were located. This example illustrates the importance of proper archiving of sampling data and publication of study findings along with information describing the locations of samples.

A historical review of fish and wildlife assemblages must also include careful documentation of common names, or terminology, used during the period being investigated. The following example shows how misinterpreted historical common names could affect management decisions. In 1982, the U.S. Fish and Wildlife Service prepared an unpublished report on the legal and administrative history of the UMR Fish and Wildlife Refuge, which extends from near Wabasha, Minnesota, in lower pool 4, to near Rock Island, Illinois, in pool 16. The report also included a review of the congressional intent of the refuge. The author of the report incorrectly interpreted the term "black bass" to mean smallmouth bass. This misinterpretation became "fact" since no follow-up was done. The purpose of the UMR Fish and Wildlife Refuge was brought up as one factor to consider when planning for habitat restoration projects in the mid-1990s. Discussions during the planning of several projects on the refuge raised the question as to which of the bass species the congressional intent of the refuge applied. At the time, the decision relied on the summary contained in the 1982 report that incorrectly indicated one purpose was for smallmouth bass. The qualitative assessment for this study documents that when the refuge was established, the term "black bass" applied to largemouth bass. Fortunately, the decision of the project planning team was to provide features that would benefit both largemouth and smallmouth bass along with a variety of other fish and wildlife species.

The need for describing the preimpoundment, or even pre-European, fish and wildlife communities of the UMR has been implied in recent publications (Sparks 1995; USGS 1999). Some of the purposes include developing bio-criteria for assessment of ecosystem health, establishing ecosystem goals and objectives aimed at restoring the UMR, and determining if current and future restoration efforts are promoting a "more natural" or "unnatural" assemblage of UMR fish and wildlife species. However, care must be taken when using historical data to define restoration objectives. A historical review will most likely lead to an unclear picture of predisturbance fish and wildlife assemblages or habitats. The sparse quantitative data will need to be augmented with qualitative information, professional judgment, and common sense.

Acknowledgments

Much of the archival search for this study could not have been done without the services provided by staff of the U.S. Fish and Wildlife Service, D.C. Booth National Historic Fish Hatchery, Spearfish, South Dakota, University of Minnesota Archives, Minnesota State Historical Society, and Indiana State University Archives. John Anfinson, National Park Service Mississippi River National Scenic and Recreation Area historian provided valuable leads to historical information sources. Brian Ickes, USGS Upper Midwest Environmental Sciences Center (UMESC), provided copies of the seining data collected by the Environmental Management Program LTRMP and reviewed a draft of the document. Don Fago, Wisconsin DNR, provided copies of the WDNR database. The GIS data were obtained from the USGS UMESC Web site.

References

Anfinson, J. O. 1993. Commerce and conservation on the upper Mississippi River. The annals of Iowa. State Historical Society of Iowa 52(4):385–417.

Anfinson, J. O. 2003. The river we have wrought: a history of the upper Mississippi. University of Minnesota Press, Minneapolis.

Atwood, E. R. 1984. A comparison of hoopnetting catches in the Illinois River from 1931 to 1979. Master's thesis. Western Illinois University, Carbondale.

Barnickol, P. G., and W. C. Starrett. 1951. Commercial and sport fishes of the upper Mississippi River between Caruthersville, Missouri and Dubuque, Iowa. Bulletin of the Illinois Natural History Survey 25(5):267–350.

Becker, G. C. 1983. Fishes of Wisconsin. The University of Wisconsin Press, Madison.

Carlander, H. B. 1954. History of fish and fishing in the upper Mississippi River. Upper Mississippi River Conservation Committee, Rock Island, Illinois.

Cavender, T. M. 1986. Review of the fossil history of North American freshwater fishes. Pages 699–724 *in* C. H. Hocutt and E. O. Wiley, editors. The zoogeography of North American freshwater fishes. Wiley, New York.

Conover, W. J. 1980. Practical nonparametric statistics. Wiley, New York.

Cox, U. S. 1897. A preliminary report on the fishes of Minnesota. Zoological series III of the geological and natural history survey of Minnesota. The Pioneer Press, St. Paul, Minnesota.

Cross, F. B., R. L. Mayde, and J. D. Stewart. 1986. Fishes in the western Mississippi basin (Missouri, Arkansas, and Red rivers). Pages 363–412 *in* C. H. Hocutt and E. O. Wiley, editors. The zoogeography of North American freshwater fishes. Wiley, New York.

Culler, C. F. 1920. Fish rescue operations. Transactions of the American Fisheries Society 50:247–250.

Eddy, S., and T. Surber. 1943. Northern fishes with special reference to the upper Mississippi valley. The University of Minnesota Press, Minneapolis.

Ellis, M. M. 1931. A survey of conditions affecting fisheries in the upper Mississippi River. U.S. Department of Commerce Bureau of Fisheries, Fishery Circular No. 5, Government Printing Office, Washington, D.C.

Fischer, J. R., and T. O. Claflin. 1995. Declines in aquatic vegetation in navigation pool No. 8, upper Mississippi River between 1975 and 1991. In K. Lubinski, J. Wiener, and N. Bhowmik, editors. Regulated Rivers – Research & Management 11(2):157–165

Fremling, C. R. and T. O. Claflin. 1984. Ecological history of the upper Mississippi River. Pages 5–24 *in* J. G. Wiener, R. V. Anderson, and D. R. McConville, editors. Contaminants in the upper Mississippi River: Proceedings of the 15th Annual Meeting of the Mississippi River Research Consortium. Butterworth Publishers, Boston.

Galstoff, P. S. 1924. Limnological observations in the upper Mississippi. U.S. Bureau of Fisheries Annual Report, Government Printing Office, Washington, D.C.

Greene, W. C. 1935. The distribution of Wisconsin fishes. State Conservation Commission of Wisconsin, Madison.

Gutreuter, S., R. Burkhardt, and K. Lubinski. 1995. Long Term Resource Monitoring Program procedures: fish monitoring. National Biological Service, Environmental Management Technical Center, LTRMP 95-P002–1, Onalaska, Wisconsin.

Gutreuter, S., and C. Theiling. 1999. Fishes. Pages 12–1 to 12–25 *in* K. Lubinski and C. Theiling, editors. Ecological status and trends of the upper Mississippi River system 1998. Upper Midwest Environmental Sciences Center, La Crosse, Wisconsin.

Hughes, R. M. 1995. Defining acceptable biological status by comparing with reference conditions. Pages 31–47 *in* W. S. Davis and T. P. Simon, editors. Biological assessment and criteria: tools for water resource planning. Lewis Publishers, Boca Raton, Florida.

Knights, B. C., B. L. Johnson, and M. B. Sandheinrich. 1995. Response of bluegill and black crappie to dissolved oxygen, temperature, and current in backwater lakes of the upper Mississippi River during winter. North American Journal of Fisheries Management 15:390–399.

Martin, L. 1916. The physical geography of Wisconsin. Wisconsin Geological and Natural History Survey, Bulletin XXXVI, Educational Series No. 4, Madison.

Merritt, R. H. 1979. Creativity, conflict & controversy: a history of the St. Paul District U.S. Army Corps of Engineers. U.S. Government Printing Office, Washington, D.C.

Pitlo, J., Jr. 1986. Mississippi River investigations: an evaluation of largemouth bass populations in the upper Mississippi River. Iowa Department of Natural Resources, Federal Aid to Fish Restoration, Project No. F-109-R, Segment 2, Des Moines.

Sheehan, R. J., W. M. Lewis, and R. L. Bodenstiener. 1990. Winter habitat requirements and overwintering of riverine fishes. Illinois Department of Conservation, Federal Aid in Sportfish Restoration, Project F-79-R, Final Report, Springfield.

Sparks, R. E. 1995. Need for ecosystem management of large rivers and their floodplains. BioScience 45(3):169–182.

State of Wisconsin. 1907. Biennial Report of the commissioners of fisheries of Wisconsin for the years 1905–1906, letter from superintendent of fisheries to commissioners, State of Wisconsin, Madison.

Surber, T. 1920. A preliminary catalogue of the fishes and fish-like vertebrates of Minnesota. Syndicate Printing Co., Minneapolis, Minnesota.

Theiling, C. H., C. Korchgen, H. De Haan, T. Fox, J. Rohweder, and L. Robinson. 2000. Habitat needs assessment for the upper Mississippi River system: technical report. U.S. Geological Survey, Upper Midwest Environmental Science Center, La Crosse, Wisconsin. Contract report prepared for the U.S. Army Corps of Engineers, St. Louis District, St. Louis, Missouri.

Tiffany, W. L. 1876. The black bass in Minnesota. Pages 374–381 *in* Bulletin of the Minnesota Academy of Natural Sciences, Minnetonka, Minnesota.

UMRCC (Upper Mississippi River Conservation Committee). 1946. Second progress report of the technical committee for fisheries. Rock Island, Illinois.

USACE (U.S. Army Corps of Engineers). 1997. Report to Congress – an evaluation of the upper Mississippi River system environmental management program. Rock Island, Illinois.

USACE (U.S. Army Corps of Engineers). 2000. Upper Mississippi River system – habitat needs assessment. U.S. Army Corps of Engineers, St. Louis, Missouri.

U.S. Congress, House, Committee on Agriculture. 1924. Mississippi River wild life and fish refuge: hearings before the committee on agriculture. House of Representatives sixty-eighth congress first session on H. R. 4088, a bill to establish the Upper Mississippi River wild life and fish refuge. U.S. Government Printing Office, Washington D.C.

U.S. Congress, Senate, Committee on Commerce. 1924. Upper Mississippi River wild life and fish refuge: hearing before the committee on commerce, United States Senate, sixty-eighth congress, first session on S. 1558, a bill to establish the upper Mississippi River wild life and fish refuge. U.S. Government Printing Office, Washington, D.C.

U.S. Department of Commerce, Bureau of Fisheries, U.S. Commissioner of Fisheries. 1903–1938. Report of the United States Commissioner of Fisheries for Fiscal Year(s) 1901–1938. U.S. Government Printing Office, Washington, D.C.

USGS (U.S. Geological Survey). 1999. Ecological status and trends of the upper Mississippi River system 1998: a report of the long term resource monitoring program. U.S. Geological Survey, Upper Midwest Environmental Sciences Center, LTRMP 99-T001, La Crosse, Wisconsin.

WEST Consultants, Inc. 2000a. Upper Mississippi River and Illinois Waterway cumulative effects study. Volume 1. Geomorphic assessment. U.S. Army Corps of Engineers, Rock Island, Illinois.

WEST Consultants, Inc. 2000b. Upper Mississippi River and Illinois Waterway cumulative effects study. Volume 2. Ecological assessment. U.S. Army Corps of Engineers, Rock Island, Ilinois.

Wilcox, D. B. 1993. An aquatic habitat classification system for the upper Mississippi River system. U.S. Fish and Wildlife Service, Environmental Management Technical Center, EMTC 93-T003, Onalaksa, Wisconsin.

Appendix A.—Comparison of fish species (listed alphabetically by common name), frequency of occurrence (%), rank of frequency of occurrence (% rank), total number of fish caught (No.), and rank of number of fish caught (No. rank) for two Upper Mississippi River (UMR) databases. Preimpoundment Wisconsin Department of Natural Resources (WDNR) seining is a summary of 81 seining events from Pools 4–12 of the UMR (source, WDNR statewide database pooled for years 1900 to 1938). Post-impoundment Long Term Resources Monitoring Program (LTRMP) seining is a summary of 945 seining events from pools 4, 8, and 13 of the UMR (source, U.S. Geological Survey Environmental Management Program [USGS, EMP-LTRMP] fisheries database pooled for years 1993 to 1999). The higher the rank, the greater the frequency of occurrence or number caught. There was significant correlation between the databases for fish species reported in both databases ($n = 75$) for ranks of frequency of occurrence and ranks of number of fish caught (Spearman's rho = 0.71, $p < 0.01$, and Spearman's rho = 0.63, $p < 0.01$, respectively).

		Preimpoundment WDNR seining				Postimpoundment LTRMP seining			
		Frequency of occurrence		Number caught		Frequency of occurrence		Number caught	
Common name	Scientific name	%	% rank	No.	No. rank	%	% rank	No.	No. rank
American brook lamprey	*Lampetra appendix*	2.5		2					
American eel	*Anguilla rostrata*	3.7		3					
banded darter	*Etheostoma zonale*					0.2		2	
bigmouth buffalo	*Ictiobus cyprinellus*	4.9	24.5	4	11.5	1.5	26	47	27
bigmouth shiner	*Notropis dorsalis*	6.2	28.5	19	37	0.7	20.5	46	26
black buffalo	*Ictiobus niger*	3.7		3					
black bullhead	*Ameiurus melas*	14.8	59	15	32.5	0.1	5.5	27	22
black crappie	*Pomoxis nigromaculatus*	23.5	71.5	293	67	12.3	57	1,020	57
blackchin shiner	*Notropis heterodon*	1.2		1					
blacknose dace (eastern)	*Rhinichthys atratulus*					0.1		1	
blacknose shiner	*Notropis heterolepis*	7.4		136					
blackside darter	*Percina maculata*					0.1		1	
blue catfish	*Ictalurus furcatus*	2.5		2					
blue sucker	*Cycleptus elongatus*	2.5	11	2	6	0.4	16.5	6	14
bluegill	*Lepomis macrochirus*	25.9	74	328	70	44.3	71	16,051	71
bluntnose minnow	*Pimephales notatus*	7.4	32.5	27	41	2.0	27.5	79	31
bowfin	*Amia calva*	2.5	11	3	8.5	0.7	20.5	7	16.5
brassy minnow	*Hybognathus hankinsoni*					0.1		1	
brook silverside	*Labidesthes sicculus*	9.9	44	198	59	25.4	64	4,123	66
brook stickleback	*Culaea inconstans*					0.3		4	
brown bullhead	*Ameiurus nebulosus*	4.9		4					

Appendix A.—Continued.

Common name	Scientific name	Preimpoundment WDNR seining: Frequency of occurrence %	Preimpoundment WDNR seining: Frequency of occurrence % rank	Preimpoundment WDNR seining: Number caught No.	Preimpoundment WDNR seining: Number caught No. rank	Postimpoundment LTRMP seining: Frequency of occurrence %	Postimpoundment LTRMP seining: Frequency of occurrence % rank	Postimpoundment LTRMP seining: Number caught No.	Postimpoundment LTRMP seining: Number caught No. rank
bullhead minnow	*Pimephales vigilax*	18.5	65	374	72	59.2	72	14,364	70
burbot	*Lota lota*	4.9		9					
central mudminnow	*Umbra limi*	1.2		1					
central stoneroller	*Campostoma anomalum*	1.2	3	1	2.5	0.1	5.5	1	4
channel catfish	*Ictalurus punctatus*	17.3	64	195	58	5.2	39	209	39
channel shiner	*Notropis wickliffi*					19.3		14,239	
chestnut lamprey	*Ichthyomyzon castaneus*	1.2	3	1	2.5	0.1	5.5	1	4
common carp	*Cyprinus carpio*	12.3	50	20	38	11.7	54	971	56
creek chub	*Semotilus atromaculatus*	3.7	19	10	29	0.1	5.5	1	4
crystal darter	*Crystallaria asprella*	2.5	11	2	6	0.4	16.5	4	11
emerald shiner	*Notropis atherinoides*	28.4	75	709	74	78.9	75	128,251	75
fantail darter	*Etheostoma flabellare*	3.7	19	6	17	0.1	5.5	1	4
fathead minnow	*Pimephales promelas*	2.5	11	18	36	3.4	34	92	32
flathead catfish	*Pylodictis olivaris*	8.6	39	7	21	0.3	13	3	9.5
freshwater drum	*Aplodinotus grunniens*	12.3	50	14	30.5	18.8	63	6,075	68
ghost shiner	*Notropis buchanani*	1.2		1					
gizzard shad	*Dorosoma cepedianum*	16.0	62	21	39.5	28.3	67	7,033	69
golden redhorse	*Moxostoma erythrurum*	4.9	24.5	5	14.5	1.3	25	12	20
golden shiner	*Notemigonus crysoleucas*	13.6	55	215	61	6.7	43	184	38
goldeye	*Hiodon alosoides*	1.2	3	1	2.5	0.1	5.5	1	4
redfin pickerel	*Esox americanus*	1.2		1					
green sunfish	*Lepomis cyanellus*					1.9		36	
highfin carpsucker	*Carpiodes velifer*	4.9	24.5	5	14.5	0.4	16.5	10	18.5
iowa darter	*Etheostoma exile*	2.5	11	4	11.5	0.1	5.5	3	9.5
johnny darter	*E. nigrum*	19.8	67.5	274	64	42.8	70	2,960	65
lake sturgeon	*Acipenser fulvescens*	6.2		5					
largemouth bass	*Micropterus salmoides*	23.5	71.5	124	54	31.9	69	4,524	67
logperch	*Percina caprodes*	13.6	55	223	62	26.7	65	1,360	58
longnose gar	*Lepisosteus osseus*	8.6	39	9	26.5	3.5	35	44	25

Appendix A.—Continued.

Common name	Scientific name	Preimpoundment WDNR seining: Frequency of occurrence %	Preimpoundment WDNR seining: Frequency of occurrence % rank	Preimpoundment WDNR seining: Number caught No.	Preimpoundment WDNR seining: Number caught No. rank	Postimpoundment LTRMP seining: Frequency of occurrence %	Postimpoundment LTRMP seining: Frequency of occurrence % rank	Postimpoundment LTRMP seining: Number caught No.	Postimpoundment LTRMP seining: Number caught No. rank
mimic shiner	*Notropis volucellus*	12.3	50	89	52	30.7	68	17,421	72
mississippi silvery minnow	*Hybognathus nuchalis*	14.8	59	362	71	3.8	36	281	44
mooneye	*Hiodon tergisus*	2.5	11	9	26.5	0.9	23	21	21
mud darter	*Etheostoma asprigene*	8.6	39	65	50	12.0	55	371	48
northern hog sucker	*Hypentelium nigricans*	3.7	19	9	26.5	0.5	19	6	14
northern pike	*Esox lucius*	13.6	55	77	51	6.8	44	115	33
orangespotted sunfish	*Lepomis humilis*	7.4	32.5	17	34.5	14.1	58	2,121	60
paddlefish	*Polyodon spathula*	3.7		3					
pallid shiner	*Hybopsis amnis*	8.6	39	7	21	0.3	13	6	14
pirate perch	*Aphredoderus sayanus*	2.5		4					
pugnose minnow	*Opsopoeodus emiliae*	3.7	19	61	49	14.6	61	1,441	59
pugnose shiner	*Notropis anogenus*	1.2		2					
pumpkinseed	*Lepomis gibbosus*	13.6	55	41	45	5.6	42	271	43
quillback	*Carpiodes cyprinus*	3.7	19	3	8.5	10.1	51	2,183	62
river carpsucker	*C. carpio*	4.9	24.5	4	11.5	9.6	49.5	2,936	64
river darter	*Percina shumardi*	12.3	50	129	55	14.2	59	470	51
river redhorse	*Moxostoma carinatum*	1.2	3	1	2.5	0.2	11	2	8
river shiner	*Notropis blennius*	21.0	70	306	69	60.2	74	41,100	74
rock bass	*Ambloplites rupestris*	4.9	24.5	17	34.5	8.2	46	250	42
rudd	*Scardinius erythrophthalmus*					0.1		2	
sand shiner	*Notropis stramineus*	8.6	39	34	44	5.3	40.5	489	52
sauger	*Sander canadensis*	8.6	39	7	21	7.4	45	120	34
shorthead redhorse	*Moxostoma macrolepidotum*	12.3	50	133	56	10.9	53	298	45
shortnose gar	*Lepisosteus platostomus*	7.4	32.5	6	17	3.2	32	48	28
shovelnose sturgeon	*Scaphirhynchus platorynchus*	4.9		4					
silver chub	*Macrhybopsis storeriana*	19.8	67.5	292	66	9.4	48	541	53
silver lamprey	*Ichthyomyzon unicuspis*	2.5		2					
silver redhorse	*Moxostoma anisurum*	7.4	32.5	9	26.5	5.3	40.5	139	36
skipjack herring	*Alosa chrysochloris*	6.2	28.5	4,222	75	0.1	5.5	1	4

Appendix A.—Continued.

		Preimpoundment WDNR seining				Postimpoundment LTRMP seining			
		Frequency of occurrence		Number caught		Frequency of occurrence		Number caught	
Common name	Scientific name	%	% rank	No.	No. rank	%	% rank	No.	No. rank
slenderhead darter	*Percina phoxocephala*	2.5	11	7	21	3.0	30	53	29
smallmouth bass	*Micropterus dolomieu*	11.1	46.5	15	32.5	14.4	60	330	46.5
smallmouth buffalo	*Ictiobus bubalus*	4.9	24.5	4	11.5	3.9	37	422	49.5
southern redbelly dace	*Phoxinus erythrogaster*					0.2		2	
speckled chub	*Macrhybopsis aestivalis*	9.9	44	278	65	2.0	27.5	249	41
spotfin shiner	*Cyprinella spiloptera*	19.8	67.5	189	57	59.4	73	34,048	73
spottail shiner	*Notropis hudsonius*	24.7	73	610	73	17.6	62	2,156	61
spotted sucker	*Minytrema melanops*	7.4	32.5	14	30.5	3.3	33	130	35
stonecat	*Noturus flavus*	1.2	3	6	17	0.1	5.5	1	4
suckermouth minnow	*Phenacobius mirabilis*	2.5	11	29	42	0.4	16.5	5	12
tadpole madtom	*Noturus gyrinus*	11.1	46.5	54	48	9.6	49.5	422	49.5
trout-perch	*Percopsis omiscomaycus*	13.6	55	107	53	2.3	29	67	30
walleye	*Sander vitreus*	8.6	39	53	47	8.7	47	154	37
warmouth	*Lepomis gulosus*	2.5	11	2	6	0.8	22	10	18.5
weed shiner	*Notropis texanus*	14.8	59	301	68	3.1	31	223	40
western sand darter	*Ammocrypta clara*	2.5	11	21	39.5	10.3	52	720	55
white bass	*Morone chrysops*	16.0	62	49	46	28.3	66	2,565	63
white crappie	*Pomoxis annularis*	16.0	62	205	60	4.2	38	330	46.5
white sucker	*Catostomus commersonii*	9.9	44	30	43	1.1	24	40	24
yellow bass	*Morone mississippiensis*	2.5	11	8	24	0.1	5.5	7	16.5
yellow bullhead	*Ameiurus natalis*	7.4	32.5	7	21	0.3	13	30	23
yellow perch	*Perca flavescens*	19.8	67.5	227	63	12.0	56	634	54

American Fisheries Society Symposium 45:345–363, 2005

Fish Assemblage Structure, Composition, and Biotic Integrity of the Wisconsin River

JOHN LYONS*

Wisconsin Department of Natural Resources, 1350 Femrite Drive, Monona, 53716, USA

Abstract.—The Wisconsin River is a lowland warmwater river located entirely within the state of Wisconsin. It is the largest river within the state, with a length of 676 km, a drainage basin of 31,800 km^2, and an estimated mean annual flow at the mouth of 292 m^3/s. The middle part of the river has been heavily modified by dams and pollution, but the lower portion is relatively nondegraded. A total of 110 native fish species have been recorded from the river, a high number for the upper Mississippi River basin. Only two alien species occur and only one, common carp *Cyprinus carpio*, is common. Five state-endangered, five state-threatened, and 10 state-special-concern (=vulnerable) fishes are known from the river. Populations of the endangered black redhorse *Moxostoma duquesnei*, threatened paddlefish *Polyodon spathula*, blue sucker *Cycleptus elongatus*, black buffalo *Ictiobus niger*, and vulnerable western sand darter *Ammocrypta clara* in the Wisconsin River are among the largest in the state for these species. Historical data are scarce, but it appears that no fish species have been eliminated from the river. However, several species have been extirpated from discrete reaches of the river. Fish species richness decreases from mouth to headwaters, but reach-specific fish assemblage structure and composition and biotic integrity vary in a more complex pattern, most likely as a consequence of habitat fragmentation and isolation by dams and industrial and municipal pollution. In particular, the Prairie du Sac Dam, lowermost on the river, prevents upstream fish movement and causes a sharp drop in species richness and biotic integrity upstream.

Introduction

Large lowland warmwater rivers are critically important ecosystems in North America, especially in the midwestern United States where they are particularly common (Sheehan and Rasmussen 1999). Their long-term movement of water and soil have shaped the landscape and created a wide variety of terrestrial, riparian, and aquatic habitats. They are major repositories of biodiversity and serve as dispersal and migration corridors for numerous organisms. In particular, they support unique, highly diverse, and economically valuable fish assemblages.

Unfortunately, midwestern warmwater rivers have been severely degraded by human activities (Karr et al. 1985). All rivers in the region have experienced dramatic changes in water, sediment, and nutrient inputs because of forest clearing, agricultural development, and urbanization in their watersheds. Nearly all have also been dammed for power production, flood control, navigation, or other human uses. Most have experienced massive pollution from cities and industries and have seen the establishment of alien flora and fauna that have fundamentally altered their biological communities.

In many regards, the Wisconsin River is typical of midwestern warmwater rivers in that it has been heavily modified and impacted by human uses. Known locally as "the hardest working river in the nation" (Goc 1993), much of the Wisconsin River has been dramatically altered by hydroelectric dams, flow regulation, municipal and industrial pollution,

* E-mail: lyonsj@dnr.state.wi.us

forest clear-cutting, agriculture, and urbanization (WDNR 1976, 2000). However, in other ways the Wisconsin River is unusual, with its nearly 150-km-long lower portion having a largely natural character and representing one of the least-degraded large river reaches remaining in the Midwest (Lyons et al. 2001). Consequently, it still retains a relatively intact fish assemblage and excellent ecosystem health. In this manuscript, I describe the fish fauna of the Wisconsin River in relation to longitudinal patterns of human impact, focusing on historical changes in occurrence and current configurations of fish diversity, assemblage structure and composition, and biotic integrity.

Methods

Study Area

The Wisconsin River is the longest and largest river within Wisconsin, flowing 676 km from its source at the outlet of Lac Vieux Desert on the Wisconsin–Michigan border to its mouth at the Mississippi River (Figure 1). It has a basin area of 31,800 km^2. Mean annual discharge at the gauge at Muscoda, 71 km upstream from the mouth, is 247 m^3/s, with daily extremes over the period 1913–2001 of 41 m^3/s (in July 1988) and 2,290 m^3/s (in September 1938) (USGS 2002). Estimated mean annual discharge at the mouth is 292 m^3/s.

The Wisconsin River basin experiences a cold-temperate continental climate. The overall mean annual air temperature is about 10°C, with a typical summer maximum near 35°C and a winter minimum of –30°C. Water temperatures in the main channel of the Wisconsin River usually range from 0–30°C, and the river is at least partially ice covered from November through March. Mean annual precipitation in the basin ranges from 71 to 86 cm, with the monthly maximum in the summer and the monthly minimum in the winter. However, because of snowmelt and reduced evapo-transpiration rates, river flows are typically highest in the spring. Minimum flows usually occur in summer because of high evapo-transpiration and in winter because of reduced precipitation that mainly falls as snow. However, short-term high flows can occur at any time of year.

I have divided the Wisconsin River into five reaches that reflect patterns of longitudinal change in natural features and human impacts (Figure 1; Table 1). The Headwaters Reach is the uppermost. Here, the river is small and meandering with a gentle gradient. The Headwaters Reach is currently in a largely natural state, with little pollution or channel modification and a primarily forested riparian zone and watershed. However, this reach was strongly impacted by extensive clear-cut logging and associated log drives during the late 1800s (Durbin 1997).

As the Wisconsin River flows through the Upper Reach, the next downstream, it grows dramatically in size and increases in gradient (Table 1). The Upper Reach, although appearing natural in some areas, has a much greater level of human modification than does the Headwaters Reach (Table 1). It too was strongly affected by clear-cut logging and log drives in the late 1800s. In addition, over half of the reach is impounded by dams. Until recently, the Upper Reach also was heavily polluted by pulp and paper mill wastes and municipal sewage discharges. During the mid 1900s, water quality was very poor below these discharges, with low dissolved oxygen, excessive nutrients, high suspended solids, noxious odors, nuisance growths of filamentous bacteria, and depressed fish abundance and diversity (CNRA 1972; WDNR 1976; Ball and Marshall 1978; Coble 1982). Beginning in the mid-1970s, massive treatment of discharges began, and since then, water quality and biotic conditions have improved dramatically (WDNR 2000, 2002; Lyons et al. 2001).

The Middle Reach is similar in natural character to the Upper Reach but has been even more modified by human activities. It has the highest population density, the greatest industrial development, and the most dams of the five reaches (Table 1). Almost all of the Middle Reach is impounded, and riverine habitat exists only in short tailwaters below the dams. Agriculture is extensive in the watershed, and agricultural runoff contributes excessive amounts of sediment and nutrients to the river. The Middle Reach has the greatest number of municipal and industrial discharges, and historically, these discharges caused major pollution problems, of a similar nature but even greater magnitude and extent than in the Upper

Figure 1.—The Wisconsin River, showing locations mentioned in the text, including reaches, dams (short lines perpendicular to the river; some lines represent more than one dam), and quantitative boomshocking sites, from 1996 to 1999 (dots; some dots represent more than one site). For clarity, no tributaries are shown.

Reach (CNRA 1972; WDNR 1976; Ball and Marshall 1978; Coble 1982). As in the Upper Reach, recent improvements in waste treatment have substantially improved water quality and biotic conditions in the Middle Reach (WDNR 1976, 2000, 2002; Lyons et al. 2001).

The character of the Wisconsin River changes dramatically in the Dells Reach (Clayton and Attig 1989). As it leaves the glacial till of the Middle Reach and enters the sandy lacustrine plain of the Dells Reach, the river becomes substantially wider and shallower, and the channel becomes braided with a shifting sand substrate and numerous sand bars and islands. A unique area, Wisconsin Dells, occurs in the middle of the reach. Here, the river flows through a 17-km-long, narrow, rocky gorge with cliffs up to 20 m high. River width, normally 300–400 m, is constricted to as little as 15 m, and water depths, usually less than 1 m, exceed 10 m. The primary human impacts in the Dells Reach are dams that impound more than half the reach and modify flows, agricultural runoff, and pollutants transported from upstream reaches.

The Lower Reach is similar in riverine habitat to the Dells Reach, with a relatively wide, shallow, braided channel of shifting sand. However, human impacts are less. No dams occur in the Lower Reach, and much of the pollution from the Upper and

TABLE 1.—Delineation and characteristics of five reaches of the Wisconsin River. Rkm = river kilometer, the distance from the mouth via the river channel.

	Reach				
Characteristic	Headwaters	Upper	Middle	Dells	Lower
Length (km)	66	183	169	109	149
Upper end	La Vieux Desert Dam (Rkm 676, source)	Otter Rapids Dam (Rkm 610)	Wausau Dam (Rkm 427)	Castle Rock Dam (Rkm 258)	Prairie du Sac Dam (Rkm 149)
Lower end	Otter Rapids Dam (Rkm 610)	Wausau Dam (Rkm 427)	Castle Rock Dam (Rkm 258)	Prairie du Sac Dam (Rkm 149)	Mississippi River (Rkm 0, mouth)
Bankfull width (m)	10–30	100–150	200–300	300–500	400–700
Mean depth (m)	<1	>1	>1	<1	<1
Substrate	Sand, gravel, and cobble	Sand, gravel, cobble and boulder in river; sand and silt in impoundments	Cobble and boulder in river; sand, silt, and gravel in impoundments	Sand, with limited gravel and cobble in river; sand and silt in impoundments	Sand, with limited gravel and cobble
Gradient (m/km)	0.3	0.8	0.5	0.3	0.3
Channel form	Meandering; pools and riffles	Relatively straight; pools and rapids (most rapids inundated by dams)	Relatively straight; pools and rapids (most rapids inundated by dams)	Braided, with numerous islands and sand bars	Braided, with numerous islands and sand bars
Conductivity (μ/cm)	60–80	70–110	110–190	170–240	210–300
Riparian land cover	Forest, wetland	Forest, localized wetland; residential on impoundments	Forest, residential, urban; residential on impoundments	Forest, wetland; residential on impoundments	Forest, wetland
Catchment land cover	Forest, wetland	Forest. wetland, limited agriculture	Agriculture, forest, limited wetlands	Agriculture, forest, wetlands	Agriculture, forest; wetlands along river
Catchment land form	Undulating glacial till plains; Sandy outwash plains	Undulating glacial till plains; Sandy outwash plains; Morainal hills	Undulating glacial till plains; Sandy outwash plains; Morainal hills; Sandy lacustrine plains	Sandy lacustrine plain; Loess-capped and deeply dissected bedrock plateau	Loess-capped and deeply dissected bedrock plateau; Sandy alluvial plain along river
Dams present	2 (including Otter Rapids)	10 (including Wausau)	13 (including Castle Rock)	2 (including Prairie du Sac)	0

TABLE 1.—Continued.

Characteristic	Reach				
	Headwaters	Upper	Middle	Dells	Lower
Percent of reach length impounded	15	56	95	58	0
Riparian cities with population > 1,000 (2000 census)	Eagle River (1,443)	Rhinelander (7,735) Tomahawk (3,770) Merill (10,146) Wausau (in part) (38,426)	Wausau (in part)(38,426) Rothschild (4,970) Schofield (2,117) Mosinee (4,063) Stevens Point (24,551) Whiting (1,760) Plover (10,520) Wisconsin Rapids (18,435) Port Edwards (1,944) Nekoosa (2,590)	Wisconsin Dells (2,418) Lake Delton (1,982) Portage (9,728)	Prairie du Sac (3,231) Sauk City (3,109) Spring Green (1,585) Muscoda (1,453) Boscobel (3,047)
Major municipal and industrial discharges	1 municipal	4 municipal 6 industrial (6 paper mills)	10 municipal 12 industrial (6 paper mills)	2 municipal	4 municipal

Middle reaches has dissipated before it reaches the Lower Reach. Agricultural runoff does occur but is localized, and water quality is generally good. Channel and riparian modifications are minimal. Overall, the Lower Reach is among the least-disturbed parts of the Wisconsin River and is one of the highest-quality large warmwater river reaches remaining in the United States (Lyons et al. 2001). It was recommended for designation as a National Scenic River and is now protected as the only Wisconsin State Riverway (WDNR 1988).

Dams are a dominant feature of the Wisconsin River. The first dams on the river were built in the late 1830s, and during the heyday of logging in the late 1800s, over 200 dams, mostly temporary low-head wooden structures, were present (Durbin 1997). Currently, there are 27 dams on the river, concentrated in the Upper and Middle reaches (Table 2). About half of the total length of the river is impounded, and river flow patterns are modified throughout the length of the river. Flow in all five of the Wisconsin River reaches is regulated in a coordinated fashion through the Wisconsin Valley Improvement Company, a consortium of pulp and paper industries and public power utilities (Goc 1993). All of the dams on the river, 24 of which

TABLE 2.—Dams currently found on the Wisconsin River. Data from Goc (1993), Durbin (1997), WDNR (2003), and Wisconsin Department of Natural Resources (unpublished data files). Note that all but the Lac Vieux Desert, Rainbow, and Lower Paper Mill dams currently produce hydroelectric power. The Lac Vieux Desert Dam increases and regulates the water level on a natural lake, Lac Vieux Desert. All of the other reservoirs on the river are impoundments produced by the dams. Reservoir area is at normal pool, and residence time is based on mean annual flow. PCP = pentachlorophenol; PCBs = polychlorinated biphenyls.

Dam name	River kilometer	Height (m)	Year first dam built/latest major modifications	Reservoir area (ha)	Reservoir residence time (days)	Known toxins in fish or sediment
Lac Vieux Desert	676	0.7	1891/1938	1,155	1,068	None
Otter Rapids	610	3.7	1878/1927	40	1	None
Rainbow	588	6.3	1937/1937	823	9	Mercury
Rhinelander	564	9.5	1876/1926	779	19	None
Hat Rapids	559	6.1	1905/1923	60	1	None
Kings	537	7.0	1910/1910	554	10	Mercury
Tomahawk	521	4.6	1888/1936	773	6	None
Grandmother	511	5.8	1923/1970	80	0.5	None
Grandfather Falls	502	9.5	1876/1939	90	0.8	None
Alexander	483	7.0	1924/1924	274	2	None
Merrill	475	4.0	1847/1913	66	0.2	PCBs, PCP
Wausau	427	8.2	1852/1921	115	0.7	PCBs
Rothschild	414	6.1	1909/1963	776	4	PCBs, PCP
Mosinee	407	6.7	1840s/1930	402	1	PCBs
DuBay	396	7.6	1943/1943	2,692	11	PCBs
Stevens Point	380	5.2	1840s/1966	959	3	PCBs
Wisconsin River Division	375	6.4	1891/1926	8	<0.1	PCBs
Lower Paper Mill	374	2.7	1893/1953	2	<0.1	PCBs
Biron	352	7.3	1830s/1933	860	3	PCBs
Wisconsin Rapids	348	9.5	1904/1932	181	0.7	PCBs
Centralia	344	4.6	1840s/1920	93	0.2	PCBs
Port Edwards	339	5.2	1840/1976	47	0.1	PCBs
Nekoosa	335	6.7	1832/1914	183	0.5	PCBs
Petenwell	278	12..2	1949/1949	9,324	47	PCBs, Dioxin
Castle Rock	258	9.8	1950/1950	6,734	13	PCBs, Dioxin
Kilbourn	222	7.6	1859/1909	809	2	PCBs
Prairie du Sac	149	11.6	1914/1933	3,642	8	PCBs

generate hydroelectric power, and 19 non-hydroelectric dams on tributaries, are operated in concert to optimize river flows for power production. During fall and winter, water levels in the reservoirs created by the dams are lowered to provide storage capacity for snowmelt and spring runoff. This leads to somewhat higher than normal flows in the fall and early winter and lower than normal reservoir levels during the winter. In spring, snowmelt and runoff is captured and stored in the reservoirs, and normally high flows in the river are dampened and reservoir levels elevated. During summer and early fall, the stored water in the reservoirs is released, leading to higher than normal flows. Throughout the year, the dams reduce flood peaks.

Analysis of Fish Data

Records of fish occurrence in the Wisconsin River were taken from general published summaries of the distribution of Wisconsin fishes (Greene 1935; Becker 1983; Fago 1992; Lyons et al. 2000), specific published studies of Wisconsin River fishes (Becker 1966; Coble 1982; Lyons 1993; Rasmussen et al. 1994; Lyons and Welke 1996; Pellett et al. 1998; Runstrom et al. 2001; Knights et al. 2002), and unpublished data files from the University of Wisconsin (UW) and the Wisconsin Department of Natural Resources (WDNR). I included data from the main channel, impoundments, side channels, sloughs, backwaters, and tributary mouths but not from the tributaries themselves or from off-channel lakes and reservoirs. I identified alien species that had been introduced from outside the Wisconsin River basin. I then classified both alien and native species as either residents, that spent all or large portions of their life within a river reach, or strays, that moved into the reach occasionally, in small numbers, for brief periods from another reach, a tributary, or the Mississippi River. I estimated current relative abundance (Appendix A) based on all available catch data since 1980. Species of high abundance were usually (>75% of time) collected, often (>50% of time) in large numbers (more than 25 fish per unit of effort), when their habitat was sampled with an appropriate technique. Species of moderate abundance were often collected but only infrequently (<25% of time) in large numbers. Species of low abundance were infrequently collected and almost never in large numbers.

Current quantitative patterns of fish species composition, abundance, and guild structure along the length of the Wisconsin River were based on standardized pulsed-DC boomshocker collections from 55 main-channel sites during 1996–1999 (Lyons et al. 2001). The 55 sites were chosen systematically from the Upper, Middle, Dells, and Lower reaches such that all existing riverine main-channel habitat types were sampled in approximate proportion to their occurrence. The Headwaters Reach was not sampled because it was too narrow and shallow for the boomshocker, and no impoundments were sampled because the technique used had not been developed for lentic environments.

Each quantitative site was sampled once; sampling took place during daylight between late May and early September, and no sampling occurred during high flows. Sites were 1.6-km long, a distance at which species richness approached an asymptote (Lyons et al. 2001). Sampling occurred along a single shoreline (chosen randomly) of the main channel, following the contours of the bank, and the distance sampled was determined from landmarks on a 1:24,000-scale topographic map. The same electroshocker design and electrical settings were used at all sites. Pulse rate was 60 Hz and duty cycle 25%, and power output to the electrodes was maintained at 3,000 W through adjustment of generator output. A single netter with a 17-mm stretch-mesh dip net attempted to collect all fish observed. This sampling technique was particularly effective for common carp *Cyprinus carpio*, catostomids, and centrarchids but tended to under-sample most cyprinids, ictalurids, and percids.

Lyons et al. (2001) used the quantitative fish data to calculate a large-river index of biotic integrity (IBI) and its constituent metrics, including species richness. I correlated the IBI and IBI metrics with the river kilometer of the collections. I also calculated mean values for each variable for each of the four reaches and compared the reaches with a one-way analysis of variance and a Scheffé multiple comparisons test. All analyses were run in SAS (1990) and considered significant if $P < 0.05$.

Results and Discussion

The Wisconsin River has a high number of native fish species (110) relative to other rivers in the Upper Mississippi River basin (Appendix A). The total in the Wisconsin River is higher than the 99 reported for the much larger Mississippi River between its source and the mouth of the Wisconsin River (1,170 km long; 175,000 km^2 basin area; Fremling et al. 1989), the 89 for the similarly sized Minnesota River (597 km; 44,300 km^2; Kavanaugh 1993), and the 84 for the smaller and particularly high-quality St. Croix River on the Wisconsin–Minnesota border (276 km; 20,020 km^2; Fago and Hatch 1993). For the entire Wisconsin River, 4 native species occur only as strays, whereas 12 species occur as strays in some reaches and are resident in others. Only two alien species have been found in the Wisconsin River, the common carp, which is common in much of the river, and the brown trout *Salmo trutta*, which occurs as a rare stray from coldwater tributaries in three reaches.

The Wisconsin River contains no federally threatened or endangered species, but at the state level has 5 endangered, 5 threatened, and 10 vulnerable (termed "special concern" by WDNR) fishes. The only extant population of endangered black redhorse *Moxostoma duquesnei* in Wisconsin occurs in the Middle Reach. Probably the largest remaining populations in the state of the threatened paddlefish *Polyodon spathula*, blue sucker *Cycleptus elongatus*, and black buffalo *Ictiobus niger*, and the vulnerable western sand darter *Ammocrypta clara*, occur in the Lower Reach (Lyons et al. 2000).

The number of species per Wisconsin River reach declines in an upstream direction (Appendix A), a pattern characteristic of many smaller rivers and streams (e.g., Lyons 1996) but not of large rivers, where species richness typically has little correlation with longitudinal position (Lyons et al. 2001). The Lower Reach has 98 native species (including 12 strays), whereas the adjacent Dells Reach has only 83 (8 strays). There is a sharp drop to 61 species (7 strays) in the Middle Reach, with more moderate declines to 49 species (3 strays) in the Upper Reach and 36 species (1 stray) in the Headwaters reach. Fifteen Wisconsin River species are known only from the Lower Reach, four only from the Headwaters Reach, one only from the Middle Reach, and none only from the Dells or Upper reaches.

The most common species also differ among reaches. Species of high abundance in the Lower Reach are (in taxonomic order) gizzard shad *Dorosoma cepedianum*, spotfin shiner *Cyprinella spiloptera*, emerald shiner *Notropis atherinoides*, sand shiner *Notropis stramineus*, quillback *Carpiodes cyprinus*, shorthead redhorse *Moxostoma macrolepidotum*, and channel catfish *Ictalurus punctatus*. The Dells Reach is similar, with high abundance of spotfin shiner, common carp, emerald shiner, sand shiner, shorthead redhorse, and bluegill *Lepomis macrochirus*. The Middle Reach has high abundance of common carp, shorthead redhorse, bluegill, smallmouth bass *Micropterus dolomieu*, and logperch *Percina caprodes*, similar to the Upper Reach, which has shorthead redhorse, bluegill, smallmouth bass, yellow perch *Perca flavescens*, and logperch. The Headwaters Reach is most distinctive, with common shiner *Luxilus cornutus*, hornyhead chub *Nocomis biguttatus*, blacknose dace *Rhinichthys atratulus*, and creek chub *Semotilus atromaculatus* in high abundance.

Long-term trends in species richness, composition, and abundance are difficult to decipher in the Wisconsin River because of a lack of historical data. Before 1925, only a few anecdotal observations on fish are available, and from most of these, the exact species cannot be determined. Greene (1935) described the first scientific collections from 1925 to 1928 but listed only 15 samples, almost all collected with small-mesh seines. In the 1960s and 1970s, UW, WDNR, and private consultants used seines, electrofishers, hoop nets, or fyke nets to collect 219 samples (Becker 1966, 1983; Coble 1982; Fago 1992; UW and WDNR, unpublished data). However, by this time, the river had been greatly modified by dams and pollution, and the samples did not represent the fish fauna prior to human impacts. Most recently, from the mid-1980s to the present, the WDNR collected 310 additional samples using seines, electrofishers, or hoop nets (Lyons 1993; Rasmussen et al. 1994; Pellett et al. 1998; Lyons et al. 2000), but again, these surveys did not provide information on historical conditions.

Although the limited historical data make any conclusion about long-term change uncertain, it appears that no resident native species have been extirpated from the entire Wisconsin River. However, resident species have been eliminated (no records in more than 50 years) from the Dells and Middle reaches (Table 1). The Dells Reach lost the state-threatened paddlefish and blue sucker and likely also the shovelnose sturgeon *Scaphirhynchus platorynchus* because of habitat fragmentation and isolation from downstream populations caused by the Prairie du Sac Dam (Lyons 1993). The Middle Reach lost the state special-concern lake sturgeon *Acipenser fulvescens* and pirate perch *Aphredoderus sayanus* and likely also the flathead catfish *Pylodictis olivaris* and slenderhead darter *Percina phoxocephala* probably because of habitat loss and fragmentation from dams and municipal and industrial pollution (CNRA 1972).

It is impossible to quantify long-term trends in the relative abundance of fish in the Wisconsin River. However, qualitative data comparisons and anecdotal accounts from anglers (WDNR, unpublished data) indicate that fish diversity and abundance in parts of the Middle and Upper reaches are now greater than they were in the mid-1900s, most likely because of improved water quality. In the 1960s, fish surveys (nonstandardized and poorly documented) conducted below major industrial and municipal discharges typically yielded only four to six species and few individuals and were often dominated by pollution-tolerant species such as common carp or bullheads *Ameiurus* spp. During the same time period, anglers complained of poor catches of sportfish species such as smallmouth bass and walleye *Sander vitreus*. At present, surveys in these same general areas typically yield 10–15 species, usually including good numbers of smallmouth bass and sometimes good numbers of walleye. Anglers also report better sportfish catch rates.

The quantitative boomshocking data from 1996 to 1999 provide a basis for a detailed comparison of current patterns of relative fish species richness and fish assemblage structure and composition among river reaches. Strong longitudinal patterns were evident within the Wisconsin River. Correlation analysis revealed decreases in an upstream direction for the number of native species (r = –0.61; $P < 0.0001$), sucker (Catostomidae) species (r = –0.53; $P < 0.0001$), intolerant species (r = –0.29; P = 0.0321), and obligate riverine species (r = –0.71; $P < 0.0001$); the proportion of individuals as riverine species (r = –0.36; P = 0.0067); and the IBI score (r = –0.48; P = 0.0002) (Table 3; Figure 2). Two metrics increased in an upstream direction: the proportion of biomass as round-bodied suckers (*Cycleptus, Hypentelium, Minytrema, Moxostoma* spp.) (r = 0.28; P = 0.0385) and the number of individuals as simple lithophilous spawners (spawning over gravel or cobble without building a nest or providing parental care) (r = 0.28; P = 0.0390). However, many of these correlations were caused by a sharp change in values between the Lower and Dells reaches rather than a consistent trend over the length of the river. The Lower Reach had a significantly higher mean number of native species, sucker species, intolerant species, and riverine species and mean IBI score than did the adjacent Dells Reach (Figure 3). If the Lower Reach was excluded and the correlations rerun for only the Dells, Middle, and Upper reaches, then there were no longitudinal trends in number of native species, number of sucker species, proportion of individuals as riverine species, proportion of individuals as simple lithophilous species, or IBI score. The correlations were reversed for number of intolerant species (r = 0.53; P = 0.0077) so that the proportion tended to increase rather than decrease in an upstream direction. Only the correlations for number of riverine species (r = –0.43; P = 0.0361) and proportion of biomass as round-bodied suckers (r = 0.48; P = 0.0167) remained significant and had the same sign as when data from the Lower Reach were included.

These results imply that habitat fragmentation and isolation caused by the Prairie du Sac Dam have had a substantial effect on fish assemblage composition and structure in the Wisconsin River. The Prairie du Sac Dam, closed in 1914, is the lowermost dam on the river and a complete barrier to upstream fish movement (Beck 1990). Before the dam was built, there were no impediments to fish movement between the Lower and Dells reaches, and early accounts (Durbin 1997) indicate that habitat in both reaches was similar (wide, shallow, braided, sandy channels). Based on what

TABLE 3.—Guild membership of Wisconsin River fishes.

Guild	Species
Intolerant	Chestnut lamprey, southern brook lamprey, silver lamprey, Mississippi silvery minnow, pallid shiner, speckled chub, blackchin shiner, spottail shiner, carmine shiner, weed shiner, highfin carpsucker, blue sucker, northern hog sucker, black buffalo, shoal sucker, black redhorse, greater redhorse, muskellunge, mottled sculpin, rock bass, smallmouth bass, western sand darter, crystal darter, rainbow darter, Iowa darter, banded darter, slenderhead darter
Obligate riverine	Chestnut lamprey, southern brook lamprey, silver lamprey, goldeye, mooneye, central stoneroller, largescale stoneroller, spotfin shiner, Mississippi silvery minnow, shoal chub, silver chub, hornyhead chub, river shiner, bigmouth shiner, carmine shiner, sand shiner, channel shiner, suckermouth minnow, bullhead minnow, blacknose dace, longnose dace, creek chub, river carpsucker, quillback, highfin carpsucker, blue sucker, northern hog sucker, smallmouth buffalo, black buffalo, spotted sucker, silver redhorse, river redhorse, black redhorse, golden redhorse, stonecat, western sand darter, crystal darter, mud darter, rainbow darter, fantail darter, banded darter, blackside darter, slenderhead darter, river darter
Simple lithophilous spawners	Lake sturgeon, shovelnose sturgeon, common shiner, emerald shiner, river shiner, carmine shiner, blacknose dace, longnose dace, white sucker, blue sucker, northern hog sucker, silver redhorse, river redhorse, black redhorse, golden redhorse, shorthead redhorse, greater redhorse, burbot, crystal darter, rainbow darter, banded darter, logperch, blackside darter, slenderhead darter, river darter, sauger, walleye

is known of current fish movement patterns and habitat use in the Lower Reach and adjacent Mississippi River, most species would have moved freely between the two reaches before the dam was in place (Pellett et al. 1998; Runstrom et al. 2001; Knights et al. 2002; WDNR, unpublished tagging data). Thus, it seems unlikely that there would have been a sharp change in fish assemblages at the boundary between the Lower and Dells reaches prior to the closing of the Prairie du Sac Dam. Apparently, the blockage of upstream fish movement by the Prairie du Sac Dam led to a decline in fish species richness and health in the Dells Reach and perhaps in reaches further upstream as well.

The probable effects of the Prairie du Sac Dam on the fish assemblage of the Wisconsin River are consistent with a theoretical model of dam impacts on rivers, the serial discontinuity concept (Ward and Stanford 1983, 1989, 1995). This model postulates that dams disrupt natural longitudinal trends in rivers, causing abrupt and large shifts in biological attributes over short distances. Differences in biological attributes between sites above and below a dam may be equivalent to those observed between sites hundreds of river kilometers apart in a river without dams. The lowermost dam often has a disproportionate influence, as it isolates the longest stretch of river. In the Wisconsin River, species richness drops an average of five species between sites downstream and upstream of the Prairie du Sac Dam (Figure 3). This drop is equivalent to the difference in richness between sites in the Lower and Upper reaches, which are 278 river kilometers apart. At no other dam on the river is the difference between upstream and downstream sites in average species richness so large.

Comparisons among reaches also suggest an effect on ecological health from the numerous dams and historical point-source pollution in the Middle Reach. The Middle Reach had significantly lower IBI scores than the Lower, Dells, and Upper reaches

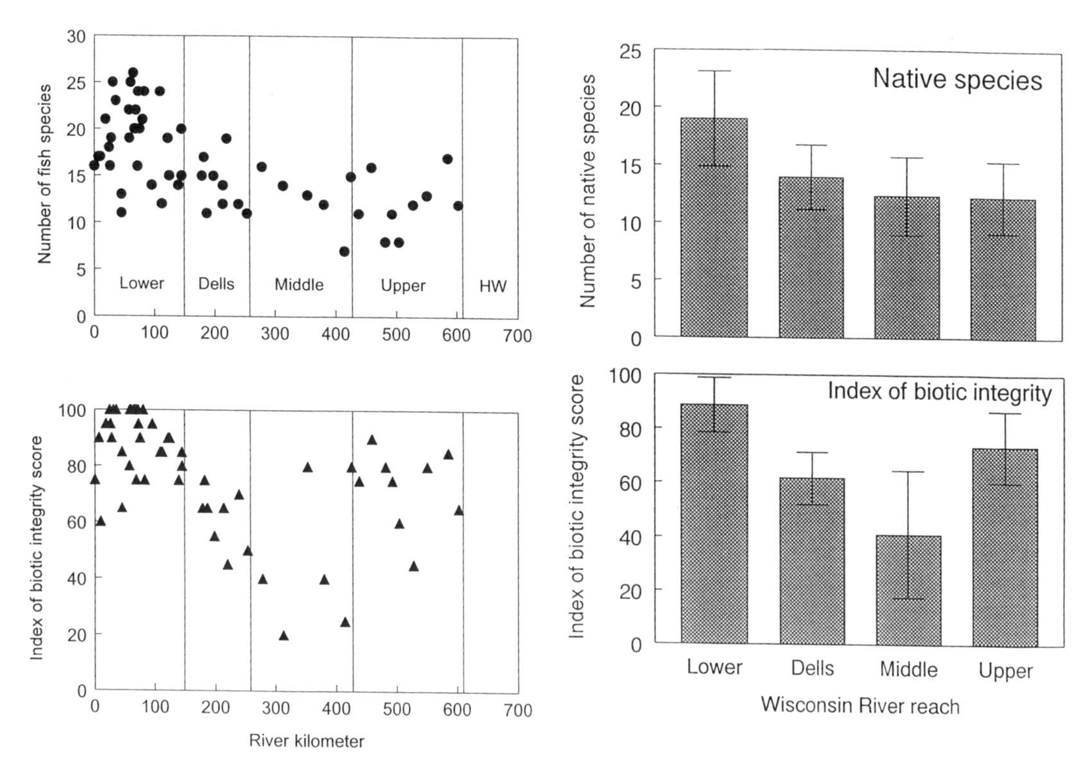

FIGURE 2.—Plot of number of native species (top) and index of botic integrity scores (bottom) versus river kilometer for the 1996–1999 quantitative sampling. The vertical lines indicate the boundaries between river reaches. HW = Headwaters Reach.

FIGURE 3.—Mean values by reach for the number of native species (top) and index of biotic integrity (IBI) scores (bottom) from the 1996–1999 quantitative sampling. The error bars indicate 1 SD.

(Figure 2) and a higher proportion of total biomass of the alien common carp than the Lower and Dells reaches (Upper Reach not included in analysis because common carp not established in most of reach) (Lower = 0.32, Dells = 0.37, Middle = 0.78; F = 18.2, $P < 0.0001$). However, anecdotal data suggest that fish assemblage condition and ecological health in the Middle Reach has improved since the 1970s because of reductions in pollution (WDNR 1976, 2000; Lyons et al. 2001). Studies on other large rivers document declines in fish assemblage condition caused by municipal and industrial pollution followed by fish assemblage recovery after treatment of discharges and improvements in water quality (e.g., Hughes and Gammon 1987; Fremling et al. 1989; Sheehan and Rasmussen 1999).

A decline in biotic integrity in an upstream direction, as observed in the Wisconsin River, was also observed in the Ohio River in Illinois, Indiana, Ohio, and Kentucky (Simon and Sanders 1999), but not in the Willamette River in Oregon (Hughes and Gammon 1987), where biotic integrity decreased in a downstream direction. This difference reflects geographic patterns of human impacts. In the Wisconsin and Ohio rivers, the greatest relative volume of pollution and the most dams were in the upper and middle portions of the river, whereas in the Willamette River, environmental degradation was highest near the mouth.

In conclusion, the Wisconsin River supports a diverse native fish fauna with 110 known species. The Lower Reach of the river has the most species, and species richness declines in an upstream direction. The Wisconsin River has been heavily modified and degraded, particularly in the Middle and Upper reaches, but no fishes have been completely eliminated from the river. However, several species have been extirpated from the Dells and Middle reaches. The Prairie du Sac Dam, the lowermost on the river, has blocked upstream fish movements and led to a decline in fish species richness and biotic integrity in the Dells Reach. Other dams further upstream also block movements and fragment habitat and probably also affect ecological health. Numerous dams and industrial and municipal pollution have reduced biotic integrity in the Middle Reach, although some recovery is evident over the last 30 years due to pollution control.

Management Implications

Efforts to restore biotic integrity and to improve fish assemblage structure and composition in the Wisconsin River must focus on the dams. These structures have fundamentally transformed the habitat and biota of the river. The dams provide important human services in terms of power production, flood control, pollution dilution, and recreation associated with the impoundments they have created (Goc 1993). Consequently, none are likely to be removed in the foreseeable future. However, modifications of the dams could benefit fish populations. In particular, development of effective upstream and downstream fish passage through the dams would eliminate barriers to fish movement and reduce effects of habitat fragmentation. At present, all of the dams prevent upstream movement. Downstream movement is possible and does occur, but fish mortality in passing through the dam (hydroelectric turbines at normal flows and floodgates at high flows) is largely unknown and could be substantial (WDNR, unpublished data). Moreover, fish that move downstream through the dam are permanently lost to upstream populations because they cannot return upstream.

Effective fish passage would be most beneficial at the Prairie du Sac Dam, and this is where fish passage is most likely to be implemented in the near future. On 11 October 2002, the U.S. Federal Energy Regulatory Commission (FERC) ordered the owner of the dam, Wisconsin Power and Light Company, to install an upstream fish passage facility within 3 years as a condition of their newly granted federal operating license (license identifier: 101 FERC at 61, 055; project number 11162-004). This passage facility is to be designed, constructed, and evaluated in consultation with the WDNR and U.S. Fish and Wildlife Service. If effective passage can be implemented, then it may be possible to reduce the barrier effects of the dam. However, the specific physical setting and construction of the Prairie du Sac Dam coupled with a lack of knowledge about the swimming capabilities and behavior of the fish species below the dam will make upstream passage challenging. An adaptive approach over many years will probably be necessary to develop an effective passage facility. The success or failure of passage at the Prairie du Sac Dam will likely determine whether passage can realistically be pursued at dams further upstream.

In addition to acting as barriers to fish movement and fragmenting habitats, dams on the Wisconsin River also greatly modify river flow patterns. In particular, dam operations reduce natural variation in river flow, dampening flood peaks and augmenting low flows. Conceivably, changes in dam operations to mimic more natural flows could benefit riverine species (Kinsolving and Bain 1993). However, such changes are unlikely to be implemented soon. Since 1995, FERC has issued 30-year operating licenses for most of the dams in the Middle and Upper reaches and affirmed the requests of dam operators to continue storing and releasing water to maximize power production according to the Wisconsin Valley Improvement Corporation guidelines. The one location where flow regulation has been changed to benefit fish and other biota is at Castle Rock Dam. This dam is the only one on the river with daily "peaking" operation, where flows are changed dramatically within each day to provide maximum hydroelectric power during periods of peak demand. The new license for the dam still allows peaking but only within a much narrower range of flows, such that daily extremes will be much less.

Although dams currently have the greatest effect on the Wisconsin River ecosystem, industrial and municipal discharges and nonpoint source pollution (agricultural and urban runoff) continue to threaten fish populations and overall ecosystem health. The WDNR and the industries and municipalities that discharge to the Wisconsin River devote considerable time and resources to ensuring that wastes are adequately treated and water quality is maintained. This effort has led to improvements in fish populations (WDNR 2000; Lyons et al. 2001) and should be continued. However, mercury, PCP (pentachlorophenol), dioxin, and PCBs (polychlorinated biphenyls) from past discharges persist in the river environment, accumulate in fish, and limit the safe consumption of fish by anglers (Table 2). Plans need to be developed and implemented to reduce contamination of sediments and fish. Also, the reservoirs behind Petenwell and Castle Rock dams are hypereutrophic because of excessive phosphorus. Sources of this phosphorus are unclear but must be identified before phosphorus loading to the river can be reduced (WDNR 2003).

At present, only one alien species, the common carp, has a strong impact on Wisconsin River fishes. Feeding and spawning by common carp are blamed for resuspending sediments and nutrients and destroying macrophyte beds in many of the impoundments in the Middle and Dells reaches (WDNR 2003). Unfortunately, other alien species may soon disrupt the Wisconsin River ecosystem. The zebra mussel *Dreissena polymorpha*, which can dramatically change water and habitat quality and biotic assemblages through its filter feeding and bottom encrusting habits, has recently been detected in the Upper Reach and is poised to spread downstream through larval drift (WDNR, unpublished data). A variety of alien fishes have become established in waters of Lake Superior and Lake Michigan since 1980 (Lyons et al. 2000) and could be brought to the Wisconsin River in bait buckets by careless anglers. Two species of Asian carp (silver carp *Hypophthalmichthys molitrix* and bighead carp *H. nobilis*) are spreading through the Mississippi River system and soon may have access to the Lower Reach. Competition, predation, or habitat and water quality modification by any one of these fishes could fundamentally alter the Wisconsin River fish assemblage.

The long, costly, and ultimately unsuccessful effort to eliminate common carp from Wisconsin waters (Becker 1983) demonstrates that once alien species are established, they will be very difficult to control. Thus, the best management strategy is to prevent alien species from entering the Wisconsin River. Restrictions on bait harvest and education of anglers, already being implemented (WDNR, unpublished data), will reduce the chances that alien species will be transported from the Great Lakes or other waters. However, it may be impossible to stop Asian carp from entering the Lower Reach. If they become established there, then any upstream passage facility at the Prairie du Sac Dam must be designed so that they cannot use it to invade the Dells Reach.

Acknowledgments

I thank Don Fago, Dave Marshall, and Nancy Nate for providing unpublished WDNR data on fish occurrences and environmental conditions in the Wisconsin River. Jay Hatch, Robert Hughes, Kyle Piller, Jeffrey Thomas, Bruce Vondracek, and an anonymous reviewer provided helpful comments on earlier drafts of this manuscript. Preparation of this manuscript was funded in part by Federal Aid in Sport Fish Restoration, Project F-95-P, Study SSOV.

References

Ball, J. R., and D. W. Marshall. 1978. Seston characterization of major Wisconsin rivers (slime survey). Wisconsin Department of Natural Resources, Technical Bulletin 109, Madison.

Beck, B. 1990. Transforming the heartland. The history of Wisconsin Power and Light Company. Wisconsin Power and Light Company, Madison.

Becker, G. C. 1966. Fishes of southwestern Wisconsin. Transactions of the Wisconsin Academy of Sciences, Arts and Letters 55:87–117.

Becker, G. C. 1983. Fishes of Wisconsin. University of Wisconsin Press, Madison.

Clayton, L., and J. W. Attig. 1989. Glacial Lake Wisconsin. Geological Society of America, Memoir 173, Boulder, Colorado.

CNRA (Citizens Natural Resources Association of Wisconsin, Inc.). 1972. The Wisconsin River: its history and a plan for restoration. CNRA, Loganville, Wisconsin.

Coble, D. W. 1982. Fish populations in relation to

dissolved oxygen in the Wisconsin River. Transactions of the American Fisheries Society 111:612–623.

Durbin, R. D. 1997. The Wisconsin River. An odyssey through time and space. Spring Freshet Press, Cross Plains, Wisconsin.

Fago, D. 1992. Distribution and relative abundance of fishes in Wisconsin. VIII. Summary report. Wisconsin Department of Natural Resources, Technical Bulletin 175, Madison.

Fago, D., and J. Hatch. 1993. Aquatic resources of the St. Croix River basin. Pages 23–56 *in* L. W. Hesse, C. B. Stalnaker, N. G. Benson, and J. R. Zuboy, editors. Proceedings of a symposium on restoration planning for the rivers of the Mississippi River ecosystem. U.S. Department of the Interior, National Biological Survey, Biological Report 19, Washington, D.C.

Fremling, C. R., J. L. Rasmussen, R. E. Sparks, S. P. Bobb, C. F. Bryan, and T. O. Claflin. 1989. Mississippi River fisheries: a case history. Pages 309–351 *in* D. P. Dodge, editor. Proceedings of the International Large River Symposium. Canadian Special Publication of Fisheries and Aquatic Sciences 106.

Goc, M. J. 1993. Stewards of the Wisconsin. Wisconsin Valley Improvement Company. New Past Press, Friendship, Wisconsin.

Greene, C. W. 1935. The distribution of Wisconsin fishes. State of Wisconsin Conservation Department, Madison.

Hughes, R. M., and J. R. Gammon. 1987. Longitudinal changes in fish assemblages and water quality in the Willamette River, Oregon. Transactions of the American Fisheries Society 116:196–209.

Karr, J. R., L. A. Toth, and D. R. Dudley. 1985. Fish communities of Midwestern rivers: a history of degradation. BioScience 35:90–95.

Kavanaugh, C. M. 1993. Minnesota River basin assessment project. Pages 5–22 *in* L. W. Hesse, C. B. Stalnaker, N. G. Benson, and J. R. Zuboy, editors. Proceedings of a symposium on restoration planning for the rivers of the Mississippi River ecosystem. U.S. Department of the Interior, National Biological Survey, Biological Report 19, Washington, D. C.

Kinsolving, A. D., and M. B. Bain. 1993. Fish assemblage recovery along a riverine disturbance gradient. Ecological Applications 3:531–544.

Knights, B. C., J. M. Vallazza, S. J. Zigler, and M. R. Dewey. 2002. Habitat and movement of lake sturgeon in the upper Mississippi River system, USA. Transactions of the American Fisheries Society 131:507–522.

Lyons, J. 1993. Status and biology of the paddlefish (*Polyodon spathula*) in the lower Wisconsin River. Transactions of the Wisconsin Academy of Sciences, Arts, and Letters 81:123–135.

Lyons, J. 1996. Patterns in the species composition of fish assemblages among Wisconsin streams. Environmental Biology of Fishes 45:329–341.

Lyons, J., P. A. Cochran, and D. Fago. 2000. Wisconsin fishes 2000. Status and distribution. University of Wisconsin Sea Grant, Madison.

Lyons, J., R. R. Piette, and K. W. Niermeyer. 2001. Development, validation, and application of a fish-based index of biotic integrity for Wisconsin's large warmwater rivers. Transactions of the American Fisheries Society 130:1077–1094.

Lyons, J., and K. Welke. 1996. Abundance and growth of young-of-year walleye (*Stizostedion vitreum*) and sauger (*S. canadense*) in Pool 10, upper Mississippi River, and at Prairie du Sac Dam, lower Wisconsin River, 1987–1994. Journal of Freshwater Ecology 11:39–50.

Pellett, T. D., G. J. Van Dyck, and J. V. Adams. 1998. Seasonal migration and homing of channel catfish in the lower Wisconsin River, Wisconsin. North American Journal of Fisheries Management 18:85–95.

Rasmussen, P. W., J. M. L. Unmuth, J. Lyons, and G. Van Dyck. 1994. A creel survey of the lower Wisconsin River, 1990–1991. Wisconsin Department of Natural Resources, Research Report 160, Madison.

Runstrom, A. L., B. Vondracek, and C. A. Jennings. 2001. Population statistics for paddlefish in the Wisconsin River. Transactions of the American Fisheries Society 130:546–556.

SAS. 1990. SAS/STAT user's guide, version 6, 4th edition. SAS Institute, Cary, North Carolina.

Sheehan, R. J., and J. L. Rasmussen. 1999. Large rivers. Pages 529–559 *in* C. C. Kohler and W. A. Hubert, editors. Inland fisheries management in North America, 2nd edition. American Fisheries Society, Bethesda, Maryland.

Simon, T. P., and R. E. Sanders. 1999. Applying an index of biotic integrity based on great river fish communities: considerations in sampling and interpretation. Pages 475–506 *in* T. P. Simon, editor. Assessing the sustainability and biological integrity of water resources using fish communities. CRC Press, Boca Raton, Florida.

USGS (U.S. Geological Survey). 2002. Water resources data. Wisconsin. Water Year 2001. USGS, Water-Data Report WI-01-1, Middleton, Wisconsin.

Ward, J. V., and J. A. Stanford. 1983. The serial discontinuity concept of lotic ecosystems. Page 29–42 *in* T. D. Fontaine and S. M. Bartell, editors. Dynamics of lotic ecosystems. Ann Arbor Science, Ann Arbor, Michigan.

Ward, J. V., and J. A. Stanford. 1989. Riverine ecosystems: the influence of man on catchment dynamics and fish ecology. Pages 56–64 *in* D. P. Dodge, editor. Proceedings of the International Large River Symposium. Canadian Special Publication of Fisheries and Aquatic Sciences 106.

Ward, J. V., and J. A. Stanford. 1995. The serial discontinuity concept: extending the model to floodplain rivers. Regulated Rivers: Research and Management 10:159–168.

WDNR (Wisconsin Department of Natural Resources). 1976. Wisconsin 1976 water quality inventory report to Congress. WDNR, Division of Environmental Standards, Madison.

WDNR (Wisconsin Department of Natural Resources). 1988. Final environmental impact statement: proposed Lower Wisconsin State Riverway. WDNR, Bureau of Environmental Assessment and Review, Madison.

WDNR (Wisconsin Department of Natural Resources). 2000. Wisconsin water quality assessment report to Congress 2000. WDNR, Bureau of Watershed Management, PUB-WT-254-2000, Madison.

WDNR (Wisconsin Department of Natural Resources). 2002. Wisconson water quality assessment report to Congress. WDNR, Water Division, Publication PUB-WT-254-2003, Madison.

WDNR (Wisconsin Department of Natural Resources). 2003. Wisconsin water quality assessment report to Congress 2002. WDNR, Bureau of Watershed Management, PUB-WT-254-2003, Madison.

Appendix A.—Fishes of the Wisconsin River by reach. R = resident; X = extirpated resident; S = stray or waif from other systems; *?* = plausible but unconfirmed; – = absent. Bold text = high abundance, roman text = moderate abundance, italics = uncommon. Alien and State of Wisconsin Endangered, Threatened, and Vulnerable (=Special Concern) species are indicated.

		Reach				
Family	Species	Lower	Dells	Middle	Upper	Headwaters
Lampreys (Petromyzontidae)	Chestnut lamprey *Ichthyomyzon castaneus*	*R*	*R*	*?*	*?*	-
	Southern brook lamprey *Ichthyomyzon gagei*	-	-	-	*S*	R
	Silver lamprey *Ichthyomyzon unicuspis*	R	*?*	-	-	-
Sturgeons (Acipenseridae)	Lake sturgeon (vulnerable) *Acipenser fulvescens*	*R*	R	*X*[a]	-	-
	Shovelnose sturgeon *Scaphirhynchus platorynchus*	R	*X*	-	-	-
Paddlefishes (Polyodontidae)	Paddlefish (threatened) *Polyodon spathula*	R	*X*	-	-	-
Gars (Lepisosteidae)	Longnose gar *Lepisosteus osseus*	R	*?*	-	-	-
	Shortnose gar *Lepisosteus platostomus*	R	-	-	-	
Bowfins (Amiidae)	Bowfin *Amia calva*	R	*R*	*R*	*R*	-
Mooneyes (Hiodontidae)	Goldeye (endangered) *Hiodon alosoides*	*R*	-	-	-	-
	Mooneye *Hiodon tergisus*	R	*R*	-	-	-
Freshwater eels (Anguillidae)	American eel (vulnerable) *Anguilla rostrata*	*R*	*?*	-	-	-
Herrings (Clupeidae)	Gizzard shad *Dorosoma cepedianum*	**R**	R	-	-	-
Minnows (Cyprinidae)	Central stoneroller *Campostoma anomalum*	*S*	*S*	-	-	-
	Largescale stoneroller *Campostoma oligolepis*	-	-	*S*	*R*	-
	Spotfin shiner *Cyprinella spiloptera*	**R**	**R**	R	-	-
	Common carp (alien) *Cyprinus carpio*	R	**R**	**R**	*R*	-
	Brassy minnow *Hybognathus hankinsoni*	*S*	*S*	-	-	*R*
	Mississippi silvery minnow *Hybognathus nuchalis*	*R*	*?*	-	-	-
	Pallid shiner (endangered) *Hybopsis amnis*	*S*	-	-	-	-
	Common shiner *Luxilus cornutus*	*R*	*R*	*R*	R	**R**
	Shoal chub (threatened) *Macrhybopsis hypstoma*	*R*	*R*	-	-	-
	Silver chub (vulnerable) *Macrhybopsis storeriana*	*R*	-	-	-	-
	Pearl dace *Margariscus margarita*	-	-	-	-	*R*
	Hornyhead chub *Nocomis biguttatus*	-	-	*S*	R	**R**

Appendix A.—Continued.

Family	Species	Reach				
		Lower	Dells	Middle	Upper	Headwaters
Minnows (Cyprinidae)	Golden shiner *Notemigonus crysoleucas*	*R*	*R*	*R*	*R*	*R*
	Emerald shiner *Notropis atherinoides*	**R**	**R**	R	R	-
	River shiner *Notropis blennius*	R	*R*	-	-	-
	Bigmouth shiner *Notropis dorsalis*	*R*	*R*	*R*	-	-
	Blackchin shiner *Notropis heterodon*	-	-	-	-	*R*
	Spottail shiner *Notropis hudsonius*	*R*	*R*	*R*	-	-
	Carmine shiner *Notropis percobromus*	-	-	*R*	*R*	*R*
	Sand shiner *Notropis stramineus*	**R**	**R**	R	-	-
	Weed shiner (vulnerable) *Notropis texanus*	*R*	*R*	-	-	-
	Mimic shiner *Notropis volucellus*	*?*	*R*	*R*	*R*	*R*
	Channel shiner *Notropis wickliffi*	R	*R*	-	-	-
	Pugnose minnow (vulnerable) *Opsopoeodus emiliae*	*R*	*R*	*?*	-	-
	Suckermouth minnow *Phenacobius mirabilis*	*S*	-	-	-	-
	Northern redbelly dace *Phoxinus eos*	-	-	-	-	*R*
	Bluntnose minnow *Pimephales notatus*	*R*	*R*	*R*	*R*	R
	Fathead minnow *Pimephales promelas*	*S*	*S*	*S*	*S*	*R*
	Bullhead minnow *Pimephales vigilax*	R	*R*	-	-	-
	Eastern blacknose dace *Rhinichthys atratulus*	-	-	*S*	*R*	**R**
	Longnose dace *Rhinichthys cataractae*	-	-	*S*	*R*	R
	Creek chub *Semotilus atromaculatus*	*S*	*S*	*S*	*R*	**R**
Suckers (Catostomidae)	River carpsucker *Carpiodes carpio*	R	*R*	-	-	-
	Quillback *Carpiodes cyprinus*	**R**	R	*R*	-	-
	Highfin carpsucker *Carpiodes velifer*	R	R	-	-	-
	White sucker *Catostomus commersonii*	*R*	*R*	R	R	R
	Blue sucker (threatened) *Cycleptus elongatus*	R	*X*	-	-	-
	Lake chubsucker (vulnerable) *Erimyzon sucetta*	*R*	-	-	-	-
	Northern hog sucker *Hypentelium nigricans*	R	*R*	*R*	R	R
	Smallmouth buffalo *Ictiobus bubalus*	R	*R*	-	-	-
	Bigmouth buffalo *Ictiobus cyprinellus*	*R*	R	*R*	-	-
	Black buffalo (threatened) *Ictiobus niger*	*R*	*R*	-	-	-

Appendix A.—Continued.

				Reach		
Suckers (Catostomidae)	Spotted sucker *Minytrema melanops*	*R*	*R*	-	-	-
Family	Species	Lower	Dells	Middle	Upper	Headwaters
	Silver redhorse *Moxostoma anisurum*	*R*	*R*	R	R	*R*
	River redhorse *Moxostoma carinatum*	*S*	-	-	-	-
	Black redhorse (endangered) *Moxostoma duquesnei*	-	-	*R*	-	-
	Golden redhorse *Moxostoma erythrurum*	*R*	*R*	R	R	*R*
	Shorthead redhorse *Moxostoma macrolepidotum*	**R**	**R**	**R**	**R**	R
	Greater redhorse (threatened) *Moxostoma valenciennesi*	*S*	-	-	*?*	*?*
Bullhead catfishes (Ictaluridae)	Black bullhead *Ameiurus melas*	*R*	*R*	R	R	*?*
	Yellow bullhead *Ameiurus natalis*	*R*	*R*	R	R	*?*
	Brown bullhead *Ameiurus nebulosus*	-	*R*	*R*	-	-
	Channel catfish *Ictalurus punctatus*	**R**	R	*R*	-	-
	Stonecat *Noturus flavus*	*S*	*S*	*R*	*R*	-
	Tadpole madtom *Noturus gyrinus*	*R*	*R*	*R*	*R*	-
	Flathead catfish *Pylodictis olivaris*	R	*R*	*X*	-	-
Pikes (Esocidae)	Grass pickerel *Esox americanus*	*R*	*R*	-	-	-
	Northern pike *Esox lucius*	*R*	*R*	*R*	*R*	*R*
	Muskellunge *Esox masquinongy*	*S*[b]	*S*[b]	*R*	R	*R*
Mudminnows (Umbridae)	Central mudminnow *Umbra limi*	*R*	*R*	*R*	*R*	R
Trouts (Salmonidae)	Brown trout (alien) *Salmo trutta*	*S*	*S*	-	-	*S*
	Brook trout *Salvelinus fontinalis*	-	-	-	-	*R*
Trout-perches (Percopsidae)	Trout-perch *Percopsis omiscomaycus*	*R*	*R*	R	*R*	*?*
Pirate perches (Aphredoderidae)	Pirate perch (vulnerable) *Aphredoderus sayanus*	*R*	*R*	*X*	-	-
Codfishes (Gadidae)	Burbot *Lota lota*	*R*	*R*	*R*	*R*	*R*
Killifishes (Fundulidae)	Starhead topminnow (endangered) *Fundulus dispar*	*R*	-	-	-	-
	Blackstripe topminnow *Fundulus notatus*	*R*	*R*	-	-	-
Silversides (Atherinidae)	Brook silverside *Labidesthes sicculus*	R	R	*R*	-	-
Sticklebacks (Gasterosteidae)	Brook stickleback *Culaea inconstans*	*S*	*S*	*S*	*R*	*R*
Sculpins (Cottidae)	Mottled sculpin *Cottus bairdii*	-	-	-	*S*	R
Temperate basses (Moronidae)	White bass *Morone chrysops*	R	R	*R*	-	-
	Yellow bass *Morone mississippiensis*	*R*	*R*	-	-	-
Sunfishes (Centrarchidae)	Rock bass *Ambloplites rupestris*	*R*	*R*	*R*	R	R
	Green sunfish *Lepomis cyanellus*	*R*	*R*	*R*	-	-
	Pumpkinseed *Lepomis gibbosus*	*R*	*R*	*R*	R	*R*
	Warmouth *Lepomis gulosus*	*R*	*R*	*R*	*R*	-
	Orangespotted sunfish *Lepomis humilis*	*R*	-	-	-	-

Appendix A.—Continued.

Family	Species	Reach				
		Lower	Dells	Middle	Upper	Headwaters
Sunfishes (Centrarchidae)	Bluegill *Lepomis macrochirus*	R	**R**	**R**	**R**	R
	Smallmouth bass *Micropterus dolomieu*	R	R	**R**	**R**	?
	Largemouth bass *Micropterus salmoides*	R	R	R	R	?
	White crappie *Pomoxis annularis*	*R*	*R*	*R*	*R*	-
	Black crappie *Pomoxis nigromaculatus*	*R*	*R*	R	R	?
Perches (Percidae)	Western sand darter (vulnerable) *Ammocrypta clara*	R	R	-	-	-
	Crystal darter (endangered) *Crystallaria asprella*	*R*	-	-	-	-
	Mud darter (vulnerable) *Etheostoma asprigene*	*R*	*R*	-	-	-
	Rainbow darter *Etheostoma caeruleum*	*R*	*R*	*R*	R	-
	Iowa darter *Etheostoma exile*	*R*	*R*	*R*	*R*	*R*
	Fantail darter *Etheostoma flabellare*	*R*	*R*	*R*	*R*	-
	Least darter (vulnerable) *Etheostoma microperca*	*R*	-	-	-	-
	Johnny darter *Etheostoma nigrum*	R	R	R	R	R
	Banded darter *Etheostoma zonale*	R	R	*R*	*R*	-
	Yellow perch *Perca flavescens*	R	R	R	**R**	R
	Logperch *Percina caprodes*	R	R	**R**	**R**	R
	Blackside darter *Percina maculata*	R	R	R	R	-
	Slenderhead darter *Percina phoxocephala*	*R*	*R*	*X*	R	*R*
	River darter *Percina shumardi*	*R*	*R*	-	-	-
	Sauger *Sander canadensis*	R	R	-	-	-
	Walleye *Sander vitreus*	R	R	R	R	*R*
Drums (Sciaenidae)	Freshwater drum *Aplodinotus grunniens*	R	R	-	-	-
Totals	Native (excluding "?")	98	83	61	49	36
	All reaches combined	110				
	Alien	2	2	1	1	1
	All reaches combined	2				

[a] Lake sturgeon have been re-introduced into the Middle Reach within the last 15 years by the Wisconsin Department of Natural Resources, but the long-term success of this stocking is uncertain.
[b] Muskellunge are currently stocked in the Lower and Dells reaches and survival has been good, but no reproduction has been observed.

American Fisheries Society Symposium 45:365–381, 2005

Wabash River Fishes from 1800 to 2000

James R. Gammon*

Department of Biological Sciences,
DePauw University, Greencastle, Indiana 46135, USA

Abstract.—The present ichthyofauna (1965–2001) of the Wabash River system is compared to that of three periods: presettlement through 1820, 1875–1900, and 1940–1950. This second largest Ohio River tributary flows freely for 350 mi. However, its environment and watershed have been altered greatly from presettlement times; two-thirds has been converted to agriculture, eliminating all prairies and most forests and wetlands. Canals, large and small dams, channelization, and effluents have extinguished 12 fish species, diminished some, and favored others. Thirteen of approximately 175 species are recent, including 3 aliens. Better municipal and industrial waste treatment has improved water quality, but excessive agricultural runoff remains detrimental to many fishes. Degraded habitats exacerbate these problems. Many sensitive species are today either absent or severely reduced in distribution and abundance compared to 50 years ago. Smallmouth bass *Micropterus dolomieu* has been replaced by largemouth bass *M. salmoides* or spotted bass *M. punctulatus,* and few visual piscivores occur except near reservoirs.

Introduction

It is probably fortuitous that the fishes of the Wabash and White rivers have received so much attention since 1800. Other medium-size rivers harbor equally diverse and interesting fishes and may be even more esthetically pleasing than the Wabash (Figure 1). Whatever the reasons, the fishes of the Wabash and its many tributaries formed the raw material used by a succession of early naturalists as they attempted to determine exactly what kinds of fish lived in our rivers. Many names are familiar only because they are appended to scientific species names—Lesueur, Rafinesque, Jordan, Copeland, and Say. Key among these early fish ecologists was David Starr Jordan, who produced another generation of talented scientists in the early 1900s: Swain, Jenkins, Gilbert, Moenkhaus, Butler, Ulrey, Forbes, Meek, Kirsch, Hay, Blatchley, Evermann, and Eigenmann.

Somewhat later, the theme broadened through the efforts of Scott, Hubbs, Lagler, Gerking, and Frey as they sought to unravel the complexities of interactions between aquatic biota and the physicochemical environment.

In recent decades, our concept of running water ecosystems, especially large rivers, has benefited

Figure 1.—Aerial view of the middle Wabash River circa 1980.

*E-mail: JRGAMMON@depauw.edu

enormously from efforts of Larimore, McReynolds, McComish, Burr, Pearson, Hamelink, Spacie, Simon, and a host of Indiana Department of Natural Resources biologists. In the future, our ecological image of rivers will continue to be better defined by yet another generation of ecologists.

This chapter broadly examines changes in distribution and abundance of the fish populations in the Wabash River system over the past 200 years. The pathway to achieving such an objective is underlain by problems galore. Should such an analysis focus primarily upon distribution or abundance or both? The problem of identifying species was the primary concern of early workers from 1800 through 1900. Rafinesque (1820) carefully detailed size, color, fin ray counts, and edibility, but sometimes included rivers inhabited, whether the species was "permanent" or seasonal, and used "common," "not uncommon," "not common," and "rare" as indications of abundance.

This kind of quasi-quantitative evaluation persisted through the Jordan era and is here assumed to refer to the relative abundance of a species within a sampled assemblage. Geographic coverage was much better from 1875 to 1900. Nevertheless, only two-thirds of Indiana's streams were sampled and the Wabash and White rivers themselves were scarcely touched.

The same abundance estimates were used when Gerking collected fishes throughout the state in the 1940s, but distributional abundance became an additional feature of terminology because of increased sampling intensity of smaller streams, but not the larger ones. The criteria for appraising relative abundance, therefore, differ between pre-1940s studies and more recent periods.

Results

Presettlement to 1820

When Marquette and Joliet reached the Mississippi River in 1673, the Iroquois had driven most previous resident agricultural cultures from the Ohio River valley (Hyde 1962). During the next decade or so, the Iroquois forced other tribes westward, but did not settle in the Ohio River valley. At the close of the 17th century, tribes from the north recolonized the Wabash and White river valleys in small, scattered villages.

In the 1700s, the French built forts and trading posts on the Wabash River. In 1779, Vincennes became the first American seat of government northwest of the Ohio River, and white settlers moved into the Wabash Valley in increasing numbers.

Until 1815, few written accounts of the rivers and streams were available, although French furtraders lived in harmony with the Indians for at least a century. In 1792, Heckewelder (McCord 1970) wrote of the river near Vincennes "...the Wabash as clear as the Monocasy, full of fish, ... "

Caleb Lownes, in a letter to Oliver Wolcott (Secretary Treasurer under Presidents Washington and Adams), wrote in 1815, (McCord 1970)

> ". . first rate lands lie on the Wabash all the way to the lakes on the most beautiful stream in my recollection—it is about 250 yards wide at this place (Vincennes) and preserves its width very nearly for 400 mi...It is a beautiful and valuable stream—the water generally perfectly clear and transparent—exhibiting a clean, gravelly bottom—It abounds with fish of various kinds—bass-pickerel, pike-perch-catfish, etc. The catfish are of every size up to 122 1/2 lb. The perch (probably smallmouth bass *Micropterus dolomieu*) are from 12 to 20 in length. . . a large white fish about 2 1/2 ft long with very little bone was yesterday caught by a gentlemen on a party said to be excellent."

By far, the finest early account of the biota is that of Thomas (1819).

> "The Wabash is four hundred yards wide at its mouth, three hundred at Vincennes, and two hundred at Fort Harrison. It is fordable in many places."

> "Whenever a high piece of land appears on one side of the river, the opposite shore is low and sunken; and from Raccoon Creek, fifteen miles above Fort Harrison to the mouth of the river, I believe there is no exception to this remark."

"All its tributary streams after a heavy shower of rain, rise above the banks; and overflow the low land adjoining, which on all, is of considerable extent. I have known it for more than four weeks at one time, that no person could get away from Union Prairie without swimming his horse, or going in a boat."

"The Wabash abounds with fish of many kinds; which, in the months of April, May and June, may be readily caught with the hook and line. The gar or bill fish is more than two feet in length."

"There are three kinds of cat-fish: the Mississippi cat, the mud cat, and the bull head. Some of the first have weighed one hundred and twenty pounds. The mud cat is covered with clouded spots and is a very homely fish. The head is very wide and flat. Some have weighed one hundred pounds."

"The real sturgeon is found in the Wabash, though the size is not large. These have been taken from twenty to sixty pounds weight."

"The shovel fish or flat nose is another species of sturgeon. It weighs about twenty pounds."

"The pond pike [northern pike *Esox lucius*] is taken in ponds from one to three feet long, but very slim."

"The river pike [muskellunge *Esox masquinongy*] is large and highly esteemed, but scarce."

"The drum or white perch weighs from one to thirty pounds. It is shaped like the sunfish."

"The black perch or bass [smallmouth bass *Micropterus dolomieu*] is excellent, and weighs from one to seven pounds."

"The streaked bass [white bass *Morone chrysops*] is scarce."

"The buffalo fish is of the sucker kind, and very common. Weight from two to thirty pounds."

"The rock mullett [smallmouth buffalo *Ictiobus bubalus*] is sometimes seen three feet long. It is slim and weighs from 10 to 15 lb."

"The red horse is also of the sucker kind. It is large and bony, weighing from five to fifteen pounds."

"The Jack pike or pickerel [walleye *Sander vitreus*] is an excellent fish, and weighs from six to twenty pounds."

"The eel is frequently taken in the Wabash, and weighs from one to three pounds."

"The freshwater clam or muscle is so plenty, as to be gathered and burnt for lime. Twenty years ago, I am told, no other kind of lime was procured."

"Craw fish, which resembles the lobster, is very common in the low lands of this country."

The great clarity of water was remarked upon by many observers. Mrs. Lydia Bacon (McCord 1970) wrote of Vincinnes on October 10, 1811,

"the local situation of the place is very pleasant, lying on a clean stream of water, which affords them a variety of fish and facilitates their intercourse with the neighboring states and territories."

"It was in the month of April (1825) when I first saw the Wabash River...Schools of fishes—salmon, bass, redhorse, and pike—swam close along the shore, catching at the blossoms of the red-bud and plum that floated on the surface of the water, which was so clear that myriads of the finny tribe could be seen darting hither and thither amid the limpid element, turning up their silvery sides as they sped out into deeper water." (Cox 1860).

McCulloch (McCord 1970) visited the Wabash River at Logansport in late May, 1833, and wrote

". . I followed an Indian trail that led along the banks of the Wabash, which had not then been deprived of any of their natural beauty by either freshets or the axe of the settler. The river was bank-full. Its water was clear, and as it sparkled in the sunlight or reflected the branches of the trees which hung over it, I thought it was more beautiful than even the Ohio...."

Fish were plentiful and easy to obtain at this time. Rafinesque (1820) stated,

"The most usual manners of catching fish...are, with seines or harpoons at night and in shallow water, with boats carrying a light, or with the hooks and line, and even with baskets."

Other methods were also used.

"At John Stitt's mill below town (Crawfordsville), on Sugar river, there is a fish-trap, and in one night we caught nine hundred fish, the first spring we were in the country (1825), most of them pike, salmon, bass, and perch. Some of the largest pike and salmon (walleye) measured from two to four feet in length, and weighed from twelve to twenty-five pounds." (Cox 1860).

At Indianapolis on the White River. . .

"I went up to McCormick's dam (just above the country club) four miles above town on the river one day and sat down at a chute . . . and I threw out with my hand eighty-seven bass, ranging in size from one pound up to five" (Dunn 1910).

"Amos Hanway says there were 'bass, salmon [walleye and/or sauger], redhorse, ordinary suckers, quillbacks, or as they were sometimes called spearbacks, perch, pike, catfish, etc. The biggest salmon I ever caught weighed sixteen pounds. I once caught a pike that measured four feet and two inches [muskellunge]; at another time a gar-fish that measured over three feet, and a blue catfish that weighed sixteen and a quarter pounds. The finest rock bass [largemouth bass] I ever took was one which weighed eight and a quarter pounds, and that was near Waverly; while the biggest river bass [smallmouth bass] I ever lifted from the water weighed six and one-fourth pounds." In Morgan County, above the Cox dam, when the fish were running "at one haul seined twelve barrels of fish, and there were thirty fish that averaged, undressed, ten pounds each . . . mostly bass and salmon, but also large redhorse, white perch, quillbacks and ordinary suckers."

Maximilian (1843) visited New Harmony (population = 600) from October 19, 1832 to March 16, 1833.

"The Wabash, a fine river, as broad as the Moselle, winds between banks which are now cultivated, but were lately covered with thick forests." Snapping, softshell, and other *emys* "are numerous." "The proteus (*Menobranchus lateralis*, Harl.) of the Ohio and of the great Canadian lakes, is found in the Wabash." This may have been either hellbender, mudpuppy or both. "There are many kinds of fish in the Wabash, on the whole the same as in the Ohio and the Mississippi. 100 lb catfish, several species of sturgeon and pikes, the horn-fish, the buffalo, . . a large fish resembling the carp, etc., paddlefish." "At places where the flat boats, laden with maize, land, the fish collect and assemble in great numbers, and fall an easy prey to the fisherman."

He commented that,

"The water of the river is clear and dark green, and the bottom, which is plainly seen, is covered with bivalve shells (*Unio*), as well as with several kinds of snails."

Great changes occurred between 1830 and 1845, as Cox (1860) writes,

". ..those prairies, for more than fifteen years past, have been like so many cultivated gardens. . . venison, wild turkeys, and fish . . . are now mostly brought from the Kankakee and the lakes."

The changes were caused by a rapidly increasing population, the establishment of extensive agriculture, and the export of agricultural goods. Navigable waters throughout the state were clearly needed, but the physical clearing of the land itself acted against this need in many of the smaller streams.

> "Unquestionably, White River is not so easily navigable now as it was ninety years ago, though probably as much water passes out through its channel in the course of a year as there did then. The flow is not so steady because of the clearing of the land and improved drainage, making the surface water pass off more rapidly. And this has increased the obstructions in the streams, for the soil, sand, and gravel wash much more easily from cleared land. Moreover, in the natural state, most of the timber that got into the river came from the undermining of banks on which it stood, and this usually did not float away, but hung by the roots where it fell. But after the axmen got to work, every freshet brought down logs and rails which formed drifts at some places. Some logs stranded as the water went down, decayed, became waterlogged, and made bases for sand and gravel bars." (Dunn 1910)

1825–1850

Canal building augmented environmental damage from agricultural erosion from 1828 to 1855. Beginning at Fort Wayne in 1828, the Wabash and Erie Canal crept steadily westward toward the mouth of the Tippecanoe River, which was considered the head of navigation for the Wabash. The canal paralleled the Wabash River on the north between Fort Wayne and Delphi, and on the south and east from Delphi to Terre Haute. At each tributary, an elevated, waterproof aqueduct had to be constructed and feeder canals channeled into the canal from tributary streams and wetlands. Tributary dams were frequently needed to ensure adequate water during the dry summer periods.

The upper portion of the Wabash and Erie Canal was in operation by July 4, 1835, but the tolls were inadequate to keep it in repair. The wooden aqueducts bridging every tributary were already rotten. The canal south of Lafayette had an insufficient supply of water throughout the 1840s, and "feeders" had to be developed from tributaries all along the canal. That same year, several sections of canal were constructed on the White River at Indianapolis and a 19-mi section on Pigeon Creek near Evansville, which, when completed, dried up completely.

The canal was instrumental in increasing the human population, and farms were enlarged by clearing and draining lands, which before were not considered worth cultivation (Anonymous 1907). More than 5,000 bushels of corn were shipped to Toledo in 1844. This increased 100 times in 1846 to 2,775,149 bushels in 1851. Water from the canal powered nine flour mills, eight saw mills, three paper mills, eight carding and fulling mills, two oil mills, and one iron foundry.

There is no record of the aggregate damage that must have been caused to the Wabash and its tributaries during this period. The zenith of canal usage was 1850 when perhaps 500 boats navigated it. By this time, repair or replacement was a regular feature of the canal which had 9 aqueducts, 37 locks, 5 dams, 71 road bridges, and 139 culverts between the state line and Perrysville alone. Water weeds clogging the canal had to be removed repeatedly by a specially invented submarine mower.

Spring floods in 1854 wrecked the Sugar Creek aqueduct and damaged another at Raccoon Creek. By this time, there were 1,300 mi of operating railroads and another 1,600 under construction. A repetition of floods, breaks, and droughts in 1857–1858 wrecked the Wildcat dam, carried away aqueducts over Wea and Shawnee creeks, and breached banks at a dozen places. Navigation was abandoned south of Terre Haute in 1860, and by 1870, little more than a succession of stagnant pools marked the site of the canal (Esarey 1912).

1875–1900

By the time David Starr Jordan arrived in 1874, the native fish populations were already altered as the result of clearing land and building dams and canals. However, he and his many students witnessed the first side effects of industrial development in Indiana. Streams and lakes in only two-thirds of Indiana were sampled during this period, but many of the fish lists produced were accompanied by general assessments of abundance (Appendix A)

Jordan assessed the Wabash River during this period as follows:

> "The upper Wabash and most of its tributaries are clear streams, . . . " but "towards its junction with the Ohio River, the Wabash becomes a large river with moderate current, the water not very clear, and the bottom covered with gravel and sand in which grow many water plants. The tributary streams are mostly sluggish and yellow with clay and mud" (Jordan 1890).

Nevertheless, he found the "fish fauna of the lower Wabash. . . to be unexpectedly rich," especially in the number of species of darters and their abundance.

The Commission of Fisheries of Indiana (1883) discussed the coming availability of a great new species that would put fish back into Indiana lakes and streams—the common carp.

During the first decade of the 1900s, Forbes and Richardson (1920) collected fish by seining five, shallow, hard-bottomed sites on the lower Wabash River between the Embarrass River and the Little Wabash River. Shortly thereafter, the problem of gross organic pollution and the human health aspects of pollution emerged.

> "Before our population was so concentrated, sewage disposal by dilution was satisfactory from a physical standpoint, but now the condition of many of our streams has become such that for a part of the year at least, the odors from them are quite obnoxious and a nuisance to the cities and to the population living along the banks, as well as a menace to their health" (Craven 1912).

Culbertson (1908) recognized the effects of deforestation on flows, stating that, "Streams that thirty years ago furnished abundant power for mills during ten months of the twelve now are even without flowing water for almost half the time."

He further stated that this "had a serious effect . . . upon the animal life of these streams."

> Rolfe (in Forbes and Richardson 1920) stated, "The waters of the Wabash are, like those of the Illinois and the Kaskaskia, commonly brown and opaque with suspended silt, never clearing even at the lowest stages; and the same is true of most of its tributary streams."

1940–1960

In the early 1940s, Gerking (1945) systematically collected fish at 412 sites scattered throughout Indiana in a comprehensive program, using a quarter-inch mesh seine, commercial gill-net captures, and verbal reports.He observed that,

> "Streams in the northern third of the state often run clear. . . ." "Creeks of the central and southern part of Indiana are usually turbid and warm." "Many of the southwestern streams are slow moving and usually carry a heavy load of suspended material."

Gerking (1945) found only 3 species of darters in the Wabash River at Delphi compared to 13 species found during Jordan's time, none compared to 4 species at Mt. Vernon, and 3 species compared to 12 at New Harmony. He attributed the decrease to increased siltation from soil erosion plus "city sewage, cannery waste, coal mine drainage, paper mill waste, and dairy-products factory waste."According to Visher (1944),

> "Deforestation has greatly increased runoff as have extensive drainage operations by open ditches and by tiles. Indiana has many miles of drainage ditches, 20,787 mi according to the 1930 Census of Drainage. Moreover,...there are...many thousand miles of small tile in addition to the 10,439 mi of large tile such as storm sewers...the extensive erosion of cultivated hillsides has carried large amounts of soil and other materials from the higher levels into the stream channels...."

Denham (1938) found poor fish assemblages in the White River downriver from Indianapolis, Indiana, where lethally low DO levels existed. In the early

1940s, Gerking (1945) found only 10 species at Indianapolis compared to 53 species found by Jordan (1878) 60 years earlier and remarked that, "The Indianapolis area is known to pollute the White River for many miles downstream." The problems of the White River continued to grow during 1950–1960, after which substantial progress was made to reduce pollution in Muncie and Indianapolis.

1965–2001

Investigations of fishes of the Wabash and White rivers were prompted by the passage of the Clean Water Act of 1967. Intensive, but restrictive, investigations near electric generating stations (EGS) (Gammon 1971, 1973, 1982) were soon expanded to systematic electrofishing in longer reaches of the river (Figure 2)(Gammon 1976, 1993, 1998). An Index of Well-Being (Iwb) was developed to provide an overall portrait of fish assemblage condition over time and space. Catch data from selected years were used to develop an Index of Biotic Integrity (IBI) (Gammon and Simon 2000).

The relative clarity of large rivers changed during the 1940s with the development of hybrid corn, which required additional fertilizer. Enhanced phytoplankton and periphyton densities facilitated movement of gizzard shad *Dorosoma cepedianum* into

Figure 2.—The Wabash River drainage basin with an enlargement of the Reaches studied of the middle Wabash river main stem.

previously unoccupied smaller tributaries. Secchi depths ranged from 25 to 35 cm because of algal densities with chlorophyll a reaching as high as 247 mg/L (Bridges et al. 1986).

Until 1984, the Wabash River fish assemblage was dominated by gizzard shad and river carpsucker *Carpiodes carpio* with contributions from a variety of catostomids, cyprinids, hyodontids, lepisosteids, and ictalurids, but very few centrarchids or percids. Visual feeding piscivores were especially scarce (catch rates = 2–3/km).

In addition to causing high turbidities during summer months, the masses of phytoplankton also produced a dissolved oxygen (DO) deficit, which imposed an additional environmental burden on the fishes. The quality of the fish assemblage as measured by the Iwb was negatively correlated to the DO deficit within 270 km of the middle Wabash.

The Iwb increased when most populations except for common carp *Cyprinus carpio* and gizzard shad improved dramatically during 1984–1992 (Figure 3). A four fold increase of predator species reduced gizzard shad to fewer and larger individuals until 1992 when reproductive success again burgeoned. Catches of channel catfish *Ictalurus punctatus*, a supposedly pollution tolerant species, increased 10-fold.

Catches of the shovelnose sturgeon *Scaphirhynchus platorynchus* doubled. The blue sucker *Cycleptus elongatus*, an Indiana species of special concern, tripled in numbers caught as populations expanded into previously unoccupied areas upriver and down. Near the junction of the White and Wabash rivers, it was recently found to be the dominant species (Stefanavage 1995). Mooneye *Hiodon tergisus*, sauger *Sander canadensis*, smallmouth bass, and spotted bass became important upriver from Lafayette, Indiana. Goldeye *Hiodon alosoides* increased further downriver.

Three factors probably influenced the sudden improvement: 1) stable summer flows leading to successful reproduction, 2) a long-term 50% reduction in biochemical oxygen demand loadings by cities and industries, and 3) the 1983 PIK (payment-in-kind) program, which reduced agricultural chemical loadings to the river by 25%.

A 1993 flood finalized this encouraging episode by eradicating early year-classes of many fishes and reducing the Iwb to extremely low levels for a 2-year period (Figure 4). Improvement of the assemblage in 1997–1998 was more dramatic than it had been in 1984–1992, especially in the lower reaches of river.

Less severe periods of high water and droughts produced shorter and more subtle effects. Diversities and abundances were lower than flanking years during droughts of 1983 and 1991, but not all river sections were affected equally.

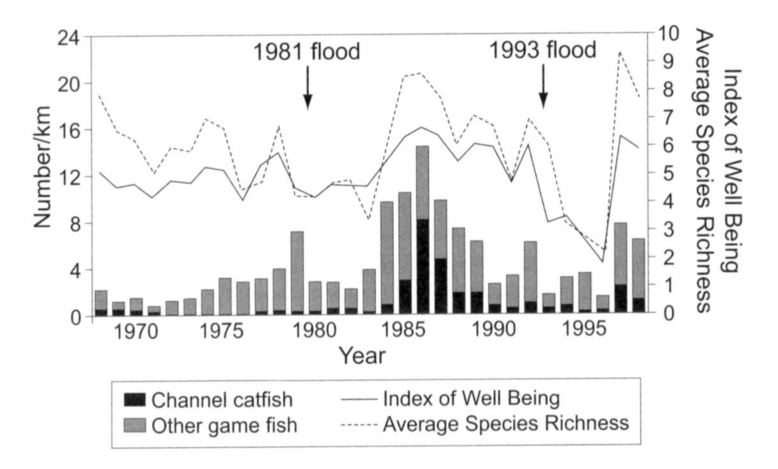

Figure 3.—Mean sport fish catch rate, number of species per kilometer, and IWB values from 1973 to 1998, all stations combined.

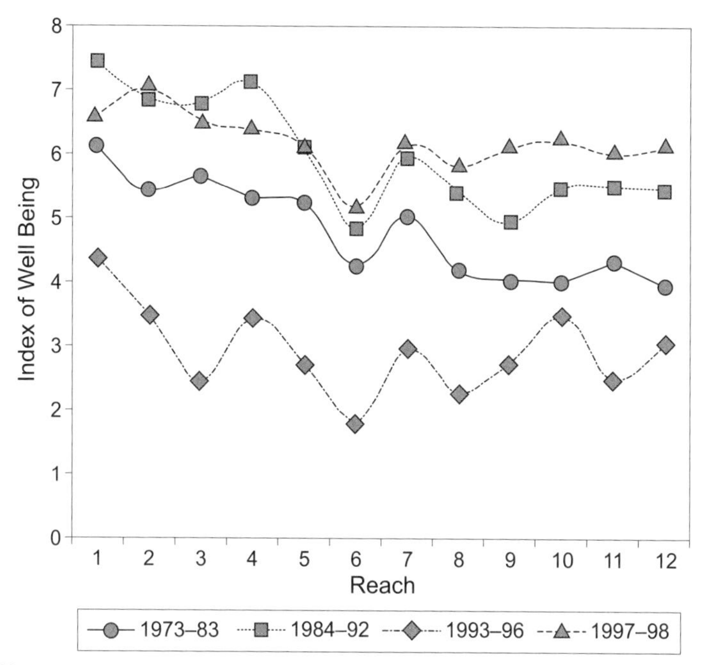

Figure 4. —Profiles over space of mean IWB values for four periods: 1973–1983, 1984–1992, 1993–1996, and 1997–1998.

Wabash and White river tributaries acted as refugia for the fish assemblages during periods of stress. In 1977, critically low DO concentrations caused fish to concentrate at the mouth of Sugar Creek (Wabash River) in 1977 (Gammon and Reidy 1981). Tributaries also served as refugia during the 1988 drought. Compared to 1987, a "normal" flow year, the 1988 IBI profile was severely depressed downriver from Lafayette/W. Lafayette except at stations located near clean tributaries (Gammon and Simon 2000).

Tributary fish populations also serve to replenish the assemblages of the Wabash River after fish kills in 1983 and 1986 produced by interactions of flow, phytoplankton, and heat additions. Phytoplankton density increased sharply near the Cayuga EGS during very low flows. The algae then settled out in the heated, lentic pool downriver, decomposed and reduced DO to very low levels (Parke and Gammon 1986; Yost and Rives 1990). This phenomenon may be common in most large rivers throughout the corn belt and undoubtedly affects local fish assemblages. It also presages the formation of the "Dead Zone" in the Gulf of Mexico.

Assessments of upper White River fish assemblages after 1960 provide direct evidence of municipal waste treatment. Sport and commercially important fishes from both forks of the White River were systematically collected by Christensen (1968), followed by more intensive studies by Kingsley (1983), Braun (1984), Simon (1992), and Foy (1996). In 1965, the only fish found immediately downriver from Indianapolis were small carp, but by 1990 a much more diverse assemblage was present. By 1999, largemouth and smallmouth bass, flathead catfish *Pilodictis olivaris*, yellow bass *Morone mississippiensis*, and other "sport" fishes comprised half of the electrofishing catch and provided excellent recreational fishing in the White River north of Indianapolis (Lewis et al. 2000).

In mid-December 1999, a toxic release of thyram from an Anderson, Indiana industry ravaged the sewage treatment plant and the combination of thyram and untreated domestic wastes flowed unabated into White River for 12 d. An estimated 187 tons of fish were killed in 50 mi of river from Anderson to Indianapolis.

The ensuing consent decree preceded current efforts to rehabilitate the fish assemblage and improve the overall instream and riparian environment. Fish from tributaries quickly colonized the main stem. Bass returned in modest numbers 2 years after the destruction. Nearly 1 million predator species were stocked by January 1, 2003, and former resident species will be reintroduced in the future. Fish in the main stem are being assessed biannually by Indiana Department of Natural Resources, Division of Fish and Wildlife. The riparian corridor and fish assemblages of tributaries are currently being appraised.

Some species have recently extended their ranges northward. Spotted bass are now as far north as the Eel River near Logansport, Indiana (Gammon and Gammon 1993). Threadfin shad *Dorosoma petenense* penetrated into the lower Wabash River. The western mosquitofish *Gambusia affinis* is now found in central Indiana's backwaters and small streams (Gammon et al. 2003).

Some species, uncollected since 1900, were rediscovered recently by electrofishing. Most of these fishes now constitute Indiana's Endangered Species List (Appendix A). The lake sturgeon *Acipenser fulvescens* was discovered in the lower White River, the only known nucleus in the entire Ohio River valley (B. Fisher, Indiana Department of Natural Resources, personal communication).

Seven species known to have been early residents of the Wabash system have not been found and are officially regarded as extirpated (Table 1). Alligator gar *Atractosteus spatula* occupied this list until a single specimen was found in the White River in 1993 at Hazelton by Baker and Frey (cited by Crawford et al. 1996)

Other species entered the Wabash system by range extension. The red shiner *Cyprinella lutrensis* entered the lower Wabash River from Illinois tributaries. Rosefin shiner *Lythrurus ardens* expanded its range from Ohio River tributaries into the East Fork of the White River. Several alien fishes have recently invaded from the Ohio River. Large grass carp *Ctenopharyngodon idella* were first found during the 1980s. The bighead carp *Hypophthalmichthys nobilis* and silver carp *H. molitrix* appeared more recently and are now reported throughout the Wabash River. Small bighead carp were observed recently in a backwater south of Terre Haute by Fisher (personal communication), indicating successful reproduction. The alien zebra mussel *Dreissena polymorpha* has also reached the Wabash River via the Tippecanoe River and northern Indiana lakes.

Table 1.—Species of fishes believed to be extirpated from the Wabash and White river drainages.

Common name	Scientific name
Alabama shad	*Alosa alabamae*
crystal darter	*Crystallaria asprella*
Ohio muskellunge	*Esox masquinongy ohiensis*
harelip sucker	*Moxostoma lacerum*
popeye shiner	*Notropis ariommus*
stargazing darter	*Percina uranidea*
saddleback darter	*P. vigil*

Summary

During the early 1800s, Wabash and White River fishes were numerous and diverse with an abundance of large sight feeding piscivores and a diversity of smaller species. Environmental conditions deteriorated thereafter because of the growth of the human population as forests and prairies were replaced by agricultural fields and canals and dams were built. From 1828 to 1856, the construction and operation of the Wabash and Erie Canal, and less extensive canals on the White River, damaged fish assemblages seriously as a result of water withdrawls.

By 1900, many prized species of fishes had diminished sufficiently so that stocking programs for native fishes were initiated and the common carp was introduced. Conversion of forests and wetlands to agriculture continued. In addition, industrial development was seriously polluting rivers near population centers such as Indianapolis.

Draining wetlands and channelizing tributary headwaters accelerated during the 1900–1950 period. For example, by 1917, the entire 402-km length of the Kankakee River, located immediately north of the upper Wabash River, had been converted to an 132-km-long straight ditch. Sections of the Wabash and White rivers and their tributaries were channelized and leveed to protect agriculture and a growing human populations from flooding. The development of hybrid corn in the 1940s, accompanied by increased use of fertilizer, herbicides, and pesticides, altered water quality physically and chemically.

Channel and flow modification continue to affect Indiana's rivers and fishes. Only 1 dam has been built on the Wabash main stem, but 17 in the upper White River block fish migrations. Many farm ponds have been created (circa 40,000) flushing small largemouth bass, bluegill *Lepomis macrochirus*, and other stocked species into headwaters after every heavy rain. Nine large, off-main-stem reservoirs, created since 1950, and uncounted numbers of smaller ones provide the same service for tributaries. The strategy of reservoir management has shifted from simply stocking bass/bluegill to stocking walleye, walleye/sauger hybrids, hybrid striped bass, and muskellunge. Escapees provide fishing opportunities that were unavailable a few years ago.

The maintenance of old channelized headwaters and the creation of new ones persists and riparian integrity is breached along almost every river. The smallmouth bass finds refuge mainly in the lower reaches of tributaries sufficiently long and distant from headwater insults (Gammon et al. 2003), but is unable to permanently occupy the lower murky main-stem waters of the Wabash and White rivers. Industries and cities sporadically pollute and cause fish kills, but they have improved to the point where nonpoint pollution clearly is the greater affliction to be addressed in future decades. The gradual replacement of older, less environmentally friendly methods of farming by no-till shows promise. Some flood-prone agricultural fields are currently being abandoned or brought into ownership by various state land trusts, thereby offering hope that continuous forested riparian borders will be reestablished. However, the effects on fish of the appreciable amounts of atrazine and other pesticides to the White and Wabash rivers are unknown (Carter et al. 1995; Crawford 1995).

Most of the ichthyofauna has survived, despite 200 years of persistent and varied abuse of the Wabash and White rivers. Twelve species of fish have been eradicated and many others are reduced in distribution and abundance. Nevertheless, 110 species of fish were found in 274 km of the middle Wabash River in 31 years of sampling (Gammon 1998). Simon (1998) collected 78 species in a single series of collections following the 1993 flood. Visual predator and darter species are uncommon. Fish assemblages are most affected by excessive nutrient and sediment runoff as are other large rivers in the corn belt (Durham 1993; Kavanaugh 1993). Although not in themselves toxic, nutrients stimulate excessive algal growth, which interferes with normal water clarity, and create DO problems, while sediment smothers spawning substrate and macroinvertebrates.

Apart from large industries and cities, almost all of the problems of the Wabash and White rivers are traceable to their tributaries, and it is here in a less formidable, more familiar setting that they must be addressed. The task of improving and protecting tributaries will be difficult, but less costly than first destroying and then rehabilitating them after understanding their value.

Acknowledgments

I extend my appreciation for the assistance of those whose familiarity with running water fish assemblages in various areas of Indiana has been acquired over decades of assiduous work. My special thanks to Harold McReynolds, Ed Braun, and Brandt Fisher.

References

Anonymous. 1907. The Wabash and Erie Canal. Indiana Magazine of History 3(3):100–107.

Braun, E. R. 1984. A fisheries investigation of the West Fork of the White River in Randolph, Delaware, and Madison counties. Indiana Department of Natural Resources, Indianapolis.

Bridges, C. L., H. L. BonHomme, G. R. Bright, D. E. Clark, J. K. Ray, J. L. Rud, and J. L. Winters. 1986. Summer primary productivity and associated data for four Indiana rivers. Proceedings of the Indiana Academy of Science 96:309–314.

Carter, D. S., M. J. Lydy, and C. G. Crawford. 1995. Water-quality assessment of the White River basin, Indiana—analysis of available information on pesticides, 1972–1992. U.S. Geological Survey, Water-Resources Investigations Report 94–4024, Indianapolis, Indiana.

Christensen, D. 1968. The distribution of fishes throughout the White River system and the effects of various environmental factors upon the commercial fishery. Indiana Department of Natural Resources, Indianapolis, .

Cox, S. C. 1860. Recollections of the early settlement of the Wabash Valley. Courier Steam Book & Job Printing House, Lafayette, Indiana.

Craven, J. 1912. The protection of our rivers from pollution. Proceedings of the Indiana Academy of Science 22:47–50.

Crawford, C. G. 1995. Occurrence of pesticides in the White River, Indiana, 1991–1995. U.S. Geological Survey, Fact Sheet 233–95, Indianapolis, Indiana.

Crawford, C. G., M. J. Lydy, and J. W. Frey. 1996. Fishes of the White River basin, Indiana. U.S. Geological Survey, Water Resources Investigations Report 96–4232, Indianapolis, Indiana.

Culbertson, G. 1908. Deforestation and its effects among the hills of southern Indiana. Proceedings of the Indiana Academy of Science 18:27–37.

Denham, S. C. 1938. A limnological investigation of the West Fork and common branch of White River. Indiana Department of Conservation, Indiana University Biological Station, Investigations of Indiana Lakes and Streams, No. 5, Indianapolis, Indiana.

Dunn, J. P. 1910. Greater Indianapolis—the history, the industries, the institutions, and the people of a city of homes. Lewis Publishing, Chicago.

Durham, L. 1993. The Embarras River. Pages 101–122 *in* L. W. Hesse, C. B. Stalnaker, N. G. Benson, and J. R. Zuboy, editors. Restoration planning for the rivers of the Mississippi River ecosystem. National Biological Survey, Biological Report 19, Washington, D.C.

Esarey, L. 1912. Internal improvements in early Indiana. Indiana Historical Society, Publication V(2), Indianapolis.

Forbes, S. A., and R. E. Richardson. 1920. The fishes of Illinois, 2nd edition. Natural History Survey of Illinois 3:1–357.

Foy, J. R. 1996. Fish population report. Pages 1–57 *in* Muncie Bureau of Water Quality. Annual Report 1995. Biological stream monitoring-fish/macroinvertebrates/mussels. Muncie Bureau of Water Quality, Muncie, Indiana.

Gammon, J. R. 1971. The response of fish populations in the Wabash River to heated effluents. Pages 513–523 *in* Proceedings of the Third National Symposium of Radioecology, Oak Ridge, Tennessee.

Gammon, J. R. 1973. The effect of thermal inputs on the populations of fish and macroinvertebrates in the Wabash River. Purdue University, Water Resources Research Center, Technical Report No. 32, Lafayette, Indiana.

Gammon, J. R. 1976. The fish populations of the middle 340 km of the middle Wabash River. Purdue University, Water Resources Research Center, Technical Report No. 86, Lafayette, Indiana.

Gammon, J. R. 1982. Changes in the fish community of the Wabash River following power plant start-up: projected and observed. Pages 350–366 *in* W. E. Bishop, R. D. Cardwell, and B. B. Heidolph, editors. Aquatic Toxicology and Hazard Assessment: Sixth Symposium. American Society for Testing and Materials STP 802, Philadelphia.

Gammon, J. R. 1993. The Wabash River: progress and promise. Pages 142–161 *in* L. W. Hesse, C. B. Stalnaker, N. G. Benson, and J. R. Zuboy, editors. Restoration planning for the rivers of the Mississippi River ecosystem. National Biological Survey, Biological Report 19, Washington, D.C.

Gammon, J. R. 1998. The Wabash River ecosystem. Indiana University Press, Bloomington.

Gammon, J. R., W. C. Faatz, and T. P. Simon. 2003. Patterns of water quality and fish assemblages in three central Indiana streams with emphasis on animal feed lot operations. Pages 373–417 *in* T. P. Simon, editor. Biological response signatures: indicator patterns using aquatic communities. CRC Press, Boca Raton, Florida.

Gammon, J. R., and C. W. Gammon. 1993. Changes in the fish community of the Eel River resulting from agriculture. Proceedings of the Indiana Academy of Science 102:67–82.

Gammon, J. R., and J. M. Reidy. 1981. The role of tributaries during an episode of low dissolved oxygen in the Wabash River. Pages 396–407 *in* L. A. Krumholz, editor. The warmwater streams symposium. American Fisheries Society, Southern Division, Lawrence, Kansas.

Gammon, J. R., and T. P. Simon. 2000. Variation in a great river index of biotic integrity over a 20-year period. Hydrobiologia 422/423:291–304.

Gerking, S. C. 1945. Distribution of the fishes of Indiana. Investigations of Indiana Lakes and Streams, Volume III:1–137. Indiana Department of Conservation and Indiana University, Indianapolis.

Hyde, G. E. 1962. Indians of the Woodlands from prehistoric times to 1725. University of Oklahoma Press, Norman.

Jordan, D. S. 1890. Report of explorations made during the summer and autumn of 1888, in the Allegheny region of Virginia, North Carolina and Tennessee and in western Indiana, with an account of the fishes found in the river basins of those regions. U.S. Fisheries Commission Bulletin 8.

Jordan, D. W. 1878. Catalogue of the fishes of Indiana. Appendix A in First Annual Report of the Commission of Fisheries of Indiana, 1883:96–103. Indianapolis Journal Company, Indianapolis, Indiana.

Kavanaugh, C. M. 1993. Minnesota River basin assessment project. Pages 5–22 *in* L. W. Hesse, C. B. Stalnaker, N. G. Benson, and J. R. Zuboy, editors. Restoration planning for the rivers of the Mississippi River ecosystem. National Biological Survey, Biological Report 19, Washington, D.C.

Kingsley, D. W. 1983. Fisheries survey of the West Fork of White River and tributaries in Marion County, 1982 stream survey report. Indiana Department of Natural Resources, Indianapolis.

Lewis, R. B., J. E. Pike, E. M. Oliver, and W. B. Taylor. 2000. Fish community analysis near the Noblesville generating station 1999. PSI Energy, Environmental Services, Technical Report EPW-069, Plainfield, Indiana.

Maximilian, Prince of Wied. 1843. Travels in the interior of North America. Ackerman and Co., London.

McCord, S. S. 1970. Travel accounts of Indiana 1679–1961. Indiana Historical Bureau, Indiana Historical Collections volume XLVII, Indianapolis.

Parke, N. J., and J. R. Gammon. 1986. An investigation of phytoplankton sedimentation in the middle Wabash River. Proceedings of the Indiana Academy of Science 95:279–288.

Rafinesque, C. S. 1820. Icthyologia Ohiensis or natural history of the fishes inhabiting the river Ohio and its tributary streams. Reprint Edition 1970 by Arno Press Inc., New York.

Simon, T. P. 1992. Biological criteria development for large rivers with an emphasis on an assessment of the White River drainage, Indiana. U.S. Environmental Protection Agency, EPA 905/R-92/006, Chicago.

Simon, T. P. 1998. Development of index of biotic integrity expectations for the Wabash River. U.S. Environmental Protection Agency, EPA 905-R-96–005, Chicago.

Stefanavage, T. C. 1995. Standing stock estimation of sport and commercial fish species in the lower one mile reach of the West Fork White River. Indiana Department of Natural Resources, Indianapolis.

Thomas, D. 1819. Travels through the western country in the summer of 1816, including notes of the natural history. David Rumsey, Auburn, New York.

Yost, J. C., and S. R. Rives. 1990. Recalibration of Wabash River dissolved oxygen model. EA Engineering, Science, and Technology, Inc., Sparks, Maryland.

Visher. 1944. Indiana floods. Proceedings of the Indiana Academy of Science 54:134–141.

Appendix A.—Relative abundance of fish species collected from the Wabash River system during four eras from 1800 to 2000, Ohio River and Lake Michigan species not included. (Code: o = not recorded, x = present, 1 = very rare, 2 = rare, 3 = not uncommon, 4 = common, 5 = abundant)

Family, common name, and scientific name	1820	1900	1945	2000
lamprey family–Petromyzontidae				
least brook lamprey *Lampetra aepyptera* (Abbott)	o	o	1	1
American brook lamprey *L. appendix* (DeKay)	x	o	2	2
silver lamprey *Ichthyomyzon unicuspis* (Hubbs & Trautman)	o	3	2	3
chestnut lamprey *I. castaneus* (Girard)	x	o	2	1
sturgeon family–Acipenseridae				
shovelnose sturgeon *Scaphirhynchus platorynchus* (Rafinesque)	4	3	3	3
lake sturgeon *Acipenser fulvescens* (Rafinesque)[ab]	x	3	o	1
paddlefish family–Polyodontidae				
paddlefish *Polyodon spathula* (Walbaum)	2	3	3	3
gar family–Lepisosteidae				
longnose gar *Lepisosteus osseus* (Linnaeus)	4	4	4	4
shortnose gar *L. platostomus* (Rafinesque)	x	o	3	3
spotted gar *L. oculatus* (Winchell)[c]	o	o	3	3
alligator gar *Atractosteus spatula*	x	1	o	1
bowfin family–Amiidae				
bowfin *Amia calva* (Linnaeus)	o	3	3	3
mooneye family–Hiodontidae				
goldeye *Hiodon alosoides* (Rafinesque)[c]	4	1	2	2
mooneye *H. tergisus* (Lesueur)[c]	4	1	1	2
freshwater eel family–Anguillidae				
American eel *Anguilla rostrata* (Lesueur)[b]	2	o	o	2
herring family–Clupeidae				
skipjack herring *Alosa chrysochloris* (Rafinesque)[c]	3	1	1	2
gizzard shad *Dorosoma cepedianum* (Lesueur)[c]	3	4	4	5
threadfin shad *D. petenense* (Gunther)	o	o	o	2
minnow family–Cyprinidae				
stoneroller *Campostoma anomalum* (Rafinesque)	5	5	5	5
goldfish *Carassius auratus* (Linnaeus)	o	o	2	2
grass carp *Ctenopharyngodon idella* (Valenciennes)	o	o	o	3
red shiner *Cyprinella lutrensis* (Baird & Girard)	o	o	o	1
spotfin shiner *C. spiloptera* (Cope)	o	5	5	5
steelcolor shiner *C. whipplei* (Girard)	o	5	4	5
common carp *Cyprinus carpio* (Linnaeus)[c]	o	2	3	5
streamline chub *Erimystax dissimilis* (Kirtland)	o	3	2	3
gravel chub *E. x–punctatus* (Hubbs & Crowe)	o	o	3	3
Mississippi silvery minnow *Hybognathus nuchalis* (Agassiz)	o	3	3	4
bigeye chub *Hybopsis amblops* (Rafinesque)	x	3	3	3
silver carp *Hypophthalmichthys molitrix* (Valenciennes)	o	o	o	3
bighead carp *H. nobilis* (Richardson)	o	o	o	3
striped shiner *Luxilus chrysocephalus* (Rafinesque)	x	3	4	5
common shiner *L. cornutus* (Mitchill)	x	x	x	5
redfin shiner *L. umbratilis* (Girard)	x	3	3	4
ribbon shiner *L. fumeus* (Evermann)	o	1	1	2
silver chub *Macrhybopsis storeriana* (Kirtland)	o	2	2	3
speckled chub *M. aestivalis* (Girard)	x	2	2	2
hornyhead chub *Nocomis biguttatus* (Kirtland)	x	3	4	4
river chub *N. micropogon* (Cope)	x	3	4	3
golden shiner *Notemigonus crysoleucus* (Mitchill)	3	3	3	4
pallid shiner *Hybopsis amnis* (Hubbs and Greene)	o	o	2	1
rosefin shiner *Lythrurus ardens* (Jordan)	o	o	3	3

Appendix A.—Continued.

Family, common name, and scientific name[c]	1820	1900	1945	2000
pugnose shiner *Notropis anogenus* (Forbes)	o	1	1	1
popeye shiner *N. ariommus* (Cope)	o	2	o	o
emerald shiner *N. atherinoides* (Rafinesque)	4	3	3	4
river shiner *N. blennius* (Girard)[c]	o	2	2	3
bigeye shiner *N. boops* (Girard)	o	1	2	2
silverjaw minnow *N. buccatus* (Cope)	o	4	4	5
ghost shiner *N. buchanani* (Meek)	o	x	x	1
ironcolor shiner *N. chalybaeus* (Abbott)	o	o	2	3
bigmouth shiner *N. dorsalis* (Agassiz)	o	o	o	1
blackchin shiner *N. heterodon* (Cope)	o	o	2	3
blacknose shiner *N. heterolepis* (Eigenmann & Eigenmann)	o	o	2	2
spottail shiner *N. hudsonius* (Clinton)	o	2	2	3
sand shiner *N. stramineus* (Cope)[c]	o	o	4	4
silver shiner *N. photogenis* (Cope)	o	1	3	3
rosyface shiner *N. rubellus* (Agassiz)	o	3	3	4
silverband shiner *N. shumardi* (Girard)	o	o	1	1
mimic shiner *N. volucellus* (Cope)	o	o	3	4
channel shiner *N. wickliffi* (Cope)	o	o	3	3
pugnose minnow *Opsopoeodus emiliae* (Hay)	o	1	1	o
suckermouth minnow *Phenacobius mirabilis* (Girard)	o	o	3	4
southern redbelly dace *Phoxinus erythrogaster* (Rafinesque)	o	3	3	3
bluntnose minnow *Pimephales notatus* (Rafinesque)	x	5	5	5
fathead minnow *P. promelas* (Rafinesque)	2	2	2	3
bullhead minnow *P. vigilax* (Baird & Girard)	o	3	3	3
creek chub *Semotilus atromaculatus* (Mitchill)	x	3	5	5
eastern blacknose dace *Rhinichthys atratulus* (Hermann)	o	3	3	4
sucker family–Catostomidae				
river carpsucker *Carpiodes carpio* (Rafinesque)	x	o	x	5
quillback *C. cyprinus* (Lesueur)	x	o	3	4
highfin carpsucker *C. velifer* (Rafinesque)	3	o	3	3
white sucker *Catostomus commersonii* (Lacepede)[c]	x	o	3	5
blue sucker *Cycleptus elongatus* (Lesueur)[c]	2	2	o	3
creek chubsucker *Erimyzon oblongus* (Mitchell)	x	o	3	3
lake chubsucker *E. sucetta* (Lacepede)	o	o	3	4
northern hog sucker *Hypentelium nigricans* (Lesueur)	x	4	4	5
smallmouth buffalo *Ictiobus bubalus* (Rafinesque)	4	3	2	4
bigmouth buffalo *I. cyprinellus* (Valenciennes)	4	3	3	4
black buffalo *I. niger* (Rafinesque)	x	1	o	2
spotted sucker *Minytrema melanops* (Rafinesque)	x	3	3	4
harelip sucker *Moxostoma lacerum* (Jordan and Brayton)	o	1	o	o
greater redhorse *M. valenciennesi* (Jordan)[a,b]	o	o	1	2
black redhorse *M. duquesnei* (Lesueur)[c]	x	2	3	4
silver redhorse *M. anisurum* (Rafinesque)	x	o	3	3
golden redhorse *M. erythrurum* (Rafinesque)	3	3	4	5
shorthead redhorse *M. macrolepidotum* (Lesueur)	o	o	3	4
river redhorse *M. carinatum* (Cope)	o	2	o	3
catfish family–Ictaluridae				
black bullhead *Ameiurus melas* (Rafinesque)[c]	x	3	3	4
yellow bullhead *A. natalis* (Lesueur)	x	3	3	4
brown bullhead *A. nebulosus* (Lesueur)	o	x	2	3
blue catfish *Ictalurus furcatus* (Lesueur)	x	o	1	3
channel catfish *I. punctatus* (Rafinesque)[c]	x	3	3	5
stonecat *Noturus flavus* (Rafinesque)	x	3	3	3
mountain madtom *N. eleutherus* (Jordan)	o	1	2	2

Appendix A.—Continued.

Family, common name, and scientific name	1820	1900	1945	2000
tadpole madtom *N. gyrinus* (Mitchill)	o	2	3	3
brindled madtom *N. miurus* (Jordan)	o	3	3	3
freckled madtom *N. nocturnus* (Jordan & Gilbert)	o	1	2	2
northern madtom *N. stigmosus* (Taylor)	o	o	o	1
flathead catfish *Pylodictis olivaris* (Rafinesque)	x	3	3	4
pike family–Esocidae				
grass pickerel *Esox americanus vermiculatus* (Leseur)	x	3	3	4
northern pike *E. lucius* (Linnaeus)	2	3	o	3
muskellunge *E. masquinongy* (Mitchell)	2	1	1	2
mudminnow family–Umbridae				
central mudminnow *Umbra limi* (Kirtland)	o	3	3	3
pirate perch family–Aphredoderidae				
pirate perch *Aphredoderus sayanus* (Gilliams)	o	3	3	3
codfish family–Gadidae				
burbot *Lota lota* (Linnaeus)	o	1	1	1
killifish family–Fundulidae				
northern studfish *Fundulus catenatus* (Storer)	o	o	1	2
banded killifish *F. diaphanus* (Lesueur)	o	2	2	3
starhead topminnow *F. dispar*	o	2	2	3
blackstripe topminnow *F. notatus* (Rafinesque)	o	3	3	3
livebearer family–Poeciliidae				
western mosquitofish *Gambusia affinis* (Baird & Girard)	o	2	2	3
silversides family–Atherinidae				
brook silverside *Labidesthes sicculus* (Cope)	o	3	3	4
stickleback family–Gasterosteidae				
brook stickleback *Culaea inconstans* (Kirtland)	o	2	1	1
temperate bass family–Moronidae				
white bass *Morone chrysops* (Rafinesque)	3	2	3	4
yellow bass *M. mississippiensis* (Jordan & Eigenmann)	o	1	2	3
striped bass *M. saxatilus* (Walbaum)	o	o	o	3
sunfish family–Centrarchidae				
rock bass *Ambloplites rupestris* (Rafinesque)	o	4	4	4
flier *Centrarchus macropterus* (Lacepede)	o	o	2	2
green sunfish *Lepomis cyanellus* (Rafinesque)	x	4	4	4
pumpkinseed *L. gibbosus* (Linnaeus)	o	3	3	4
orangespotted sunfish *L. humilis* (Girard)	o	3	3	3
longear sunfish *L. megalotis* (Rafinesque)	o	5	5	5
warmouth *L. gulosus* (Cuvier)	o	3	3	3
bluegill *L. macrochirus* (Rafinesque)	x	3	4	5
redear sunfish *L. microlophus* (Gunther)	o	2	3	3
spotted sunfish *L. punctatus* (Valenciennes)	o	x	o	2
bantam sunfish *L. symmetricus* (Forbes)[a, b]	o	o	o	1
smallmouth bass *Micropterus dolomieu* (Lacepede)	4	4	4	4
spotted bass *M. punctulatus* (Rafinesque)	4	3	3	4
largemouth bass *M. salmoides* (Lacepede)	4	4	4	5
white crappie *Pomoxis annularis* (Rafinesque)	3	3	3	4
black crappie *P. nigromaculatus* (Lesueur)	o	3	3	3
perch family–Percidae				
crystal darter *Crystallaria asprella* (Jordan)	o	2	o	o
western sand darter *Ammocrypta clara* (Jordan & Meek)	o	o	1	2
eastern sand darter *Ammocrypta pellucida* (Putnam)	o	3	3	3
mud darter *Etheostoma asprigene* (Forbes)	o	2	2	2
greenside darter *E. blennioides* (Rafinesque)	4	3	4	4

Appendix A.—Continued.

Family, common name, and scientific name	1820	1900	1945	2000
rainbow darter *E. caeruleum* (Storer)	o	4	4	4
bluebreast darter *E. camurum* (Cope)[b]	o	2	2	3
bluntnose darter *E. chlorosomum* (Hay)	2	2	2	2
fantail darter *E. flabellare* (Rafinesque)	4	3	3	3
slough darter *E. gracile* (Girard)	o	2	2	2
harlequin darter *E. histrio* (Jordan & Gilbert)[b]	o	1	o	1
spotted darter *E. maculatum* (Kirtland)[b]	o	2	o	2
least darter *E. microperca* (Jordan & Gilbert)	o	2	2	3
johnny darter *E. nigrum* (Rafinesque)[c]	o	3	4	5
orangethroat darter *E. spectabile* (Agassiz)	o	3	3	4
spottail darter *E. squamiceps* (Jordan)[b]	o	2	o	3
Tippecanoe darter *E. tippecanoe* (Jordan & Evermann)[b]	o	2	2	3
variegate darter *E. variatum* (Kirtland)[a, b]	o	2	2	2
banded darter *E. zonale* (Cope)	o	2	2	2
yellow perch *Perca flavescens* (Mitchill)	o	3	3	4
logperch *Percina caprodes* (Rafinesque)[c]	4	4	4	5
channel darter *P. copelandi* (Jordan)[a]				
gilt darter *P. evides* (Jordan & Copeland)[a, b]	o	2	1	1
blackside darter *P. maculata* (Girard)	o	3	3	4
slenderhead darter *P. phoxocephala* (Nelson)	o	3	3	3
dusky darter *P. sciera* (Swain)	o	3	3	3
river darter *P. shumardi* (Girard)	o	2	2	2
stargazing darter *P. uranidea* (Jordan & Gilbert)	o	1	o	o
saddleback darter *P. vigil* (Hay)	o	x	o	o
sauger *Sander canadensis* (Smith)	o	3	2	3
walleye *S. vitreus* (Mitchill)	3	2	2	3
sculpin family–Cottidae				
mottled sculpin *Cottus bairdii* (Girard)	o	3	3	4
banded sculpin *C. carolinae* (Gill)	o	3	3	3
drum family–Sciaenidae				
freshwater drum *Aplodinotus grunniens* (Rafinesque)[c]	5	3	3	5

[a] Indiana endangered fishes.
[b] Recently rediscovered fishes.
[c] Species of fishes increasing from 1940–1950 to 2000.

American Fisheries Society Symposium 45:383–398, 2005

Changes in the Biological Integrity of Fish Assemblages in the Patoka River Drainage as a Result of Anthropogenic Disturbance from 1888 to 2001

THOMAS P. SIMON*

U.S. Fish and Wildlife Service, 620 South Walker Street, Bloomington, Indiana 47403-2121, USA

RONDA L. DUFOUR

Dufour Consultants, 5609 Michigan Road, Indianapolis, Indiana 46201, USA

BRANT E. FISHER

Indiana Department of Natural Resources, Atterbury Fish and Wildlife Area, 7970 South Rowe Street, Post Office Box 3000, Edinburgh, Indiana 46124-3000, USA

Abstract.—The Patoka River drainage is a lowland-gradient watershed of the Wabash River lowlands in southwestern Indiana. During the late 18th century, the river was part of an extensive riparian floodplain wetland that connected the White River with the lower Wabash River. Through anthropogenic changes as a result of ditching, channelization, levee creation, coal extraction, and oil and gas exploration, the Patoka River drainage has been highly altered. These changes have resulted in a loss of site-specific biological diversity and integrity, causing drainage-wide biological diversity decline. Extirpations in the watershed have resulted in the local loss of 12.7% of the fish fauna during the last century. The local extirpations of six species included central mudminnow *Umbra limi*, black redhorse *Moxostoma duquesnei*, brindled madtom *Noturus miurus*, bluebreast darter *Etheostoma camurum*, slenderhead darter *Percina phoxocephala*, and saddleback darter *P. vigil*. Black redhorse, bluebreast darter, slenderhead darter, and saddleback darter were only known from pre-1900, while brindled madtom and central mudminnow were known until the early 1940s. These species may have been rare to begin with in the Patoka River drainage, but since they are widespread elsewhere, it seems more probable that they disappeared as a result of the land-use changes. Sensitive species of darters and minnows have declined in abundance, but recent sampling has shown that they remain in the watershed at low abundance. Based on a probability sample, less than 12% of the channels represented reference least-disturbed conditions, while 61% exhibited degraded conditions.

Introduction

The nationwide modification of streams as a result of extensive land-use changes at the turn of the 19th century has resulted in the loss of floodplain wetlands (Gammon 1998). Loss of floodplain wetlands has reduced species richness and has made other species either rare or imperiled (Delleur 1994; Gammon 1994). With the loss of riparian wetland habitats came an increased dependence on stream permanence; however, the increased activity of humans within headwater

* Corresponding author: Thomas_Simon@fws.gov

streams has decreased low flows, and desiccated streams (Exl and Simon 2002; Simon and Exl 2002). To compensate for the lack of water permanence, aquatic life in the Patoka River drainage was forced into main channel habitats that were eventually channelized and leveed (Simon et al. 1994).

The history of degradation in the Patoka River drainage is essentially a history of all streams in the eastern coal field of southwestern Indiana and southeastern Illinois. The area shares many of the same environmental impacts as a result of acid mine drainage (Corbett 1969; Gray 1979; Renn 1989). Mineral extraction and oil and gas exploration have left an indelible image on the Patoka River watershed (Allen et al. 1978). Prior to 1890, little effort was made to document the fishes present in the Patoka River (Simon et al. 1994). Investigations were mainly concerned with the presence of fish as food to entice settlers to move into the Indiana territory (Gammon 1998). After the expansion, numerous investigations of the Patoka River drainage occurred. Jordan (1890) investigated the Patoka River near the town of Patoka (Gibson county) and documented the species of fishes found. Moenkhaus (1896) collected from two sites near Huntingburg (Dubois county), and Hubbs and Lagler (1942) and Lagler and Ricker (1942) collected from Foots Pond (Gibson county). Despite Foots Pond being outside the immediate Patoka River watershed, it represents species of fish that would probably have been present in the many oxbow ponds that occurred along the Patoka River. Further efforts by Gerking (1945) during 1940–1944 at eight sites included four sites on the Patoka River main stem. Additional investigations by the Indiana Department of Natural Resources (DNR) (Hottell 1978; Stefanavage 1993) during 1970–1991 documented species found in the river and its tributaries. Hottell evaluated 12 main-stem sites on the Patoka River, Houchins Ditch, and 12 tributaries, while Stefanavage collected from 8 main-stem locations. Simon et al. (1994) documented the present and historic distribution of fish in the Patoka River watershed based on 13 main-stem and 36 tributary sites supplemented with unpublished information from Lawrence M. Page, Illinois Natural History Survey (11 tributary sites) and the Indiana Department of Transportation (2 main-stem and 13 tributary sites). Carnahan (1997a, 1997b) surveyed five oxbow lakes and a single wetland in the Patoka River drainage. During additional sampling in these oxbows, he found paddlefish *Polyodon spathula* in 1997 while collecting fish for the state fair display. Additional, unpublished survey data (T. P. Simon and B. E. Fisher, unpublished data), including information from the Indiana Department of Environmental Management (IDEM) are reported from over 203 locations.

The purpose of this chapter is to document the changes in the fish assemblage and the biological integrity of the Patoka River based on historical collection information from the literature, museum records (University of Michigan, National Museum of Natural History, Illinois Natural History Survey), and unpublished data (T. P. Simon and B. E. Fisher, unpublished data) collected post-1994.

Methods

Study Area.—The Patoka River drains the Wabash River lowland and historically was part of a vast riverine floodplain that included extensive forested wetlands (Figure 1). The Patoka River is a large river tributary of the Wabash River, originating near the town of Valeene, Orange County, Indiana. The river flows west for 260.7 km across the Crawford upland and the Wabash lowland physiographic units (Schneider 1966). The Patoka River watershed drains 2,232.6 km^2 (223,257 ha) in eight counties of southwestern Indiana. The basin contains areas of extensive lacustrine deposits and minimum glacial deposits. At the upper eastern end of the watershed, the Patoka River flows rapidly through a narrow floodplain of forested uplands and has a gradient of 3.7 m per kilometer (U.S. Fish and Wildlife Service 1994). West of the city of Jasper, Indiana, the velocity of the Patoka quickly diminishes as the river's gradient decreases to 0.304 m per kilometer. The predominant land use changes

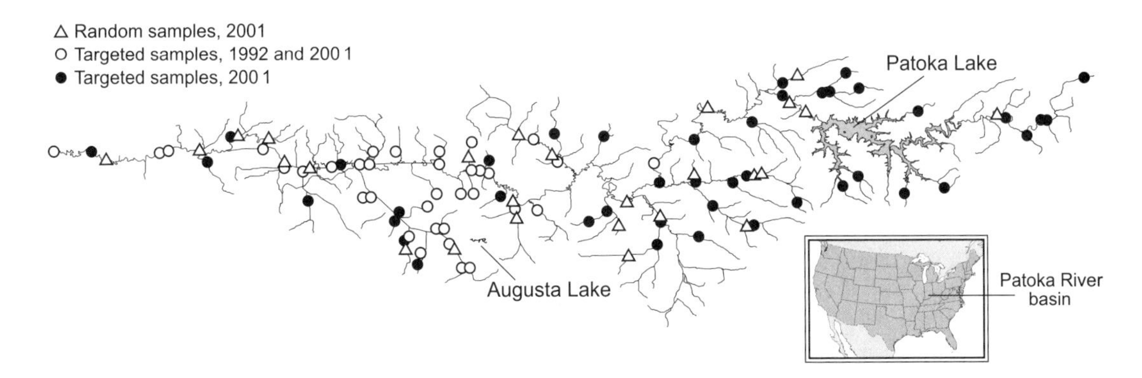

Figure 1.—The Patoka River drainage showing principal tributary streams and sites sampled during 1992 and 2001.

from forested uplands in the headwaters to agriculture in the lowlands downstream.

Several large public land holdings include federal and state properties that protect the ecological integrity of the Patoka River. The U.S. Forest Service owns large tracts of land in the Patoka River unit of Hoosier National Forest, which includes large well-preserved upland habitats. This area contains much of the headwaters of the river before forming Patoka Lake. The Army Corps of Engineers possesses the land surrounding Patoka Lake, which is a man-made impoundment of the Patoka River. Both of these areas are in the Patoka River headwaters. The U.S. Fish and Wildlife Service maintains a national wildlife refuge in Pike and Gibson counties. The Patoka National Wildlife Refuge (PNWR) was established in 1994 and consists of 8,903.1 ha, including 2,751.9 ha set aside as wildlife management areas. The project encompasses one of the last remaining stretches of bottomland hardwood wetlands, uplands, migratory bird habitat, and threatened or endangered species habitat in the state of Indiana. The refuge is located in the slow, meandering, lowland stretch of the river, which includes a wide floodplain, numerous oxbows, and low rolling uplands. In addition, the Indiana DNR manages the Sugar Ridge Fish and Wildlife Area. This area is a reclaimed coal strip mine.

A drainage project in the 1920s and the U.S. Army Corps of Engineers' construction of Patoka Lake in the late 1970s affected current flows in the Patoka River. In the early 1920s, nearly 40,468.6 ha of forested wetlands were drained in the floodplain of the lower reaches of the river to advance farming (U.S. Fish and Wildlife Service 1994). The drainage project changed 57.94 km of meandering river into an approximately 27.4 km, dredged, channelized ditch. Nearly 30.6 km of natural river meanders were isolated from the main channel, creating oxbow ponds. Water exchanges within these man-made oxbows are now limited to floods. Heavy sediment loads carried during these high water periods resulted in increased sediment deposition in the oxbows. Consequently, the hydrology of these ecologically important units is continually being altered. Major tributaries to the Patoka River include the South Fork Patoka River and Flat Creek, both of which were channelized during the early part of the 20th century.

History of Land Use Change in the Patoka River Drainage

Period of Discovery between 1700 and 1800.—The first explorers of the Indiana territory found the Patoka River to be a sinuous stream flowing westward through hardwood forested wetlands (National Park Service, George Rogers Clark Museum, unpublished diary). The water was tannin stained as a result of extensive wetlands. George Rogers Clark and troops indicated that it took them several days to cross the Wabash River floodplain as they marched from the area near New Albany to the present day St. Louis due to the extensive forested riparian wetland. With the settlement of the Indiana territory, the watershed was extensively cleared, drained, and many of the tributaries and main stem were channelized to enable agriculture and grazing opportunities for the westward expansion. As the extensive hardwood forest was cleared and land was plowed, there is no record of the changes that occurred in the aquatic assemblages (U.S. Fish and Wildlife Service 1994). However, based on current knowledge, there had to be an increase in ambient turbidity and temperature with the removal of the canopy cover and loss of bank stability. An extensive series of ditches was excavated to drain the wetlands and put more land into crop production.

Mineral Extraction Period between 1800 and 1900.—As the land was cleared and rapidly changed into cropland, the predominant crops growing in the corn belt plain further north were unable to produce the same yields in the interior river lowland. As a result, the land became pastureland and was of little use, since the extensive wetlands and hardwood forests were problematic for agriculture (U.S. Fish and Wildlife Service 1994). Despite the poor yields, the land was still used for subsistence farming. During early portions of this period, the land remained unsettled with mostly French traders and merchants settling in Vincennes and Terre Haute (Gammon 1998). Although much of the land was cleared, areas along the river bottoms were not logged because of the rolling topography and the agriculturally unproductive land.

The exploration and discovery of the eastern coal field was the next large change (U.S. Fish and Wildlife Service 1994). This vast area extended from West Virginia to Illinois and provided coal for the westward expansion. The invention of the steam locomotive created a need for coal as the power for the boilers. Coal was extensively mined beginning in the late 1800s and continuing to the present. Numerous piles of coal were removed from the landscape and large amounts of overburden waste coal were dumped. The overburden leached acid and sulfite that caused many of the streams to become acidic. The loss of stream assimilative capacity and the leaching of heavy metals caused acute toxicity in areas surrounding gob piles and abandoned coal mines (U.S. Fish and Wildlife Service, unpublished data).

The Industrial Revolution Period 1920–1950.—With the invention of the automobile, oil exploration in the eastern coal field ensued. As oil and gas fields and exploratory derricks multiplied across the land, there was little regulation or concern over the increasing acidic coal gob piles or the runoff of oil and oil brines into streams (U.S. Fish and Wildlife Service 1994). As subsurface coal mining diminished, greater efforts were put into surface mining. These extraction changes caused groundwater and surface water table changes in the area. Small headwater streams went dry with increased frequency. The loss of all aquatic life in the south fork of the Patoka River resulted from acid mine drainage (Corbett 1969). Deep pits littered the landscape and filled with water too acidic to support aquatic life. Little was done to reclaim the land or to change mineral extraction practices.

Post World War II to Present 1950–2002.—There were additional declines in environmental quality as servicemen returned from war. The Works Project Administration (WPA) built dams, bridges, and channelized extensive portions of major drainages to reduce flooding and provide more agricultural land (U.S. Fish and Wildlife Service 1994). The Patoka River was extensively straightened, and instead of a sinuous stream, changes were made so that flows could be shunted through Houchins Ditch to keep the Patoka River bottoms from flooding. The draining of Big Bottoms, a large floodplain wetland lake, resulted in a small ditched remnant stream flowing through agricultural fields.

The degradation of the environment reached catastrophic proportions until the early 1970s (Corbett 1969). The advent of the Clean Water Act and formation of the U.S. Environmental Protection Agency improved point-source protection and regulated the production of coal and waste leachate. As important strides were made in the regulatory arena and more environmental advocates raised questions regarding the extraction of coal, oil, and gas, further efforts to restore the Patoka River watershed were occurring. The State of Indiana Department of Environmental Management was formed in 1986 and since has reported on the condition of the Patoka River drainage. The watershed was not meeting designated uses for a substantial portion of its river miles as a result of acid mine drainage. With economic prosperity and the need for recreation and leisure, the U.S. Army Corps of Engineers constructed a large reservoir near the headwaters of the Patoka River, forming Patoka Lake. This 3,237.5-ha lake provided drinking water, recreation, and leisure for residents in southwestern Indiana. The lake also reduced main-stem migration of species and inundated streams in the Hoosier National Forest. This resulted in the isolation of species in the lake headwaters into various tributary units.

Another important public land acquisition was the formation of the Patoka River National Wildlife Refuge and Sugar Ridge State Fish and Wildlife Area in the middle to lower portion of the watershed (U.S. Fish and Wildlife Service 1994). The occurrence of Hoosier National Forest in the headwaters, Patoka Lake in the upper river, and the Patoka National Wildlife Refuge and Sugar Ridge Fish and Wildlife Area in the mid-lower river provide extensive public lands.

Field Sampling and Study Design

Fish assemblages were sampled at reaches that were based on previous water and sediment sampling stations above and below point-source discharges in the Patoka River (Simon et al. 1994). Reaches were evaluated between early June 1992 (66 sites) and late September 2001 to quantify fish assemblage condition at 125 sites (Figure 1).

Every attempt was made to collect at historical locations conducted by Jordan (1890), Moenkhaus (1896), Hubbs and Lagler (1942), Lagler and Ricker (1942), Hottell (1978), Simon et al. (1994), and Stefanavage (1993) so that comparisons of fish composition changes could be documented. Sampling at 66 of the original sites sampled in 1992 by Simon et al. (1994) was repeated during 2001 to compare changes in biological integrity over a decade. In addition to the 125 hand-picked reaches, a probability sample of 26 sites was surveyed by the IDEM during 2001 to compare changes in the watershed between 1992 and 2001 (Stacy Sobat, IDEM, unpublished data).

Fish species composition and relative abundance (the number of fish/min of electrofishing effort) data were gathered by electrofishing surveys at reaches using appropriate electrofishing gear. Sampling gear included a model 6A Smith-Root boat-mounted electrofisher in the main-stem river and Smith-Root backpack and longline systems in tributaries. Electrofishing surveys included systematic sampling of representative habitat within reaches, including the thalweg or deepest point in the cross sectional profile, usually for distances of 500 m for a minimum of 1,800 s. Captured fish were placed in an onboard holding tank until a sampling event was completed. Data recorded for each survey event included species' identifications and weights, number of fish caught, examination for external disease and anomalies (DELTs), and sample and habitat conditions.

Results

Status of Water Quality

The Patoka River watershed is affected by acid mine drainage (Corbett 1969; Renn 1989). From the 1800s to the early 1900s, almost all the coal produced in Indiana was from underground mines. Then, surface mining steadily increased until it became the only current mining practice (Powell 1972). From 1941 to 1980, 16% of Pike county's acres were disturbed by surface mining (Renn 1989). A significant portion (62%) of Pike county was mined prior to Indiana's comprehensive reclamation law of 1968, resulting in 75.3 ha of refuse piles, 52.2 ha of slurry ponds, 1,259.8 ha of land with less than 75% vegetation cover, and 151.8 ha of surface water impoundments affected by coal-mine drainage (Allen et al. 1978).

The detrimental effects of coal production on water quality in the watershed were first documented in the early 1960s (Corbett 1969). Acid mine drainage from underground and surface mining has altered water quality in the Patoka watershed. Corbett (1969) found a range in pH from 2.6 to 5.5, in alkalinity from 0 to 4 mg/L, in acidity from 38 to 554 mg/L, and in sulfate from 138 to 4,400 mg/L based on 37 water samples from the South Fork Patoka River at the SR 57 bridge during the period from April 9, 1962, to July 8, 1968. In addition to coal mining, oil-well-waste pollution has increased. Corbett (1969) found that during low flows, the Patoka River (in the vicinity of Wheeling) had high chloride levels. In 1969, chloride concentrations ranged from 3 to 952 ppm with an average of 47 ppm (sample size = 110). The mean chloride concentration in river water in North America is 8 ppm (Livingstone 1963). In 1989, Renn reported that acid mine drainage continued to impair water quality in the South Fork Patoka River. Surface water screening toxicity test results, invertebrate assemblage assessments, and water quality data show that acid mine drainage still impairs water quality in the watershed (U.S. Fish and Wildlife Service, unpublished data).

Changes in Species Richness and Composition

The Patoka River currently and historically supported 97 fish species (Simon et al. 1994; Appendix A). This represents 46.6% of the fish fauna of Indiana (Simon et al. 2002). Overall, the biological diversity of the Patoka River drainage has remained stable with slight increases in species richness as a result of increased sampling during the last two decades. Local extirpations and site specific biological integrity changes have declined dramatically since 1890.

Historical distributions (Pre-1900).—The Patoka River, prior to 1900, contained several species characteristic of clear streams of the Wabash River lowlands. Data from the turn of the century are limited, primarily based on three localities (Jordan 1890; Moenkhaus 1896); however, these three collections documented 38 fish species (Appendix A). Species such as *Noturus miurus*, *Etheostoma camurum*, *Etheostoma histrio*, *Percina phoxocephala*, *P. sciera*, *Moxostoma duquesnei*, and *P. vigil* were present in the drainage. Based on museum specimens, sites on the Patoka River were sampled frequently and the numbers of specimens are adequate to document the majority of species previously occurring in the lower Patoka River. The only species not represented by multiple collections is *E. camurum* (Simon et al. 1994). Based on historical information (Jordan 1890), *E. camurum* was probably not a significant component of the Patoka River fauna. *Etheostoma camurum* is usually associated with gravel-cobble riffle habitats in moderate to large rivers (Trautman 1981) and usually requires the clean interstitial spaces of elevated cobble and boulders to clump eggs during reproduction (Page and Simon 1988). Typically, *E. camurum* occurs in the glacial till areas of the last Wisconsian advance. The main-stem Patoka River probably never had significant amounts of this microhabitat because it was dominated by lowland habitats and shale outcrops; however, a portion of the lower river was glaciated during the Wisconsian advance. The presence of the other six species, which are characteristic of clear streams, is consistent with the expected fauna of the headwater tributaries of the Wabash River lowlands.

Moxostoma macrolepidotum, *Carpiodes velifer*, *M. duquesnei*, *Cyprinella whipplei*, *Dorosoma cepedianum*, *Micropterus salmoides*, *P. vigil*, *E. histrio*, *P. phoxocephala*, *E. nigrum*, and *P. maculata* are now included in the pre-1900 list. These species should have been included in Simon et al. (1994); however, most of these species were not included as a result of nomenclature changes that clouded the identity of species. For example, Moenkhaus (1896) included *Moxostoma aureolum* from both the Patoka River and Short Creek. Neither Gerking (1945) nor Simon et al. (1994) included this species in their lists since this name previously was used for *M. macrolepidotum, M. erythrurum,* and *M. duquesnei*. Smith (1979) identified *M. aureolum* from Indiana collections as *M. macrolepidotum*. Gerking (1945) also did not include any records prior to 1927 for

Micropterus salmoides. Prior to this time, the species included both *M. punctulatus* and *M. salmoides*. Both of these species have been collected from the Patoka River. Finally, Moenkhaus (1896) reported the presence of *Notropis microstomus,* which may have been either *N. stramineus, N. volucellus,* or *N. wickliffi*. We speculated that the species was probably both *N. stramineus* and *N. volucellus* based on the location of the collection and the types of species occurring in the main-stem Patoka River. Changes in Appendix A reflect an updated understanding of Patoka River species based on investigations of nomenclature and errors and omissions from the original literature.

Changes between 1940 and 1950.—Hubbs and Lagler (1942), Lagler and Ricker (1942), and Gerking (1945) made nine collections (eight in the Patoka drainage and a single collection in Foots Pond) during this period and documented 71 species (Appendix A). The loss of *E. camurum, E. histrio, P. sciera, P. phoxocephala, P. vigil, M. duquesnei,* and *M. macrolepidotum* prior to this time suggests that significant changes were occurring. Species such as *Noturus miurus, E. caeruleum,* and *Umbra limi* were collected for the last time during this period. Additional species documented by these investigators included *Polyodon spathula, Notropis blennius, Morone mississippiensis, Amblop-lites rupestris, Micropterus dolomieu, Etheostoma gracile, E. chlorosoma,* and *Cottus carolinae*. The clearing of land and channelization of streams had increased sediment loads and turbidity. Paddlefish and yellow bass were collected from Foots Pond, which was never connected to the Patoka River. However, their presence in the Wabash River lowlands suggests that these species may have occurred in the Patoka River drainage, but gear methodology would have been incapable of capture.

Ambloplites rupestris, M. dolomieu, and *C. carolinae* are all species characteristic of upland, coolwater streams. The change from clear, well-oxygenated waters to more turbid, warmer waters may have led to the demise of these species. All of these species were collected for the first time during this period and have been found since.

Assemblage changes between 1960 and 1991.—Indiana DNR biologists evaluated 31 stations in the main-stem Patoka River and tributaries and documented 63 species (Hottell 1978; Stefanavage 1993; Appendix A). During this period, four fish species were added to the faunal list, including *Lepomis microlophus, L. miniatus, Esox lucius,* and *Etheostoma spectabile*. After the formation of Patoka Lake and the creation of extensive lacustrine habitat, numerous species were stocked by the Indiana DNR, including *L. microlophus* and *Esox lucius*. Hottell's (1978) *E. lucius* specimens were undoubtedly stocked specimens; neither previous nor subsequent investigators have confirmed their persistence.

Assemblage changes from 1992 to 2002.—Five times as much collection effort has been expended in the watershed since 1992 than had previously occurred over the last two centuries. Simon et al. (1994) documented the increase in species diversity as a result of increased sampling intensity. They documented 14 new species in the watershed, including *Dorosoma petenense, Hiodon alosoides, Hybognathus hayi, Lythrurus fumeus, Macrhybopsis hyostoma, N. wickliffi,* pallid shiner *Hybopsis amnis, Phoxinus erythrogaster,* fathead minnow *Pimephales promelas, Rhinichthys atratulus,* lake chubsucker *Erimyzon sucetta, Moxostoma anisurum, Noturus flavus,* and *Fundulus dispar*. The range extension of *L. fumeus* may have been a result of misidentification since *L. umbratilis* does occur in the upper portion of the watershed. The *L. fumeus* was previously known from only a few small streams in southwestern Indiana (Gerking 1945). Bait-bucket release of fathead minnow *Pimephales promelas* into the watershed was speculated by Simon et al. (1994), while the presence of *Dorosoma petenense* was probably a result of immigration from upstream reservoir habitats. *F. dispar* suggests that water quality conditions were improving, because this species is considered sensitive to acidity and turbidity (Simon 1991).

With additional sampling since 1993, we add *Erimyzon sucetta* and the first authenticated record for *Polyodon spathula* (verification of species presence by Dan Carnahan, Indiana DNR, personal communication) in oxbow lakes from the floodplain wetlands. Species that were rediscovered include *Notropis blennius*, sand shiner *N. stramineus*, *Pimephales vigilax*, *Moxostoma macrolepidotum*, *Labidesthes sicculus*, *Morone mississippiensis*, *M. dolomeiu*, *E. gracile*, *P. maculata*, *and C. carolinae*. The increase in the number of new species records is due to increased sampling. The rediscovery of *Etheostoma histrio* and *Percina sciera* was also documented upstream to the mouth of South Fork Patoka River. *Erimyzon sucetta*, *Polyodon spathula*, *Ambloplites rupestris*, *Etheostoma gracile*, *E. histrio*, *Percina maculata*, *P. sciera*, and *C. carolinae* are all species sensitive to siltation and acidity. The increased occurrence of these species in the watershed may be an environmental indicator of improved water quality.

Authenticity of records.—Simon et al. (1994) questioned the authenticity of several records. Weed shiner *Notropis texanus* was reported by Hottell (1978) and Stefanavage (1993) from two Patoka River main-stem sites, Cup Creek, and Hunley Creek. *Notropis texanus* typically occurs in habitats with an abundance of aquatic vegetation. This shiner is similar to *Opsopoeodus emiliae* in appearance and is only known in Indiana from the Kankakee River drainage. During this study, no *N. texanus* were collected. Since the species has never been collected since, we speculate that the specimens were misidentified. Hottell (1978) reported *O. emiliae* to be sympatric with *N. texanus* at Hunley Creek. *Morone chrysops* was reported from the mouth of the Patoka River by Hottell (1978). Although this record is probable, *M. chrysops* is probably not a significant resident of the Patoka River; however, the species was also reported by Carnahan (1997a) in oxbows. *Morone chrysops* has been reported as far upstream as Patoka River km 53.9 and from Foots Pond. We speculate that the species may have been captured as a result of escape from adjacent oxbow lakes. *Polyodon spathula* reported by Carnahan from oxbow lakes of the Patoka also included reports from conservation officers of anglers snagging paddlefish below the dam at Patoka Lake. This suggests that this species seasonally migrates up the Patoka River and historically enters the oxbow lakes. The entrapment of *P. spathula* in oxbow wetlands during high water may have been a seasonal occurrence; however, the high levees along the river virtually eliminate seasonal use of this habitat. *Polyodon spathula* cannot escape back into the Wabash River except during high flows when the oxbow floodplain wetlands are connected. Some may be running up the river to Patoka Lake and may remain in the river for long durations.

Biological Diversity Changes in the Patoka River Drainage

The species of fish occurring in the Patoka River has slightly increased over the past 120 years, probably as a result of more intensive collection effort and the use of more efficient collection gear. Current and historic fish records shows that 97 species have occurred in the watershed; however, 86 species are currently documented (Appendix A).

Changes in fish assemblage structure at the town of Patoka have been substantial; 22 species of fish were documented in 1888 (Jordan, 1890), 8 species in 1992 (Simon et al. 1994), while 11 species were collected during 2001 (T. P. Simon and B. E. Fisher, unpublished data). The Patoka River near Huntingburg has been sampled between 1896 and 2001 and 22 species were collected in 1893 (Moenkhaus 1896), which declined to 4 species in 1978 (Hottell 1978), rising to 12 species in 1992 (Simon et al. 1994), then 22 species in 2001 (T. P. Simon and B. E. Fisher, unpublished data). Despite this loss and subsequent recovery of site specific biological diversity during 1950–1991, overall watershed diversity has declined. Extirpations in the watershed have steadily risen, and loss of 12.7% of the fauna has occurred in the last century. Many species are represented by small remnant populations. More extensive effort is required to find the same number of species that were easily found in 1893.

Despite the lack of alien species in the Patoka River, bighead carp *Hypopthalmichthys nobilis*, silver carp *H. molitrix*, and grass carp *Ctenopharyngodon idella* are found in the lower Wabash River watershed. Currently, none of these species have been collected from the Patoka River drainage, but they have been collected in the Wabash and lower White River. We speculate that these alien species will advance into the lower Patoka River.

Changes in Biological Condition

Comparison of historical biological integrity shows changes in the Patoka River near the town of Patoka declined 50% in species richness and biological integrity, while Huntingburg showed the most dramatic changes with declines in species richness from 22 species to 4 fish species and then recovery to 22 species in the main-stem river.

Based on targeted least-impacted sampling at 34 least-disturbed sites between 1992 and 2001, biological condition of the Patoka River has declined slightly (Figure 2). During 1992, sampled fish assemblages had IBI scores above 45, while during 2001, sampled fish assemblages did not exceed IBI scores of 40. The mean assemblage score did not decline substantially between the two periods, but more sites scored in the very poor range of integrity during 2001 than during 1992 (Figure 2). Over this time period, prolonged drought conditions possibly reduced nonpoint source runoff of toxic materials and acidity into streams, while groundwater infiltration has potentially enabled some species to recolonize areas that had been decimated in the past (U.S. Geological Survey Water Resources data, 1992–2001). Unfortunately, the lack of water also caused declines in species richness and changes in trophic dynamics.

We chose an unbiased approach to verify our understanding of overall biological condition during 2001. We sampled 87 least-disturbed targeted sites during 2001 and another 26 random sites using an equally weighted probability-based sampling design (Stacy Sobat, IDEM, unpublished data). Based on index of biotic integrity (IBI) scores for the watershed, the two designs had similar results (Figure 3). The random sites had slightly higher IBI scores than the targeted, least-

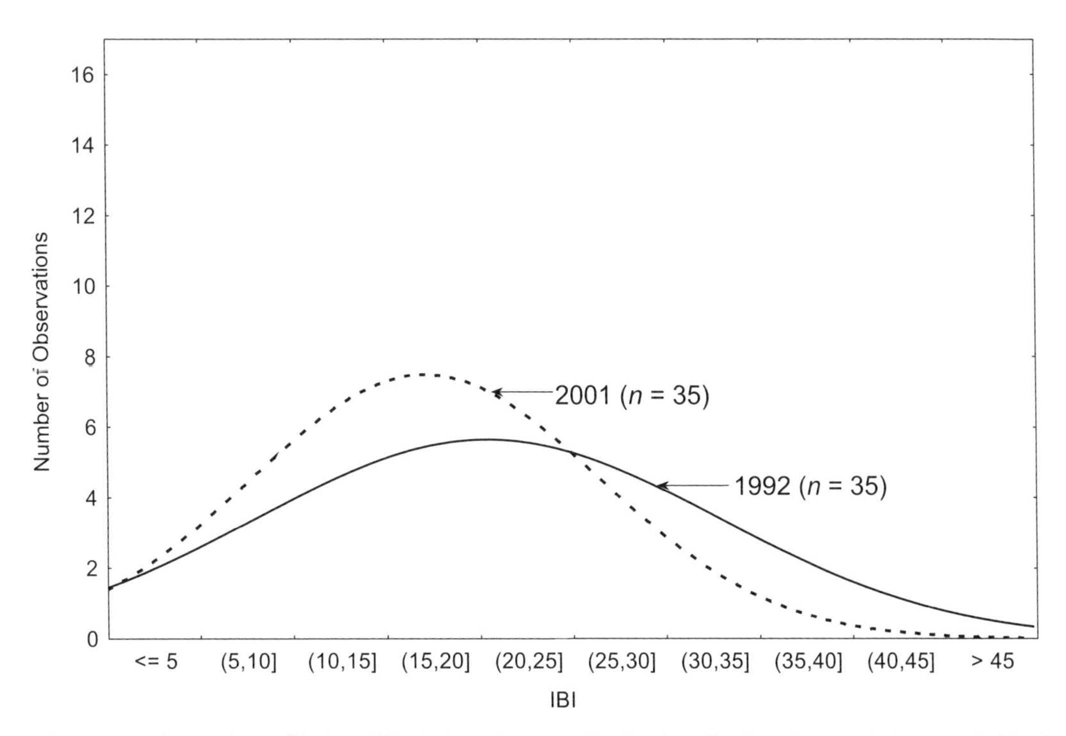

Figure 2.—Comparison of index of biotic integrity score distributions for the 35 targeted sites sampled in the Patoka River during 1992 and repeat sampled in 2001.

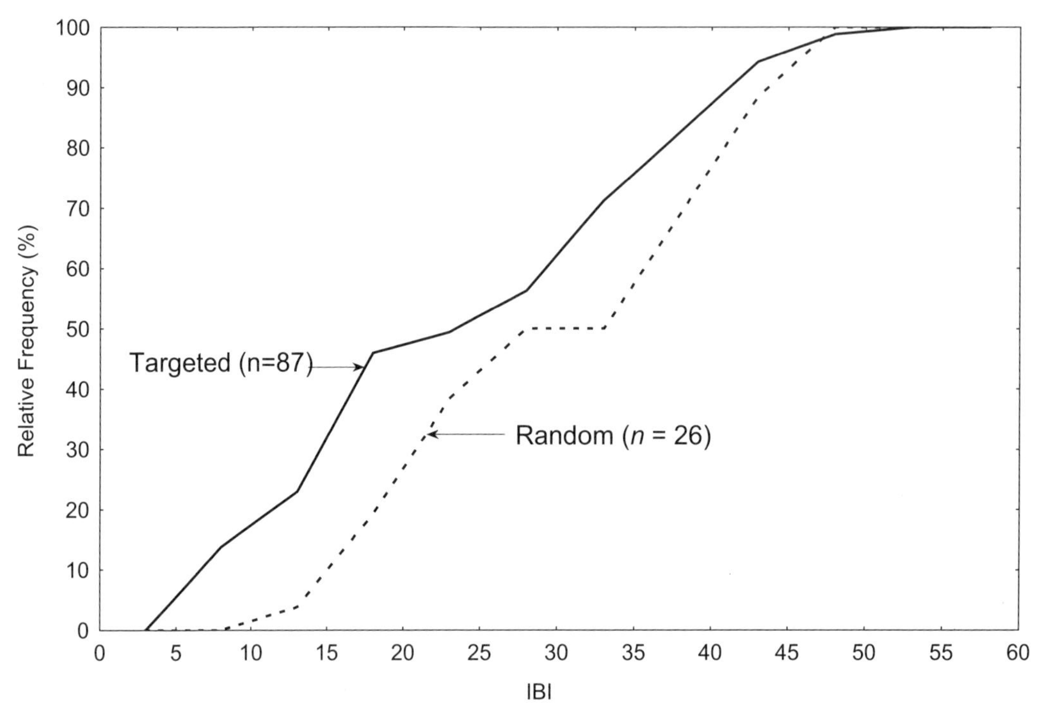

Figure 3.—Cumulative frequency distributions for 87 targeted sites and 26 random sites sampled during 2001 in the Patoka River drainage.

disturbed sites. The two sample designs showed that fishless sites were similar in proportion.

Discussion

The Patoka River watershed has endured substantial environmental changes and has exhibited significant losses in biological integrity, yet despite the local extirpation of 12.7% of the fish fauna, the watershed still maintains remnant populations of many fish species, providing a similar species composition list over the last 120 years (Simon et al. 1994). Despite the widespread habitat loss from the draining of riparian forested wetlands, channelization and ditching of wetlands, straightening of the main channel river, and damming of the headwaters to form Patoka Lake, remnant portions of the watershed are still least-disturbed. Areas remaining least-disturbed include portions of the Hoosier National Forest and several of the streams maintained in the Patoka Lake watershed by the Army Corps of Engineers. However, the extraction of minerals, oil, and gas has changed the landscape so that few areas have not been disturbed (Allen et al. 1978), including areas that are now part of the Sugar Ridge Fish and Wildlife Area and the Patoka National Wildlife Refuge.

Extensive changes in the fish assemblage as a result of watershed disturbance have caused the local extirpation of species such as *Noturus miurus*, *Percina phoxocephala*, and *Umbra limi*. These species were collected for the last time during the 1940–1950 period, with the exception of *P. phoxocephala,* which was actually collected for the last time pre-1900. These changes in the Patoka River mirror much more widespread losses within the Wabash River lowlands. The draining of wetlands has resulted in the extirpations of alligator gar *Atractosteus spatula*, banded pygmy sunfish *Elassoma zonatum*, and bantam sunfish *Lepomis symmetricus* from the lower Wabash River (Simon et al. 2002). These species were never documented from the Patoka River, but are dependent on floodplain forested wetlands for reproduction.

Historically, 97 species of fish cumulatively resided in the Patoka River drainage (Appendix A). Significant land-use changes over the last 120 years have reduced the number of residents since 1991–1985 species. With increased sampling efforts, 14 new species have been documented in the watershed since 1991. Local species extinctions and rare occurrence of some taxa suggest that the Patoka River bottomlands and adjacent habitats are important for maintaining and preserving stable biological trophic function and biodiversity. Local extirpations have caused a decline in site-specific biological diversity.

The permanence of headwater streams shows that many of the first-order streams of the Patoka River are dry during a portion of the year. Simon et al. (1994) had shown that Hog Branch, an unnamed tributary at Sugar Ridge Road, Wheeler Creek, the old Wabash and Erie Canal, and Lick Creek all were dry during 1993. These same streams were also dry during 2000 and 2001. Additional stream sites have also been dry and reduced flows in many other first and second order streams surrounding Patoka Lake have created stagnant standing pools where fish are trapped (Simon, personal observation). These isolated pools do not have subsurface flow as do many of the interior plateau streams further east. The stable flows are a result of extensive floodplain wetlands, which mitigated these low flow periods. However, with the drainage of those wetlands during the middle of the 19th century came reduced standing water on the landscape.

Biological integrity of streams in the Patoka River watershed has certainly declined since the 1890s. With the initial clearing of the land during the early 1800s, baseline efforts of Jordan and Moenkhaus documented widespread changes that had already occurred. Our sampling efforts during 1992–2001 have shown that few high quality sites remain in the watershed. It would be difficult to collect 38 species today at any three sites in the watershed as Jordan and Moenkhaus did in the late 1890s. However, based on a random probability sample collected by the IDEM during 2001, less than 12% of the streams represent least-disturbed reference conditions, while 61% are degraded below the state average (Stacy Sobat, IDEM, unpublished data).

Some signs of recovery are beginning to be observed in areas previously degraded by acid mine drainage. For example, Augusta Lake, a headwater impoundment of Mill Creek on the Sugar Ridge State Wildlife Refuge property, for the first time had aquatic life during 2000 and the presence of darters in the stream downstream. This lake had water of pH 3.7 over the spillway in 1992, which caused loss of aquatic life downstream. Through intensive mitigation and construction of an anoxic limestone drain, the lake water has improved to where aquatic macroin-vertebrates now inhabit the lake (Comer et al. 2000). Habitat reclamation in the South Fork Patoka River has enabled some aquatic life to return to the uppermost stretches of the stream (Simon 1999). The return of *C. carolinae*, *Polyodon spathula*, *Etheostoma histrio*, and *Percina sciera* in the river indicates marked improvements. These species are considered sensitive indicators and should continue to improve with improving water quality.

Acknowledgments

The authors thank Scott Sobiech, Stacy Sobat, and Anthony Branam for their efforts in documenting the fish fauna of the Patoka River. Special thanks to Bill McCoy for his support and management of the Patoka River National Wildlife Refuge. Although this study was partly funded by the U.S. Fish and Wildlife Service, no endorsement by that agency should be inferred. This study was funded by an off-refuge grant to TPS through contaminants grant #31440–1261–3N29.

References

Allen, J. H., T. C. Thomas, and R. R. Kelly. 1978. A survey to determine the extent and environmental effect of derelict lands resulting from the surface extraction of coal. Indiana Department of Natural Resources, P. L. 62, Acts of 1976, Indianapolis.

Carnahan, D. P. 1997a. Standing stock estimation of sport and commercial fish species in two, one mile reaches of the Patoka River. Indiana Department of Natural Resources, Division of Fish and Wildlife, Fisheries Section, Indianapolis.

Carnahan, D. P. 1997b. Initial fisheries and habitat evaluation of five Patoka River oxbow lakes and one wetland in Gibson and Pike counties. Indiana Department of Natural Resources, Division of Fish and Wildlife, Fisheries Section, Indianapolis.

Comer, J. B., R. T. Smith, M. V. Ennis, T. D. Branam, L. R. Welp, N. E. Morales, S. A. Sobiech, and T. P. Simon. 2000. Effects of passive reclamation on water quality in the northeastern drainage of Augusta Lake, Pike County, Indiana. Indiana Geological Survey, Open File Study 00–2, Bloomington.

Corbett, D. M. 1969. Acid mine-drainage problems of the Patoka River watershed, southwestern Indiana. Indiana University, Water Resources Research Center, Number 4, Bloomington.

Delleur, J. W. 1994. Indiana's wetlands: past, present, and future. Proceedings of the Indiana Academy of Science 103:139–142.

Exl, J. A., and T. P. Simon. 2002. Biological integrity of several streams located in northeastern Minnesota's northern lakes and forest ecoregion (USA) with emphasis on differentiating natural from anthropogenic disturbances. Dimensions of Pollution 1:1–25.

Gammon, J. R. 1994. The status of riparian wetlands in west-central Indiana streams. Proceedings of the Indiana Academy of Science 103:195–214.

Gammon, J. R. 1998. The Wabash River ecosystem. Indiana University Press, Bloomington.

Gerking, S. D. 1945. The distribution of the fishes of Indiana. Investigations of Indiana Lakes and Streams 3:1–37.

Gray, H. H. 1979. Indiana. U.S. Geological Survey, Professional Papers 1110:K1–K20, Indianapolis, Indiana.

Hottell, H. E. 1978. Fisheries survey of the Patoka watershed above, in, and below Patoka Lake. Indiana Department of Natural Resources, Division of Fish and Wildlife, Fisheries Section, Indianapolis.

Hubbs, C. L., and K. F. Lagler. 1942. Annotated list of fishes of Foots Pond, Gibson County, Indiana. Investigations of Indiana Lakes and Streams 2:73–83.

Jordan, D. S. 1890. Report of explorations made during the summer and autumns of 1888, in the Allegheny region of Virginia, North Carolina, and Tennessee, and in western Indiana, with an account of the fishes found in each of the river basins. Bulletin United States Fisheries Commission 8:97–173.

Lagler, K. F., and W. E. Ricker. 1942. Biological fisheries investigations of Foots Pond, Gibson county, Indiana. Investigations of Indiana Lakes and Streams 2:47–72.

Livingstone, D. A. 1963. Chemical composition of rivers and lakes. Chapter 6. Data on biochemistry. 6th edition. U.S. Geological Survey, Professional Papers 440-G, Indianapolis, Indiana.

Moenkhaus, W. J. 1896. Notes on a collection of fishes of Dubois county, Indiana. Proceedings of the Indiana Academy of Science 11:159–162.

Page, L. M., and T. P. Simon. 1988. Observations on the reproductive behavior and eggs of four species of darters, with comments on *Etheostoma tippecanoe* and *E. camurum.* Transactions of the Illinois State Academy of Science 81:205–210.

Powell, R. L. 1972. Coal strip-mined land in Indiana. Environmental Data Service, Asheville, North Carolina 89(12):1–15.

Renn, D. E. 1989. Streamflow and stream quality in the coal mine region. Patoka River basin, southwestern Indiana, 1983–1985. U.S. Geological Survey, Water Resources Investigation Report 88–4150, Indianapolis, Indiana.

Schneider, A. F. 1966. Physiography. Pages 40–56 *in* A. A. Lindsey, editor. Natural features of Indiana. Indiana Academy of Science, Indianapolis.

Simon, T. P. 1991. Development of index of biotic integrity expectations for the ecoregions of Indiana. I. Central corn belt plain. U.S. Environmental Protection Agency, EPA 905/9–91/025, Chicago.

Simon, T. P. 1999. Evaluation of habitat improvement structures in the South Fork Patoka River: baseline November 1998. U.S. Fish and Wildlife Service, Bloomington, Indiana.

Simon, T. P., and J. A. Exl. 2002. Effects of silviculture on indices of biotic integrity for benthic macroinvertebrate and fish assemblages in northeastern Minnesota's northern lakes and forest ecoregion (USA). In T. P. Simon, editor. Biological response signatures: patterns in aquatic assemblages. CRC Press, Boca Raton, Florida.

Simon, T. P., S. A. Sobiech, T. H. Cervone, and N. E. Morales. 1994. Historical and present distribution of fishes in the Patoka River basin: Pike, Gibson, and Dubois counties, Indiana. Proceedings of the Indiana Academy of Science 104:193–206.

Simon, T. P., J. O. Whitaker, Jr., J. S. Castrale, and S. A. Minton. 2002. Revised checklist of the vertebrates of Indiana. Proceedings of the Indiana Academy of Science 111:182–214.

Smith, P. W. 1979. The fishes of Illinois. The University of Illinois Press, Champaign.

Stefanavage, T. C. 1993. Fisheries survey of the Patoka River in Gibson and Pike counties. Indiana Department of Natural Resources. Division of Fish & Wildlife, Fisheries Section, Indianapolis.

Trautman, M. B. 1981. The fishes of Ohio. Ohio State University Press, Columbus.

U.S. Fish and Wildlife Service. 1994. Patoka River national wetlands project environmental impact statement. U.S. Fish and Wildlife Service, Fort Snelling, Minnesota.

Appendix 1.—Fish species collected from the Patoka River drainage from four different time periods. Collections pre-1900 were made by Jordan (1890) and Moenkhaus (1896); 1940–1950 were made by Gerking (1945), Hubbs and Lagler (1942), and Lagler and Ricker (1942); 1960–1991 were made by the Indiana Department of Natural Resources (Hottell 1978; Stefanavage 1993); and 1992–2002 by Simon et al. (1994), Carnahan (1997a, 1997b), Brant Fisher (unpublished data, 1996–2001 [*N* = 7 sites]), T. P. Simon (unpublished data, 2000–2001 [*N* = 87 sites]), and Indiana Department of Environmental Management (unpublished data, 2001 [*N* = 26 sites]). (?) = species not included by Gerking (1945), but listed by Jordan (1890) and/or Moenkhaus (1896) based on other nomenclature; P = probable occurrence in the Patoka River oxbows based on occurrence in Foots Pond collections by Hubbs and Lagler (1942) and Lagler and Ricker (1942).

Species	Pre-1900	1940–1950	1960–1991	1992–2002
Lepisosteidae, gar				
Lepisosteus platostomus, shortnose gar		X	X	X
L. osseus, longnose gar		P	X	X
L. oculatus, spotted gar		X	X	X
Amiidae, bowfin				
Amia calva, bowfin		X	X	X
Polyodontidae, paddlefish				
Polyodon spathula, paddlefish		P		X
Esocidae, pikes				
Esox americanus, grass pickerel	X	X	X	X
E. lucius, northern pike			X	
Umbridae, mudminnows				
Umbra limi, central mudminnow		X		
Clupeidae, shad and herrings				
Dorosoma cepedianum, gizzard shad	X	X	X	X
D. petenense, threadfin shad				X
Hiodontidae, mooneyes				
Hiodon alosoides, goldeye				X
Cyprinidae, minnows				
Campostoma anomalum, central stoneroller		X	X	X
Cyprinella spiloptera, spotfin shiner		X	X	X
C. whipplei, steelcolor shiner	X(?)	X	X	X
Cyprinus carpio, common carp		X	X	X
Hybognathus hayi, cypress minnow				X
H. nuchalis, Mississippi silvery minnow	X	X	X	X
Hybopsis amnis, pallid shiner				X
Luxilus chrysocephalus, striped shiner	X	X	X	X
Lythrurus fumeus, ribbon shiner				X
L. umbratilis, redfin shiner	X	X	X	X
Macrhybopsis hyostoma, shoal chub				X
Notemigonus crysoleucas, golden shiner	X	X	X	X
Notropis atherinoides, emerald shiner	X	P	X	X
N. blennius, river shiner		P		X
Notropis buccatus, silverjaw minnow		X	X	X
N. stramineus, sand shiner	X(?)	X		X
N. volucellus, mimic shiner	X(?)	X	X	X
N. wickliffi, channel shiner				X
Opsopoeodus emiliae, pugnose minnow	X	X	X	
Phenacobius mirabilis, suckermouth minnow		X	X	X
Phoxinus erythrogaster, southern redbelly dace				X
Pimephales notatus, bluntnose minnow	X	X	X	X
P. promelas, fathead minnow				X
P. vigilax, bullhead minnow	X	X		X
Rhinichthys atratulus, eastern blacknose dace				X
Semotilus atromaculatus, creek chub		X	X	X

Appendix A.—Continued.

Species	Pre-1900	1940–1950	1960–1991	1992–2002
Catostomidae, suckers				
Carpiodes carpio, river carpsucker		P	X	X
C. cyprinus, quillback		X	X	X
C. velifer, highfin carpsucker	X(?)		X	X
Catostomus commersonii, white sucker		X	X	X
Erimyzon oblongus, creek chubsucker		X	X	X
E. sucetta, lake chubsucker				X
Ictiobus bubalus, smallmouth buffalo		P	X	X
I. cyprinellus, bigmouth buffalo		X	X	X
I. niger, black buffalo		P	X	X
Minytrema melanops, spotted sucker		X	X	X
Moxostoma anisurum, silver redhorse				X
M. duquesnei, black redhorse	X(?)			
M. erythrurum, golden redhorse		X	X	X
M. macrolepidotum, shorthead redhorse	X(?)			X
Ictaluridae, catfishes				
Ameiurus melas, black bullhead	X	X	X	X
A. natalis, yellow bullhead		X	X	X
A. nebulosus, brown bullhead		X	X	X
Ictalurus punctatus, channel catfish	X	X	X	X
Noturus flavus, stonecat				X
N. gyrinus, tadpole madtom		P	X	X
N. miurus, brindled madtom	X	X		
Pylodictis olivaris, flathead catfish	X	P	X	X
Aphredoderidae, pirate perch				
Aphredoderus sayanus, pirate perch	X	X	X	X
Fundulidae, topminnows				
Fundulus dispar, northern starhead topminnow				X
F. notatus, blackstripe topminnow	X	X	X	X
Atherinidae, silversides				
Labidesthes sicculus, brook silverside	X	P		X
Poeciliidae, mosquitofish				
Gambusia affinis, western mosquitofish		P	X	X
Moronidae, temperate bass				
Morone chrysops, white bass		P	X	X
M. mississippiensis, yellow bass		P		X
Centrarchidae, sunfish and black bass				
Ambloplites rupestris, rock bass		X		X
Centrarchus macropterus, flier		X	X	X
Lepomis cyanellus, green sunfish		X	X	X
L. gulosus, warmouth	X	X	X	X
L. humilus, orangespotted sunfish		X	X	X
L. macrochirus, bluegill	X	X	X	X
L. megalotis, longear sunfish	X	X	X	X
L. microlophus, redear sunfish			X	X
L. miniatus, redspotted sunfish			X	X
Micropterus dolomieu, smallmouth bass		X		X
M. punctulatus, spotted bass		X	X	X
M. salmoides, largemouth bass	X(?)	P	X	X
Pomoxis annularis, white crappie	X	X	X	X
P. nigromaculatus, black crappie	X	X	X	X
Percidae, perch and darters				
Etheostoma asprigene, mud darter		P	X	X
E. caeruleum, rainbow darter		X		

Appendix A.—Continued.

Species	Pre-1900	1940–1950	1960–1991	1992–2002
E. camurum, bluebreast darter	X			
E. chlorosoma, bluntnose darter		X	X	
E. gracile, slough darter		X		X
E. histrio, harlequin darter	X			X
E. nigrum, johnny darter	X	X	X	X
E. spectabile, orangethroat darter			X	X
Percina caprodes, logperch	X	X	X	X
P. maculata, blackside darter	X	X		X
P. phoxocephala, slenderhead darter	X			
P. sciera, dusky darter	X			X
P. shumardi, river darter		P	X	
P. vigil, saddleback darter	X			
Sander canadensis, sauger		P	X	X
Sciaenidae, drums				
Aplodinotus grunniens, freshwater drum	X	X	X	X
Cottidae, sculpins				
Cottus carolinae, banded sculpin		X		X
Total number of species	38	71	62	86
Number of collection site visits	3	9	31	203

American Fisheries Society Symposium 45:399–429, 2005

Changes in Fish Assemblage Status in Ohio's Nonwadeable Rivers and Streams over Two Decades

CHRIS O. YODER* AND EDWARD T. RANKIN

Midwest Biodiversity Institute and Center for Applied Bioassessment and Biocriteria Post Office Box 21561, Columbus, Ohio 43221–0561, USA

MARC A. SMITH, BRIAN C. ALSDORF, DAVID J. ALTFATER, CHARLES E. BOUCHER, ROBERT J. MILTNER, DENNIS E. MISHNE, RANDALL E. SANDERS, AND ROGER F. THOMA

Ohio Environmental Protection Agency, 4675 Homer Ohio Lane, Groveport, Ohio 43125, USA

Abstract.—A systematic, standardized approach to monitor fish assemblages has been applied in Ohio's rivers since 1979. A primary objective is the assessment of changes in response to water pollution abatement and other water quality management programs. All major, nonwadeable rivers were intensively sampled using standardized electrofishing methods and a summer–early fall index period. Most rivers were sampled two or three times, before and after implementation of pollution controls at major point source discharges and best management practices for nonpoint sources. A modified and calibrated index of biotic integrity (IBI) was used to demonstrate and evaluate changes at multiple sampling locations in major river segments. An area of degradation value (ADV) and an area of attainment value (AAV) were also calculated from IBI results to demonstrate the magnitude and extent of changes in fish assemblage condition along segments and between sampling years. Positive responses in the IBI and the ADV/AAV were observed 4 to 5 years after implementing improved municipal wastewater treatment. Positive responses were much less apparent in rivers predominantly influenced by complex industrial sources, agricultural nonpoint sources, and extensive hydrologic modifications. The ADV/AAV showed incremental improvements in river fish assemblages, unlike pass/fail IBI thresholds, and tiered IBI biocriteria provided more appropriate benchmarks than chemical, physical, or qualitative biological criteria. The results show the value of standardized and intensive fish assemblage monitoring and the use of tools that reveal the extent and severity of impairments to determine the effectiveness of water pollution control programs.

Introduction

The Ohio Environmental Protection Agency (EPA) initiated a comprehensive and standardized assessment of the fish assemblages in Ohio rivers in 1979. Its purpose was to provide information for establishing water quality standards (WQS), developing permit terms and conditions for pollutant dischargers, awarding grants for pollution control projects, and planning water quality projects (Yoder and Rankin 1995a; Yoder and Smith 1999). Fish assemblage assessments are one part of Ohio EPA biological and water quality surveys. These surveys are interdisciplinary monitoring efforts planned, coordinated, and conducted on specific water bodies and individual watersheds as part of a comprehensive statewide monitoring strategy. In main-stem rivers, these involve entire reaches, multiple and overlapping stressors, and tens of sampling sites.

* Corresponding author: mbi@rrohio.com

The aggregate database has supported research programs and projects, including the original work on ecoregions and regionalization (Hughes et al. 1986; Larsen et al. 1986; Omernik 1987), biological criteria (Larsen 1995; Yoder and Rankin 1995a, 1995b; Sanders et al. 1999; Thoma 1999; Barbour and Yoder 2000), water quality criteria (Ohio EPA 1999; Miltner and Rankin 1998), analysis of land use impacts (Yoder et al. 2000; Miltner et al. 2004), diagnosis of biological responses (Yoder and Rankin 1995b; Norton et al. 2000; Yoder and DeShon 2003), defining risk and application to management programs (Yoder 1998; Yoder and Rankin 1998; Cormier et al. 1999b; Barbour et al. 2000; NRC 2001; Karr and Yoder 2004; Erekson et al. 2005), and research to develop and validate physiological and genetic indicators (Silbiger et al. 1998; Cormier et al. 1999a).

Ohio EPA annually conducts biological surveys at 400–600 stream and river sampling sites. To date, fish assemblage sampling has included more than 8,000 sites in 1,750 rivers and streams since 1979. More than 100 reports, applied research papers, and technical publications have been developed by Ohio EPA and others (most reports are available at http://www.epa.state.oh.us/dsw/document_index/psdindx.html).

Standardized biological, chemical, and physical monitoring and assessment techniques are used to satisfy three baseline water quality management objectives: 1) determine the extent to which water body classifications assigned in the Ohio WQS are either attained or not attained; 2) determine if the classifications assigned to a given water body are appropriate and attainable; and 3) determine if any changes in ambient biological, chemical, or physical indicators have taken place over time, particularly before and after implementation of mandatory point source pollution controls or voluntary best management practices. Underlying all of the objectives is the identification of the causes and sources associated with impairments or threats identified by an integrated assessment (Yoder and DeShon 2003).

While there is not a single definition of a nonwadeable river, it functionally includes rivers that cannot be sampled effectively by wading techniques (Ohio EPA 1989a). The development of biological assessment tools for nonwadeable rivers, particularly those focused on assessments of condition and status, has lagged behind the development of wadeable methods in the United States. Biological assessments of great and large rivers have been conducted since the late 1940s, but few of these early efforts included fish. The routine assessment of fish assemblages is a comparatively recent addition and followed the development of effective electrofishing technologies. Single-gear electrofishing assessments include the pioneering work by Gammon (1973, 1976, 1980) and Gammon et al. (1981) in the Wabash River of Indiana. Other efforts followed and many were associated with studies of thermal effluents in response to Section 316[a] of the Clean Water Act (CWA) conducted mostly in the 1970s. Many of these studies lacked a conceptual framework for analyzing data and producing meaningful and consistent assessments. The development of the index of biotic integrity (IBI; Karr 1981; Fausch et al. 1984; Karr et al. 1986) provided a conceptual framework for the development of fish assemblage assessment approaches that can be applied to nonwadeable rivers. In this chapter, we report on changes in the fish assemblages of nonwadeable rivers in Ohio, before and after the implementation of pollution controls, using an IBI modified and calibrated for Ohio rivers. This was accomplished by assessing changes in the IBI and the area of degradation value (ADV; Yoder and Rankin 1995b) and an area of attainment value (AAV).

Methods

All methods for capturing, identifying, and processing electrofishing samples follow procedures developed by Ohio EPA (Ohio EPA 1980, 1989a). The rationale for the development of these procedures is described in greater detail in Ohio EPA (1987a, 1987b, 1989a) and Yoder and Smith (1999).

Assessment Design

The Ohio EPA rotating basin approach consists of surveys of specific river basins repeated at varying intervals depending on the need for information about spatial and temporal changes, but generally within 5–10 years. A spatially intensive design is employed to sample fish assemblages in a river reach to comprehensively assess major disturbances. The design requires multiple sampling sites in spatial proximity to suspected sources so that results can be analyzed and displayed in a longitudinal context. Sampling sites are established to represent all habitats including pools, runs, riffles, shoals, and backwaters as available. The sampling design and results interpretation relative to disturbance sources is based on the concepts originally described by Bartsch (1948) and Doudoroff and Warren (1951) to facilitate detection and quantification of varying pollution influences along a reach (i.e., pollution zones). Typically, we sample reaches upstream from major sources of disturbance, in areas of immediate impact and potentially acute effects, through zones of increasing and lessening degradation, and zones of recovery. We attempt to determine the role of specific sources as well as cumulative effects of multiple sources. Large rivers are treated as a single study unit to understand how changes take place along a longitudinal continuum with respect to both natural and anthropogenic influences. Important in the delineation of these study units are natural features and transitional boundaries (e.g., ecological and geological boundaries) and clusters of anthropogenic sources (e.g., major urban/industrial areas, impoundments, etc.). Some study reaches are up to 160 km long to capture all relevant influences, include zones of impact and recovery, and provide context for interpreting results within a localized reach or at a given location (e.g., Figure 1). This design yields a detailed assessment of status, the extent and severity of indicator responses in a particular river reach, and temporal changes (Figure 1). It produces assessments of the severity (departure from the desired state) and extent (lineal extent of the departures) of biological impairments in the various rivers.

When the assessment process described here

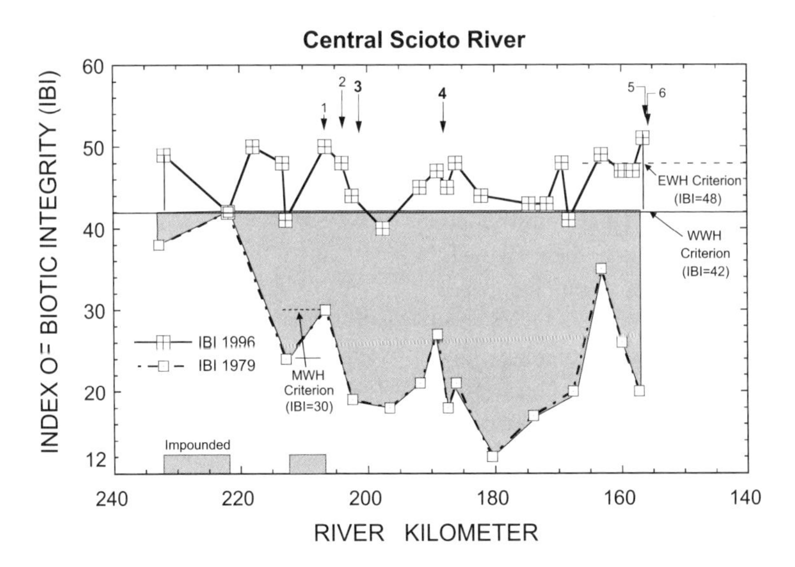

Figure 1.—IBI results from 2 years of electrofishing at multiple locations in a 64-km-long segment of the central Scioto River between Columbus and Circleville, Ohio. The biological criteria for the Warmwater Habitat (WWH), Modified Warmwater Habitat (MWH), and Exceptional Warmwater Habitat (EWH) use designations and major pollution sources are shown (1 = Whittier Street Combined Sewer Overflow; 2 = Techneglass; 3 = Jackson Pike Wastewater Treatment Plant; 4 = Columbus Southerly Wastewater Treatment Plant; 5 = Jefferson Smurfitt Corporation; 6 = Circleville Wastewater Treatment Plant; different font sizes indicate proportional pollutant loading). The shaded area below the WWH biocriterion yields the area of degradation value (ADV) and the unshaded area above the WWH biocriterion yields the area of attainment value (AAV).

was initiated in 1979, management programs were principally focused on major point source discharges such as municipal wastewater treatment plants (WWTPs) and heavy industries (steel making, petroleum refining, metal finishing, paper mills, and major manufacturing). This represented the focus of water quality management at that time, the administration of NPDES permits (National Pollution Discharge Elimination System, Section 402 of the Clean Water Act) and the municipal construction grants program. However, to meet the major objectives of the biological assessment program, we included the influence of other sources in the monitoring and assessment design. These included physical habitat alterations (dams, flow alterations, channelization), nonpoint sources (row crop agriculture, urbanization), and stressors resulting from land use changes (silt, nutrients, flow alterations). We characterized river status and apportioned impacts from various sources regardless of regulatory priority or status, precluding assessment bias towards a particular stressor. We focused on documenting changes in fish assemblages in Ohio rivers before and after implementation of mandatory wastewater treatment upgrades required to meet chemical water quality criteria and voluntary best management practices for the abatement of nonpoint sources of pollution. We chose these two time periods to demonstrate pre- to postpollution control changes, primarily upgrades in wastewater treatment at major municipal and industrial point sources. The principal regulatory driver was the EPA National Municipal Policy, which required wastewater treatment plants to attain effluent quality consistent with instream water quality criteria by July 1, 1988. The year 1990 was chosen as the boundary between "before" and "after" to reflect changes in the fish assemblage that allowed a 2–3-year period for achieving operational stability and initial biological recovery. The IBI and ADV/AAV were used to demonstrate the direction and degree of such changes in the subject rivers and to rank each in terms of their existing and potential quality and recovery status.

Training and Logistics

Ohio EPA samples fish with field crews composed of one full time, professionally qualified crew leader and two technicians. Crew leaders have taxonomic and field sampling skills and are tested for this expertise during the interview and hiring process. New crew leaders undergo an apprenticeship with an experienced crew supervisor prior to performing crew leader duties on a routine basis (Ohio EPA 1989b). Technicians receive basic safety and sampling training prior to the field season.

Field Methods

Fish are collected via standardized boat-mounted electrofishing (Ohio EPA 1987b, 1989b). Johnboats 3.6–4.8 m long are rigged with a hinged aluminum boom mounted on a bow platform. Four 6.25 mm diameter by 0.9 m long woven steel cables serve as anodes and are spaced evenly on a cross-member extending 2.0–2.75 m in front of the bow. Four 25 mm diameter by 1.8 m flexible stainless steel cathodes hang directly from the bow. Pulsed direct current is produced and transformed by Smith-Root type VI-A, 2.5, 3.5, and 5.0 GPP electrofishing units. Power output is varied depending on relative conductivity to produce 12–20 A.

Electrofishing is conducted during daylight June 15–October 15 each year, and for 0.5 km near shore (Ohio EPA 1989a; Yoder and Smith 1999). Time is recorded and a *minimum* of 2,000–2,500 s is specified to ensure sufficient intensity of sampling effort. All habitats (pools, runs, riffles, shoals, and backwaters) are thoroughly electrofished. The boat is maneuvered by outboard motors (9–15 hp) and/or manually by pushing in shallow water where motoring is not possible. Stunned and immobilized fish are netted by one person standing on the bow platform and placed in an aerated live well. This method of sampling is effective for a wide spectrum of river fish, including smaller benthic and riffle dwelling species, large bottom dwellers, pool dwellers, and deep, fast water inhabitants. Our sampling increased known ranges for numerous Ohio river fish species beyond those documented in Trautman (1981). Each sample typically produces 20–35 species (maximum = 50) and 250–500 fish (maximum > 1,000), provided chemical quality and physical habitat are not limiting.

Captured fish are enumerated by species, weighed, examined for external anomalies, and released or preserved in 10% formalin. Lengths are taken on selected species, otherwise species are classified as adults, 1+, or 0+. Fish less than 20–25 mm total length generally are not included in the samples, following the recommendation of Angermeier and Karr (1986).

A qualitative habitat assessment is conducted during each sampling pass using the Qualitative Habitat Evaluation Index (QHEI; Ohio EPA 1989a; Rankin 1989, 1995). The QHEI is a visual estimate of the quality, composition, amount, and extent of substrate, cover, channel, riparian, flow, pool/run/riffle, and gradient variables. The QHEI corresponds to key attributes of fish assemblage quality (Rankin 1989, 1995) and is an important tool in determining the appropriate and attainable use classification for Ohio rivers and streams (Rankin 1995; Yoder and Rankin 1995a). Those data are also entered and stored in the Ohio ECOS data management system.

Chemical and physical water quality data are collected near each biological site by separate field crews dedicated to this type of sampling. Core parameters collected at each site include field measurements (temperature, dissolved oxygen, conductivity, and pH) and a baseline set of conventional parameters (nitrogen series, phosphorus, biochemical oxygen demand, suspended and dissolved solids, chlorides, sulfates, and hardness), and common heavy metals. Additional parameters (other heavy metals, organic chemicals) are added if these contaminants are suspected. Most sampling consists of grab samples collected 3–8 times from the water column during the summer–early fall index period. Composite samples, continuous data, and analyses of bottom sediments (metals and organics) are included as the complexity of the situation dictates. This provides the essential data on stressors and exposures against which the response by the fish assemblage is interpreted (Yoder and Rankin 1998; Yoder and DeShon 2003.)

Data Entry and Data Analysis

Data are recorded on field sheets, and after verifying voucher specimens, entered into the Ohio ECOS data management system. Each crew leader and a data entry analyst proofread all entries before the data are considered valid. We use a modified IBI that was developed and calibrated for boat-mounted electro-fishing (Ohio EPA 1987b; Yoder and Rankin 1995a; Yoder and Smith 1999). The IBI consists of 12 metrics (Table 1) and generally adheres to the original guidance of Karr et al. (1986). The IBI values are calculated for individual sampling passes by a program in the Ohio ECOS routine following the procedures in Ohio EPA (1987b, 1989b). Data are analyzed with box-and-whisker plots and graphical routines for IBI scores and ADV values. The IBI is used to determine the proportion of sampled reaches that attain designated aquatic life uses (Ohio Administrative Code 3745–1). A site is considered impaired if the sample result is more than four IBI units below the criterion. The statistical properties of this IBI were described by Ohio EPA (1987b), Rankin and Yoder (1990), and Fore et al. (1993).

The ADV (Yoder and Rankin 1995b) was originally developed to quantify the extent and severity of departures from biocriteria within a defined river reach. We have added an Area of Attainment Value (AAV) that quantifies the extent to which minimum attainment criteria are surpassed. The ADV/AAV correspond to the area of the polygon formed by the longitudinal profile of IBI scores and the straight line boundary formed by a criterion, the ADV below and the AAV above (Figure 1). The computational formula (after Yoder and Rankin 1995b) is

$$ADV/AAV = \sum [(aIBIa + aIBIb) - (pIBIa + pIBIb)] * (RMa - RMb), \text{ for } a = 1 \text{ to } n, \text{ where}$$

$aIBI_a$ = actual IBI at river mile a,

$aIBI_b$ = actual IBI at river mile b,

$pIBI_a$ = IBI biocriterion at river mile a,

$pIBI_b$ = IBI biocriterion at river mile b,

RM_a = upstream most river mile,

RM_b = downstream most river mile, and

n = number of samples.

The average of two contiguous sampling sites is assumed to integrate fish assemblage status for the distance between the points. The intensive survey

Table 1.—Index of biotic integrity metrics and scoring criteria based on fish assemblage data collected with boat electrofishers at nonwadeable sites in Ohio (after Ohio EPA 1987b). All percent metrics are based on fish numbers. Species metric assignments are available in Ohio EPA (1987b).

Metric	Scoring criteria 5	3	1
Native species[a]	>20	10–20	<10
% Round-bodied suckers[b]	>38	19–38	<19
Sunfish species[c]	>3	2–3	<2
Sucker species	>5	3–5	<3
Intolerant species	>3	2–3	<2
% tolerant	<15	15–27	>27
% omnivores	<16	16–28	>28
% insectivores	>54	27–54	<27
% top carnivores	>10	5–10	<5
% simple lithophils			
≤1,560 ha (600 mi^2)	>50	25–50	<25
>1,560 ha (600 mi^2)		Varies by drainage area	
% DELT anomalies	<0.5[d]	0.5–3.0[e]	>3.0
Fish numbers (no./km)[f]	>450	200–450	<200

[a] Excludes all introduced and alien fish species.
[b] Includes *Moxostoma, Hypentelium, Minytrema, Erimyzon,* and *Cycleptus*; excludes white sucker *Catostomus commersonii.*
[c] Excludes black basses (*Micropterus* sp.).
[d] Or > 1 individual at sites with < 200 total fish.
[e] Or > 2 individuals at sites with < 200 total fish.
[f] Excludes tolerant and alien species and all hybrids; metric scoring adjustments are made at < 50 and 50–200 fish/km (Yoder and Smith 1999).

design typically positions sites in close enough proximity to sources of stress and along probable zones of impact and recovery so that meaningful changes are adequately captured. We have observed fish assemblages as portrayed by the IBI to change predictably in proximity to major sources and types of pollution in numerous instances (Ohio EPA 1987a; Yoder and Rankin 1995b; Yoder and Smith 1999). Thus, the longitudinal connection of contiguous sampling points produces a reasonably accurate portrayal of the extent and severity of impairment in a specified river reach as reflected by the IBI (Yoder and Rankin 1995a). The total ADV/AAV for a specified river segment is normalized to ADV/AAV units/km for making comparisons between years and rivers.

The ADV is calculated as a negative (below the biocriterion) expression; the AAV is calculated as a positive (above the biocriterion) expression. Each depicts the extent and degree of impairment (ADV) and attainment (AAV) of a biological criterion, which provides a more quantitative depiction of quality than pass/fail descriptors. It also allows the visualization of incremental changes in condition that may not alter the pass/fail status, but are nonetheless meaningful in terms of quantitative change over space and time. In our analyses, the Warmwater Habitat (WWH) biocriterion for the IBI, which varies by use designation and ecoregion (Table 2), is used as the threshold for calculating the ADV and AAV. The WWH use designation represents the minimum goal required by the Clean Water Act for the protection and propagation of aquatic life, thus it is used as a standard benchmark for ADV/AAV analyses.

Integrated Assessments

Data and information are analyzed in accordance with a stress-exposure-response sequence (Figure 2; Yoder and Rankin 1998; Karr and Yoder 2004). The fish assemblage data are used to characterize and quantify the biological response to accompanying chemical/physical data and disturbance data

Table 2.—Biological criteria for the index of biotic integrity that are applicable to boat electrofishing sites in Ohio (Ohio Administrative Code Chapter 3745-1).

Ecoregion	Modified Warmwater Habitat (MWH)[a]	Warmwater Habitat (WWH)	Exceptional Warmwater Habitat (EWH)
HELP – Huron/Erie Lake Plain	20/22	34	48
EOLP – Erie/Ontario Lake Plain	24/30	40	48
IP - Interior Plateau	24/30	38	48
ECBP – East Corn Belt Plains	24/30	42	48
WAP – West Allegheny Plateau	24/30	40	48

[a] MWH biocriteria for channelized/impounded sites.

such as pollutant loadings, spills, land uses, and other indicators of human activity. This process, first developed by U.S. EPA (1990, 1995), has been extensively described by Yoder and Rankin (1998), Yoder and Smith (1999), Yoder and Kulik (2003), and Karr and Yoder (2004) and it is routinely employed by Ohio EPA. Key to the process is the accurate identification and quantification of biological impairments and the association of these impairments with relevant chemical and physical indicator thresholds and criteria (Yoder and Rankin 1995b; Yoder and DeShon 2003).

Results

Changes in Fish Assemblages

We analyzed data from 1979 through 2001, when we collected 135 of the 172 fish species recorded for Ohio (Trautman 1981; Appendix A). Many species occurred sporadically and rarely (58 species occurred in >10% of samples after 1990, versus 43 before 1990). Most species increased in relative abundance after 1990, and 36 species more than doubled in abundance. Of these, 14 are considered

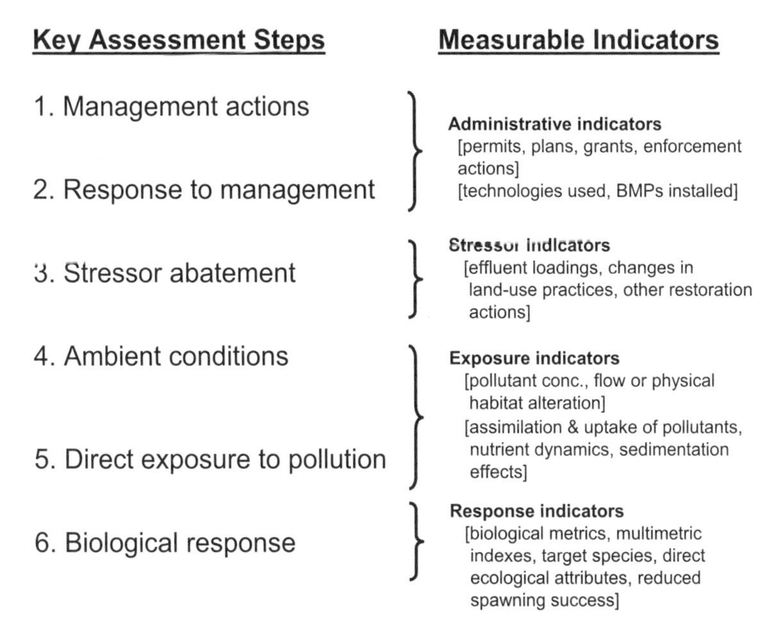

Figure 2.—Linkages between key steps of an adequate monitoring and assessment process, including their measurable indicators; the key assessment steps are sequential in a descending manner and comprise a feedback loop among and between steps (modified from U.S. EPA 1995, Yoder and Rankin 1998; and Karr and Yoder 2004).

highly intolerant, 5 are round-bodied suckers, 10 are darters, 5 are large river riffle species, and 5 are obligate large river species. One highly tolerant species, blacknose dace *Rhinichthys atratulus*, doubled in abundance but was rare in nonwadeable rivers and occurred in less than 5% of samples. Species declining at least 25% after 1990 were goldfish *Carassius auratus*, brown bullhead *Ameiurus nebulosus*, green sunfish *Lepomis cyanellus*, and white perch *Morone americana*, the first three of which are considered tolerant of poor water quality. Three other tolerant species declined slightly and six tolerant species increased slightly, but only one, bluntnose minnow *Pimephales notatus*, increased substantially. Our results indicated increased abundance and distribution of many Ohio fish species, with range extensions of several large river species collected by boat electrofishing (e.g., river redhorse *Moxostoma carinatum*, greater redhorse *M. valenciennesi*, smallmouth buffalo *Ictiobus bubalus*, black buffalo *I. niger*, sauger *Sander canadensis*, and gravel chub *Erimystax x-punctatus*). However, many other species remain well below their historical abundances and distributional ranges.

We ranked the status of 45 of Ohio's nonwadeable rivers including all major main-stem rivers and their largest tributaries draining more than 390 ha (Figure 3) using box-and-whisker plots of pre- and post-1990 IBI results. Rivers were ranked according to their post-1990 75th percentile IBI values, which better indicate assemblage condition *potential* than do medians or averages, and minimized the influence of reaches and sites that have not fully responded to pollution abatement practices and/or where those efforts were incomplete (Figure 4). Twelve of the top ranked 14 rivers were either wholly or partially classified as Exceptional Warmwater Habitat (EWH), which reflects a quality consistent with the 75th percentile of Ohio's least disturbed reference sites (Yoder and Rankin 1995a). The median IBI for 30 rivers met the Warmwater Habitat (WWH) IBI criterion, which is set at the 25th percentile of least disturbed reference sites and is the minimum restoration goal of the Clean Water Act in Ohio. Least disturbed ecoregional sites represent attainable background conditions and were used as reference sites for developing the Ohio biocriteria (Ohio EPA 1987b; Yoder and Rankin 1995a). Typically, this includes nonwadeable river reaches that are upstream or outside of the immediate influence of point sources and major habitat modifications. As such, these sites reflect attainable biological condition in terms of the concept of regional reference envisioned by Hughes et al. (1986).

We used changes in the median IBI and changes in the ADV/AAV to express the degree and significance of temporal changes for 38 river segments (Appendices B–D). Positive changes in median IBI values were significant (i.e., >4 units; Ohio EPA 1987b; Rankin and Yoder 1990) in 28 rivers, with changes of greater or equal to 10 or more units in 17 rivers (Appendix B). There were no significant declines in the median IBI in any of the rivers. The percent change in median IBI values was positive in all but three rivers, exceeding 100% in three rivers (central Scioto, middle Great Miami, and Cuyahoga rivers) and more than 25% in 15 others (Appendix C). Changes in the ADV/AAV were expressed as a real change in impairment (negative ADV), a real change in attainment (AAV), and the net change in both the ADV and AAV (Appendix C). Reduced impairment occurred in all but seven rivers, and gains in attainment occurred in all but seven rivers. Net gains in ADV/AAV were observed in all but five rivers.

General Disturbance Types

In terms of the relative ranking of the rivers using the 75th percentile IBI (Appendix B), the highest ranked rivers received effluents from municipal WWTPs and runoff from agricultural nonpoint sources, followed by industrial, urban, and other sources. Seven rivers predominantly disturbed by complex toxics and acid mine drainage ranked at the bottom, even though incremental improvements were observed (Appendix B). They reflected residual impacts from uncontrolled toxic impacts and contaminated benthic sediments that are not as amenable to conventional abatement practices.

The overall distribution of IBIs and changes in the median IBIs between rivers predominantly disturbed by agricultural nonpoint sources, municipal WWTPs, and all other sources (industrial, urban, complex toxic, and acid mine drainage) showed

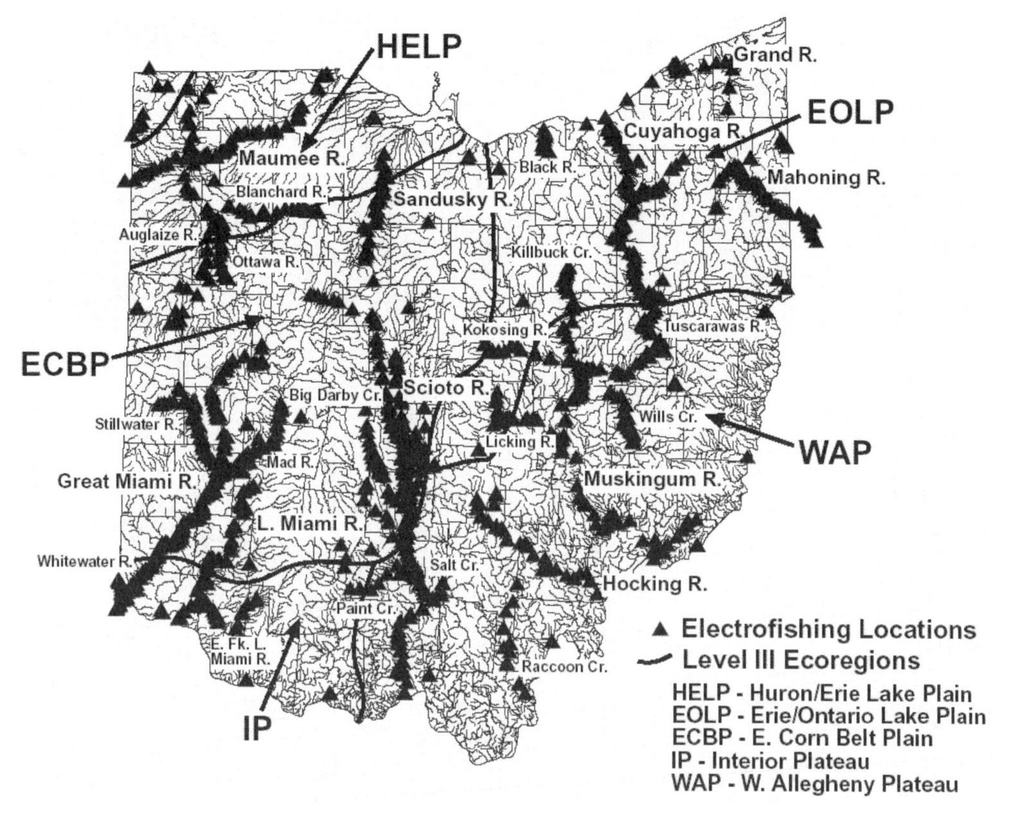

Figure 3.—Ohio EPA nonwadeable electrofishing locations sampled between 1979 and 2001. All sites drain greater or equal to 390 ha and were sampled with boat electrofishers per Ohio EPA (1989b). Sampled main-stem rivers and significant tributaries are labeled, and level-III ecoregion boundaries are shown.

improvements after 1990 (Figure 5). This improvement reflected the relative severity of each disturbance type, with the lowest IBI values associated with complex toxic and industrial sources, followed by WWTP and agricultural nonpoint sources. In terms of changes in median IBI values, this order was reversed. The greatest gains in the 25th–75th percentiles occurred with the complex toxic and industrial disturbance types, followed by WWTPs and agricultural nonpoint sources. This reflects, in part, the greater severity of these impacts and hence more potential for improvement. Nevertheless, half the rivers disturbed by complex toxic and industrial sources had IBIs that attained the WWH biocriterion after 1990, compared with no values meeting the biocriterion before 1990. No IBIs attained the EWH biocriterion for complex toxic and industrial sources before 1990 and only a few sites did so after 1990. Rivers with the least or no positive changes in the ADV or AAV were disturbed by complex toxics and acid mine drainage (Appendix C). Agricultural nonpoint sources showed the least improvement in terms of net ADV + AAV increases, although the median IBI improved from marginal WWH to EWH quality. The WWTP impacted rivers showed slightly more improvement.

The largest increases in median IBIs consistently occurred in rivers that were predominantly impacted by municipal WWTPs and included the central Scioto (+26), upper Hocking (+22), Licking (+16), upper Great Miami (+14), and lower Great Miami (+12) rivers (Appendix B). In each of these areas, agricultural row cropping was the predominant land use upriver of a major urban area that also included combined sewer overflows in all except one river. Areas of impairment associated with combined sew-

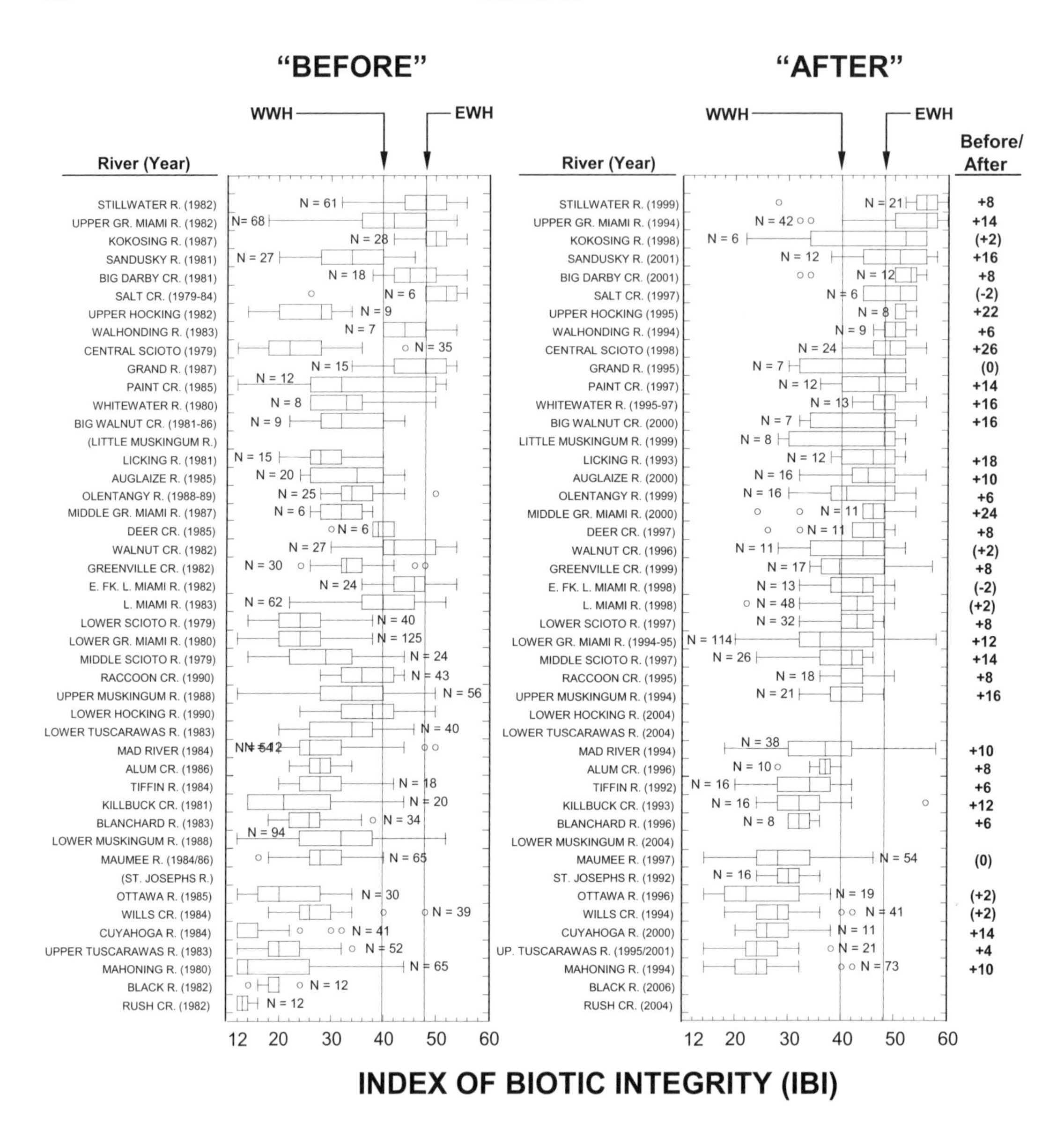

Figure 4.—Box-and-whisker plots of index of biotic integrity (IBI) results from 45 Ohio river segments sampled by boat electrofishers. Results from 1979 to 1990 appear in the left figure box and are labeled "Before." Results after 1990 appear in the right figure box and are labeled "After." Rivers are ranked according to the 75th percentile IBI value in the "After" period. The net change (±) in the median IBI from "Before" to "After" is listed under ΔBefore/After column; nonsignificant changes (< 4 IBI units) are listed in parentheses. The biological criteria for the Warmwater Habitat (WWH) and Exceptional Warmwater Habitat (EWH) use designations are indicated by vertical lines. N is sample size.

ers were observed in some urban areas, but were localized compared to the larger reaches affected by the WWTP effluents.

Notable improvements also occurred in some rivers disturbed by industrial, complex toxic, urban, and mine drainage sources, and included the lower Scioto River (industrial; +18), Big Walnut Creek (urban; +16), Paint Creek (industrial; +14), Cuyahoga River (complex toxic; +14), and middle Scioto River (industrial; +14). Improvements of a

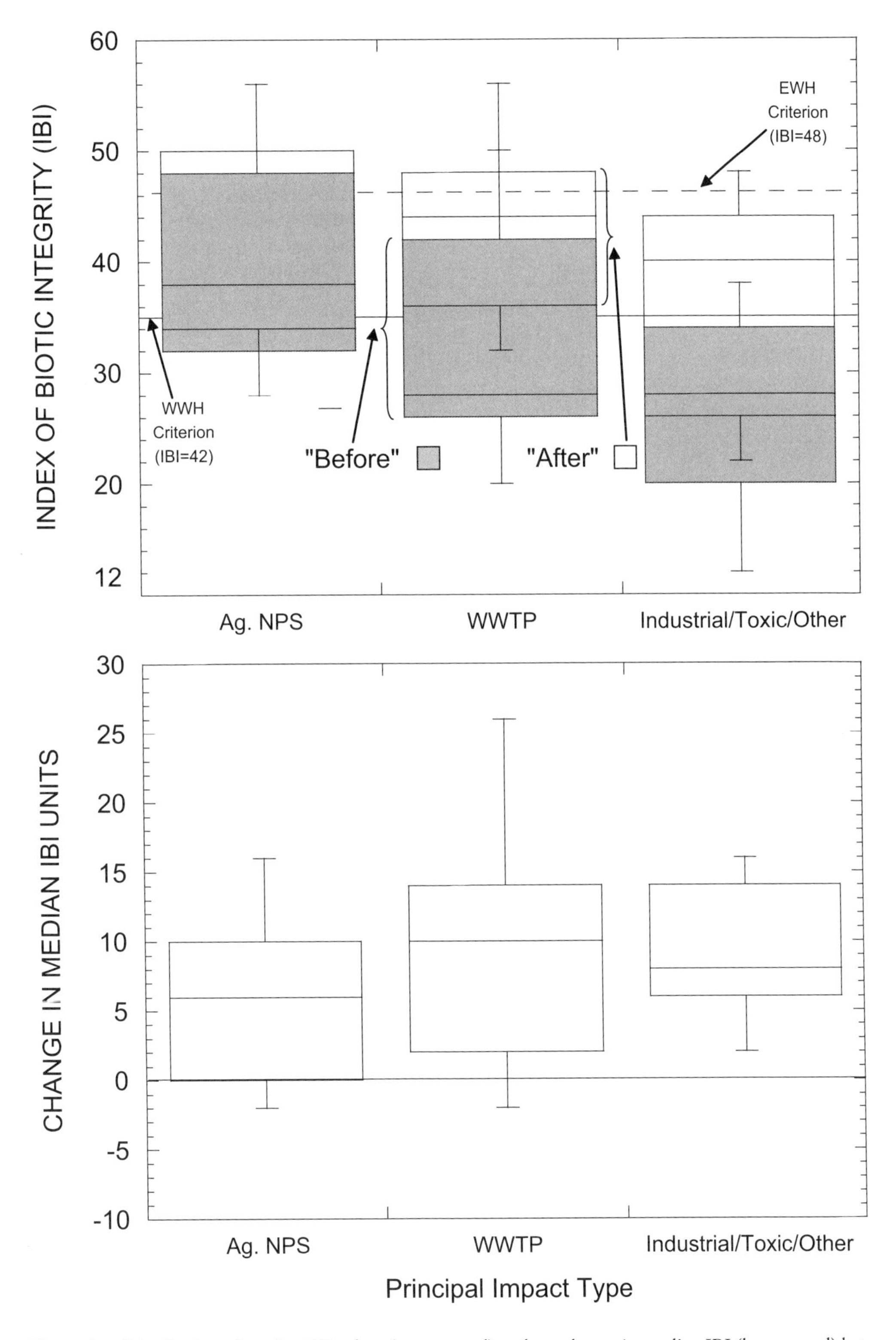

Figure 5.—Distribution of median IBI values (upper panel) and net change in median IBI (lower panel) between the "Before" and "After" periods aggregated by the three predominant disturbance types of Yoder and Rankin (1995b).

similar magnitude were less common in rivers receiving agricultural nonpoint sources, but included comparable changes in the Sandusky River (+16), Whitewater River (+16), and Auglaize River (+10). Many of the highest quality rivers in Ohio were predominantly disturbed by agricultural nonpoint sources and exhibited very good to exceptional quality before 1990. As a result, none of these rivers exhibited significant increases in median IBIs after 1990. Declines in median IBIs occurred in two rivers (East Fork Little Miami River and Salt Creek), but the changes were not significant (i.e., <4 IBI units). However, both are designated EWH and any decline is noteworthy. The East Fork Little Miami River was disturbed by municipal WWTPs that have recently approached treatment capacity. Salt Creek is disturbed by nonpoint sources, especially by row cropping that has increasingly encroached on the riparian zone.

Magnitude and Extent of Changes

A comparison of the ranking of selected rivers for the median IBI and ADV/AAV statistics (Figure 4; Appendix D) shows consistency in some, but not all rivers. The most impaired rivers before 1990 included the Cuyahoga, central Scioto, Mahoning, Ottawa, upper Tuscarawas, lower Great Miami, Whitewater, and lower Scioto rivers, all with ADV greater than 100 units/km and zero AAV/km. After 1990, impairment was substantially reduced in the central Scioto (ADV/km = 0), Whitewater (–0.3), lower Scioto (–0.3), and lower Great Miami (–17.8) rivers, but remained high in the Mahoning (–126) and Ottawa (–116) rivers. The reduced impairments were associated with improved wastewater treatment and included all rivers with large gains in median IBI values (Appendix C). Those rivers with the greatest impairments and least improved IBIs after 1990 are disturbed by complex toxics.

Only 12 rivers had pre-1990 AAVs greater than 10 units/km, and 17 had values of zero. AAVs increased markedly after 1990 with three rivers exceeding 100 units/km and only three rivers with values of zero (Appendix C). Rivers showing the largest gains included the upper Hocking River (recovery from WWTP and complex toxic impacts), middle Great Miami River (recovery from acute thermal impacts), Auglaize River (recovery from agricultural nonpoint source impacts), Stillwater River (existing high quality), and upper Great Miami River (recovery from WWTP impacts). In terms of net gains made in both ADVs and AAVs, 7 of the top 10 rivers were disturbed by WWTPs and increased as much as three- to fourfold over other sources (Appendix D).

The greatest reduction in impairment occurred for the mix of other sources (industrial, urban, complex toxic, and acid mine drainage) and WWTP. Reductions were less for agricultural sources, primarily because the IBI departures before 1990 were less severe (Figure 6). Agricultural nonpoint sources and WWTPs showed greater changes in the AAV, although some rivers with other pollution sources showed the greatest gains.

Case Studies

It is useful to examine three rivers in different natural settings that had differing disturbance and rehabilitation histories, both to better understand historical changes and future possibilities.

Central Scioto River (Management of Municipal Wastewater Pollution).—The central Scioto River includes the main stem between Columbus and Circleville, Ohio, a distance of approximately 75 km. The main-stem Scioto River flows through the greater Columbus metropolitan area, with a human population exceeding one million. Wastewater treatment is provided by the Jackson Pike and Southerly WWTPs, with daily capacities of 283,875 and 454,200 m^3 of treated wastewater, respectively. These discharges collectively comprise 90–95% of the normal summer–fall flow in the Scioto River down river from Columbus. At the turn of the 20th century, the Scioto River was grossly polluted and only six fish species were found (Trautman 1981). By the 1960s, the acute impacts from point sources extended from Columbus to the Ohio River, a distance of 200 km. The landscape and environmental setting is typical of many other municipalities throughout Ohio, which includes an agricultural watershed, an urbanized area including combined sewer overflows, run-of-river impoundments, and WWTPs that dominate the summer–fall flow regime down river from the city.

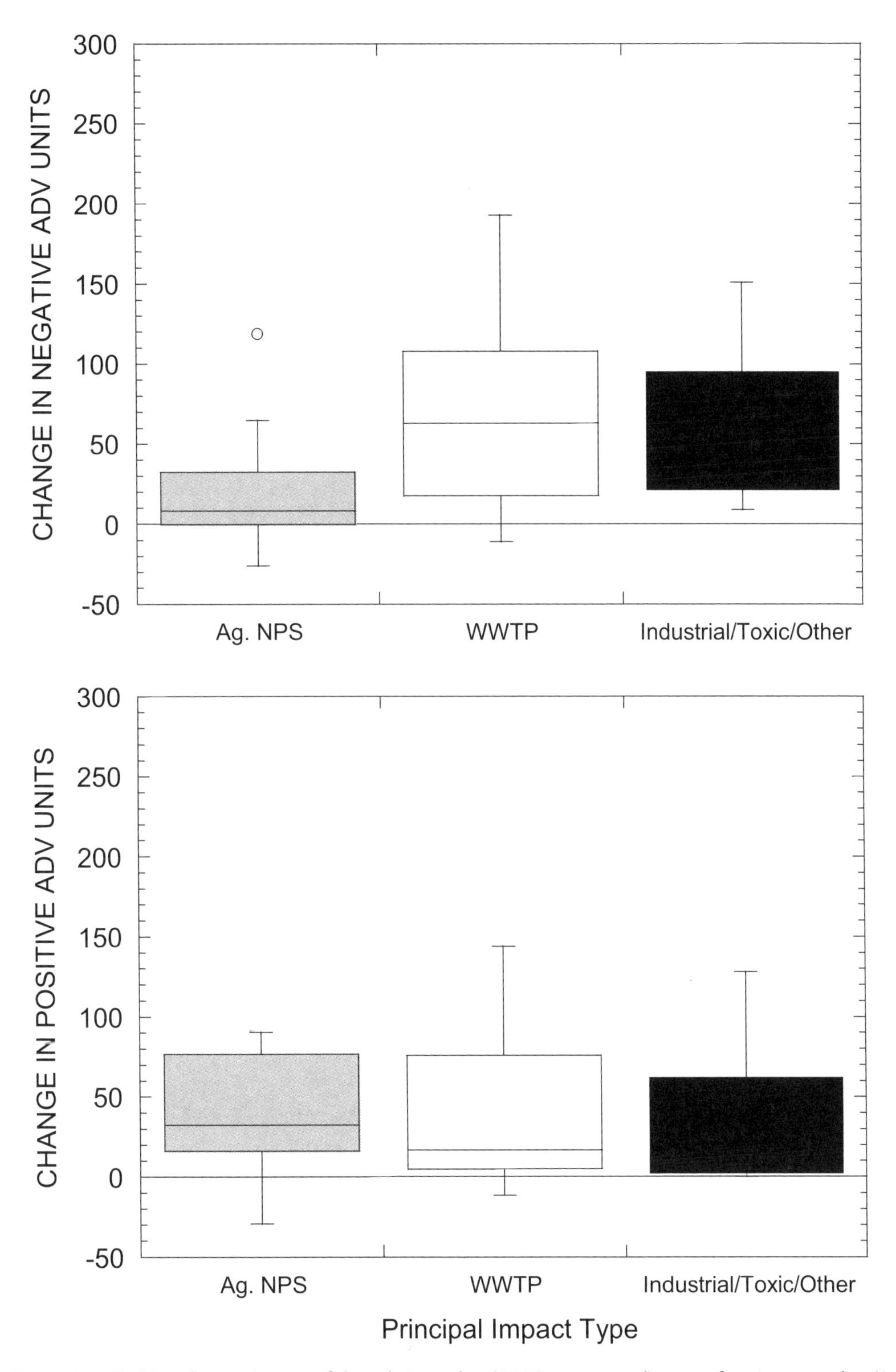

Figure 6. —Positive changes in area of degradation value (ADV, upper panel), area of attainment value (AAV, middle panel), and net gain (lower panel) between "Before" and "After" periods and aggregated by the three predominant disturbance types.

Fish assemblage data were collected 1979–1981 and annually since 1985 to better understand year-to-year changes and the management of large pollution sources (Figure 7). The ADV reflected incremental changes corresponding to the full suite of pollution controls implemented at the Columbus WWTPs since the early 1980s. The longitudinal IBI results show that improvements occurred through time and represent incremental recovery at most sampling locations impacted by the WWTPs. The 1996 results show most sites above the WWH biocriterion and some sites attaining the EWH biocriterion, indicating that a change in classification of the lower 13 km to EWH is warranted.

The improved fish assemblage corresponded with reduced pollutant loads and lower pollutant concentrations in the river (Figure 8). For example, ammonia-N loads were reduced from 1,000s of kg/d to less than 25–50 kg/d following implementation of advanced wastewater treatment (AWT) in 1988. These improvements in effluent and chemical water quality were followed by improvements in biological quality as indicated by the fish assemblage results (Figure 7). This exemplifies the process outlined in Figure 2 in which management actions and responses were followed by changes in chemical/physical indicators that produced a positive biological response in the receiving environment. Similar results have since been documented in other rivers disturbed by WWTP discharges that have been subjected to the same sequence of pollution controls.

Upper Great Miami River (Response of a High Quality Resource).—The upper Great Miami River is located in western Ohio between Indian Lake and Dayton, a distance of 130 km. Land use is predominantly row crop agriculture, and there are major WWTPs at Sidney, Piqua, and Troy. While none of these dominate the flow of the main stem, each requires water quality-based effluent limitations to meet water quality criteria for common pollutants such as ammonia-N and biochemical oxygen demand. In 1982, the IBI results indicated attainment of the WWH biocriterion for the majority of the river reach (Figure 9). Localized zones of degradation and impairment were observed downstream from the WWTPs, and abatement measures were based on meeting water quality criteria for the

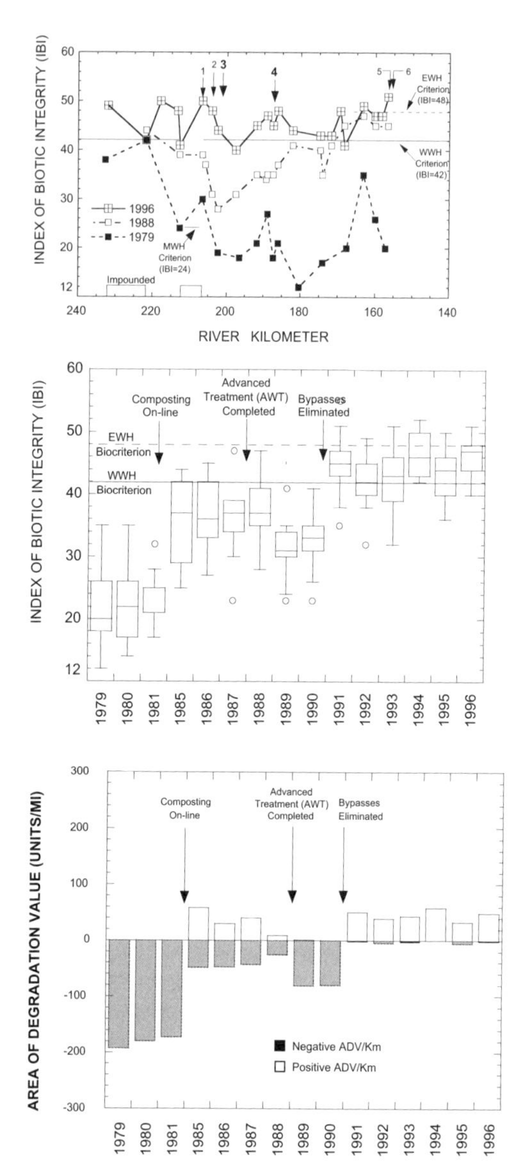

Figure 7.—Longitudinal profile of IBI scores in the central Scioto River main stem in and downstream from Columbus, Ohio in 1979, 1988, and 1996 (upper panel; WWH = Warmwater Habitat; EWH = Exceptional Warmwater Habitat; 1 = Whittier Street Combined Sewer Overflow; 2 = Techneglass; 3 = Jackson Pike Wastewater Treatment Plant; 4 = Columbus Southerly Wastewater Treatment Plant; 5 = Jefferson Smurfitt Corporation; 6 = Circleville Wastewater Treatment Plant). Annual IBI results from the central Scioto River main stem between 1979 and 1996 (middle panel), and ADV and AAV/km during the same period (lower panel). Significant changes in the operation of the Columbus sewage treatment system are noted on the lower two panels.

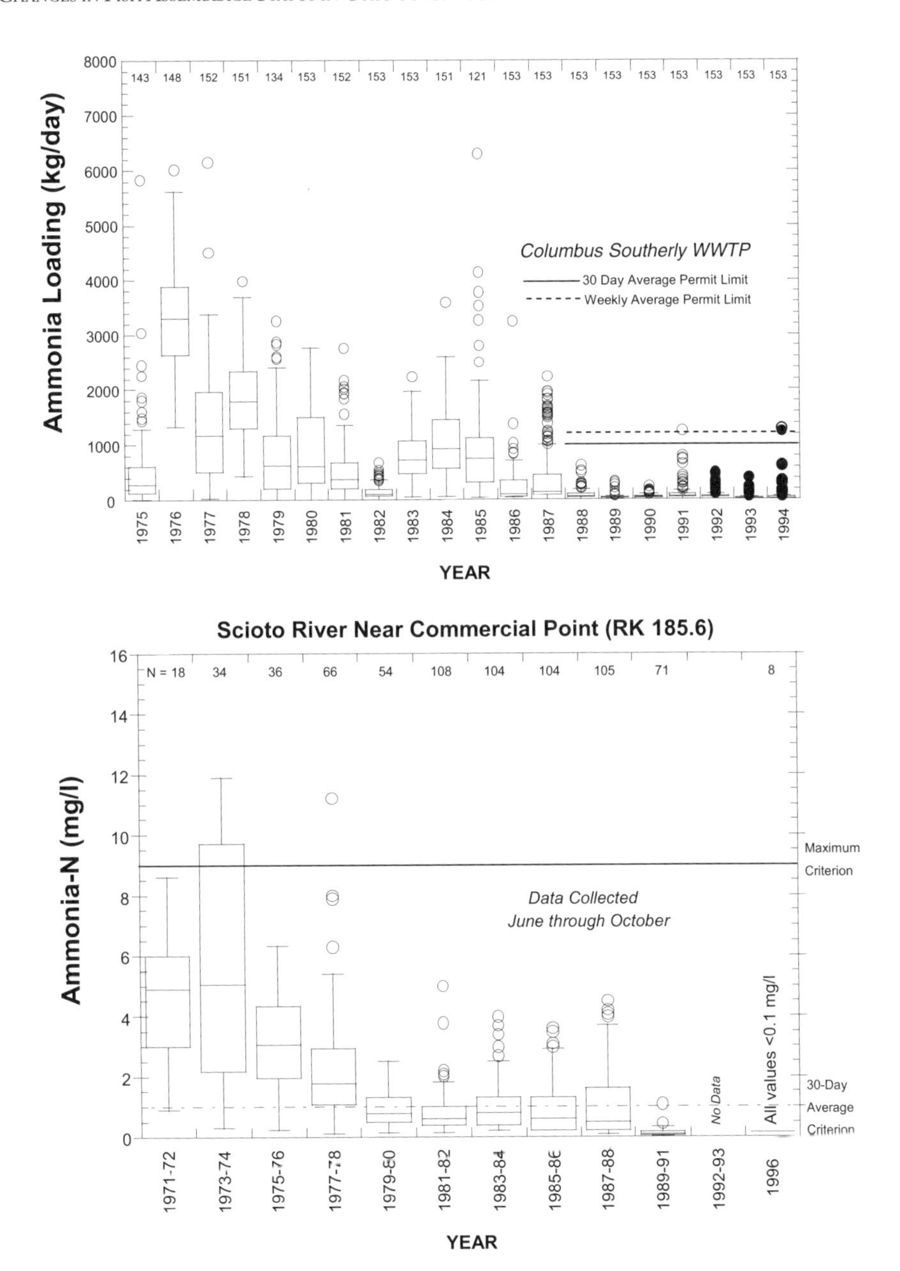

Figure 8.—Box-and-whisker plots of ammonia-N loads (kg/d) discharged annually by the Columbus Southerly WWTP between 1975 and 1994 based on frequent measurements of flow and ammonia-N concentration (mg/L) taken from the final effluent (upper panel). June–October concentrations of ammonia-N (mg/l) measured in the Scioto River 4.6 km (2.9 mi.) downstream from the Columbus Southerly WWTP between 1971 and 1996 (lower panel). Effluent and instream water quality criteria are indicated in each panel, and sample size is indicated at the top of each panel.

WWH classification. The implementation of advanced wastewater treatment common to municipal WWTPs with water quality-based effluent limitations resulted in significant reductions in ammonia-N loads. Loads of greater than 200–300 kg/d were reduced to less than 10–20 kg/d following implementation of advanced wastewater treatment in 1987 and 1988 (Figure 9). This was followed by

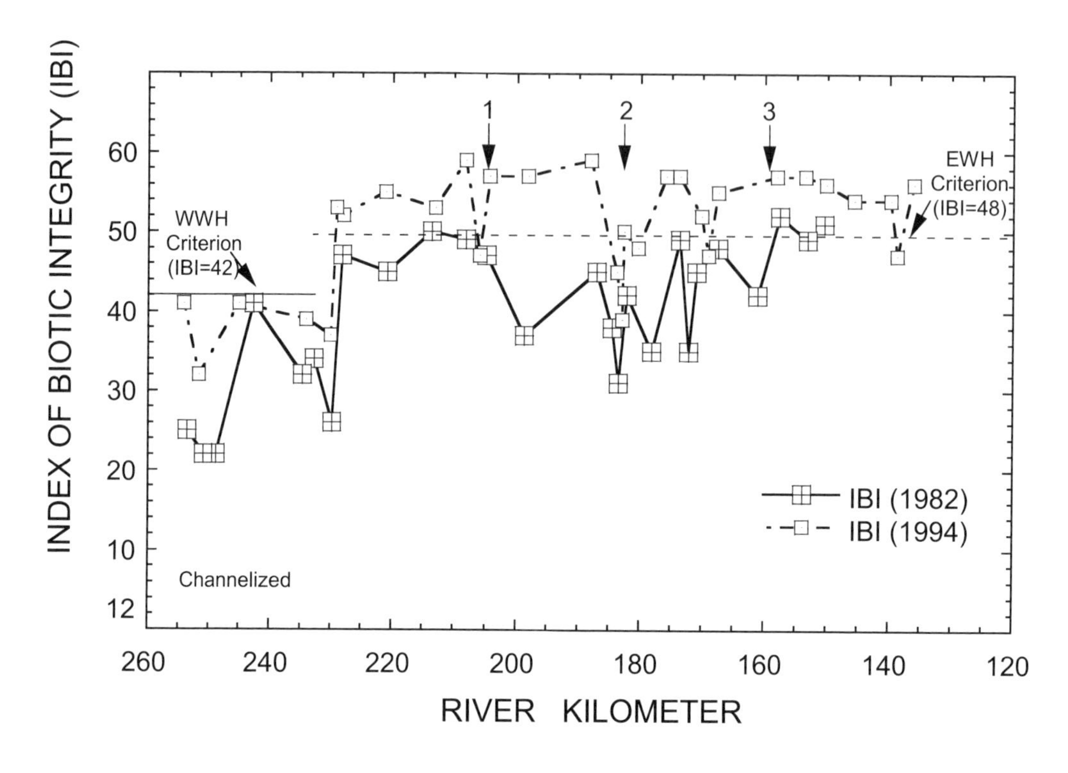

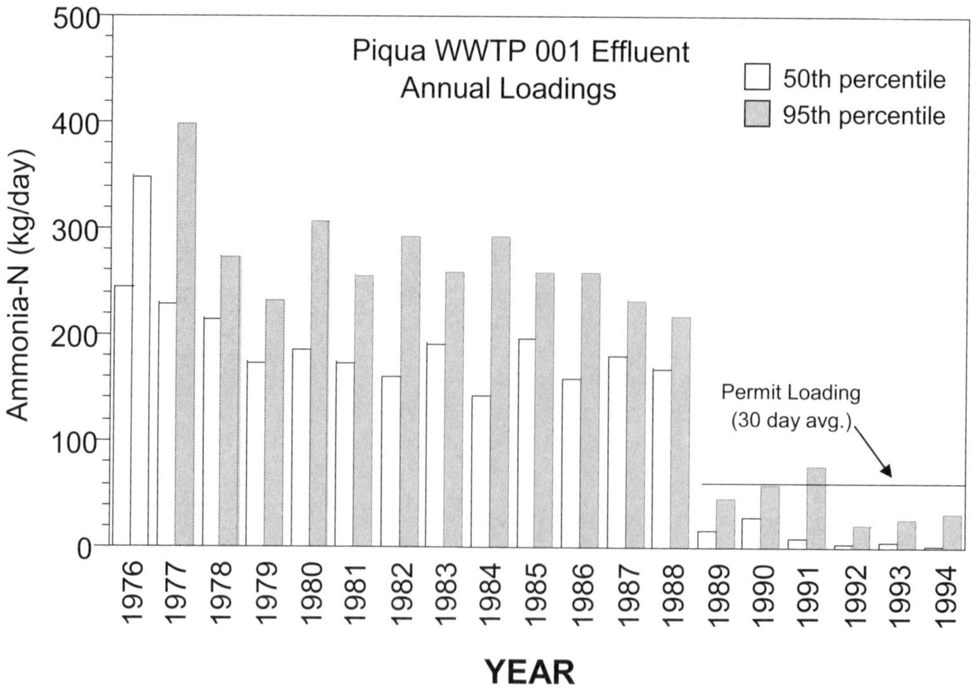

Figure 9.—Longitudinal IBI profile in the upper Great Miami River main stem between Indian Lake and Dayton, Ohio in 1982 and 1994 (upper panel; WWH = Warmwater Habitat; EWH = Exceptional Warmwater Habitat; 1 = Sidney WWTP, 2 = Piqua WWTP, 3 = Troy WWTP). Annual median and 95th percentile loads (kg/d) of ammonia-N discharged by the Piqua WWTP between 1976 and 1994 (lower panel).

IBI improvements in 1994 that surpassed the minimum biocriterion for the EWH classification. This resulted in a redesignation of the upper main stem between Sidney and Dayton to EWH, thus increasing the level of protection for one of the highest quality river segments in Ohio. The channelized area in the upper portion of this reach showed insufficient improvement between 1982 and 1994, and this segment remained classified as WWH.

This example is particularly noteworthy in that the implementation of pollution controls resulted in improvements that went beyond the minimum goal of the CWA. Furthermore, the existence of a system of tiered use classifications in the Ohio WQS resulted in a higher level of protection for this reach in accordance with its ecological attributes. Most state WQS consist of single uses that do not include the multiple classification tiers demonstrated here. Such WQS leave high quality waters vulnerable to unintended degradation or lower levels of protection. That would have been especially true in this case since this reach lacks many of the ecological attributes (i.e., rare, threatened, or endangered species) that are usually required to draw special attention in single use systems. The exceptional IBIs were sufficient evidence of an outstanding and exceptional resource that merited a higher level of protection than CWA minima. The problems with single use classifications were highlighted by the NRC (2001) and extensively described by Karr and Yoder (2004).

Auglaize River (Best Management Practices for Nonpoint Sources).—The Auglaize River is a major tributary in the Maumee drainage of northwestern Ohio. The sampled nonwadeable segment extends for approximately 64 km from near Wapakoneta to Ft. Jennings and is disturbed primarily by row crop agriculture. Most small tributaries and the headwaters were channelized to enhance surface and subsurface drainage, which makes row cropping sustainable in this glaciated lake plain. The 2000 biological survey revealed significant increases in IBI values in this reach compared to 1985 and 1991 (Figure 10). The increase in the 2000 IBI was not explained by improvements in WWTP effluent quality as this was not a major disturbance in this river. Myers et al. (2000) noted that the implementation of tillage practices that leave crop residues on the soil surface increased markedly in this region in the mid 1980s through the 1990s (Figure 10) and corresponded with reduced concentrations of suspended sediments in the Auglaize River near Ft. Jennings. These reductions in suspended and deposited sediments coincided with higher IBI scores. The metrics most involved in the improved IBI scores (increased proportion of intolerant species, insectivores, simple lithophils and reduced tolerant species and omnivores) are those expected to respond to reduced fine sediments.

The improvement in IBI scores did not adequately communicate the quantity of improvement in the fish assemblage, thus the ADV and AAV were used to better document these changes (Figure 10). Much of the gain made in the Auglaize River fish assemblage was in the extent of attainment (as shown by the AAV), much more so than declines in impairment (shown by the ADV). This would not have been apparent in a conventional focus on simple pass/fail thresholds.

Discussion

Ohio's nonwadeable rivers have been subjected to multiple impacts from human activities ranging from untreated discharges of human and industrial waste, deforestation, extensive changes in land use, and major hydrological modifications during the past 200 years. These impacts began in the early 19th century and peaked in the early to mid-20th century. Some rivers were so polluted that fish were virtually absent for significant distances (Trautman 1981). The Clean Water Act amendments passed in 1972 required improved wastewater treatment and basin planning, both aimed at reducing the gross impacts of decades of pollution. The Ohio EPA biological monitoring and assessment program was initiated to directly determine the effectiveness of CWA pollution control programs and to develop new and improved tools to better protect and manage water resources. As the gross pollution problems of the preceding decades were better controlled, new and less well understood issues were identified and required new and improved technologies and management practices. In addition, the costs of treatment and man-

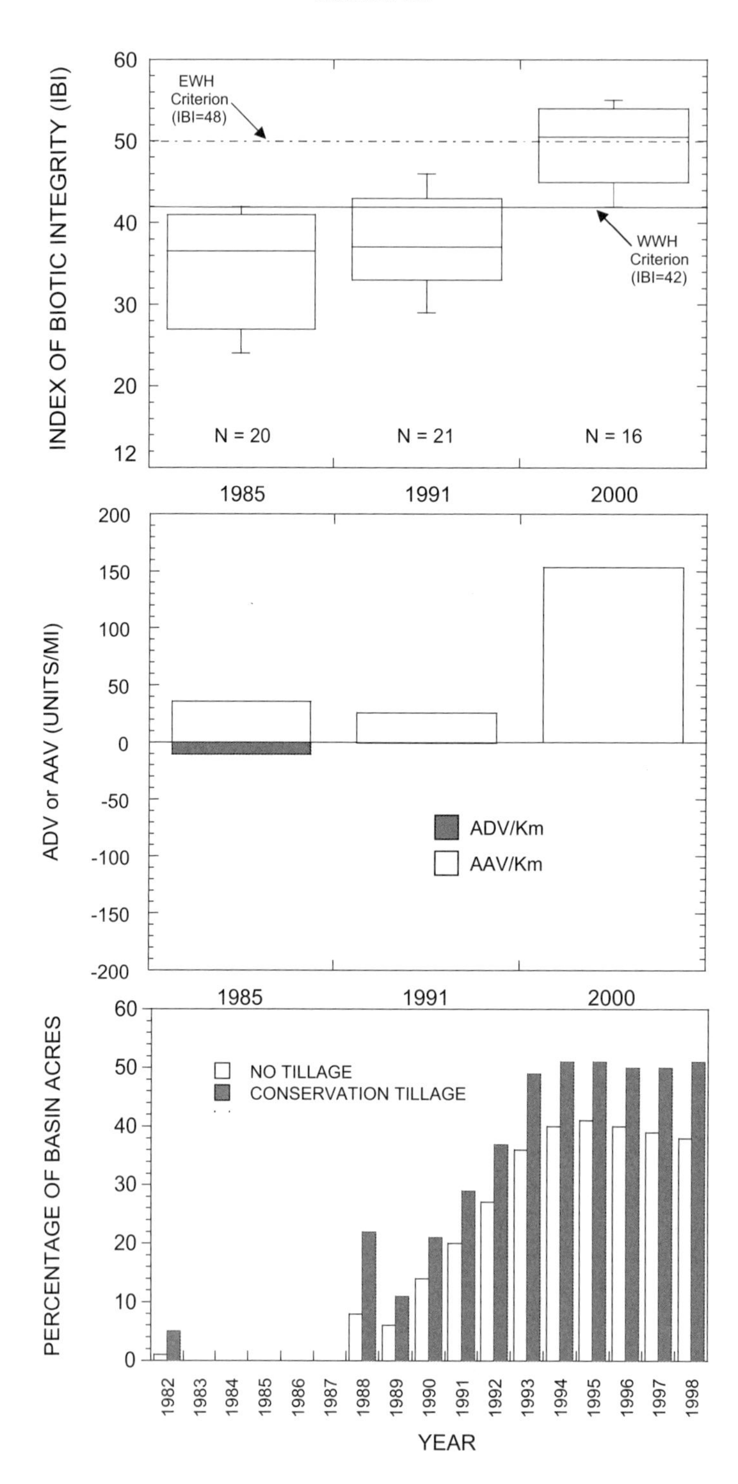

Figure 10. —Box-and-whisker plot of IBI values in the Auglaize River main stem between Wapakoneta and Ft. Jennings, Ohio in 1985, 1991, and 2000 (upper panel; WWH = Warmwater Habitat; EWH = Exceptional Warmwater Habitat; N = number of samples). ADV and AAV/km for the same segment and years (middle panel). Percent of conservation tillage and no till acres in northwestern Ohio between 1982 and 1998 (lower panel; modified from Myers et al. 2000).

agement programs made documentation of improvements in ecological and water quality resources more important (GAO 2000). Biological assessment such as that described here became increasingly important to document the ecological outcomes of water quality management programs (Karr and Yoder 2004).

All variables we used portray changes in fish assemblages before and after 1990 and indicated marked improvements in most Ohio rivers. In contrast, the declines observed in some rivers were comparatively small, but nonetheless meaningful. Lack of improvement or smaller than those observed elsewhere indicates remaining or increased pollution. The three case studies clarified the linkages between management programs and biological responses, the values of tiered uses, and the need for considering incremental improvements beyond pass/fail criteria.

We have demonstrated the value and benefits of operating a systematic, standardized approach to assess fish assemblage quality on a statewide basis. Not only is such an approach cost-effective, it is essential for accurate and proportionate assessments of surface water quality and management program effectiveness. By using a tool like the IBI, we have fulfilled one of its important purposes to "monitor biotic integrity at specific sites... screening a large number of sites in order to identify those that require attention, and for assessing trends over time"...and "...to interpret large amounts of data from complex fish communities when the objective is to assess biotic integrity" (Karr et al. 1986).

Sampling Ohio's nonwadeable rivers for more than 20 years has demonstrated improvements in fish assemblage quality that can be related to efforts to manage and improve water quality through CWA driven regulation and policy. In most states, waters are typically assessed for Clean Water Act purposes by comparing chemical, physical, and biological sampling results to chemical and physical criteria in simple pass/fail assessments. The availability of information rich tools like the IBI makes more sophisticated, quantitative, and meaningful assessments possible. When used within an appropriate survey design, sufficient biological data are generated to more quantitatively assess aggregate changes in biological quality and condition over spatially meaningful areas and relate them directly to water pollution abatement efforts. The pass/fail system presently emphasized in various EPA reporting venues (e.g., impaired waters listings, integrated reports) does not sufficiently recognize incremental changes that do not result in full compliance with pass/fail benchmarks and criteria. Nor does it recognize the extent and severity of departures below such benchmarks or by how much they are surpassed. When used with tools like the ADV and AAV, the approach documented here can produce incremental assessments and potentially provide a more integrative method of tracking progress than the conventional approaches that rely on surrogate endpoints such as chemical concentrations and pollutant loadings. Incremental changes not recognized by pass/fail approaches can be quantified and used as feedback for documenting, validating, and improving pollution abatement practices.

The IBI is not satisfactory for all aquatic resource management applications, such as single species management (Lyons et al. 2001). However, in terms of assessing progress towards important water quality management goals (Karr and Yoder 2004), it has performed well and represents an appropriate tool (Yoder and Kulik 2003), particularly where the administrative burden of managing hundreds and thousands of sources over wide geographical areas is required. More detailed assessment of specific sites, sources, and reaches is possible (and without the need to collect extensive new data) within the approach outlined here and is practiced routinely as part of the Ohio EPA integrated assessment program. However, success in this approach requires a robust derivation and calibration database, which requires several years to develop. It is therefore imperative that state biological monitoring programs incorporate quality assurance evaluations of sampling, sample processing, and data analyses; reference condition assessments; and index development as part of baseline efforts. Without these initial investments in data collection and data analyses, a poorly derived and calibrated indicator will result, leading to erroneous assessments.

In our experience the calibrated and modified IBI performed as described by Steedman (1988) as being "based on simple, definable ecological relationships that are quantitative as ordinal, if not linear measures, that respond in an intuitively correct manner to known environmental gradients. Further, when incorporated with mapping, monitoring, and modeling information it is invaluable in determining management and restoration requirements." In Ohio, we have been able to document which pollution control practices have worked and which sources and rivers have seen the most success in terms of attaining the goal of the CWA, the protection and propagation of balanced, indigenous populations of fish, shellfish, and wildlife. The two decades of effort described above provide a baseline for judging future changes and modifying and adapting existing practices in response to these changes.

Acknowledgments

The active support of past and present Ohio EPA managers of surface water programs was essential for sustaining the systematic collection of electrofishing data over the past two decades. These visionaries included Gary Martin, Pat Abrams, Linda Friedman, Ava Hottman, Lisa Morris, and Jeff DeShon. Charles Staudt was instrumental in developing much of the Ohio ECOS programming. We acknowledge the support of the USEPA national biocriteria program, including Suzanne Marcy, Susan Jackson, and Bill Sweitlik, and the USEPA Region V, including Jim Bland, Wayne Davis, Tom Simon, and Ed Hammer. Phil Larsen, Jim Omernik, and Bob Hughes of the USEPA Western Ecology Laboratory were instrumental in the development of the Ohio EPA biological assessment and biological criteria program throughout its existence. We greatly appreciate the continuing advice and counsel of James Karr with IBI development issues. Finally the assistance of more than 300 summer technicians, too numerous to name individually, made data collection possible. Much of this work was partially supported by USEPA grants.

References

Angermeier, P. L., and J. R. Karr. 1986. Applying an index of biotic integrity based on stream-fish communities: considerations in sampling and interpretation. North American Journal of Fisheries Management 6:418–427.

Barbour, M. T., W. F. Swietlik, S. K. Jackson, D. L. Courtemanch, S. P. Davies, and C. O. Yoder. 2000. Measuring the attainment of biological integrity in the USA: a critical element of ecological integrity. Hydrobiologia 423:453–464.

Barbour, M. T., and C. O. Yoder. 2000. The multimetric approach to bioassessment, as used in the United States of America. Pages 281–292 *in* J. F. Wright, D. W. Sutcliffe, and M. T. Furse, editors. Assessing the biological quality of fresh waters. RIVPACS and similar techniques. Freshwater Biological Association, Ambleside, UK.

Bartsch, A. F. 1948. Biological aspects of stream pollution. Sewage Works Journal 20(1948):292–302.

Cormier, S. M., E. L. C. Lin, M. R. Millward, M. K. Schubauer-Berigan, D. E. Williams, B. Subramanian, R. Sanders, B. Counts, and D. Altfater. 1999a. Using regional exposure criteria and upstream reference data to characterize spatial and temporal exposures to chemical contaminants. Environmental Toxicology and Chemistry 19(4):1127–1135.

Cormier, S. M., M. Smith, S. Norton, and T. Niehiesel. 1999b. Assessing ecological risk in watersheds: a case study of problem formulation in the Big Darby Creek watershed, Ohio, USA. Environmental Toxicology and Chemistry 19(4):1082–1096.

Doudoroff, P., and C. E. Warren. 1951. Biological indices of water pollution with special reference to fish populations. Pages 144–153 *in* Biological problems in water pollution. U.S. Public Health Service, Cincinnati, Ohio.

Erekson, O. H., O. L. Loucks, S. R. Elliot, D. S. McCollum, M. Smith, and R. J. F. Bruins. 2005. Evaluating development alternatives for a high-quality stream threatened by urbanization: Big Darby Creek watershed. Pages 227–247 *in* R. F. J. Bruins and M. T. Heberling, editors. Economics and ecological risk assessment: applications to watershed management. CRC Press, Boca Raton, Florida.

Emery, E. B., T. P. Simon, F. H. McCormick, P. A. Angermier, J. E. DeShon, C. O. Yoder, R. E. Sanders, W. D. Pearson, G. D. Hickman, R. J. Reash, and J. A. Thomas. 2003. Development of a multimetric index for assessing the biological condition of the Ohio River. Transactions of the American Fisheries Society 132:791–808.

Fausch, K. D., J. R. Karr, and P. R. Yant. 1984. Regional application of an index of biotic integrity based on stream fish communities. Transactions of the American Fisheries Society 113:39–55.

Fore, L. S., J. R. Karr, and L. L. Conquest. 1993. Statistical properties of an index of biotic integrity used to evaluate water resources. Canadian Journal of Fisheries and Aquatic Sciences 51:1077–1087.

Gammon, J. R. 1973. The effect of thermal inputs on the populations of fish and macro-invertebrates in the Wabash River. Purdue University, Water Resources Research Center Technical Report 32, West Lafayette, Indiana.

Gammon, J. R. 1976. The fish populations of the middle 340 km of the Wabash River, Purdue University, Water Resources Research Center Technical Report 86, West Lafayette, Indiana.

Gammon, J. R. 1980. The use of community parameters derived from electrofishing catches of river fish as indicators of environmental quality. Pages 335–363 *in* Seminar on water quality management trade-offs (point source vs. diffuse source pollution). U.S. Environmental Protection Agency, EPA-905/9-80-009, Washington, D.C.

Gammon, J. R., A. Spacie, J. L. Hamelink, and R. L. Kaesler. 1981. Role of electrofishing in assessing environmental quality of the Wabash River. Pages 307–24 *in* J. M. Bates and C. I. Weber, editors. Ecological assessments of effluent impacts on communities of indigenous aquatic organisms. American Society of Testing and Materials, ASTM STP 730, Philadelphia.

GAO (General Accounting Office). 2000. Water quality – key EPA and state decisions are limited by inconsistent and incomplete data. U.S. General Accounting Office, GAO/RCED-00-54, Washington, D.C.

GAO (General Accounting Office). 2003. Water quality: improved EPA guidance and support can help states develop standards that better target cleanup efforts. U.S. General Accounting Office, GAO-03-308, Washington, D.C.

Hughes, R. M., D. P. Larsen, and J. M. Omernik. 1986. Regional reference sites: a method for assessing stream pollution. Environmental Management 10:629–635.

ITFM (Intergovernmental Task Force on Monitoring Water Quality). 1992. Ambient water quality monitoring in the United States: first year review, evaluation, and recommendations. Interagency Advisory Committee on Water Data, Washington, D.C.

ITFM (Intergovernmental Task Force on Monitoring Water Quality). 1995. The strategy for improving water-quality monitoring in the United States. Final report of the Intergovernmental Task Force on Monitoring Water Quality + Appendices. Interagency Advisory Committee on Water Data, Washington, D.C.

Karr, J. R. 1981. Assessment of biotic integrity using fish communities. Fisheries 6(6): 21–27.

Karr, J. R. 1991. Biological integrity: a long-neglected aspect of water resource management. Ecological Applications 1(1):66–84.

Karr, J. R., K. D. Fausch, P. L. Angermier, P. R. Yant, and I. J. Schlosser. 1986. Assessing biological integrity in running waters: a method and its rationale. Illinois Natural History Survey Special Publication 5, Champaign.

Karr, J.R., and C.O. Yoder. 2004. Biological assessment and criteria improve TMDL planning and decision making. Journal of Environmental Engineering 130(6):594–604.

Larsen, D. P. 1995. The role of ecological sample surveys in the implementation of biocriteria. Pages 287–302 *in* W. S. Davis and T. P. Simon, editors. Biological assessment and criteria: tools for water resource planning and decision making. Lewis Publishers, Boca Raton, Florida.

Larsen, D. P., J. M. Omernik, R. M. Hughes, C. M. Rohm, T. R. Whittier, A. J. Kinney, A. L. Gallant, and D. R. Dudley. 1986. The correspondence between spatial patterns in fish assemblages in Ohio streams and aquatic ecoregions. Environmental Management 10:815–828.

Lyons, J., R. R. Piette, and K. W. Niermeyer. 2001. Development, validation, and application of a fish-based index of biotic integrity for Wisconsin's large warmwater rivers. Transactions of the American Fisheries Society 130:1077–1094.

Miltner, R. J., and Rankin, E. T. 1998. Primary nutrients and the biotic integrity of rivers and streams. Freshwater Biology 40:145-158.

Miltner, R. J., D. White, and C. Yoder. 2004. The biotic integrity of streams in urban and suburbanizing landscapes. Landscape and Urban Planning 69:87–100.

Myers, D. N., K. D. Metzker, and S. Davis. 2000. Status and trends in suspended-sediment discharges, soil erosion, and conservation tillage in the Maumee River basin; Ohio, Michigan, and

Indiana. U.S. Geological Survey, WRI Report 00-4091, Washington, D.C.

Norton, S. B., S. M. Cormier, M. Smith, and R. C. Jones. 2000. Can biological assessments discriminate among types of stress? A case study from the eastern corn belt plains ecoregion. Environmental Toxicology and Chemistry 19:1113–1119.

NRC (National Research Council). 2001. Assessing the TMDL approach to water quality management. National Academy Press, Washington, D.C.

Ohio EPA (Environmental Protection Agency). 1980. Manual of surveillance methods and quality assurance practices. Office of Wastewater Pollution Control, Columbus, Ohio.

Ohio EPA (Environmental Protection Agency). 1987a. Biological criteria for the protection of aquatic life: volume I. The role of biological data in water quality assessment. Division of Water Quality Monitoring and Assessment, Surface Water Section, Columbus, Ohio.

Ohio EPA (Environmental Protection Agency). 1987b. Biological criteria for the protection of aquatic life: volume II. Users manual for biological field assessment of Ohio surface waters. Division of Water Quality Monitoring and Assessment, Surface Water Section, Columbus, Ohio.

Ohio EPA (Environmental Protection Agency). 1989a. Biological criteria for the protection of aquatic life. volume III: standardized biological field sampling and laboratory methods for assessing fish and macroinvertebrate communities. Division of Water Quality Monitoring and Assessment, Columbus, Ohio.

Ohio EPA (Environmental Protection Agency). 1989b. Addendum to biological criteria for the protection of aquatic life. volume II: users manual for biological field assessment of Ohio surface waters. Division of Water Quality Planning and Assessment, Surface Water Section, Columbus, Ohio.

Ohio EPA (Environmental Protection Agency). 1997. Biological and water quality study of the middle Scioto River and Alum Creek Franklin, Delaware, Morrow, and Pickaway counties, Ohio. Ohio EPA Technical Report MAS/1997-12-12, Columbus.

Ohio EPA (Environmental Protection Agency). 1999. Associations between nutrients, habitat, and the aquatic biota of Ohio's rivers and streams. Division of Surface Water, Monitoring and Assessment Section, Technical Bulletin MAS/1999-1-1, Columbus, Ohio.

Omernik, J. M. 1987. Ecoregions of the conterminous United States. Annals of the Association of American Geographers 77(1):118–125.

Rankin, E. T. 1989. The qualitative habitat evaluation index (QHEI), rationale, methods, and application. Ohio EPA, Division of Water Quality Planning and Assessment, Ecological Assessment Section, Columbus.

Rankin, E. T. 1995. The use of habitat assessments in water resource management programs. Pages 181–208 *in* W. S. Davis and T. P. Simon, editors. Biological assessment and criteria: tools for water resource planning and decision making. Lewis Publishers, Boca Raton, Florida.

Rankin, E. T., and C. O. Yoder. 1990. The nature of sampling variability in the index of biotic integrity (IBI) in Ohio streams. Pages 9–18 *in* W. S. Davis, editor. Proceedings of the 1990 midwest pollution control biologists conference. U.S. EPA, Region V, Environmental Sciences Division, EPA-905-9-90/005, Chicago.

Sanders, R. S., R. J. Miltner, C. O. Yoder, and E. T. Rankin. 1999. The use of external deformities, erosions, lesions, and tumors (DELT anomalies) in fish assemblages for characterizing aquatic resources: a case study of seven Ohio streams. Pages 225–248 *in* T. P. Simon, editor. Assessing the sustainability and biological integrity of water resources using fish communities. CRC Press, Boca Raton, Florida.

Silbiger, R. N., S. A. Christ, A. C. Leonard, M. Garg, D. L. Lattier, S. Dawes, P. Dimsoski, F. McCormick, T. Wessendarp, D. A. Gordon, A. C. Roth, M. Smith, and K. G. P. Toth. 1998. Preliminary studies on population genetics of the central stoneroller (*Campostoma anomalum*) from the Great Miami River basin, Ohio. Environmental Monitoring and Assessment 51:81–495.

Steedman, R. J. 1988. Modification and assessment of an index of biotic integrity to quantify stream quality in southern Ontario. Canadian Journal of Fisheries and Aquatic Sciences 45:492–501.

Thoma, R. F. 1999. Biological monitoring and an index of biotic integrity for Lake Erie's nearshore waters. Pages 417–462 *in* T. P. Simon, editor. Assessing the sustainability and biological integrity of water resources using fish communities. CRC Press, Boca Raton, Florida.

Trautman, M. B. 1981. The fishes of Ohio. The Ohio State University Press, Columbus.

U.S. EPA (Environmental Protection Agency). 1990. Feasibility report on environmental indicators for surface water programs. U.S. EPA, Office of Water Regulations and Standards, Office of Policy, Planning, and Evaluation, Washington, D.C.

U.S. EPA (Environmental Protection Agency). 1995. A conceptual framework to support development and use of environmental information in decision-making. Office of Policy, Planning, and Evaluation, EPA 239-R-95-012, Washington, D.C.

Yoder, C. O. 1998. Important concepts and elements of an adequate state watershed monitoring and assessment program. Proceedings of the NWQMC National Conference Monitoring: Critical Foundations to Protecting Our Waters. U.S. Environmental Protection Agency, Washington, D.C.

Yoder, C. O., and DeShon, J. E. 2003. Using biological response signatures within a framework of multiple indicators to assess and diagnose causes and sources of impairments to aquatic assemblages in selected Ohio rivers and streams. Pages 23–81 *in* T. P. Simon, editor. Biological response signatures: indicator patterns using aquatic communities. CRC Press, Boca Raton, Florida.

Yoder, C. O. and B. H. Kulik. 2003. The development and application of multimetric biological assessment tools for the assessment of impacts to aquatic assemblages in large, non-wadeable rivers: a review of current science and applications. Canadian Journal of Water Resources 28(2):1–28.

Yoder, C. O., R. J. Miltner, and D. White. 2000. Using biological criteria to assess and classify urban streams and develop improved landscape indicators. National conference on tools for urban water resource management and protection. Pages 32–44 *in* S. Minamyer, J. Dye, and S. Wilson, editors. U.S. Environmental Protection Agency, EPA/625/R-00/001, Cincinnati, Ohio.

Yoder, C. O., and E. T. Rankin. 1995a. Biological criteria program development and implementation in Ohio. Pages 109–144 *in* W. S. Davis and T. P. Simon, editors. Biological assessment and criteria: tools for water resource planning and decision making. Lewis Publishers, Boca Raton, Florida.

Yoder, C. O., and E. T. Rankin. 1995b. Biological response signatures and the area of degradation value: new tools for interpreting multimetric data. Pages 263–286 *in* W. S. Davis and T. P. Simon, editors. Biological assessment and criteria: tools for water resource planning and decision making. Lewis Publishers, Boca Raton, Florida.

Yoder, C. O., and E. T. Rankin. 1998. The role of biological indicators in a state water quality management process. Environmental Monitoring and Assessment 51:61–88.

Yoder, C. O. and M. A Smith. 1999. Using fish assemblages in a state biological assessment and criteria program: essential concepts and considerations. Pages 17–56 *in* T. P. Simon, editor. Assessing the sustainability and biological integrity of water resources using fish communities. CRC Press, Boca Raton, Florida.

Appendix A.— Fish species and their average relative abundance (numbers/km) at boat electrofishing sites draining >390 ha (150 mi.[2]) before (n = 3471 samples) and after 1990 (n = 2176 samples). Species are ranked by post-1990 relative abundance; the percentage of sampling locations at which each occurred is also shown.

	Before 1990		After 1990	
Species	No./km	% occurrence	No./km	% occurrence
gizzard shad *Dorosoma cepedianum*	63.9	74.0	70.8	77.8
spotfin shiner *Cyprinella spiloptera*	33.7	71.2	48.3	80.9
golden redhorse *Moxostoma erythrurum*	21.9	59.7	37.5	70.2
bluntnose minnow *Pimephales notatus*	19.9	58.1	36.2	74.8
northern hog sucker *Hypentelium nigricans*	10.3	42.4	26.5	65.3
common carp* *Cyprinus carpio*	24.3	91.4	25.7	94.5
bluegill *Lepomis macrochirus*	16.2	64.7	24.3	71.2
emerald shiner *Notropis atherinoides*	13.5	23.1	23.8	37.5
longear sunfish *Lepomis megalotis*	18.7	47.2	22.6	57.8
white sucker *Catostomus commersonii*	16.6	45.5	16.3	35.8
smallmouth redhorse *Moxostoma breviceps*	2.83	23.8	15.5	37.0
smallmouth bass *Micropterus dolomieu*	8.67	50.4	15.3	67.0
central stoneroller *Campostoma anomalum*	4.90	20.7	14.8	40.9
green sunfish *Lepomis cyanellus*	22.7	64.2	14.2	60.8
sand shiner *Notropis stramineus*	2.97	15.7	10.2	36.8
freshwater drum *Aplodinotus grunniens*	3.46	29.5	9.54	48.7
black redhorse *Moxostoma duquesnei*	4.47	22.2	9.27	30.0
channel catfish *Ictalurus punctatus*	5.23	41.9	9.24	62.7
suckermouth minnow *Phenacobius mirabilis*	1.27	11.1	8.50	24.2
river carpsucker *Carpiodes carpio*	4.22	24.9	7.39	31.5
gravel chub *Erimystax x-punctatus*	0.16	2.4	6.82	19.7
striped shiner *Luxilis chrysocephalus*	3.92	20.0	6.26	28.7
rock bass *Ambloplites rupestris*	6.38	46.2	5.97	41.2
greenside darter *Etheostoma blennioides*	0.91	12.7	5.27	36.2
silver shiner *Notropis photogenis*	4.45	16.1	5.05	22.2
silver redhorse *Moxostoma anisurum*	3.45	30.6	4.98	47.9
logperch *Percina caprodes*	1.31	16.1	4.78	38.7
orangespotted sunfish *Lepomis humilis*	4.27	27.1	4.45	24.3
largemouth bass *Micropterus salmoides*	4.81	48.1	4.42	43.6
creek chub *Semotilus atromaculatus*	4.79	17.0	4.37	14.5
spotted sucker *Minytrema melanops*	2.32	22.8	4.28	23.1
quillback *Carpiodes cyprinus*	2.50	30.0	3.97	44.5
steelcolor shiner *Cyprinella whipplei*	1.29	11.0	3.96	21.6
spotted bass *Micropterus punctulatus*	3.98	20.6	3.62	30.2
common shiner *Luxilis cornutus*	0.81	3.97	3.41	3.91
pumpkinseed *Lepomis gibbosus*	3.76	23.2	3.39	21.0
mottled sculpin *Cottus bairdii*	0.18	1.30	3.28	4.73
river chub *Nocomis micropogon*	0.78	5.07	3.17	14.8
bullhead minnow *Pimephales vigilax*	0.65	5.56	2.91	19.9
smallmouth buffalo *Ictiobus bubalus*	0.71	1.34	2.88	27.8
banded darter *Etheostoma zonale*	0.35	5.01	2.61	21.8
white crappie *Pomoxis annularis*	2.64	30.6	2.43	26.3
golden shiner *Notemigonus crysoleucas*	2.20	18.5	2.14	11.9
black crappie *Pomoxis nigromaculatus*	1.27	20.2	1.81	26.6
rainbow darter *Etheostoma caeruleum*	0.20	3.60	1.75	17.3
rosyface shiner *Notropis rubellus*	0.66	3.63	1.58	11.5
goldfish* *Carrasius auratus*	3.84	18.1	1.49	12.1
white bass *Morone chrysops*	1.34	16.5	1.44	21.4
yellow bullhead *Ameiurus natalis*	0.97	19.4	1.78	21.5
eastern blacknose dace *Rhinichthys atratulus*	0.20	1.15	1.15	4.04

Appendix A.—Continued.

Species	Before 1990 No./km	Before 1990 %Occurrence	After 1990 No./km	After 1990 %Occurrence
sauger *Sander canadensis*	0.53	12.9	1.15	21.9
fathead minnow *Pimephales promelas*	0.89	4.47	1.02	4.91
flathead catfish *Pylodictis olivaris*	0.57	12.6	1.02	23.7
river redhorse *Moxostoma carinatum*	0.34	6.71	0.92	13.6
mimic shiner *Notropis volucellus*	0.24	2.45	0.89	6.71
black buffalo *Ictiobus niger*	0.08	2.25	0.84	13.1
brook silverside *Labidesthes sicculus*	0.48	6.31	0.83	11.2
yellow perch *Perca flavescens*	0.55	7.63	0.73	9.10
white perch *Morone americana*	0.98	5.24	0.70	3.35
blackside darter *Percina maculata*	0.31	6.91	0.68	10.3
shorthead redhorse *Moxostoma macrolepidotum*	0.52	4.02	0.67	3.89
variegate darter *Etheostoma variatum*	0.06	1.01	0.66	6.00
ghost shiner *Notropis buchanani*	0.08	1.12	0.61	2.30
stonecat *Noturus flavus*	0.10	2.56	0.60	13.1
johnny darter *Etheostoma nigrum*	0.37	6.17	0.59	11.9
spottail shiner *Notropis hudsonius*	0.28	2.62	0.59	2.67
longnose gar *Lepisosteus osseus*	0.28	8.04	0.55	14.8
streamline chub *Erimystax dissimilis*	0.11	1.50	0.48	4.23
redfin shiner *Lythrurus umbratilis*	0.44	3.63	0.47	4.60
silverjaw minnow *Notropis buccatus*	0.19	2.59	0.39	2.80
black bullhead *Ameiurus melas*	0.24	4.75	0.39	3.54
bigmouth buffalo *Ictiobus cyprinellus*	0.21	5.47	0.39	7.12
skipjack herring *Alosa chrysochloris*	0.05	1.09	0.38	5.19
bluebreast darter *Etheostoma camurum*	0.01	0.20	0.37	2.90
highfin carpsucker *Carpiodes velifer*	0.21	5.50	0.35	6.85
brown bullhead *Ameiurus nebulosus*	0.71	10.9	0.32	4.41
greater redhorse *Moxostoma valenciennesi*	0.07	0.98	0.30	3.45
hornyhead chub *Nocomis biguttatus*	0.05	0.63	0.30	1.33
trout-perch *Percopsis omiscomaycus*	0.36	4.49	0.29	3.03
warmouth *Lepomis gulosus*	0.45	5.82	0.29	6.11
dusky darter *Percina sciera*	0.03	0.86	0.27	4.37
blackstripe topminnow *Fundulus notatus*	0.16	3.26	0.24	4.43
central mudminnow *Umbra limi*	0.02	0.72	0.23	0.51
walleye *Sander vitreus*	0.14	3.66	0.23	5.24
slenderhead darter *Percina phoxocephala*	0.05	0.89	0.22	5.88
fantail darter *Etheostoma flabellare*	0.10	1.53	0.22	5.79
grass pickerel *Esox americanus*	0.42	9.39	0.20	5.15
Tippecanoe darter *Etheostoma tippecanoe*	<0.01	0.09	0.19	2.25
brown trout* *Salmo trutta*	0.07	0.40	0.15	0.74
redear sunfish *Lepomis microlophus*	0.02	0.54	0.14	1.93
channel shiner *Notropis wickliffi*	0	0	0.14	0.60
northern pike *Esox lucius*	0.12	4.12	0.13	4.46
mooneye *Hiodon tergisus*	0.05	1.47	0.12	3.26
rainbow trout* *Oncorynchus mykiss*	0.20	0.72	0.11	1.47
bowfin *Amia calva*	0.05	1.35	0.09	1.75
scarlet shiner *Lythrurus fasciolaris*	0.09	1.73	0.06	1.19
tonguetied minnow *Exoglossum laurae*	<0.01	0.14	0.05	0.65
mountain madtom *Noturus eleutherus*	0	0	0.05	0.46
silver lamprey *Ichthyomyzon unicuspis*	0.01	0.46	0.05	2.21
silver chub *Macrhybopsis storeriana*	0.04	0.75	0.05	1.42
bigeye chub *Hybopsis amblops*	0.02	0.32	0.05	0.41
muskellunge *Esox masquinongy*	<0.01	0.20	0.05	1.75

Appendix A.—Continued.

Species	Before 1990		After 1990	
	No./km	%Occurrence	No./km	%Occurrence
orangethroat darter *Etheostoma spectabile*	<0.01	0.09	0.04	1.56
tadpole madtom *Noturus gyrinus*	0.03	0.66	0.04	0.87
American brook lamprey *Lampetra appendix*	0.04	0.84	0.04	0.92
brindled madtom *Noturus miurus*	0.02	0.84	0.04	1.61
eastern sand darter *Ammocrypta pellucida*	0.01	0.20	0.03	0.74
grass carp* *Ctenopharyngodon idella*	<0.01	0.03	0.02	0.74
popeye shiner *Notropis ariommus*	0	0	0.02	0.10
blue sucker *Cycleptus elongatus*	<0.01	0.03	0.02	0.60
least brook lamprey *Lampetra aepyptera*	0.01	0.46	0.02	0.60
river shiner *Notropis blennius*	0.02	0.49	0.01	0.14
alewife* *Alosa pseudoharengus*	<0.01	0.12	0.01	0.32
threadfin shad *Dorosoma petenense*	0	0	0.01	0.23
striped bass* *Morone saxatilis*	0	0	0.01	0.18
northern madtom *Noturus stigmosus*	0	0	0.01	0.18
redside dace *Clinostomus elongatus*	<0.01	0.03	<0.01	0.14
shortnose gar *Lepisosteus platostomus*	<0.01	0.12	<0.01	0.18
spotted darter *Etheostoma maculatum*	0	0	<0.01	0.10
paddlefish *Polyodon spathula*	0	0	<0.01	0.10
American eel *Anguilla rostrata*	<0.01	0.12	<0.01	0.10
river darter *Percina shumardi*	<0.01	0.09	<0.01	0.10
coho salmon* *Oncorhynchus kisutch*	<0.01	0.14	<0.01	0.05
round goby* *Neogobius melanstomus*	0	0	<0.01	0.05
white catfish* *Ameirus catus*	0	0	<0.01	0.05
rainbow smelt* *Osmerus mordax*	0	0	<0.01	0.05
goldeye *Hiodon alosoides*	<0.01	0.06	<0.01	0.05
brook stickleback *Culaea inconstans*	<0.01	0.06	0	0
Iowa darter *Etheostoma exile*	<0.01	0.09	0	0
western mosquitofish* *Gambusia affinis*	<0.01	0.03	0	0
banded killifish *Fundulus diaphanus*	0.2	0.43	0	0
pugnose minnow *Opsopoedus emiliae*	<0.01	0.03	0	0
creek chubsucker *Erimyzon oblongus*	<0.01	0.03	0	0
sea lamprey* *Petromyzon marinus*	<0.01	0.06	0	0

* alien to Ohio

Appendix B.— Fish assemblage status for boat electrofishing in 45 Ohio rivers comparing index of biotic integrity results for the earliest and latest Ohio EPA surveys during the period 1979 through 2001; rivers are ranked by their 75th percentile value for the latest ("After") period. Results prior to 1988–1990 ("Before") reflect conditions prior to the upgrades of major municipal wastewater treatment facilities and most industrial discharges; conditions after 1990 ("After") reflect reduced impacts. The IBI statistics are for the entire river segment included in the survey; maximum and minimum values exclude statistical outliers (>2 interquartile ranges [IQR] beyond the median). Principal disturbance types are listed in order of importance. Rivers in boldface are classified Exceptional Warmwater Habitat (EWH) over most of the surveyed segment, the remaining rivers are classified Warmwater Habitat (WWH). The IBI values that fail to meet numeric criteria for the IBI in each river are underlined; poor and very poor values are also in boldface.

	Earliest ("Before")						Latest ("After")						Median	
River	Year	N	Med. (±2IQR)	Max.	75%ile	Min.	Year	N	Med. (±2IQR)	Max.	75%ile	Min.	Change (±)	Disturbance type(s)
Stillwater R.	1982	61	48 (1.34)	56	52	32	1999	21	56 (3.96)	60	58	52	+8	Ag. NPS, WWTPs
Upper Gr. Miami	1982	68	42 (2.00)	54	48	40	1994	42	56 (2.00)	60	58	40	+14	WWTPs, Ag. NPS, Channelized
Kokosing R.	1987	28	50 (1.46)	54	52	18	1998	6	52 (10.0)	56	56	22	(+2)	WWTPs, Ag. NPS
Sandusky R.	1981	27	34 (3.02)	46	40	20	2001	12	50 (3.66)	58	56	38	+16	Ag. NPS, WWTPs, Impounded
Big Darby Cr.	1981	18	44 ((2.68)	56	50	38	2001	12	52 (4.64)	56	54	44	+8	Ag. NPS, WWTPs
Salt Cr.	1979–1984	6	52 (9.12)	56	54	48	1992	6	50 (4.06)	54	54	44	(-2)	Ag. NPS, Silviculture
Upper Hocking	1982	9	28 (4.58)	34	30	14	1995	8	50 (1.06)	54	52	50	+22	WWTPs, Complex toxic, Chan.
Walhonding R.	1983	7	44 (3.78)	54	48	40	1994	9	50 (1.70)	54	52	46	+6	Ag. NPS, Flow regulation
Central Scioto R.	1979	35	22 (2.46)	44	28	12	1998	24	48 (1.52)	56	52	40	+26	WWTPs, CSOs, Urban, Ag. NPS
Grand R.	1987	15	48 (3.22)	54	52	34	1995	7	48 (7.26)	52	52	30	(0)	Ag. NPS, WWTPs
Paint Cr.	1985	12	32 (7.24)	52	50	12	1997	12	46 (3.44)	54	52	36	+14	Industrial, Ag. NPS
Whitewater R.	1980	8	32 (5.28)	50	35	26	1995–1997	13	48 (2.18)	56	50	42	+16	Ag. NPS, Flow reg.
Big Walnut Cr.	1981–1986	9	32 (5.02)	44	40	22	2000	7	48 (6.30)	54	50	32	+16	Urban, WWTPs, Ag. NPS
L. Muskingum R.	(No survey before 1990)						1999	8	48 (6.66)	52	50	28		Silviculture, Ag. NPS
Licking R.	1981	15	28 (2.60)	40	32	20	1993	12	46 (2.70)	52	50	38	+18	WWTPs, Ag. NPS, CSO, Urban
Auglaize R.	1985	20	34 (2.90)	44	40	24	2000	16	44 (3.46)	56	50	32	+10	Ag. NPS, WWTPs
Olentangy R.	1988–1989	25	34 (2.42)	44	38	28	1999	16	40 (3.48)	54	50	30	+6	Urban, CSO, Impounded
Mid. Gr. Miami	1988	27	22 (4.36)	44	28	16	2000	11	46 (5.22)	54	48	44	+14	Thermal, WWTPs, Urban
Deer Cr.	1985	6	38 (3.62)	42	42	38	1997	11	46 (5.50)	50	48	42	+8	Ag. NPS, WWTPs
Walnut Cr.	1982	27	42 (2.26)	54	50	30	1996	11	44 (4.92)	52	48	28	(+2)	WWTPs, Ag. NPS
Greenville Cr.	1982	30	32 (2.08)	42	36	26	1999	17	40 (3.52)	56	48	32	+8	WWTPs, Ag. NPS
E. Fk. L. Miami	1982	24	46 (1.76)	54	48	36	1998	13	44 (2.98)	50	46	32	(-2)	WWTPs, Flow reg.
L. Miami R.	1983	62	40 (1.78)	52	46	22	1998	48	42 (1.54)	50	46	32	(+2)	WWTPs, Ag. NPS, CSO
Lower Scioto R.	1979	40	24 (1.86)	38	28	14	1997	32	42 (1.54)	48	46	32	+18	Industrial, Ag. NPS, Silviculture

Appendix B.—Continued

River	Earliest ("Before") Year	N	Med. (±2IQR)	Max.	75%ile	Min.	Latest ("After") Year	N	Med. (±2IQR)	Max.	75%ile	Min.	Median Change (±)	Principal impact type(s)
Lower Gr. Miami	1980	125	24 (1.04)	38	28	12	1994–1995	114	36 (1.62)	58	46	20	+12	WWTPs, Industrial, Urban
Middle Scioto R.	1979	24	28 (2.70)	44	34	14	1997	26	42 (2.00)	46	44	24	+14	Industrial, Ag. NPS, Chan.
Raccoon Cr.	1990	10	36 (3.54)	44	42	28	1994–1995	18	44 (1.78)	50	44	36	+6	AMD, Silviculture
Upper Muskingum	1988	43	34 (2.84)	50	40	12	1994	21	40 (2.12)	48	42	32	+16	Industrial, Thermal, WWTPs
Lower Hocking R.	1990	56	38 (1.66)	50	42	24	(First survey after 1990 planned for 2004)							AMD, WWTPs, Ag. NPS
Lower Tuscarawas	1983	40	34 (2.20)	46	38	20	(First survey after 1990 planned for 2004)							Industrial, Ag. NPS, WWTPs
Mad R.	1984	54	26 (1.90)	44	30	20	1994	38	36 (2.88)	58	42	18	+10	WWTPs, Ag. NPS, Channelized
Alum Cr.	1986	12	28 (1.90)	34	30	22	1996	10	36 (2.10)	40	38	34	+8	Urban, CSOs, WWTPs, Flow reg.
Tiffin R.	1984	18	28 (2.96)	42	32	20	1992	16	34 (3.24)	42	38	20	+6	Ag. NPS, Channelized
Killbuck Cr.	1981	20	20 (3.68)	44	30	14	1993	16	32 (3.78)	42	36	24	+12	WWTP, Ag. NPS, Channelized
Blanchard R.	1983	34	26 (1.60)	36	28	18	1996	8	32 (1.52)	36	34	30	+6	WWTP, Ag. NPS, CSO
Lower Muskingum	1988	54	32 (2.06)	52	38	12	(First survey after 1990 planned for 2004)							Industrial, Thermal, Impounded
Maumee R.	1984–1986	65	28 (1.24)	40	32	18	1997	54	28 (2.02)	46	34	14	(0)	Ag. NPS, WWTPs, Impounded
St. Josephs R.	(No survey before 1990)						1992	16	30 (1.70)	36	32	24		Ag. NPS, WWTPs, Channelized
Ottawa R.	1985	30	20 (2.14)	34	30	12	1996	19	22 (3.50)	38	32	14	(+2)	Complex toxic, Ind./WWTP, CSO
Wills Cr.	1984	39	26 (1.82)	34	30	18	1994	41	28 (1.60)	36	30	18	(+2)	AMD, WWTP, Ag. NPS, Silvic.
Cuyahoga R.	1984	41	12 (1.54)	22	16	12	2000	11	26 (3.14)	38	30	20	+14	Complex toxic, WWTP/Ind., CSO
Upper Tuscarawas	1983	52	20 (1.50)	32	24	12	1995/01	21	24 (2.34)	32	28	14	+4	Comp. toxic, WWTP/Ind., Chan.
Mahoning R.	1980	65	14 (2.18)	44	26	12	1994	73	24 (1.38)	32	26	14	+10	Com. Tox., WWTP/Ind., CSO, Imp.
Black R.	1982	12	20 (1.42)	20	20	16	(First survey after 1990 planned for 2006)							Complex toxic, WWTP/Ind., CSO
Rush Cr.	1982	12	12 (0.90)	16	14	12	(First survey after 1990 planned for 2004)							AMD, Ag. NPS

Impact type abbreviations: WWTPs = Municipal wastewater treatment facilities; Industrial = Industrial wastewater treatment facilities; Ag. NPS = Agricultural nonpoint sources (mainly row cropping); Chan. = channelization of main channel; Imp. = impoundment of main channel by run-of-river low head dams; Silv = Silvicultural practices; Complex Toxic = Complex mixture of toxic sources including industrial, WWTP, and NPS; CSO = Combined sewer overflows; Flow = flow regulation; AMD = Acidic mine drainage.

Appendix C.—Fish assemblage results from earliest and latest Ohio EPA surveys for 38 river segments comparing changes in median index of biotic integrity (IBI) values, area of degradation value (ADV) and the area of attainment value (AAV). Segments are arranged in the same order as Figure 4. Changes are expressed as the increase (+) or decrease (–) in the % median IBI, ADV/km, AAV/km, and net change in ADVand/or AAV/km. The IBI values that fail to meet numeric criteria for the IBI in each river are underlined; poor and very poor values are also in boldface. Declining net DADV+ AAV results are in boldface and underlined.

	"Before"				"After"				Changes			
River	Year	Median IBI (±2IQR)	ADV	AAV	Year	Median IBI (±2IQR)	ADV	AAV	%ΔMed. IBI (±)	Δ ADV	Δ AAV	Net Δ ADV + AAV
Stillwater R.	1982	48 (1.34)	–2.0	+41.1	1999	56 (3.96)	–4.6	+118	+17%	**–2.6**	+76.9	+74.3
Upper Gr. Miami	1982	42 (2.00)	–17.8	+25.5	1994	56 (2.00)	–0.1	+115	+33%	+17.7	+89.5	+158
Kokosing R.	1987	50 (1.46)	0	+90.7	1998	52 (10.0)	0	+88.7	+4%	0	–2.0	**–2.0**
Sandusky R.	1981	34 (3.02)	–32.4	+6.7	1999	50 (3.66)	0	+85.9	+47%	+32.4	+79.2	+112
Big Darby Cr.	1981	44 (2.68)	0	+71.1	2001	52 (4.64)	0	+90.3	+18%	0	+19.2	+19.2
Salt Cr.	1984	52 (9.12)	0	+80.0	1992	50 (4.06)	–0.2	+50.6	-3.8%	**–0.2**	**–29.4**	**–29.6**
Upper Hocking R.	1982	28 (4.58)	–108	0	1995	50 (1.06)	0	+144	+79%	+108	+144	+252
Walhonding River	1983	44 (3.78)	–5.5	+3.2	1994	50 (1.70)	0	+70.8	+14%	+5.5	+67.6	+73.1
Central Scioto R.	1979	22 (2.46)	–193	0	1998	48 (1.52)	0	+76	+118%	+193	+76	+269
Grand River	1987	48 (3.22)	–32.9	+39.6	1995	48 (7.26)	–3.9	+72.0	0%	+29	+32.4	+61.4
Paint Cr.	1985	32 (7.24)	–61.6	0	1997	46 (3.44)	0	+56	+44%	+61.6	+56	+118
Whitewater River	1980	32 (5.28)	–119.5	0	1995	48 (2.18)	–0.3	+51.3	+50%	+119	+51.3	+171
Big Walnut Cr.	1981–1986	32 (5.02)	–14.9	0	2000	48 (6.30)	–2.5	+70.9	+50%	+12.4	+70.9	+83.3
Licking R.	1981	28 (2.60)	–84.4	0	1993	46 (2.70)	0	+101	+64%	+84.4	+101	+185
Auglaize R.	1985	34 (2.90)	–10.4	+35.7	2000	44 (3.46)	–1.9	+126	+29%	+8.5	+90.3	+98.8
Olentangy R.	1988–1989	34 (2.42)	–35.0	+11.3	1999	40 (3.48)	–0.4	+67.8	+18%	+34.6	+56.5	+91.1
Mid. Gr. Miami R.	1988	22 (4.36)	–147	+1.5	2000	46 (5.22)	0	+129.1	+100%	+147	+128	+275
Deer Cr.	1985	38 (3.62)	–69.6	0	1997	46 (5.50)	–4.7	+16.5	+21%	+64.9	+16.5	+81.4
Walnut Cr.	1982	42 (2.26)	–0.7	+55.3	1996	44 (4.92)	–11.5	+54.4	+4.7%	**–10.8**	**–1.1**	**–11.9**
Greenville Cr.	1982	32 (2.08)	–85.2	+1.0	1999[1]	40 (3.52)	–22.3	+8.6	+25%	+62.9	+7.6	+70.5
E. Fk. L. Miami	1982	46 (1.76)	–6.9	+17.1	1998	44 (2.98)	–11.3	+5.8	-4.3%	**–4.4**	**–11.3**	**–15.7**
L. Miami R.	1983	40 (1.78)	–54.5	+3.1	1998	42 (1.54)	–22.8	+8.3	+4.7%	+31.7	+5.2	+36.9
Lower Scioto R.	1979	24 (1.86)	–117	0	1997	42 (1.54)	–0.3	+61.7	+75%	+116	+61.7	+178
Lower Gr. Miami	1980	24 (1.04)	–131	0	1994-5	36 (1.62)	–17.8	+25.5	+50%	+113	+25.5	+139
Middle Scioto R.	1979	28 (2.70)	–104	0	1997	42 (2.00)	–9.2	+31.1	+50%	+94.8	+31.1	+126
Raccoon Cr.	1990	36 (3.54)	–10.0	0	1994-5	44 (1.78)	0	+62.8	+22%	+10.0	+62.8	+72.8
Upper Muskingum R.	1988	34 (2.84)	–12.2	+13.4	1994	40 (2.12)	–3.0	+31.2	+18%	+9.2	+17.8	+27.0
Mad R.	1984	**26** (1.90)	–109	+1.7	1994	36 (2.88)	–37.1	+18.5	+38%	+71.9	+16.8	+88.7
Alum Cr.	1986	28 (1.90)	–103	0	1996	36 (2.10)	–40.7	0	+29%	+62.3	0	+62.3

Appendix C.—Continued.

River	Earliest Year	Earliest Median IBI ($\pm$2IQR)	Earliest ADV	Earliest AAV	Latest Year	Latest Median IBI ($\pm$2IQR)	Latest ADV	Latest AAV	Changes %ΔMed. IBI ($\pm$)	Changes Δ ADV	Changes Δ AAV	Net Δ ADV + AAV
Tiffin R.	1984	28 (2.96)	–21.8	+5.0	1992	34 (3.24)	–6.6	+25.4	+21%	+15.2	+20.4	+35.6
Killbuck Cr.	1981	20 (3.68)	–136	0	1993	32 (3.78)	–20.6	+40.1	+60%	+115	+40.1	+155
Blanchard R.	1983	26 (1.60)	–46.0	0	1996	32 (1.52)	0	+17.0	+23%	+46.0	+17.0	+63.0
Maumee R.	1984–1986	28 (1.24)	–7.2	+23.8	1997	28 (2.02)	–33.2	+36.9	0	-26.0	+13.1	–12.9
Ottawa R.	1985	20 (2.14)	–146	+0.8	1996	22 (3.50)	–116.0	+3.5	+10%	+29.8	+2.7	+32.5
Wills Cr.	1984	26 (1.82)	–94.0	0	1994	28 (1.60)	–72.0	0	+7.6	+22.0	0	+22.0
Cuyahoga R.	1984	12 (1.54)	–207	0	2000	26 (3.14)	–56.3	+3.9	+100%	+151	+3.9	+155
Upper Tuscarawas	1983	14 (1.50)	–131	0	1995/01	24 (2.34)	–36.4	+4.0	+20%	+94.6	+4.0	+98.6
Mahoning R.	1980	24 (2.18)	–185	+0.2	1994	24 (1.38)	–126	0	0	+59.0	–0.2	+58.8

[a] Data collected with wading methods; included for comparison purposes only.

Appendix D.—Rank order of selected rivers for before and after changes (Δ) in median index of biotic integrity (IBI), % median IBI, area of degradation value (ADV), area of attainment value (AAV), and net ADV + AAV.

ΔMedian IBI	%ΔMedian IBI	Δ ADV	ΔAAV	ΔNet ADV + AAV
Central Scioto R. (+26)	Central Scioto R. (+118)	Central Scioto R. (+193)	Upper Hocking R. (+144)	Middle Gr. Miami R. (+275)
Upper Hocking R. (+22)	Middle Gr. Miami R. (+100)	Cuyahoga R. (+151)	Middle Gr. Miami R. (+128)	Central Scioto R. (+269)
Lower Scioto R. (+18)	Cuyahoga R. (+100)	Middle Gr. Miami R. (+147)	Licking R. (+101)	Upper Hocking R. (+252)
Sandusky R. (+16)	Upper Hocking R. (+79)	Whitewater R. (+119)	Auglaize R. (+90.3)	Licking R. (+185)
Big Walnut Cr. (+16)	Lower Scioto R. (+75)	Lower Scioto R. (+116)	Upper Gr. Miami R. (+89.5)	Lower Scioto R. (+178)
Licking R. (+16)	Licking R. (+64)	Killbuck Cr. (+115)	Sandusky R. (+79.2)	Whitewater R. (+171)
Upper Gr. Miami R. (+14)	Killbuck Cr. (+60)	Lower Gr. Miami R. (+113)	Stillwater R. (+76.9)	Upper Gr. Miami R. (+158)
Paint Cr. (+14)	Whitewater R. (+50)	Upper Hocking (+108)	Central Scioto R. (+76)	Cuyahoga R. (+155)
Whitewater R. (+14)	Big Walnut Cr. (+50)	Middle Scioto R. (+94.8)	Big Walnut Cr. (+70.9)	Killbuck Cr. (+155)
Middle Gr. Miami R. (+14)	Lower Gr. Miami R. (+50)	Upper Tuscarawas R. (+94.6)	Walhonding R. (+67.6)	Lower Gr. Miami R. (+139)
Cuyahoga R. (+14)	Middle Scioto R. (+50)	Licking R. (+84.4)	Raccoon Cr. (+62.8)	Middle Scioto R. (+126)
Lower Great Miami R. (+12)	Sandusky R. (+47)	Mad R. (+71.9)	Lower Scioto R. (+61.7)	Paint Cr. (+118)
Middle Scioto R. (+12)	Paint Cr. (+44)	Deer Cr. (+64.9)	Olentangy R. (+56.5)	Sandusky R. (+112)
Auglaize R. (+10)	Mad R. (+38)	Greenville Cr. (+62.9)	Paint Cr. (+56)	Auglaize R. (+98.8)
Greenville Cr. (+10)	Upper Gr. Miami R. (+33)	Alum Cr. (+62.3)	Whitewater R. (+51.3)	Upper Tuscarawas R. (+98.6)
Mad R. (+10)	Auglaize R. (+29)	Paint Cr. (+61.6)	Killbuck Cr. (+40.1)	Olentangy R. (+91.1)
Killbuck Cr. (+10)	Alum Cr. (+29)	Mahoning R. (+59)	Grand R. (+32.4)	Mad R. (+88.7)
Mahoning R. (+10)	Blanchard R. (+23)	Blanchard R. (+46)	Middle Scioto R. (+31.1)	Big Walnut Cr. (+83.3)
Stillwater R. (+8)	Raccoon Cr. (+22)	Olentangy R. (+34.6)	Lower Gr. Miami R. (+25.5)	Deer Cr. (+81.4)
Big Darby Cr. (+8)	Deer Cr. (+21)	Sandusky R. (+32.4)	Tiffin R. (+20.4)	Stillwater R. (+74.3)
Raccoon Cr. (+8)	Tiffin R. (+21)	L. Miami R. (+31.7)	Big Darby Cr. (+19.2)	Walhonding R. (+73.1)
Alum Cr. (+8)	Upper Tuscarawas R. (+20)	Ottawa R. (+29.8)	Up. Muskingum R. (+17.8)	Raccoon Cr. (+72.8)
Walhonding R. (+6)	Big Darby Cr. (+18)	Grand R. (+29)	Blanchard R. (+17)	Greenville Cr. (+63.0)
Olentangy R. (+6)	Olentangy R. (+18)	Wills Cr. (+22)	Mad R. (+16.8)	Blanchard R. (+62.3)
Deer Cr. (+6)	Upper Muskingum R. (+18)	Upper Gr. Miami R. (+17.7)	Deer Cr. (+16.5)	Alum Cr. (+72.8)
Upper Muskingum R. (+6)	Stillwater R. (+17)	Tiffin R. (+15.2)	Maumee R. (+13.1)	Grand R. (+61.4)
Tiffin R. (+6)	Walhonding R. (+14)	Raccoon Cr. (+10)	Greenville Cr. (+7.6)	Mahoning R.(+58.8)

American Fisheries Society Symposium 45:431–449, 2005

Detection of Temporal Trends in Ohio River Fish Assemblages Based on Lockchamber Surveys (1957–2001)

JEFF A. THOMAS* AND ERICH B. EMERY
Ohio River Valley Water Sanitation Commission (ORSANCO), 5735 Kellogg Avenue, Cincinnati, Ohio 45228, USA

FRANK H. MCCORMICK
USDA Forest Service, Pacific Northwest Research Station, Olympia Forestry Sciences Laboratory Olympia, Washington 98512, USA

Abstract.—The Ohio River Valley Water Sanitation Commission (ORSANCO), along with cooperating state and federal agencies, sampled fish assemblages from the lockchambers of Ohio River navigational dams from 1957 to 2001. To date, 377 lockchamber rotenone events have been conducted, resulting in the collection of nearly three million fishes, representing 116 taxa, including 7 hybrids, in 19 families. We observed significant temporal trends in Ohio River fish riverwide at the assemblage, guild, and species levels. Modified index of well-being (MIWB) scores and changes in guild structure indicated significantly ($p < 0.05$) improving fish assemblages throughout the Ohio River. Quantile regression of the abundance of individual species by year revealed significant declines ($p < 0.05$) in populations of several pollution-tolerant species (e.g., *Ameiurus* spp., goldfish *Carassius auratus*) with time, while some intolerant species (e.g., smallmouth redhorse *Moxostoma breviceps,* smallmouth bass *Micropterus dolomieu,* and mooneye *Hiodon tergisus*) have increased in recent years. In all, 40 of the 116 taxa collected in the lockchamber surveys changed significantly over time. Sixteen species did not change. Sixty species could not be analyzed either because of incomplete data or insufficient abundance. Fish assemblage metrics that would be expected to decrease with improving conditions in the Ohio River (percent tolerant individuals, percent nonindigenous individuals, and percent detritivore individuals) also declined ($p < 0.05$). These changes coincide with marked improvement of the water quality in the Ohio River over the last 50 years, particularly in the aftermath of the Clean Water Act (1972). Some species and metric responses may also be due to the replacement of the 50 wicket dams by the construction of 18 high-lift dams.

Introduction

The Ohio River basin (Figure 1) has undergone dramatic changes since the early settlement period (1781–1803). The first Europeans found a beautiful, free-flowing, clear river buffered by hardwood forests and an abundance of wetlands (Trautman 1981; Pearson and Pearson 1989). Sycamores 3–5 m in circumference at 1 m high were common in the upper section of the river (Cramer 1824). As the land was settled over the next 100 years, the entire basin was heavily logged and many of the wetlands along the river were drained, causing increased nutrients, siltation, and turbidity. As the

* Corresponding author: jthomas@orsanco.org

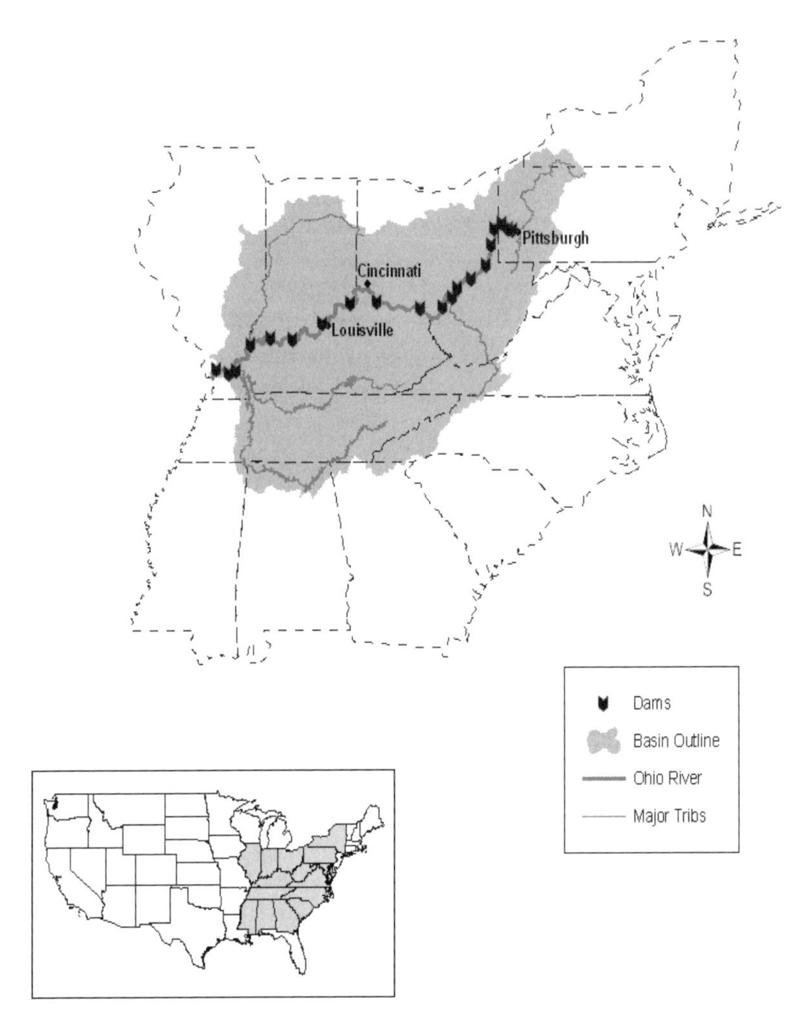

Figure 1.—The Ohio River showing the location of the 20 navigational dams, major tributaries, and three major metropolitan areas (Pittsburgh, Cincinnati, and Louisville).

population increased in the Ohio River basin, greater amounts of municipal, industrial, agricultural, and mine effluents were directed into the system. Today, 48.3% of the Ohio River watershed is agricultural, 4.1% is urban, and 1% is mines, while only 36.3% is forested (ORSANCO 1994). The Ohio River has over 600 National Pollutant Discharge and Elimination System (NPDES) permitted discharges to its waters, including those from industry, power generating facilities, and municipalities. One fourth of the total number of NPDES discharges occurs in the upper 160 river kilometers (ORSANCO 1994).

In the early 1800s, there was already interest in altering the channel of the Ohio River to allow for year round navigation (Cramer 1824). While easily navigable in times of high flow, the numerous islands throughout the river resulted in many shoals and sandbars at times of low water, greatly impeding navigation. Interest in removing "ripples" and blowing out rocks from low areas was considerable in order to allow for economic growth to all parts of the river. Also, construction of a "lock-canal" around the Falls of the Ohio at Louisville, Kentucky, was already under consideration (Cramer 1824).

In 1885, the Ohio River underwent the first in a progression of channel modifications that drastically altered the river when the U.S. Army Corps of Engineers (USACE) installed a series of low-lift navigation dams (Pearson and Pearson 1989). Eventually, 50 such "wicket" dams were erected, providing a 3-m-deep navigational channel throughout the year. Starting in 1921, USACE began to replace the low-lift dams with fewer high-lift dams (ORSANCO 1994). Today, there are 18 high-lift dams distributed from Ohio River kilometer 10 (ORKm 10; or 10 km below Pittsburgh) to ORKm 1,478. Below the last high-lift dam are the only two remaining wicket dams, Lock and Dam 52 (L&D 52) and L&D 53 (ORKm 1,549). These two locks and dams are scheduled to be replaced by a high-lift dam currently under construction in the area of Olmsted, Illinois, just downstream of L&D 53. The final 29.6 km of the Ohio River flow freely to the Mississippi River (ORSANCO 1994).

These dams and impoundments have profoundly altered the channel morphology and aquatic habitats of the Ohio River and greatly influenced its fish assemblage (Pearson and Pearson 1989; Lowman 2000). The natural flow of the river was controlled, in effect creating a series of lakes, increasing sedimentation, and altering the temperature regime (Lowman 2000). Fine sediments covered the river's naturally coarse substrate, adversely affecting many lithophilic fish species (Pearson and Pearson 1989). The high lift dams also created physical barriers for many migratory fish, such as American eel *Anguilla rostrata*, greatly limiting their distribution in the Ohio River (Pearson and Krumholz 1984).

Since sections of the Ohio River have been continually impounded for over 100 years, most of the changes observed since 1957 are attributable to changes in water quality (Cavanaugh and Mitsch 1989; Pearson and Pearson 1989). Water pollution affects aquatic life when toxicity results in fish kills, when sublethal effects alter behavior, impede growth, and reduce disease resistance, and when cumulative effects render fish unsafe for consumption (Welcomme 1985). By the middle of the 19th century, Ohio River water quality declined as a result of extensive logging, agriculture, mining, and sewage discharge (Trautman 1981; Taylor 1989). The coal mining and steel industries were major polluters through the 19th and 20th centuries. Acid mine drainage was responsible for pH values of less than 4.0 in the upper 160 km of the Ohio River before 1950 (Pearson and Krumholz 1984). One steel mill effluent with a pH of 3.4 was reported by Williams et al. (1961) to contribute a deep yellow substance to the river that was observable for 100 m downstream. Input of untreated sewage was another major factor in water quality decline. One tributary in the upper part of the river was described as "unpleasant" in appearance and as having an odor "characteristic of sewage-laden streams" (Williams et al. 1961). At least three of 159 species reported from the Ohio River have been extirpated (lake sturgeon *Acipenser fulvescens*, Alabama shad *Alosa alabamae*, and crystal darter *Crystallaria asprella*), primarily as a result of pollution (Pearson and Pearson 1989).

Sporadic attempts to reduce water pollution in the Ohio River valley were being made in the early 1900s (ORSANCO 1962). Significant improvements in Ohio River water quality, especially those related to dissolved oxygen, pH, and total and dissolved metals, have been made since 1950, particularly in the upper 160-km section of the river (Pearson and Krumholz 1984; Reash and Van Hassel 1988; Van Hassel et al. 1988; Cavanaugh and Mitsch 1989). Coal mining and steel operations have been reduced throughout the basin and regulations on discharges have been tightened. Primary sewage treatment was achieved for 90% of the basin's residents by 1960 and secondary treatment facilities were in place by the middle 1970s (Pearson and Krumholz 1984). In the late 1970s, the U.S. Environmental Protection Agency (USEPA) issued the National Municipal Policy that declared all wastewater treatment facilities in the country should have secondary treatment in place by July 1, 1988. While not all facilities accomplished this goal, most in the Ohio River basin met the deadline. These water quality improvements over the last 50 years have improved the quality of the aquatic life of the river.

Rafinesque (1820) conducted the earliest extensive collection of fishes from the Ohio River in 1818 and described over 100 new species of fish, including 26 of 30 species whose type locality is the Ohio River (Pearson and Pearson 1989). Taxonomic revisions reduced the total, but Pearson and Krumholz (1984) concluded that Rafinesque had seen 52 species. Trautman (1981) reported 93 species in 1950 and 103 species in 1980.

The Ohio River Valley Water Sanitation Commission (ORSANCO), an interstate water pollution control agency, was created in 1948 to mitigate water pollution throughout the Ohio River basin. Since the passage of the Clean Water Act (1972), populations of many Ohio River fish species have increased (Emery et al. 1998). Subsequently, a positive shift in fish species composition occurred as pollution tolerant species were replaced by more pollution sensitive species (Cavanaugh and Mitsch 1989). In 1957, ORSANCO, along with cooperating state agencies and academic institutions, conducted the first lockchamber survey of the navigational dams along the Ohio River (ORSANCO 1962). The purpose of this chapter is to document fish assemblage responses to changes in water quality and riverine habitats in the Ohio River over the last 45 years through the analysis of data collected in a consistent manner since 1957.

Methods

Fish Collection Techniques

From 1957 to 2001, 377 fish collections were made using rotenone in lockchambers of 59 different Ohio River navigational dams, including all 18 current high-lift dams and 41 of the 50 original low-lift dams (Table 1; ORSANCO 1992). Most of the collections (90%) were made during the same index period (July–October) as ORSANCO's electrofishing surveys. The lower lock gate was left open for 24 h prior to collection to allow fish to move upstream into the chamber. Field crews using 4–6-m johnboats entered the lockchamber from the downstream side and the lower gate was closed. One crew dispensed rotenone to achieve a concentration of 1 ppm (based on lockchamber size and water depth at time of sampling), making several passes along the chamber.

As they surfaced, fish were netted and placed in containers on the boats. This continued until no more fish surfaced. Fish were identified, sorted into 3-cm size-classes, and weighed. Any small fish not easily identified in the field were preserved in 10% formalin and returned to the laboratory for identification. Most fish identified on site were disposed of according to the regulations of the state in which they were processed. Some specimens were retained as vouchers and are currently stored in museum collections.

Gaps exist in the data used in this analysis because of inconsistencies in the level of identification or reporting of small fishes, particularly species in the families Cyprinidae, Atherinidae, and Fundulidae, and members of the genera *Etheostoma, Percina,* and *Noturus*. At some sampling events, the majority of which occurred in the 1950s and 1960s, small fish either were not identified to species or the data were not added to the database (Pearson and Krumholz 1984). In some cases, minnows were not identified past the family level. Instead, the total number of cyprinids was estimated and batch weighed. Events in which the complete identification of all fish could not be ensured were excluded from analyses of assemblage condition and guild structure, limiting the samples used in these analyses to 157. Also, the numbers for any species considered troublesome to identify were excluded from the percent catch analysis for each individual species. To improve our ability to detect temporal trends, we merged data from the low-lift dams with those from the high lift dams that replaced them.

Water Quality

Temperature, dissolved oxygen, pH, and conductivity were recorded hourly year round from 1961 to 1986 with the use of robot monitoring stations at several Ohio River locations. Since the majority of the samples occurred in September, average values for that month for these four parameters were calculated for these years from five stations (64.6, 489.9, 744.8, 966.6, 1,273.8) to determine water quality trends.

Data Analysis

Fish were assigned to taxonomic, habitat, trophic, and reproductive guilds according to Simon and Emery (1995) and Emery et al. (2002). The modi-

Table 1.—Location, number, and year of lockchamber sampling events.

Location	River kilometer	Dam type	Number of events	Year range
Emsworth	10.0	high lift	4	1958–1992
Dashields	21.3	high lift	17	1958–1991
Montgomery	51.0	high lift	16	1957–2001
Lock #8	74.7	wicket	1	1958
New Cumberland	87.6	high lift	12	1968–2000
Lock #9	90.3	wicket	1	1958
Lock #10	106.6	wicket	1	1958
Lock #11	123.8	wicket	1	1958
Pike Island	135.6	high lift	20	1967–1999
Lock #12	140.7	wicket	1	1958
Lock #13	154.7	wicket	1	1960
Lock #14	183.5	wicket	4	1958–1959
Hannibal	203.5	high lift	13	1976–2000
Lock #15	207.9	wicket	4	1958–1970
Lock #16	235.9	wicket	1	1958
Willow Island	260.3	high lift	11	1980–1999
Lock #17	269.7	wicket	2	1958–1967
Lock #18	289.6	wicket	1	1958
Lock #19	309.4	wicket	1	1958
Lock #20	326.0	wicket	1	1958
Belleville	328.3	high lift	15	1968–1991
Lock #21	345.5	wicket	1	1958
Lock #22	355.6	wicket	2	1958–1959
Lock #23	372.6	wicket	3	1958–1968
Racine	382.4	high lift	11	1981–2000
R.C. Byrd	449.5	high lift	18	1958–2000
Lock #27	484.6	wicket	1	1958
Lock #28	501.7	wicket	2	1958–1959
Lock #29	515.0	wicket	4	1958–1960
Lock #30	546.4	wicket	4	1957–1959
Greenup	549.0	high lift	12	1967–1999
Lock #31	578.5	wicket	3	1958–1960
Lock #32	616.0	wicket	3	1957–1959
Lock #33	652.2	wicket	2	1958–1959
Lock #34	698.9	wicket	2	1958–1959
Meldahl	702.3	high lift	14	1967–1999
Lock #35	725.9	wicket	3	1957–1959
Lock #36	742.0	wicket	2	1958–1959
Lock #37	778.0	wicket	5	1957–1960
Lock #38	810.3	wicket	2	1958–1959
Markland	855.7	high lift	14	1968–2001
Lock #39	856.0	wicket	4	1957–1959
Mcalpine	976.9	high lift	14	1960–1991
Lock #41	977.3	wicket	28	1957–1959
Lock #43	1,019.5	wicket	7	1957–1970
Lock #44	1,067.8	wicket	3	1957–1959
Lock #45	1,131.8	wicket	3	1957–1959
Cannelton	1,160.3	high lift	15	1968–2001
Lock #46	1,219.3	wicket	5	1957–1974
Newburgh	1,249.5	high lift	9	1968–1991
Lock #47	1,252.1	wicket	4	1957–1959
Lock #48	1,303.5	wicket	3	1957–1959
Lock #49	1,360.5	wicket	4	1957–1960
J.T. Myers	1,362.1	high lift	17	1969–2001

Table 1.—Continued.

Location	River kilometer	Dam type	Number of events	Year range
Lock #50	1,411.6	wicket	6	1957–1970
Lock #51	1,454.0	wicket	3	1957–1959
Smithland	1,478.8	high lift	9	1976–2001
Lock #52	1,511.6	wicket	3	1957–1959
Lock #53	1,549.8	wicket	4	1957–1981

fied index of well-being (MIWB) was used to determine changes in assemblage condition in the Ohio River over time. The MIWB, a measure of overall fish assemblage health developed by Gammon (1976) and modified by Ohio EPA to exclude tolerant individuals, sums numbers of individuals and their biomass, as well as Shannon diversity indices based on abundance and weight (Ohio EPA 1987; ORSANCO 1992). We used a linear model form of quantile regression analysis (Terrell et al. 1996; Dunham et al. 2002) to identify temporal trends in species richness, individual species abundance, and guild structure. Water quality data were analyzed by principal components analysis. Spearman correlations were calculated to detect significant relationships among fish assemblage (e.g., MIWB) and water quality variables on the first principal component axis. All analyses were conducted using SAS v. 8.0 (SAS Institute, Cary, North Carolina).

Results

Species Composition

Since 1957, 116 taxa (including 7 hybrids) representing 19 families were identified from the Ohio River lockchamber surveys (Appendix A). The five most common species based on abundance during these surveys were *Dorosoma petenense, D. cepedianum, Ictalurus punctatus, Aplodinotus grunniens, and Notropis atherinoides.* Only 6 of the 116 taxa recorded in this survey are alien to the Ohio River (Appendix A).

Assemblage, Guild, and Species Level Trends

The MIWB increased significantly over time (r^2 = 0.60; $p < 0.005$, Figure 2). Similar trends were observed in individual lockchambers from which there were multiple samples. The MIWB scores for events from 1988 to 2001 were significantly ($p < 0.05$) higher than events sampled before 1988 (Figure 3). Secondary treatment of municipal wastewater was established practice in the Ohio River by 1988. At the guild level, several metrics showed significant ($p < 0.05$) temporal trends. Number of native species (Figure 4) and six other metrics showed positive trends with time (Table 2). Percent tolerant individuals (Figure 4), percent alien individuals, percent invertivore individuals, percent simple lithophil individuals, and percent detritivore individuals declined with time (Table 2). The positive correlation of percent individuals as great river species with time likely reflects increasing populations of skipjack herring and blue catfish (Table 2). The biomass of all individuals excluding tolerants also increased significantly ($p < 0.05$) over time (Table 2).

Of the 116 taxa, 60 could not be analyzed because taxonomic identification was incomplete or because the range of their abundance was small (0–2). Thirty-five taxa (30%) showed a significant increase with time (e.g., Figure 4), while three pollution-tolerant species (2.6%) declined since 1957, *Carassius auratus, Ameiurus melas,* and *A. nebulosus* (Appendix A). Five pollution sensitive species, *Micropterus dolomieu, Moxostoma breviceps, Percina caprodes, Hiodon tergisus,* and *Polyodon spathula* (Appendix A), increased in abundance since 1957; only two tolerant species, *Cyprinus carpio* and *Lepomis cyanellus,* increased; and only one intolerant species, *Hiodon alosoides,* decreased. Sixteen species showed no significant changes over time.

Nonindigenous and Hybrid Species

Only 6 of the 116 species recorded in these surveys are alien to the Ohio River. Twenty-one alewife were

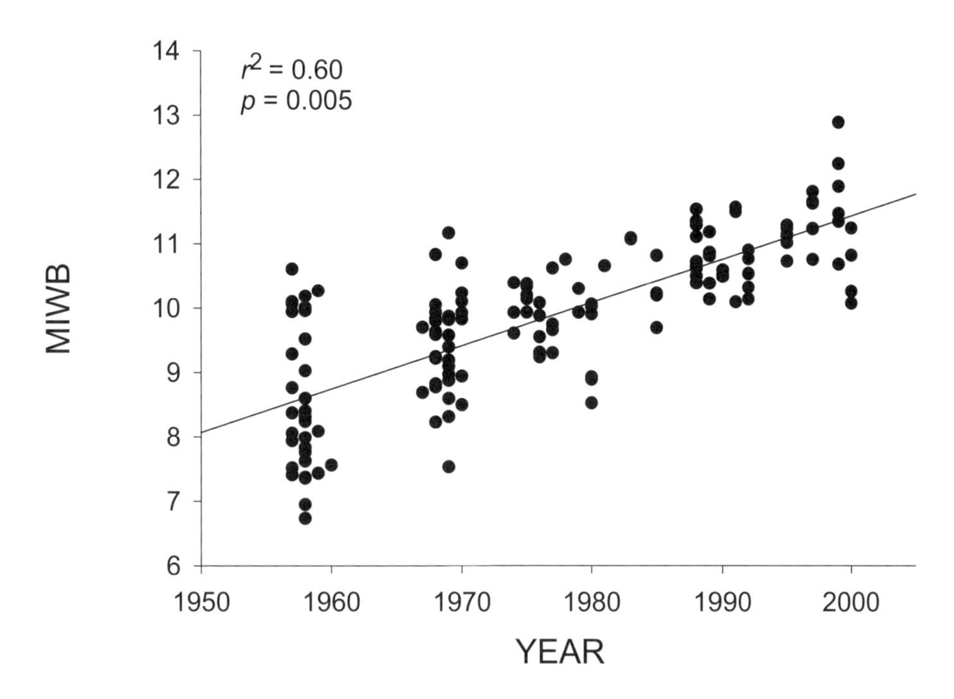

Figure 2.— Modified index of well-being (MIWB) scores from lockchamber data versus year.

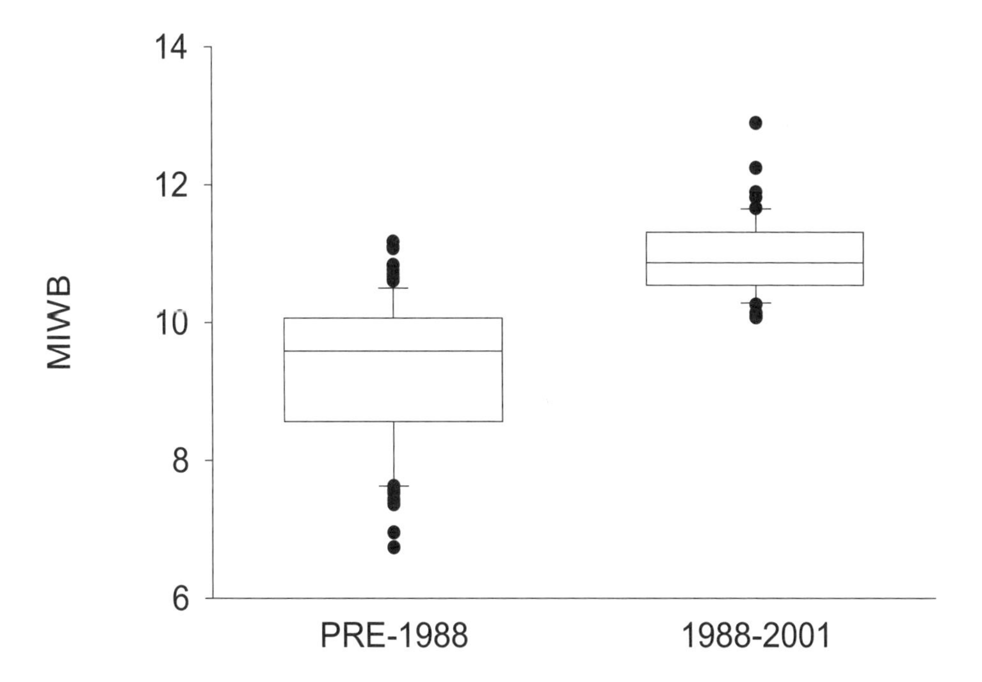

Figure 3.—Modified index of well-being (MIWB) scores before 1988 and after 1988, showing median value, 75th and 25th percentiles (box), 90th and 10th percentiles (whiskers), and outlier values. Secondary treatment of municipal wastewater was established practice in the Ohio River by 1988.

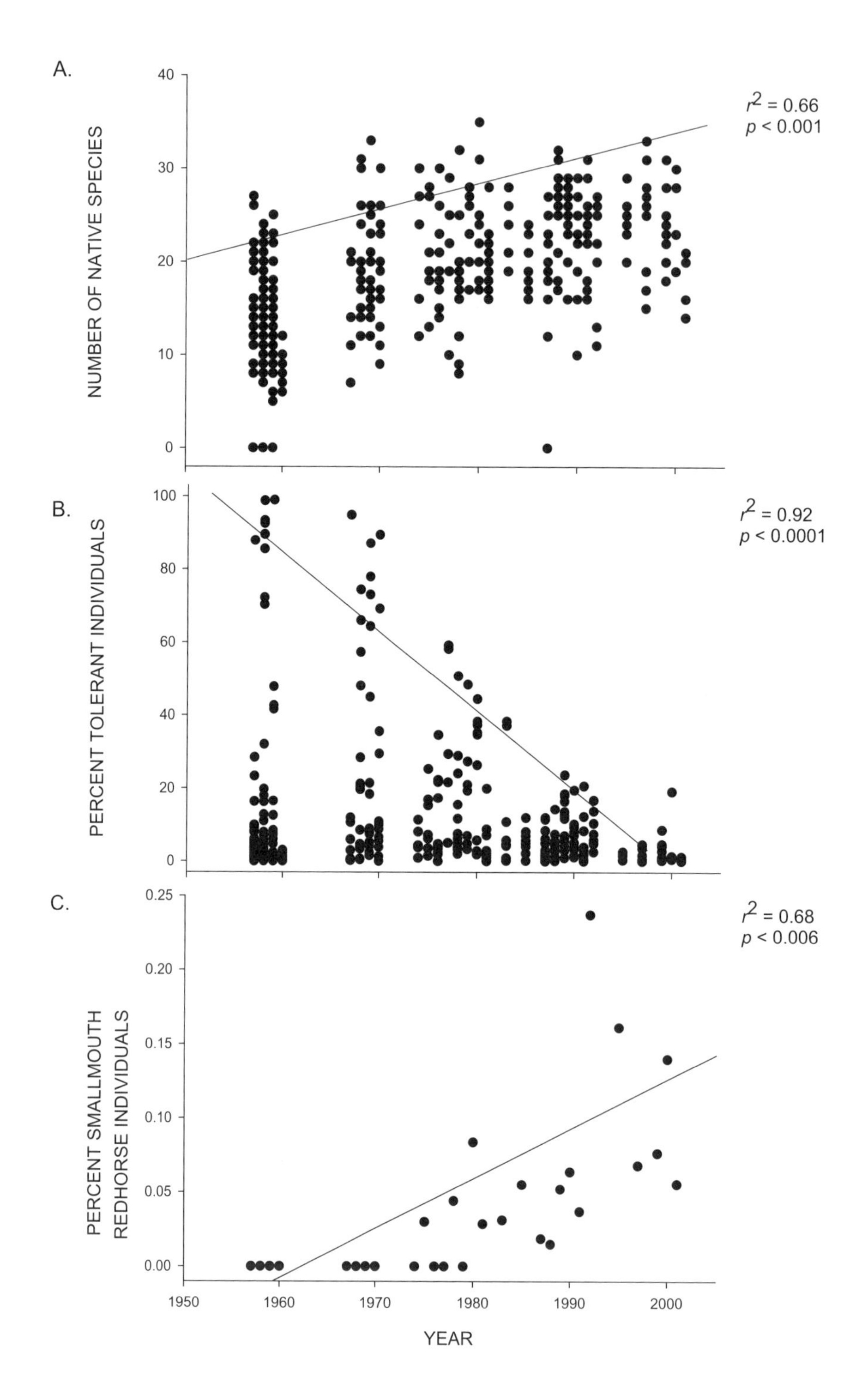

Figure 4.—Quantile regression analyses showing changes in (A) number of species, (B) percent of tolerant individuals, and (C) percent smallmouth redhorse *Moxostoma breviceps* in the Ohio River lockchamber surveys from 1957 to 2001.

Table 2.—Results of quantile regression analysis of fish assemblage metrics showing significant correlations of metrics with year. NS = not significant.

Metric	R^2	Trend
Number of species	0.31	+
Number of catostmid spp.	0.34	+
Number of centrarchid spp.	NS	
Number of intolerant spp.	0.49	+
Number of great river spp.	0.31	+
Percent individuals as great river spp.	0.20	+
Percent individuals as tolerant spp.	0.83	–
Percent individuals as nonindigenous spp.	0.43	–
Percent individuals as simple lithophils	0.41	–
Percent individuals as detritivores	0.21	–
Percent individuals as invertivores	0.56	–
Percent individuals as piscivores	0.18	+
Catch per unit effort	NS	
Total biomass, excluding tolerant individuals	0.35	+

collected since 1970. These fish most likely are the result of bait releases and stocking in private lakes (Pearson and Krumholz 1984). Common carp has been a major part of the fish fauna of the Ohio River since the early 1900s (Pearson and Krumholz 1984) and has increased in abundance in lock-chamber surveys. Since 1957, the species comprised about 1% of the total catch (ninth most abundant taxon). The first record of bighead carp was reported in 1997 at the J.T. Myers lock-chamber. Bighead carp has been stocked in ponds in Arkansas since 1973 to control algae growth (Robison and Buchanan 1988). It was first found in the natural waters of Arkansas in 1986 (Robison and Buchanan 1988), and since then it has been taken with some regularity from the Mississippi in Missouri (Etnier and Starnes 1993). The forty-seven goldfish collected between 1957 and 1989 most likely represent aquarium releases. It has declined in abundance and has not been collected from the lockchambers in 12 years. It has, however, been collected occasionally since 1991 in electrofishing samples conducted by ORSANCO (unpublished data). White catfish was commonly collected at Ohio River lock-chambers from 1968 until 1981 above ORKm 450. It has only been recorded once since 1981 at ORKm 1160 in 1990. Striped bass was the 14th most commonly collected species between 1957 and 2001, comprising a little more than 0.3% of the total catch. Striped bass is routinely stocked in the Ohio River and is most likely reproducing (D. Henley, Kentucky Division of Fish and Wildlife Services, personal communication). While several other alien species have been recorded in other Ohio River surveys (ORSANCO unpublished data; Pearson and Krumholz 1984; Burr et al. 1996), these six species represent the only aliens collected in the Ohio River lockchamber surveys. The percent of alien individuals metric has declined significantly with time (Table 2).

Of the seven hybrid taxa, only two occurred in sufficient numbers to be analyzed. Both have increased significantly ($p < 0.05$) since 1957. Saugeye *Sander canadensis* × *S. vitreus* and hybrid striped bass *Morone saxatilis* × *M. chrysops* were caught more frequently in recent lock chamber samples than historically. Karr (1981) indicated that increases in hybrid populations may indicate degraded conditions, these two hybrids are common game fish stocked in the river.

Sensitive Species

Of the 104 native fish species collected in the lockchamber surveys, 39 have special conservation status in the states bordering the Ohio River (Table 2). Twenty-one species represented in Ohio River lockchamber collections are listed as endangered in at least one of the border states, 13 are considered threatened in at least one state, and 11 have a status of special concern in one or more of those states. There is some overlap with these numbers since one species may have different designations in different states. Eleven species listed as special concern, threatened, or endangered in one or more of the Ohio River border states are significantly increasing in abundance based on the percent catch of each species (Appendix A). Conversely, two state-listed species have significantly decreased in abundance in the surveys (Appendix A). Black bullhead is listed as endangered in Pennsylvania. Goldeye, threatened in Pennsylvania and endangered in Ohio, is less common than it was 25 years ago. The number of sensitive species has increased over time (Table 2).

Water Quality Trends

Four water quality parameters (temperature, conductivity, pH, and dissolved oxygen) were highly correlated with the first principal components axis (PCA1), which accounted for 56% of the variability, and showed a significant ($p < 0.05$) increase in water quality over time (Figure 5). Higher dissolved oxygen and higher pH contributed the most to PCA1. A bivariate plot of PCA1 against time for the upper 160 kilometers of the Ohio River, which historically contained the densest concentration of industrial and municipal point sources, showed a significant response in improved water quality with time (Figure 5) and a significant improvement in assemblage condition (MIWB) with improved water quality (Figure 6).

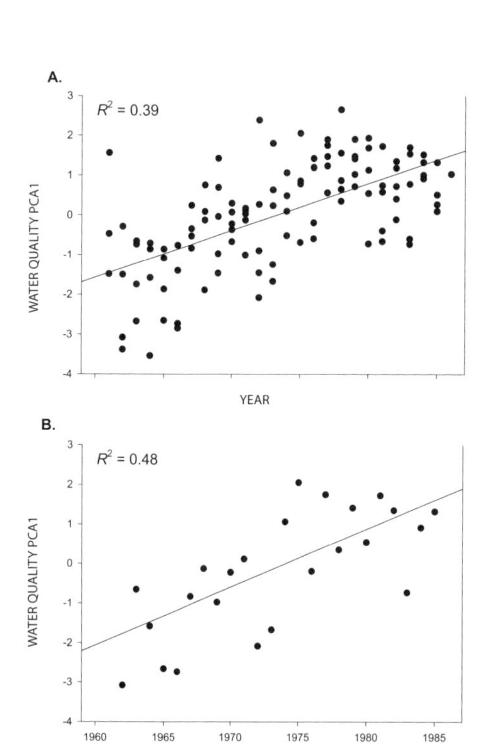

Figure 5.—The first principal component axis of water quality variables (56% of variance explained) for (A) all Ohio River monitoring stations and (B) Ohio River km 0–160 from 1961 through 1986. Positive end of the water quality axis reflects higher dissolved oxygen levels and higher pH.

Discussion

In earlier analyses of this lockchamber data, Pearson and Krumholz (1984) found that nearly all species of fish above ORKm 160 increased in density from 1957 to 1980. They also observed declines in the populations of certain tolerant species, such as bullheads *Ameiurus* spp., throughout the river. Emery et al. (1998) reported replacement of tolerant sucker species such as white sucker by more sensitive species such as redhorses *Moxostoma* spp. in the transitional period from the 1960s to the 1990s. The trends observed in these two analyses indicate that the Ohio River fish assemblage reacted positively to reduced pollution loads. We observed evidence of these same trends in our analysis of the dataset.

Temporal changes in assemblage condition, guild structure, and species structure indicate an improvement in Ohio River lockchamber fish populations (and presumably those of the river) from 1957 to 2001. Increases in MIWB scores, decreased percentages of tolerant individuals, declines in pollution-tolerant species, and recovery of pollution-sensitive species reflect improved water quality as observed by Reash and Van Hassel (1988), Van Hassel et al. (1988), and Cavanaugh and Mitsch (1989). Secondary sewage treatment and reduced toxics discharges are partially responsible for the change in MIWB scores before and after 1988. Increasing species richness, particularly sucker, great river, and intolerant species, indicate general improvement in Ohio River fish assemblages as reported by Pearson and Krumholz (1984) and Emery et al. (1998).

The improvement in the condition of the fish assemblages of the Ohio River is consistent with reports of improvement in its major tributaries and other large floodplain rivers. Gammon (1993) reported significant improvements in the condition of the Wabash River fish assemblage (based on his index of well being) from 1973 to 1992 and attributed the improvement to reduced pollution loads, improvement in wastewater treatment, and reductions in nonpoint source pollution from agricultural sources. Yoder and Rankin (1995) attribute comparable improvements in fish and macroinvertebrate assemblages in the Scioto River in Ohio to improvements in wastewater

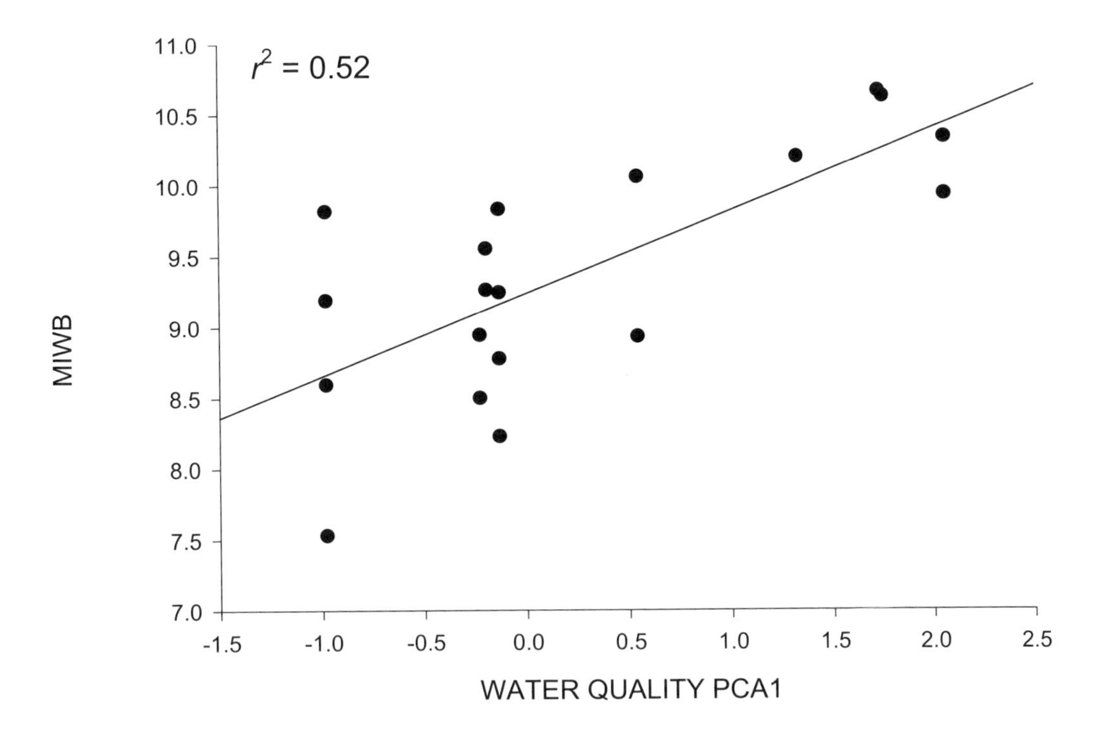

Figure 6—Plot of modified index of well-being (MIWB) versus first principal component axis of water quality variables (56% of variance explained) for Ohio River km 0–160 from 1961 through 1986. Positive end of the water quality axis reflects higher dissolved oxygen levels and higher pH.

treatment. Hughes and Gammon (1987) cited improvement in Willamette River fish assemblage integrity as a result of improvements in water quality.

However, riverine fish assemblages continue to be threatened by habitat alteration, pollution, hybridization, and invasion by alien species (Warren et al. 2002). Waite and Carpenter (2000) found that impaired fish assemblages were still associated with poor habitat and water quality in Willamette basin streams. In a review of the serial discontinuity concept (SDC) using nine large floodplain rivers, Stanford and Ward (2001) found that downstream recovery of fish assemblages below dams was overwhelmed by water quality degradation. Sparks et al. (1990) cited the disconnected flood plain, sediment loading, and the cumulative effects of point source and diffuse pollution as impediments to the rehabilitation of the Illinois River.

Dams alter a river's most important ecological processes, including impacts to the flow of water, sediments, nutrients, energy, and biota (Ligon et al. 1995; Poff et al. 1997). Alteration of the natural flow regime may further affect fish species with habitat-specific requirements for lotic environments because of altered temperature regimes and barriers to fish migration (Edwards 1978; Winston et al. 1991; De Jalon et al. 1994; Kriz 2000). Nicola et al. (1996) found that migratory species were particularly vulnerable; species of eel, lamprey, sturgeon, and shad were found to be extirpated from wide areas. Fish species diversity was significantly reduced in regulated catchments in the Murray–Darling river system in Australia (Gehrke et al. 1995). In the Vltava River, Czech Republic, dams and pollution regulation were major influences in fish density increases (Kubecka and Vostradovsky 1995). Some of the increases in Ohio River fish species may be attributed to changes in the hydrologic regime from lotic to lentic conditions that favor slackwater species.

The improvement of Ohio River water quality in the last 50 years, particularly in the upper section of the river, has had positive influences on the fish assemblage of the river. These measurable improvements in the fish populations tangibly reflect the gradual abatement of water pollution throughout the Ohio River watershed. While the navigational dams throughout the river will continue to be a major factor affecting fish assemblages, nonpoint source pollution from agricultural and urban runoff, and invasive aquatic species remain greater long-term threats to the health of the Ohio River fish assemblage.

This study shows the value of long-term, consistent monitoring data in detecting trends in fish assemblages and the importance of continuing the lockchamber surveys. The study also indicates that managing for the entire fish assemblage by improving water quality throughout the system may be an important aspect in maintaining a healthy fisheries for recreational purposes.

Acknowledgments

We would like to thank the many people from state and federal wildlife and water pollution control agencies for their assistance in conducting the surveys. The U.S. Army Corps of Engineers provided access to their facilities and manipulated the lock chambers to optimize sampling efficiency. We thank them. K. Blocksom provided advice on statistical analyses. G. Suskauer provided the map. K. Blocksom, R. Preston, W. Pearson, R. Hughes, and an anonymous reviewer provided helpful comments that greatly improved this manuscript.

References

Burr, B. M., D. J. Eisenhour, K. M. Cook, C. A. Taylor, G. Seegert, R. W. Sauer, E. R. Atwood. 1996. Nonnative fishes in Illinois waters: what do the records reveal? Transactions of the Illinois Academy of Science 89:73–91.

Cavanaugh, T. M., and W. J. Mitsch. 1989. Water quality trends of the upper Ohio River from 1977 to 1987. Ohio Journal of Science 89:153–163.

Cramer, Z. 1824. The Navigator. 12th edition. Cramer and Spear, Pittsburgh, Pennsylvania.

De Jalon, D. G., P. Sanchez, and J. A. Camargo. 1994. Downstream effects of a new hydropower impoundment on macrophyte, macroinvertebrate, and fish communities. Regulated Rivers: Research and Management 9:253–261.

Dunham, J. B., B. S. Cade, and J. W. Terrell. 2002. Influences of spatial and temporal variation on fish-habitat relationships defined by regression quantiles. Transactions of the American Fisheries Society 131:86–98.

Edwards, R. J. 1978. The effect of hypolimnion reservoir releases on fish distribution and species diversity. Transactions of the American Fisheries Society 107:71–77.

Emery, E. B., F. H. McCormick, and T. P. Simon. 2002. Response patterns of great river fish assemblage metrics to outfall effects from point source dischargers. Pages 473–485 *in* T. P. Simon, editor. Biological response signatures: indicator patterns using aquatic communities. CRC Press, Boca Raton, Florida.

Emery, E. B., T. P. Simon, and R. Ovies. 1998. Influence of the family Catostomidae on the metrics developed for a great rivers index of biotic integrity. Pages 203–224 *in* T. P. Simon, editor. Assessing the sustainability and biological integrity of water resources using fish communities. CRC Press, Boca Raton, Florida.

Etnier, D. A., and W. C. Starnes. 1993. The fishes of Tennessee. The University of Tennessee Press, Knoxville.

Gammon, J. R. 1976. The fish populations of the middle 340 km of the Wabash River. Purdue University, Water Resources Research Center, Technical Report 86, West Lafayette, Indiana.

Gammon, J. R. 1993. The Wabash River: progress and promise. Pages 141–161 *in* Restoration planning for the rivers of the Mississippi River ecosystem. United States Department of the Interior, National Biological Survey, Biological Report 19, Washington, D.C.

Gehrke, P. C., P. Brown, C. B. Schiller, D. B. Moffatt, and A. M. Bruce. 1995. River regulation and fish communities in the Murray–Darling river system, Australia. Regulated Rivers: Research and Management 11:363–375.

Hughes, R. M., and J. R. Gammon. 1987. Longitudinal changes in fish assemblages and water quality in the Willamette River, Oregon. Transactions of the American Fisheries Society 116:196–209.

Karr, J. R. 1981. Assessment of biotic integrity using fish communities. Fisheries 6(6):21–27.

Kriz, M. 2000. Shakedown at Snake River. National Journal (1/8/00):92–97.

Kubecka, J., and J. Vostradovsky. 1995. Effects of dams, regulation, and pollution on fish stocks in the Vltava River in Prague. Regulated Rivers: Research and Management 10:93–98.

Ligon, F. K., W. E. Dietrich, and W. J. Thrush. 1995. Downstream ecological effects of dams. BioScience 45:183–192.

Lowman, B. 2000. Changes among Ohio River fish populations due to water quality improvements and high-lift dams. Master's thesis. Marshall University, Huntington, West Virginia.

Nicola, G. G., B. Elvira, and A. Almodovar. 1996. Dams and fish passage facilities in the large rivers of Spain: effects on migratory species. Archiv fur Hydrobiologie Supplement 113:375–379.

Ohio EPA (Environmental Protection Agency). 1987. Biological criteria for the protection of aquatic life: volume II: users manual for biological field assessment of Ohio surface waters. Ohio Environmental Protection Agency, Division of Water Quality Planning and Assessment, Ecological Assessment Section, Columbus.

ORSANCO (Ohio River Valley Water Sanitation Commission). 1962. Aquatic life resources of the Ohio River. ORSANCO, Cincinnati, Ohio.

ORSANCO (Ohio River Valley Water Sanitation Commission). 1992. Assessment of ORSANCO fish population data using the modified index of well being (MIWB). ORSANCO, Cincinnati, Ohio.

ORSANCO (Ohio River Valley Water Sanitation Commission). 1994. The Ohio River fact book. ORSANCO, Cincinnati, Ohio.

Pearson, W. D., and L. A. Krumholz. 1984. Distribution and status of Ohio River fishes. Oak Ridge National Laboratory, ORNL/SUB/79–7831/1, Oak Ridge, Tennessee.

Pearson, W. D., and B. J. Pearson. 1989. Fishes of the Ohio River. The Ohio Journal of Science 89(5):181–187.

Poff, N. L., N. L. Allan, M. B. Bain, J. R. Karr, K. L. Prestagaard, B. D. Richter, R. E. Sparks, and J. C. Stromberg. 1997. The natural flow regime: a paradigm for river conservation. BioScience 47:769–784.

Rafinesque, C. S. 1820. Ichthyologia Ohiensis, or natural history of the fishes inhabiting the river and its tributary streams, preceded by a physical description of the Ohio and its branches. W. G. Hunt, Lexington, Kentucky.

Reash, R. J., and J. H. Van Hassel. 1988. Distribution of upper and middle Ohio River fishes, 1973–1985: II. Influence of zoogeographic and physicochemical tolerance factors. Journal of Freshwater Ecology 4:459–476.

Robison, H. W., and T. M. Buchanan. 1988. Fishes of Arkansas. The University of Arkansas Press, Fayetteville.

Simon, T. P., and E. B. Emery. 1995. Modification and assessment of an index of biotic integrity to quantify water resource quality in great rivers. Regulated Rivers: Research and Management 11:283–298.

Sparks, R. E., P. B. Bayley, S. L. Kohler, and L. L. Osborne. 1990. Disturbance and recovery of large floodplain rivers. Environmental Management 14:699–709.

Stanford, J. A., and J. V. Ward. 2001. Revisiting the serial discontinuity concept. Regulated Rivers: Research and Management 17:303–310.

Taylor, R. W. 1989. Changes in freshwater mussel populations of the Ohio River 1000 BP to recent times. Ohio Journal of Science 89(5):188–191.

Terrell, J. W., B. S. Cade, J. Carpenter, and J. M. Thompson. 1996. Modeling stream fish habitat limitations from wedge-shaped patterns of variation in standing stock. Transactions of the American Fisheries Society 125:104–117.

Trautman, M. B. 1981. The Fishes of Ohio. Ohio State University Press, Columbus.

Van Hassel, J. H., R. J. Reash, H. W. Brown, J. L. Thomas, and R. C. Mathews, Jr. 1988. Distribution of upper and middle Ohio River fishes, 1973–1985: I. Associations with water quality and ecological variables. Journal of Freshwater Ecology 4:441–458.

Waite, I. R., and K. D. Carpenter. 2000. Associations among fish assemblage structure and environmental variables in Willamette basin streams, Oregon. Transactions of the American Fisheries Society 129:754–770.

Warren, M. L., B. M. Burr, S. J. Walsh, H. L. Bart, R. C. Cashner, D. A. Etnier, B. J. Freeman, B. R. Kuhajda, R. L. Mayden, H. W. Robison, S. T. Ross, and W. C. Starnes. 2002. Diversity, distribution, and conservation status of the native freshwater fishes of the southern United States. Fisheries 25(10):7–29.

Welcomme, R. L. 1985. River fisheries. FAO Fisheries Technical Papers 262. Food and Agriculture Organization of the United Nations, Rome.

Williams, J. C., J. E. Hannegan, and W. M. Clay. 1961. ORSANCO – University of Louisville surveillance study of the Ohio River, 1960 – 1961 Final Report. The Potamological Institute, Louisville, Kentucky.

Winston, M. R., C. M. Taylor, and J. Pigg. 1991. Upstream extirpation of four minnow species due to damming of a prairie stream. Transactions of the American Fisheries Society 120:98–105.

Yoder, C. O., and E. T. Rankin. 1995. Biological response signatures and the area of degradation value: new tools for interpreting multimetric data. Pages 263–286 *in* W. S. Davis, and T. P. Simon, editors. Biological assessment and criteria: tools for water resource planning and decision making. CRC Press, Boca Raton, Florida.

Appendix A.—Species collected from Ohio River lockchamber surveys from 1957 to 2001 with guild assignments. R-square values are for significant ($p < 0.05$) quantile regressions of species abundance in individual lockchamber surveys vs. year. (NA = not analyzed, NS = not significant). For GRS, X = great river species. For SL, X = simple lithophil. For Tol, I = intolerant, and T = tolerant. For the five state columns, E = endangered, T = threatened, and V = vulnerable.

Species	Count	R^2	Trend	GRS	Trophic	SL	Tol	NIS	PA	OH	KY	IN	IL
Ohio lamprey													
Ichthyomyzon bdellium	3	NA			piscivore		I		V	E			
silver lamprey													
I. unicuspis	78	0.80	+		piscivore								
Lampetra sp.	33	NA			detritivore								
paddlefish													
Polyodon spathula	319	0.49	+	X	planktivore	X	I			T			
spotted gar													
Lepisosteus oculatus	39	NS			piscivore				E	E			
longnose gar													
L. osseus	699	0.27	+		piscivore				V				
shortnose gar													
L. platostomus	58	0.30	+	X	piscivore					E			
bowfin													
Amia calva	16	NS			piscivore				V				
American eel													
Anguilla rostrata	299	0.57	+	X	piscivore					T			
skipjack herring													
Alosa chrysochloris	46,514	0.49	+	X	piscivore				T				
alewife													
A. pseudoharengus	21	0.44	+		planktivore			X					
gizzard shad													
Dorosoma cepedianum	1,084,684	0.42	+		herbivore								
threadfin shad													
D. petenense	62,843	0.56	+		planktivore								
goldeye													
Hiodon alosoides	411	0.13	–	X	invertivore	X	I		T	E			
mooneye													
H. tergisus	939	0.18	+	X	invertivore	X	I		T				
northern pike													
Esox lucius	1	NA			piscivore								
muskellenge													
E. masquinongy	6	NA			piscivore					V		V	
E. masquinongy x lucius	5	NA			piscivore								
central stoneroller													
Campostoma anomalum	2	NA			herbivore								
red shiner													
Cyprinella lutrensis	1	NA			invertivore								
spotfin shiner													
C. spiloptera	98	NA			omnivore								
steelcolor shiner													
C. whipplei	3	NA			invertivore								
silverjaw minnow													
Notropis buccatus	2	NA			invertivore								
Mississippi silvery minnow													
Hybognathus nuchalis	704	NA		X	detritivore					E			V
bigeye chub													
Hybopsis amblops	29	NA			invertivore	X	I						E

Appendix A.—Continued.

Species	Count	R^2	Trend	GRS	Trophic	SL	Tol	NIS	PA	OH	KY	IN	IL
common shiner *Luxilus cornutus*	225	NA			invertivore								
speckled chub *Macrhybopsis aestivalis*	385	NA		X	invertivore					E			
silver chub *M. storeriana*	11,083	NA		X	invertivore	X			E				
golden shiner *Notemigonus chrysoleucas*	52	NA			omnivore		T						
emerald shiner *Notropis atherinoides*	543,151	NA			planktivore								
river shiner *N. blennius*	765	NA		X	invertivore	X			E				
bigeye shiner *N. boops*	8	NA			invertivore					T			E
ghost shiner *N. buchanani*	3,835	NA		X	invertivore				E				
spottail shiner *N. hudsonius*	98	NA			omnivore						V		
rosyface shiner *N. rubellus*	148	NA			invertivore		I						
sand shiner *N. stramineus*	5,367	NA			invertivore								
mimic shiner *N. volucellus*	62,098	NA			invertivore		I						
channel shiner *N. wickliffi*	11,378	NA		X	invertivore								
pugnose shiner *Opsopoeodus emiliae*	3	NA			detritivore								
suckermouth minnow *Phenacobius mirabilis*	4	NA			invertivore								
bluntnose minnow *Pimephales notatus*	5,028	NA			detritivore		T						
fathead minnow *P. promelas*	38	NA			detritivore		T						
bullhead minnow *P. vigilax*	61	NA			omnivore								
creek chub *Semotilus atromaculatus*	4	NA			invertivore		T						
goldfish *Carassius auratus*	47	0.79	–		detritivore		T	X					
common carp *Cyprinus carpio*	29,594	0.31	+		detritivore		T	X					
C. carpio x *Carassius auratus*	3	NA			detritivore		T						
bighead carp *Hypophthalmichthys nobilis*	1	NA			omnivore		T	X					
river carpsucker *Carpiodes carpio*	2,794	NS			detritivore								
quillback *C. cyprinus*	971	0.51	+		detritivore								
highfin carpsucker *C. velifer*	136	0.64	+		detritivore								
white sucker *Catostomus commersonii*	152	NS			detritivore	X	T						

Appendix A.—Continued.

Species	Count	R^2	Trend	GRS	Trophic	SL	Tol	NIS	PA	OH	KY	IN	IL
blue sucker													
Cycleptus elongatus	16	NA		X	invertivore	X	I			E		V	
smallmouth buffalo													
Ictiobus bubalus	8,124	0.62	+		detritivore				T				
bigmouth buffalo													
I. cyprinellus	566	0.89	+		detritivore				E				
black buffalo													
I. niger	105	0.92	+		detritivore						V		
spotted sucker													
Minytrema melanops	280	0.19	+		invertivore				T				
silver redhorse													
Moxostoma anisurum	23	0.41	+		invertivore	X							
river redhorse													
M. carinatum	14	NA			invertivore	X	I		V	V		V	T
black redhorse													
M. duquesnei	48	NS			invertivore	X	I						
golden redhorse													
M. erythrurum	129	0.32	+		invertivore	X							
smallmouth redhorse													
M. breviceps	237	0.51	+		invertivore	X	I						
white catfish													
Ameiurus catus	781	NS			omnivore			X					
yellow bullhead													
A. natalis	702	0.16	–		omnivore		T						
black bullhead													
A. melas	6,738	0.37	–		omnivore				E				
brown bullhead													
A. nebulosus	8,156	0.13	–		omnivore		T						
blue catfish													
Ictalurus furcatus	22,170	NS		X	omnivore					E			
channel catfish													
I. punctatus	93,927	NS			omnivore								
mountain madtom													
Noturus eleutherus	6	NA			invertivore				E	E			
slender madtom													
N. exilis	2	NA			invertivore						E		
stonecat													
N. flavus	14	NA			invertivore		I						
tadpole madtom													
N. gyrinus	18	NA			invertivore				E				
brindled madtom													
N. miurus	1	NA			invertivore				T				
freckled madtom													
N. nocturnus	23	NA			invertivore								
flathead catfish													
Pylodictis olivaris	5,730	NS			piscivore								
trout-perch													
Percopsis omiscomaycus	70	NA			invertivore						V		
pirate perch													
Aphredoderus sayanus	78	NA			invertivore					E			
brook silverside													
Labidesthes sicculus	14	NA			invertivore		I		V				
banded killifish													
Fundulus diaphanus	5	NA			invertivore		T						T

Appendix A.—Continued.

Species	Count	R^2	Trend	GRS	Trophic	SL	Tol	NIS	PA	OH	KY	IN	IL
white perch													
Morone americana	1	NA			invertivore								
white bass													
M. chrysops	17,199	0.22	+		piscivore								
yellow bass													
M. mississippiensis	215	0.52	+		invertivore								
striped bass													
M. saxatilis	9,172	NS			piscivore			X					
M. saxatilis x *M. chrysops*	1,529	0.31	+		piscivore								
rock bass													
Ambloplites rupestris	178	0.63	+		piscivore								
green sunfish													
Lepomis cyanellus	552	0.39	+		invertivore		T						
pumpkinseed													
L. gibbosus	342	NS			invertivore								
L. gibbosus x *L. cyanellus*	2	NA			invertivore								
warmouth													
L. gulosus	457	NS			invertivore				E				
orangespotted sunfish													
L. humilis	375	NS			invertivore								
bluegill													
L. macrochirus	6,417	0.33	+		invertivore								
L. macrochirus x *L. cyanellus*	5	NA			invertivore								
L. macrochirus x *L. megalotis*	10	NA			invertivore								
longear sunfish													
L. megalotis	1,185	0.37	+		invertivore				E				
redear sunfish													
L. microlophus	70	NS			invertivore								
smallmouth bass													
Micropterus dolomieu	324	0.29	+		piscivore		I						
spotted bass													
M. punctulatus	1,735	0.11	+		piscivore								
largemouth bass													
M. salmoides	464	0.34	+		piscivore								
white crappie													
Pomoxis annularis	6,773	0.13	+		piscivore								
black crappie													
P. nigromaculatus	1,970	NS			invertivore								
mud darter													
Etheostoma asprigene	1	NA			invertivore								
greenside darter													
E. blennioides	1	NA			invertivore		I						
rainbow darter													
E. caeruleum	1	NA			invertivore	X							
stripetail darter													
E. kennicotti	13	NA			invertivore								
johnny darter													
E. nigrum	1	NA			invertivore								
banded darter													
E. zonale	1	NA			invertivore		I						
yellow perch													
Perca flavescens	71	NS			invertivore								
logperch													
Percina caprodes	172	0.37	+		invertivore	X	I						

Appendix A.—Continued.

Species	Count	R^2	Trend	GRS	Trophic	SL	Tol	NIS	PA	OH	KY	IN	IL
channel darter													
P. copelandi	3	NA		X	invertivore	X	I		T	T			
blackside darter													
P. maculata	9	NA			invertivore	X							
slenderhead darter													
P. phoxocephala	2	NA			invertivore	X	I						
dusky darter													
P. sciera	14	NA			invertivore	X	I						
river darter													
P. shumardi	10	NA		X	invertivore	X				T			
sauger													
Sander canadensis	5,598	0.38	+		piscivore	X							
S. canadensis x vitreus	280	0.07	+		piscivore								
walleye													
S. vitreus	622	0.27	+		piscivore	X							
freshwater drum													
Aplodinotus grunniens	207,963	NS			invertivore								

American Fisheries Society Symposium 45:451–470, 2005

The Susquehanna River Fish Assemblage: Surveys, Composition, and Changes

Blaine D. Snyder*

Tetra Tech, Inc., 400 Red Brook Boulevard, Suite 200, Owings Mills, Maryland 21117, USA

Abstract.—The Susquehanna River drains portions of New York, Pennsylvania, and Maryland, and is the 18th largest river (by discharge) in the United States. Although relatively undeveloped (i.e., 63% of the basin is forested, whereas 9% is urban), the river and its fish assemblage have experienced stresses associated with coal mining, logging, electric power generation, population growth, and agricultural and industrial operations. Surveys of Susquehanna River fishes have a rich history, with the qualitative surveys of 19th century naturalists giving way to the quantitative studies of 20th century environmental impact assessment specialists. Ichthyofaunal surveys of the Susquehanna drainage were compiled and summarized herein to examine species composition, losses, and additions. Collection records indicate that the Susquehanna River drainage supports a diverse and relatively stable assemblage of 60 native species (or 51% of all species), 33 (28%) alien species, 22 (19%) euryhaline or diadromous fishes, and 2 (2%) extirpated or extinct species. Stocking efforts, bait-bucket releases, range extensions, and new species descriptions accounted for most contemporary species additions. Overall reduction in species richness has been limited to one cyprinid that has not been collected since 1862, and one darter species that has not been collected since 1987. Construction of four large hydroelectric dams on the lower Susquehanna (in the early 20th century) eliminated 98% of historic anadromous fish habitat, leading to notable reductions in commercial/recreational clupeid stocks. Recent increases in the occurrence and abundance of anadromous fish in the Susquehanna River are a credit to an extensive restoration program that began with fish trap and transfer operations in 1972, included fish culture programs, and led to the installation of fish passage technologies at each of the four dams.

Whereas an abundance of general survey and species account information is available for the Susquehanna drainage, assessments of temporal change in fish assemblage composition and abundance have been limited. The extensive quantitative fish monitoring data collected at three nuclear power stations during the 1970s and 1980s offer benchmarks for trend studies of Susquehanna fishes. The greatest challenge will be generating interest and funding to support such trend assessments of Susquehanna River fish resources.

Introduction

Many large rivers have a sparse historical record concerning ichthyofaunal composition and temporal changes, and in many cases serious study of large rivers began only after widespread anthropogenic modifications had occurred (Reash 1999). The sparse record is due, at least in part, to the physical challenges (e.g., width, depth, and discharge) associated with large river access and sampling. Simon and Sanders (1999) noted that large river survey and assessment techniques have been among the last to be developed, due to the enormous size of rivers, the complexity of their sampling requirements, and their hypothesized intricate biological relationships.

* E-mail: Blaine.Snyder@tetratech.com

Recent advances in the development of assessment methods for streams and wadeable rivers (Gibson et al. 1996; Barbour et al. 1999; Karr and Chu 1999), coupled with concern for the condition of large rivers (Graf 2001) has resulted in renewed interest in river studies (past and present) and kindled interest in the continued development of bioassessment methods for nonwadeable rivers (Ohio EPA 1987; Simon and Sanders 1999; Schiemer 2000; Schmutz et al. 2000; Peck et al. 2001; Hughes et al. 2002; Emery et al. 2003; Mebane et al. 2003).

The Susquehanna River is one of the largest river basins on the east coast of the United States, and studies of Susquehanna fishes date back to the late 1800s. Many of the ichthyofaunal surveys were spatially and/or temporally limited, and comprehensive summaries of survey results are few. Denoncourt and Cooper (1975) and McKeown (1984) noted the lack of a comprehensive account of Susquehanna ichthyofauna. Denoncourt et al. (1975) emphasized the importance of maintaining a complete list of Susquehanna fish species, identifying the native fauna, monitoring the status of rare species, and documenting the presence of alien species. Ichthyofaunal studies in the Susquehanna River drainage were compiled to prepare this chapter, and are summarized herein to catalog the chronology and extent of surveys. The specific objectives of this chapter are to present a comprehensive list of Susquehanna fishes; examine fish species composition, losses, and additions; and explore the major anthropogenic disturbances within the river and its watershed.

River Description

The Susquehanna River is the nation's 18th largest river (based on an average discharge >1,080 m^3/s) (Kammerer 1990), with a 71,251 km^2 basin, and a watershed of 50,200 total stream km (USEPA 1993). It drains portions of New York, Pennsylvania, and Maryland (Figure 1), and contributes over half of the freshwater inflow to the Chesapeake Bay. The Susquehanna originates at Otsego Lake (near Cooperstown) in New York state, and flows 714 km to the Chesapeake Bay at Havre de Grace, Maryland (Hoffman 2002). The North Branch (508 km in length) and the West Branch (367 km) join at Northumberland/Sunbury, Pennsylvania, to form the lower (or 206 km main stem) Susquehanna (Brubaker 2000). Considering its large size, the Susquehanna is unusually shallow and rocky. Over 90% of the river basin is underlain by sedimentary rock that is largely undisturbed in the Appalachian Plateau, but convoluted and eroded in the Valley and Ridge Province (Hoffman 2002). Johnson (1931) and Thompson (1949) described the lower Susquehanna as a superimposed river that has occupied its lower basin for some time, possibly through much of the Tertiary Period (1.8–65 million years ago). Williams (1895) and Leverett (1934) noted extensive glacial deposits throughout the drainage, confirming that the river system was well established during the Illinoian Glacial Stage (130,000–300,000 years ago). The Susquehanna derived fauna (including elements from the northern Mississippi basin) via glacial outlets into its headwaters (Gilbert 1980; Hocutt et al. 1986). The extended lower Susquehanna (now drowned by the Chesapeake Bay) allowed exchange of fishes among the Susquehanna, Potomac, Rappahannock, York, and James drainages. Written records of harvest of Susquehanna fishes date back to the 1600s (Gerstell 1998). Commercial shad and herring fisheries were common by the 1800s (St. Pierre 2002), and important recreational fisheries for both introduced and native game species persist today.

The Susquehanna River borders major population centers, and although relatively undeveloped, it has experienced the stresses of overuse and water pollution (Hoffman 2002). Throughout the 19th century, water quality was diminished by coal mining (prevalent in the North and West branches) and erosion/siltation from extensive logging (particularly in the West Branch) (Stranahan 1993). Acid mine drainage has had devastating effects (Heard et al. 1997) on fishes and invertebrates of the West Branch and its tributaries, since the river has little natural buffering capacity; however, recent improvements have been observed (Lovich and Lovich 1996). Early in the 20th century, coal operations in the upper anthracite region began using water in their cleaning process, carrying tons of coal particles to the river and creating "river

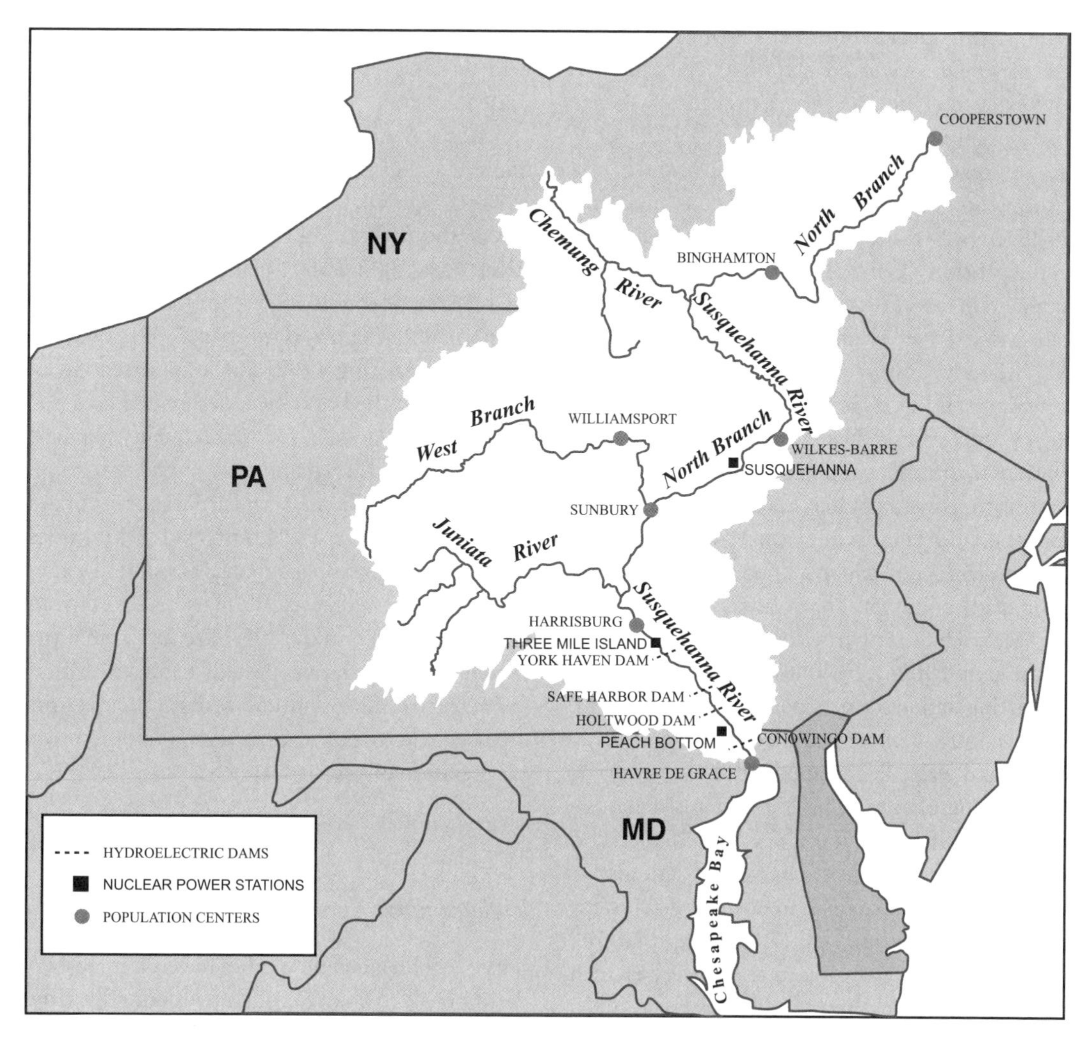

Figure 1.—Susquehanna River basin, showing major tributaries, population centers, hydroelectric dams, and nuclear power facilities.

coal" (Stranahan 1993). Once the value and abundance of the Susquehanna's depositional coal particles was recognized, a cottage industry developed. River coal dredging (via barge-mounted bucket dredges and pumps) became a common practice in the Susquehanna during the early 1900s. Additional stresses throughout the 19th and 20th centuries resulted from agricultural (e.g., nutrient enrichment, siltation, and loss of riparian vegetation) and industrial development (e.g., point source discharges) in the lower Susquehanna, and municipal discharges from growing riverside communities.

Early 20th century development along the river also included construction of electric generating facilities that used waters of the Susquehanna for hydropower production or for cooling purposes at steam electric stations. The steep river slope and the narrow valley of the lower Susquehanna subbasin provided areas for hydropower development (McMorran 1986b). Four hydroelectric dams were built on the lower Susquehanna between 1904 and 1932 (Gerstell 1998). The dams altered the flow regime, interrupted fish migrations, and eliminated many commercial and recreational fisheries for anadromous American shad *Alosa sapidissima* and river herring species (blueback herring *A. aestivalis* and alewife *A. pseudoharengus*) that had been active and extensive since the early 19th century (Myers 1991). Holtwood Dam was built in 1910, just 40 km upstream from the mouth of the Susquehanna. Completion of Conowingo Dam construction in 1928 left only the lowest 16 river km open to migratory fishes (Gerstell 1998). A program was launched in the 1970s to restore anadromous American shad and river herring to the Susquehanna River. The intensive restoration program included construction of fish passage facilities, culture and stocking of American shad larvae, and trap/transfer operations to introduce adult American shad and river herring to spawning waters above dams. Fish elevators and ladders built between 1991 and 2000 at the four hydropower dams opened 338 km of historic shad habitat (St. Pierre 2002).

Susquehanna hydroelectric dams constructed in the early 1900s were followed by steam electric generating station construction in the mid- and late-20th century. Concern over cooling water intake effects and waste-heat disposal into waterways began in earnest in the 1960s and 1970s (Richards et al. 1977; USEPA 1977). The Clean Water Act of 1972 required thermal dischargers to demonstrate that their effluents resulted in no appreciable detrimental impact to aquatic biota, and required cooling water intake structures to reflect the best technology available for minimizing adverse environmental impact. Potential impacts from cooling water intakes included the entrainment of fish eggs and larvae through a steam electric facility's cooling system, subjecting them to mechanical, pressure differential, thermal, and chemical stressors (USEPA 1977). Also of concern was the impingement of young, juvenile, and adult fish on intake screening or other cooling water system protection structures. The largest intake volumes of Susquehanna waters were required for the operation of nuclear power stations. Peach Bottom Atomic Power Station, located upstream from Conowingo Dam, began commercial operation in 1967 (Exelon Corporation 2002a). Operation of the Three Mile Island Nuclear Station (located just upstream from the York Haven Dam) began in 1974 (Exelon Corporation 2002b). The Susquehanna Steam Electric Station began operations in 1983, and is located near Berwick, Pennsylvania (PPL 2001). All three facilities continue to operate.

Recent land use estimates for the Susquehanna drainage indicated that 63% of the basin is forested; 19% is cropland; 9% is urban; 7% is pasture; and 2% is water (Ott et al. 1991). The Susquehanna River Basin 2002 Water Quality Inventory and Assessment Report to Congress (or Clean Water Act Section 305(b) report) (Hoffman 2002) noted that 91% (representing 1,740 km) of assessed stream kilometers in the basin meet designated uses (i.e., water uses identified in state water quality standards that must be achieved and maintained as required under the Clean Water Act). Partial support of designated uses was reported for 6% (111 km) of assessed streams, and nonsupport was reported for 3% (60 km) of assessed waters. Primary causes of stream impairment cited in the 305(b) report were nutrient enrichment, siltation, physical habitat alterations (primarily from agricultural activities), acid mine drainage, removal of riparian vegetation, and physical alteration of stream channels.

Methods

Ichthyofaunal Surveys

Surveys of Susquehanna fishes have a long history, dating back to the 1800s. Early reports by Cope (1862 and 1881) and Bean (1892), and updates by Fowler (1919, 1927, and 1940), documented fish species collected throughout Pennsylvania. Cooper (1983) published a book of Pennsylvania fish species accounts, documenting decades of sampling experience that included the Susquehanna drainage. Numerous ichthyofaunal surveys included Susquehanna River and/or tributary samples (Table 1). Greeley (1936) provided an annotated list of species specifically for the Susquehanna and Delaware River basins, and Bielo (1963) completed a fishery investigation of the Susquehanna River as a Master's thesis. Denoncourt and Cooper (1975) summarized historical and recent fish collections from the Susquehanna drainage above Conowingo Dam; Webster (1980) considered historical fish distribution in the upper Susquehanna; and McKeown

Table 1.—Ichthyofaunal surveys conducted in the Susquehanna River and its tributaries (years represent year of publication).

Author(s)	Year	Survey location
Cope	1862	Susquehanna River drainage
Cope	1881	Susquehanna River drainage
Bean	1892	Susquehanna River drainage
Fowler	1919	Susquehanna River drainage
Fowler	1927	Susquehanna River drainage
Greeley	1936	Susquehanna River drainage
Fowler	1940	Susquehanna River drainage
Bielo	1963	Susquehanna River drainage
PA DER and USEPA	1964	Susquehanna River drainage
Kneib	1972	Little Pine Creek drainage
Denoncourt and Stambaugh	1974	Codorus Creek drainage
Denoncourt	1975	Susquehanna River drainage (above Conowingo Dam)
Denoncourt and Cooper	1975	Susquehanna River drainage (above Conowingo Dam)
Hocutt and Stauffer	1975	Conowingo Creek drainage
Hocutt and Drawas	1975	Susquehanna River, West Branch
Stambaugh and Denoncourt	1975	Conewago Creek drainage
Lee et al.	1976	Susquehanna River drainage
Malick and Potter	1976	Fishing Creek drainage
Potter et al.	1976	Swatara Creek drainage
Ritson et al.	1977	Conowingo Creek drainage
Denoncourt et al.	1979	Chillisquaque Creek drainage
Zuck and Denoncourt	1979	Yellow Breeches Creek drainage
Brezina et al.	1980	Lower Susquehanna basin
Lee et al.	1980	Susquehanna River drainage
Barila et al.	1981	Susquehanna River, Raystown Branch
Lee et al.	1981	Susquehanna River drainage
Stauffer et al.	1982	Central and northern Appalachian Mountain drainages
Cooper	1983	Susquehanna River drainage
McKeown	1984	Susquehanna River drainage (below Conowingo Dam)
Smith	1985	Susquehanna River drainage (NY state)
Hocutt et al.	1986	Susquehanna River drainage
McMorran	1986a	Juniata River sub-basin
McMorran	1986b	Lower Susquehanna basin
Umstead and Denoncourt	1991	Middle Creek drainage
Argent et al.	1998a	Susquehanna River drainage
Argent et al.	1998b	Susquehanna River drainage
Bilger and Brightbill	1998	Lower Susquehanna basin
Snyder et al.	1999	Codorus Creek drainage

(1984) summarized Susquehanna survey results from Conowingo Dam to the upper Chesapeake Bay. Argent et al. (1998a, 1998b, and 2000) recently compiled Pennsylvania fish collection records (including the Susquehanna drainage), mapped species distributions based on those records, and developed a method to characterize rare or vulnerable fishes in the state.

Early and wide-ranging natural history surveys gave way to more detailed, localized studies of fish abundance and distribution (Table 1). In addition to the published studies summarized in Table 1, unpublished synoptic ecological studies were conducted at nuclear power stations on the Susquehanna River (primarily during the 1970s and 1980s). All reports and volumes are too numerous to mention here; however, examples include Ichthyological Associates (1974), Robbins and Mathur (1974 and 1975), Potter and Associates (1975), Nardacci and Associates (1977), Jacobsen and Associates (1979), and EA Engineering, Science, and Technology (1987). These were detailed, quantitative aquatic monitoring studies of the Susquehanna River in the vicinity of the Susquehanna, Three Mile Island, and Peach Bottom nuclear power stations. In most cases, the reports were mandated and reviewed by the Nuclear Regulatory Commission, as part of each facility's operational or environmental technical specifications. Fishes were collected from the river via seine, boat electrofisher, trapnets, gill nets, and ichthyoplankton nets. Samples were collected over extended index periods for (in many cases) several years, as part of impact assessment programs. In addition to the ambient river surveys, multi-year studies of fish entrainment and impingement were conducted at facility intakes (Mathur et al. 1977).

Management Surveys

An important (but often unpublished) source of additional information on Susquehanna River fishes can be found in fisheries management reports. Surveys associated with fisheries management are often species-specific, but may encompass large study areas and extended time frames. Hoopes et al. (1984 and 1985) and Hoopes and Cooper (1985) conducted evaluations of riverine smallmouth bass *Micropterus dolomieu* populations from 1974 to 1984, and the Pennsylvania Fish and Boat Commission continues to compile annual fisheries management field reports to assess smallmouth bass and walleye *Sander vitreus* stocks. In addition, creel surveys have been conducted in the Susquehanna River to evaluate the extent and success of sportfishing in reference to heated discharges from nuclear power stations (Euston and Mathur 1979; Snyder 1986). Extensive surveys have also been sponsored and conducted by the Susquehanna River Anadromous Fish Restoration Committee (formerly the Susquehanna Shad Advisory Committee), as part of ongoing anadromous fish research, restoration, and management programs (St. Pierre 2002).

Survey Compilation

Susquehanna River ichthyofaunal survey (Table 1) and management survey results were compiled to provide a comprehensive list of fishes for the Susquehanna River drainage. Recent species additions and losses (since the earliest natural history surveys) were noted. Trophic and tolerance designations were summarized for both native Susquehanna River fishes and introduced species. Feeding guild and pollution tolerance designations were patterned after Karr et al. (1986) and Ohio EPA (1987).

Results

Species Compilation

Survey accounts found in the literature were used to compile a comprehensive ichthyofaunal list of 117 species for the Susquehanna River drainage (Appendix A). Species richness in the drainage is dominated by minnows (Cyprinidae) and sunfishes (Centrarchidae) (Figure 2). Species totals include 60 native fishes (or 51% of all species), 33 (28%) introduced species, 22 (19%) euryhaline or diadromous fishes, 1 (1%) extirpated species, and 1 (1%) species that is believed to be extinct. The Maryland darter *Etheostoma sellare* is possibly extinct, but is currently federally listed as endangered throughout its range. Nearly 70% of fishes native to the Susquehanna are insectivorous species, whereas the majority (52%) of introduced fishes are piscivores (Figure 3). Species with an intermediate tolerance to anthropogenic stress dominate the Susquehanna River fish assemblage (Figure 4) . A total of 25% of native species are characterized as sensitive/intolerant species, whereas only 9% of introduced fishes could be classified as intolerant species.

Species Additions

Additions to the Susquehanna River ichthyofauna (since the surveys of the late 19th and early 20th centuries) were the result of stocking, bait-bucket introductions, range extensions, and new species descriptions (Table 2). Both game and forage species were stocked, and suspected bait-bucket introductions included minnow, killifish, and stickleback species. Range extensions of selected species into the Susquehanna drainage may have been caused (or accelerated) by hurricane activity and resulting

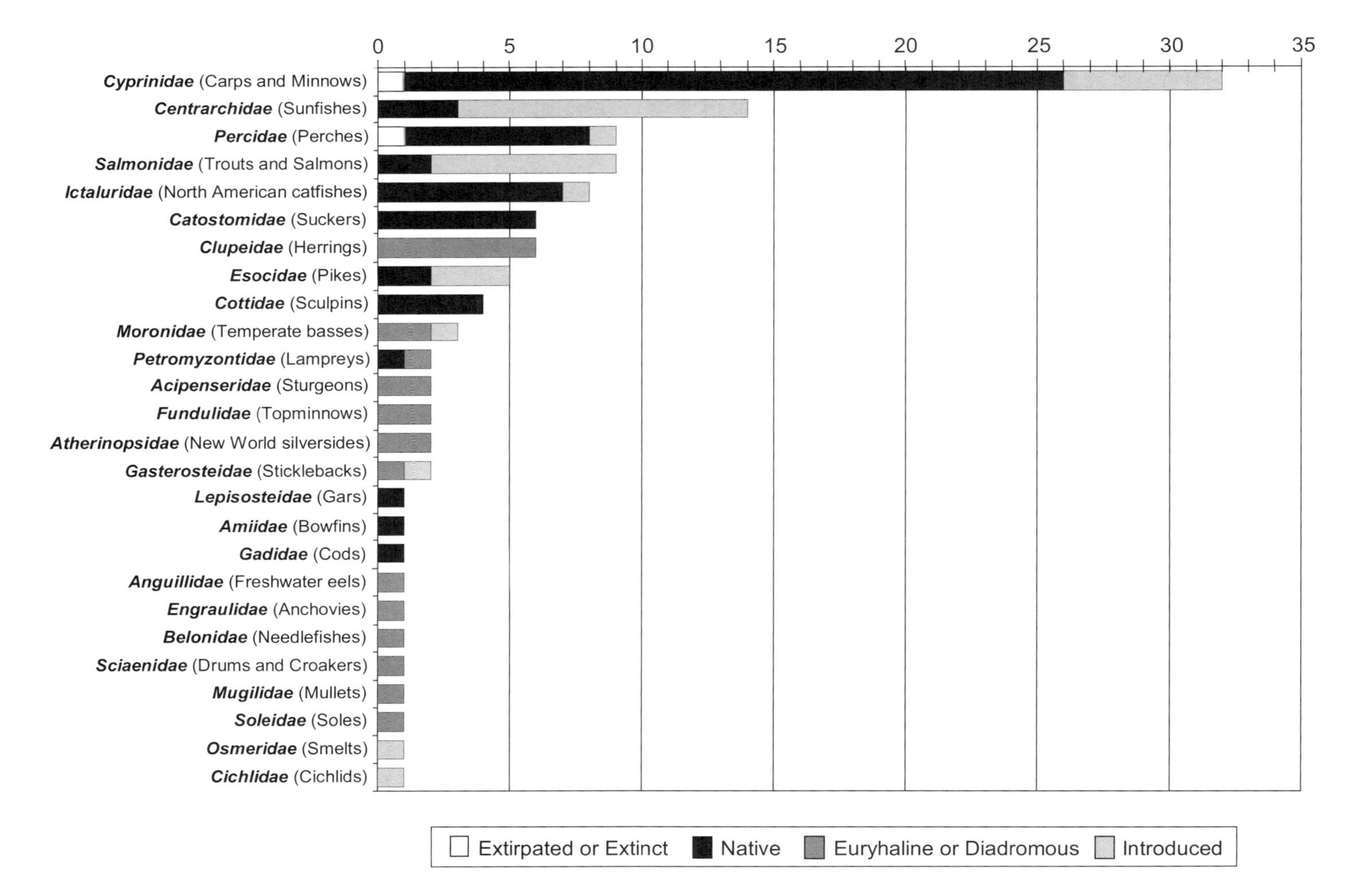

Figure 2.—The number of native, introduced, diadromous, euryhaline, and extirpated species reported in ichthyofaunal surveys of the Susquehanna River drainage.

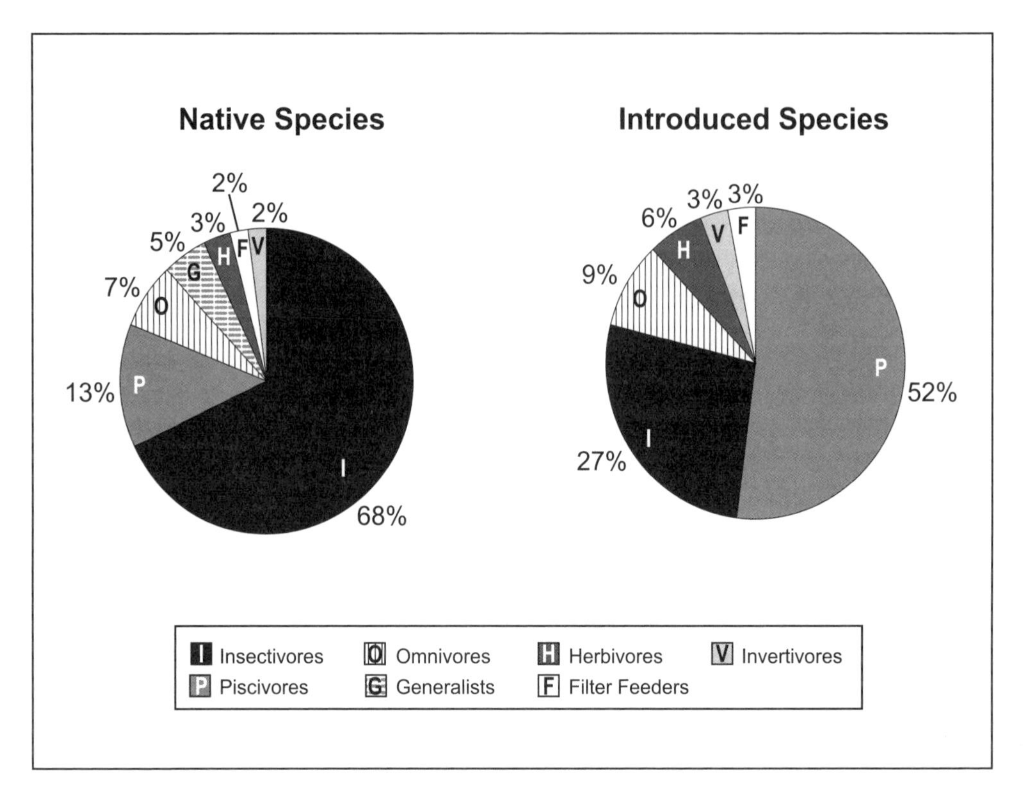

Figure 3.—Feeding guilds of native and introduced Susquehanna River fish species.

floods. Range extensions for some species may also be attributed to stream capture from the Allegheny and Potomac river drainages. Ichthyofaunal additions also included blue tilapia *Oreochromis aureus*, an alien species that escaped from an aquaculture facility in the vicinity of the Susquehanna River, and the recent appearance of the flathead catfish *Pylodictis olivaris* from an unknown source of origin (possibly transplanted by anglers).

Species Losses

Argent et al. (1998a) noted that Pennsylvania fish species of the families Petromyzontidae, Cyprinidae, Catostomidae, Ictaluridae, Centrarchidae, and Percidae experienced reductions in distribution over the past 100 years. Eleven species (shortnose sturgeon *Acipenser brevirostrum*, Atlantic sturgeon *A. oxyrinchus*, hickory shad *Alosa mediocris*, cisco *Coregonus artedi*, bridle shiner *Notropis bifrenatus*, blackchin shiner *N. heterodon*, tadpole madtom *Noturus gyrinus*, burbot *Lota lota*, banded sunfish *Enneacanthus obesus*, warmouth *Lepomis gulosus*, and longear sunfish *L. megalotis*) known to have occurred in the Susquehanna River drainage are listed as endangered by the state of Pennsylvania. Three of the species are euryhaline or anadromous, four are native to the Susquehanna, and four are introduced species.

Only two Susquehanna species—the northern redbelly dace *Phoxinus eos* and the Maryland darter—are believed to be extirpated or extinct. The northern redbelly dace was reported in the Susquehanna River drainage of Pennsylvania (Meshoppen Creek), and was last collected by Cope in 1862. It was not found in a 1935 survey of the Susquehanna (Greeley 1936), has not been collected recently anywhere in the drainage, and is believed to be extirpated (Cooper 1983). Its limited and disjunct distribution in the drainage may have made it susceptible to extirpation resulting from local natural or anthropogenic perturbations. The Maryland

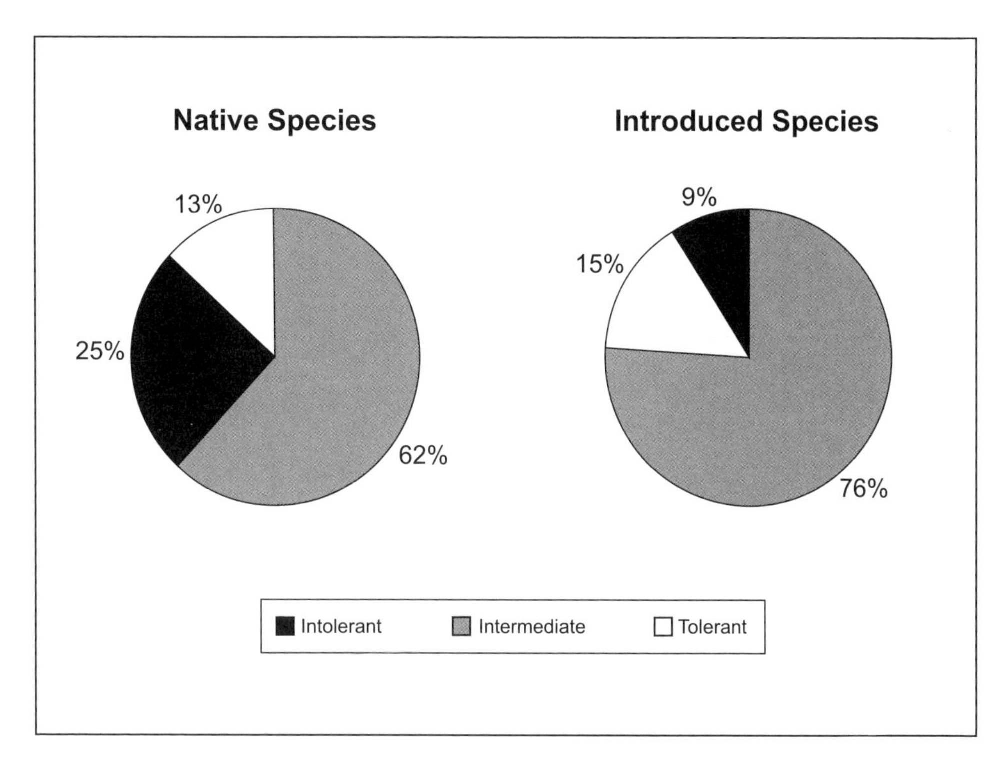

Figure 4.—Pollution tolerance designations for native and introduced Susquehanna River fish species.

darter was last observed in the lower Susquehanna River basin (Deer Creek) in 1987, and is possibly extinct (USEPA 1999). Completion of the Conowingo Dam in 1928 may have restricted the range of the Maryland darter, and the limited, discontinuous, and disjunct populations were susceptible to extinction. Their extremely specialized habitat requirements contributed to their scarcity, and increased silt loads and water withdrawals (e.g., for irrigation and potable water uses) may have contributed to their loss (USEPA 1999).

Species Conservation and Recovery

Susquehanna River diadromous fish stocks experienced at least local extirpations due to poor water quality, overfishing, canal and dam construction, and industrial development throughout the 19th century. Recovery efforts may be too late for some species such as the shortnose sturgeon, whereas other species (e.g., American shad) show promising signs of recovery from recent restoration efforts. Shortnose sturgeon have been reported only sporadically in the upper Chesapeake Bay and lower Susquehanna River for more than a century (Blankenship 1999). It is unclear whether any native Chesapeake Bay shortnose sturgeon remain (NMFS 1998). Individuals can move between the Delaware Bay and Chesapeake Bay by traversing the Chesapeake and Delaware Canal (Welsh et al. 2002). Genetic testing of a limited number of specimens collected in recent years from the upper Chesapeake Bay indicated a Delaware Bay point of origin (Blankenship 1999).

In the early 20th century, construction of four large hydroelectric dams in the lower Susquehanna eliminated 98% of historic American shad habitat (Gerstell 1998). All shad fisheries in Maryland were closed in 1980, in response to the severe decline in American shad and river herring numbers. Prior to the completion of Conowingo Dam in 1928, the annual harvest of American eel *Anguilla rostrata* in the Susquehanna River was nearly 1 million pounds (Myers 1991). Since then, the harvest has been zero, and until fish passage facilities were recently installed,

Table 2.—A summary of contemporary additions to Susquehanna River ichthyofauna (refer to Appendix A for scientific names of species).

Author(s)	**Species**	Comments
Kneib (1972)	banded darter	First to report pressure in drainage. Denoncourt et al. (1975b), Denoncourt and Stauffer (1976), and Robbins and Mathur (1975) showed range extension to Conowingo Dam. Flooding by Hurricane Agnes in 1972 may have accelerated the range extension
Denoncourt et al. (1975a)	Amur pike	Introduced into Glendale Lake by PA Fish and Boat Commission. More may have escaped from state hatchery
Denoncourt et al. (1975a)	lake cisco sockeye salmon lake trout	Introduced into Harvey's Lake in 1967 by PA Fish and Boat Commission
Denoncourt et al. (1975a)	Atlantic salmon	Introduced into Laural Lake
Denoncourt et al. (1975a)	rainbow smelt	Introduced into Harvey's Lake in 1952 by PA Fish and Boat Commission
Denoncourt et al. (1975a)	redear sunfish	Introduced to many locations by the PA Fish and Boat Commission
Denoncourt et al. (1975b)	silverjaw minnow	Range extension may be the result of bait-bucket transfers and/or 1972 flooding by Hurricane Agnes
Denoncourt et al. (1975a)	brook stickleback	Collected in the Susquehanna drainage but origin unknown
Denoncourt et al. (1975a)	bluespotted sunfish	Reproducing population found in Mountain Creek. May represent relict population that was widely distributed in interglacial periods
Denoncourt et al. (1976)	white bass	Only one specimen collected from Susquehanna River. Stocked in Middle Creek Lake in 1970
Knapp (1976)	Maryland darter	May have been more widespread in lower Susquehanna before construction of Conowingo Dam. Known to exist only in lower reaches of Deer Creek. Last observed in Deer Creek watershed in 1987, and believed to be extinct (USEPA 1999)
Denoncourt et al. (1977)	greenside darter	Suggested dispersal to Susquehanna River via stream capture from the Allegheny River. Range extensions in the West Branch Susquehanna and its tributaries may have been limited by acid mine drainage

Table 2.—Continued.

Author(s)	Species	Comments
Denoncourt et al. (1978)	mummichog	Freshwater population found in Letort Spring Run. Suspected bait-bucket introduction
Malick et al. (1978)	eastern silvery minnow	Reported for first time from Susquehanna tributaries (Little Muncy Creek and Beaverdam Creek). Beaverdam Creek specimens may have entered Susquehanna via stream capture from the Potomac drainage
Malick (1978)	mimic shiner	Reported for first time in Susquehanna River and Sherman Creek. Previously known only from Ohio drainage in Pennsylvania. Suspected bait-bucket introduction or stream capture from Allegheny River
Strauss (1980)	Potomac sculpin	Localized in the Conodoguinet System. Probably derived from the Potomac River drainage
Cooper (1983)	fourspine stickleback	Breeding populations found in Harvey's Lake and Big Spring. Suspected bait-bucket introduction
Skinner (1984) Skinner (1986) Stauffer et al. (1988) Courtenay and Williams (1992)	blue tilapia	Escaped from Pennsylvania Power and Light Company's Brunner Island Aquaculture Facility sometime after 1982. Specimens were collected from the Susquehanna River as far as 48 miles downstream from the Brunner Island site
Gutowski and Raesly (1993)	tadpole madtom	First recorded occurrence in the Susquehanna drainage
Kinziger et al. (2000)	Blue Ridge sculpin	A member of the *Cottus bairdii* Complex, recently described as a distinct species. Found in Atlantic slope drainages from Roanoke River north to Lower Susquehanna
PFBC (2002)	flathead catfish	First specimens in the Susquehanna River caught via hook and line near Safe Harbor Dam

American eel were rarely observed above the dam. Anadromous fish restoration efforts included fish trap and transfer operations beginning in 1972 at Conowingo Dam, construction of an additional fish lift at Conowingo in 1991, installation of fish elevators at Holtwood and Safe Harbor dams in 1997, and completion of fish ladder construction at York Haven Dam in 2000 (Gerstell 1998; St. Pierre 2002). Numbers of American shad returning to the Susquehanna have grown from only a few hundred each year from 1972 through 1984, to an average of 150,000 annually during 2000 through 2002 (St. Pierre 2002).

Discussion

Collection records indicate that the Susquehanna River supports a diverse and relatively stable assemblage of 60 native fishes, 33 introduced species, and 22 euryhaline or diadromous species. More than one quarter of all Susquehanna fishes are introduced species. Interdrainage comparisons of 21 major basins in the central Appalachians and central Atlantic Coastal Plain by Hocutt et al. (1986) indicated that the Susquehanna had the third highest percentage of introduced species (exceeded only by the Kanawha River above the fall line and the Potomac River). The majority of the introduced fishes are piscivores, whereas most native Susquehanna species are insectivores. Additions to the ichthyofaunal list since the earliest studies of Susquehanna fishes are due to alien species introductions; species range expansions; the addition of euryhaline fishes known to occur in the tidal section of the river; increased sampling effort among fisheries scientists; and improvements in sampling equipment. Recent increases in the occurrence and abundance of anadromous fish are a credit to an extensive restoration program (including fish culture, fish trap/transport, and fish passage efforts).

Overall reduction in species richness has been limited to a single cyprinid species (believed to be extirpated) that has not been collected since 1862, and a single darter species (believed to be extinct) that has not been collected since 1987. At least eleven other species (two acipenserids, one clupeid, one salmonid, two cyprinids, one ictalurid, one gadid, and three centrarchids) are rare (representing 9.4% of all Susquehanna fishes), and two species (the shortnose sturgeon and the Maryland darter) are federally listed as endangered (1.7% of all Susquehanna fishes). Argent et al. (1998a) identified species that have experienced declines in Pennsylvania drainages and documented reduction in the distributions of 29 species in the Susquehanna, compared to 36 in the Ohio River, and 25 in the Delaware River drainages.

Accounts of Susquehanna River fishes have a long and rich history, with the qualitative studies of natural historians giving way to the quantitative studies of environmental impact assessment specialists. Whereas an abundance of ichthyofaunal survey information is available for the Susquehanna, assessment of fish assemblage composition (and abundance) through time has been limited. The compilation presented herein was intended to summarize the current knowledge of Susquehanna River species composition and change. Studies specifically linking historical changes in the ichthyofauna to anthropogenic stressors are few. Unfortunately, in the case of the Susquehanna River, the faunal changes that have resulted from the hydrological modifications associated with the addition of dams may mask underlying effects of land use changes or water quality impairment. Important resources for future study that should not be overlooked are the unpublished reports detailing the synoptic ecological studies at Susquehanna River nuclear power plants. These facilities are in the upper, middle and lower portions of the river, and their associated ecological studies (in most cases) span the decades of the 1970s and 1980s in defined portions of the river. Their extensive, quantitative fish monitoring results are suitable for use as benchmarks for trend studies of Susquehanna fishes. A legitimate concern (since they were not published or widely distributed) is that the reports may be lost in the archives of the utility companies or regulatory agencies with the passage of time. The greatest challenge, however, will continue to be the generation of sufficient interest and funding to support the important task of assessing trends in the condition of the invaluable fish resources of the Susquehanna River.

Acknowledgments

This chapter is dedicated to ichthyologist, educator, mentor, and friend Robert F. Denoncourt, recognizing his many years of study of Susquehanna River fishes. Jennifer Pitt, Brenda Decker, Kristen Pavlik, and James Choe greatly assisted in the preparation of this chapter. The assistance of Richard St. Pierre is also gratefully acknowledged. This manuscript would not have been possible without the efforts of numerous professionals from academia, water resource agencies, and private environmental companies who dedicated at least a portion of their careers to the study of Susquehanna River ichthyofauna.

References

Argent, D. G., R. F. Carline, and J. R. Stauffer, Jr. 1998a. Changes in the distribution of Pennsylvania fishes: the last 100 years. Journal of the Pennsylvania Academy of Science 72(1):32–37.

Argent, D. G., R. F. Carline, and J. R. Stauffer, Jr. 2000. A method to identify and conserve rare fishes in Pennsylvania. Journal of the Pennsylvania Academy of Science 74(1):3–12.

Argent, D. G., J. R. Stauffer, Jr., R. F. Carline, C. Paola Ferreri, and A. L. Shiels. 1998b. Fishes: Review of status in Pennsylvania *in* J. D. Hassinger, R. J. Hill, G. L. Storm and R. H. Yahner, editors. Inventory and monitoring of biotic resources in Pennsylvania. Pennsylvania Biological Survey, Harrisburg.

Barbour, M. T., J. Gerritsen, B. D. Snyder, and J. B. Stribling. 1999. Rapid bioassessment protocols for use in streams and wadeable rivers: periphyton, benthic macroinvertebrates and fish, 2nd edition. U.S. Environmental Protection Agency, EPA 841-B-99–002, Office of Water, Washington, D.C.

Barila, T. Y., R. D. Williams, and J. R. Stauffer, Jr. 1981. The influence of stream order and selected streambed parameters on fish diversity in Raystown Branch, Susquehanna River drainage, Pennsylvania. Journal of Applied Ecology 18(1):125–132.

Bean, T. H. 1892. The fishes of Pennsylvania, with descriptions of the species and notes on their common names, distribution habitats, reproduction, rate of growth and mode of capture. Pages 1–149 *in* Report to the Pennsylvania Commission of Fisheries (1889–91), Harrisburg, Pennsylvania.

Bielo, R. J. 1963. A fishery investigation of the Susquehanna River in Pennsylvania. Master's thesis. University of Delaware, Norfolk.

Bilger, M. D., and R. A. Brightbill. 1998. Fish communities and their relation to physical and chemical characteristics of streams from selected environmental settings in the lower Susquehanna River basin, 1993–95. U.S. Geological Survey, National Water Quality Assessment Program, Water-Resources, Investigations Report 98–4004, Lemoyne, Pennsylvania.

Blankenship, K. 1999. Shortnose sturgeon. Pennsylvania Angler and Boater 68(6):49–51.

Brezina, E. R., K. K. Shaeffer, J. T. Ulanoski, M. J. Arnold, R. Hughey, and T. P. Clista. 1980. Lower Susquehanna River basin water quality. Pennsylvania Department of Environmental Resources, Bureau of Water Quality Management, Publication Number 54, Harrisburg, Pennsylvania.

Brubaker, J. 2000. Down the Susquehanna to the Chesapeake. Pennsylvania State University Press, University Park, Pennsylvania.

Cooper, E. L. 1983. Fishes of Pennsylvania and the northeastern United States. Pennsylvania State Press, University Park, Pennsylvania.

Cope, E. D. 1862. Observations on certain cyprinoid fish in Pennsylvania. Proceedings of the Academy of Natural Sciences of Philadelphia 13:522–524.

Cope, E. D. 1881. The fishes of Pennsylvania. Pages 59–145 *in* Report to the Pennsylvania Commission of Fisheries (1879–1880), Harrisburg, Pennsylvania.

Courtenay, W. R., Jr., and J. D. Williams. 1992. Dispersal of exotic species from aquaculture sources, with emphasis on freshwater fishes. Pages 49–81 *in* A. Rosenfield, and R. Mann, editors. Dispersal of living organisms into aquatic ecosystems. Maryland Sea Grant Publication, College Park, Maryland.

Denoncourt, R. F. 1975. Key to the families and genera of Pennsylvania freshwater fishes and the species of freshwater fishes of the Susquehanna River drainage above Conowingo Dam. Proceedings of the Pennsylvania Academy of Science 49:82–88.

Denoncourt, R. F. , and E. L. Cooper. 1975. A review of the literature and checklist of fishes of the Susquehanna River drainage above Conowingo Dam. Proceedings of the Pennsylvania Academy of Science 49:121–125.

Denoncourt, R. F., J. C. Fisher, and K. M. Rapp. 1978. A freshwater population of the mummichog, *Fundulus heteroclitus*, from the Susquehanna River drainage in. Pennsylvania Estuaries 1(4):269–272.

Denoncourt, R. F., C. H. Hocutt, and J. R. Stauffer, Jr. 1975a. Additions to the Pennsylvania ichthyofauna of the Susquehanna River drainage. Proceedings of the Academy of Natural Sciences of Philadelphia 127(9):67–69.

Denoncourt, R. F., C. H. Hocutt, and J. R. Stauffer, Jr. 1975b. Extensions to the known ranges of *Ericymba buccata* Cope and *Etheostoma zonale* (Cope) in the Susquehanna River drainage. Proceedings of the Pennsylvania Academy of Science 49(1):45–46.

Denoncourt, R. F., W. A. Potter, and D. W. Daniels. 1976. A record of the white bass, *Morone chrysops*, from the Susquehanna River. Chesapeake Science 17(4):304–305.

Denoncourt, R. F., W. A. Potter, and J. R. Stauffer, Jr. 1977. Records of the greenside darter, *Etheostoma blennioides* from the Susquehanna River drainage in Pennsylvania. The Ohio Journal of Science 77(1):38–42.

Denoncourt, R. F., T. W. Robbins, and R. Hesser. 1975. Recent introductions and reintroductions to the Pennsylvania fish fauna of the Susquehanna River drainage above Conowingo Dam. Proceedings of the Pennsylvania Academy of Science 49:57–58.

Denoncourt, R. F., W. F. Skinner, and B. D. Snyder. 1979. Fishes of the Chillisquaque Creek drainage, Montour County, Pennsylvania. Proceedings of the Pennsylvania Academy of Science 53:145–150.

Denoncourt, R. F., and J. W. Stambaugh. 1974. An ichthyofaunal survey and discussion of fish species as an indicator of water quality, Codorus Creek drainage, York County, Pennsylvania. Proceedings of the Pennsylvania Academy of Science 48:71–78.

Denoncourt, R. F., and J. R. Stauffer, Jr. 1976. A taxonomic study of recently introduced populations of the banded darter, *Etheostoma zonale* (Cope), in the Susquehanna River. Chesapeake Science 17(4):303–304.

Emery, E. B., T. P. Simon, F. H. McCormick, P. L. Angermeier, J. E. Deshon, C. O. Yoder, R. E. Sanders, W. D. Pearson, G. D. Hickman, R. J. Reash, and J. A. Thomas. 2003. Development of a multimetric index for assessing the biological condition of the Ohio River. Transactions of the American Fisheries Society 132:791–808.

EA Engineering, Science and Technology, Inc. 1987. An ecological study of the Susquehanna River near the Three Mile Island Nuclear Station. Annual Report for 1986. EA Engineering, Science and Technology, Inc., Sparks, Maryland.

Euston, E. T., and D. Mathur. 1979. Effects of heated discharges on the winter fishery in Conowingo Pond, Pennsylvania. Proceedings of the Pennsylvania Academy of Science 53(2):156–160.

Exelon Corporation. 2002a. Power Generation: Peach Bottom Atomic Power Station. Available: http://www. exeloncorp. com/generation/nuclear/gn_peach. shtml. (November 18, 2002).

Exelon Corporation. 2002b. Power Generation: Three Mile Island Unit - 1. Available: http://www. exeloncorp. com/generation/nuclear/gn_three_mile. shtml. (November 18, 2002).

Fowler, H. W. 1919. A list of the fishes of Pennsylvania. Proceedings of the Biological Society of Washington 32:49–74.

Fowler, H. W. 1927. Notes on Pennsylvania fishes. Fish Culturist 7(3):191–192.

Fowler, H. W. 1940. A list of the fishes recorded from Pennsylvania. Commonwealth of Pennsylvania Board of Fish Commissioners, Bulletin 7:1–25.

Gerstell, R. 1998. American shad in the Susquehanna River basin: a three-hundred-year history. The Pennsylvania State University Press, University Park, Pennsylvania.

Gibson, G. R., M. T. Barbour, J. B. Stribling, J. Gerritsen, and J. R. Karr. 1996. Biological criteria: technical guidance for streams and small rivers. U.S. Environmental Protection Agency, Office of Water, EPA 822-B-96–001, Washington, D.C.

Gilbert, C. R. 1980. Zoogeographic factors in relation to biological monitoring of fish. Pages 309–355 *in* C. H. Hocutt and J. R. Stauffer, Jr., editors. Biological monitoring of fish. D.C. Heath and Company, Lexington, Massachusetts.

Graf, W. L. 2001. Damage control: restoring the

physical integrity of America's waters. Annals of the Association of American Geographers 91:1–27.

Greeley, J. R. 1936. Fishes of the area with annotated list. Pages 45–88 *in* A biological survey of the Delaware and Susquehanna watersheds, Supplement to 25th Annual Report. New York State Conservation Department, Albany, New York.

Gutowski, M. J. and R. L. Raesly. 1993. Distributional records of madtom catfishes (Ictaluridae: *Noturus*) in Pennsylvania. Journal of the Pennsylvania Academy of Science 67(2)79–84.

Heard, R. M., W. E. Sharpe, R. F. Carline, and W. G. Kimmel. 1997. Episodic acidification and changes in fish diversity in Pennsylvania headwater streams. Transactions of the American Fisheries Society 126:977–984.

Hocutt, C. H., and N. M. Drawas. 1975. An ecological reconnaissance of the west branch of the Susquehanna River, Clearfield County, Pennsylvania. Environmental Research and Compliance Report M-0087. Pennsylvania Electric Company, Akron, Ohio.

Hocutt, C. H., R. E. Jenkins, and J. R. Stauffer, Jr. 1986. Zoogeography of the fishes of the central Appalachians and central Atlantic Coastal Plain. Pages 161–211 *in* C. H. Hocutt and E. O. Wiley, editors. The zoogeography of North American freshwater fishes. Wiley, New York.

Hocutt, C. H., and J. R. Stauffer, Jr. 1975. Influence of gradient on the distribution of fishes in Conowingo Creek, Maryland and Pennsylvania. Chesapeake Science 16(2):143–147.

Hoffman, J. L. R. 2002. The 2002 Susquehanna River basin water quality assessment 305(b) report. Susquehanna River Basin Commission, Publication 220, Harrisburg, Pennsylvania.

Hoopes, R. L., R. Carline, and C. C. Cooper. 1985. Fisheries management report: riverine smallmouth bass population evaluations. Pennsylvania Fish Commission, Pleasant Gap, Pennsylvania.

Hoopes, R. L., and C. C. Cooper. 1985. Fish management report: angler use of the smallmouth fishery at Lake Frederic, a Susquehanna River hydropower impoundment. Pennsylvania Fish Commission, Pleasant Gap, Pennsylvania.

Hoopes, R. L., R. A. Lahr, and C. C. Cooper. 1984. Fisheries management report: Riverine smallmouth bass population evaluations. Pennsylvania Fish Commission, Pleasant Gap, Pennsylvania.

Hughes, R. M., P. R. Kaufmann, A. T. Herlihy, S. S. Intelmann, S. C. Corbett, M. C. Arbogast, and R. C. Hjort. 2002. Electrofishing distance needed to estimate fish species richness in raftable Oregon rivers. North American Journal of Fisheries Management 22:1229–1240.

Ichthyological Associates. 1974. An ecological study of the North Branch Susquehanna River in the vicinity of Berwick, Pennsylvania. Ichthyological Associates, Inc., Ithaca, New York.

Jacobsen, T. V., and Associates. 1979. Ecological studies of the Susquehanna River in the vicinity of the Susquehanna Steam Electric Station. Annual Report for 1978. Ichthyological Associates, Inc., Ithaca, New York.

Johnson, D. W. 1931. Stream sculpture on the Atlantic slope, a study in the evolution of Appalachian rivers. Columbia University Press, New York.

Kammerer, J. C. 1990. Largest rivers in the United States. U.S. Geological Survey Fact Sheet, Open File Report 87–242, Reston, Virginia.

Karr, J. R., and E. W. Chu. 1999. Restoring life in running waters: better biological monitoring. Island Press, Washington, D.C.

Karr, J. R., K. D. Fausch, P. L. Angermeier, P. R. Yant, and I. J. Schlosser. 1986. Assessing biological integrity in running waters: a method and its rationale. Illinois Natural History Survey, Special Publication 5, Champaign, Illinois.

Kinziger, A. P., R. L. Raesly, and D. A. Neely. 2000. New species of *Cottus* (Teleostei: Cottidae) from the middle Atlantic, eastern United States. Copeia 2000(4):1007–1018.

Knapp, L. W. 1976. Re-description relationships and status of the Maryland darter, *Etheostoma sellare*, an endangered species. Proceedings of the Biological Society of Washington 89(6):99–118.

Kneib, R. T. 1972. The effects of man's activity on the distribution of five stream fishes in Little Pine Creek, Pennsylvania. Proceedings of the Pennsylvania Academy of Science 46:49–51.

Lee, D. S., C. R. Gilbert, C. H. Hocutt, R. E. Jenkins, D. E. McAllister, and J. R. Stauffer, Jr. 1980. Atlas of North American freshwater fishes. North Carolina Museum of Natural History, Raleigh, North Carolina.

Lee, D. S., A. Norden, C. R. Gilbert, and R. Franz. 1976. A list of the freshwater fishes of Maryland and Delaware. Chesapeake Science 17(3):205–211.

Lee, D. S., S. P. Platania, C. R. Gilbert, R. Franz, and A. Norden. 1981. A revised list of the freshwater fishes of Maryland and Delaware. Proceedings of the Southeastern Fishes Council 3:1–10.

Leverett, F. 1934. Glacial deposits outside the Wisconsin terminal moraine in Pennsylvania. Fourth Series Bulletin G7:1–123. Pennsylvania Geological Survey, Harrisburg, Pennsylvania.

Lovich, J. E., and R. E. Lovich. 1996. The decline of native brook trout (*Salvelinus fontinalis*) populations along the upper west branch of the Susquehanna River: canaries outside the coal mine. Journal of the Pennsylvania Academy of Science 70(2):55–60.

Malick, R. W., Jr. 1978. The mimic shiner, *Notropis volucellus* (Cope) in the Susquehanna River drainage of Pennsylvania. Proceedings of the Pennsylvania Academy of Science 52(2):199–200.

Malick, R. W., Jr., and W. A. Potter. 1976. The distribution and relative abundance of the ichthyofauna of Fishing Creek, York County, Pennsylvania. Proceedings of the Pennsylvania Academy of Science 50:96–100.

Malick, R. W., Jr., P. C. Ritson, and J. L. Polk. 1978. First records of the silvery minnow, *Hybognathus nuchalis* Agassiz, in the Susquehanna River drainage of Pennsylvania. Proceedings of the Academy of Natural Sciences in Philadelphia 129(5):83–85.

Mathur, D., P. G. Heisey, and N. C. Magnusson. 1977. Impingement of fishes at Peach Bottom Atomic Power Station, Pennsylvania. Transactions of the American Fisheries Society 106(3):258–267.

McKeown, P. E. 1984. Additions to ichthyofauna of the Susquehanna River with a checklist of fishes of the Susquehanna River drainage below Conowingo Dam. Proceedings of the Pennsylvania Academy of Science 58:187–192.

McMorran, C. P. 1986a. Water quality and biological survey of the Juniata River subbasin. Susquehanna River Basin Commission, Publication Number 103, Harrisburg, Pennsylvania.

McMorran, C. P. 1986b. Water quality and biological survey of the lower Susquehanna subbasin. Susquehanna River Basin Commission, Publication Number 104, Harrisburg, Pennsylvania.

Mebane, C. A., T. R. Maret, and R. M. Hughes. 2003. An index of biological integrity (IBI) for Pacific Northwest Rivers. Transactions of the American Fisheries Society 132:239–261.

Myers, C. E. 1991. The history of shad fishing on the Susquehanna River and current efforts to restore the species. Wilkes University Press, Wilkes-Barre, Pennsylvania.

Nardacci, G. A., and Associates. 1977. An ecological study of the Susquehanna River in the vicinity of the Three Mile Island Nuclear Station. Annual Report for 1976. Ichthyological Associates, Inc., Ithaca, New York.

Nelson, J. S., E. J. Crossman, H. Espinosa-Perez, L.T. Findley, C. R. Gilbert, R. N. Lea, and J. D. Williams. 2004. Common and scientific names of fishes from the United States, Canada, and Mexico. American Fisheries Society, Special Publication 29, Bethesda, Maryland.

NMFS (National Marine Fisheries Service). 1998. Recovery plan for the shortnose sturgeon (*Acipenser brevirostrum*). Prepared by the Shortnose Sturgeon Recovery Team for the National Marine Fisheries Service, Silver Spring, Maryland.

Ohio EPA (Ohio Environmental Protection Agency). 1987. Biological criteria for the protection of aquatic life. Volumes I–III. Ohio EPA, Division of Water Quality Monitoring and Assessment, Surface Water Section, Columbus, Ohio.

Ott, A. N., C. S. Takita, R. E. Edwards, and S. W. Bollinger. 1991. Loads and yields of nutrients and suspended sediment transported in the Susquehanna River basin, 1985–1989. Susquehanna River Basin Commission, Publication No. 136, Harrisburg, Pennsylvania.

Peck, D. V., D. K. Averill, D. L. Klemm, and J. M. Lazorchak. 2001. Field operations manual for non-wadeable streams and rivers. U.S. Environmental Protection Agency, Corvallis, Oregon.

Pennsylvania Department of Environmental Resources (PADER), and U.S. Environmental Protection Agency (USEPA). 1964. Biological survey of the Susquehanna River and its tributaries between Danville, Pennsylvania and Conowingo, Maryland. Federal Water Pollution Control Administration, Washington, D.C.

PFBC (Pennsylvania Fish and Boat Commission). 2002. Information Paper: Flathead catfish, *Pylodictis olivaris*. Pennsylvania Fish and Boat Commission, Harrisburg, Pennsylvania.

PPL (PPL Corporation). 2001. Who we are: PPL Susquehanna. Available: http://www. pplweb.com/who_we_are/powerplants/susquehanna.html. (November 18, 2002).

Potter, W. A., and Associates. 1975. An ecological study of the Susquehanna River in the vicinity of the Three Mile Island Nuclear Station. Annual Report for 1974. Ichthyological Associates, Inc., Ithaca, New York.

Potter, W. A., R. W. Malick, Jr., and J. L. Polk. 1976. The composition and distribution of fishes of the Swatara Creek drainage, Pennsylvania. Proceedings of the Pennsylvania Academy of Science 50:136–140.

Reash, R. J. 1999. Considerations for characterizing midwestern large-river habitats. Pages 463–505 *in* T. P. Simon, editor. Assessing the sustainability and biological integrity of water resources using fish communities. CRC Press, Boca Raton, Florida.

Richards, F. P., W. W. Reynolds, and R. W. McCauley. 1977. Temperature preference studies in environmental impact assessments: an overview with procedural recommendations. Journal of the Fisheries Research Board of Canada 34(5):728–761.

Ritson, P. C., J. L. Polk, and R. W. Malick, Jr. 1977. Fishes of the Conewago Creek drainage, York and Adams counties, Pennsylvania. Proceedings of the Pennsylvania Academy of Science 51:59–66.

Robbins, T. W., and D. Mathur. 1974. Peach Bottom Atomic Power Station preoperational report No. 1 on the ecology of Conowingo Pond for Units No. 2 and 3. Ichthyological Associates, Inc., Ithaca, New York.

Robbins, T. W., and D. Mathur. 1975. Peach Bottom Atomic Power Station postoperational report No. 2 on the ecology of Conowingo Pond for Units No. 2 and 3. Ichthyological Associates, Inc., Ithaca, New York.

Schiemer, F. 2000. Fish as indicators for the assessment of the ecological integrity of large rivers. Hydrobiologia 423:271–278.

Schmutz, S., M. Kauffman, B. Vogel, M. Jungwirth, and S. Muhar. 2000. A multi-level concept for fish-based, river-type-specific assessment of ecological integrity. Hydrobiologia 423:279–289.

Simon, T. P., and R. E. Sanders. 1999. Applying an index of biotic integrity based on Great River fish communities: considerations in sampling and interpretation. Pages 475–506 *in* T. P. Simon, editor. Assessing the sustainability and biological integrity of water resources using fish communities. CRC Press, Boca Raton, Florida.

Skinner, W. F. 1984. *Oreochromis aureus* (Steindachner; Cichlidae), an exotic fish species, accidentally introduced to the lower Susquehanna River, Pennsylvania. Proceedings of the Pennsylvania Academy of Science 58:99–100.

Skinner, W. F. 1986. Susquehanna River tilapia. Fisheries 11(4):56–57.

Smith, C. L. 1985. The inland fishes of New York State. New York State Department of Environmental Conservation, Albany, New York.

Snyder, B. D. 1986. A ten-year sportfishing survey of the lower Susquehanna River in the vicinity of Three Mile Island Nuclear Station. Northeast Fish and Wildlife Conference, Hershey, Pennsylvania.

Snyder, B. D., J. B. Stribling, M. T. Barbour, and C. L. Missimer. 1999. Integrating assessments of fish and macroinvertebrate assemblages and physical habitat condition. Pages 639–652 *in* T. P. Simon, editor. Assessing the sustainability and biological integrity of water resources using fish communities. CRC Press, Boca Raton, Florida.

Stambaugh, J. W., Jr., and R. F. Denoncourt. 1975. A preliminary report on the Conewago Creek faunal survey, Lancaster County, Pennsylvania. Proceedings of the Pennsylvania Academy of Science 48:50–60.

Stauffer, J. R., Jr., S. E. Boltz, and J. M. Boltz. 1988. Cold shock susceptibility of blue tilapia from the Susquehanna River, Pennsylvania. North American Journal of Fisheries Management 8:329–332.

Stauffer, J. R., Jr., B. M. Burr, C. H. Hocutt, and R. E. Jenkins. 1982. Checklist of the fishes of the central and northern Appalachian Mountains. Proceedings of the Biological Society of Washington 95(1):27–47.

St. Pierre, R. 2002. History of the American shad restoration program on the Susquehanna River. U.S. Fish and Wildlife Service, Harrisburg, Pennsylvania.

Stranahan, S. Q. 1993. Susquehanna, river of dreams. The Johns Hopkins University Press, Baltimore, Maryland.

Strauss, R. E. 1980. Genetic and morphometric variation and the systematic relationships of eastern North American sculpins (Pisces: Cottidae). Doctoral dissertation. Pennsylvania State University, University Park, Pennsylvania.

Thompson, H. D. 1949. Drainage evolution in the Appalachians of Pennsylvania. Annals of the New York Academy of Sciences 52:31–62.

Umstead, T. M., and R. F. Denoncourt. 1991. Fishes of the Middle Creek drainage, Snyder County, Pennsylvania. Journal of the Pennsylvania Academy of Science 65(1):10–16.

USEPA (United States Environmental Protection Agency). 1977. Draft guidance for evaluating the adverse impact of cooling water intake structures on the aquatic environment: Section 316(b) P. L. 92–500. U.S. Environmental Protection Agency, Washington, D.C.

USEPA (United States Environmental Protection Agency). 1993. Total waters estimates for United States streams and lakes: total waters database and reporting program. U.S. Environmental Protection Agency, Washington, D.C.

USEPA (United States Environmental Protection Agency). 1999. From the mountains to the sea: the state of Maryland's freshwater streams. USEPA Office of Research and Development, EPA/903/R-99/023, Washington, D.C.

Webster, D. A. 1980. DeWitt Clinton's "...Fishes of the western waters of the state of New York" reexamined. Fisheries 5(2):5–12.

Welsh, S. A., M. F. Mangold, J. E. Skjeveland, and A. J. Spells. 2002. Distribution and movement of shortnose sturgeon (*Acipenser brevirostrum*) in the Chesapeake Bay. Estuaries 25(1):101–104.

Williams, E. H., Jr. 1895. Notes on the southern ice limit in eastern Pennsylvania. American Journal of Science 49:174–185.

Zuck, W. H., and R. F. Denoncourt. 1979. Fishes of the Yellow Breeches Creek drainage, Cumberland and York counties, Pennsylvania. Proceedings of the Pennsylvania Academy of Science 53:161–168.

Appendix A.—Fishes collected in the Susquehanna River drainage (Nomenclature follows Nelson et al. 2004).

Scientific name	Common name
Petromyzontidae	**Lampreys**
Lampetra aepyptera	least brook lamprey
Petromyzon marinus	sea lamprey[d]
Acipenseridae	**Sturgeons**
Acipenser brevirostrum	shortnose sturgeon[d]
A. oxyrinchus	Atlantic sturgeon[d]
Lepisosteidae	**Gars**
Lepisosteus osseus	longnose gar
Amiidae	**Bowfins**
Amia calva	bowfin
Anguillidae	**Freshwater eels**
Anguilla rostrata	American eel[d]
Engraulidae	**Anchovies**
Anchoa mitchilli	bay anchovy[c]
Clupeidae	**Herrings**
Alosa aestivalis	blueback herring[d]
A. mediocris	hickory shad[d]
A. pseudoharengus	alewife[d]
A. sapidissima	American shad[d]
Brevoortia tyrannus	Atlantic menhaden[c]
Dorosoma cepedianum	gizzard shad[c]
Cyprinidae	**Carps and minnows**
Campostoma anomalum	central stoneroller
Carassius auratus	goldfish[a]
Clinostomus elongatus	redside dace
C. funduloides	rosyside dace
Ctenopharyngodon idella	grass carp[a]
Cyprinella analostana	satinfin shiner
C. spiloptera	spotfin shiner
Cyprinus carpio	common carp[a]
Exoglossum maxillingua	cutlip minnow
Hybognathus regius	eastern silvery minnow
Luxilus cornutus	common shiner
Margariscus margarita	pearl dace
Nocomis biguttatus	hornyhead chub
N. micropogon	river chub
Notemigonus crysoleucas	golden shiner
Notropis amoenus	comely shiner
N. atherinoides	emerald shiner[a]
N. bifrenatus	bridle shiner
N. buccatus	silverjaw minnow
N. heterodon	blackchin shiner
N. heterolepis	blacknose shiner
N. hudsonius	spottail shiner
N. procne	swallowtail shiner
N. rubellus	rosyface shiner
N. volucellus	mimic shiner[a]
Phoxinus eos	northern redbelly dace[b]
Pimephales notatus	bluntnose minnow
P. promelas	fathead minnow[a]
Rhinichthys atratulus	blacknose dace
R. cataractae	longnose dace
Semotilus atromaculatus	creek chub
S. corporalis	fallfish
Catostomidae	**Suckers**
Carpiodes cyprinus	quillback
Catostomus commersonii	white sucker
Erimyzon oblongus	creek chubsucker
Hypentelium nigricans	northern hog sucker
Moxostoma anisurum	silver redhorse
M. macrolepidotum	shorthead redhorse
Ictaluridae	**North American catfishes**
Ameiurus catus	white catfish
A. natalis	yellow bullhead
A. nebulosus	brown bullhead
Ictalurus punctatus	channel catfish
Noturus flavus	stonecat
N. gyrinus	tadpole madtom
N. insignis	margined madtom
Pylodictis olivaris	flathead catfish[a]
Esocidae	**Pikes**
Esox americanus	redfin pickerel
E. lucius	northern pike[a]
E. masquinongy	muskellunge[a]
E. niger	chain pickerel
E. reicherti	Amur pike[a]
Osmeridae	**Smelts**
Osmerus mordax	rainbow smelt[a]
Salmonidae	**Trouts and salmons**
Coregonus artedi	cisco[a]
C. clupeaformis	lake whitefish
Oncorhynchus kisutch	coho salmon[a]
O. mykiss	rainbow trout[a]
O. nerka	sockeye salmon[a]
Salmo salar	Atlantic salmon[a]
S. trutta	brown trout[a]
Salvelinus fontinalis	brook trout
S. namaycush	lake trout[a]
Gadidae	**Cods**
Lota lota	burbot
Mugilidae	**Mullets**
Mugil curema	white mullet[c]
Atherinopsidae	**New World Silversides**
Menidia beryllina	inland silverside[c]
M. menidia	Atlantic silverside[c]
Belonidae	**Needlefishes**
Strongylura marina	Atlantic needlefish[c]
Fundulidae	**Topminnows**
Fundulus diaphanus	banded killifish[c]
F. heteroclitus	mummichog[c]
Gasterosteidae	**Sticklebacks**
Apeltes quadracus	fourspine stickleback[c]
Culaea inconstans	brook stickleback[a]
Cottidae	**Sculpins**
Cottus bairdii	mottled sculpin
C. caeruleomentum	Blue Ridge sculpin
C. cognatus	slimy sculpin
C. girardi	Potomac sculpin

Appendix A.—Continued.

Scientific name	Common name	Scientific name	Common name
Moronidae	**Temperate basses**	**Percidae**	**Perches**
Morone americana	white perch[c]	*Etheostoma blennioides*	greenside darter
M. chrysops	white bass[a]	*E. flabellare*	fantail darter
M. saxatilis	striped bass[c]	*E. olmstedi*	tessellated darter
Centrarchidae	**Sunfishes**	*E. sellare*	Maryland darter[b]
Ambloplites rupestris	rock bass[a]	*E. zonale*	banded darter[a]
Enneacanthus gloriosus	bluespotted sunfish	*Perca flavescens*	yellow perch
E. obesus	banded sunfish[a]	*Percina caprodes*	logperch
Lepomis auritus	redbreast sunfish	*P. peltata*	shield darter
L. cyanellus	green sunfish[a]	*Sander vitreus*	walleye
L. gibbosus	pumpkinseed	**Sciaenidae**	**Drums and croakers**
L. gulosus	warmouth[a]	*Leiostomus xanthurus*	spot[c]
L. macrochirus	bluegill[a]	**Cichlidae**	**Cichlids**
L. megalotis	longear sunfish[a]	*Oreochromis aureus*	blue tilapia[a]
L. microlophus	redear sunfish[a]	**Soleidae**	**Soles**
Micropterus dolomieu	smallmouth bass[a]	*Trinectes maculatus*	hogchoker[c]
M. salmoides	largemouth bass[a]		
Pomoxis annularis	white crappie[a]		
P. nigromaculatus	black crappie[a]		

[a] Introduced.
[b] Extirpated or extinct.
[c] Euryhaline.
[d] Diadromous.

American Fisheries Society Symposium 45:471–503, 2005

Changes in Fish Assemblages in the Tidal Hudson River, New York

ROBERT A. DANIELS*
New York State Museum, Cultural Education Center 3140, Albany, New York 12230, USA

KARIN E. LIMBURG
SUNY College of Environmental Science and Forestry, Syracuse, New York 13210, USA

ROBERT E. SCHMIDT
Simon's Rock College of Bard, 84 Alford Road, Great Barrington, Massachusetts 01230, USA

DAVID L. STRAYER
Institute of Ecosystem Studies, Box AB, Millbrook, New York 12545-0219, USA

R. CHRISTOPHER CHAMBERS
National Marine Fisheries Service, Coastal Ecology Branch,
74 Magruder Road, Highlands, New Jersey 07732, USA

Abstract.—The main channel of the Hudson River is a tidal estuary from its mouth in New York Harbor to Troy, New York, 247 km upstream. It drains about 35,000 km^2 and is an important navigational, commercial, and recreational system. Since the arrival of European settlers over 400 years ago, it has undergone numerous environmental changes. These changes have included channel maintenance by dredging, wholesale dumping of industrial and domestic wastes, scattered in-basin urbanization and shoreline development, deforestation of the watershed and an increase in agriculture, and water removal for commercial, industrial, and agricultural needs. In addition, the biota of the river has supported commercial and recreational harvesting, exotic species have become established, and habitats have become fragmented, replaced, changed in extent, or isolated. The tidal portion of the Hudson River is among the most-studied water bodies on Earth. We use data from surveys conducted in 1936, the 1970s, the 1980s, and the 1990s to examine changes in fish assemblages and from other sources dating back to 1842. The surveys are synoptic but use a variety of gears and techniques and were conducted by different researchers with different study goals. The scale of our assessment is necessarily coarse. Over 200 species of fish are reported from the drainage, including freshwater and diadromous species, estuarine forms, certain life history stages of primarily marine species, and marine strays. The tidal Hudson River fish assemblages have responded to the environmental changes of the last century in several ways. Several important native species appear to be in decline (e.g., rainbow smelt *Osmerus mordax* and Atlantic tomcod *Microgadus tomcod*), others, once in decline, have rebounded (e.g., striped bass *Morone saxatilis*), and populations of some species seem stable (e.g., spottail shiner *Notropis hudsonius*). No native species is extirpated from the system, and only one, shortnose sturgeon

* Corresponding author: rdaniels@mail.nysed.gov

Acipenser brevirostrum, is listed as endangered. The recent establishment of the exotic zebra mussel *Dreissena polymorpha* may be shifting the fish assemblage away from open-water fishes (e.g., *Alosa*) and toward species associated with vegetation (e.g., centrarchids). In general, the Hudson River has seen an increase in the number and importance of alien species and a change in dominant species.

Introduction

Fish assemblages in large rivers are among the most altered of any in northern temperate regions (Tallman, this volume). The extent of change is difficult to assess in any given river because most such rivers were extensively modified before synoptic studies of the fish were undertaken. Norris and Hawkins (2000) argue that the best way to assess change when descriptions of the original assemblage or system are absent is to examine trends in the status and condition of individual species. Others (e.g., Karr et al. 1986; Daniels et al. 2002) suggest that the entire species assemblage best reflects system-wide change. All recognize the need to use several types of data, including historical distribution of fishes, in order to establish reference conditions with which to compare change (e.g., Hughes 1995). However, in most rivers the historical record is too short to provide information adequate for distinguishing reference conditions. Northeastern rivers have long records, but alterations were extensive before these records began, again limiting their value in establishing reference conditions. For the Hudson River, a series of reports detailing fish distribution extends to the early years of the 19th century and offers an unusually, if not uniquely, long data series on fish assemblages in the river. Although a 200-year record of fish distribution is useful in assessing change in the fish assemblage, it is important to recognize that this river was already affected by environmental change as a direct result of European settlement before fish distribution within the river was assessed.

Modifications within the tidal Hudson River are extensive and include impacts that range from local to regional to global. Limburg et al. (1986) and Barnthouse et al. (1988) note examples of modifications related to pollution, urbanization and shoreline development, water removal, and power plant operations. Agriculture has been an important activity within the basin for more than two centuries, and 14% of the land in the basin currently supports agricultural activities (Wall and Phillips 1998). The banks of the tidal Hudson River also are urbanized. Runoff from both agricultural and urban areas brings pesticides and other pollutants into the river, although the type and amount of contaminants entering the system vary seasonally and regionally (Wall and Phillips 1998; Phillips et al. 2002). Urbanization has increased the amount of sewage entering the river as well. As late as 1970, the "Albany Pool," an area from the Troy Dam to about 45 km downstream, was devoid of dissolved oxygen and had extremely high coliform densities (Boyle 1979). The condition described by Boyle (1979) was long in the making. Faigenbaum (1935, 1937) noted the extent and type of pollutants entering the river in the 1930s; his numbers are only slightly different from those reported by Boyle (1979). In addition to raw sewage entering the river, Faigenbaum (1935, 1937) reported wholesale dumping of wastes from paper production, tanneries, textile production, and canneries. Sewage problems were severe until the implementation of the Clean Water Act revisions of 1972, but still exist in the New York Harbor vicinity (Stanne et al. 1996). Toxic substances were also discharged for decades in the Hudson (Limburg et al. 1986), and the large load of polychlorinated biphenyls (PCBs) discharged from several General Electric factories above the Troy Dam continues to pollute downriver food webs (Baker et al. 2001). The presence of PCBs in fish tissue has led to restrictions on, and closures of, commercial and sport fisheries.

The channel has been modified over the last two centuries by episodic dredging for navigation (beginning in 1834), alteration of shoreline wetlands related, in part, to construction of railroads (Squires 1992), and the construction of dams in the upper drainage and on tributaries in the lower drainage. Schmidt and Cooper (1996) noted that dams exist on 30 of the 62 larger tributaries to the tidal Hudson River. These dams not only store significant amounts of floodwater, but also alter flow regimes by reducing peak discharges downstream and mete out flow over time to maintain minimum depths for navigation (e.g., Lumia 1998). Dams in the Hudson River drainage, as elsewhere, restrict fish movement between the main channel and tributaries and among tributaries, effectively fragmenting the watershed (Schmidt and Cooper 1996). The most extreme land-use alterations came in the form of deforestation, primarily for agriculture, timbering, and tanning in the mid-1800s to late 1800s (McMartin 1992) and were noted as serious problems for fisheries by the end of the 19th century (Stevenson 1899). Although agricultural activity is considerably less widespread in the watershed today, problem areas still exist (Phillips et al. 2002). These and other types of environmental change have affected the composition of the Hudson River fish assemblage and the abundance of species in them.

Commercial fishing also has affected fish abundance and assemblage structure. Striped bass *Morone saxatilis*, Atlantic tomcod *Microgadus tomcod*, American shad *Alosa sapidissima*, and sturgeon *Acipenser* spp. currently or formerly supported commercial harvesting. The history of the Atlantic sturgeon *A. oxyrhyinchus* fishery is typical of commercial operations in the Hudson River. In the 19th century, sturgeon, called Albany beef (Lossing 1876), was shipped across the country. This harvesting depleted stocks drastically; the population had a long and slow recovery, and there was a short period of harvesting in the 1980s and 1990s, followed by a second decline and fishing closure in 1996 (Bain et al. 2000).

The role of alien species in the development and composition of the current assemblage is also important. Mills et al. (1997) discuss changes related to the introduction of exotic aquatic species and list 65 animals and 55 plants. The effect of exotic species on native fishes varies. For example, the invasion of the Hudson River by zebra mussel *Dreissena polymorpha* may have had a pervasive effect on the native fish assemblage. Zebra mussels first appeared in the Hudson in 1991 and, since September 1992, have constituted more than half of heterotrophic biomass in the freshwater tidal section of the river (Strayer et al. 1996). Consequently, in the freshwater tidal Hudson, river kilometer (rkm) 100–247 (rkm 0 is at the southern tip of Manhattan Island), biomass of phytoplankton and small zooplankton declined 80–90%, whereas biomass of planktonic bacteria, water clarity, and concentrations of dissolved nutrients rose substantially (Caraco et al. 1997; Findlay et al. 1998; Pace et al. 1998). Most directly relevant to fish, overall biomass of forage invertebrates (zooplankton plus macrobenthos, excluding large bivalves) fell by 50% (Pace et al. 1998; Strayer and Smith 2001). This change in the forage base was distributed unevenly through the habitats of the Hudson: biomass of zooplankton and deepwater macrobenthos fell sharply, whereas biomass of macrobenthos in the vegetated shallows actually rose. Strayer et al. (2004) hypothesized that the zebra mussel invasion should have caused growth and population size of open-water fishes to fall, and the abundance of these species to increase downriver into brackish-water reaches where zebra mussels are scarce or absent. Littoral fishes were hypothesized to show opposite trends in growth, abundance, and distribution.

Here, we examine change in the tidal Hudson River fish assemblage by comparing species composition and richness of the assemblage over time and by examining trends in the abundance and distribution of selected species. We attempt to relate these changes to human-related changes occurring within the river channel, its tributaries, and surrounding landscapes.

Methods

Study Area

The Hudson River arises from Lake Tear of the Clouds in the Adirondack Mountains of upstate New York, on the flanks of Mount Marcy, the highest point in the state. From there it flows southward some 507 km and drains into the Atlantic Ocean at Manhattan. The drainage area is 34,680 km^2 (Limburg et al. 1986). A dam at Troy, New York (rkm 247), effectively bisects the river into nontidal upriver and tidal downriver sections. Additionally, the Mohawk River, a major tributary, enters the Hudson River just above Troy (Figure 1). The Hudson River and its Mohawk River tributary are closely linked to an extant, and historically more extensive, canal network that tied the Hudson River system to all other drainages in the state (Daniels 2001).

Below Federal Dam at Troy, the Hudson is a drowned river valley that becomes an estuary, with only a 1.5-m drop in elevation from Troy to the estuary mouth (Helsinger and Friedman 1982). The upper tidal reach is fresh, averages 8 m in depth, and has a bottom substrate composed of mud and sand. Below Poughkeepsie, the tidal river widens into Newburgh Bay (rkm 93–103), the first of three embayments in the southern, downriver reach. Just below Newburgh Bay, the river becomes a fjord, deepening and passing over a series of sills. This is the reach of the dramatic Hudson River Highlands, where the Continental Army stretched a great chain across the river from Garrison to West Point beginning in 1778 to prevent upriver movement of British ships. South of the fjord, the Hudson opens into Haverstraw Bay (rkm 50–64) and the Tappan Zee (rkm 39–50). These are both important nursery areas for many fish species. The river deepens and narrows again as it passes the Palisades, a basalt formation of columnar cliffs rising above the river's western shore, and remains moderately deep to where it empties into New York Harbor. Nevertheless, sediment transport necessitates dredging both the harbor and many upriver areas to maintain a 9-m to 11-m shipping channel (McFadden et al. 1978).

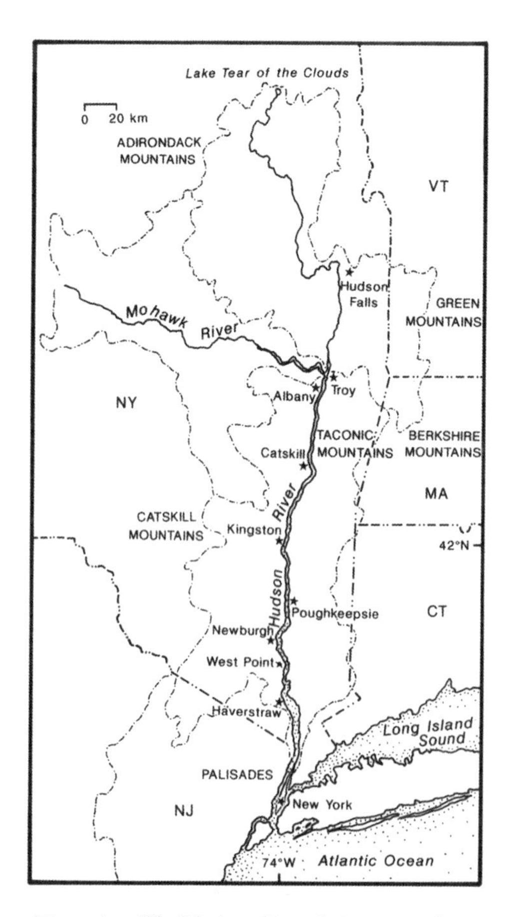

Figure 1.—The Hudson River drains most of eastern New York and parts of Vermont, Massachusetts, and Connecticut. It flows southerly into New York Bay from headwaters in the Adirondack Mountains.

Freshwater discharge averages 390 m^3/s at Green Island, which is near the Troy dam (U.S. Geological Survey records, 1948–1994), but varies seasonally, with highest flows in spring freshets driven by snowmelt within the watershed and lowest flows in July and August (Figure 2). Freshwater flows in the lower Hudson River are 538–567 m^3/s (Central Hudson Gas and Electric 1977). The flushing time, estimated as the ratio of water volume to mean freshwater discharge, is 126 d (Simpson et al. 1974), which makes the Hudson one of the fastest flushed of the East Coast estuaries.

Salinities in the tidal Hudson vary with season and freshwater discharge. The river above rkm 97 remains fresh except in drought; thus, the city of Poughkeepsie at rkm 121 draws its drinking water from the Hudson, and New York City maintains an

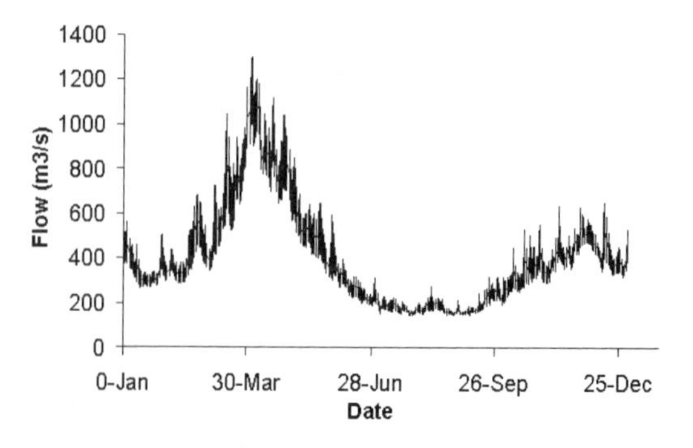

Figure 2.—Freshwater discharge based on daily averages at Green Island, New York, 1948–1994.

emergency pump station at Chelsea (rkm 108). Haverstraw Bay and the Tappan Zee vary from oligohaline (0–5 practical salinity units [psu]) in higher flows to mesohaline (5–18 psu) in summertime. The salt front (defined as 0.5 psu) generally remains below Newburgh, although in times of drought it has been recorded as far north as Kingston (rkm 145; Butch et al. 2001). The Hudson is polyhaline (18–30 psu) in its lower reaches.

Vertical salinity and temperature measurements show that the tidal Hudson is mostly a well-mixed system, due in large part to its semidiurnal tides. This is particularly true in the tidal freshwater zone (Cole et al. 1992; Howarth et al. 1996). When discharges are high, a layer of freshwater can ride over the denser brackish water in the lower Hudson (Busby and Darmer 1970). In summertime low flows, water residence time increases, and the residence time of water (which differs from flushing time by including the influence of tidal influxes of seawater) in the photic zone (where primary production occurs) can be on the order of 7–10 d (Howarth et al. 2000). At that time, some thermal stratification can occur in the relatively shallow bays of Haverstraw and the Tappan Zee.

Tidal flow is 10–100 times greater than flows resulting from upper basin and tributary input (Cooper et al. 1988; Firda et al. 1994) and can exceed 14,000 m^3/s. In general, tidal flow reverses the current direction twice a day, but vertical and horizontal distribution of flow is not uniform due to the morphometry of the channel and effects of the salt front. Shoreline points and bends in the channel produce eddies that affect the flow regime. Tidal amplitude is greatest at Albany (rkm 240), with a mean of 1.6 m, and least at West Point (rkm 83), with a mean of 0.8 m (Giese and Barr 1967).

Data Collection

We used data obtained from many diverse sources. DeKay (1842) and Bean (1903) reviewed the fish fauna of New York. DeKay did not describe his data collection methods, but they appear to have been a mixture of reports, sightings, and collections from local fish markets. For our analysis, we regard the species as present in the Hudson River if DeKay (1842) specifically identified the species as occurring in the Hudson River or if he noted its presence in all waters in the state. Bean (1903) relied on the results of synoptic surveys (e.g., Bean 1899, 1900; Scott 1902), but he also gleaned information from markets, anglers, and commercial fishers. He included a table of fish distribution by drainage and, for our analysis, we accept his tabulation. DeKay (1842) and Bean (1903) only noted the presence of the species in the drainage; species distribution and abundances within the drainage were not detailed.

The first detailed, synoptic survey of fishes undertaken in the tidal Hudson River was completed in 1936 (Greeley 1937). All fish were identified, the abundance of each species was assessed (usually by an actual count, sometimes by a relative abundance estimate), and many specimens were vouchered for later study. Greeley (1937) collected fish via seines, gill nets, and angling. Beginning in 1970, interest in the fishes of the Hudson River increased and both long- and short-term surveys have been conducted by several private and government agencies during the last three decades. We include information from surveys using seines, bottom trawls, and midwater trawls. Because the sampling equipment and effort varied among the studies, we use only presence-absence information when making comparisons among studies. If comparing information within a study, we use all the data.

Some of the data presented here were taken from the Hudson River Estuary Monitoring Program (HRMP), an annual monitoring program sponsored by the utility companies on the Hudson River. Rainbow smelt *Osmerus mordax* and gizzard shad *Dorosoma cepedianum* larval and adult Atlantic tomcod distribution information came from the HRMP's Long River Ichthyoplankton Survey and the catfish young-of-year data from its trawl survey. These two surveys are river-wide and cover early and late seasons, respectively. They have been performed continuously since 1974 with few changes in methodology. The details of the sampling methods and survey design are in Klauda et al. (1988b).

Pop nets (see Serafy et al. 1988) were deployed in dense water chestnut *Trapa natans* beds in Tivoli South Bay (rkm 156) during several studies (Pelczarski and Schmidt 1990; Hankin and Schmidt 1991; Gilchrest and Schmidt 1997) following the protocol outlined in Pelczarski and

Schmidt (1990). Twenty-two samples were taken between 1989 and 2002.

To examine changes in abundance of nongame fishes and to test for zebra mussel effects on fish populations, we used data collected by the electric utilities and the New York State Department of Environmental Conservation (NYS DEC). To test for zebra mussel effects on fish populations, Strayer et al. (2004) divided the fish assemblage into two groups: open-water species such as American shad, blueback herring *Alosa aestivalis*, alewife *A. pseudoharengus*, gizzard shad, white perch *Morone americana*, and striped bass, and littoral species, such as spottail shiner *Notropis hudsonius*, common carp *Cyprinus carpio*, banded killifish *Fundulus diaphanus*, fourspine stickleback *Apeltes quadracus*, redbreast sunfish *Lepomis auritus*, pumpkinseed *L. gibbosus*, bluegill *L. macrochirus*, smallmouth bass *Micropterus dolomieu*, largemouth bass *M. salmoides*, and tessellated darter *Etheostoma olmstedi*. Their analyses were restricted to young-of-year fish. Strayer et al. (2004) should be consulted for further methodological details.

Results

The rich Hudson River fish fauna comprises a mixture of freshwater, diadromous, estuarine, and marine species. To date, 210 species have been reported from the Hudson River drainage (Appendix A). Of these, 129 species are found in the main channel of the tidal portion of the river; the remaining 81 are confined to tributaries of the lower Hudson River or are reported only from the upper Hudson River or Mohawk River systems (Table 1). Of the species present in the tidal portion of the river, 49 are primarily marine visitors and 80 species are either resident freshwater or diadromous forms.

The number of species reported from the Hudson River drainage has increased since first tabulated in 1842 (Table 1). The trend is consistent: species richness has increased over time in the lower drainage, the main channel, and in the freshwater component. There has been a 1.5 to 2-fold increase in the number of species in the drainage since the 1930s (Table 1). The increase is due partially to an increase in the effort spent in collecting and reporting species. It is also due, in part, to the number of exotic species now known to occur in the lower drainage. Although the number of native, freshwater species has increased about 21% in the past 70 years, the number of alien, freshwater species has increased 130%. The presence of two alien species (common carp and rock bass *Ambloplites rupestris*) in the lower Hudson River was noted by DeKay (1842), indicating that they gained access to the system before any assessment of the fauna was undertaken. The number of alien fishes in the system has increased during each sample period, doubling since the 1930s and increasing by seven species in the last decade. Moreover, the extent of the range of alien species and the size of their populations are increasing in the tidal Hudson River.

The effect of alien species on the native assemblage varies across species. The effect of long-established species, such as those noted as alien by DeKay (1842) is impossible to evaluate. Sufficient data exist to assess the status of several of the new arrivals. Gizzard shad has been collected consistently in the Hudson River since 1974, and a single specimen was reported from a pound net on Long Island in 1975 (Hickey and Lester 1976). Reports of gizzard shad in the Connecticut River (O'Leary and Smith 1987), Merrimack River (Hartel et al. 2002), and Kennebec River suggest that this species is expanding its range northward through a combination of marine migrations and accidental stockings. Marine migration may have been the mechanism of introduction of the Hudson River population. A second hypothesis explaining the presence of gizzard shad in the Hudson River is it that it may be derived from inland populations, as suggested by Carl George (Smith 1985).

Gizzard shad is becoming more abundant in the tidal Hudson River (Figure 3). It is now common in large and small tributaries in the spring and summer, although there are no reports of it spawning in these areas. Commercial fishers have reported that catches of gizzard shad in American shad nets have reached nuisance status. Based on the results of the HRMP, gizzard shad are spawning in the river, although the numbers of larvae remain modest and are concentrated in the estuary around Albany. It is possible that spawning occurs in the Mohawk River and larvae moving downriver may enhance the population in the lower river.

Table 1.—Species richness of fish in the Hudson River, 1842–2002. Data are from DeKay (1842), Bean (1903), Greeley (1935, 1937), and Beebe and Savidge (1988). The numbers for the 1990s represent information from researchers actively working with the Hudson River and from museum specimens. Percentage change is computed between values from the 1930s and 1990s.

	1842	1902	1930s	1980s	1990s	% increase
Species in drainage			114		202	77
Species in lower Hudson system			99	140	187	89
Species in main channel	38	44	68		123	81
Freshwater species in channel	31	37	54	71	80	48
Native freshwater species in channel	29	32	41	48	50	21
Alien freshwater species in channel	2	5	13	23	30	130

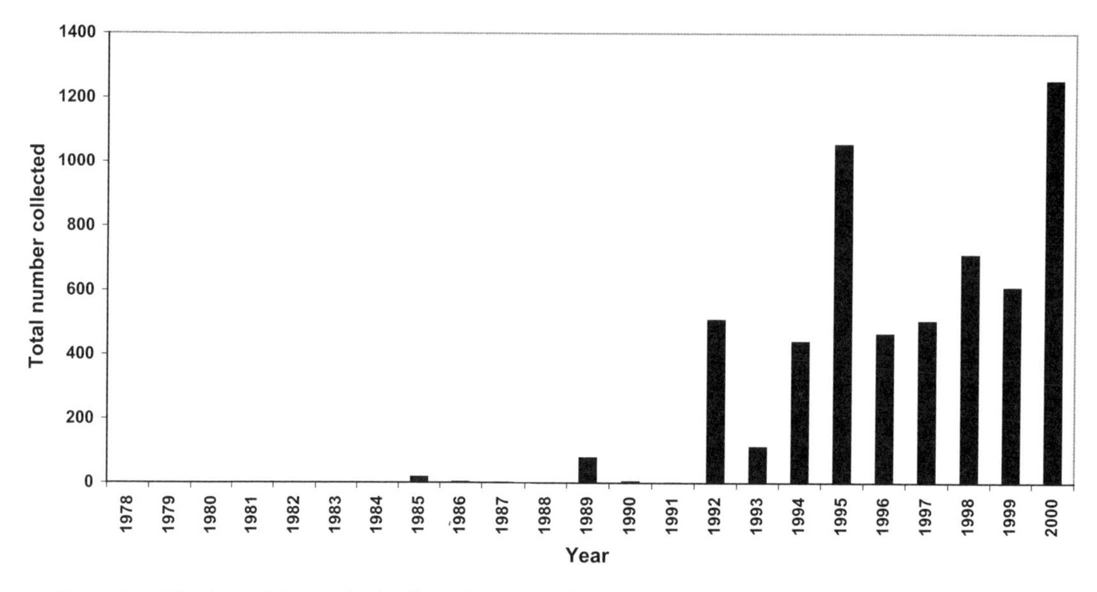

Figure 3.—Number of gizzard shad yolk sac larvae caught in ichthyoplankton tows in the Hudson River, New York, 1974–2000. Data are from the utilities sponsored Long River Ichthyoplankton survey.

Gizzard shad are small-particle feeders and facultative detritivores, and as such, large populations could affect the phytoplankton and microzooplankton populations in the Hudson River Estuary, but we are unaware of any effort to document the effect of the spread of this species.

Channel catfish has become increasingly common in the river during the last decade. A few specimens were taken before 1985, and the earliest Hudson River capture was in 1974 (Smith and Lake 1990). The origin of this species in the Hudson River is not known, but it is available from commercial fish suppliers. Beginning 10 years ago, channel catfish were common in the Sturgeon Pool and contiguous areas in the Wallkill River, a major, mid-basin tributary (R. Pierce, NYS DEC, personal communication); this pattern suggests that channel catfish was illegally released into the drainage before then. Such releases could easily explain the establishment of this species in the Hudson River. A second possible explanation is that channel catfish moved into the system from the Saint Lawrence River drainage, via the Barge Canal.

We (R.E.S., R.A.D.) have collected channel catfish in marshes and tidal tributaries, mostly in the northern part of the estuary. Channel catfish grows to a larger size than the native white catfish and could potentially displace the latter. Trawl surveys show increasing numbers of channel catfish (Jordan et al. 2004), which ultimately may lead to a decline in the native white catfish, a situation that has been observed in the Delaware and Connecticut rivers.

Although not reported in Beebe and Savidge (1988), freshwater drum have been taken from the river since 1978. Reports of this species have become more common in recent years, primarily from anglers in bass tournaments. These anglers have reported 2–4-kg fish taking jigs fished for bass. The abundance of this species may be underreported if it inhabits deep, difficult to sample areas of the estuary. Small numbers of early life stages were collected as early as 1985 in the Long River Ichthyoplankton survey, but based on these tows, spawning in the river appears limited. The invasion of zebra mussel provided an abundant food source for this molluscivorous fish, and the presence of this new forage base may account for increasing numbers of freshwater drum.

Northern pike *Esox lucius*, norlunge (*E. lucius* × muskellunge *E. masquinongy*), and walleye *Sander vitreus* have been stocked in the Hudson River or its tributaries to establish a sport fishery. These large, piscivorous fish have become increasingly abundant in the estuary. The esocids have been taken at several upriver localities, and R.E.S. caught a spawning pair of northern pike at the head of tide in Coxsackie Creek in April 1999. Northern pike young of year have been reported from the estuary in recent years (K. Hattala, NYS DEC, personal communication.). We have a single report of walleye young of year in the tidal Hudson River (W. Gilchrest, Norrie Point Environmental Laboratory, personal communication), but we have no evidence that walleye spawn in the tidal Hudson.

Fathead minnow *Pimephales promelas* is becoming increasingly common in the watershed. This species is widely sold as bait and as forage for largemouth bass stocked in farm ponds. Although no adults have been reported from the main channel, larvae are common in drift samples taken at the mouths of tributaries. Fathead minnow was absent from 16 tributary mouths in 1988 (Schmidt and Limburg 1989), but was abundant in Stockport Creek 5 years later (Schmidt and Stillman 1994) and in Moordener Kill and Coxsackie Creek (Schmidt and Lake 2000).

Alien fishes affect the native fish assemblage directly by interspecific interactions, such as competition and predation. Other alien components of the community also affect the fish assemblage. Water chestnut was first reported in the Hudson River drainage in 1884 (Mills et al. 1997). The large beds that currently dominate river shallows from the Troy Dam to Iona Island (rkm 71) have altered the habitat available to fishes by increasing the amount of cover and spatial complexity throughout the littoral zone and have affected the dynamics of dissolved oxygen and nutrients (Caraco and Cole 2002).

The arrival of the zebra mussel is the best illustration of the effect of an alien species on the Hudson River fish assemblage (Strayer et al. 2004). The zebra mussel invasion was associated with large, pervasive changes in young-of-year fish in the Hudson (Figure 4). Abundance of many littoral species rose, with populations of several species more than doubling. In contrast, populations of openwater species showed no pervasive changes, although numbers of postyolk sac larvae of *Alosa* spp. declined sharply (Strayer et al. 2004). The distribution of fish within the Hudson also shifted following the zebra mussel invasion. As hypothesized, populations of open-water species generally shifted downriver at the same time that populations of littoral species shifted upriver. Many of these shifts were large (more than twofold). Finally, apparent growth rates of almost all open-water fish species fell after the zebra mussel invasion. Apparent growth rates rose for

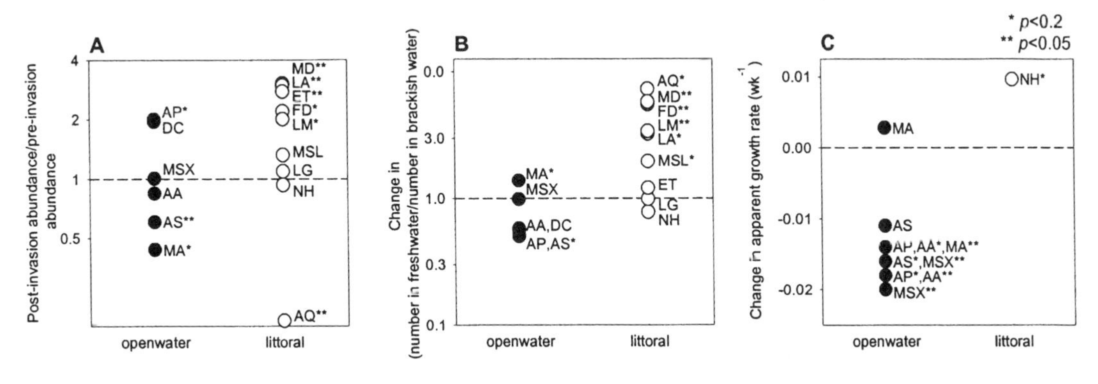

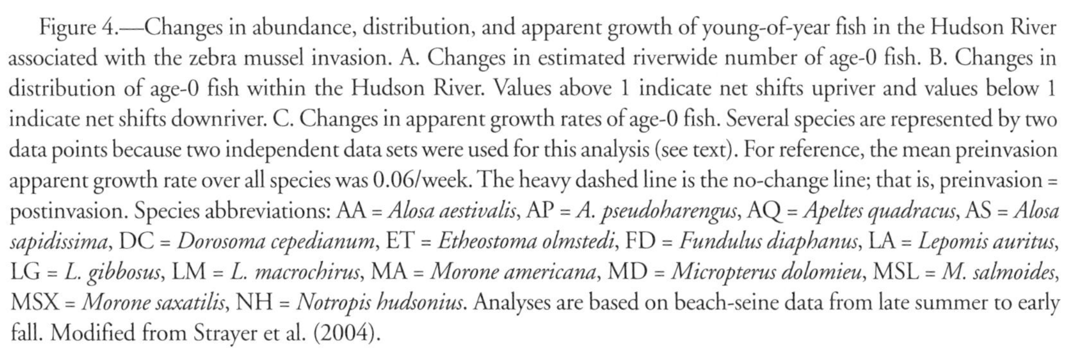

Figure 4.—Changes in abundance, distribution, and apparent growth of young-of-year fish in the Hudson River associated with the zebra mussel invasion. A. Changes in estimated riverwide number of age-0 fish. B. Changes in distribution of age-0 fish within the Hudson River. Values above 1 indicate net shifts upriver and values below 1 indicate net shifts downriver. C. Changes in apparent growth rates of age-0 fish. Several species are represented by two data points because two independent data sets were used for this analysis (see text). For reference, the mean preinvasion apparent growth rate over all species was 0.06/week. The heavy dashed line is the no-change line; that is, preinvasion = postinvasion. Species abbreviations: AA = *Alosa aestivalis*, AP = *A. pseudoharengus*, AQ = *Apeltes quadracus*, AS = *Alosa sapidissima*, DC = *Dorosoma cepedianum*, ET = *Etheostoma olmstedi*, FD = *Fundulus diaphanus*, LA = *Lepomis auritus*, LG = *L. gibbosus*, LM = *L. macrochirus*, MA = *Morone americana*, MD = *Micropterus dolomieu*, MSL = *M. salmoides*, MSX = *Morone saxatilis*, NH = *Notropis hudsonius*. Analyses are based on beach-seine data from late summer to early fall. Modified from Strayer et al. (2004).

spottail shiner, the only littoral species for which sufficient data were available. Changes in apparent growth rates were large (>25%) compared to preinvasion growth rates in many cases.

Related to the observations of Strayer et al. (2004), ongoing studies of the trophic effects of zebra mussels on juvenile blueback herring show a dramatic change in the role of the latter in the Hudson food web following mussel invasion. In the 1980s the diet of juvenile blueback herring consisted almost entirely of pelagic zooplankton (Limburg and Strayer 1987; Grabe 1996). Today, their diet contains nearly no *Bosmina leydmani* (formerly a primary component) and littoral/benthic macroinvertebrates dominate the diet (K.E.L., unpublished data). The composition of dietary items mirrors the findings of Strayer et al. (2004), suggesting that blueback herring are compensating somewhat for the loss of pelagic prey by foraging in macrophyte-associated habitats. This is suggestive of a possible regime change in production pathways in the Hudson, mediated by the strength of zebra mussel filtration of the water column.

Changes in the abundance and macrodistribution of native fishes have also been reported. Rainbow smelt is anadromous in the Hudson River and there are historical records of spawning runs in many tributaries (Smith 1985). Rose (1993), based on ichthyoplankton surveys, found no evidence of a declining population between 1974 and 1991, although Daniels (1995) showed a decline in adult catches between 1974 and 1989.

The last tributary run of rainbow smelt that we (R.E.S., K.E.L.) observed was in Rondout Creek in 1988. Rose (1993) suggested that rainbow smelt enter tributaries during high adult-population years, but spawn every year in the main channel of the river. A crude analysis of the ichthyoplankton data, extending Rose's (1993) assessment to the end of the century, shows a very different picture (Figure 5). After 1995, smelt essentially disappeared from the Hudson River ichthyoplankton. We show data for the post-

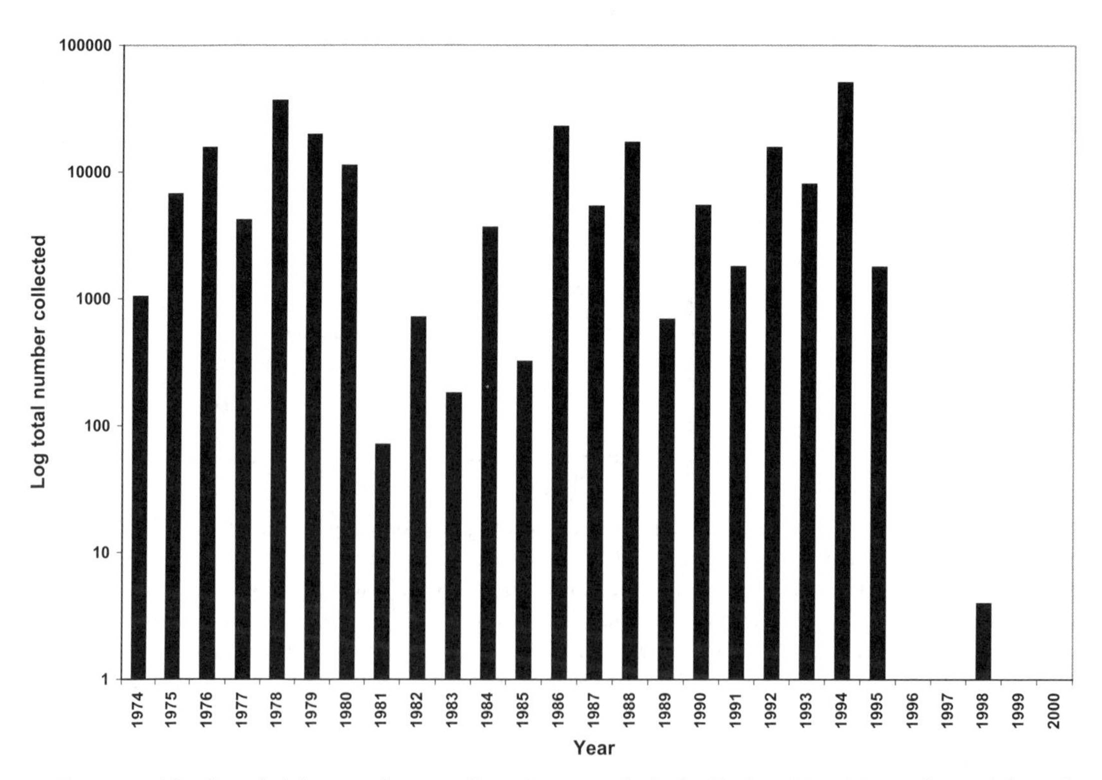

Figure 5.—Number of rainbow smelt post-yolk-sac larvae caught in the Hudson River, New York, 1974 through 2000. Data are from the utilities-sponsored Long River Ichthyoplankton survey.

yolk sac life history stage, but the other stages (egg, yolk sac, young of year, older) show the same pattern. Although there has been variation in the number of samples taken in this survey (1,561–3,684 tows/year), there were 2,329–2,437 tows/year in the last 5 years of the data set, so this disappearance is not an artifact of a change in effort.

The Hudson River is home to the only anadromous member of the family Gadidae on the North American Atlantic Coast. A population of Atlantic tomcod is largely contained in the lower tidal portions of the river, surrounding bays of the lower estuary, and in the outer bay and coastal habitats. Historically, tomcod was reported as far south as Virginia (Bigelow and Schroeder 1953), but there are no recent reports of spawning in any drainage south of the Hudson River (Stewart and Auster 1987). The fact that Hudson River tomcod are at the southernmost boundary of the species' spawning distribution may foretell future reductions in its population with warming climate.

Collections of Hudson River tomcod from 1974 to 2000 made by the HRMP, which entail estimates of abundance for all life stages, suggest cyclical change in tomcod abundance (Figure 6). It is noteworthy that the HRMP data for the most recent years (1997–2000) show tomcod abundance, quantified here as an index for the feeding age 0 stages, to be on a protracted decline and with the lowest values in the 27-year time series occurring in 2000. More recent collections by us (R.C.C.) of adult tomcod from spawning areas near Garrison, New York (rkm 82) in the winters of 2000–2001 and 2001–2002, and of juveniles in the mid to lower reaches of the tidal Hudson River during the summers of 2000–2002, reveal a continuation of extraordinarily low numbers.

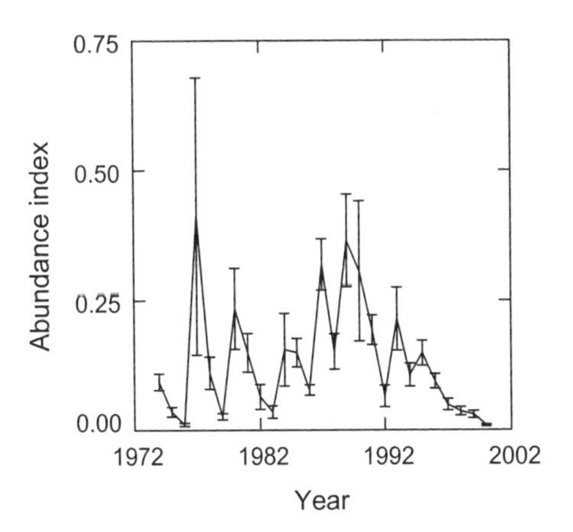

Figure 6.—Abundance index (mean ± SE) for post-yolk-sac larval and young-of-year juvenile Atlantic tomcod in the Hudson River from 1974 through 2000. The abundance index is derived by collapsing the weekly standing crop indices for the regions of the estuary that were sampled (Battery to Albany) over the period from approximately May through early July for each year. Source of data is the utilities-sponsored Long River survey.

Fourspine stickleback populations appear to be declining in the tidal Hudson River (Figure 7). Historically, this estuarine species was found upriver into the nontidal stretches upstream of the Troy Dam, where it was described as locally common (Greeley 1935). More recent collections were confined to the tidal portion downstream of Catskill Creek (rkm 177; Smith 1985). Pop-net samples in water chestnut beds in Tivoli South Bay showed this species to be one of two dominant fishes (Pelczarski and Schmidt 1990; Hankin and Schmidt 1991; Gilchrest and Schmidt 1997). Water chestnut provides cover, food items of appropriate size, and structure for nest building. This plant became abundant beginning in the late 1970s when chemical control efforts ceased (Hankin and Schmidt 1991). The pop-net surveys show a decline in the number of fourspine stickleback from 75% of the catch in 1989 to 1% of the catch in 2002 (Figure 7). In addition, Strayer et al. (2004) documented a 99% decline in the abundance of fourspine stickleback between 1974 and 1999 in the utility-sponsored beach seine survey.

The abundance pattern displayed by goldfish *Carassius auratus* is one of decline and re-establishment. Goldfish flourished in the Hudson River and supported a commercial fishery until 1979–1980 when an epidemic of furunculosis led to a catastrophic decline in numbers (Smith 1985). Recent pop-net surveys indicate that numbers are rising again (Figure 7).

Abundance of other species seems to be little changed, but other aspects of fish life history appear to have been affected by environmental changes of the last half-century. Spottail shiner has

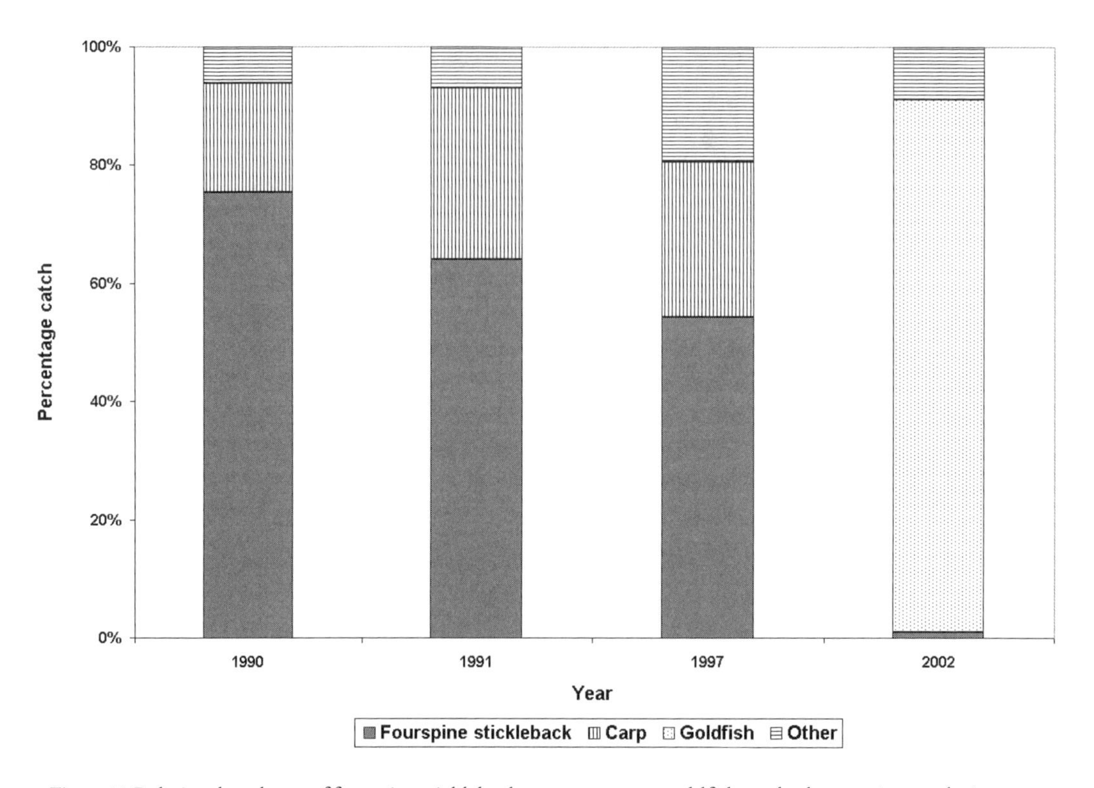

Figure 7. Relative abundance of fourspine stickleback, common carp, goldfish, and other species caught in pop nets from Tivoli South Bay, Hudson River, New York, in 1990, 1991, 1997, and 2002.

always been one of the most abundant species in the tidal Hudson River and we have no evidence that its numbers have changed. Instead, age structure of the population appears to have changed. In the late 1980s, R.E.S. and Tom Lake (NYS DEC, personal communication) independently began looking for rainbow smelt in tributaries using small-mesh gill nets (1.2-cm bar). Efforts to catch rainbow smelt in Hudson River tributaries were unsuccessful, but runs of large spottail shiners adults (total length [TL] > 110 mm) were common in the spring. Lake and Schmidt (1997) documented a substantial spawning run of these large spottail shiner in Quassaic Creek (rkm 97) and conservatively estimated that 2,800 individuals were observed. No spawning run was observed in a similar study conducted the following year in the same creek (Lake and Schmidt 1998). We have seen no further runs of large spottail shiner even though we continue to catch larvae in the drift (Schmidt and Lake 2000). A change in age structure in spottail shiner may negatively affect the status of this species in the Hudson River.

Discussion

The Hudson River fish assemblage is rich, diverse, and dynamic. Although the marine species that enter the river are an interesting component of the assemblage, many of these are listed based on a single collection. In contrast, the resident freshwater and estuarine species and the diadromous species are the key components to the Hudson River assemblage. Changes in the abundance, distribution, or life history of these species have the greatest effect on the overall assemblage. Nothing is known about the character of the pre-1800 fish assemblage. Mitchill's (1815) treatment of New York fish unfortunately fails to include consistent information on distribution within the state. DeKay (1842) is the first to include zoogeographic information.

It is difficult to accurately assess the change in the number of indigenous freshwater or diadromous species in the Hudson River since the middle years of the 19th century, although there appears to be nearly a twofold increase (Table 1). However, the 29 native species identified by DeKay (1842) is a conservative number; it is based on species identified as present in the Hudson River by DeKay (1842) who did not always provide information on the distribution of species. Undoubtedly, some species present in the Hudson River were not included in DeKay's count, simply because he did not mention the Hudson River in his commentary. Changes in nomenclature and taxonomy also affect DeKay's (1842) count; many of the minnow species not listed by DeKay (1842) had yet to be described. Bean (1903) provided more information on distribution and the taxonomy of fishes was better developed. Still, his count was less than all later 20th century counts. There also are some curious omissions from the lists published by DeKay (1842) and Bean (1903) (Table 1). DeKay (1842) did not include blueback herring, bay anchovy *Anchoa michelli*, white perch, and redbreast sunfish, and he included only five species of native minnow. Bean (1903) failed to list blueback herring, American shad, bay anchovy, rainbow smelt, and tessellated darter. These species were included in all later 20th century lists; several of these species are among the most frequently encountered species in the river in the 20th century and were obviously present in the 19th century; for example, there was an active commercial fishery for American shad in the Hudson River before 1900. Despite the differences in actual numbers, the native freshwater and diadromous species in the Hudson River make up a core assemblage that arguably has varied little since the earliest records, dating back almost 200 years.

What has changed is the number of freshwater alien species now established in the tidal Hudson River (Table 1). DeKay (1842) noted only two aliens: common carp and rock bass. Since then, there has been a 15-fold increase in the number of established alien species in the river and a doubling in the last six decades alone. Many of the new arrivals are large, predatory species, like channel catfish, northern pike, smallmouth and largemouth bass, and walleye. The establishment of several of these predatory species coincides with the decline of the small, tributary species (e.g., bridle shiner *Notropis bifrenatus*, common shiner *Luxilus cornutus*, satinfin shiner *Cyprinella analostana*) that once inhabited the nearshore areas of the main channel (Daniels 1995). Based on presence-absence information, the assemblage of the late 20th century differs from that of the early 19th century. The size and composition of the alien component of the assemblage has increased, but it is not the only change affecting the assemblage.

Large, widespread changes in fish populations were associated with the zebra mussel invasion in the Hudson. These changes were consistent with what was expected from observed changes in the Hudson's forage base, with large losses of open-water forage (zooplankton and deepwater macrobenthos) and simultaneous increases in littoral forage (Pace et al. 1998; Strayer and Smith 2001). Consequently, open-water fish (especially *Alosa* spp.) declined in abundance, apparently grew more slowly, and shifted in distribution downriver into brackish sections of the Hudson where zebra mussels are scarce or absent (Figure 4; Strayer et al. 2004). At the same time, littoral fish (especially the centrarchids) became more abundant, apparently grew more quickly, and shifted in distribution upriver into the sections of the Hudson most affected by zebra mussels. Many of these changes were large (>50%).

Abiotic factors also have affected the assemblage, although it is often difficult to identify the extent of these impacts based on available information. Habitat modification in the tidal Hudson River has been extensive: maintenance of a 10-m deep ship channel by dredging, filling in of shorelines by dredge spoils and as a result of railroad construction, creation of wetlands by shoreline railroad construction, and the stabilization of flows by dam construction upstream have all affected the abundance and distribution of fish in the river (Jackson et al. in press). Most of these changes occurred before synoptic surveys of fish were undertaken; however, the changes have been so drastic that changes in the relative importance of species is likely. The small forms that are still common in the tributaries, but are now rare in the main channel are likely to be particularly affected by these changes. Species like bridle shiner, cutlip minnow *Exoglossum maxillingua*, and blacknose dace *Rhinichthys atratulus* may not find suitable habitat in the main channel to sustain prolonged residence or through which to migrate to other tributaries. There is a negative correlation between the number of alewife larvae exiting Hudson River tributaries and the degree of watershed urbanization (Limburg and Schmidt 1990). Overfishing of stocks has led to the decline of once abundant commercially important species (e.g., Bain et al. 2000; Limburg et al. in press).

The Hudson River population of rainbow smelt is at the southern extreme of the reproductive range (Lee et al. 1980), although historically it occurred farther south (Smith 1985). The abrupt decline in rainbow smelt early life stages in the ichthyoplankton may result from global warming. Ashizawa and Cole (1994) documented the trend of slowly increasing water temperature in the Hudson River. The rainbow smelt runs in the coastal streams of western Connecticut have drastically declined or disappeared simultaneously with the decline in the Hudson River population (S. Gephard, Connecticut Department of Environmental Protection, personal communication).

Perhaps the most important explanation of the recent decline of Atlantic tomcod in the Hudson River is the thermal environment it has experienced. Atlantic tomcod is at the southern extreme of its geographic range in the Hudson River (Lee et al. 1980). Grabe (1978), Klauda et al. (1988a), McClaren et al. (1988), and the data collected by the HRMP suggest that growth of tomcod is reduced as temperatures warm in the summer. Experimental work has confirmed a temperature-dependent growth and condition response of juveniles as temperatures surpass 22°C (R.C.C., D. A. Witting, unpublished data). Atlantic tomcod experiences such temperatures regularly in July and August in the Hudson River. The degree and extent of critical temperatures have been more limiting during the last several summers, which have been exceptionally warm and dry. The decrease in frequency of reports, during the last century, of tomcod from bays to the south of the Hudson River is consistent with a retreating southern range boundary, although the contribution of habitat alterations to the decline of tomcod in these systems cannot be dismissed.

Alternatively, rather than one factor directly affecting smelt or tomcod populations, an interaction of factors may be at play in what appears to be a significant reduction in the abundance of both species in the Hudson River. For example, contaminants may be important in understanding the reduction of tomcod abundance in the Hudson River. Congener-specific analyses have demonstrated that levels of PCBs, dioxins, and furans in livers and eggs of tomcod from the Hudson River are much higher than in conspecifics from elsewhere (Wirgin et al. 1992; Courtenay et al. 1999; Roy et al. 2001; Yuan et al. 2001). Further work is needed in order to determine the degree to which molecular- and individual-level effects of toxic substances are manifested in population- and community-level responses.

We present data that indicate that the Hudson River fish assemblage is rich and dynamic. Given the number and intensity of environmental change in the drainage, the fish assemblage has demonstrated itself to be remarkably resilient. We argue that the causes affecting change are varied and include an increase in the number of alien fishes in the drainage, the arrival and establishment of alien invertebrates and abiotic factors associated with land-use practices, urbanization, nonpoint source pollution, and global warming.

Of particular interest are two factors that may foster an increase in richness in the Hudson River system. Global warming may trigger increases in the number of marine strays entering the estuary. Some of these species (e.g., Atlantic croaker and black drum *Pogonias cromis*), which now occur in estuaries that are more southern, but were once more common in New York (Bean 1903), may be able to utilize the Hudson River Estuary as a nursery or as temporary feeding grounds. Other species, such as the striped shiner *Luxilus chrysocephalus*, may use the Erie or Champlain canals to enter the Hudson River drainage. Daniels (2001) argued that the modern canals provide suitable migratory routes since they are large, flowing systems that actually include the river channels for most of their courses. Marsden et al. (2000) reviewed the status of brook silverside *Labidesthes sicculus* and suggested that it used the canal system to gain access to the lower Hudson River and Lake Champlain. Documentation of changes in ranges of species resulting from changing temperature and movement through canals should be an important research goal in future studies of river fishes. Continued monitoring of changes in the Hudson River fish assemblage will allow more rigorous testing of many of the observations and relationships examined here.

Acknowledgments

We thank Bill Dey and John Young for providing information from utility-sponsored surveys. We also thank the Hudson River Foundation for support of several projects of each of the authors over the years. The five reviewers provided useful information and comments. Our thanks. KL acknowledges support by NSF project number 0238121.

References

Ashizawa, D., and J. J. Cole. 1994. Long term temperature trends of the Hudson River: a study of the historical data. Estuaries 17:166–171.

Bain, M., N. Haley, D. Peterson, J. R. Waldman, and K. Arend. 2000. Harvest and habitats of Atlantic sturgeon *Acipenser oxyrinchus* Mitchill, 1815, in the Hudson River estuary: lessons for sturgeon conservation. Boletin Instituto Espanol de Oceanografia 16:43–53.

Baker, J. E., W. F. Bohlen, R. Bopp, B. Brownawell, T. K. Collier, K. J. Farley, W. R. Geyer, and R. Nairn. 2001. PCBs in the upper Hudson River: the science behind the dredging controversy. Report to the Hudson River Foundation, New York.

Barnthouse, L. W., R. J. Klauda, D. S. Vaughan, and R. L. Kendall, editors. 1988. Science, law, and Hudson River power plants: a case study in environmental impact assessment. American Fisheries Society, Monograph 4, Bethesda, Maryland.

Bean, T. H. 1899. Fishes of the south shore of Long Island. Science 9(211):52–55.

Bean, T. H. 1900. Report on the fishes of Long Island collected in the summer of 1898. New York State Museum Report of the Director 52:r92–r111, Albany.

Bean, T. H. 1903. Catalogue of the fishes of New York. New York State Museum, Bulletin 60, Albany.

Beebe, C. A., and I. R. Savidge. 1988. Historical perspective on fish species composition and distribution in the Hudson River estuary. Pages 25–36 *in* L. W. Barnthouse, R. J. Klauda, D. S. Vaughan, and R. L. Kendall, editors. Science, law, and Hudson River power plants: a case study in environmental impact assessment. American Fisheries Society, Monograph 4, Bethesda, Maryland.

Bigelow, H. B., and W. C. Schroeder. 1953. Fishes of the Gulf of Maine. U.S. Fish and Wildlife Service Fishery Bulletin 53:1–577.

Boyle, R. H. 1979. The Hudson River. A natural and unnatural history. Expanded edition. W. W. Norton and Company, New York.

Busby, M. W., and K. I. Darmer. 1970. A look at the Hudson estuary. Water Resources Bulletin 6:802–812.

Butch, G. K., P. M. Murray, J. A. Robideau, and J. A. Gardiner II. 2001. Water Resources Data, New York, Water Year 2001, Volume 1: eastern New York. U.S. Geological Survey Water Data Report NY-01–1, New York.

Caraco, N. F., and J. J. Cole. 2002. Contrasting impacts of a native and alien macrophyte on dissolved oxygen in a large river. Ecological Applications 12:1469–1509.

Caraco, N. F., J. J. Cole, P. A. Raymond, D. L. Strayer, M. L. Pace, S. E. G. Findlay, and D. T. Fischer. 1997. Zebra mussel invasion in a large, turbid river: phytoplankton response to increased grazing. Ecology 78:588–602.

Carlson D. M., and R. A. Daniels. 2004. Status of fishes in New York: increases, declines and homogenization of watersheds. American Midland Naturalist 152:104–139.

Central Hudson Gas and Electric. 1977. Roseton Generation Station. Near-field effects of once-through cooling system operation on Hudson River biota. Central Hudson Gas and Electric Corporation, Poughkeepsie, New York.

Cole, J. J., N. F. Caraco, and B. L. Peierls. 1992. Can phytoplankton maintain a positive carbon balance in a turbid, freshwater, tidal estuary? Limnology and Oceanography 37:1608–1617.

Cooper, J. C., F. R. Cantelmo, and C. E. Newton. 1988. Overview of the Hudson River estuary. Pages 11–24 *in* L. W. Barnthouse, R. J. Klauda, D. S. Vaughan, and R. L. Kendall, editors. Science, law, and Hudson River power plants: a case study in environmental impact assessment. American Fisheries Society, Monograph 4, Bethesda, Maryland.

Courtenay, S., C. Grunwald, G.-L. Kreamer, W. L. Fairchild, J. T. Arsenault, M. Ikonomou, and I. Wirgin. 1999. A comparison of the dose and time response of CYP1A1 mRNA induction in chemically treated Atlantic tomcod from two populations. Aquatic Toxicology 47:43–69.

Daniels, R. A. 1995. Long-term change in the near-shore fish assemblage of the tidal Hudson River. Pages 260–263 *in* E. T. LaRoe, G. S. Farris, C. E. Puckett, P. D. Doran, and M. J. Mac, editors. Our living resources: a report to the nation on the distribution, abundance, and health of U.S. plants, animals, and ecosystems. U.S. Department of the Interior, National Biological Service, Washington, D.C.

Daniels, R. A. 2001. Untested assumptions: the role of canals in the dispersal of sea lamprey, alewife, and other fishes in the eastern United States. Environmental Biology of Fishes 60:309–329.

Daniels, R. A., K. Riva-Murray, D. B. Halliwell, D. L. Vana-Miller, and M. D. Bilger. 2002. An index of biological integrity for northern mid-Atlantic slope drainages. Transactions of the American Fisheries Society 131:1044–1060.

DeKay, J. E. 1842. Zoology of New York or the New York fauna, Part IV. Fishes. W. and A. White and J. Visscher. Albany, New York.

Faigenbaum, H. M. 1935. Chemical investigation of the Mohawk-Hudson watershed. Pages 160–213 *in* E. Moore, editor. A biological survey of the Mohawk-Hudson watershed. New York State Conservation Department, Supplement to the twenty-fourth annual report, Albany.

Faigenbaum, H. M. 1937. Chemical investigation of the lower Hudson area. Pages 146–216 *in* E. Moore, editor. A biological survey of the lower Hudson watershed. New York State Conservation Department, Supplement to the twenty-sixth annual report, Albany.

Findlay, S., M. L. Pace, and D. T. Fischer. 1998. Response of heterotrophic planktonic bacteria to the zebra mussel invasion of the tidal freshwater Hudson River. Microbial Ecology 36:131–140.

Firda, G. D., P. M. Murry, and W. O. Freeman. 1994. Water Resources Data, New York, Water Year 1994. Volume 1: eastern New York excluding Long Island. U. S. Geological Survey Water-Data Report. NY-94–1, New York.

Giese, G. L., and J. W. Barr. 1967. The Hudson River estuary, a preliminary investigation of flow and water characteristics. New York State Water Commission, Bulletin 61, Albany.

Gilchrest, W. R., and R. E. Schmidt. 1997. Comparison of fish communities in open and occluded freshwater tidal wetlands in the Hudson River estuary. Section IX *in* J. R. Waldman and W. C. Nieder, editors. Final reports of the Tibor T. Polgar Fellowship Program 1997. Hudson River Foundation, New York.

Grabe, S. A. 1978. Food and feeding habits of juvenile Atlantic tomcod, *Microgadus tomcod*, from Haverstraw Bay, Hudson River, New York. U.S. National Marine Fisheries Service Fishery Bulletin 76:89–94.

Grabe, S. A. 1996. Feeding chronology and habits of *Alosa* spp. (Clupeidae) juveniles from the lower Hudson River estuary, New York. Environmental Biology of Fishes 47:321–326.

Greeley, J. R. 1935. Fishes of the area with an annotated list. Pages 63–101 *in* E. Moore, editor. A biological survey of the Mohawk-Hudson watershed. New York State Conservation Department, Supplement to the twenty-fourth annual report, Albany.

Greeley, J. R. 1937. Fishes of the area with an annotated list. Pages 45–103 *in* E. Moore, editor. A biological survey of the lower Hudson watershed. New York State Conservation Department, Supplement to the twenty-sixth annual report, Albany.

Hankin, N., and R. E. Schmidt. 1991. Standing crop of fishes in water celery beds in the tidal Hudson River. Section VIII *in* J. R. Waldman and E. A. Blair, editors. Final reports of the Tibor T. Polgar Fellowship Program 1991. Hudson River Foundation and Hudson River National Estuarine Research Reserve, New York.

Hartel, K. E., D. B. Halliwell, and A. E. Launer. 2002. Inland fishes of Massachusetts. Massachusetts Audubon Society, Lincoln.

Helsinger, M. H., and G. M. Friedman. 1982. Distribution and incorporation of trace elements in the bottom sediments of the Hudson River and its tributaries. Northeastern Environmental Science 1:33–47.

Hickey, C. R., Jr., and T. E. Lester. 1976. First record of the gizzard shad from Long Island, New York. New York Fish and Game Journal 23:188–189.

Howarth, R. W., R. Schneider, and D. Swaney. 1996. Metabolism and organic carbon fluxes in the tidal, freshwater Hudson River. Estuaries 19:848–865.

Howarth, R. W., D. Swaney, T. J. Butler, and R. Marino. 2000. Climatic control on eutrophication of the Hudson River estuary. Ecosystems 3:210–215.

Hughes, R. M. 1995. Defining acceptable biological status by comparing with reference conditions. Pages 31–47 *in* W. S. Davis and T. P. Simon, editors. Biological assessment and criteria: tools for water resource planning and decision making. CRC Press, Boca Raton, Florida.

Jackson, J. K., A. D. Huryn, D. L. Strayer, D. Courtemanch, and B. W. Sweeney. In press. Atlantic rivers—Northeastern states. In A. C. Benke and C. E. Cushing, editors. Rivers of North America. Academic Press, San Diego, California.

Jordan, S. M., R. M. Newmann, and E. T. Schultz. 2004. Distribution, habitat use, growth, and condition of a native and introduced catfish species in the Hudson River estuary. Journal of Freshwater Ecology. 19:59–67.

Karr, J. R., K. D. Fausch, P. L. Angermeier, P. R. Yant, and I. J. Schlosser. 1986. Assessing biological integrity in running waters: a method and its rationale. Illinois Natural History Survey, Special Publication 5, Champaign.

Klauda, R. J., R. E. Moos, and R. E. Schmidt. 1988a. Distribution and movements of Atlantic tomcod in the Hudson River estuary and adjacent waters. Pages 219–251 *in* C. L. Smith, editor. Fisheries research in the Hudson River. State University of New York Press, Albany.

Klauda, R. J., P. H. Muessig, and J. A. Matousek. 1988b. Fisheries data sets compiled by utility-sponsored research in the Hudson River estuary. Pages 7–85 *in* C. L. Smith, editor. Fisheries research in the Hudson River. State University of New York Press, Albany.

Lake, T. R., and R. E. Schmidt. 1997. Seasonal presence and movement of fish populations in the tidal reach of Quassaic Creek, a Hudson River tributary (HRM 60): documentation of potamodromy, anadromy, and residential components. Section VII *in* W. C. Nieder and J. R. Waldman, editors. Final reports of the Tibor T. Polgar Fellowship Program 1996. Hudson River Foundation, New York.

Lake, T. R., and R. E. Schmidt. 1998. The relation between fecundity of an alewife (*Alosa pseudoharengus*) spawning population and egg productivity in Quassaic Creek, a Hudson River tributary (HRM 60) in Orange County, New York. Section II *in* W. C. Nieder and J. R. Waldman, editors. Final reports of the Tibor T. Polgar Fellowship Program 1997. Hudson River Foundation, New York.

Lee, D. S., C. R. Gilbert, C. H. Hocutt, R. E. Jenkins, D. E. McAllister and J. R. Stauffer, Jr. 1980. Atlas of North American freshwater fishes. North Carolina State Museum of Natural History, Raleigh.

Limburg, K. E., M. A. Moran, and W. H. McDowell. 1986. The Hudson River ecosystem. Springer-Verlag, New York.

Limburg, K. E., K. A. Hattala, A. W. Kahnle, and J. R. Waldman. In press. Fisheries of the Hudson River. In J. Levinton and J. R. Waldman, editors. The Hudson River ecosystem.

Limburg, K. E., and R. E. Schmidt. 1990. Patterns of fish spawning in Hudson River tributaries: response to an urban gradient? Ecology 71:1238–1245.

Limburg, K. E., and D. L. Strayer. 1987. Studies of young-of-the-year river herring and American shad in the Tivoli bays, Hudson River, New York. Chapter VII *in* E. A. Blair and J. R Waldman, editors. Final reports to Polgar Fellowship Program 1987. Hudson River Foundation, New York.

Lossing, B. J. 1876. The Hudson from the wilderness to the sea. Virtue and Yorston, New York.

Lumia, R. 1998. Flood of January 19–20, 1996 in New York State. U.S. Geological Survey, Water-Resources Investigations Report 97–4252. Albany, New York.

Marsden, J. E., R. W. Langdon, and S. P. Good. 2000. First occurrence of the brook silverside (*Labidesthes sicculus*) in Lake Champlain, Vermont. Northeastern Naturalist 7:248–254.

McClaren, J. B., T. H. Peck, W. P. Dey, and M. Gardinier. 1988. Biology of Atlantic tomcod in the Hudson River estuary. Pages 102–112 *in* L. W. Barnthouse, R. J. Klauda, D. S. Vaughan, and R. L. Kendall, editors. Science, law, and Hudson River power plants: a case study in environmental impact assessment. American Fisheries Society, Monograph 4, Bethesda, Maryland.

McFadden, J. T., Texas Instruments Inc., and Lawler, Matusky and Skelly, Engineers. 1978. Influence of the proposed Cornwall pumped storage project and steam electric generating plants on the Hudson River estuary, with emphasis on striped bass and other fish populations. Revised. Consolidated Edison of New York, New York.

McMartin, B. 1992. Hides, hemlocks and Adirondack history: how the tanning industry influenced the region's growth. North Country Books, Utica, New York.

Mills, E. L., M. D. Scheuerell, J. T. Carlton, and D. L. Strayer. 1997. Biological invasions in the Hudson River basin: an inventory and historical analysis. New York State Museum Circular 57, Albany.

Mitchill, S. L. 1815. Fishes of New York. Transactions of the Literary and Philosophical Society 1:355–492.

Norris, R. H., and C. P. Hawkins. 2000. Monitoring river health. Hydrobiologia 435:5–17.

O'Leary, J., and D. G. Smith. 1987. Occurrence of the first freshwater migration of the gizzard shad, *Dorosoma cepedianum*, in the Connecticut River, Massachusetts. Fisheries Bulletin 85:380–383.

Pace, M. L., S. E. G. Findlay, and D. Fischer. 1998. Effects of an invasive bivalve on the zooplankton community of the Hudson River. Freshwater Biology 39:103–116.

Pelczarski, K., and R. E. Schmidt. 1990. Evaluation of a pop net for sampling fishes from water-chestnut beds in the tidal Hudson River. Section V *in* E. A. Blair and J. R. Waldman, editors. Final reports of the Tibor T. Polgar Fellowship Program 1990. Hudson River Foundation and Hudson River National Estuarine Research Reserve, New York.

Phillips, P. J., D. A. Eckhardt, D. A. Freehafer, G. R. Wall, and H. H. Ingleston. 2002. Regional patterns of pesticide concentrations in surface waters of New York in 1997. Journal of the American Water Resources Association 38:731–745.

Rose, F. P. 1993. Have all the smelt gone somewhere? Assessing changes in population size of the rainbow smelt (*Osmerus mordax*) in the Hudson River estuary. M.S. thesis. Bard College, Annandale-on-Hudson, New York.

Roy, N. K., S. C. Courtenay, G. Maxwell, M. Ikonomou, and I. Wirgin. 2001. An evaluation of the etiology of reduced CYP1A1 mRNA expression in the Atlantic tomcod from the Hudson River, New York, USA, using reverse transcriptase polymerase chain reaction analysis. Environmental Toxicology and Chemistry 20:1022–1030.

Schmidt, R. E., and S. Cooper. 1996. A catalog of barriers to upstream movement of migratory fishes in Hudson River tributaries. Final report to Hudson River Foundation, New York.

Schmidt, R. E., and T. R. Lake. 2000. Alewives in Hudson River tributaries, two years of sampling. Report to the Hudson River Foundation, New York.

Schmidt, R. E., and K. E. Limburg. 1989. Fishes spawning in non-tidal portions of Hudson River tributaries. Report to the Hudson River Foundation, New York.

Schmidt, R. E., and T. Stillman. 1994. Drift of early life stages of fishes in Stockport Creek and significance of the phenomenon to the Hudson River estuary. Report to the Hudson River Foundation, New York.

Scott, G. G. 1902. Notes on the marine food fishes of Long Island and a biologic reconnaissance of Cold Springs Harbor. New York State Museum Report of the Director 54:r214–r229, Albany.

Serafy, J. E., R. M. Harrell, and J. C. Stevenson. 1988. Quantitative sampling of small fishes in dense vegetation: design and field testing of portable "pop-nets." Journal of Applied Ichthyology 4:149–157.

Simpson, H. J., R. Bopp, and D. Thurber. 1974. Salt movement patterns in the Hudson. Paper 9 *in* Hudson River ecology, 3rd Symposium of the Hudson River Environmental Society, Bronx, New York.

Smith, C. L. 1985. Inland fishes of New York State. Department of Environmental Conservation, Albany, New York.

Smith, C. L., and T. R. Lake. 1990. Documentation of the Hudson River fish fauna. American Museum Novitates, Number 2981.

Squires, D. F. 1992. Quantifying anthropogenic shoreline modification of the Hudson River and estuary from European contact to modern time. Coastal Management 20:343–354.

Stanne, S. P., R. G. Panetta, and B. E. Forist. 1996. The Hudson. An illustrated guide to the living river. Rutgers University Press, New Brunswick, New Jersey.

Stevenson, C. H. 1899. The shad fisheries of the Atlantic Coast of the United States. Pages 101–269 *in* U.S. Commission of Fish and Fisheries, Part XXIV. Report of the Commissioner for the year ending June 30, 1898. Government Printing Office, Washington, D.C.

Stewart, L. L., and P. J. Auster. 1987. Species profiles: life history and environmental requirements of coast fishes and invertebrates (North Atlantic): Atlantic tomcod. U.S. Fish and Wildlife Service Biological Report 82(11.76). U.S. Army Corps of Engineers, RT EL-82–4, Washington D.C., Vicksburg, Mississippi.

Strayer, D. L., K. Hattala, and A. Kahnle. 2004. Effects of an invasive bivalve (*Dreissena polymorpha*) on fish populations in the Hudson River estuary. Canadian Journal of Fisheries and Aquatic Sciences. 61:924–941.

Strayer, D. L., J. Powell, P. Ambrose, L. C. Smith, M. L. Pace, and D. T. Fischer. 1996. Arrival, spread, and early dynamics of a zebra mussel (*Dreissena polymorpha*) population in the Hudson River estuary. Canadian Journal of Fisheries and Aquatic Sciences 53:1143–1149.

Strayer, D. L., and L. C. Smith. 2001. The zoobenthos of the freshwater tidal Hudson River and its response to the zebra mussel (*Dreissena polymorpha*) invasion. Archiv fuer Hydrobiologie Supplement (Monographic Studies) 131:1–52.

Wall, G. R., and P. J. Phillips. 1998. Pesticides in the Hudson River basin, 1994–96. Northeastern Geology and Environmental Sciences 20:299–307.

Wirgin, I. I., G.-L., Kreamer, C., Grunwald, K., Squibb, S. J., Garte, and S. Courtenay. 1992. Effects of prior exposure history on cytochrome P450IA mRNA induction by PCB congener 77 in Atlantic tomcod. Marine Environmental Research 34:103–108.

Yuan, Z., M. Wirgin, S. Courtenay, M. Ikonomou, and I. Wirgin. 2001. Is hepatic cytochrome P4501A1 expression predictive of hepatic burdens of dioxins, furans, and PCBs in Atlantic tomcod from the Hudson River estuary? Aquatic Toxicology 54:217–230.

Appendix A.—Fishes reported from the Hudson River drainage, New York. The drainage is divided into three units: the lower Hudson is tidal and runs from Troy, New York to the Battery at the tip of Manhattan Island; the Mohawk is a tributary system that enters the Hudson just upstream of Troy; the upper watershed is that portion upstream of Troy, excluding the Mohawk River and its tributaries. Presence (denoted by x) in 1842 is based on DeKay (1842), in 1902 on Bean (1903), and in 1936 on Greeley (1937). A date in the early specimen column refers to the oldest museum specimen collected prior to 1925; a ? denotes an early, undated specimen exists. We searched for specimens in NYSM, AMNH, and USNM. An x in the 1970–2002 column signifies that the species has been verified from the main channel of the lower Hudson River during those years. An x in the distribution columns indicates that the species has been reported from the watershed. Three species have been reported from the drainage (see Carlson and Daniels, 2004), but no specimens have been accessioned into any collection to allow verification. If established in the drainage, *Acipenser fulvescens*, *Ameiurus melas*, and *Esox masquinongy* are alien.

Common and						Distribution within drainage						
scientific names	1842	1902	1936	Early specimen	1970–2003	Lower	Upper	Mohawk	Life history	Abundance status	Origin, if alien	Comments
silver lamprey												
Ichthyomyzon unicuspis						x	x		freshwater	rare	St. Lawrence	first record 1937
American brook lamprey												
Lampetra appendix	x			1909		x			freshwater	rare		
sea lamprey												
Petromyzon marinus	x								anadromous	common		rare in upper drainage
smooth dogfish												
Mustelus canis						x			marine	rare		no voucher specimen
spiny dogfish												
Squalus acanthias			x			x			marine	rare		
little skate												
Raja erinacea						x			marine	rare		single specimen, 1987
barndoor skate												
R. laevis			x			x			marine	rare		single specimen, 1932
shortnose sturgeon												
Acipenser brevirostrum	x								anadromous	common		federal endangered
Atlantic sturgeon												
A. oxyrinchus	x		x	1878	x	x			anadromous	rare, increasing		
longnose gar												
Lepisosteus osseus					x	x			freshwater	rare	unknown	first record 1989
bowfin												
Amia calva						x			freshwater	rare	unknown	first record 1988
ladyfish												
Elops saurus						x			marine	rare		
bonefish												
Albula vulpes						x			marine	rare		
American eel												
Anguilla rostrata	x	x	x		x	x	x	x	catadromous	common		

Appendix A.—continued

Common and Scientific names	1842	1902	1936	Early specimen	1970–2003	Distribution within drainage: Lower	Upper	Mohawk	Life history	Abundance status	Origin, if exotic	Comments
conger eel *Conger oceanicus*					x	x			marine	rare		transforming leptocephali
blueback herring *Alosa aestivalis*			x		x	x		x	anadromous	abundant		
hickory shad *A. mediocris*					x	x			marine	rare		
alewife *A. pseudoharengus*	x								anadromous	common		rare in Mohawk
American shad *A. sapidissima*	x								anadromous	common		rare in Mohawk, extirpated from upper drainage by 1850s
Atlantic menhaden *Brevoortia tyrannus*			x		x	x			marine	episodically common		
Atlantic herring *Clupea harengus*					x	x			marine	rare		
gizzard shad *Dorosoma cepedianum*					x	x		x	freshwater	common, increasing	unknown	first record in lower drainage 1972, recent range expansion
round herring *Etrumeus teres*						x			marine	rare		
striped anchovy *Anchoa hepsetus*					x	x			marine	rare		
bay anchovy *A. mitchilli*			x		x	x			estuarine	abundant		
central stoneroller *Campostoma anomalum*								x	freshwater	common	Mississippi	widely distributed in Mohawk and central tributaries, recent range expansion

Appendix A.—continued

Common and scientific names	1842	1902	1936	Early specimen	1970–2003	Distribution within drainage: Lower	Upper	Mohawk	Life history	Abundance status	Origin, if alien	Comments
goldfish												
Carassius auratus		x	x		x	x	x	x	freshwater	once common, increasing	Europe	first record 1842, earlier introduction
redside dace												
Clinostomus elongatus								x	freshwater	rare, declining		tributaries
lake chub												
Couesius plumbeus							x	x	freshwater	rare		
grass carp												
Ctenopharyngodon idella					x	x			freshwater	rare	Asia	escapees from introduction, 1988 no reproduction
satinfin shiner												
Cyprinella analostana						x		x	freshwater	rare		tributaries
spotfin shiner												
C. spiloptera			x		x	x	x	x	freshwater	common		tributaries
common carp												
Cyprinus carpio	x	x	x		x	x	x	x	freshwater	common	Europe	first record 1840s
cutlip minnow												
Exoglossum maxillingua		x		1921	x	x	x	x	freshwater	common		
brassy minnow												
Hybognathus hankinsoni					x	x	x	x	freshwater	rare		tributaries
eastern silvery minnow												
H. regius		x	x		x	x	x	x	freshwater	common		
common shiner												
Luxilus cornutus	x	x	x	1883	x	x	x	x	freshwater	abundant		tributaries
pearl dace												
Margariscus margarita						x	x	x	freshwater	rare		tributaries
hornyhead chub												
Nocomis biguttatus								x	freshwater	rare	St. Lawrence	first record 1900s
golden shiner												
Notemigonus crysoleucas	x	x	x	1909	x	x	x	x	freshwater	common		
comely shiner												
Notropis amoenus		x	x		x	x			freshwater	rare	Delaware	first record 1930s

Appendix A.—continued

Common and scientific names	1842	1902	1936	Early specimen	1970–2003	Distribution within drainage: Lower	Upper	Mohawk	Life history	Abundance status	Origin, if alien	Comments
emerald shiner												
N. atherinoides					x	x	x	x	freshwater	common	St. Lawrence	first record 1930s
bridle shiner												
N. bifrenatus			x		x	x	x	x	freshwater	rare, declining		
blackchin shiner												
N. heterodon							x		freshwater	rare, declining		
blacknose shiner												
N. heterolepis							x	x	freshwater	rare, declining		
spottail shiner												
N. hudsonius	x	x	x	1875	x	x	x	x	freshwater	abundant		
sand shiner												
N. stramineus						x			freshwater	rare	St. Lawrence	tributaries, first record 1985
rosyface shiner												
N. rubellus					x	x	x	x	freshwater	common		
northern redbelly dace												
Phoxinus eos						x	x	x	freshwater	rare		
finescale dace												
P. neogaeus							x		freshwater	rare		
bluntnose minnow												
Pimephales notatus					x	x	x	x	freshwater	abundant		
fathead minnow												
P. promelas					x	x	x	x	freshwater	abundant, increasing	unknown	first record in lower drainage, 1930s
blacknose dace												
Rhinichthys atratulus	x	x	x	1883	x	x	x	x	freshwater	abundant		tributaries
longnose dace												
R. cataractae			x		x	x	x	x	freshwater	abundant		tributaries
bitterling												
Rhodeus sericeus						x			freshwater	rare, may be extirpated	Europe	first record 1923
rudd												
Scardinius erythrophthalmus					x	x			freshwater	rare, increasing	Europe	first record 1930s
creek chub												
Semotilus atromaculatus		x	x	1881	x	x	x	x	freshwater	abundant		tributaries

Appendix A.—continued

Common and scientific names	1842	1902	1936	Early specimen	1970–2003	Distribution within drainage			Life history	Abundance status	Origin, if alien	Comments
						Lower	Upper	Mohawk				
fallfish												
S. corporalis	x	x	x	1921	x	x	x	x	freshwater	common		
longnose sucker												
Catostomus catostomus						x	x	x	freshwater	rare, declining		tributaries
white sucker												
C. commersonii	x	x	x	1924	x	x	x	x	freshwater	abundant		tributaries
creek chubsucker												
Erimyzon oblongus	x	x	x			x		x	freshwater	rare		tributaries
northern hog sucker												
Hypentelium nigricans			x		x	x		x	freshwater	rare		tributaries
shorthead redhorse												
Moxostoma macrolepidotum		x			x	x		x	freshwater	rare		first record in lower 1999, expanding range
pirapatinga												
Piaractus brachypomus						x		x	freshwater	rare	aquarium release	first record 1990s
white catfish												
Ameiurus catus	x	x	x		x	x			freshwater	common		
yellow bullhead												
A. natalis					x	x	x	x	freshwater	rare		
brown bullhead												
A. nebulosus	x	x	x	1902	x	x	x	x	freshwater	abundant		
channel catfish												
Ictalurus punctatus					x	x			freshwater	rare, increasing	stocked	first record 1976
stonecat												
Noturus flavus								x	freshwater	common		
tadpole madtom												
N. gyrinus						x	x	x	freshwater	rare		tributaries
margined madtom												
N. insignis							x	x	freshwater	rare	Delaware	first record 1930s
brindled madtom												
N. miurus								x	freshwater	rare	St. Lawrence	first record 1980s
redfin pickerel												
Esox americanus		x	x	1913	x	x	x	x	freshwater	common		
northern pike												
E. lucius					x	x	x	x	freshwater	rare, increasing	St. Lawrence	first record 1840s, and later introductions

Appendix A.—continued

Common and scientific names	1842	1902	1936	Early specimen	1970–2003	Distribution within drainage: Lower	Upper	Mohawk	Life history	Abundance status	Origin, if alien	Comments
chain pickerel *E. niger*	x		x	1920	x	x	x	x	freshwater	common		
central mudminnow *Umbra limi*					x	x		x	freshwater	common	unknown	first record 1932, range expansion
eastern mudminnow *U. pygmaea*	x	x		1855	x	x			freshwater	common		tributaries
rainbow smelt *Osmerus mordax*	x								anadromous	rare, declining		stocked in upper drainage, Mohawk
cisco *Coregonus artedi*							x	x	freshwater	rare		stocked in Mohawk, native in upper
lake whitefish *C. clupeaformis*						x	x	x	freshwater	rare		stocked in lower and Mohawk, native in upper, first record in lower drainage 1936
rainbow trout *Oncorhynchus mykiss*						x	x	x	freshwater	common	Pacific rim	stocked, tributaries
sockeye salmon *O. nerka*					x	x	x		freshwater	rare	Pacific rim	first record 1974
Chinook salmon *O. tshawytscha*						x			freshwater	rare	Pacific rim	first record 1988
round whitefish *Prosopium cylindraceum*							x	x	freshwater	rare		stocked in Mohawk, native in upper, state endangered in NY
Atlantic salmon *Salmo salar*	x		x			x	x	x	freshwater	rare	stocked	numerous stockings beginning in 1880s, tributaries, lakes, alien in upper, Mohawk, possibly native in lower

Appendix A— continued

Common and scientific names	1842	1902	1936	Early specimen	1970–2003	Distribution within drainage: Lower	Upper	Mohawk	Life history	Abundance status	Origin, if alien	Comments
brown trout												
S. trutta			x		x	x	x	x	freshwater	common	Europe	stocked, tributaries
brook trout												
Salvelinus fontinalis	x		x	1902		x	x	x	freshwater	common		tributaries
lake trout												
S. namaycush						x	x	x	freshwater	rare	unknown	stocked widely
inshore lizardfish												
Synodus foetens					x	x			marine	rare		
trout-perch												
Percopsis omiscomaycus		x			x	x	x	x	freshwater	rare		
fourbeard rockling												
Enchelyopus cimbrius					x	x			marine	rare		
Atlantic cod												
Gadus morhua						x			marine	rare		
burbot												
Lota lota							x		freshwater	extirpated	St. Lawrence	single record, reported 1842
silver hake												
Merluccius bilinearis					x	x			marine	rare		
Atlantic tomcod												
Microgadus tomcod	x	x	x		x	x			estuarine	common, declining		
pollock												
Pollachius virens						x			marine	rare		
red hake												
Urophycis chuss					x	x			marine	rare		
spotted hake												
U. regia					x	x			marine	common		
white hake												
U. tenuis						x			marine	rare		
striped cusk-eel												
Ophidion marginatum					x	x			marine	rare		
oyster toadfish												
Opsanus tau						x			marine	rare		
goosefish												
Lophius americanus						x			marine	rare		

Appendix A.—continued

Common and scientific names	1842	1902	1936	Early specimen	1970–2003	Distribution within drainage: Lower	Upper	Mohawk	Life history	Abundance status	Origin, if alien	Comments
Atlantic needlefish *Strongylura marina*			x						anadromous	common		
houndfish *Tylosurus crocodilus*						x			marine	rare		single record, 1989
sheepshead minnow *Cyprinodon variegatus*						x			marine	rare		
banded killifish *Fundulus diaphanus*	x	x	x	1859	x	x	x	x	freshwater	common		
mummichog *F. heteroclitus*			x	1883	x	x			estuarine	common		
spotfin killifish *F. luciae*						x			estuarine	rare		first noted 2000
striped killifish *F. majalis*					x	x			marine	common		
western mosquitofish *Gambusia affinis*						x			freshwater	rare	unknown	first record 1992
brook silverside *Labidesthes sicculus*					x	x		x	freshwater	common, increasing	St. Lawrence	first record 1998 in lower, 1930s in Mohawk
rough silverside *Membras martinica*					x	x			marine	rare		common in areas of high salinity
inland silverside *Menidia beryllina*			x		x	x			estuarine	rare		common in areas of high salinity
Atlantic silverside *M. menidia*			x	1858	x	x			marine	rare		
fourspine stickleback *Apeltes quadracus*	x		x		x	x			estuarine	declining		
brook stickleback *Culaea inconstans*				1909		x		x	freshwater	common		tributaries

Appendix A.—continued

Common and scientific names	1842	1902	1936	Early specimen	1970–2003	Distribution within drainage			Life history	Abundance status	Origin, if alien	Comments
						Lower	Upper	Mohawk				
threespine stickleback												
Gasterosteus aculeatus	x	x	x		x	x			anadromous?	rare		
ninespine stickleback												
Pungitius pungitius			x			x			anadromous?	rare		
bluespotted cornetfish												
Fistularia tabacaria					x	x			marine	rare		
lined seahorse												
Hippocampus erectus	x				x	x			marine	rare		
northern pipefish												
Syngnathus fuscus	x				x	x			marine	common		
flying gurnard												
Dactylopterus volitans						x			marine	rare		
northern searobin												
Prionotus carolinus					x	x			marine	rare		
striped searobin												
P. evolans					x	x			marine	rare		
slimy sculptin												
Cottus cognatus				1903		x	x	x	freshwater	common		tributaries
sea raven												
Hemitripterus americanus						x			marine	rare		
grubby												
Myoxocephalus aenaeus					x	x			marine	rare		abundant in areas of high salinity
longhorn sculpin												
M. octodecemspinosus					x				marine	rare		
lumpfish												
Cyclopterus lumpus						x			marine	rare		
Atlantic seasnail												
Liparis atlanticus					x	x			marine	rare		
white perch												
Morone americana		x	x	1921	x	x		x	estuarine	abundant		
white bass												
M. chrysops					x	x		x	freshwater	rare	Mississippi	first record 1975
striped bass												
M. saxatilis	x	x	x		x	x			anadromous	common, increasing		

Appendix A.—continued

Common and scientific names	1842	1902	1936	Early specimen	1970–2003	Distribution within drainage Lower	Upper	Mohawk	Life history	Abundance status	Origin, if alien	Comments
black seabass *Centropristis striata*					x	x			marine	rare		
gag *Mycteroperca microlepis*						x			marine	rare		
rock bass *Ambloplites rupestris*	x		x		x	x	x	x	freshwater	common	St. Lawrence	first record 1840s
bluespotted sunfish *Enneacanthus gloriosus*						x			freshwater	rare		
banded sunfish *E. obesus*							x			freshwater	rare	state threatened in NY, single record 1936
redbreast sunfish *Lepomis auritus*		x	x	?	x	x	x	x	freshwater	common		
green sunfish *L. cyanellus*						x		x	freshwater	rare, increasing	stocked	first record 1936
warmouth *L. gulosus*					x	x			freshwater	rare	stocked?	first record 1936
pumpkinseed *L. gibbosus*x		x	x		x	x	x	x	freshwater	abundant		
bluegill *L. macrochirus*			x	1855	x	x	x	x	freshwater	common	stocked	
smallmouth bass *Micropterus dolomieu*		x	x		x	x	x	x	freshwater	common	stocked	first record 1830s
largemouth bass *M. salmoides*		x	x	1882	x	x	x	x	freshwater	common	stocked	first record 1830s
white crappie *Pomoxis annularis*					x	x		x	freshwater	rare	stocked	
black crappie *P. nigromaculatus*			x		x	x	x	x	freshwater	common	stocked	
greenside darter *Etheostoma blennioides*								x	freshwater	common		

Appendix A.—continued

Common and scientific names	1842	1902	1936	Early specimen	1970–2003	Distribution within drainage: Lower	Upper	Mohawk	Life history	Abundance status	Origin, if alien	Comments
fantail darter												
E. flabellare								x	freshwater	common		
tessellated darter												
E. olmstedi	x		x	1903	x	x	x	x	freshwater	abundant		
yellow perch												
Perca flavescens	x	x	x	1902	x	x	x	x	freshwater	common		
log perch												
Percina caprodes			x		x	x	x	x	freshwater	common	St. Lawrence	first record 1932
shield darter												
P. peltata			x		x	x			freshwater	rare	Delaware	first record 1936
walleye												
Sander vitreus			x		x	x	x	x	freshwater	common	stocked	first record 1893
short bigeye												
Pristigenys alta						x			marine	rare		
bluefish												
Pomatomus saltatrix	x	x	x	1881	x	x			marine	common		seasonal
cobia												
Rachycentron canadum		x	x			x			marine	rare		single record 1876
sharksucker												
Echeneis naucrates	x			1854	x	x			marine	rare		
crevalle jack												
Caranx hippos	x		x		x	x			marine	episodically common		
Atlantic moonfish												
Selene setapinnis					x	x			marine	rare		
lookdown												
S. vomer					x	x			marine	rare		
permit												
Trachinotus falcatus						x			marine	rare		
gray snapper												
Lutjanus griseus						x			marine	rare		
spotfin mojarra												
Eucinostomus argenteus						x			marine	rare		
pigfish												
Orthopristis chrysoptera						x			marine	rare		

Appendix A.—continued

Common and scientific names	1842	1902	1936	Early specimen	1970–2003	Distribution within drainage Lower	Upper	Mohawk	Life history	Abundance status	Origin, if alien	Comments
pinfish												
Lagodon rhomboides						x			marine	rare		
scup												
Stenotomus chrysops					x	x			marine	rare		
freshwater drum												
Aplodinotus grunniens					x	x			freshwater	rare, increasing	St. Lawrence	
silver perch												
Bairdiella chrysoura					x	x			marine	rare		
weakfish												
Cynoscion regalis					x	x			marine	episodically abundant		
spot												
Leiostomus xanthurus	x		x	1917	x	x			marine	episodically abundant		
northern kingfish												
Menticirrhus saxatilis					x	x			marine	rare		
Atlantic croaker												
Micropogonias undulatus					x	x			marine	rare		
foureye butterflyfish												
Chaetodon capistratus						x			marine	rare		single record
spotfin butterflyfish												
C. ocellatus						x			marine	rare		
striped mullet												
Mugil cephalus					x	x			marine	rare		
white mullet												
M. curema					x	x			marine	rare		
northern sennet												
Sphyraena borealis						x			marine	rare		
guaguanche												
S. guachancho						x			marine	rare		
tautog												
Tautoga onitis					x	x			marine	rare		
cunner												
Tautogolabrus adspersus					x	x			marine	rare		
rock gunnel												
Pholis gunnellus					x	x			marine	rare		

Appendix A.— continued

Common and scientific names	1842	1902	1936	Early specimen	1970–2003	Distribution within drainage			Life history	Abundance status	Origin, if alien	Comments
						Lower	Upper	Mohawk				
northern stargazer												
Astroscopus guttatus					x	x			marine	rare		
feather blenny												
Hypsoblennius hentz						x			marine	rare		
freckled blenny												
H. ionthas						x			marine	rare		
American sand lance												
Ammodytes americanus					x	x			marine	rare		larvae common early in season
fat sleeper												
Dormitator maculatus					x	x			marine	rare		
highfin goby												
Gobionellus oceanicus						x			marine	rare		single record 2000
naked goby												
Gobiosoma bosc					x	x			marine	common		
seaboard goby												
G. ginsburgi					x	x			marine	common		
Atlantic cutlassfish												
Trichiurus lepturus						x			marine	rare		
Atlantic makerel												
Scomber scombrus					x	x			marine	rare		
Spanish mackerel												
Scomberomorus maculatus					x	x			marine	rare		
butterfish												
Peprilus triacanthus			x		x	x			marine	rare		
Gulf Stream flounder												
Citharichthys arctifrons					x	x			marine	rare		
smallmouth flounder												
Etropus microstomus					x	x			marine	rare		
summer flounder												
Paralichthys dentatus			x		x	x			marine	common		
fourspot flounder												
P. oblongus					x	x			marine	rare		
windowpane												
Scophthalmus aquosus					x	x			marine	rare		

Appendix A—continued

Common and						Distribution within drainage						
scientific names	1842	1902	1936	Early specimen	1970–2003	Lower	Upper	Mohawk	Life history	Abundance status	Origin, if alien	Comments
winter flounder												
Pseudopleuronectes americanus			x		x	x			marine	common		
yellowtail flounder												
Pleuronectes ferrugineus					x	x			marine	rare		
northern tonguefish												
Symphurus pusillus						x			marine	rare		
hogchocker												
Trinectes maculatus	x								estuarine	common		
orange filefish												
Aluterus schoepfi						x			marine	rare		
planehead filefish												
Monacanthus hispidus						x			marine	rare		
scrawled cowfish												
Acanthostracion quadricornis					x	x			marine	rare		single specimen, 1999
striped burrfish												
Chilomycterus schoepfi						x			marine	rare		
smooth puffer												
Lagocephalus laevigatus				1848		x			marine	rare		single specimen, 1848
northern puffer												
Sphoeroides maculatus					x	x			marine	rare		

American Fisheries Society Symposium 45:505–521, 2005

Fish Assemblage Structure in Relation to Multiple Stressors along the Saint John River, New Brunswick, Canada

R. ALLEN CURRY*

Canadian Rivers Institute, New Brunswick Cooperative Fish and Wildlife Research Unit, Department of Biology, University of New Brunswick, Fredericton, New Brunswick E3B 6E1, Canada

KELLY R. MUNKITTRICK

Canadian Rivers Institute, Department of Biology, University of New Brunswick, Saint John, New Brunswick E2L 4L5, Canada

Abstract.—The Saint John River is located on the mainland of eastern North America, forming in northern Maine–southeastern Quebec, and flows east and south through New Brunswick. Fish collections were conducted at sites located from 135 to 625 km above the mouth in 2000 and 2001. Methods development trials demonstrated that the highest success was achieved with a standardized netting protocol consisting of a combination of dusk seining, nighttime electrofishing, and short-term gill net sets. A total of 36 species of fish were collected, with the greatest diversity occurring at the farthest downstream site. Upstream migration of anadromous species is restricted by the absence or poor performance of fish passage facilities at the five hydroelectric dams along the river system. The downstream migration of introduced muskellunge *Esox masquinongy* and upstream range expansions of introduced smallmouth bass *Micropterus dolomieu* and rainbow trout *Oncorhynchus mykiss* were observed. Fish species and abundances varied along the river, but the cumulative effects of human activities were not easily identified within the fish assemblage. There may have been a critical threshold within the fish assemblage defined by an accumulation of 20 anthropogenic developments. Cyprinid species declined in abundance and yellow perch *Perca flavescens* and brown bullhead *Ameiurus nebulosus* increased in abundance downstream of this apparent threshold.

Introduction

Understanding the responses of an ecosystem to perturbations begins with establishing an ecological foundation from which to assess change. This is typically achieved by comparisons to reference ecosystems presumed to represent a natural state. To assess the reference systems as appropriate models, we rely on established knowledge derived from both experimental (empirical) and theoretical information.

In the case of spatially expansive ecosystems like large rivers, it is difficult to assess ecosystem-level changes because there are few comprehensive studies from which to develop theoretical foundations. The most often cited proposed theory of riverine ecosystem dynamics is the River Continuum Concept (RCC; Vannote et al. 1980). The RCC was developed from research on smaller, low order river ecosystems and was initially assumed to be applicable

* Corresponding author: racurry@unb.ca

in a cumulative pattern as watercourses progressed downstream in the watershed; that is, the ecosystem processes of downstream reaches were driven by longitudinal energy flows generated from upstream sources. Later, the serial discontinuity concept recognized how human activities such as dam construction would alter the pattern predicted by the RCC (Ward and Stanford 1995). The flood pulse concept addressed the addition of energy (nutrients) from seasonal flooding or lateral inputs to river systems (Junk et al. 1989; Sedell et al. 1989). More recent research suggests significant autochthonous production and inputs from the riparian zone regulate processes in downstream reaches (Thorp and Delong 1994; Thorp et al. 1998). The four concepts have been applied with success along reaches of large rivers; however, no one concept defines the complete ecosystem of large rivers (Cummins et al. 1995; Dettmers et al. 2001).

The Saint John River of eastern North America is typical of large river ecosystems associated with human habitation. It flows approximately 700 km from pristine headwaters through agricultural, industrial, and urban areas and offers an opportunity to begin to improve our understanding of the spatial and temporal dynamics of large river ecosystems and how humans affect these natural systems, thus beginning our process of determining appropriate reference conditions for assessing change in this large-scale environment.

To date, much of our knowledge about the Saint John River ecosystem comes from assessments of the dramatic decline in populations of Atlantic salmon *Salmo salar*. Their declining numbers, from tens of thousands to a few thousand in recent years, are strongly correlated with the construction of dams, particularly the farthest downstream dam which allows no free passage of fishes (Dominy 1973). A major study of the biology and socioeconomics of the river occurred in the late 1960s and early 1970s, but the surveys of the fish assemblage were incomplete (Meth 1973).

New threats to the biodiversity of fishes have arisen in recent years. The absence of salmon angling opportunities has resulted in an increased importance of the fishery for smallmouth bass *Micropterus dolomieu* as the species slowly expands its range northward in the Saint John River system, both naturally and from unauthorized human introductions. These fish have become a dominant top predator in the system (Arndt 1996). Muskellunge *Esox masquinongy* were introduced to a headwater lake during the early 1970s and have subsequently expanded dramatically throughout most of the upper and middle basin (Stocek et al. 1999). Our studies also indicate that new species have been introduced upstream of the impassable Grand Falls (smallmouth bass and rainbow trout *Oncorhynchus mykiss*).

We have initiated a number of studies along the river and have been developing an overall assessment of the impact of human activities on the fish populations. The fish assemblage research reported here serves to initiate our understanding of fish distributions and habitats in relation to human activities along the river, as well as develop information required to select sentinel species for localized and basin-wide monitoring programs (see Gray et al. 2002; Galloway et al. 2003). In this paper, we explore the fish assemblage distributions from open, headwater reaches downstream through multiple urban and industrial discharges, hydroelectric dams, and intense potato cultivation areas.

Methods

Saint John River

The Saint John River is located on the mainland of eastern North America, just south of the estuary of the St. Lawrence River (Figure 1). Its main stem begins to form in northern Maine (36% of the total drainage area) and southeastern Quebec and flows east and south through New Brunswick, Canada. The main stem is close to 700 km long, and the total watershed encompasses 55,000 km^2. At more than 400 km upstream of the river mouth at the Bay of Fundy, river width and depth in the main stem average 50 m and 2 m, respectively, and summer and winter low discharge averages 135 m^3/s. At Fredericton, (river km 135), low water discharge averages 250 m^3/s (average width = 750 m and depth = 3 m).

The river flows through six distinct geologic zones that are primarily sedimentary in origin and ranging in age from 200 to 300 million years

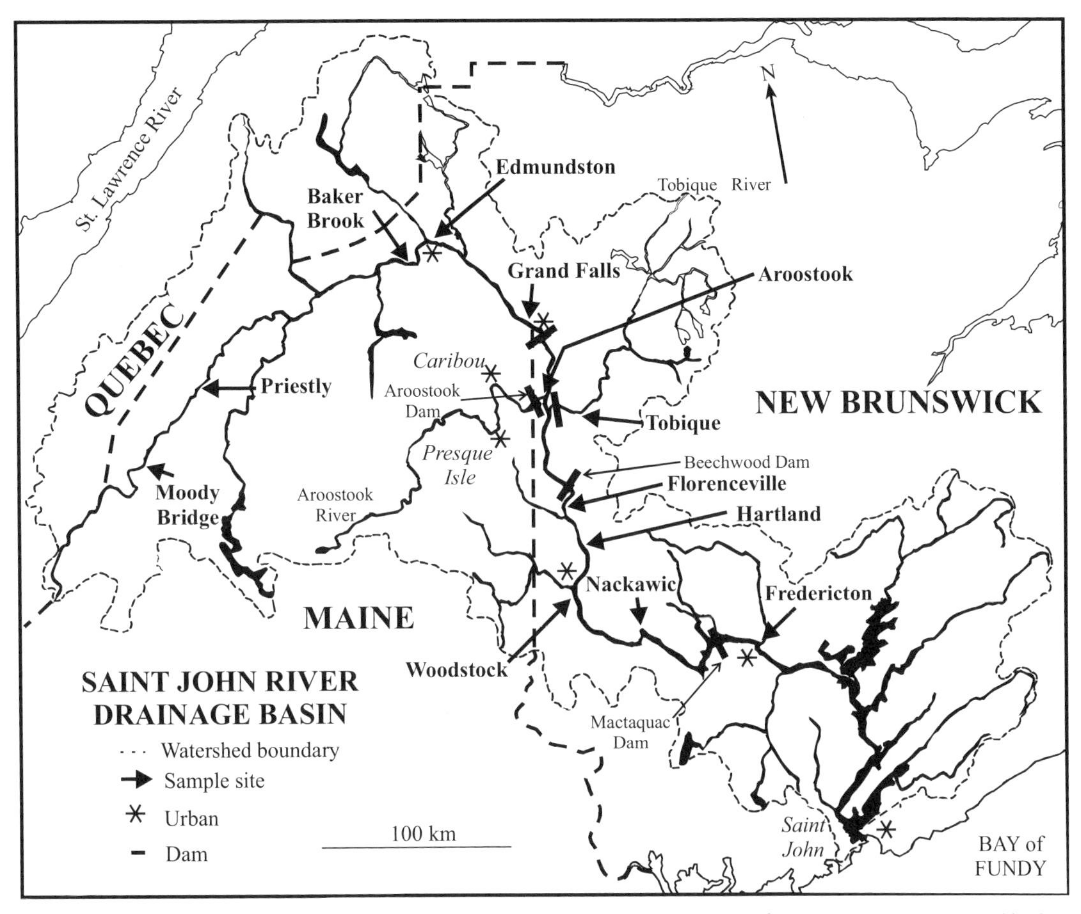

FIGURE 1.—The Saint John River basin, showing sampling sites and locations of major urbanized areas and hydroelectric facilities.

(Fensome and Williams 2001). The river would be characterized as existing in a relatively confined valley with a generalized substrate of cobbles varying to sand from Fredericton upstream. There are some outcroppings of bedrock and few larger boulders; most of these were likely removed during earlier log driving days of forestry operations (pre-1960). In the relatively nonimpacted reaches upstream of the city of Edmundston (km 420), water chemistry medians are pH = 7.0, color = 35 total color units, total dissovled solids = 35 mg/L, total P = 0.03 mg/L, and total N = 0.2 mg/L.

Sampling Protocols

Sampling of large rivers is complicated by the spatial separation and complexity of habitats as well as the challenges of collecting fishes in large volumes of flowing water. We developed our fish sampling protocol by first testing the effectiveness of a series of gear types in capturing all species in the river near Fredericton (Figure 1).

Twelve sites along the main stem were selected for study (Figure 1). They spanned a multitude of human-impacted reaches and included areas with reservoirs, forestry operations, dam and flow regulation, potato farming, pulp and paper mills, urban sewage discharge (varying treatments) and runoff, food processing plants, and pig or poultry production (Table 1). A relative index of anthropogenic activities was created where each activity (hydroelectric dam, extensive agricultural development, food processing industry, etc.) received a score of 1, and these scores were summed to create

TABLE 1.—Site locations for the assemblage collections on the Saint John River.

Site	Type	Stressors	Comments	Location (km upstream)	Elevation (m a.s.l.)
Priestly, Moody Bridge	Main stem	Forestry, recreation		>475–625	338
Baker Brook	Main stem	Sewage, poultry processing		447	295
Edmundston	Main stem	Pulp mill, paper mill, sewage treatment plants, piggeries		420	136
Grand Falls	Main stem(reservoir)	Potato agriculture, potato processing	Hydroelectric dam, no fish passage	360	135
Tobique	Tributary(reservoir)	Forestry	Hydroelectric dam, with fish passage	325	100
Aroostook	Main stem	Potato farming and processing, sewage treatment facilities, de-commissioned air force base	Hydroelectric dam, no fish passage	329	80
Florenceville	Main stem	Two food processing plants, potato farming, and sewage treatment facilities	Hydroelectric dam	275	55
Hartland	Main stem	Presque Isle tributary has starch production industries and potato production fields		255	50
Woodstock	Main stem	Large municipal sewage treatment facility, intense potato production, Meduxnekeag River drains urban and potato production	Head of the flooded reservoir created by the Mactaquac Hydroelectric Dam	233	48
Nackawic	Main stem(reservoir)	Pulp mill	At the head of the lacustrine reach of the reservoir (Mactaquac Lake)	185	45
Fredericton	Main stem		Downstream of the Mactaquac Dam, no fish passage	135	9

a qualitative index of cumulative development along the river basin.

The sites at Florenceville, Hartland, and Fredericton experience substantial fluctuations in water levels, owing to the hydroelectric facilities. There is no fish passage at the Mactaquac Dam, where anadromous species are captured in a trap, sorted to some degree, and selectively transported upstream of the dam (species targeted for transport are Atlantic salmon, alewife *Alosa pseudoharengus*, and blueback herring *A. aestivalis*; species targeted for incidental transport are American eel *Anguilla rostrata*, striped bass *Morone saxatilis*, white perch *M. americana*, and American shad *Alosa sapidissima*). The dam is the upstream limit for at least 10 anadromous species of fishes.

At each site, the standard sampling protocol was undertaken over 2 years (late July to early August 2000 and 2001). If new species continued to be captured and relative abundances continued to change, then sampling effort was increased. Fishes were counted, measured for length (nearest 1 mm), and a proportion measured for weight (nearest 1 g). Ten individuals of each species at each site were targeted for lethal sampling to obtain 1-g samples of epaxial muscle tissue for stable isotope analysis. Whole fish were used for small-bodied species.

Tissue samples for isotope analyses were dried at 50°C for 48 h and ground into a fine powder with a mortar and pestle. Approximately 0.200 mg aliquots were packed into 3-mm × 5-mm tin cups. Samples were combusted in a Carolo Erba NC2500 elemental analyzer (Thermo Electron, Milan, Italy) and resultant gases delivered via a continuous flow to a Thermo Finnigan MAT Delta Plus mass spectrometer (Thermo Finnigan, Bremen, Germany). Results of ^{13}C:^{12}C and ^{15}N:^{14}N isotope ratios are reported as:

$$dX = [\,(R_{sample}/R_{standard}) - 1\,] \times 1{,}000,$$

where $X = {}^{15}N$ or ${}^{13}C$ and $R = {}^{15}N/{}^{14}N$ or ${}^{13}C/{}^{12}C$. Replicates of commercially available isotope standards yielded results that were both accurate and precise (International Atomic Energy Agency, unpublished). Ten percent of the samples were analyzed in duplicate.

Because different efforts were applied at each site, a method of comparing among sites using relative abundance was developed. For each gear type, catches were normalized among sites and adjusted based on the lowest negative value (deviation) equaling zero. The result was a series of relative catches ranging from zero to the largest positive deviation for each gear type. These were then averaged among gear types within a site to give a species relative abundance by site. The total catch was also determined in the same manner for the total catch among sites. Measures of variability could not be generated.

Comparisons among sites relied on graphical interpretations. An analysis of covariance was attempted for comparing growth rates among sites for white sucker *Catostomus commersonii* and yellow perch *Perca flavescens* (largest sample sizes among species). Significant interactions forced an analysis of variance based on comparison of condition factor ($K = 100 \times \text{weight} \times \text{length}^{-3}$), and young of the year were excluded ($\log_{10}$ transformed data, $\alpha = 0.05$). Comparisons were followed by a Tukey pairwise comparison ($a = 0.05$). In addition, a multidimensional scaling (MDS) plot was applied to the species relative abundance data (Clarke and Warwick 1994).

Results

During the protocol development, we tested gill nets, seines, trap nets, minnow traps, and backpack and boat electrofishing during night and day (Figure 2). We observed that the number of species captured averaged 70% (SD = 11%) of the total species captured by each gear type after three sampling efforts. We observed an expected difference among gear related to body size (e.g., most cyprinids were not captured in gill nets) and between night and day sampling. We considered these results and the requirement to complete the fish assemblage survey within a 2–3-week period to reduce seasonal variability (e.g., effects of migration on presence–absence) and selected a standard sampling protocol of three night seines (15 × 1.5 m, 5 mm mesh, fishing 20–30 m of shoreline), three periods of ≥ 500 s of either boat or backpack electrofishing at night (total of ≥ 1,500 s), and three sets of a gill-net gang at night (20-min sets; 1 gang = 3, 50' panels: 40

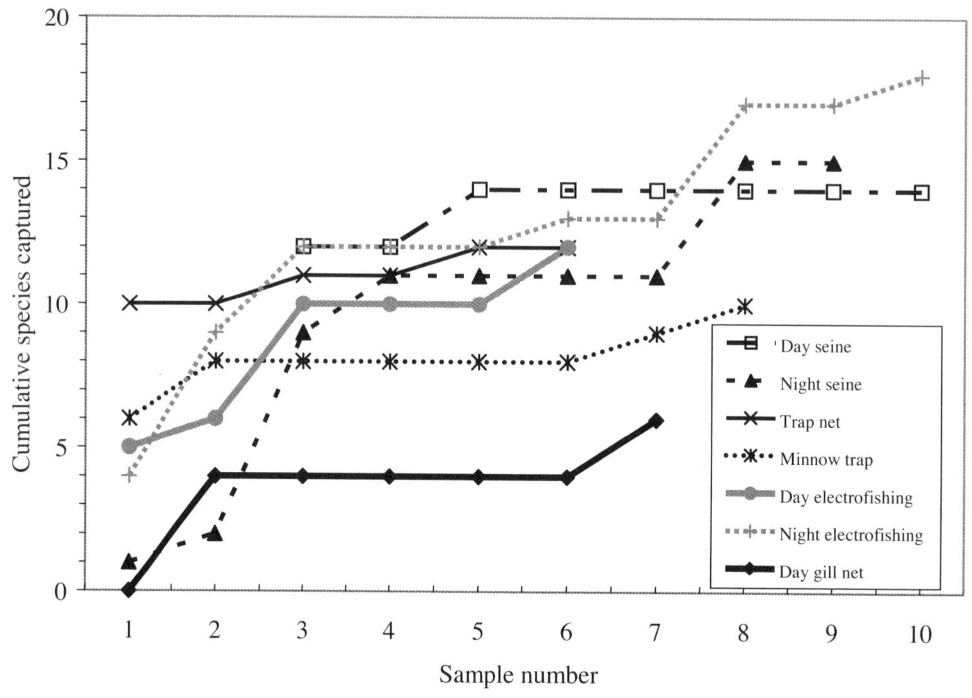

FIGURE 2.—Cumulative catch as a function of sampling effort for common species using various capture methods to determine a standard sampling protocol. Location is Fredericton, June 2000 (Figure 1).

mm, 50 mm, and 65 mm stretch measure monofilament).

A total of 36 species are known to occur in the river; however, only 30 species were captured among the sites (Appendix A). Several species captured during protocol development trials were missing: American shad, Atlantic sturgeon *Acipenser oxyrinchus*, brown trout *Salmo trutta*, muskellunge, rainbow smelt, sea lamprey *Petromyzon marinus*, and striped bass.

Common shiners and white suckers were the most widely distributed species along the river, occurring at all 12 sites, although only young of the year were captured at Moody Bridge (Appendix A). Young of the year of eight species were observed: smallmouth bass, white sucker, chain pickerel, gaspereau (alewife or blueback herring), yellow perch, common shiner, fallfish, and lake chub, although all of the species except striped bass are known to spawn in the system.

The greatest diversity occurred at Fredericton (21 species) and declined upriver to 7–8 species upstream of Grand Falls (Figure 3; Appendix A). Blacknose dace, slimy sculpin, white sucker and burbot, with smaller numbers of creek chub, common shiner, lake chub, and fallfish, dominated the upstream fish communities. This assemblage also included very small numbers of large predators like brook trout and muskellunge, but they were not captured during these surveys (these species are known from other surveys, Curry, unpublished data). The farthest downstream site was dominated by sticklebacks Gasterosteidae (three species), banded killifish *Fundulus diaphanus*, blacknose shiner, and alewife (Appendix A). The fish assemblage downstream also had relatively large numbers of predators, such as smallmouth bass, American eel, yellow perch, and chain pickerel; smaller numbers of lake whitefish, burbot, lake chub, white sucker, shortnose sturgeon; and very small numbers of fallfish, white perch, golden shiner, pumpkinseed, and brown bullhead.

Several species are only found in the middle reaches, including longnose sucker and northern redbelly dace. Several species exhibit a trend that decreases with distance upstream, including species that are anadromous (American eel, alewife,

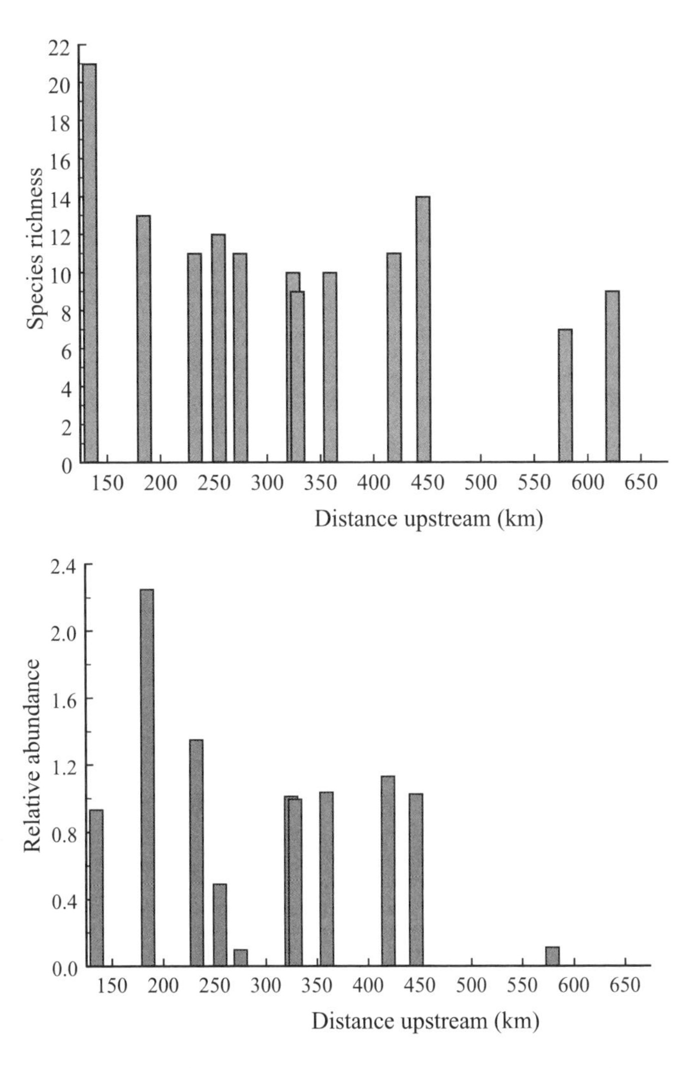

FIGURE 3.—Fish species richness (top) and relative abundance (bottom) at sites along the Saint John River during the summers of 2000 and 2001.

shortnose sturgeon, white perch), species that have been introduced and extending their range (smallmouth bass, chain pickerel, and possibly pumpkinseed), and other small-bodied species, such as banded killifish and golden shiner. Several species are relatively new introductions found occasionally, including rainbow trout and brown trout.

The relative abundance of fishes fluctuated along the river, with fewer fish in the upstream reaches (Priestly and Moody Bridge) and a reduction in numbers downstream of the hydroelectric facilities at Aroostook and Beechwood (Florenceville, Hartland; Figure 3). Individual species followed similar patterns of change along the river (Appendix A) as indicated for three representative species (Figure 4). Correlations between location in the river (distance upstream) and abundance were typically weak (rs < 0.50, Ps > 0.10; e.g., the common shiner is a small-bodied cyprinid that had its lowest abundance in the downstream reaches, and insectivorous/piscivorous yellow perch were reduced in abundance upstream of Grand Falls. The number of species did not appear to influence abundances (rs < 0.50, Ps > 0.10). We checked for possible differences between riverine (n = 9) and reservoir (n = 3) effects on the fish assemblage. Almost all differences in parameters

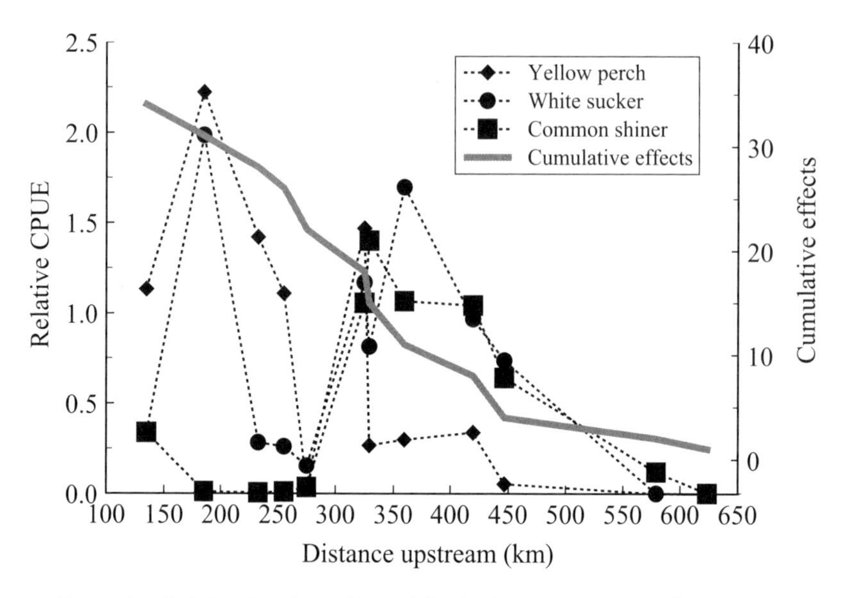

FIGURE 4.—Relative abundance (CPUE) for the three most common fish species along the Saint John River during the summers of 2000 and 2001. The indicator for cumulative effects is a subjective sum of the total, large-scale human activites along the river.

between these river habitats were not statistically significant (T-test, $P > 0.10$). White sucker relative abundance was significantly greater in the reservoirs ($P = 0.01$, $n = 9$ and $n = 3$ for riverine and reservoir sites, respectively) and smallmouth bass relative abundance was reduced in the reservoir ($P = 0.02$, $n = 5$ and $n = 2$ for riverine and reservoir sites, respectively).

The upper reaches of the Saint John River beyond km 475 are isolated with limited development other than industrial forestry. As the river progresses downstream, there is a substantial increase in anthropogenic developments. The cumulative human impact scores ranged from a low of 1 in the upper reaches more than 500 km upstream (forestry operations) to an accumulated 34 effects at Fredericton (km 135; Figure 4). Excluding reservoir sites suggested two trends in abundance (Figure 5). First, cyprinid species (common shiners, fallfish, and lake chub) had high abundance in the upper basin where there were lower levels of cumulative activities. The insectivorous/piscivorous or predatory yellow perch and brown bullheads increased in abundance with increasing cumulative activities, with declines in abundance observed at Fredericton. Multidimensional scaling plots based on species' abundances grouped the central reaches together, where cumulative activities were concentrated, and separated headwater sites (Priestly and Moody Bridge; Figure 6).

Yellow perch, white sucker, and common shiner occurred along the river with sufficient population sizes to be useful as indicators at present. Both length and weight of white suckers and yellow perch varied substantially among sites along the river. This may have been related to activities, but no aging of fish was undertaken to determine age effects on observed measures. The condition factor for both species was similar among sites (Figure 7; $P = 0.14$ and 0.32 for white sucker and perch, respectively).

Changes in carbon isotope levels measured as $d^{13}C$ within species and, on average, among species were minimal upstream of Florenceville (km 275; Figure 8a). Levels were reduced in the Tobique reservoir, which is a tributary to the main stem (km 325), elevated at Hartland (km 255), and then generally reduced downstream to Fredericton (km 135). Nitrogen levels measured as $d^{13}N$ generally became enriched in a progression downstream (Figure 8b) and, again, the Tobique reservoir displayed a unique, lower level. Smallmouth bass and yellow perch (downstream of Woodstock at least) were minimally

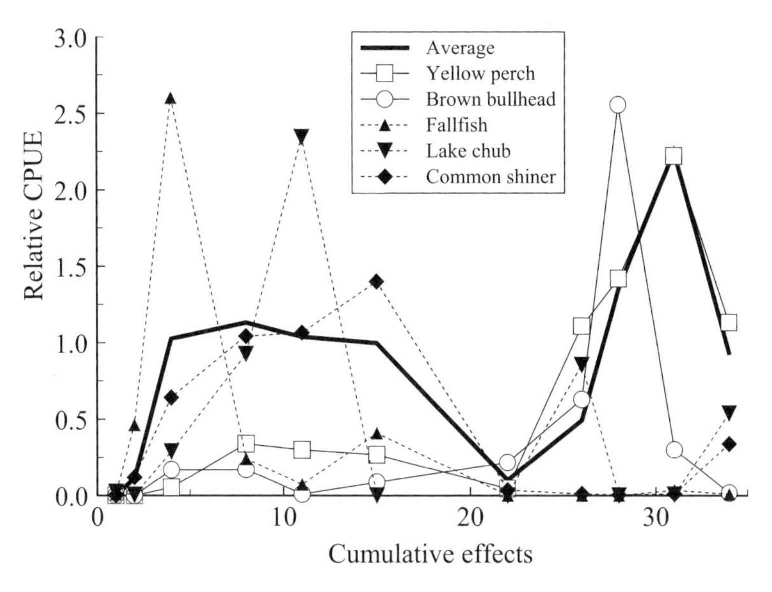

FIGURE 5.—Relative abundance (CPUE) of common fish species in the Saint John River during the summers 2000 and 2001. The indicator for cumulative effects is a subjective sum of the total, large-scale human activites along the river.

reduced in dN^{15} and enhanced in $d^{13}C$ levels downstream of Tobique, with $d^{13}C$ levels declining farther downstream from Woodstock (km 230). A comparison of riverine (n = 9) and reservoir (n = 3) sites indicated $d^{13}C$ levels were generally reduced in the reservoirs (average ± SD = -–26.9 ± 1.0 and 25.1 ± 2.4 for reservoir and rivers sites, respectively). $d^{13}N$ was similar between the river and reservoirs (10.7 ± 1.1 and 10.6 ± 1.1, respectively).

Discussion

The fish assemblage of the Saint John River of eastern Canada along its main stem from km 185 to km 624 exemplifies the complexity of large river ecosystems and particularly those in areas of human habitation. Fish distributions are predicted to be related to the postglacial, reinvasion routes into the region about 10,000 years ago, which appear to be via the upper Saint John River watershed and upstream from the Bay of Fundy (Dadswell 1974). The fish assemblage upstream of Grand Falls, the one natural barrier on the main stem, probably most resembles the historical freshwater assemblage with the exceptions of introductions of muskellunge, rainbow trout, unknown species from baitfish use in Maine, and, most recently, smallmouth bass (Curry, unpublished data). There have also been historical introductions of Atlantic salmon upstream of Grand Falls, although we have not encountered any in the main stem of the river. Human-created barriers now exist at the Mactaquac and Beechwood dams on the main stem of the river (km 135 and 280, respectively), on the Aroostook and Tobique tributaries (km 329 and 325, respectively), and at several smaller dams located in the various tributaries. These barriers are clearly affecting the dispersal and, therefore, structure of the modern assemblage of fishes in the Saint John River.

Downstream at Fredericton, the fish assemblage is most probably stable with respect to adaptations to the regulated flow environment that began in 1967. The exception is the anadromous species as most notably depicted by the significant decline in abundance of Atlantic salmon (DFO 2003) and possibly other migratory species that have been less studied such as American eel and striped bass (e.g., Jessop 1975; Smith and Clugston 1997). Four anadromous species were observed, and an additional six anadromous species were not captured at Fredericton, The returning alewife, blueback her-

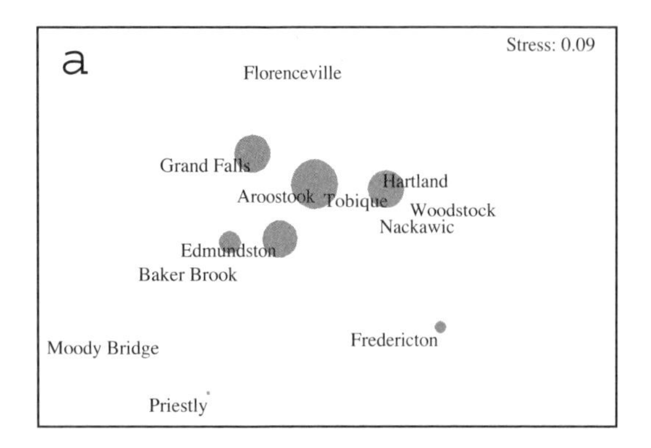

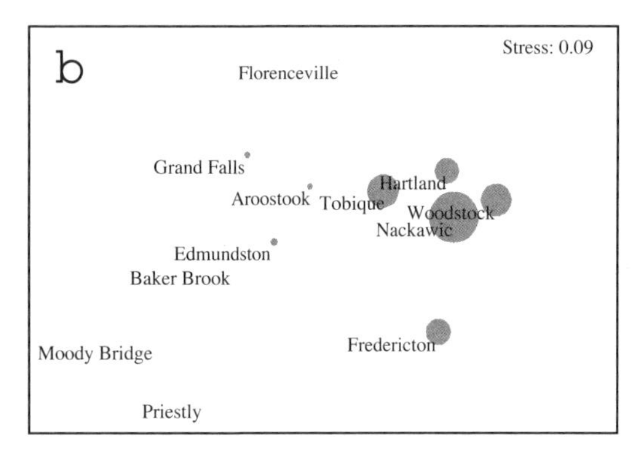

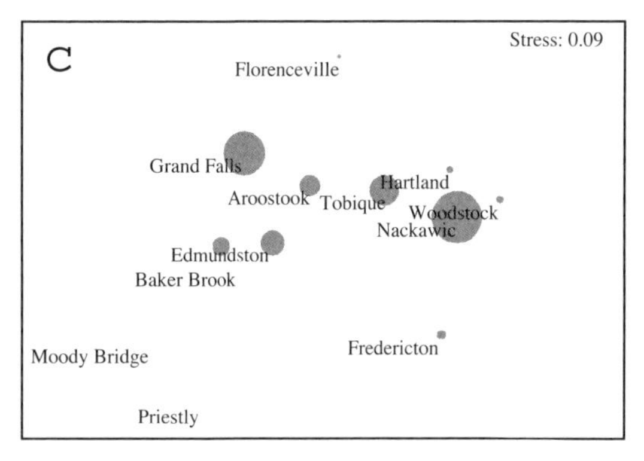

FIGURE 6.—Multidimensional scaling plot for the fish assemblage of the Saint John River during the summers of 2000 and 2001 indicating relative abundance (CPUE) for (a) common shiner, (b) yellow perch, and (c) white sucker.

ring, and Atlantic salmon are captured and trucked to various locations upstream of the dam (and there are incidental transfers of other species). Atlantic salmon are transported as far upstream and past the Beechwood and Tobique dams. Alewife and herring are typically transferred directly above the dam. These species are managed because they support fisheries deemed important by the Department of Fisheries and Oceans, Canada, although the Atlantic salmon commercial fishery has been closed for many years and the recreational fishery has been closed since 1997. Fewer than 4,000 adult salmon return to the dam each year. Approximately 1.2 million alewife and blueback herring spawners are released into the Mactaquac reservoir annually in support of a downstream commercial fishery (Jessop 1990). Prior to the dam's opening in 1968, the escapement upstream was substantially less, on the order of tens of thousands (Jessop 2001). Large numbers of alewife young of the year can have positive and negative impacts on fish communities (Tisa and Ney 1991; Mason and Brandt 1996), and there is increasing evidence that this management practice is affecting the fish assemblage (Hanson and Curry, in press).

Abundance was low for all species at Florenceville (km 275). Florenceville is subject to irregular fluctuations in river discharge because the Beechwood hydroelectric dam and the majority of the riverbed can be exposed on the low flow cycle. Fluctuations in water levels and, therefore, available habitat below hydroelectric dams are well known for having negative impacts on fish populations (e.g., Holden 1979). In addition, there is a major food processing plant discharging treated effluent to the river that elevates nutrient levels and the biological oxygen demand at Florenceville (Curry, unpublished data). We are continuing work to determine how these cumulative activities are related to the fish and other biota at this site.

In the upstream areas where only industrial forestry occurs, the main stem has low species richness and abundances (>km 475). Farther downstream, as the potential human activities begin to accumulate (Baker Brook and Edmundston, km 450) and increase through the middle reaches, species richness and abundance increase but are also highly variable. This pattern is most closely aligned with the predictions of the Serial Discontinuity Concept (Ward and Stanford 1995). However, the discontinuity effects (e.g., cumulative human impacts and potential tributary effects along the river) were poorly correlated to changes in fish assemblage.

The most apparent effect was the Tobique River and reservoir, where distinctly reduced d^{15}N and

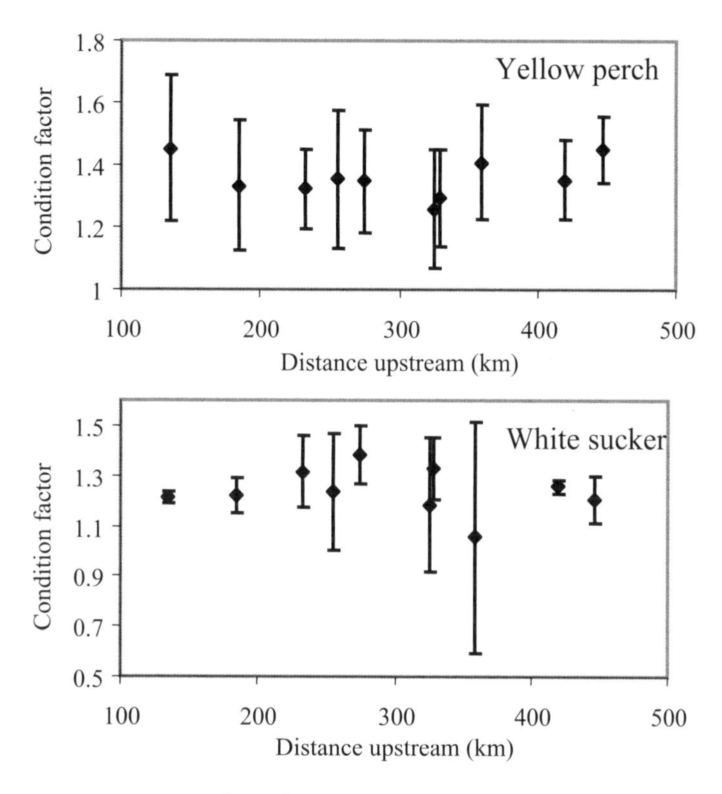

FIGURE 7.—Condition factor ($K = 100 \times \text{weight} \times \text{length}^{-3}$) for yellow perch and white sucker at sites along the Saint John River in summer 2000 (average + SE).

$d^{13}C$ levels occurred in the fishes. There were suggestions these changes in carbon and nitrogen signatures persisted in the main stem of the river downstream of the Tobique River, at least within the species that are benthic omnivores; however, this middle stretch of river is also affected by the greatest occurrence of human activities, which can affect isotope signatures (e.g., agricultural runoff, food processing effluent, sewage treatment and urban runoff, and flow regulation and reservoir effects from a major hydroelectric facility; Harrington et al. 1998; Lake et al. 2001). This stretch of the main river had the greatest variability among parameters measured. Fish abundance in the river was greatest at Nackawic and Woodstock and lowest at Florenceville. The former were located at the head of the Mactaquac Dam reservoir and downstream of urban and pulp and paper discharges; the latter was located downstream of the Beechwood Dam, where water levels fluctuate substantially as per user demands versus natural, in-stream biota requirements. Florenceville separates distinctly from all other sites in terms of species richness and abundance (MDS plots). Downstream of the Mactaquac Dam at Fredericton, also clustered with Florencville, however, the fish communities were different between these two sites, which reminds us that other factors, individually or cumulatively, also influence the fish communities along these large and complex rivers.

The cumulative effects of human activities along the entire river were not easily identified within the fish assemblage. There were clearly site-specific changes that altered patterns in the assemblage structure, but correlated changes with cumulative activities were weak. There may have been a critical threshold of an accumulation of 20 individual effects. Cyprinid species declined in abundance, and yellow perch and brown bullheads increased in abundance downstream of this threshold. This may be related to the increasing impacts of human ac-

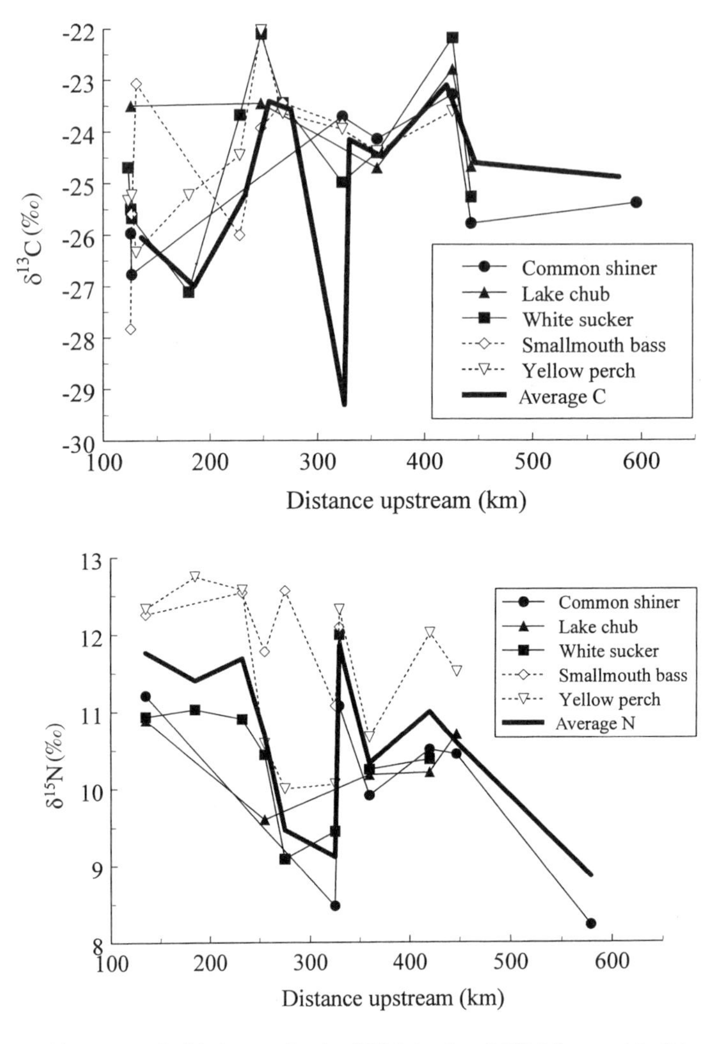

FIGURE 8.—Stable isotope levels of $\delta^{13}C$ (top) and $\delta^{15}N$ (bottom) in fishes along the Saint John River during the summers of 2000 and 2001.

tivities on specific feeding guilds, or it may be an autocorrelation with downstream distance (i.e., a natural large river response). We are testing this result in other rivers where fewer human activities occur.

Against the backdrop of originally different communities, new barriers, and new species introductions, it is difficult to detect changes related to the impacts of human activities on aquatic communities in this complex river. It is apparent that the barriers restrict access for anadromous species, but further interpretation is difficult. We are continuing our studies of the fishes and aquatic ecosystems in general in the Saint John River and other eastern rivers in search of potential indicators of change. Within specific reaches, we have a number of studies using sentinel species that are aimed at developing a better understanding of the localized effects of point source discharges and developments.

Summary

1. A standardized method of simple fish collections and analyses was developed to compare sites within and among rivers.

2. The historic fish assemblages of the river were best represented at sites upstream of Edmundston (km 450), where human influences were the lowest, and downstream at Fredericton (km 135), where the first barrier to fish migration occurs. Between these reaches, site-specific changes in the fish assemblage were apparent; however, there was little support for predictions from theoretical concepts of riverine ecology and no strong links between fish assemblage changes and the cumulative activities of humans.

3. Certain changes in the fish assemblage are clearly linked to humans (i.e., the introduction of three species and the management of the alewife and blueback herring commercial fishery in the Mactaquac Reservoir). The longer-term impacts of these activities remain to be determined.

Acknowledgments

This study was supported by funding from the New Brunswick Wildlife Trust Fund, New Brunswick Environmental Trust Fund, and the Toxic Substances Research Initiative (TSRI) of Health Canada (TSRI Project 205). Field collections were conducted by staff of the New Brunswick Cooperative Fish and Wildlife Research Unit and Canadian Rivers Institute. We are particularly indebted to S. Currie, M. Gautreau, and C. Doherty. Comments by M. Dadswell and an anonymous reviewer greatly improved the manuscript.

References

Arndt, S. K. A. 1996. A baseline study of smallmouth bass biology in New Brunswick. New Brunswick Cooperative Fish and Wildlife Research Unit, University of New Brunswick, Fredericton.

Clarke, K., and R. Warwick. 1994. Change in marine communities: an approach to statistical analysis and interpretation. Plymouth Marine Laboratory, Plymouth, UK.

Cummins, K. W., C. E. Cushing, and G. W. Minshall. 1995. Introduction: an overview of stream ecosystems. Pages 1–8 *in* C. E. Cushing, K. W. Cummins, and G. W. Minshall, editors. Ecosystems of the world 22. Rivers and stream ecosystems. Elsevier, New York.

Dadswell, M. J. 1974. Distribution, ecology and postglacial dispersal of certain crustaceans and fishes in eastern North America. Publication of the Zoological National Museum of Natural Sciences, National Museum of Canada 11:1–110.

Dettmers, J. M., D. H. Wahl, D. A. Soluk, and S. Gutreuter. 2001. Life in the fast lane: fish and foodweb structure in the main channel of large rivers. Journal of the North American Benthological Society 20:255–265.

DFO (Department of Fisheries and Oceans, Canada). 2003. Atlantic salmon Maritime Provinces overview for 2002. DFO Science, Stock status report 2003/026. Available: *www.dfo-mpo.gc.ca/CSAS/CSAS/status/2003/SSR2003_026_E.pdf* (May 2004).

Dominy, C. L. 1973. Recent changes in Atlantic salmon (*Salmo salar*) runs in the light of environmental changes in the Saint John River, New Brunswick, Canada. Biological Conservation 5:105–113.

Fensome, R. A., and G. L. Williams, editors. 2001. The last billion years: a geological history of the Maritime Provinces of Canada. Atlantic Geoscience Society, Special Publication 15, Halifax, Nova Scotia.

Galloway, B. J., K. R. Munkittrick, S. Currie, M. A. Gray, R. A. Curry, and C. S. Wood. 2003. Examination of the responses of slimy sculpin (*Cottus cognatus*) and white sucker (*Catostomus commersoni*) collected on the Saint John River (Canada) downstream of pulp mill, paper mill, and sewage discharges. Environmental Toxicology and Chemistry 22:2898–2907.

Gray, M. A., R. A. Curry, and K. R. Munkittrick. 2002. Non-lethal sampling methods for assessing environmental impacts using a small-bodied sentinel fish species. Water Quality Research Journal of Canada 37:195–211.

Hanson, S. D. 2002. Size structure and trophic interactions of age-0 river herring and age-0 smallmouth bass in the Mactaquac reservoir and Oromocto Lake, New Brunswick. Master's thesis. University of New Brunswick, Fredericton.

Hanson, S. D., and R. A. Curry. In press. Effects of river herring management in the Saint John River, New Brunswick on trophic interactions with age-0 smallmouth bass. Transactions of the American Fisheries Society.

Harrington, R. R., B. P. Kennedy, C. P. Chamberlain, J. D. Blum, and C. L. Folt. 1998. 15N enrichment in agricultural catchments: field patterns and applications to tracking Atlantic salmon (*Salmo salar*). Chemical Geology 147:281–294.

Holden, P. 1979. Ecology of riverine fishes in regulated stream systems with emphasis on the Colorado River. Pages 57–74 *in* J. Ward and J. Stanford, editors. The ecology of regulated rivers. Plenum, New York.

Jessop, B. M. 1975. A review of the American shad (*Alosa sapidissima*) stocks of the Saint John River, New Brunswick, with particular reference to the adverse effects of hydroelectric developments. Department of Fisheries and Oceans, Canada, Fisheries and Marine Service, Resource Development Branch, Technical Report Series MAR/T 75-6, Dartmouth, Nova Scotia.

Jessop, B. M. 1990. Stock–recruitment relationships of alewives and blueback herring returning to the Mactaquac Dam, Saint John River, New Brunswick. North American Journal Fisheries Management 10:19–32.

Jessop, B. M. 2001. Stock status report of the alewife and blueback herring returning to the Mactaquac Dam, Saint John River, N.B. Department of Fisheries and Oceans, Canada, Canadian Science Advisory Secretariat, Research Document 2001/059, Dartmouth, Nova Scotia.

Junk, W., P. Bayley, and R. Sparks. 1989. The flood pulse concept in river-floodplain systems. Canadian Special Publication of Fisheries and Aquatic Sciences 106:89–109.

Lake, J. L., R. A. McKinney, R. A. Osterman, R. J. Pruell, J. Kiddon, S. A. Ryba, and A. D. Libby. 2001. Stable nitrogen isotopes as indicators of anthropogenic activities in small freshwater systems. Canadian Journal of Fisheries and Aquatic Sciences 58:870–878.

Mason, D. M., and S. B. Brandt. 1996. Effect of alewife predation on survival of larval yellow perch in an embayment of Lake Ontario. Canadian Journal of Fisheries and Aquatic Sciences 53:1609–1617.

Meth, F. F. 1973. Fishes of the upper and middle Saint John River. Saint John River Basin Board, Report 7c, Fredericton, New Brunswick.

Sedell, J. R., J. E. Richey, and F. J. Swanson. 1989. The river continuum concept: a basis for the expected ecosystem behavior of very large rivers? Canadian Special Publication of Fisheries and Aquatic Sciences 106:49–55.

Smith, T. I. J., and J. P. Clugston. 1997. Status and management of Atlantic sturgeon, *Acipenser oxyrinchus*, in North America. Environmental Biology of Fishes 48:335–346.

Stocek, R. F., P. J. Cronin, and P. D. Seymour. 1999. The muskellunge, *Esox masquinongy*, distribution and biology of a recent addition to the ichthyofauna of New Brunswick. Canadian Field Naturalist 113:230–234.

Thorp, J., and M. Delong. 1994. The riverine productivity model: an heuristic view of carbon sources and organic. Oikos 70:305–308.

Thorp, J., M. Delong, K. Greenwood, and A. Casper. 1998. Isotopic analysis of three food web theories in constricted and floodplain regions of a large river. Oecologia 117:551–563.

Tisa, M. S., and J. J. Ney. 1991. Compatibility of alewives and gizzard shad as reservoir forage fish. Transactions of the American Fisheries Society 120:157–165.

Vannote, R., G. Minshall, K. Cummins, J. Sedell, and C. Cushing. 1980. The river continuum concept. Canadian Journal of Fisheries and Aquatic Sciences 3:130–137.

Ward, J., and J. Stanford. 1995. The serial discontinuity concept: extending the model to floodplain rivers. Regulated Rivers 10:159–168.

Appendix A.—Fish assemblage of the Saint John River indicating site locations (km upstream of the river mouth), species richness, and relative abundances (see text for description). YOY = young of year.

Species	Sites	%	Freder-icton	Nacka-wic	Wood-stock	Hart-land	Florence-ville	Tobique	Aroo-stook	Grand Falls	Edmunds-ton	Baker Brook	Priestly	Moody Bridge
Three spine stickle-back *Gasterosteus aculeatus*	2	17	1.55525	0.00000	0.00000	0.00000	0.00000	0.00000	0.00000	0.00000	0.00000	0.59076	0.00000	0.00000
Four spine stickle-back *Apeltes quadracus*	1	8	2.56121	0.00000	0.00000	0.00000	0.00000	0.00000	0.00000	0.00000	0.00000	0.00000	0.00000	0.00000
Nine spine stickle-back *Pungitius pungitius*	2	17	1.26823	0.00000	0.00000	0.00000	0.00000	0.00000	0.00000	0.00000	0.00000	0.80440	0.00000	0.00000
Atlantic salmon *Salmo salar*	1	8	0.00000	0.00000	0.00000	0.00000	0.86603	0.00000	0.00000	0.00000	0.00000	0.00000	0.00000	0.00000
Brown bullhead *Amerius nebulosus*	10	83	0.01737	0.29970	2.55604	0.62829	0.21716	1.08802	0.08687	0.00980	0.17057	0.16943	0.00000	0.00000
Banded killifish *Fundulus diaphanous*	6	50	2.57828	1.16349	0.00000	0.01351	0.01351	0.00000	0.06065	0.00000	0.04814	0.00000	0.00000	0.00000
Eastern blacknose dace *Rhinichthys atratulus*	4	33	0.00000	0.00000	0.00000	0.00000	0.00000	0.00000	0.00000	0.00000	1.19149	1.74137	0.25929	0.10641
Blacknose shiner *Notropis heterolepis*	1	8	2.59808	0.00000	0.00000	0.00000	0.00000	0.00000	0.00000	0.00000	0.00000	0.00000	0.00000	0.00000
Brook trout *Salvalinus fontinalis*	2	17	0.00000	0.00000	0.00000	0.00000	0.86603	0.00000	0.00000	0.00000	0.00000	1.73526	0.00000	0.00000
Burbot *Lota lota*	3	25	0.32656	0.00000	0.00000	0.00000	0.00000	0.00000	0.00000	0.00000	0.00000	0.00000	0.56348	0.68577
Creek chub *Semotilus atromaculatus*	7	58	0.00000	0.00000	0.00000	0.00000	0.02979	0.97460	0.33480	1.35466	0.36451	0.86923	0.00000	0.02651
chain pickerel *Esox niger* (introduced)	2	17	0.46459	0.00000	0.77432	0.00000	0.00000	0.00000	0.00000	0.00000	0.00000	0.00000	0.00000	0.00000
YOY	1	8												

Appendix A.—Continued.

Species	Sites	%	Fredericton	Nackawic	Woodstock	Hartland	Florenceville	Tobique	Aroostook	Grand Falls	Edmundston	Baker Brook	Priestly	Moody Bridge
Common shiner *Luxilus cornutus*	12	100	0.33913	0.01063	0.00453	0.01051	0.03504	1.05720	1.40100	1.06665	1.04369	0.64152	0.12066	0.00245
YOY	3	25												
Northern redbelly dace *Phoxinus eos*	2	17	0.00000	0.00000	0.00000	0.17289	0.86446	0.00000	0.00000	0.00000	0.00000	0.00000	0.00000	0.00000
American eel *Anguilla rostrata*	5	42	1.73205	0.31871	0.84988	0.10624	0.00000	0.10624	0.00000	0.00000	0.00000	0.00000	0.00000	0.00000
Fallfish *Semotilus corporalis*	9	75	0.00913	0.03326	0.00000	0.00000	0.00000	1.64529	0.40876	0.07215	0.24027	2.60170	0.46161	0.00822
YOY	1	8												
Alewife *Alosa pseudoharengus*	3	25	2.88810	0.84828	0.17289	0.00000	0.00000	0.00000	0.00000	0.00000	0.00000	0.00000	0.00000	0.00000
YOY	1	8												
Blueback herring *Alosa aestivalis*	2	17												
Golden shiner *Notemigonus chrysoleucas*	5	42	0.14107	3.00386	0.92548	0.04760	0.00000	0.33087	0.00000	0.00000	0.00000	0.00000	0.00000	0.00000
Lake chub *Couesius plumbeus*	8	67	0.53850	0.00626	0.00000	0.85322	0.00000	0.00000	0.00000	2.34465	0.92460	0.28825	0.00389	0.02255
YOY	1	8												
Longnose sucker *Catostomus catostomus*	4	33	0.00000	0.00000	0.00000	0.00000	0.00000	0.00000	0.70249	0.55902	1.51811	0.00000	0.00000	0.63226
Lake whitefish *Coregonus clupeaformis*	2	17	0.86603	0.00000	0.00000	0.00000	0.00000	0.00000	0.00000	0.00000	0.00000	0.86603	0.00000	0.00000
Rainbow trout *Oncorhynchus mykiss* (introduced)	3	25	0.00000	0.00000	0.00000	0.00000	0.86923	0.00000	0.00000	0.86603	0.00000	0.86603	0.00000	0.00000
Pumpkinseed *Lepomis gibbosus*	5	42	0.11586	0.19492	1.10884	0.49025	0.00000	1.67465	0.00000	0.00000	0.00000	0.00000	0.00000	0.00000
Smallmouth bass *Micropterus dolomieu* (introduced)	7	58	1.95269	0.20944	1.68362	0.59668	0.26367	0.16029	1.98918	0.00000	0.00000	0.00000	0.00000	0.00000
YOY	5	42												

Appendix A.—Continued.

Species	Sites	%	Freder-icton	Nackawic	Wood-stock	Hart-land	Florence-ville	Tobique	Aroo-stook	Grand Falls	Edmunds-ton	Baker Brook	Priestly	Moody Bridge
Rainbow smelt *Osmerus mordax*	1	8	0.00000	0.00000	0.00000	0.00000	0.00000	0.00000	0.00000	0.86603	0.00000	0.00000	0.00000	0.00000
Slimy sculpin *Cottus cognatus*	4	33	0.00000	0.00000	0.00000	0.00000	0.00000	0.00000	0.00000	0.00000	0.07928	1.74366	0.14738	0.08968
Shortnose sturgeon *Acipenser brevirostrum*	1	8	0.86603	0.00000	0.00000	0.00000	0.00000	0.00000	0.00000	0.00000	0.00000	0.00000	0.00000	0.00000
White perch *Morone saxatilis*	4	33	0.05895	2.91681	0.40014	0.94071	0.00000	0.00000	0.00000	0.00000	0.00000	0.00000	0.00000	0.00000
White sucker *Catostomus commersonii*	12	100	0.34317	1.98498	0.28416	0.26206	0.15484	1.16999	0.81398	1.69767	0.96905	0.73743	0.00082	0.00003
YOY	6	50												
Yellow perch *Perca flavescens*	10	83	1.13294	2.22206	1.41949	1.10887	0.04444	1.46865	0.26875	0.30008	0.33917	0.05281	0.00000	0.00000
YOY	3	25												

American Fisheries Society Symposium 45:523–556, 2005

Changes in the Fish Fauna of the Kissimmee River Basin, Peninsular Florida: Nonnative Additions

LEO G. NICO*
U.S. Geological Survey, Center for Aquatic Resources Studies, 7920 NW 71st Street, Gainesville, Florida 32653, USA

Abstract.—Recent decades have seen substantial changes in fish assemblages in rivers of peninsular Florida. The most striking change has involved the addition of nonnative fishes, including taxa from Asia, Africa, and Central and South America. I review recent and historical records of fishes occurring in the Kissimmee River basin (7,800 km^2), a low-gradient drainage with 47 extant native fishes (one possibly the result of an early transplant), at least 7 foreign fishes (most of which are widely established), and a stocked hybrid. Kissimmee assemblages include fewer marine fishes than the nearby Peace and Caloosahatchee rivers, and fewer introduced foreign fishes than south Florida canals. Fish assemblages of the Kissimmee and other subtropical Florida rivers are dynamic, due to new introductions, range expansions of nonnative fishes already present, and periodic declines in nonnative fish populations during occasional harsh winters. The addition, dispersal, and abundance of nonnative fishes in the basin is linked to many factors, including habitat disturbance, a subtropical climate, and the fact that the basin is centrally located in a region where drainage boundaries are blurred and introductions of foreign fishes commonplace. The first appearance of foreign fishes in the basin coincided with the complete channelization of the Kissimmee River in the 1970s. Although not a causal factor, artificial waterways connecting the upper lakes and channelization of the Kissimmee River have facilitated dispersal. With one possible exception, there have been no basin-wide losses of native fishes. When assessing change in peninsular Florida waters, extinction or extirpation of fishes appears to be a poor measure of impact. No endemic species are known from peninsular Florida (although some endemic subspecies have been noted). Most native freshwater fishes are themselves descended from recent invaders that reached the peninsula from the main continent. These invasions likely were associated with major fluctuations in sea level since the original mid-Oligocene emergence of the Florida Platform. As opportunistic invaders, most native freshwater fishes in peninsular Florida are resilient, widespread, and common. At this early stage, it is not possible to predict the long-term consequences caused by the introduction of foreign fishes. We know a few details about the unusual trophic roles and other aspects of the life histories of certain nonnatives. Still, the ecological outcome may take decades to unfold.

Introduction

The Kissimmee River basin, draining an area about 7,800 km^2, is situated in peninsular Florida just south of Orlando (Figure 1) (Toth et al. 1998). The basin is the principal tributary to Lake Okeechobee and is the only major river in peninsular Florida without a direct outlet to the ocean. Since the late 1800s, the basin has been subjected to an array of environmental stresses that have modified the aquatic environment and altered its hydrology (Kushlan 1991). Changes include excavation of canals, ditches, ponds, and other artificial waterways

* E-mail: leo_nico@usgs.gov

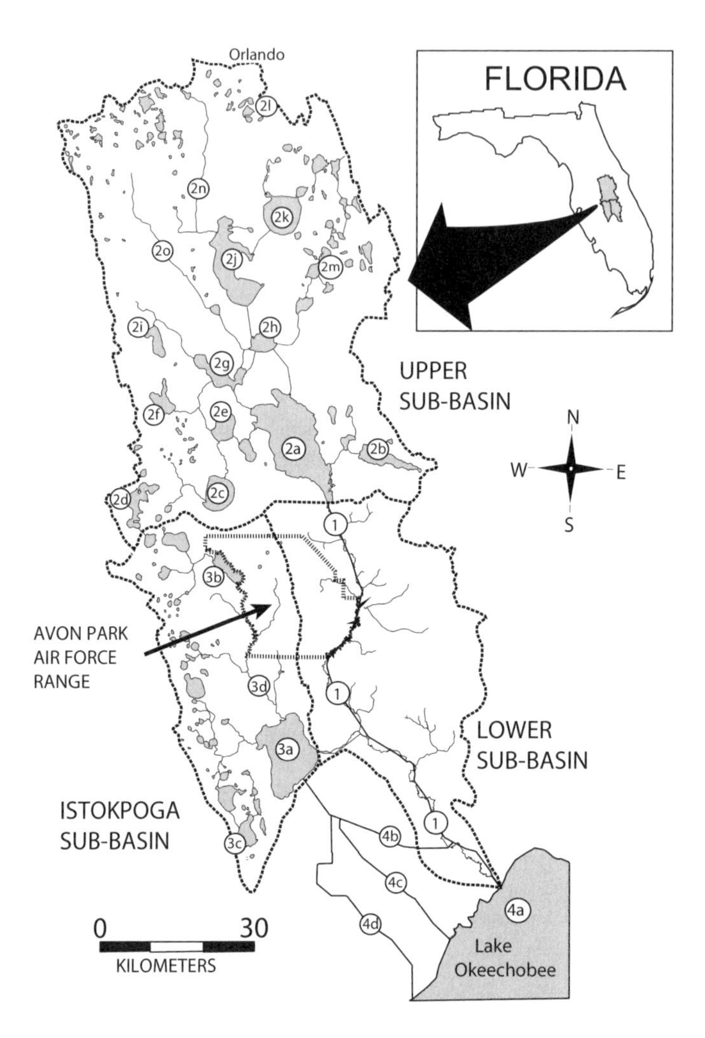

Figure 1.—The Kissimmee River basin, showing the boundaries of the three major subbasins (boundaries indicated by dotted lines) and location of Avon Park Air Force Range (boundary indicated by parallel dashes). Names of major water bodies are as follows: Lower Sub-Basin—1, Kissimmee River/Canal C-38. Upper Sub-basin—2a, Lake Kissimmee; 2b, Lake Marian; 2c, Lake Weohyakapka; 2d, Crooked Lake; 2e, Lake Rosalie; 2f, Lake Pierce; 2g, Lake Hatchineha; 2h, Cypress Lake; 2i, Lake Marion; 2j, Lake Tohopekaliga; 2k, East Lake Tohopekaliga; 2l, Lake Conway; 2m, Alligator Lake; 2n, Shingle Creek; 2o, Reedy Creek. Istokpoga Sub-basin—3a, Lake Istokpoga; 3b, Lake Arbuckle; 3c, Lake Placid; 3d, Arbuckle Creek. Other—4a, Lake Okeechobee (partial); 4b, Canal C-41a; 4c, Indian Prairie Canal; 4d, Harney Pond Canal.

(beginning in the 1800s), wetland drainage, roadway construction, widespread cattle ranching and other forms of agriculture, and extensive urbanization, particularly along the basin's northern and western boundaries. Of marked significance was the construction of a massive flood-control project during the period 1962–1971 (Toth et al. 1997, 1998). As a result of that project, major lakes in the upper part of the basin were partitioned into flood-storage reservoirs and connected via artificial canals. In the southern half of the basin, the Kissimmee River channel and adjacent wetlands were greatly modified by excavation of a 90-km long drainage canal slicing through the river-floodplain ecosystem, which straightened a meandering river (Toth et al. 1998). During the past two decades, the Kissimmee River has gained fame because of attempts to restore historical flows and channel patterns by removal or modification of selected water control structures, backfilling canals, and changes in water flow regulation. Various symposia (e.g., Loftin et al. 1990) and an entire issue of the journal *Restoration Ecology* (1995, 3[3]) are devoted to the subject.

Several recent authors have compiled lists of fish species from the Kissimmee River basin. Counts range from 22 to 48 species, with differences in total numbers largely due to some authors clearly limiting their analysis to the lower Kissimmee subbasin or portions thereof or basing their list on collections from a relatively short time period. Published and unpublished compilations appear in Dineen et al. (1974) [32 natives], Montalbano et al. (1979a) [39 natives], Perrin et al. (1982) [22 natives and 2 alien], Bass (1983) [43 natives and 1 alien], and Trexler (1995) [44 natives and 4 alien]. Without elaboration, Bass (1991) indicated that the system contains 39 native fishes. In their zoogeographical analysis of southeastern freshwater fishes, Swift et al. (1986) combined the Kissimmee River and Lake Okeechobee into a single drainage unit and reported a total of 39 species; however, their summary Appendix and species list in the same chapter cataloged a total of 52 species, including 38 native freshwater fishes, 10 euryhaline species, and an additional 4 species identified as "suspected records" (i.e., possibly valid but without known museum records or other substantiation). Miller (1990) and Toth (1993) list 39 fishes found in the prechannelization (1956–1957) Kissimmee River ecosystem, citing lower-river data appearing in Florida Game and Fresh Water Fish Commission (FGFWFC) (1957). I examined the original FGFWFC (1957) report and discovered that both recent authors are in error concerning two species, mistakenly including dollar sunfish *Lepomis marginatus*, a species not collected by FGFFWC in the 1950s, and not listing mullet, a fish taken by FGFWFC during the early study. Trexler (1995) presents the most comprehensive compilation of Kissimmee basin fishes. Based largely on a review of the literature, he reported a total of 48 fish species. Of these, Trexler categorized 34 species as native freshwater fishes, 10 euryhaline species, and 4 alien species. His focus was on the lower Kissimmee subbasin, apparently excluding Lake Istokpoga and its drainage network. Trexler did not include certain native fishes now known to be present. Furthermore, since publication of his paper, there have been additions to the basin's nonnative fish fauna.

During the period January 1995 through October 1997, colleagues and I conducted fieldwork to inventory fishes and other aquatic fauna in the Avon Park Air Force Range (APR), Florida. The APR is a federal military reservation occupying 43,142 ha within the Kissimmee River basin. To better understand observed distribution patterns of fish found within APR, I reviewed historical and recent records of fishes in the basin. For a variety of reasons, gathering and analyzing this information was a challenge. Despite the fact that there have been numerous surveys, most after 1970, conducted in the upper lakes and in portions of the lower Kissimmee River, there is relatively little information on Kissimmee River fishes in the primary scientific literature. Most data on distribution and abundance of fishes from the basin only exist in unpublished reports or file documents.

As discussed by Collette (1990), gray literature raises questions about credibility of information provided. I have attempted to address these and other issues in the present review. Unfortunately, prior to our APR project, relatively few fish specimens from the Kissimmee River basin were preserved and deposited in museum collections. This is a surprising situation given the size and importance of the basin. Shortage of museum material from the basin is evident in published errors on the geographic range of certain species. For instance, fish distribution maps appearing in Lee et al. (1980) and Page and Burr (1991), generated largely from museum material, erroneously show a number of Florida natives as absent or having a limited distribution in the Kissimmee basin. In reality, longnose gar *Lepisosteus osseus*, gizzard shad *Dorosoma cepedianum*, and redbreast sunfish *Lepomis auritus*, among others, are broadly distributed in the basin. Conversely, a map for the sailfin shiner *Pteronotropis hypselopterus* given by Page and Burr (1991) shows its range as including a portion of the Kissimmee basin, but this species has never been taken in the basin.

Even more critical are potential errors in species identifications. It is obvious that some fish names used in early reports on the Kissimmee River are in error or should be considered suspect. Uncertainty of identification is sometimes unavoidable, simply because it reflects changes in our knowledge of systematics and taxonomy. In a number of cases, however, researchers unfamiliar with certain nongame species or juvenile sportfishes simply used higher taxonomic groupings or lumped several different fishes into broad categories. As a result, fishes mentioned in many reports (and even some museum specimens) are not identified to species (e.g., sunfishes, minnows, *Notropis* sp., *Lepomis* sp., ictalurid). Without voucher material, the accuracy of fish names referenced in past reports cannot be verified.

The purpose of this chapter is to provide an overview and historical perspective of the fish fauna of the Kissimmee River basin based on original fieldwork and on surveys and collections of others. Existing information is examined to provide an updated inventory of the ichthyofauna of the entire basin and to assess changes in fish fauna composition over time. Emphasis is given to native fishes considered uncommon or rare as well as any whose abundance and distribution may have been altered recently. Special attention is given to documenting the occurrence and distribution of introduced fishes, because the most obvious change in fish composition in the basin has been the addition of nonnatives.

Study Area

The Kissimmee River basin, occupying parts of seven Florida counties, is one of several major river basins in peninsular Florida. The basin is subtropical, having its most northern boundary at about 28°33'N latitude. Its most southern point is where the Kissimmee River feeds into Lake Okeechobee, about 27°08'40'N. The U.S. Geological Survey stream gauge data for the Kissimmee River near its mouth indicate that annual mean flow for the period of record 1929–2000 has ranged from a low of 3 m^3/s for 1981 to a high of 166 m^3/s for 1960 (but see FGFWFC 1957). Prior to channelization (1929–1961), mean annual flow averaged about 62 m^3/s, but that amount is now reduced (USCOE 1991). Topography of the basin is relatively flat and low, ranging from about 4 to 35 m above sea level. The headwaters of the basin are in upland plains, a region known as the Osceola Plain, but near Lake Okeechobee the basin consists of coastal lowlands known as the Okeechobee Plain (Fernald 1981; Kushlan 1991; USCOE 1991). In the north, the basin is becoming increasingly urbanized, largely because of expansion of the city of Orlando, but the character of the basin gradually changes downstream to rangeland for cattle and cropland and then to wetland (Paulson et al. 1993). According to Kushlan (1990), undulating topography within the Kissimmee Valley has created many isolated swale marshes that blend into drier grasslands known as the Kissimmee Prairie.

The Kissimmee River basin is bounded on the north by lakes of the Orlando area, on the west by the Peace River basin, on the east by the upper St. Johns River basin, and to the south by Lake Okeechobee. Like many other rivers in peninsular Florida, the exact boundary of the basin is difficult to delineate in many places because of the overall flat topography. As a consequence, adjacent basins (as well as subbasins within the Kissimmee basin) frequently have surface water connections during high water periods. These periodic water connections include both natural and artificial waterways, mostly marshes, swamps, ditches, and small canals or channels. In contrast, a number of small lakes within the basin are relatively isolated and rarely have surface connections to other water bodies. Unlike most basins in central Florida, major springs are noticeably absent.

The Kissimmee River basin is part of a much larger and complex drainage unit, the 77,000 km^2 Kissimmee–Okeechobee–Everglades system, often referred to as the Greater Everglades Ecosystem (Kushlan 1991; Toth et al. 1997). Prior to drainage and installation of water control structures, the entire system was connected hydrologically. Historically, the unregulated Kissimmee River discharged freely into Lake Okeechobee, and during wet cycles, the lake would overflow its south bank, providing additional flow of freshwater to the Everglades (Light and Dineen 1994). For analytical purposes, I divide the Kissimmee River basin into three major subbasins: (1) lower Kissimmee, (2) upper Kissimmee, and (3) Lake Istokpoga (Figure 1).

Lower Kissimmee Subbasin

The lower Kissimmee Subbasin includes the Kissimmee River from its beginning at the southern end of Lake Kissimmee to its mouth on the north side of Lake Okeechobee (Figure 1). Drainage area estimates vary, ranging from 1,772 km^2 (Scarlatos et al. 1990) to 1,963 km^2 (FGFWFC 1957; USCOE 1991). The subbasin includes portions of five counties: Glades, Okeechobee, Highlands, Polk, and Osceola. The historic Kissimmee River was a low-gradient, highly meandering stream, about 166-km long, within a 1.5- to 3-km-wide floodplain. More than 20 lateral tributary sloughs are known (Toth 1996). Because of its many meanders, early boatmen commonly referred to the Kissimmee as a long and crooked river (Will 1965). The river averaged only about 1.2 m deep and flows were highly variable (Kushlan 1991; Toth 1995), and the natural drop from its outlet at Lake Kissimmee to the mouth was 11 m (USCOE 1991).

Prior to the 1960s, only portions of the Kissimmee River had been extensively modified. For instance, at the river's lower end a 5-km-long canal known as Government Cut was excavated that diverted flow and created an isolated remnant of the river known as Paradise Run. Nevertheless, the most drastic change was the creation of Canal C-38, between 1962 and 1971, which involved cutting through the river floodplain from Lake Kissimmee southward, effectively reducing the 160-km-long Kissimmee River main channel to a 90-km-long box-cut canal 9 m deep and 64 to 105 m wide (USCOE 1991; Toth et al. 1997, 1998). The remaining 70 km or so of natural meandering river were converted either into isolated oxbows or side-channels (often referred to as remnant river channels) with either one or both ends connected to the main canal. The project divided the artificial canal into a series of five impoundments (Pools A, B, C, D, and E) separated by a series of six water control structures (i.e., gated spillways with locks for boat passage) to regulate water levels and essentially eliminating seasonal and annual fluctuation in water levels. The project also resulted in the loss of more than half of the floodplain marsh. According to Bass (1983), the river has a gradient of 0.07 m/km. The mean pH is about 6.9 (Bass and Cox 1985). During the period 1980–1998, U.S. Geological Survey water temperature readings for the Kissimmee River near its mouth ranged from a low of 12°C (January 1981) to a high of 35°C (July 1996), with a mean of about 25°C. The main river/canal, however, is relatively warm and winter temperatures rarely drop below 15°C or16°C (L. Glenn, South Florida Water Management District [SFWMD], unpublished data).

In an effort to reverse negative environmental effects resulting from canal construction, portions of the river are in various stages of restoration under a project known as the Kissimmee River Restoration Project (Loftin et al. 1990; USCOE 1991; Toth et al. 1997). Restoration of the Kissimmee River began in 1994 (Toth et al. 1998). Although earlier proposals called for filling a 46-km length of the canal and diverting discharge to historical channels and floodplains, later plans subsequently were modified and scaled back. According to L. Toth (SFWMD, personal communication), the current restoration project includes removal of two of the dam-like water control structures along the canal. One of the dams was removed with explosives in early summer 2000.

Upper Kissimmee Subbasin

The upper Kissimmee or "Headwaters" area includes all water bodies from Lake Kissimmee upstream. It includes portions of Orange, Osceola, and Polk counties. The upper Kissimmee is the largest of the three subbasins, with drainage area estimates ranging from 4,150 km^2 (Scarlatos et al. 1990; Toth et al. 1997) to as much as 4,229 km^2 (FGFWFC 1957; USCOE 1991). State Road 60 marks its southern boundary where the subbasin's largest lake, Lake Kissimmee, discharges into the Kissimmee River (USCOE 1991). Although the subbasin includes several streams, it is dominated by an extensive network of at least 26 natural freshwater lakes, known as the Kissimmee Chain of Lakes, ranging in size from a few acres to 144 km^2 (Toth et al. 1997). In addition to Lake Kissimmee, major lakes in the upper Kissimmee include Cypress, Jackson, Hatchineha, Marian, Marion, Pierce, Rosalie, Russell, Tiger, Tohopekaliga, Weohyakapka, and 10 or more major lakes in what is known as the Alligator Lake Chain, including Alligator, Brick, Center,

Coon, East Lake Tohopekaliga, Gentry, Hart, Lizzie, Mary Jane, and Trout (Moyer et al. 1985a). During historic times, all or most of these lakes were interconnected by marsh and possibly by small natural channels. Nevertheless, a series of channels excavated in the 1880s and later enlarged now exist that permanently connect the larger lakes (USCOE 1991). Many of these interconnecting canals have water control structures. Major streams in the upper Kissimmee subbasin include Reedy Creek, Shingle Creek, Boggy Creek, and Lake Marion Creek.

Lake Istokpoga Subbasin

The third part of the Kissimmee River basin includes Lake Istokpoga (11,207 ha) and its tributary streams and lakes (USCOE 1991:5). The entire subbasin covers an area estimated to be somewhere between 1,611 and 1,684 km^2 in Highlands and southern Polk counties (FGFWFC 1957; Milleson 1978; USCOE 1991). According to Milleson (1978), the Istokpoga subbasin originates on the Highlands Ridge in the vicinity of Lake Clinch to the north and Lake Annie to the south. More than forty lakes are located within the subbasin, and many of these are connected by natural streams or by excavated channels and canals. At least 10 lakes are considered isolated (e.g., Lake Clinch). According to Kushlan (1990), Lake Istokpoga is part of the Kissimmee marsh complex and once was covered by shallow marsh, embedded with numerous deeper marshes. Other lakes in the Istokpoga subbasin greater than 1,000 ha surface area include Jackson, June-in-Winter, and Placid (formerly Lake Childs) in Highlands County, and Arbuckle and Reedy in Polk County. Major streams in the Istokpoga drainage include Reedy, Arbuckle (and its main tributaries Bonnet, Carter, and Morgan Hole creeks), and Josephine creeks. Although historically a part of the Kissimmee River basin, Lake Istokpoga and its drainages now provide only a portion of their former flows to the Kissimmee River (USCOE 1991). Until the late 1940s, the only surface water outlet from Lake Istokpoga was Istokpoga Creek, a natural lowland stream that flowed east, feeding directly into the Kissimmee River. Historically, during high water periods, Lake Istokpoga overflowed its southern banks and drained towards the southeast. To control flooding and provide agricultural irrigation, a canal parallel to the creek (Istokpoga Canal) with a water control structure (S-67) was constructed in 1949. In 1962, an additional canal system (C-41A) and several water control structures were installed on the southeast shore. As a result, major regulatory releases from Lake Istokpoga now are controlled by water control structure S-68 with water flow to Lake Okeechobee via C-41A, Indian Prairie Canal, and Harney Pond Canal. Canal C-41A flows into the lower Kissimmee Canal; however, Indian Prairie and Harney Pond canals discharge directly into Lake Okeechobee (Milleson 1978; USCOE 1991; Moxley et al. 1993). Because of the artificial changes to its drainage pattern, many recent authors do not include the Lake Istokpoga drainage as part of the Kissimmee River basin (e.g., Scarlatos et al. 1990; USCOE 1991; Koebel 1995).

Methods

Information on the fish fauna of the Kissimmee River basin has been drawn from a variety of sources: (1) fish samples by U.S. Geological Survey (USGS) personnel from in and around APR between January 19, 1995 and October 31, 1997, as well as personal knowledge and field experience in other parts of the basin; (2) vouchered specimen records in museum collections; (3) the primary literature; (4) the gray literature, including results of fish surveys and state stocking records; (5) personal communications with fishery biologists active in the basin; and (6) specimens obtained from anglers and others. All of the above sources are cited in Appendix A. Museum abbreviations are defined by Leviton et al. (1985).

The 1995–1997 sampling of APR included a total of 79 collections made at more than 60 different stations representing all major aquatic and wetland habitats. Sampling yielded more than 14,500 specimens and 44 species. In shallow-water areas, fishes were sampled using various types of small seines (typically 1.8 × 3 m dimensions with mesh sizes of 3.2 mm, 4.8 mm, or 6.4 mm), long-handled dip nets (3.2-mm and 6.4-mm mesh size), and backpack electrofishers. In deeper waters, a boat electrofisher was used to capture fishes. Other gear occasionally used included gill nets and small minnow traps. In most cases, sampling was done during daylight hours and was qualitative rather than quantitative. Sites sampled in the basin, outside APR, by my colleagues and me included lakes Marian and Istokpoga, and the Kissimmee River canal and remnant channels in pools B and E. The majority of voucher material from all the above collections is deposited at the Florida Museum of Natural History at the University of Florida (UF). A few voucher specimens are deposited at the North Carolina State Museum of Natural History (NCSM) and the University of Alabama (UAIC).

To locate voucher material collected by others, searches of museum databases were conducted via the Internet and by direct contact with curators. In many cases, museum holdings included collections not referenced in earlier reports or publications. Most include relatively small collections made by groups of ichthyologists or other individuals visiting the basin. In a few cases, where catalogued names were suspect, material was borrowed to verify identifications.

Most gray literature consists of reports providing results of surveys and stockings conducted by the Florida Fish and Wildlife Conservation Commission (FFWCC, until recently known as the Florida Game and Fresh Water Fish Commission). Many FFWCC surveys were large-scale efforts focusing on larger lakes and selected river and canal reaches. A wide variety of gear was used (e.g., block nets, rotenone, electrofishing, gill nets) and often resulted in the capture of many thousands of fishes. Nevertheless, no vouchers exist for the vast majority of specimens and for most species. I have excluded most unidentified material from the present analysis, except where noted. In addition, there are other taxonomic issues that had to be addressed. Most changes with respect to names are shown in Appendix A. For instance, following Gilbert et al. (1992a), all Kissimmee basin fish previously identified as *Fundulus cingulatus* are listed here as *F. rubrifrons*. Also, all past references to *Gambusia affinis* are treated here as *G. holbrooki*. In contrast, *Elassoma* listed in most reports and publications generally are referenced as *E. evergladei*, although I realize that some may have been *E. okefenokee*, now known to naturally occur in the basin (see later discussion concerning this species).

In addition to potential errors in nomenclature and identification, interpreting historical changes in fish assemblages has traditionally been suspect due to other problems. For many early collections, amount of effort and specific gear type at a given locality are largely unknown. Although many recent surveys have been quantitative, different surveys generally involved the sampling of differing types of habitat and relied on different types of methods and gear. Thus, results do not allow catch-per-unit-effort comparisons, either over time or among subbasins and major habitats. On the positive side, collecting effort and variety of gear used suggest that fishes of the basin have been well sampled, providing substantial information on species presence-absence and, to a lesser extent, on relative abundance. In this report (see Appendices A, B), relative abundances of different fishes within each subbasin are estimated based on frequency of capture and on how widespread the species appears to be based on selected recent sampling effort.

Most stocking records used in this report are based on information provided by the state-operated Richloam Fish Hatchery (Sumter County) for the years 1979–2002. Two other facilities in the state, Welaka National Fish Hatchery (Putnam County), in operation since 1926, and Blackwater Hatchery (Santa Rosa County) exist and also probably stocked fish in the Kissimmee basin on occasion. Most stocking records from the Welaka Hatchery do not include site-specific information. Early records from Blackwater Hatchery are not available.

Historical Perspective

Prior to 1900, and particularly during the late 1800s, a number of well-known researchers were actively describing and documenting inland fishes found in the southeastern United States (Bass and Cox 1985). Few scientists collected or examined fishes taken from the Kissimmee River basin during this period. The only recorded scientific observations before 1900 on fishes inhabiting the basin were by the Swedish naturalist and scientist Einar Lönnberg (1865–1942). From fall 1892 to summer 1893, Lönnberg traveled in Florida, a trip that included small collections of fishes at sites in and around Orlando and the town of Kissimmee. As part of several publications on Florida and its wildlife, Lönnberg (1894) included observations of Florida fishes. Lönnberg's contributions are of special interest because many of the specimens that he collected and preserved still survive in two major museums in Sweden, including material not mentioned in his 1894 publication.

In addition to the records of Lönnberg, there are a few published accounts providing anecdotal information on sport and commercially valuable fish found in the Kissimmee basin during the period before 1900. In a report for the U.S. Fish Commission, Smiley (1885) included a letter dated February 2, 1885, from a Mr. H. R. Clarke, who, while passing through the town of Kissimmee, noted that natives using trawling tackle take "large-mouth bass" and "croppies" (some weighing up to 16 lb) from area lakes as well as down the river to Kissimmee Lake and "Okechobee" (and elsewhere). Similarly, Daugé (1886) published a short article describing a boat trip by tourists venturing from Kissimmee City into nearby lakes during the 1880s. Daugé described the capture of a "trout" (probably largemouth bass) weighing about 5 kg. Will (1965) provided information on commercial fishermen who worked Lake Okeechobee and parts of the Kissimmee River during the late 1800s and early 1900s. Unfortunately, data on fishes are limited to a few minor details, mostly concerning catfishes and bass. Will gives the common names for a few food fishes, some apparently of local origin.

In 1886, Angelo Heilprin, a scientist associated with the Academy of Natural Sciences of Philadelphia (ANSP), traveled aboard the schooner "Rambler" through parts of Florida. According to Heilprin (1887), during 6 d on Lake Okeechobee his party made a small collection of fishes "some distance out from the mouth of the Kissimmee River" where they took "black-bass" (*Micropterus salmoides*) and what Heilprin described as a new catfish species, *Ictalurus okeechobeensis* (a species later determined to be a synonym of *Ameiurus catus*). Eschmeyer (1998) and Gilbert (1998) erroneously give the type locality of *I. okeechobeensis* as the Kissimmee River. In his 1887 publication, Heilprin makes no mention of entering the Kissimmee and his fish collections were obviously in the lake itself and not within the river mouth. Part of the confusion likely resulted from the locality data filed with the type specimens in Philadelphia, given in an abbreviated fashion as "Lake Okeechobee, Kissimmee River."

During the period 1900–1950, various groups of researchers passed through the Kissimmee River basin and made small collections of fishes. Most information from this time period is based on museum records. In the lower Kissimmee subbasin, fishes were collected by L. Giovannoli in 1931 and C. J. Goodnight in 1940. In the upper Kissimmee subbasin, museum specimens survive from collections of Edgar A. Mearns in 1901; F. Harper in 1917; O. C. Van Hyning in 1928; G. V. Harry and A. C. Bauman in 1939; C. Hemphill, S. M. Brown, and H. Trapido in 1940; V. Walters and P. Taleporos in 1945; and Ford, Parkes, Hilton, and Van Etten in 1949. In the Istokpoga subbasin, the only known fish collections were made by J. E. Hill and H. L. Hill in 1943. In 1948–1949, the FFWCC conducted studies in several lakes and connecting canals in the upper Kissimmee River basin. Lakes sampled included Kissimmee, Tohopekaliga, East Tohopekaliga, Hatchineha, and Cypress (FGFWFC 1957). They sampled fishes using haul seines (76-mm stretch mesh) and reported that the lakes supported a large game fish population, with black crappie being the most common. Lists of species taken are included in FGFWFC (1957); nevertheless, there are no known voucher specimens.

The period since 1950 has seen the most intensive sampling of the Kissimmee basin. Prior to the 1995–1998 sampling in APR, museum voucher specimens from this time period consisted largely of collections made by small groups of scientists, including E. C. Raney, C. R. Robins, R. H. Backus, and R. W. Crawford in 1952; J. M. Barkuloo and crew in 1960; N. R. Foster and party in 1962; C. C. Swift and R. W. Hastings in 1966; R. W. Yerger and crew in 1974; and D. A. Hensley and R. S. Dial in 1979. Various state and federal agencies conducted large- and small-scale investigations of fishes in the basin. In 1955–1956, the U.S. Fish and Wildlife Service (1958) gathered and reviewed fisheries data pertaining to the Kissimmee River basin. Although my copy of the report may be incomplete, it contains little information of value on fishes. In 1956–1957, the former Florida Game and Fresh Water Fish Commission sampled fishes in two areas within the lower Kissimmee River. The resulting report provides the only baseline data on prechannelization fish populations (FGFWFC 1957). Since 1970, there have been many investigations, of various focus and scope, on fishes in different parts of the drainage. The FFWCC has conducted numerous surveys, primarily focusing on fishes inhabiting larger lakes and main river and canal habitats. Except for a few studies conducted in the upper Kissimmee subbasin (e.g., Wegener et al. 1974; Guillory 1979; Guillory et al. 1979; Moyer et al. 1995), resulting data from most FFWCC investigations remain unpublished. Biologists of the South Florida Water Management District (SFWMD) and FFWCC, often in conjunction with university researchers, have conducted studies of fish populations in the lower Kissimmee subbasin. The thrust of these studies has been to assess the different stages of the Kissimmee River Restoration Project (e.g., Wullschleger et al. 1990a, 1990b; Bull et al. 1991; Miller 1997). The U.S. Geological Survey investigators, as part of the agency's National Water-Quality Assessment (NAWQA) program, annually sampled portions of the lower Kissimmee River from 1995 to 1998. Additional studies of consequence since 1950 include Yerger (1975)and Montalbano et al. (1979a, 1979b), among others.

Results

Faunal Composition

A total of 59 fish taxa (58 species and 1 hybrid), representing 23 families, have been recorded from the Kissimmee River basin from the late 1800s to 2002 (Appendices A and B). Of these, 48 species are native and one of these is known only from a single historical record. The remaining 11 taxa consist of introduced fishes, including foreign species, transplantation of fishes native to other parts of the continent, and an artificial hybrid. For some of the introduced forms there is no evidence of reproduction. Most species are permanent freshwater fishes, but a few are euryhaline. None of the Kissimmee's extant species are listed as endangered or threatened by federal or state authorities. Moreover, the basin has no endemic fish species (although the Florida peninsula itself apparently does harbor several endemic subspecies; see Gilbert 1987). All strictly freshwater fishes native to the basin are derived from North American temperate waters and most are relatively widespread in Florida and other parts of the southeastern United States. Considering natives, the four dominant families are Centrachidae (9 species), Cyprindae (5 species), Ictaluridae (5 species), and Fundulidae (5 species); in total they comprise 51% of the 47 extant native fishes. The smallest fishes in the basin are the pygmy killifish and the least killifish, both less than about 30 mm total length. The largest is the longnose gar (>1.5 m). The eastern mosquitofish is the most abundant and widespread fish. Native fishes in the basin that are relatively uncommon or rare include *Lepisosteus osseus, Anguilla rostrata, Notropis chalybaeus, N. petersoni, Aphredoderus sayanus, Strongylura marina, Menidia beryllina, Leptolucania ommata, Elassoma okefenokee, Lepomis auritus, Percina nigrofasciata*, and *Mugil cephalus*. There are no records of *Lepisosteus osseus, Aphredoderus sayanus, Menidia beryllina*, or *Lepomis auritus* from the Istokpoga subbasin, even though all four are either occasionally or regularly taken in the other two subbasins. The other extant native species have been recorded in all three subbasins.

Changes in the Native Fish Fauna

Human modification of the Kissimmee River basin undoubtedly resulted in changes in the relative abundance of many species. Nevertheless, there has been no system-wide loss of a native, with one possible exception. During examination of accession records of the Swedish Museum of Natural History (NRM), I discovered a collection of 22 *Cyprinodon variegatus* (NRM 30025, 10–25 mm SL) collected on 15 February 1893 by E. Lönnberg from a site identified as "Kissimmee Creek" in Osceola county. This is the only known record of *Cyprinodon* from the Kissimmee basin. Without explanation, Lönnberg (1894) makes no mention of this collection in his publication on Florida fishes, but did report on other samples taken near the town of

Kissimmee. Colleagues and I examined the NRM specimens and confirmed their identification as *C. variegatus*, but are uncertain if they represent the subspecies *C. v. hubbsi*, an inland form currently known only from sites in the adjacent St. Johns River basin. The more common *C. v. variegatus* has been recorded from several sites along the northern shore of Lake Okeechobee (Gilbert et al. 1992b). If the NRM locality is accurate, the disappearance of *Cyprinodon* from the Kissimmee basin represents the only record of a recent fish becoming extirpated from the system.

Based on comparisons of fish survey data from pre- and postchannelization of the Kissimmee River below Lake Kissimmee, some have argued that the 1962–1971 channel modification project caused the decline, and possible disappearance, of certain native fishes. For instance, Perrin et al. (1982; also see USCOE 1991) remarked on the absence of six indigenous fish species in collections made in the Kissimmee River/C-38 system after channelization, including three euryhaline fishes, *Anguilla rostrata*, *Menidia beryllina*, and *Mugil cephalus*, and three primary freshwater fishes, *Opsopoeodus emiliae*, *Notropis petersoni*, and *Percina nigrofasciata*. Elimination or reduction in numbers of the three euryhaline species was believed to have been the result of installation of control structures in the lower canal; absence of the other three species was attributed to alteration in stream flows due to channelization. Subsequent sampling in the subbasin revealed that three of the six species mentioned by Perrin et al. (1982) were still present. Consequently, Bull et al. (1991), repeated by Trexler (1995), reevaluated the list of species considered to possibly be disappearing. They reported *Percina nigrofasciata*, *Notropis petersoni*, and *Menidia beryllina* as absent in postchannelization collections within and around the Kissimmee River/ C-38 system.

In addition to localized loss of several species, other changes in fisheries resources have been noted. Based on 1950s sampling conducted within Government Cut and the unchannelized main channel, it was reported that total fish biomass in the channelized portions was significantly less than that of the historical Kissimmee River (Miller 1997). Later, following the 1962–1971 channelization of the entire river south of Lake Kissimmee, it was reported that the largemouth bass fishery had substantially decreased and that forage fish populations, also had declined (Perrin et al. 1982; but see Trexler 1995). Changes in fish populations were attributed to a combination of factors, including low dissolved oxygen levels in the canal and remnant channels, loss of wetland habitat, and lack of river habitat diversity in the canal (Perrin et al. 1982; Miller 1997). Decline or disappearance of some species, in particular *P. nigrofasciata* and possibly *N. petersoni*, was most likely the result of decreased habitat, especially large shallow areas with flowing, relatively clear water. Bull et al. (1991) concluded that the channelization project had degraded habitats in the Kissimmee River and resulted in the dominance of more tolerant species, particularly *Lepisosteus platyrhincus* and *Amia calva*. For instance, the investigators noted that these two species accounted for more than 60% of the total fish biomass in Pool B during their 1990 samples. Nevertheless, these species were fairly common in the Kissimmee prior to channelization. In the estimate of the USCOE (1991), the "rough fish" (gar and bowfin) to game fish ratio taken during the 1950s census is believed to have been about two to one.

Uncommon and Rare Fishes

At least 12 native fishes can be considered uncommon or rare in the basin (Appendix A). In addition, several fishes have been mentioned as possibly disappearing or declining following the 1962–1971 channelization work, whereas others are only known from postchannelization samples.

Lepisosteus osseus.—There is only a single record of longnose gar in earlier reports of Kissimmee basin fishes (FGFWFC 1957: Appendix B), but the species has been regularly taken during recent surveys, particularly in the upper Kissimmee subbasin. Because of the shortage of data, it is unclear if the overall population in the basin has actually changed to any significant extent. Perrin et al. (1982) and Toth (1995), based on data in FGFWFC (1957), listed *L. platyrhincus* as the only gar collected in the mid-1950s during prechannelization surveys in the lower Kissimmee subbasin. The 1957 report dealing with fishes from the lower river, however, used the general term "gar" and it is conceivable that any captured *L. osseus* were simply lumped under that general heading along with the more abundant *L. platyrhincus*. In their distribution maps for *L. osseus*, Wiley (1980) and Page and Burr (1991) show no records of this species in the Kissimmee River basin (or in other parts of the southern half of the Florida peninsula), indicating the lack of any museum material. Although locally uncommon, the species may be more abundant in the Kissimmee and other parts of central Florida than museum records and other accounts suggest. As is often the case, large fish specimens are much less likely to be preserved and subsequently deposited in museum collections. Differences in sampling techniques over the years also probably contributed to the fact that few, if any, specimens were taken by early investigators.

Anguilla rostrata.—Records suggest that American eel has never been common in the Kissimmee basin, even before extensive canalization of the river. Collections by FGFWFC (1957) and Bull et al. (1991) in the lower Kissimmee represent the only records of this species for the entire basin. FGFWFC (1957) reported capturing 11 eels, all taken in wire traps set in stations in the lower river, including several from the artificial channel known as Government Cut. No additional *A. rostrata* were recorded from the basin until Bull et al. (1991) reported taking a single specimen while electrofishing in a remnant river channel in Pool E in 1989. There is not enough information to determine if the species has truly declined. None of the recent studies has used traps specific for the capture of eels. Nevertheless, it is likely that construction of water control structures in the lower Kissimmee River has hindered upstream migration of the eel.

Notropis chalybaeus.—In the Kissimmee basin, the ironcolor shiner is known only from a few streams in the upper Kissimmee and Istokpoga subbasins. In 1952, specimens were taken from Josephine Creek in the Istokpoga subbasin (Cornell collection, CU 22095). Yerger (1975) reported taking 83 individuals from the upper Kissimmee subbasin at stations in the upper and middle reaches of Shingle Creek during a 1974 survey. During our 1995–1998 survey of APR, we collected a total of 167 specimens, all from two sites in the lower reaches of Arbuckle Creek, a blackwater tributary of Istokpoga Lake. Based on observations made in Florida, *N. chalybaeus* occurs in a diversity of stream habitats, but it prefers quiet pools in sluggish streams (Marshall 1947). McLane (1955) noted that the species is most frequently found in small streams flowing through bay and gum hammocks receiving drainage from pine flatwoods land. In many parts of its range, *N. chalybaeus* is rare or uncommon and apparently is on the decline in some areas (Albanese and Slack 1998). Despite its apparent rarity, the species is not listed as threatened or endangered by federal or state authorities, most likely because of its broad distribution. C. Gilbert and H. Jelks examined a series of preserved *Notropis chalybaeus* and *N. petersoni* specimens from the Kissimmee basin and confirmed identifications. It is difficult to distinguish between the two species, particularly those from the Kissimmee basin. Anal ray counts do not always provide a clear separation (1 *N. chalybaeus* examined has 7 rays, and 3 of 11 *N. petersoni* taken from the study area have 8 anal-fin rays). According to H. Jelks (U.S. Geological Survey, personal communication, 2000), *N. chalybaeus* taken from APR are less deep-bodied than those from the Florida panhandle. Specimens taken by Yerger (1975) from Shingle Creek in the upper Kissimmee subbasin are similar in appearance to specimens from the panhandle in terms of body shape. One trait sometimes used to distinguish *N. chalybaeus* from a number of other *Notropis* species is the presence of black pigment inside the mouth. The *N. petersoni* from the study area also have pigmentation inside the mouth, but it is typically slight. Both species usually have a pharyngeal teeth formula of 2,4–4,2. Given their similarity, it is possible that some minnows identified as *N. petersoni* taken during other surveys in the basin were actually *N. chalybaeus*. The absence of *N. chalybaeus* in the lower Kissimmee subbasin may reflect a lack of suitable habitat.

Notropis petersoni.—The coastal shiner has been reported from all three of the major subbasins of the Kissimmee River basin. The species is widespread and locally common in certain streams and lakes in the Istokpoga subbasin (Moxley et al. 1988, 1993; museum records) and in some lakes, canals, and streams in the upper Kissimmee subbasin (e.g., Yerger 1975; Guillory et al. 1979; Moyer et al. 1985i). Until recently, the only records from the lower Kissimmee were based on early collections made by FGFWFC (1957: Tables 37, 43). One specimen was taken in 1956 in a wire trap set along the main channel of the lower Kissimmee River adjacent to Government Cut, and two others were collected in 1957 during rotenone sampling of a marsh associated with the lower river. Unfortunately, there are no known vouchers to support these records. Apparently *N. petersoni* was never very common in the lower Kissimmee subbasin, and channelization may have eliminated it from parts of the basin. The only recent records of *N. petersoni* from the lower Kissimmee subbasin are based on larval stages collected in Pools A and C during 1997–1998. According to L. Glenn (SFWMD, unpublished data), although larvae of *N. petersoni* were collected regularly during that period with plankton nets, no adults were taken during electrofishing surveys (1992–1994) conducted in the same areas. If the larval identifications are correct, the results suggest the species has now apparently recolonized the lower Kissimmee subbasin, although it is possible that a small localized population has persisted since the 1950s.

Aphredoderus sayanus.—The pirate perch is known from the upper and lower Kissimmee subbasins, where it is occasionally taken in samples. It is surprising that there are no records of the species in the Istokpoga subbasin and few records in Florida south of the Kissimmee River. Little is known about its life history in Florida and what factors determine the species' distribution pattern in the state. Loftus (2000) could not confirm its presence in Everglades National Park, but we captured one specimen in Tamiami Canal near the northern boundary of Everglades National Park in October 2000. In addition, university researchers have taken pirate perch from sites in the Florida Everglades, all since early 2001 (J. C. Trexler, Florida International University, personal communication). It is conceivable that the species only recently invaded the southern part of the peninsula, likely dispersing via artificial canals.

Strongylura marina.—The Atlantic needlefish occurs in all three subbasins. In the Istokpoga subbasin it is known only from Lake Istokpoga (Moxley et al. 1988, 1993). This species was not collected during prechannelization surveys in the lower Kissimmee subbasin (FGFWFC 1957). There is, however, a record supported by a museum specimen (UF 68833), part of a 1966 collection from what was then (presumably) an unchannelized reach of the lower Kissimmee River. Since about 1980, *S. marina* has been more regularly taken in samples, mostly in the upper Kissimmee lakes. Overall, it probably is more common in the basin than sampling indicates, particularly in the larger water bodies. Absence of this species in most collections may be a sampling artifact. Ager (1971) reported *S. marina* reproducing in Lake Okeechobee. Given its common occurrence in some lakes, the species likely also is breeding in the upper Kissimmee basin.

Menidia beryllina.—The inland silverside, a marine species that ascends rivers, is relatively rare in the Kissimmee basin. A few records exist for the upper and lower Kissimmee subbasins, but there are none from Lake Istokpoga or its tributaries. Prior to about 1970, the only evidence of its presence in the upper Kissimmee is existence of a few museum specimens, three collections from 1939 (UMMZ fish collection) and two from the mid-1960s (UF fish collection). Recent upper-basin records, all from Lake Kissimmee, are provided by Williams et al. (1979) and Tugend (2001). In the lower-Kissimmee subbasin, *M. beryllina* was taken by the FGFWFC (1957) during the mid-1950s. The only other records from the lower Kissimmee are collections by Bull et al. (1994) and Furse and Davis (1996) from the 1990s.

Leptolucania ommata.—Early sampling suggested that the pygmy killifish was quite rare in the Kissimmee basin. During 1995–1998 surveys of APR, however, we collected it at 17 different sites and found the fish to be fairly common and widespread in shallow backwater habitats. Surprisingly, the APR collections document the only records of its existence in the lower-Kissimmee and Istokpoga subbasins. The pygmy killifish has been taken from a few sites in the upper Kissimmee, including a 1961 record from a tributary of Crooked Lake (ANSP collection) and in collections made during the 1980s at lakes Hart and Mary Jane (Moyer et al. 1985h, Moyer et al. 1985i). Distribution maps of Gilbert and Burgess (1980) and Page and Burr (1991) mistakenly show this species as absent from all or most of the Kissimmee basin. Similarly, Trexler (1995) did not include it in his list of Kissimmee River fishes. The shortage of data are most likely due to the fact that *L. ommata* occurs in habitats less sampled by fishery biologists, most often in places distant from main channels and other deeper bodies of water. In APR, we encountered the species in marshes, swamps, sluggish small streams, ditches, and shallow vegetated margins of lakes and ponds. Because of its small size, it likely is missed by most sampling gear.

Elassoma okefenokee.—There are only a handful of records of Okefenokee pygmy sunfish in the Kissimmee basin and only one or a few records from each of the three major subbasins. Gilbert (1987) and Trexler (1995) believed it was absent from the Kissimmee system. In 1893, E. Lönnberg collected specimens (which he later described as *Elassoma evergladei orlandicum*) from Fern Creek, Orange county, and from Tohopekaliga and other waters around the town of Kissimmee, Osceola county (Lönnberg 1894; Gilbert 1998). In 1945, V. Walters and P. Taleporos captured *E. okefenokee* at a site in Shingle Creek, Osceola County, in the upper basin (Cornell collection, CU 68482 [5]; identification verified by the ichthyologist R. M. Bailey). In 1931, L. Giovannoli collected four specimens (UF 2487) from near the mouth of the Kissimmee River (identification verified by C. R. Gilbert). Bull et al. (1994) reported taking *E. okefenokee* from two remnant channels of the Kissimmee River, including specimens from Micco Bluff Run and Montsdeoca Run in Pool C in 1993 and one specimen from Pine Island Slough in Pool B in 1994. Fortunately, vouchers exist to support their discovery (UF 96451 and 96452). Unaware of the Giovannoli specimens, Bull et al. (1994) mistakenly assumed that their specimens were the first record of the species in the Kissimmee River. Subsequently, Furse and Davis (1996) captured two additional specimens from Pool A in June 1995, and L. Glenn (SFWMD, unpublished data) reports capturing a few *E. okefenokee* with throw traps in "broadleaf marsh" in Pools A and C and within "woody shrub" in Pools A and D during 1996–1999. We captured a single specimen from Arbuckle Creek in APR in 1996, the only record of *E. okefenokee* for the Istokpoga subbasin. Although the species probably has never been common in the Kissimmee basin, its distribution and abundance in the region are probably underestimated. During the past few decades, state personnel reported capturing large numbers of *Elassoma* during lake surveys in the upper Kissimmee subbasin (e.g., Moyer et al. 1985b, 1985c, 1985d, 1985e, 1985f, 1985g, 1985h, 1985i, 1990, 1991). All *Elassoma* were reported as *E. evergladei*, but it is conceivable that some may actually have been *E. okefenokee*. Unfortunately, there are no known vouchers to verify identifications.

Lepomis auritus.—In the Kissimmee basin, the redbreast sunfish is known only from the upper and lower subbasins. The species is most common along the main river channel. In APR, we captured a total of 15 *L. aurtitus* at six different sites, all from the Kissimmee River canal and nearby remnant channels. Distribution maps of Lee et al. (1980) and Page and Burr (1991) erroneously show this species as absent from the Kissimmee basin. Based on my own sampling, it appears that the Kissimmee basin is at or near the southern boundary of its range. While sampling the lower Kissimmee River canal in 1997, we collected one specimen that we tentatively identified as a hybrid *L. auritus* × *L. macrochirus.*

Percina nigrofasciata.—The Kissimmee basin is unique in terms of the natural distribution of the blackbanded darter. The Kissimmee forms the far southern limits of the darters' range and also is the only basin where the species occurs within the entire southern half of the Florida peninsula (Gilbert 1987). Due to various factors, including its need for flowing waters, *P. nigrofasciata* apparently has never been particularly common or widespread in the basin. Only one or a few records exist from each of the three major Kissimmee River subbasins. In reference to the lower Kissimmee subbasin, Bull et al. (1991) and Trexler (1995) stated that *P. nigrofasciata* has not been collected in the Kissimmee since channelization. The FGFWFC (1957) report taking a total of 11 specimens with wire traps from several locations, typically sand-bottom sites, in the lower river. Included were two sites in the artificial canal Government Cut. The only other record from the subbasin is a 1966 collection that resulted in two museum specimens of *Percina* (UF 98827), from what was presumably still an unchannelized reach of the lower Kissimmee River. During our 1995–1998 survey of APR, we captured a total of 38 *P. nigrofasciata* at four stream sites in the Arbuckle Creek drainage, Istokpoga subbasin. Other museum material includes specimens taken from stream sites in the Arbuckle Creek drainage in 1966 and 1979. The only record from the upper Kissimmee subbasin is a collection in Shingle Creek from the 1970s (Yerger 1975).

Mugil cephalus.—Striped mullet has been recorded from inland waters throughout much of peninsular Florida (Burgess 1980b; Loftus 2000); however, it is relatively uncommon and restricted in distribution in the Kissimmee basin. All basin records are from the lower Kissimmee subbasin. The only prechannelization record appears in FGFWFC (1957), a report of a single "mullet" from a May 1957 gill-net sample. All other records are based on post-1980 surveys (Wullschleger et al. 1990a; Bull et al. 1991, 1994; Furse and Davis 1996). Because no museum voucher material existed for the basin, the distribution map of Burgess (1980b) shows the species as absent in the Kissimmee basin. Since July 1996, a ban on inshore use of commercial gill nets has been in effect and striped mullet numbers appear to be increasing in many areas. There is, however, no information as to whether this species is expanding its range or increasing its numbers in the Kissimmee basin.

Two additional species, *Fundulus lineolatus* and *Lepomis marginatus*, were not recorded during prechannelization surveys in the lower Kissimmee subbasin, but recent samples suggest both fishes are relatively common in their appropriate habitats, particularly in other parts of the basin (Appendix A). The reason for their absence in earlier surveys is unclear, but may be attributable to bias based on gear-types used and habitats sampled.

Introduced Foreign Fishes

At least seven alien fishes have been recorded for the Kissimmee River basin and most of these are established (Appendix B). The first documented cases are from the 1970s (Figure 2). Three recent species to become established date from the late 1990s. All were probably present in the basin several years or more before presence was initially confirmed. Several additional foreign fishes are reported from Lake Okeechobee, but there are no confirmed records of their occurrence in the Kissimmee basin.

Cyprinus carpio.—The common carp, a native of Eurasia, has been widely introduced in the United States (Fuller et al. 1999). Although common in much of temperate North America, the species is uncommon in most of Florida. Although Trexler (1995) presumably considered the species established in the Kissimmee basin, there is little or no

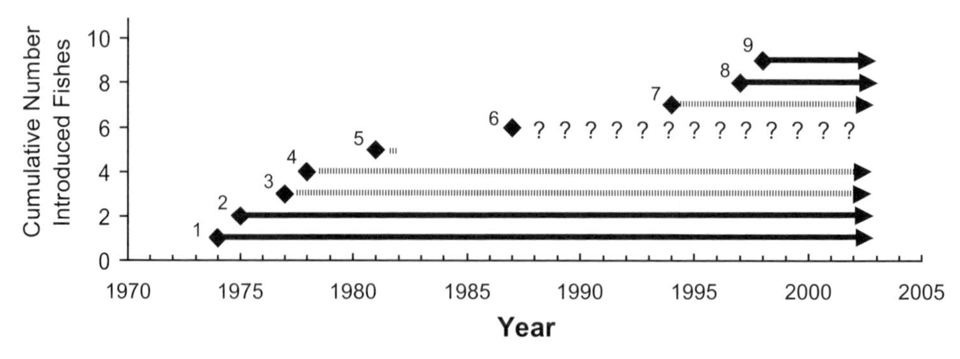

Figure 2.—Time line showing earliest reported occurrence and subsequent persistence of nonnative fishes introduced into the Kissimmee River basin. 1 = *Oreochromis aureus*; 2 = *Clarias batrachus*; 3 = *Ctenopharyngodon idella;* 4 = *Morone chrysops* × *M. saxatilis*; 5 = *Sander vitreus*; 6 = *Cyprinus carpio*; 7 = *Astronotus ocellatus*; 8 = *Pterygoplichthys cf. disjunctivus*; 9 = *Hoplosternum littorale.* Solid line = established (reproducing) population; segmented line = nonreproducing or reproductive status unknown; ? = uncertain. (See text for source references and discussion of *Dorosoma petenense*).

evidence of breeding populations in the basin or elsewhere in much of peninsular Florida. The only record of *C. carpio* in the Kissimmee basin is from the lower Kissimmee subbasin. Davis et al. (1990) reported taking a single specimen from Paradise Run while electrofishing in November 1987. Nevertheless, it is probable that populations of ornamental strains (e.g., koi) are present in urban ponds within the basin.

Ctenopharyngodon idella.—The Asian grass carp has been widely introduced throughout much of the United States (Fuller et al. 1999). There are records of grass carp, either being stocked as vegetation control agents or as being sighted or recaptured during fish surveys, for all three major subbasins. Richloam Fish Hatchery records, for the period July 1974–October 2002, indicate the state has stocked more than 155,000 grass carp in public lakes and ponds in the upper Kissimmee and Istokpoga subbasins since the late 1970s (Thomas 1994). More than 125,000 triploids were released into Lake Istokpoga during the period 1989–1993 (Starling 1990, 1991, 1993; Thomas 1994). The majority of grass carp stocked in the Kissimmee River basin were triploids, but some earlier stockings in the upper Kissimmee subbasin involved use of diploids and hybrids. For instance, 291 diploid grass carp were stocked into Lake Wales during the period 1979–1980, and approximately 7,000 hybrid grass carp (female) × bighead carp *Hypophthalmichthys nobilis* (male) were stocked in the Orlando area during the early 1980s (Starling 1981–1984; C. Starling, FFWCC, personal communication). Lake Conway was stocked with 7,000 grass carp in September 1977 (Keown and Russell 1982). Although the ploidy of these fish was not reported, the Lake Conway stocking may have involved diploids or possibly hybrids. An estimated 125,300 triploid grass carp were stocked into Lake Istokpoga during 1989–1993. To date, there is no evidence of reproduction and establishment of this species in the Kissimmee basin is considered questionable. Nevertheless, stocked individuals may live well over 10 years. The state of Florida restricts import of grass carp, but does allow citizens to purchase and stock triploid grass carp through a permit program.

Clarias batrachus.—The walking catfish is an Asian species first introduced into peninsular Florida in the mid-1960s. Soon after initial introduction, the species rapidly became established and spread northward from southeast Florida. It invaded the basin, entering the lower river and canal about 1975, possibly earlier (Courtenay 1975). For instance, Courtenay and Miley (1975) noted the appearance of walking catfish in the nearby Indian Prairie Canal as early as 1971. The catfish most likely entered the Kissimmee River via Lake Okeechobee, having been present near the river's mouth since 1972 (Courtenay 1978). By 1979, the species had spread throughout much of the lower Kissimmee and Lake Istokpoga (Courtenay 1979: Figure 1). Currently, the walking catfish is firmly established in all three subbasins, although periodic cold winters reduce its numbers.

Pterygoplichthys cf. *disjunctivus.*—This South American armored sailfin catfish has been present in Florida for more than a decade, first reported from the Hillsborough River drainage (Fuller et al. 1999). The first confirmed record for the Kissimmee basin is from lakes in the upper Kissimmee subbasin. A fish camp operating at Lake Marian reported catching hundreds of "plecos" in cast nets during the late 1990s. Subsequently, we collected two specimens in Lake Marian on 11 April 1997. Juvenile sailfin catfish, probably this species, were taken from Lake Kissimmee by researchers from the University of Florida in July 1999. The species may have originally entered the basin from the northwest by the natural spread of populations already present in the adjacent Peace River (where it has been present since about 1995). Nevertheless, intentional transport by humans cannot be ruled out. Its abundance, population size structure, and distribution in the basin provide evidence of its establishment. The spread of *Pterygoplichthys* into the lower Kissimmee subbasin has been confirmed based on a specimen taken from MacArthur Run in April 2001 (L. Glenn, SFWMD, unpublished data). Currently I am unaware of reports of its occurrence in the Istokpoga subbasin, but I suspect the species is present in at least part of that drainage. During recent years, large numbers of *Pterygoplichthys* cf. *disjunctivus* have been found in and around Lake Okeechobee (L. G. Nico, personal observation). The origin of this population may represent colonization by individuals moving down the Kissimmee canal. Existence of seemingly abnormal or intermediate abdominal color patterns in some Florida specimens suggests the possibility of introgressive hybridization (Nico and Martin 2000). Some Florida *P.* cf. *disjunctivus* are indistinguishable from the nominal *P. anisitsi* (Eigenmann and Kennedy 1903), giving rise to questions of species identity and whether the Kissimmee basin includes one or two species, or possibly a hybrid mixture. Genetic work, in progress, may help resolve the identification problem. Another member of the genus, *P. multiradiatus*, is established in southern Florida, but this species is not known to have spread north into Lake Okeechobee. A large die-off, apparently cold related, involving many foreign and native fishes occurred during the 2000–2001 winter in peninsular Florida. In January 2001, I counted 89 carcasses of *P.* cf. *disjunctivus* along 500 m of shoreline of the St. Lucie Canal at its junction with Lake Okeechobee. There also were reports of cold-related deaths of *Pterygoplichthys* and *Hoplosternum* in the Kissimmee basin. Nevertheless, surviving catfish populations apparently still thrive and young juveniles, presumably young of the year, were taken during the latter part of 2001. Unknown is whether there is any natural selection for individuals more tolerant of cooler temperatures.

Hoplosternum littorale.—The brown hoplo is native to lowland areas of South America east of the Andes. In Florida, it was first discovered in 1995 in the Indian River Lagoon drainage (Nico et al. 1996). Earliest records from the Kissimmee basin date to May 1998, when the species was taken in the lower Kissimmee subbasin. First reports included captures by anglers fishing the Kissimmee River near Avon Park Air Force Range (L. G. Nico, unpublished data) and several specimens taken by state researchers in two remnant river channels (L. Glenn, SFWMD, unpublished data). *Hoplosternum* probably invaded the basin from the east, apparently spreading into the basin from the Indian River via waters of the upper St. Johns River. Within a few years, *Hoplosternum* has spread rapidly through much of the Kissimmee basin and records now exist for all three major subbasins. As with *Pterygoplichthys*, it is conceivable that spread of this fish has been augmented by human transport. In some lakes, this catfish is now abundant. In addition to preserved material from APR within the lower Kissimmee subbasin, voucher specimens are available from several upper Kissimmee sites, including a marsh adjacent to Lake Kissimmee and lakes Cypress, Hatchineha, and Tohopekaliga, and from a site in the Istokpoga subbasin in the vicinity of Archbold Biological Station. Researchers captured 18 *Hoplosternum* from several remnant channels in the lower Kissimmee subbasin during the period 1998–2002 (L. Glenn, SFWMD, unpublished data). Recently, Nico and Muench (2004) described the nests and nest habitats of *H. littorale* populating Lake Tohopekaliga.

Oreochromis aureus.—The blue tilapia is an African cichlid first brought into Florida in 1961 by the former Florida Game and Fresh Water Fish Commission (Buntz and Manooch 1969). Subsequently, the species became established throughout the central part of the state. Based on the distribution map of Courtenay et al. (1974), as early as 1974, *O. aureus* was already present in many parts of the Kissimmee basin, including most of the upper Kissimmee subbasin and portions of the other two subbasins. More than likely, it invaded the basin several years before, apparently the result of natural spread from basins to the northeast, possibly in combination with intentional releases. Since the mid-1970s, the species has extended its range throughout the Kissimmee River main channel and associated backwaters.

Astronotus ocellatus.—The cichlid genus *Astronotus* is native to South America. The oscar *A. ocellatus*, or a closely related species, has been present in Florida since the 1950s (Courtenay et al. 1974). Introduced populations are firmly established in peninsular Florida, but most confirmed records are from areas south of Lake Okeechobee. The only reports of *Astronotus* in the Kissimmee River basin are from the lower Kissimmee subbasin and are based on creel survey records of the mid-1990s. A total of three oscar were reported from the Pool D–E area for the period from winter 1994 to spring 1996 (Furse and Davis 1995:10, Table 4; Furse and Davis 1996: Tables 3, 4). According to D. Fox (FFWCC, personal communication), the oscar has been in the lower Kissimmee River for a number of years and is frequently taken by anglers. There are no museum voucher specimens from the basin and some questions remain concerning validity of the Kissimmee records.

Unconfirmed foreign fishes.—Several other foreign fishes may occur in the Kissimmee River basin, but to date, there is insufficient evidence to confirm their presence. As many as three cichlid species reportedly reside and are possibly established in or around the mouth of the Kissimmee River or in northern portions of Lake Okeechobee. These include the black acara *Cichlasoma bimaculatum*, Mayan cichlid *C. urophthalmus*, and spotted tilapia *Tilapia mariae*. All three are firmly established in south Florida, and recent records suggest dispersal northward. Still, there are no voucher specimens or valid records from the Kissimmee basin itself. Recently, I obtained voucher specimens of *C. urophthalmus* and *T. mariae* captured in a backwater area along the northeast side of Lake Okeechobee near 27°11'N latitude. D. Fox (personal communication) informed me that *C. bimaculatum* has been present in or around the mouth of the Kissimmee River for 10 years or longer, but no voucher specimens exist. Fox also remarked on the occasional capture of pacus (*Piaractus* or *Colossoma*) in the Lake Okeechobee area and mentioned a recent report of an angler having transported and released peacock cichlid *Cichla* sp. from south Florida into the lake. There is, however, no evidence of reproduction and no reports confirming that either pacu or *Cichla* are established in the Kissimmee basin. Nevertheless, pacus, close-relatives of piranhas, are common aquarium species and probably released periodically into urban ponds within the basin. Goldfish *Carassius auratus*, as with ornamental carp, also are probably stocked in urban ponds. In fact, Hensley and Courtenay (1980a) show goldfish to be distributed throughout the Florida p0eninsula. I have not, however, uncovered any confirmed records for the Kissimmee basin.

Transplanted and Stocked Natives

Several North American fishes have been stocked in the basin, including some not native to the basin (Appendix B; Figure 2).

Sander vitreus.—The only known occurrence of this species in the Kissimmee basin is a stocking of 2,000 walleye into Lake Emerald, Orange County, in the upper Kissimmee subbasin during the period 1981–1982 (Starling 1982; Chapman et al. 1983; C. Starling, personal communication). The fish were originally produced in Kentucky. The stocked population reportedly did not survive more than a year (C. Starling and P. Chapman, FFWCC, personal communication).

Morone chrysops × *M. saxatilis hybrids.*—Names for the different "hybrid striper" crosses include both sunshine bass (i.e., white bass *Morone chrysops*, female × striped bass *M. saxatilis*, male) and palmetto bass (i.e., *M. chrysops*, male × *M. saxatilis*, female). Based on records from Richloam Fish Hatchery, the FFWCC has stocked more than 350,000 hatchery-produced hybrid *M. chrysops* × *M. saxatilis* (including both sunshine bass and palmetto bass) in public lakes in the upper Kissimmee and Istokpoga subbasins since the late 1970s (Starling 1974–2002). In the upper Kissimmee subbasin, stocked lakes have included Cherokee, Clear, Emerald, Eva, Moody, and Wales. In the Istokpoga subbasin, stocked lakes have included Bonnet, Francis, Huntley, Jackson, June-in-Winter, Red Beach, Reedy, and probably others. According to Champeau (1984), about 39,000 hybrid striped bass were stocked in Reedy Lake during the period 1980–1983. In total, approximately 129,600 sunshine bass-type hybrids were stocked in Reedy Lake during the period 1979–1986 (Champeau 1984; Starling 1974–2002). To date, there are no reports of these hybrids moving into the lower Kissimmee subbasin.

Dorosoma petenense.—Threadfin shad has been widely introduced as a forage species, making an accurate delineation of its native range difficult (Burgess 1980a; Fuller et al. 1999). Most publications treat this species as a Florida native. Nevertheless, C. R. Gilbert (Florida Museum of Natural History, personal communication) believes there is a considerable amount of circumstantial evidence to indicate that the species is not native east of the Mississippi River basin, but was introduced as a forage fish beginning in the early part of the 1900s (also see Ravenel 1901; Fuller et al. 1999:48).

Stocked natives.—In addition to the above mentioned stocked fishes, several fishes native to the Kissimmee basin have had their populations supplemented by stocking of hatchery-produced individuals. Based on Richloam Fish Hatchery records for the period 1974–2002, selected lakes and ponds in the Kissimmee basin have been stocked with more than 81,000 channel catfish, more than 93,000 largemouth bass, more than 32,000 black crappie, and more than 6,000 bream (a category that includes both bluegill and redear sunfish) (Starling 1974–2002). I do not have information concerning the genetic background of the stocked populations. Welaka National Fish Hatchery has distributed thousands of native fishes, including *I. punctatus*, *M. salmoides*, *L. macrochirus*, and *L. microlophus*, to places in central Florida. File records from the period 1976–1992, most relating to an earlier farm pond program, indicate that many live fishes were given to private interests at distribution points in the towns of Kissimmee, Sebring, and Okeechobee. Unfortunately, files do not provide information on the locations and names of water bodies in the Kissimmee basin, except for a 1993 record of 8,200 juvenile *M. salmoides* stocked into Clear Lake (Orlando).

Discussion

In large part, peninsular Florida's sparse freshwater fish fauna may be attributed to its young geological history (Gilbert 1987; but see Trexler 1995). Most native freshwater fishes are themselves descended from recent invaders that reached the peninsula from the main continent. Multiple invasions likely were associated with major fluctuations in sea level since the original mid-Oligocene (30 million years before present) emergence of the Florida Platform. Because of their supposed origin as opportunistic invaders, most native freshwater fishes in peninsular Florida are relatively resilient, widespread, and common. The present-day Kissimmee River fauna consists of native and nonnative fishes, mostly primary freshwater species. Unlike other major Florida rivers, the Kissimmee does not drain directly into the ocean. Therefore, its assemblages include fewer marine fishes than the nearby Peace and Caloosahatchee rivers. With one possible exception (i.e., *Cyprinodon variegatus*), there have been no basin-wide losses of native fishes from the Kissimmee. Consequently, when assessing change in peninsular Florida waters, extinction or extirpation of fishes appears to be a poor measure of habitat disturbance.

During recent decades, fish assemblages of the Kissimmee and Florida's other subtropical rivers have been relatively dynamic, due to new introductions, range expansions of nonnative fishes already present, and periodic declines in nonnative fish populations during occasional harsh winters. Currently, more than 30 nonnative fishes are established in Florida (Fuller et al. 1999; Nico and Fuller 1999). The addition, spread, and abundance of nonnative fishes, particularly foreign species, in the Kissimmee basin likely is linked to many factors, including widespread habitat disturbance, a subtropical climate, as well as the fact that the basin is centrally located within a region where basin boundaries are indistinct and introductions of foreign fishes commonplace. The basin has fewer introduced foreign fishes than south Florida canals, but the number of foreign species established in the basin continues to increase at a fairly rapid pace, apparently aided by dispersal via canals (Courtenay and Miley 1975).

According to Gilbert (1987), certain native fishes initially colonized central peninsular Florida by way of the Kissimmee. Recent evidence indicates nonnative fishes also are using the river as a dispersal pathway. The first appearance of foreign fishes in the basin coincided with the complete channelization of the Kissimmee River in the 1970s. Although the massive modification of the aquatic system is not linked to the initial introduction of nonnative fishes, artificial waterways connecting the upper lakes and channelization of the Kissimmee River have facilitated their rapid dispersal. Originally, a natural meandering river 160 km long, the post-1970 channel is a deep, relatively straight canal 90 km in length. Available evidence indicates that *Oreochromis aureus*, and probably *Pterygoplichthys* cf. *disjunctivus* and *Hoplosternum littorale* made their first appearance in the basin in the north and, within just a few years, invaded the southern portions of the basin and Lake Okeechobee. Downstream spread was most likely aided by passage through the main canal. *Oreochromis* and *Pterygoplichthys*, in particular, are often abundant in artificial waterways and their dispersal southward was most certainly by way of the main Kissimmee Canal. *Hoplosternum* typically inhabits backwater areas. Nevertheless, the earliest record of this catfish in the lower Kissimmee subbasin is from the main Kissimmee Canal. The catfish's rapid rate of colonization of the basin and its early occurrence in the main channel provide supporting evidence that *Hoplosternum* is using artificial waterways as dispersal routes.

In contrast to the above-named foreign species, *Clarias batrachus* invaded the Kissimmee basin from the south. Although canals south of Lake Okeechobee certainly promoted the species' northward expansion into the lake, it is not clear to what extent the channelized Kissimmee aided the catfish in colonizing more northern parts of the basin. Likely, *Clarias* invaded the Lake Istokpoga subbasin via artificial canals that directly connect Istokpoga to Lake Okeechobee. A number of other foreign fishes are present in or around the mouth of the Kissimmee River and in Lake Okeechobee. Although many of the foreign fishes now present in peninsular Florida are largely of tropical origin, some species (e.g., *Hoplosternum*) have native ranges that extend into temperate zones. Fortunately, most foreign fishes established in Florida are strictly tropical forms with low tolerance of cold temperatures, and thus are unlikely to migrate from Lake Okeechobee very far north into the Kissimmee basin.

Many of the more successful and widespread foreign fishes introduced to peninsular Florida share certain life history attributes; some of these characteristics may prove useful in predicting future invasions (Nico and Fuller 1999). In the Kissimmee basin, all established foreign species are of medium- to large body size, and among the largest in terms of body weight and length of all fishes found in the basin (Figure 3). A larger body size provides an advantage because of reduced risk of individuals being preyed on by other animals. In addition, nearly all established foreign fishes in peninsular Florida exhibit high levels of parental care, including all cichlids and the catfishes *Pterygoplichthys* and *Hoplosternum*, among others. Trexler (1995) categorized native and introduced Kissimmee basin fishes according to their tolerance for low oxygen conditions, reproductive mode, and trophic level. His classification scheme is valuable in evaluating the more recent introductions. Nearly all foreign fishes established in peninsular Florida are tolerant species; almost all are able to survive in extremely disturbed habitats. In fact, several widespread introduced catfishes are air breathers (e.g., *Clarias*, *Pterygoplichthys*, and *Hoplosternum*).

All foreign fishes now established in the Kissimmee basin are unique in terms of diet preferences, and even more so if body size also is taken into consideration (Figure 3). In assigning trophic position, Trexler (1995) categorized *Ctenopharyngodon* as an herbivore, *Cyprinus* as an omnivore, and *Oreochromis aureus* as a mixed herbivore/detritivore. *Ctenopharyngodon* is unique in the Kissimmee in that it feeds heavily on aquatic macrophytes. In addition to *O. aureus*, two of the more recent foreign invaders feed heavily on detritus. These include *Pterygoplichthys*, an algae-grazing catfish that I categorize as a mixed herbivore/detritivore (although it also consumes benthic invertebrates), and the benthic-feeding *Hoplosternum*, predominantly an invertivore/detritivore (Figure 3). The only large native fish that feeds on detritus to any significant extent is *Mugil cephalus*, a mixed herbivore/detritivore (Trexler 1995). The benthic-dwelling *Erimyzon sucetta* consumes detritus on occasion (e.g., Shireman et al. 1978), but diet of wild populations in Florida consists largely of invertebrates (W. F. Loftus, U.S. Geological Survey, personal communication). The significance of the addition of *Oreochromis*, *Ctenopharyngodon*, *Pterygoplichthys*, and *Hoplosternum* to the Kissimmee basin aquatic food web is unknown, but their high abundances in certain areas are likely correlated with food availability. During the past 100 years, many aquatic habitats in Florida have exhibited increased levels of planktonic and attached algae and large accumulations of organic- and nutrient-rich sediments, changes generally attributed to increased nutrient loading from agricultural (including cattle ranch-

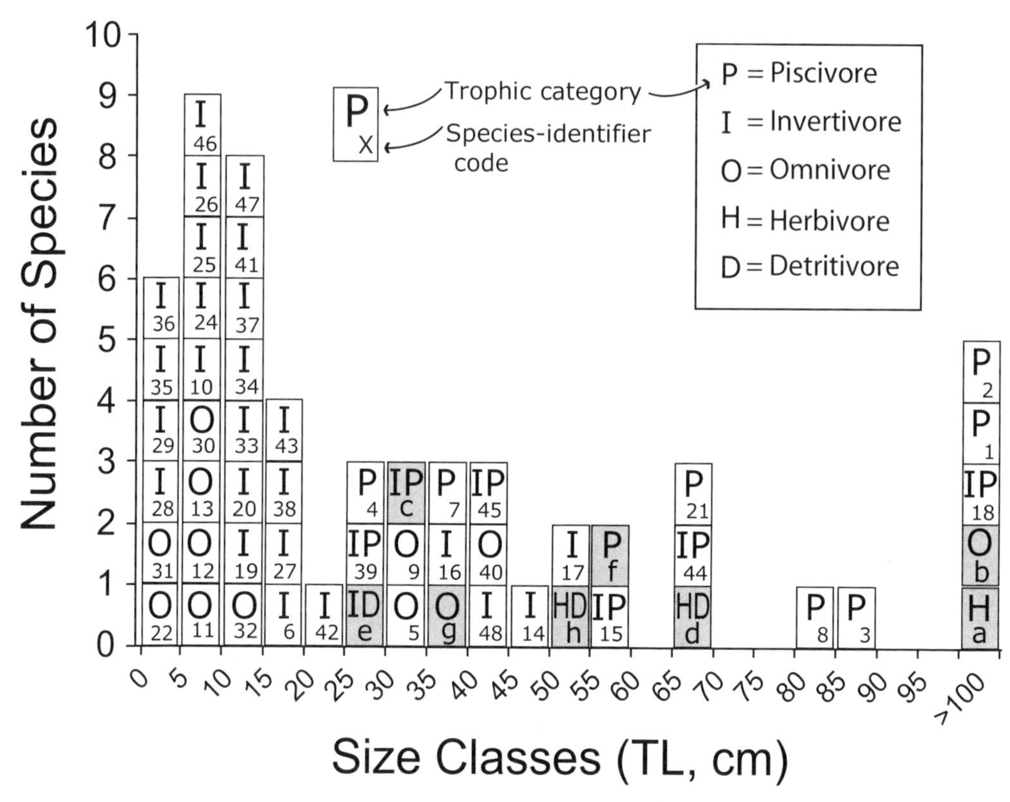

Figure 3.—Histogram showing extant Kissimmee basin fishes grouped by maximum body size (i.e., total length [TL]), with tropic category identified. Each box represents a different species. Maximum total lengths derived from Lee et al. (1980), except for certain introduced species. Most trophic category designations are taken from Trexler (1995). Species identification codes for natives (1–48) correspond to species sequence/number of Appendix A. Introduced species represented by shaded boxes. Identification codes for introduced fishes (a–h) are as follows: a = *Ctenopharyngodon idella;* b = *Cyprinus carpio*; c = *Clarias batrachus*; d = *Pterygoplichthys cf. disjunctivus*; e = *Hoplosternum littorale*; f = *Morone chrysops* × *M. saxatilis*; g = *Astronotus ocellatus*; h = *Oreochromis aureus*. Refer to Appendices A and B for status of species.

ing), residential, and urban sources, and also from hydrologic modifications and aquatic plant control (e.g., Moyer et al. 1995; Brenner et al. 1999). Thus, there appears to be no shortage of food resources to limit the numbers of detritus- and algae-feeding fishes. Unfortunately, interactions between introduced and native fishes are poorly understood. The possibility that introduced catfishes prey to any extent on the eggs of other species has not yet been determined. The increase in spawning and juvenile fish habitats potentially provided by the Kissimmee restoration may be offset by increased egg and juvenile mortality from invaders.

The future of the Kissimmee River and its ichthyofauna is unclear, but will probably continue changing. As in the recent past, the resilient native fauna will likely persist, but environmental modification may continue altering their patterns of distribution and abundance. The most obvious changes are expected to result from increases in the numbers of nonnative fishes already present and from invasions by new and different nonnative fishes in the future. Foreign nonnative fishes first appeared in the Kissimmee in the 1970s; most are much more recent. At this early stage, it is not possible to predict the long-term consequences caused by these introductions. We know a few details about the unusual trophic roles and other aspects of the life histories of certain nonnatives. Still, the ecological outcome may take decades to unfold, shaped by a complex interplay between established foreign fishes and their new environment.

Acknowledgments

This project was partially funded by the U.S. Air Force. I thank all the individuals of the Avon Park Air Force Range who provided assistance, in particular, Patrick Walsh, Paul Ebersbach, Robert Progulske, and Margaret Margosian. For help with field sampling, I thank Jon Brookshire, Gary Hill, Tim Hogan, Ann-Marie Holmes, Dan Johnson, Ricardo Lattimore, Gary Meffe, Linda S. Nico, Dave Reynolds, Kevin Schuck, William F. Smith-Vaniz, Johnna Thackston, and P. Walsh. James Williams provided critical help in sampling APR. Robert H. Robins was indispensable in both the field and laboratory, collecting, sorting, and identifying specimens. Carter R. Gilbert and Howard L. Jelks identified selected small fishes. Jason Evert assisted in data analysis. I also acknowledge the contributions of Albrey Arrington, Eva Hohausova, Ann Marie Muench, Roy Smith, Horace W. Barr, Jr., and the E. Keeler family, among others, for providing important fish voucher specimens. For contributing unpublished data, I thank Richard Owen, Frank Jordan (Loyola University, New Orleans), Michael Allen and Kim Tugend (University of Florida), and Lawrence Glenn and Louis Toth (SFWMD). The following FFWCC employees shared information: Thomas Champeau, Phil Chapman, Larry Davis, Don Fox, Robert Hujik, Carolyn Kendrick, Patrick Pence, Gary Warren, and Charles Starling and his coworkers at the Richloam Fish Hatchery. Mark Hoyer (University of Florida), James Milleson (formerly with SFWMD), and Allan Brown (U.S. Fish and Wildlife Service) provided additional help. The following museum curators and staff generously replied to queries or loaned specimens: R. H. Robins and George Burgess (Florida Museum of Natural History; University of Florida); Sven O. Kullander (Swedish Museum of Natural History, Stockholm, Sweden); Doug W. Nelson (Museum of Zoology, University of Michigan); Scott Schaefer, Barbara Brown, and Xenia Freilich (American Museum of Natural History); Jonathan W. Armbruster (Auburn University), John Friel and Charles Dardia (Cornell University Vertebrate Collections); Mats Eriksson (Uppsala University Museum of Zoology, Uppsala, Sweden); Hank Bart, Nelsen Rios, and Michael S. Taylor (Tulane University Museum of Natural History); Jeffrey T. Williams (National Museum of Natural History, Smithsonian Institution); and William Saul (Academy of Natural Sciences, Philadelphia). For comments, discussion, and suggestions on earlier versions of this manuscript, I thank Walter R. Courtenay, Jr., C. R. Gilbert, L. Glenn, H. L. Jelks, W. F. Smith-Vaniz, and Joel C. Trexler. Lisa Jelks kindly provided editorial comments.

References

Ager, L. A. 1971. The fishes of Lake Okeechobee, Florida. Quarterly Journal of the Florida Academy of Sciencc 34:53–62.

Albanese, B., and W. T. Slack. 1998. Status of the ironcolor shiner, *Notropis chalybaeus*, in Mississippi. Southeastern Fishes Council Proceedings 37:1–6.

Bass, D. G., Jr. 1983. Rivers of Florida and their fishes. Completion report 1980–1983 for Investigations Project, Dingell-Johnson Project F-36, North Florida Streams Research Project Study III. Florida Game and Fresh Water Fish Commission, Tallahassee.

Bass, D. G., Jr. 1991. Riverine fishes of Florida. Pages 65–83 *in* R. L. Livingston, editor. The rivers of Florida. Springer-Verlag, New York.

Bass, D. G., Jr., and D. T. Cox. 1985. River habitat and fishery resources of Florida. Pages 121–187 *in* W. Seaman, Jr., editor. Florida aquatic habitat and fishery resources. American Fisheries Society, Florida Chapter, Kissimmee, Florida.

Brenner, M., L. W. Keenan, S. J. Miller, and C. L. Schelske. 1999. Spatial and temporal patterns of sediment and nutrient accumulation in shallow lakes of the upper St. Johns River basin, Florida. Wetlands Ecology and Management 6:221–240.

Bull, L. A., L. J. Davis, and J. B. Furse. 1991. Lake Okeechobee-Kissimmee River-Everglades resource evaluation 1986–1991: Study I–Kissimmee River fisheries survey. Federal Wallop-Breaux completion report (F-52–5). Florida Game and Fresh Water Fish Commission, Tallahassee.

Bull, L. A., L. J. Davis, and J. B. Furse. 1994. Lake

Okeechobee-Kissimmee River-Everglades resource evaluation 1991–1994: Study III–Kissimmee River fish population surveys. Federal Wallop-Breaux completion report (F-52). Florida Game and Fresh Water Fish Commission, Tallahassee.

Buntz, J. and C. S. Manooch. 1969. *Tilapia aurea* (Steindachner), a rapidly spreading exotic in south central Florida. Proceedings of the Annual Conference of Southeastern Game and Fish Commissioners 22(1968):495–501.

Burgess, G. H. 1980a. *Dorosoma petenense* (Günther), threadfin shad. Page 70 *in* D. S. Lee, C. R. Gilbert, C. H. Hocutt, R. E. Jenkins, D. E. McAllister, and J. R. Stauffer, Jr., editors. Atlas of North American freshwater fishes. North Carolina State Museum Natural History, Raleigh.

Burgess, G. H. 1980b. *Mugil cephalus* Linnaeus, striped mullet. Page 779 *in* D. S. Lee, C. R. Gilbert, C. H. Hocutt, R. E. Jenkins, D. E. McAllister, and J. R. Stauffer, Jr., editors. Atlas of North American freshwater fishes. North Carolina State Museum Natural History, Raleigh.

Canfield, D. E., Jr., and M. V. Hoyer. 1992. Aquatic macrophytes and their relation to the limnology of Florida lakes. Final report submitted by the Department of Fisheries and Aquaculture Center for Aquatic Plants, University of Florida, Gainesville, to the Bureau of Aquatic Plant Management, Florida Department of Natural Resources, Tallahassee.

Champeau, T. R. 1984. Survival of hybrid striped bass in central Florida. Proceedings of the Annual Conference Southeastern Association of Fish and Wildlife Agencies 38(1984):446–449.

Chapman, P., F. Cross, and W. Fish. 1983. Sportfish introduction project, completion report (1982–1983), Study No. IV: walleye investigation. Florida Game and Fresh Water Fish Commission, Lakeland.

Collette, B. B. 1990. Problems with gray literature in fishery science. Pages 27–31 *in* J. Hunter, editor. Writing for fishery journals. American Fisheries Society, Bethesda, Maryland.

Courtenay, W. R., Jr. 1975. Status of introduced walking catfish in Florida. Bulletin of the International Union for Conservation of Nature and Natural Resources 6(6):23–24.

Courtenay, W. R., Jr. 1978. Additional range expansion in Florida of the introduced walking catfish. Environmental Conservation 5(4):273–275.

Courtenay, W. R., Jr. 1979. Continued range expansion in Florida of the walking catfish. Environmental Conservation 6(1):20.

Courtenay, W. R., Jr., and W. W. Miley, II. 1975. Range expansion and environmental impress of the introduced walking catfish in the United States. Environmental Conservation 2(2):145–148.

Courtenay, W. R., Jr., H. F. Sahlman, W. W. Miley, and D. J. Herrema. 1974. Exotic fishes in fresh and brackish waters of Florida. Biological Conservation 6:292–302.

Daugé, H. 1886. Mr. Weg's party on the Kissimmee. Harper's New Monthly Magazine. Reprinted 1987, pages 313–322 *in* F. Oppel and T. Meisel, editors. Tales of old Florida. Castle, Secaucus, New Jersey.

Davis, L. J., S. J. Miller, and J. G. Wullschleger. 1990. Paradise Run fisheries investigations. Pages 149–159 *in* M. K. Loftin, L. A. Toth, and J. Obeysekera, editors. Proceedings of the Kissimmee River Restoration Symposium, Orlando, Florida, October 1988. South Florida Water Management District, West Palm Beach.

Dineen, J. W., R. L. Goodrick, D. W. Hallett, and J. F. Milleson. 1974. The Kissimmee River revisited. In Depth Report, Central and South Florida Flood Control District (West Palm Beach, Florida) 2(2):1–12.

Eschmeyer, W. N., editor. 1998. Catalogue of fishes. Three volumes. California Academy of Sciences, San Francisco.

Fernald, E. A., editor. 1981. Atlas of Florida. Florida State University, Institute of Science and Public Affairs, Tallahassee, Florida.

FGFWFC (Florida Game and Fresh Water Fish Commission). 1957. Recommended program for Kissimmee River basin. Unpublished report: A contribution of Federal Aid Projects F-8-R, W-19-R, and W-39-R (August 1957). Florida Game and Fresh Water Fish Commission, Tallahassee.

Fuller, P. L., L. G. Nico, and J. D. Williams. 1999. Nonindigenous fishes introduced into inland waters of the United States. American Fisheries Society, Special Publication 27, Bethesda, Maryland.

Furse, J. B., and L. J. Davis. 1995. Kissimmee River-Lake Okeechobee-Everglades resource evaluation report 1994–1995: Study V–Kissimmee River fish population surveys. Florida Game and Fresh Water Fish Commission, Federal Wallop-Breaux F-52–9 performance report, Tallahassee.

Furse, J. B., and L. J. Davis. 1996. Kissimmee River-Lake Okeechobee-Everglades resource evaluation report 1994–1996: Study V–Kissimmee River fish population surveys. Florida Game and Fresh Water Fish Commission, Federal Wallop-Breaux F-52 completion report, Tallahassee.

Gilbert, C. R. 1987. Zoogeography of the freshwater fish fauna of southern Georgia and peninsular Florida. Brimleyana 13:25–54.

Gilbert, C. R. 1998. Type catalogue of recent and fossil North American freshwater fishes: families Cyprinidae, Catostomidae, Ictaluridae, Centrarchidae, and Elassomatidae. Florida Museum of Natural History, Special Publication no. 1., Gainesville.

Gilbert, C. R., and R. M. Bailey. 1972. Systematics and zoogeography of the American cyprinid fish *Notropis* (*Opsopoeodus*) *emiliae.* Occasional Papers of the Museum of Zoology, University of Michigan 664:1–35.

Gilbert, C. R., and G. H. Burgess. 1980. *Leptolucania ommata* (Jordan), pygmy killifish. Page 533 *in* D. S. Lee, C. R. Gilbert, C. H. Hocutt, R. E. Jenkins, D. E. McAllister, and J. R. Stauffer, Jr., editors. Atlas of North American freshwater fishes. North Carolina State Museum Natural History, Raleigh.

Gilbert, C. R., R. C. Cashner, and E. O. Wiley. 1992a. Taxonomic and nomenclatural status of the banded topminnow, *Fundulus cingulatus* (Cyprinodontiformes: Cyprinodontidae). Copeia 1992:747–759.

Gilbert, C. R., W. E. Johnson, and F. F. Snelson, Jr. 1992b. Lake Eustis pupfish *Cyprinodon variegatus hubbsi.* Pages 194–199 *in* C. R. Gilbert, editor. Rare and endangered biota of Florida. Volume II: fishes. University Press of Florida, Gainesville.

Guillory, V. 1979. A comparison of fish communities in vegetated and beach habitats. Florida Scientist 42:113–122.

Guillory, V., M. D. Jones, and M. Rebel. 1979. Species assemblages of fish in Lake Conway. Florida Scientist 42:158–162.

Heilprin, A. 1887. Explorations on the west coast of Florida and in the Okeechobee wilderness. Wagner Free Institute of Science, Philadelphia.

Hensley, D. A., and W. R. Courtenay, Jr. 1980a. *Carassius auratus* (Linnaeus), goldfish. Page 147 *in* D. S. Lee, C. R. Gilbert, C. H. Hocutt, R. E. Jenkins, D. E. McAllister, and J. R. Stauffer, Jr., editors. Atlas of North American freshwater fishes. North Carolina State Museum Natural History, Raleigh.

Hensley, D. A., and W. R. Courtenay, Jr. 1980b. *Clarias batrachus* (Linnaeus), walking catfish. Page 475 *in* D. S. Lee, C. R. Gilbert, C. H. Hocutt, R. E. Jenkins, D. E. McAllister, and J. R. Stauffer, Jr., editors. Atlas of North American freshwater fishes. North Carolina State Museum Natural History, Raleigh.

Hoyer, M. V., and D. E. Canfield, Jr. 1994. Handbook of common freshwater fish in Florida lakes. University of Florida, Florida Cooperative Extension Service and Institute of Food and Agricultural Sciences, Gainesville.

Hulon, M. W., and V. P. Williams. 1982. A preliminary evaluation of blue tilapia population expansion in Lake Tohopekaliga, Florida. Proceedings of the Annual Conference Southeastern Association of Fish and Wildlife Agencies 36(1982):264–271.

Keown, M. P., and R. M. Russell, Jr. 1982. Radiotelemetry tracking at Lake Conway, Florida. U.S. Army Engineer Waterways Experiment Station, Technical Report A-82-4, Vicksburg, Mississippi.

Koebel, J. W., Jr. 1995. An historical perspective on the Kissimmee River restoration project. Restoration Ecology 3:149–159.

Kushlan, J. A. 1990. Freshwater marshes. Pages 324–363 *in* R. L. Myers and J. J. Ewel, editors. Ecosystems of Florida. University Presses of Florida, Gainesville.

Kushlan, J. A. 1991. The Everglades. Pages 121–142 *in* R. J. Livingston, editor. The rivers of Florida. Springer-Verlag, New York.

Lee, D. S., C. R Gilbert, C. H. Hocutt, R. E. Jenkins, D. E. McAllister, and J. R Stauffer, Jr. 1980. Atlas of North American freshwater fishes. North Carolina State Museum Natural History, Raleigh.

Leviton, A. E., J. R. H. Gibbs, E. Heal, and C. E. Dawson. 1985. Standards in herpetology and ichthyology: Part I. Standard symbolic codes for institutional resource collections in herpetology and ichthyology. Copeia 1985:802–832.

Light, S. S., and J. W. Dineen. 1994. Water control in the Everglades: a historical perspective. Pages 47–84 *in* S. M. Davis and J. C. Ogden, editors. Everglades: the ecosystem and its restoration. St. Lucie Press, Boca Raton, Florida.

Loftin, M. K., L. A. Toth, and J. Obeysekera, editors. 1990. Proceedings of the Kissimmee River Restoration Symposium, Orlando, Florida,

October 1988. South Florida Water Management District, West Palm Beach.

Loftus, W. F. 2000. Inventory of fishes of Everglades National Park. Florida Scientist 63:27–47.

Lönnberg, E. 1894. List of fishes observed and collected in South-Florida. Oversigt af Kongl. Vetenskaps-Akademiens Forhandlingar, Stockholm 51(3):109–131.

Marshall, N. 1947. Studies on the life history and ecology of *Notropis chalybaeus* (Cope). Quarterly Journal of Florida Academy of Sciences 9(1946):163–188.

McLane, W. M. 1955. The fishes of the St. Johns River system. Doctoral dissertation. University of Florida, Gainesville.

Miller, D. E. 1997. Evaluation of fish populations and relative abundance within Pools A, B, and C of the Kissimmee River: possible responses to the Pool B Demonstration Project. Master's thesis. Florida Atlantic University, Boca Raton.

Miller, S. J. 1990. Kissimmee River fisheries: a historical perspective. Pages 31–42 *in* M. K. Loftin, L. A. Toth, and J. Obeysekera, editors. Proceedings of the Kissimmee River Restoration Symposium, Orlando, Florida, October 1988. South Florida Water Management District, West Palm Beach.

Milleson, J. F. 1976. Environmental responses to marshland reflooding in the Kissimmee River basin. Central and Southern Florida Flood Control District, Technical Publication 76–3 (September 1976), West Palm Beach.

Milleson, J. F. 1978. Limnological investigations of seven lakes in the Istokpoga drainage basin. South Florida Water Management District, Technical Publication 78–1, West Palm Beach.

Montalbano, F., III, K. J. Foote, M. W. Olinde, and L. S. Perrin. 1979a. Summary of selected fish and wildlife population data and associated recreational opportunities for the Kissimmee River valley. A report to the U.S. Army Corp of Engineers (May 1979). Florida Game and Fresh Water Fish Commission, Tallahassee.

Montalbano, F., III, K. J. Foote, L. S. Perrin, and M. W. Olinde. 1979b. Fish and wildlife populations and habitat parameters on upland detention/retention sites. Interim report for the Upland Detention/Retention Demonstration Project (July 1979). Florida Game and Fresh Water Fish Commission, Tallahassee.

Moxley, D., L. Ager, D. Gleckler, and T. Rosegger. 1993. Statewide resource restoration project 1987–1988 completion report for Study II: Lake Istokpoga improvement. Florida Game and Fresh Water Fish Commission, Tallahassee.

Moxley, D., T. Rosegger, and V. Williams. 1988. Statewide resource restoration project 1987–1988 annual report for Study III: Lake Istokpoga improvement. Florida Game and Fresh Water Fish Commission, Tallahassee.

Moyer, E. J., M. W. Hulon, J. Buntz, R. S. Butler, R. W. Hujik, C. S. Michael, and D. C. Arwood. 1991. Kissimmee Chain of Lakes studies annual progress report (1990–1991) for Study I—Lake Tohopekaliga investigations; Study II–Lake Kissimmee investigations; Study IV—Reedy Creek/Lake Russell investigations; Study V—Lakes Cypress and Hatchineha investigations. Florida Game and Fresh Water Fish Commission, Tallahassee.

Moyer, E. J., M. W. Hulon, R. S. Butler, D. C. Arwood, and C. Michael. 1987. Kissimmee Chain of Lakes studies annual progress report (1987) for Study II–Lake Kissimmee investigations; Study III—Boggy Creek/East Lake Tohopekaliga investigations; Study IV—Reedy Creek/Lake Russell investigations; Study V—Lakes Cypress and Hatchineha investigations. Florida Game and Fresh Water Fish Commission, Tallahassee.

Moyer, E. J., M. W. Hulon, R. S. Butler, D. C. Arwood, and C. Michael. 1988. Kissimmee Chain of Lakes studies annual progress report (1987–1988) for Study I—Lake Tohopekaliga investigations; Study III—Boggy Creek/East Lake Tohopekaliga investigations; Study IV—Reedy Creek/Lake Russell investigations; Study V—Lakes Cypress and Hatchineha investigations. Florida Game and Fresh Water Fish Commission, Tallahassee.

Moyer, E. J., M. W. Hulon, R. S. Butler, R. W. Hujik, D. C. Arwood, and C. S. Michael. 1989. Kissimmee Chain of Lakes studies annual progress report (1988–1989) for Study I—Lake Tohopekaliga investigations; Study II–Lake Kissimmee investigations; Study III—Boggy Creek/East Lake Tohopekaliga investigations; Study IV—Reedy Creek/Lake Russell investigations; Study V—Lakes Cypress and Hatchineha investigations. Florida Game and Fresh Water Fish Commission, Tallahassee.

Moyer, E. J., M. W. Hulon, J. Buntz, R. W. Hujik, J. J. Sweatman, C. Michael, and D. C. Arwood.

1992. Kissimmee Chain of Lakes studies annual progress report (1991–1992) for Study II–Lake Kissimmee investigations; Study III—Boggy Creek/East Lake Tohopekaliga investigations; Study IV—Reedy Creek/Lake Russell investigations; Study V—Lakes Cypress and Hatchineha investigations. Florida Game and Fresh Water Fish Commission, Tallahassee.

Moyer, E. J., M. W. Hulon, J. Buntz, R. W. Hujik, J. J. Sweatman, C. S. Michael, and D. C. Arwood. 1993. Kissimmee Chain of Lakes studies annual progress report (1992–1993) for Study I—Lake Tohopekaliga investigations; Study III—Boggy Creek/East Lake Tohopekaliga investigations; Study IV—Reedy Creek/Lake Russell investigations. Florida Game and Fresh Water Fish Commission, Tallahassee.

Moyer, E. J., M. W. Hulon, J. Buntz, R. W. Hujik, J. J. Sweatman, A. S. Furukawa, and D. C. Arwood. 1994. Kissimmee Chain of Lakes studies annual progress report (1993–1994) for Study I—Lake Tohopekaliga investigations; Study II–Lake Kissimmee investigations; Study III—Boggy Creek/East Lake Tohopekaliga investigations; Study IV—Reedy Creek/Lake Russell investigations. Florida Game and Fresh Water Fish Commission, Tallahassee.

Moyer, E. J., M. W. Hulon, J. J. Sweatman, R. S. Butler, and V. P. Williams. 1995. Fishery responses to habitat restoration in Lake Tohopekaliga, Florida. North American Journal of Fisheries Management 15:591–595.

Moyer, E. J., M. W. Hulon, G. L. Zuhl, D. C. Arwood, and W. H. Kurtz. 1985a. Kissimmee Chain of Lakes studies completion report for Study No. II: fisheries management plan for the Alligator Lake Chain. Florida Game and Fresh Water Fish Commission, Tallahassee.

Moyer, E. J., M. W. Hulon, G. L. Zuhl, D. C. Arwood, and W. H. Kurtz. 1985b. Kissimmee Chain of Lakes studies completion report (1983–1985) for Study No. III: Lake Gentry fish population investigations. Florida Game and Fresh Water Fish Commission, Tallahassee.

Moyer, E. J., M. W. Hulon, G. L. Zuhl, D. C. Arwood, and W. H. Kurtz. 1985c. Kissimmee Chain of Lakes studies completion report (1983–1985) for Study No. IV: Brick Lake fish population investigations. Florida Game and Fresh Water Fish Commission, Tallahassee.

Moyer, E. J., M. W. Hulon, G. L. Zuhl, D. C. Arwood, and W. H. Kurtz. 1985d. Kissimmee Chain of Lakes studies completion report (1983–1985) for Study No. V: Lake Center fish population investigations. Florida Game and Fresh Water Fish Commission, Tallahassee.

Moyer, E. J., M. W. Hulon, G. L. Zuhl, D. C. Arwood, and W. H. Kurtz. 1985e. Kissimmee Chain of Lakes studies completion report (1983–1985) for Study No. VI: Coon Lake fish population investigations. Florida Game and Fresh Water Fish Commission, Tallahassee.

Moyer, E. J., M. W. Hulon, G. L. Zuhl, D. C. Arwood, and W. H. Kurtz. 1985f. Kissimmee Chain of Lakes studies completion report (1983–1985) for Study No. VII: Lake Lizzie fish population investigations. Florida Game and Fresh Water Fish Commission, Tallahassee.

Moyer, E. J., M. W. Hulon, G. L. Zuhl, D. C. Arwood, and W. H. Kurtz. 1985g. Kissimmee Chain of Lakes studies completion report (1983–1985) for Study No. VIII: Trout Lake fish population investigations. Florida Game and Fresh Water Fish Commission, Tallahassee.

Moyer, E. J., M. W. Hulon, G. L. Zuhl, D. C. Arwood, and W. H. Kurtz. 1985h. Kissimmee Chain of Lakes studies completion report (1983–1985) for Study No. IX: Lake Hart fish population investigations. Florida Game and Fresh Water Fish Commission, Tallahassee.

Moyer, E. J., M. W. Hulon, G. L. Zuhl, D. C. Arwood, and W. H. Kurtz. 1985i. Kissimmee Chain of Lakes studies completion report (1983–1985) for Study No. X: Lake Mary Jane fish population investigations. Florida Game and Fresh Water Fish Commission, Tallahassee.

Moyer, E. J., M. W. Hulon, G. L. Zuhl, D. C. Arwood, and C. Michael. 1986. Kissimmee Chain of Lakes studies annual progress report (1986) for Study I—Lake Tohopekaliga investigations; Study II–Lake Kissimmee investigations; Study III—Boggy Creek/East Lake Tohopekaliga investigations; Study IV—Reedy Creek/Lake Russell investigations; Study V—Lakes Cypress and Hatchineha investigations. Florida Game and Fresh Water Fish Commission, Tallahassee.

Moyer, E. J., and unnamed coauthors. 1990. Kissimmee Chain of Lakes studies annual performance report (1989–1990) for Study I—Lake Tohopekaliga investigations. Florida Game and Fresh Water Fish Commission, Tallahassee.

Nico, L. G., and P. L. Fuller. 1999. Spatial and temporal patterns of nonindigenous fish introductions in the United States. Fisheries 24(1):16–27.

Nico, L. G., and R. T. Martin. 2000. The South American suckermouth armored catfish *Pterygoplichthys anisitsi* (Pisces: Loricariidae) in Texas, with comments on foreign fish introductions in the American Southwest. Southwestern Naturalist 46:98–104.

Nico, L. G., and A. M. Muench. 2004. Nests and nest habitats of the invasive catfish *Hoplosternum littorale* in Lake Tohopekaliga, Florida: a novel association with non-native *Hydrilla verticillata*. Southeastern Naturalist 3:451–466.

Nico, L. G., S. J. Walsh, and R. H. Robins. 1996. An introduced population of the South American callichthyid catfish *Hoplosternum littorale* in the Indian River Lagoon system, Florida. Florida Scientist 59:189–200.

Overdorf, T. R. 1999. Fishes of the Kissimmee River floodplain, prior to proposed restoration. Master's thesis. Florida Atlantic University, Boca Raton.

Page, L. M., and B. M. Burr. 1991. A field guide to freshwater fishes of North America north of Mexico. Houghton Mifflin, Boston.

Paulson, R. W., E. B. Chase, J. S. Williams, and D. W. Moody (compilers). 1993. National water summary 1990–1991: hydrologic events and stream water quality. U.S. Geological Survey, Water-Supply Paper 2400, Reston, Virginia.

Perrin, L. S., M. J. Allen, L. A. Rowse, F. Montalbano, K. J. Foote, and M. W. Olinde. 1982. A report of fish and wildlife studies in the Kissimmee River basin and recommendations for restoration. Florida Game and Fresh Water Fish Commission, Okeechobee.

Ravenel, W. C. 1901. Report on the propagation and distribution of food-fishes. Pages 25–118 *in* Report of the Commissioner for the year ending June 30, 1900, Part XXVI. U.S. Commission of Fish and Fisheries, Washington, D.C.

Scarlatos, P. D., M. K. Loftin, and J. Obeysekera. 1990. Restoration hydraulics: hydraulic energy gradients along pre-project and revitalized river sections. Pages 223–237 *in* M. K. Loftin, L. A. Toth, and J. Obeysekera, editors. Proceedings of the Kissimmee River Restoration Symposium, Orlando, Florida, October 1988. South Florida Water Management District, West Palm Beach.

Shireman, J. V., R. L. Stetler, and D. E. Colle. 1978. Possible use of the lake chubsucker as a baitfish. Progressive Fish-Culturist 40:33–34.

Smiley, C. W. 1885. Notes upon fish and the fisheries. Bulletin of the United States Fish Commission for 1885, 5(22):337–352.

Starling, C. C. 1979–1985. Fish hatcheries: review and annual progress reports for 1978–1985. Florida Game and Fresh Water Fish Commission, Tallahassee.

Starling, C. C. 1986–1989. Fish hatcheries annual progress reports for 1985–1989. Florida Game and Fresh Water Fish Commission, Tallahassee.

Starling, C. C. 1990–2002. Richloam Fish Hatchery annual progress reports for 1989–2002. Florida Game and Fresh Water Fish Commission, Tallahassee.

Swift, C. C., C. R. Gilbert, S. A. Bortone, G. H. Burgess, and R. W. Yerger. 1986. Zoogeography of the freshwater fishes of the southeastern United States: Savannah River to Lake Pontchartrain. Pages 213–265 *in* C. H. Hocutt and E. O. Wiley, editors. The zoogeography of North American freshwater fishes. Wiley, New York.

Thomas, P. W. 1994. The use of grass carp in Florida: past, present, and future. Program Issue and Abstracts, 14th Annual Symposium of the North American Lake Management Society, Orlando, Florida, 31 Oct–5 Nov 1994. Lake and Reservoir Management (Washington, DC) 9(2):119.

Toth, L. A. 1993. The ecological basis for the Kissimmee River restoration plan. Florida Scientist 56:25–51.

Toth, L. A. 1995. Principles and guidelines for restoration of river/floodplain ecosystems—Kissimmee River, Florida. Pages 49–73 *in* J. Cairns, Jr., editor. Rehabilitating damaged ecosystems. CRC Press, Boca Raton, Florida.

Toth, L. A. 1996. Restoring the hydrogeomorphology of the channelized Kissimmee River. Pages 369–383 *in* A. Brookes and F. D. Shields, editors. River channel restoration: guiding principles for sustainable projects. Wiley, New York.

Toth, L. A., D. A. Arrington, and G. Begue. 1997. Headwater restoration and reestablishment of natural flow regimes: Kissimmee River of Florida. Pages 425–442 *in* J. E. Williams, C. A. Wood, and M. P. Dombeck, editors. Watershed restoration: principles and practices. American Fisheries Society, Bethesda, Maryland.

Toth, L. A., S. L. Melvin, D. A. Arrington, and J. Chamberlain. 1998. Hydrologic manipulations of the channelized Kissimmee River. BioScience 48:757–764.

Toth, L. A., S. J. Miller, and M. K. Loftin. 1990. September 1988 Kissimmee River fish kill. Pages 241–247 *in* M. K. Loftin, L. A. Toth, and J. Obeysekera, editors. Proceedings of the Kissimmee River Restoration Symposium, Orlando, Florida, October 1988. South Florida Water Management District, West Palm Beach.

Trexler, J. C. 1995. Restoration of the Kissimmee River: a conceptual model of past and present fish communities and its consequences for evaluating restoration success. Restoration Ecology 3:195–210.

Tugend, K. I. 2001. Changes in the plant and fish communities in enhanced littoral areas of Lake Kissimmee, Florida, following a major habitat enhancement. Master's thesis. University of Florida, Gainesville.

USCOE (U.S. Army Corps of Engineers). 1991. Central and Southern Florida Project. Final integrated feasibility report and environmental impact statement: environmental restoration of the Kissimmee River, Florida. U.S. Corps of Engineers, Jacksonville, Florida.

U.S. Fish and Wildlife Service. 1958. Kissimmee River basin, Florida: a detailed report of the fish and wildlife resources in relation to the Corps of Engineers' plan of development. U.S. Fish and Wildlife Service, Bureau of Sport Fisheries and Wildlife, Atlanta.

Wegener, W., D. Holcomb, and V. Williams. 1974. Sampling shallow water fish populations using the Wegener ring. Proceedings of the Annual Conference Southeastern Association of Game and Fish Commissioners 27(1973):663–673.

Wegener, W., and V. Williams. 1975. Fish population responses to improved lake habitat utilizing an extreme drawdown. Proceedings of the Annual Conference Southeastern Association of Game and Fish Commissioners 28(1974):144–161.

Wiley, E. O. 1980. *Lepisosteus osseus* Linnaeus, longnose gar. Page 49 *in* D. S. Lee, C. R. Gilbert, C. H. Hocutt, R. E. Jenkins, D. E. McAllister, and J. R. Stauffer, Jr., editors. Atlas of North American freshwater fishes. North Carolina State Museum Natural History, Raleigh.

Will, L. E. 1965. Okeechobee catfishing. Great Outdoors Publishing Co., St. Petersburg, Florida.

Williams, V. P., E. J. Moyer, and M. W. Hulon. 1979. Water level manipulation project 1974–1979: Study III—Lower Kissimmee basin study. Dingell-Johnson Project, completion report (F-29–8). Florida Game and Fresh Water Fish Commission, Tallahassee.

Wullschleger, J. G., S. J. Miller, and L. J. Davis. 1990a. An evaluation of the effects of the restoration demonstration project on Kissimmee River fishes. Pages 67–81 *in* M. K. Loftin, L. A. Toth, and J. Obeysekera, editors. Proceedings of the Kissimmee River Restoration Symposium, Orlando, Florida, October 1988. South Florida Water Management District, West Palm Beach.

Wullschleger, J. G., S. J. Miller, and L. J. Davis. 1990b. A survey of fish communities in Kissimmee River oxbows scheduled for Phase II restoration. Pages 143–148 *in* M. K. Loftin, L. A. Toth, and J. Obeysekera, editors. Proceedings of the Kissimmee River Restoration Symposium, Orlando, Florida, October 1988. South Florida Water Management District, West Palm Beach.

Yerger, R. W. 1975. Aquatic vertebrate fauna of the Kissimmee River-Lake Okeechobee watershed. Technical Series 1(6) (December):1–39, Florida Department of Environmental Regulation, Tallahassee.

Appendix A.—Native fishes of the Kissimmee River basin by major subbasin, including upper Kissimmee (A, Upper Kiss); lower Kissimmee (B, Lower Kiss); and Lake Istokpoga. (C, Istokpoga) * = recorded during U.S. Geological Survey survey of Avon Park Air Force Range and adjacent areas; [v] = presence substantiated by preserved voucher specimen; ? = uncertain.

Family/species	Status	A Upper Kiss	B Lower Kiss	C Istok-poga	Citation
Lepisosteidae					
1. *Lepisosteus osseus* longnose gar	U	A	B		**A:** 17; 19; 21; 22b; 31; 36b–e, 36g–i; 37; 38; 39; 40; 43; 44; 44a; 47; 48. **B:** 26; 27; 28; 29; 32.
2. *L. platyrhincus* Florida gar [v]	C	A	B*	C*	**A:** 4; 6; 9; 17; 19; 21; 22; 22b; 51; 31; 36b–i; 37; 38; 39; 40; 41?; 42; 43; 44; 44a; 47; 48; 57. **B:** 1; 4; 23; 26; 27; 28; 31 [as "gar"]; 32; 33; 45; 50; 51; 58; 60. **C:** 1; 2; 34; 35; 59.
Amiidae					
3. *Amia calva* bowfin	C	A	B*	C*	**A:** 6; 9; 17; 19; 21; 22; 22b; 31; 36b–e, 36g–h; 37; 38; 39; 40; 42; 43; 44; 44a; 47; 48; 51. **B:** 1; 4; 7; 23; 26; 27; 28; 29; 31; 32; 45; 50; 51; 60. **C:** 1; 7; 34; 35; 59.
Anguillidae					
4. *Anguilla rostrata* American eel	R		B		**B:** 28; 31.
Clupeidae					
5. *Dorosoma cepedianum* gizzard shad [v]	C	A	B*	C	**A:** 7; 17; 19; 21; 22; 22b; 31; 36a–i; 37; 38; 39; 40; 42; 43; 44; 44a; 47; 48; 51; 57. **B:** 1; 7; 23; 25; 26; 27; 28; 29; 31; 32; 45; 50; 51; 59; 60. **C:** 34; 35.
6. *D. petenense* threadfin shad [v]	N?, C	A	B*	C	**A:** 7; 17; 19; 21; 22; 22b; 36e; 36g; 37; 38; 39; 40; 42; 43; 44; 47; 48; 51; 57. **B:** 1; 7; 23; 26; 28; 29; 31; 32; 58. **C:** 34; 35.
Esocidae					
7. *Esox americanus* redfin pickerel [v]	C	A	B*	C*	**A:** 4; 9; 19; 36b–c; 36e–h; 37; 38; 39; 40; 44; 44a; 48; 51. **B:** 1; 26; 27; 28; 31; 32. **C:** 1; 7.
8. *E. niger* chain pickerel [v]	C	A	B*	C*	**A:** 9 [as *Esox reticulatus*]; 17; 19; 21; 31; 36b–c, 36f–i; 37; 38; 39; 40; 41?; 42; 43; 44; 44a; 47; 48; 51; 57. **B:** 1; 7; 25; 26; 27; 28; 29; 31; 32; 45; 51; 58; 60. **C:** 1; 34; 35.
Cyprinidae					
9. *Notemigonus crysoleucas* golden shiner [v]	C	A	B*	C*	**A:** 4; 7; 9?; 17; 19; 21; 22; 22b; 31; 36b, d–i; 37; 38; 39; 40; 42; 43; 44; 44a; 47; 48; 51; 57. **B:** 1; 7; 23; 25; 26; 27; 28; 29; 31; 32; 33; 45; 50; 51; 58. **C:** 1; 4; 34; 35; 52.

Appendix A.—Continued

Family/species	Status	A Upper Kiss	B Lower Kiss	C Istok-poga	Citation
10. *Notropis chalybaeus* ironcolor shiner [v]	U	A		C*	**A:** 7; 51. **C:** 1; 4.
11. *N. maculatus* taillight shiner [v]	C	A	B*	C	**A:** 7; 8; 19; 21; 22; 22b; 36b; 37; 38; 39; 40; 41?; 42; 43; 44; 44a; 47; 48; 57; 59. **B:** 1; 4; 7; 26; 28; 29; 31; 32; 51; 58. **C:** 34; 35.
12. *N. petersoni* coastal shiner [v]	U	A	B	C*	**A:** 4; 7; 8; 17; 36i; 37; 38; 47; 51; 57. **B:** 31; 55. **C:** 1; 4; 7; 34; 35 [as "weed shiner"].
13. *Opsopoeodus emiliae* pugnose minnow [v]	C	A	B*	C	**A:** 4; 6; 7; 8; 16; 36b, 36e–f; 38; 39; 47; 57. **B:** 1; 28; 29; 31; 32; 51. **C:** 4; 7; 16; 59.
Catostomidae					
14. *Erimyzon sucetta* lake chubsucker [v]	C	A	B*	C*	**A:** 4; 5; 6; 10; 17; 19; 20; 21; 22; 31; 36b–i; 37; 38; 39; 40; 41?; 42; 43; 44; 44a; 47; 48; 51; 57. **B:** 1; 4; 7; 23; 25; 26; 27; 28; 29; 31; 32; 33; 45; 50; 51; 58; 60. **C:** 1; 2; 4; 7; 34; 35; 59.
Ictaluridae					
15. *Ameiurus catus* [v] white catfish	C	A	B	C*	**A:** 8; 19; 21; 31; 36b, i; 37; 38; 39; 40; 43; 44; 44a; 47; 48; 51. **B:** 8; 25; 26; 28; 29; 31; 32; 45; 50; 51; 58; 60; 61. **C:** 1; 7.
16. *A. natalis* yellow bullhead[v]	C	A	B*	C*	**A:** 3; 4; 9; 10; 17; 19; 20 [page 113]; 21; 31; 36b–i; 37; 38; 39; 40; 42; 43; 47; 48; 51; 57. **B:** 1; 4; 7; 8; 19; 23; 25; 26; 27; 28; 29; 31; 32; 33; 45; 50; 51; 58; 60. **C:** 1; 2; 7; 34; 35.
17. *A. nebulosus* brown bullhead [v]	C	A	B	C*	**A:** 6; 17; 19; 21; 22; 31; 36b–f, h–i; 37; 38; 39; 40; 41?; 42; 43; 44; 44a; 47; 48; 51; 57. **B:** 4; 23; 25; 26; 27; 28; 29; 31; 32; 33; 45; 50; 51; 58; 60. **C:** 1; 2; 34; 35; 59. [some as "speckled catfish"]
18. *Ictalurus punctatus* channel catfish	N?, C, s	A	B	C*	**A:** 21; 31; 36b, d–e, g–i; 37; 39; 40; 43; 46a, o, t, u, x; 47. **B:** 25; 26; 28; 31; 32; 45; 50; 58. **C:** 1; 34; 35.
19. *Noturus gyrinus* tadpole madtom [v]	C	A	B*	C*	**A:** 3; 4; 8; 10; 19; 20 [page 113, Fern Creek near Orlando]; 21; 36b–d, f–g; 37; 38; 39; 40; 42; 44; 47; 48; 51; 57. **B:** 1; 4 [as *N. leptacanthus*]; 7; 28; 31 [as "madtom"]; 32; 51; 58. **C:** 1; 2; 4; 7; 34; 35. [Some as *Schilbeodes mollis*]

Appendix A.—Continued

Family/species	Status	A Upper Kiss	B Lower Kiss	C Istok-poga	Citation
Aphredoderidae					
20. *Aphredoderus sayanus* pirate perch [v]	U	A	B*		**A:** 4; 21; 36c–i; 37; 38; 39; 40; 47; 51. **B:** 1; 4; 7; 28; 29; 31; 32; 50; 51; 58; 61.
Belonidae					
21. *Strongylura marina* Atlantic needlefish [v]	U	A	B	C	**A:** 7; 21; 47; 36b, 36g, i; 37; 38; 39; 40; 42; 43; 44; 57. **B:** 7; 11; 26; 28; 32. **C:** 34; 35.
Cyprinodontidae					
22. *Jordanella floridae* flagfish [v]	C	A	B*	C*	**A:** 4; 5; 8; 9; 17; 20; 21; 22; 36b, e–f, h; 37; 38; 39; 40; 41?; 42; 43; 44; 44a; 47; 51; 57. **B:** 1; 4; 7; 8; 28; 29; 31; 32; 33; 45; 50; 51; 58; 61. **C:** 1; 34; 35.
23. *Cyprinodon variegatus* sheepshead minnow [v]	H/Ex	A			**A:** 5.
Fundulidae					
24. *Fundulus chrysotus* golden topminnow [v]	C	A	B*	C*	**A:** 4; 8; 9; 10 [as *Zygonectes henshalli,* mispelled as *henskelli*]; 17; 19; 20 [page 116, in "Ferncreek" as *Zygonectes henshalli*]; 21; 22; 36b–i; 37; 38; 39; 40; 41?; 42; 43; 44; 44a; 47; 48; 51; 57. **B:** 1; 4; 7; 8; 28; 29; 31; 32; 33; 45; 50; 51; 58; 61. **C:** 1; 2; 4; 7; 34; 35; 59.
25. *F. lineolatus* lined topminnow [v]	C	A	B*	C*	**A:** 4; 8; 19; 36b–c, 36e–i; 37; 39; 40; 43; 44; 44a; 47; 48. **B:** 1; 28. **C:** 1; 2; 4. [Often reported as *F. notti* or as starhead topminnow]
26. *F. rubrifrons* redface topminnow [v]	C	A	B*	C*	**A:** 8 [as *F. confluentus*]; 57. **B:** 1; 9; 28; 33; 51; 1. **C:** 1; 2 [incorrectly identified as *F. confluentus*]; 7. [Many reported as *F. cingulatus*]
27. *F. seminolis* Seminole killifish [v]	C	A	B*	C*	**A:** 4; 7; 8; 17; 19; 21; 22; 22b; 36b, e; 37; 38; 39; 40; 42; 43; 44; 44a; 47; 48; 51; 57. **B:** 1; 7; 23; 28; 29; 31; 32; 33; 58. **C:** 1; 4; 34; 35; 59.
28. *Leptolucania ommata* pygmy killifish [v]	U	A	B*	C*	**A:** 3; 36h–i. **B:** 1. **C:** 1.
29. *Lucania goodei* bluefin killifish [v]	C	A	B*	C*	**A:** 4; 5; 6; 7; 8; 9; 19; 20 [page 118]; 21; 22; 51; 17; 36b, e–i; 37; 38; 39; 40; 41?; 42; 43; 44; 44a; 47; 48; 57. **B:** 1; 4; 7; 23; 28; 29; 31; 32; 33; 45; 50; 51; 58. **C:** 1; 4; 7; 34; 35; 59.

Appendix A.—Continued

Family/species	Status	A Upper Kiss	B Lower Kiss	C Istok-poga	Citation
Poeciliidae					
30. *Gambusia holbrooki* eastern mosquitofish [v]	C	A	B*	C*	**A:** 3; 4; 5; 6; 7; 8; 9; 10 [as *Gambusia patinclis*]; 17; 19; 20 [page 117, as *Gambusia patruelis*]; 21; 22; 36b–i; 37; 38; 39; 40; 41?; 42; 43; 44; 44a; 47; 48; 51; 57. **B:** 1; 4; 7; 8; 23; 28; 29; 31; 32; 33; 45; 50; 51; 58. **C:** 1; 2; 4; 7; 52; 59; 61. [Often reported as *Gambusia affinis* or mosquitofish]
31. *Heterandria formosa* least killifish [v]	C	A	B*	C*	**A:** 3; 4; 5; 6; 8; 9; 17; 20; 22; 36b, d–i; 37; 38; 39; 40; 41?; 42; 43; 44; 44a; 47; 51; 57. **B:** 1; 4; 7; 8; 28; 29; 31; 32; 33; 45; 50; 51; 58; 61. **C:** 1; 2; 4; 7; 34; 35; 59.
32. *Poecilia latipinna* sailfin molly [v]	C	A	B*	C*	**A:** 4; 5; 7; 9; 20 [page 118]; 21; 22; 36b; 37; 39; 40; 42; 43; 44; 44a; 47; 51. **B:** 1; 4; 7; 23; 28; 29; 31; 33; 50; 51; 57; 58; 61. **C:** 1; 4; 34; 59. [Often reported as *Mollienesia latipinna*].
Atherinidae					
33. *Labidesthes sicculus* brook silverside [v]	C	A	B*	C*	**A:** 4; 8; 6; 7; 17; 19; 21; 22; 36b–g, i; 37; 39; 40; 42; 43; 44; 44a; 47; 48; 51; 57. **B:** 1; 4; 7; 28; 29; 31 [as *Labedesthes*]; 32; 51; 58. **C:** 1; 4; 7; 34; 35; 52; 59.
34. *Menidia beryllina* inland silverside [v]	R	A	B		**A:** 7; 8; 47; 57. **B:** 29, 31 [as *Menidia*], 32.
Elassomatidae					
35. *Elassoma evergladei* Everglades pygmy sunfish [v]	C	A	B*	C*	**A:** 3; 4; 5 [possibly part of material identified as *E. 'orlandicum'* by Lönnberg]; 8; 9; 22; 36b–i; 37; 38; 39; 40; 41?; 42; 43; 44; 44a; 47; 51. **B:** 1; 4; 7; 8; 28; 31 [as "pigmy sunfish"]; 32; 33; 50; 51; 61. **C:** 1; 2; 7.
36. *E. okefenokee* Okefenokee pygmy sunfish [v]	R?	A	B	C*	**A:** 4; 5; 20 [*E. evergladei orlandicum*, see Gilbert [1998:254–255]; 57. **B:** 7; 29; 55. **C:** 1.
Centrarchidae			3		
37. *Enneacanthus gloriosus* bluespotted sunfish [v]	C	A	B*	C*	**A:** 3; 4; 6; 8; 9; 10 [as *Enneacanthus* sp.]; 17; 19; 20 [as *Enneacanthus simulans*]; 21; 22; 36b–i; 37; 38; 39; 40; 41?; 42; 43; 44; 44a; 47; 48; 51; 57. **B:** 1; 4; 7; 8; 28; 29; 31; 32; 33; 45; 50; 51; 58. **C:** 1; 2; 4; 7; 34; 35; 59.

Appendix A.—Continued

Family/species	Status	A Upper Kiss	B Lower Kiss	C Istok-poga	Citation
38. *Lepomis auritus* redbreast sunfish [v]	U	A	B*		**A:** 31; 36c; 51. **B:** 1; 25; 26; 28; 29; 31 [as "longear sunfish"]; 32; 58; 60.
39. *L. gulosus* warmouth [v]	C	A	B*	C*	**A:** 4; 8; 9; 10 [as *Pomoxis gulosus*]; 17; 19; 20 [page 124, "Ferncreek" near Orlando]; 21; 22; 36b–i; 37; 38; 39; 40; 41?; 42; 43; 44; 44a; 47; 48; 51; 57. **B:** 1; 4; 7; 8; 23; 25; 26; 27; 28; 29; 31; 32; 33; 45; 50; 51; 58; 61. **C:** 1; 2; 4; 7; 34; 35; 59; 60. [Some reported as *Chaenobryttus coronarius*].
40. *L. macrochirus* bluegill [v]	C, s	A	B*	C*	**A:** 4; 7; 5; 8; 17; 19; 21; 22; 22b; 31; 36b, 36d–i; 37; 38; 39; 40; 41?; 42; 43; 44; 44a; 46x; 47; 51; 48; 57. **B:** 1; 4; 7; 23; 25; 26; 27; 28; 29; 31; 32; 33; 45; 50; 51; 58; 60. **C:** 1; 2; 4; 7; 34; 35; 46i; 59.
41. *L. marginatus* dollar sunfish [v]	C	A	B*	C*	**A:** 4; 8; 17; 19; 21; 22; 36b–i; 37; 38; 39; 40; 41?; 42; 43; 44; 44a; 47; 48; 51; 57. **B:** 1; 4; 7; 23; 26; 27; 28; 29; 32; 33; 45; 51; 58. **C:** 1; 2; 7; 34; 35.
42. *L. microlophus* redear sunfish [v]	C, s	A	B*	C*	**A:** 7; 8; 17; 19; 21; 22; 22b; 31; 36b–i; 37; 38; 39; 40; 41?; 42; 43; 44; 44a; 47; 48; 51; 57. **B:** 1; 4; 7; 23; 25; 26; 27; 28; 29; 31 [as "shellcracker"]; 32; 33; 45; 50; 51; 58; 60. **C:** 1; 7; 34; 35; 46i; 59.
43. *L. punctatus* spotted sunfish [v]	C	A	B*	C*	**A:** 4; 8; 17; 19; 21; 22; 36d–i; 39; 40; 42; 43; 44; 44a; 47; 48; 51; 57. **B:** 1; 4; 7; 23; 25; 26; 27; 28; 29; 31; 32; 33; 50; 51; 57; 58; 60; 61. **C:** 1; 2; 7; 34; 35.
44. *Micropterus salmoides* largemouth bass [v]	C, s	A	B*	C*	**A:** 1; 3; 4; 7; 6; 8. 17; 19; 21; 22; 22b; 31; 36b–i; 37; 38; 39; 40; 41?; 42; 43; 44; 44a; 46a–b, o, s, t, x; 47; 48; 51; 57. **B:** 1; 4; 7; 23; 25; 26; 27; 28; 29; 31; 32; 33; 45; 50; 51; 58; 60. **C:** 1; 2; 4; 7; 34; 35; 46c, i; 59.
45. *Pomoxis nigromaculatus* black crappie [v]	C, s	A	B*	C*	**A:** 6; 7; 17; 19; 21; 22; 22b; 31; 36b–i; 37; 38; 39; 40; 42; 43; 44; 44a; 47; 48; 51; 57. **B:** 1; 7; 23; 25; 26; 27; 28; 29; 31; 32; 45; 50; 51; 58; 60. **C:** 1; 7; 34; 35; 46w; 52; 59.
Percidae					
46. *Etheostoma fusiforme* swamp darter [v]	C	A	B*	C*	**A:** 4 [some as *Boleichthys fusiformis*, some as *Etheostoma barratti*]; 5 [as *Etheostoma* sp.]; 8 [some as *Hololepis barratti*]; 10 [as *Etheostoma quiescens*]; 17; 19; 20 [page 126, as *Etheostoma quiescens*]; 21; 22; 36b–i; 37; 38; 39; 40; 41?; 42; 43; 44; 44a; 47; 48; 51; 57. **B:** 1; 4; 7; 10a; 28; 29; 32; 31 [as *Etheostoma barratti*]; 33; 45; 50; 51; 58. **C:** 1; 2; 4 [as *Etheostoma barratti* or as *E. edwini*]; 7; 34; 35.

Appendix A.—Continued

Family/species	Status	A Upper Kiss	B Lower Kiss	C Istok-poga	Citation
47. *Percina nigrofasciata* blackbanded darter [v]	R?	A	B	C*	**A:** 51. **B:** 7; 31 [Table 43, as *Hadropterus*]. **C:** 1; 7.
Mugilidae					
48. *Mugil cephalus* striped mullet	U		B		**B:** 26; 28; 29; 31 [as "mullet"]; 32.

Status: C = common (frequently observed or collected); U = uncommon (irregularly observed or collected); R = rare (only known from a few samples); H/Ex = historical/extirpated; N? = status as native, somewhat uncertain; s = supplemental stocking of hatchery-produced populations.

Citation codes are as follows: 1 = present study (USGS survey of Avon Park Air Force Range and adjacent water bodies).

Museum Collections (2–10a).—2 = AMNH; 3 = ANSP; 4 = CU; 5 = NRM; 6 = TU; 7 = UF; 8 = UMMZ; 9 = USNM; 10 = ZMUU; 10a = INHS.

Publications (11–22a).—11 = Burgess (1980b); 12 = Champeau (1984); 13 = Courtenay (1978); 14 = Courtenay (1979); 14a = Courtenay et al. (1974); 15 = Gilbert (1998:254–255); 16 = Gilbert and Bailey (1972); 17 = Guillory et al. (1979); 18 = Hensley and Courtenay (1980b); 19 = Hoyer and Canfield (1994); 20 = Lönnberg (1894); 21 = Wegener and Williams (1975); 22 = Wegener et al. (1974); 22a = Hulon and Williams (1982); 22b = Moyer et al. (1995).

Symposium Proceedings and Abstracts (23–27).—23 = Davis et al. (1990); 24 = Thomas (1994); 25 = Toth et al. (1990); 26 = Wullschleger et al. (1990a); 27 = Wullschleger et al. (1990b). *Reports of Florida Fish and Wildlife Conservation Commission (28–47).*—28 = Bull et al. (1991); 29 = Bull et al. (1994); 30 = Chapman et al. (1983); 31 = FGFWFC (1957); 32 = Furse and Davis (1996); 33 = Montalbano et al. (1979b); 34 = Moxley et al. (1988); 35 = Moxley et al. (1993); 36 = Moyer et al. (1985a–i); 37 = Moyer et al. (1986); 38 = Moyer et al. (1987); 39 = Moyer et al. (1988); 40 = Moyer et al. (1989); 41 = Moyer et al. (1990); 42 = Moyer et al. (1991); 43 = Moyer et al. (1992); 44 = Moyer et al. (1993); 44a = Moyer et al. (1994); 45 = Perrin et al. (1982); 46a = Starling (1979); 46b = Starling (1980); 46c = Starling (1981), 46d = Starling (1982); 46e = Starling (1983); 46f = Starling (1984); 46g = Starling (1985), 46h = Starling (1986); 46i = Starling (1987); 46j = Starling (1988); 46k = Starling (1989), 46l = Starling (1990); 46m = Starling (1991); 46n = Starling (1992); 46o = Starling (1993); 46p = Starling (1994); 46q = Starling (1995); 46r = Starling (1996); 46s = Starling (1997); 46t = Starling (1998); 46u = Starling (1999); 46v = Starling (2000); 46w = Starling (2001); 46x = Starling (2002); 47 = Williams et al. (1979).

Other unpublished reports (48–51).—48 = Canfield and Hoyer (1992); 49 = Keown and Russell (1982); 50 = Milleson (1976); 51 = Yerger (1975).

Other sources (52–62).—52 = C. R. Robins (University of Miami, unpublished field notes); 53 = C. Starling (FFWCC, personal communication); 54 = D. Fox (FFWCC, personal communication); 55 = L. Glenn (SFWMD, unpublished data); 56 = L. G. Nico (unpublished data); 57 = Tugend (2001); 58 = U.S. Geological Survey collection (NAWQA project); 59 = preserved voucher specimen(s), uncatalogued (U.S. Geological Survey, Gainesville); 60 = Miller (1997); 61 = Overdorf (1999); 62 = A. M. Muench (University of Florida, personal communication).

Appendix B.—Non-native fishes introduced into the Kissimmee River Basin by major sub-basin, including: upper Kissimmee (A, Upper Kiss); lower Kissimmee (B, Lower Kiss); and Lake Istokpoga (C, Istokpoga). * = recorded during USGS survey of Avon Park Air Force Range and adjacent areas; [v] = presence substantiated by preserved voucher specimen; ? = uncertain. See text for additional information. Citation codes are identified in Appendix A.

Family/Species	Status	A Upper Kiss	B Lower Kiss	C Istok-poga	Citation
Cyprinidae					
1. *Cyprinus carpio*					
common carp	F, E?, R		B		**B:** 23.
2. *Ctenopharyngodon idella*					
grass carp	F, E?, U, s	A	B	C*	**A:** 39:I-3 [reported as "triploids"]; 46c–n, r–v [early stockings reported as hybrids, post-1984 reported as triploids]; 49; 53 [diploids stocked in Lake Wales in 1979–1980]. **B:** 26; 28 [pages 95, 104]; 32; 58. **C:** 1; 46k–m, o–p, u–v; 24;56.
Clariidae					
3. *Clarias batrachus*					
walking catfish [v]	F, E, C?	A	B*	C*	**A:** 7; 37; 38; 40; 43; 59. **B:** 1; 13; 14; 18; 23; 26; 27; 28; 29; 32; 45; 58; 61. **C:** 1; 7; 13; 14; 18; 35.
Loricariidae					
4. *Pterygoplichthys* cf. *disjunctivus*					
vermiculated sailfin catfish [v]	F, E, U?	A	B		**A:** 7; 57; 59; 62. **B:** 55.
Callichthyidae					
5. *Hoplosternum littorale*					
brown hoplo [v]	F, E, C	A	B*	C	**A:** 7; 57; 59; 62. **B:** 1; 55. **C:** 59.
Moronidae					
6. *hybrid Morone chrysops x M. saxatilis*					
sunshine bass/palmetto bass	Hy, U, s	A		C	**A:** 19; 46b–c, e–h, j–k, m–n, p–q, s, u, w; 48. **C:** 12; 46a–j, o.
Percidae					
7. *Sander vitreus*					
walleye	I/T, Ex	A			**A:** 30; 46d; 53. [All pertain to a stocking in Emerald Lake, Orange County]
Cichlidae					
8. *Astronotus ocellatus*					
oscar	F, E?, R		B		**B:** 29; 32; 54.
9. *Oreochromis aureus*					
blue tilapia [v]	F, E, C	A	B*	C	**A:** 14a; 19; 22a, 22b; 37; 38; 39; 40; 41?; 42; 43; 44; 44a;;48; 56; 57. **B:** 1; 14a; 25; 26; 27; 28; 29; 32; 45; 55; 60. **C:** 14a. [Note: listed as "tilapia" in many state reports].

Status: C = common (frequently observed or collected); U = uncommon (irregularly observed or collected); R = rare (only known from a few samples); F = introduced, foreign origin; T = introduced, transplanted native; Hy = introduced, artificial hybrid;

American Fisheries Society Symposium 45:557–585, 2005

Status and Conservation of the Fish Fauna of the Alabama River System

MARY C. FREEMAN*
U.S. Geological Survey, Patuxent Wildlife Research Center, University of Georgia, Athens, Georgia 30602, USA

ELISE R. IRWIN
U.S. Geological Survey, Alabama Cooperative Fish and Wildlife Research Unit, 108 M. White Smith Hall, Auburn University, Auburn, Alabama 36849, USA

NOEL M. BURKHEAD
U.S. Geological Survey, Florida and Caribbean Science Center, 7920 Northwest 71st Street, Gainesville, Florida 32653, USA

BYRON J. FREEMAN
Institute of Ecology, University of Georgia and Georgia Museum of Natural History, Athens, Georgia 30602, USA

HENRY L. BART, JR.
Tulane University Museum of Natural History, Belle Chasse, Louisiana 70037, USA

Abstract.—The Alabama River system, comprising the Alabama, Coosa, and Tallapoosa subsystems, forms the eastern portion of the Mobile River drainage. Physiographic diversity and geologic history have fostered development in the Alabama River system of globally significant levels of aquatic faunal diversity and endemism. At least 184 fishes are native to the system, including at least 33 endemic species. During the past century, dam construction for hydropower generation and navigation resulted in 16 reservoirs that inundate 44% of the length of the Alabama River system main stems. This extensive physical and hydrologic alteration has affected the fish fauna in three major ways. Diadromous and migratory species have declined precipitously. Fish assemblages persisting downstream from large main-stem dams have been simplified by loss of species unable to cope with altered flow and water quality regimes. Fish populations persisting in the headwaters and in tributaries to the main-stem reservoirs are now isolated and subjected to effects of physical and chemical habitat degradation. Ten fishes in the Alabama River system (including seven endemic species) are federally listed as threatened or endangered. Regional experts consider at least 28 additional species to be vulnerable, threatened, or endangered with extinction. Conserving the Alabama River system fish fauna will require innovative dam management, protection of streams from effects of urbanization and water supply development, and control of alien species dispersal. Failure to manage aggressively for integrity of remaining unimpounded portions of the Alabama River system will result in reduced quality of natural resources for future generations, continued assemblage simplification, and species extinctions.

* Corresponding author: mary_freeman@usgs.gov

Introduction

The Alabama River system, comprising the Alabama, Coosa, and Tallapoosa rivers and their tributaries (Figure 1), forms the eastern portion of the Mobile River drainage, one of the most biologically rich aquatic ecosystems in North America (Lydeard and Mayden 1995; Mettee et al. 1996; Neves et al. 1997). Physiographic diversity and a geologic history of relative isolation and protection from Pleis-

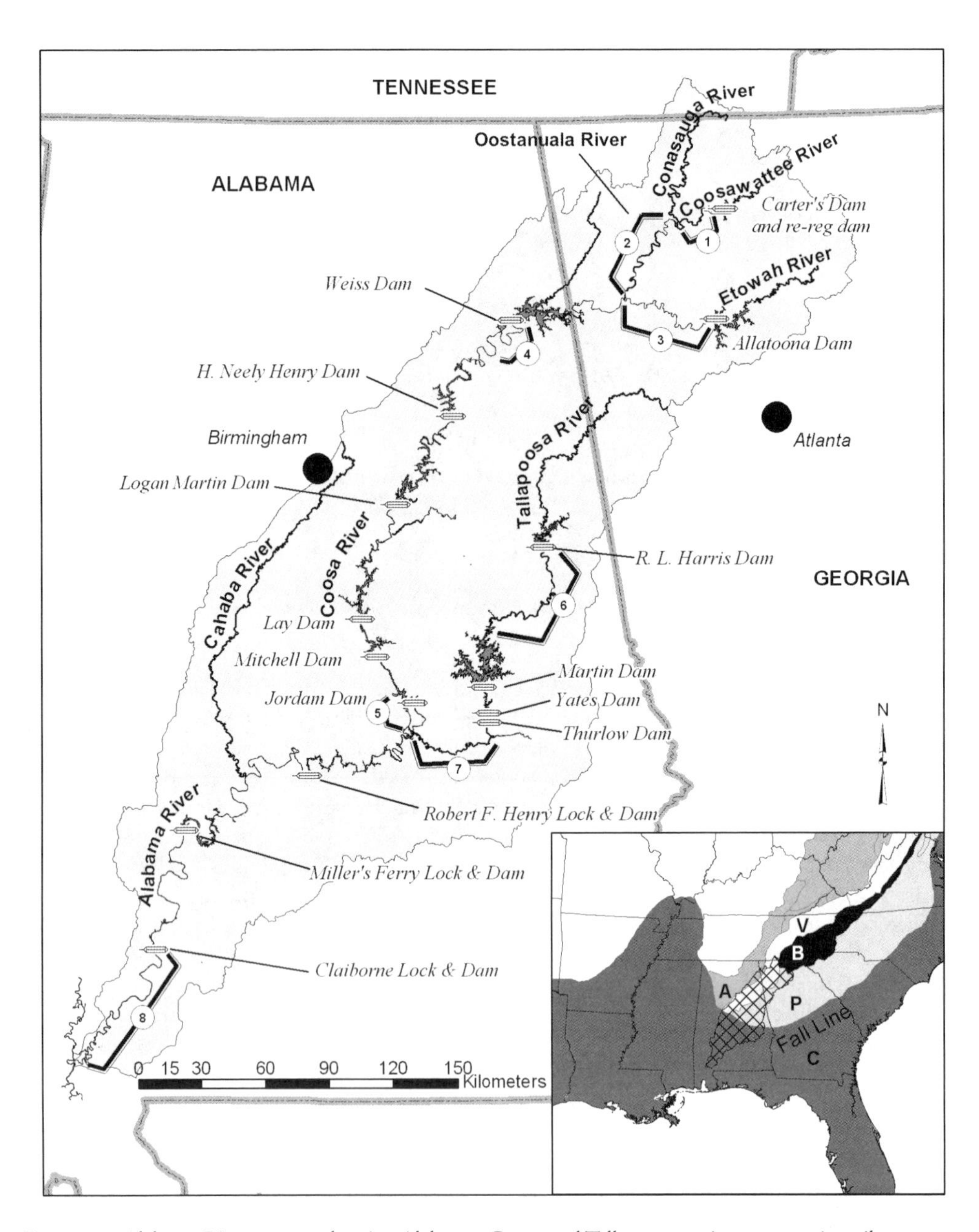

Figure 1.—Alabama River system, showing Alabama, Coosa and Tallapoosa main stems, major tributary systems, and locations of main-stem dams. Flow-regulated reaches are numbered to correspond with Table 1. Inset shows drainage (cross-hatched) overlain on physiographic provinces: Appalachian Plateau (A), Valley and Ridge (V), Blue Ridge (B), Piedmont (P), and coastal plain (C). The fall line separates the coastal plain from upland provinces.

tocene glaciation have fostered development in the Alabama River system of some of the highest levels of aquatic faunal diversity and endemism recorded in temperate freshwaters. At least 184 fishes are native to the system, including at least 33 endemic species (Appendix A).

Our knowledge of the Alabama River ichthyofauna began with the eclectic discovery and description of fishes in the mid-19th century, rapidly transitioned to more rigorous, systematic investigations by scholars at newly established national and regional natural history museums, and continues as diverse biological investigations conducted at state universities, natural history research collections, and by state and federal agencies. Boschung (1992) provides a brief synopsis of contributions by 19th century natural historians who initiated early ichthyofaunal exploration of the Alabama River. The list includes luminaries associated with ichthyological exploration of North America, but three scientists stand out by virtue of describing more than half the fishes known from the Alabama River. The anatomist Jean Louis Rodolphe Agassiz (1807–1873) described the first species from the Mobile River drainage, the blackbanded darter (as *Hadropterus nigrofasciatus*) and 28 other valid forms. The renowned scholar and humanist David Starr Jordan (1851–1931), and his student and colleague Charles Henry Gilbert (1859–1928), collectively described more than a hundred species. The first endemic species described from the Alabama River system, the blue shiner was described as being common by Jordan (1877); it is now a federally listed threatened species.

The rich fauna of Alabama has attracted many students of fishes in the 20th century, far too many to list. However, two southeastern colleagues, Herbert T. Boschung (Professor Emeritus, Department of Biology, University of Alabama) and Royal D. Suttkus (Professor Emeritus, Tulane University, Museum of Natural History), and several decades of their graduate students, made outstanding contributions to our knowledge of the composition and distribution of the Alabama River ichthyofauna. Collectively with their graduate students, Suttkus and Boschung described 36 species from Alabama, 31 of which occur in or are endemic to the Alabama River system. Boschung published the first annotated list of Coosa River fishes (Boschung 1961) and a catalog of freshwater and marine fishes (Boschung 1992). His student, James D. Williams, surveyed the Tallapoosa River for his Master's thesis (Williams 1965). William Smith-Vaniz (1968) authored the first book on the fishes of Alabama, primarily an illustrated key accompanied by black and white photographs. More recently, Maurice F. Mettee and Patrick E. O'Neil (both graduate students of Boschung) and J. Malcolm Pierson authored a beautifully illustrated book on the fishes of Alabama and the Mobile basin (Mettee et al. 1996). A third book on Alabama's fishes, by Boschung and Richard L. Mayden and illustrated by Joseph R. Tomelleri (Boschung and Mayden 2004), provides a comprehensive treatment of the systematics, distribution, biology and conservation status of the state's ichthyofauna.

Remarkably, new species discovery continues in the system (e.g., 12 fish species have been described since 1990), partly driven by recognition of cryptic species (Suttkus and Etnier 1991; Wood and Mayden 1993; Suttkus et al. 1994a, 1994b; Bauer et al. 1995; Thompson 1997a, 1997b; Burr and Mayden 1999). Our objectives are to describe how the rivers and fish assemblages of the Alabama River system have changed over the past century and a half, to highlight conservation issues, and to discuss management activities that are either in place or are needed to prevent undesirable faunal changes in the future.

The Alabama River system drains approximately 59,000 km^2, including substantial portions of northwest Georgia, east-central Alabama, and a small area of southeastern Tennessee. Physiographic diversity of the system creates a mosaic of lotic habitats that, prior to construction of large dams, formed a fluvial continuum from the mountains to the Gulf. The northernmost headwaters of the Coosa River system dissect the southern terminus of the Blue Ridge, Valley and Ridge, and upland Piedmont along the southern bend of Appalachia. These headwater rivers derive their distinctiveness from the varied lithography and soil horizons of these provinces in northwestern Georgia and northeastern Alabama (Wharton 1978). The main stem of the Coosa River

(460 km in length) originates in the relatively open Great Valley subsection of the Valley and Ridge, at Rome, Georgia. The lower third of the Coosa River main stem historically cascaded through a series of large virtually unnavigable bedrock shoals (Jackson 1995). The shoals abruptly disappear just below the fall line where the Alabama River is formed by the junction of the Coosa River with the Tallapoosa River near Montgomery, Alabama. The Tallapoosa River has similar physiographic diversity, flowing 415 km from Piedmont uplands in west Georgia and east Alabama, crossing the fall line in another set of large falls (i.e., prior to impoundment), and continuing across the coastal plain to join the Coosa River. The Cahaba River forms the western-most large tributary of the Alabama River and is the system's longest unregulated river (Figure 1). The Cahaba flows over 300 km from the Valley and Ridge province, across the fall line, onto the coastal plain and into the Alabama River near Selma, Alabama. The Alabama River main stem winds 500 km across the coastal plain, joining with the Tombigbee River approximately 72 km from Mobile Bay.

Rainfall is abundant in most years in the Alabama River system, averaging from about 127–142 cm/year across most of the basin, to more than 150 cm/year in the Coosa system headwaters and in the lower Alabama River. Natural flow regimes include seasonally highest flows in February, March, and April and lowest flows in September, October, and November. Streams on the coastal plain typically experience spring high flows that inundate riparian habitats. Annual flow in the lower Alabama River averages about 950 m^3/s; seasonal high flows (e.g., exceeding 2,250 m^3/s) spread into historically forested floodplain areas for up to 50% of days from March through September (Irwin et al. 1999).

Brief History of Alteration in the Alabama River System

Channel alteration of the main stems.—Early changes to the system's rivers brought about by European settlers include construction of small dams to power grist and textile mills and efforts to improve the rivers for navigation. Examples of the navigation improvements include rock removal, construction of rock dikes, and channel straightening in reaches of the upper Coosa system in the late 19th century (Corps of Engineers reports to the U.S. Congress, compiled by Bill Frazier, Decatur, Georgia). Steep gradient and numerous, large shoals prevented access by barges to the lower Coosa River, but the lower-gradient Alabama River provided a major corridor of river-borne commerce through the 19th century (Jackson 1995). In 1878, Congress authorized maintenance of a 1.2 × 61 m navigation channel on the length of the Alabama River, largely achieved by construction of jetties and dykes and by snag removal (Jackson 1995). The U.S. Army Corps of Engineers (USACE) presently dredges sand and gravel from shoals to facilitate travel in low-flow periods.

Beginning in 1914 and continuing to the 1980s, dam construction for hydropower generation and navigation resulted in 16 reservoirs in the Alabama River system. The first hydropower dams were constructed on the Coosa and Tallapoosa rivers in the vicinity of the fall line, where steeper gradients erode beds down to former continental shelf bedrock. Eventually, 12 hydropower projects (10 private, 2 federal) were constructed in the Coosa and Tallapoosa systems (Figure 1). The USACE constructed three additional lock and dam projects on the Alabama River main stem between 1963 and 1972 for purposes of hydropower generation and to provide a 2.7 × 61 m navigational channel. These projects resulted in extensive alteration of free-flowing, large river habitat in the Alabama River system; approximately 74% of the length of the Alabama River main stem, 87% of the Coosa River main stem and 29% of the Tallapoosa River main stem were eventually impounded by the pools behind navigation and hydropower dams.

Free-flowing riverine habitat in the Alabama River system now consists of unimpounded main-stem sections below dams and major tributary streams (many of which are now truncated by main-stem impoundments). The main-stem dams regulate flow regimes in nearly all remaining large-river habitat, with three segments experiencing the hourly flow fluctuations produced by peaking hydropower operations (Table 1). Water releases during nonpower generation periods have been as low as

Table 1.—Flow-regulated segments of the Alabama River system, showing year of initial flow regulation, segment length, and flow regime characteristics (hydropower releases and base flow levels). Numbers in parentheses correspond to numbered segments in Figure 1.

River segment	Year regulated	Length (km)	Flow regime characteristics
Coosawattee River below Carter's Dam and reregulation dam (1)	1975	40	Hydropeaking, releases buffered by the reregulation dam; base flow = 20% mean annual flow.
Oostanala River below Carter's Dam and rereg dam (2)	1975	76	Hydropeaking, releases buffered by the reregulation dam and by Conasauga River inflow.
Etowah River below Allatoona Dam (3)	1950	77	Hydropeaking; base flow = 13% mean annual flow.
Coosa River bypass, below Weiss Dam (4)	1962	36	Bypassed; flow supplied by tributaries. Flow reversed in the lower portion of the reach during power generation.
Coosa River below Jordan Dam (5)	1929[a]	13	Hydropeaking (1929–1967, 1975–1980) or bypassed (1967–1975; 1980–1990); Presently, seasonally varied baseflows with periodic hydro power releases.
Tallapoosa River below Harris Dam (6)	1982	78	Hydropeaking; base flow = 2% mean annual flow.
Tallapoosa River below Thurlow Dam (7)	1930	80	Hydropeaking; base flow increased from leakage to 25% mean annual flow in 1989.
Alabama River below Claiborne Lock and Dam (8)	1969	132	Navigation lock and dam; upstream hydropower dams operated to maintain mean daily flow = approx. 20% mean annual flow.

[a] Flow in the lower Coosa River initially was altered by construction of Lay Dam (1914) and Mitchell Dam (1923), both located upstream of Jordan Dam (1929).

flow leakage from the dams in the Coosa River below Jordan Dam and the Tallapoosa River downstream from Thurlow Dam, and remain low in the Tallapoosa River below Harris Dam (Table 1). Two segments of the Coosa River are bypassed, with flows for power generation usually released through artificial channels. Downstream from Jordan Dam, the Coosa River presently is afforded continuous, seasonally varied flows. Flow in the bypassed section below Weiss Dam is entirely from tributary inflow except during power generation, when flow is reversed as a portion of the released water is forced back upstream through the bypassed channel. Carters Dam and reregulation dam (in the upper Coosa system) operate as a pump-storage peaking project; water is released through Carters Dam to generate power during high-demand periods and is pumped back upstream from the reregulation pool when demand is low. The reregulation dam dampens the effects of peaking releases on the downstream segments of the Coosawattee and Oostanaula rivers.

Alterations from watershed activities.—Expanding agriculture in the 19th century brought extensive conversion of forests to agricultural fields and consequently enormous increases in sediment loading to streams and rivers. Mining activities, (e.g., for gold in the upper Coosa system, coal in the Cahaba system) also added increased bedload and contaminants to river segments (Ward et al. 1992; Leigh 1994; Shepard et al. 1994). Tributaries and main-stem sections were channelized to improve drainage from the eroding agricultural landscape, a practice that began a century ago and continues to present. Tributary systems have also been shortened and fragmented by construction of thousands of farm ponds and watershed dams.

Ecological integrity of the Alabama River system presently is threatened by human population expansion and urbanization. Through the 1990s, the Atlanta metropolitan area (which extends northward and westward into the Coosa and Tallapoosa systems) was included among the fastest growing counties in the United States (U.S. Census Bureau, http://www.census.gov/population/www/cen2000/phc-t4.html). Urban growth in and surrounding Birmingham also affects flows, water quality, and biological integrity in the Cahaba River system (Shepard et al. 1994). The expanding human population is placing increasing pressures on the Alabama River system for water supply. The states of Alabama and Georgia have been struggling to resolve a plan for sharing waters in the system for over a decade through joint study of water availability and demands, and since 1997, through formal negotiations under an interstate compact (USACE 1998). Symptomatic of Georgia's increasing water needs, at least six new water-supply reservoirs are proposed for streams in the upper Coosa and Tallapoosa systems.

Methods

To describe the status of the Alabama-Coosa-Tallapoosa (ACT) River basin fish fauna, we examined evidence of species imperilment and extirpations, the occurrence of alien fishes, and the condition of assemblages persisting in the longest flow-regulated main-stem segments. We initially listed all freshwater and diadromous fish species known from eight major ACT subsystems (Appendix A), based on historic records plotted by Mettee et al. (1996) and Walters (1997), supplemented with observations by Jordan (1877) for the Oostanaula and Etowah systems, and additional recent records for the upper Coosa (N. M. Burkhead and B. J. Freeman, unpublished data) and Tallapoosa (E. R. Irwin and M. C. Freeman, unpublished data) systems. We also listed fishes that are primarily marine but that commonly occur in the downstream portion of the Alabama River system. The establishment of alien species was assessed from Mettee et al. (1996) and Fuller et al. (1999).

We followed Warren et al. (2000) in listing the conservation status of each taxon, except for species listed as threatened or endangered under the Endangered Species Act (ESA), in which case we listed the species' status under the ESA. To illustrate trends in faunal conservation status, we compared number of species by conservation category in earlier assessments (Miller 1972; Deacon et al. 1979; Williams et al. 1989) with the more recent assessment by Warren et al. (2000). We also examined changes in imperilment of the Alabama River system fauna in comparison to changes for the entire fish fauna of the southeastern United States. For these comparisons, we followed Warren et al. (2000) in equating the "special concern" category used in the earlier assessments with "vulnerable" (i.e., may become threatened or endangered as a result of relatively minor habitat disturbances).

The lack of extensive faunal surveys prior to dam construction and other major habitat disturbances, along with the difficulty of sampling species that may be rare or elusive (e.g., in deep water habitats), hampers our ability to conclude that species have been extirpated from a particular reach or river system. The strongest evidence of species extirpation consists of records of past occurrence along with failure to collect a species over a prolonged period of sampling; the hypothesis of extirpation is further strengthened if the species' habitat has been severely altered or otherwise made inaccessible. Nonetheless, we recognize that rarity and difficulty in sampling can result in "rediscovery" of fishes presumed extinct or extirpated for decades (Mayden and Kuhajda 1996), and so our conclusions must be tempered with the possibility of future discoveries. For diadromous species, we presumed extirpation from those portions of the range made inaccessible by downstream dams that lack provisions for fish passage. For other species, we presumed extirpation if the species had not been observed in at least two decades. We counted a limited number of species as extirpated from flow-regulated reaches for which we lack historical records but hypothesize that the species likely occurred on the basis of habitat characteristics and proximity to areas with known occurrences.

We assessed the condition of fish assemblages in flow-regulated segments by examining evidence of species persistence for three taxonomic groups that commonly occur in wadeable habitats: sunfishes and basses (Centrarchidae), minnows (Cyprinidae), and darters (Percidae: Etheostomatini). Focusing on these three groups allowed us to include a large portion of the native fish diversity in the Alabama River system (Appendix A), while avoiding biases attributable to inefficient sampling in deep water. Examining extant diversity of these three groups also allowed us to compare persistence of habitat-generalists species that are tolerant of lotic and lentic conditions (i.e., sunfishes and basses) with that of fishes primarily adapted to flowing-water habitats (i.e., the darters and the majority of the native riverine minnows; Etnier and Starnes 1993; Jenkins and Burkhead 1994). We estimated expected native richness for these taxonomic groups in the flow-regulated reaches on the basis of known species occurrences either within the reach or in similar mainstem or large tributary habitats within the subsystem.

We used information from 11 studies (Table 2), collected over varying time periods and by different researchers, to estimate the numbers of native sunfish and basses, minnows, and darters persisting in each flow-regulated segment. In four of the flow-regulated reaches, prepositioned area electrofishers (PAEs: 1.5 × 6 m in size; Bain et al. 1985) have been used to sample fishes in wadeable habitats using similar effort on multiple occasions (i.e., over 2 to 5 years; Table 2). We used presence-absence data from PAE samples from sequential years to estimate native species detectability and richness for each site and taxonomic group, to examine the possibility that low observed species richness in some reaches resulted from low detectability. Detectability and species richness estimates were made using the jackknife estimator for closed-populations with heterogeneous detectabilities among species (model M_h), computed using program CAPTURE (Williams et al. 2002). Additional species occurrence data for flow-regulated reaches were obtained by boat electrofishing, backpack electrofishing and seining, and collection with rotenone (Table 2). Taken together, these studies provided data on species persistence for those faunal groups that were vulnerable to at least one of the sampling methods employed in each flow-regulated reach. For comparison, we estimated percent of native species persisting in the unregulated portions of the Conasauga, Etowah, and Tallapoosa systems based on Walters (1997) and our unpublished collection records.

Last, to understand how impoundments and other navigation-related changes to the Alabama River main stem have altered the fish assemblages, we summarized the results of previous studies (Buckley 1995; Buckley and Bart 1996), which used data from a long-term, fish-monitoring survey conducted by Royal D. Suttkus and the late Gerald E. Gunning to examine trends in fish species richness and abundance over time in the impounded reach of the river. Suttkus and Gunning initiated a semiannual, fish-monitoring survey of eight stations, and an annual survey at 10 stations, along a 100 km stretch of the Alabama River in 1967, continuing through the 1990s. The start of the Alabama River Fish Monitoring Survey roughly coincides with the period of intensive modification of the Alabama River for navigation. Work on the two dams that encompasses most of the Suttkus and Gunning survey area (Miller's Ferry Lock and Dam and Claiborne Lock and Dam) was initiated in 1965 and completed in 1972. Dredging of the river for maintenance of the navigation channel occurred periodically throughout the survey.

Specimens and data from the Alabama River Fish Survey are archived in the Royal D. Suttkus Fish Collection in the Tulane University Museum of Natural History. In summarizing the data, Buckley (1995) and Buckley and Bart (1996) used records from the museum database as well as information from the personal field notes of R. D. Suttkus. Sampling gear remained constant during the survey, consisting of 3.3 × 2 m, 0.5-cm mesh seines and (rarely) trammel nets. Initially, sampling was conducted at night, but changed to mostly daylight hours starting in 1983. Early collection efforts generally lasted for 45 min to 1 h, whereas later collections (after 1985) were from 15 to 30 min.

The overall species richness trend is based on pooled data for all collections within a given year.

Table 2.—Studies used to assess species occurrences in flow-regulated main-stem reaches of the Alabama River system, showing years over which studies were conducted, methods, and references.

Segment	Year(s)	Methods	Reference(s)
Coosawattee and Oostanaula rivers below Carter's Dam and rereg dam	1993–1998	Backpack or boat electrofished 14 sites; also compiled historical records (pre-dam, 1936–1962: 6 sites; post-dam, 1977–1984: 18 sites, including 1 rotenone sample).	Freeman 1998
Etowah River below Allatoona Dam	1992–1998	Backpack or boat electrofished 16 sites; also compiled historical data (pre-dam, 1949, 1 site; post-dam, 1959–1979: 4 sites).	Burkhead et al. 1997; Freeman 1998
Coosa River bypass below Weiss Dam	1999–2000	Boat and backpack electrofishing collections at 12 sites distributed throughout reach, in 5 seasons.	Stewig 2001
	1999–2000	260 pre-positioned area electrofisher (PAE) samples at 2 sites in upper third of reach, 3 seasons.	Irwin et al. 2001
Coosa River below Jordan Dam	1992–1997	PAE samples collected in a shoal complex in upper half of reach, monthly (1992–1996, 833 samples) or annually (1997–1999, 300 samples).	Peyton and Irwin 1997; E.R. Irwin, unpublished data
Tallapoosa River below Harris Dam	1990–1992, 1994–1997	Samples at 3 sites by seasonal boat electrofishing (1990–1992) and PAE sampling in spring and summer, (1990–1992: 307 samples, and 1994–1997: 791 samples).	Travnichek and Maceina 1994; Bowen et al. 1998; Freeman et al. 2001
Tallapoosa River below Thurlow Dam	1990–1992, 1994–1995, 1997	Samples at 2 sites in upper half of reach by seasonal boat electrofishing (1990–1992); PAE sampling in spring and summer, (1990–1992: 177 samples, and 1994–1995: 400 samples); rotenone (1990, 1992, 1997).	Travnichek and Maceina 1994; Travnichek et al. 1995; Bowen et al. 1998; Alabama Game and Fish Division, unpublished data

Species abundance trends are based on sampling time-adjusted data (minutes of sampling effort) and thus account for the decrease in sampling effort of collections after 1985. The data are summarized in 6-year time blocks.

Results

The Alabama River system contains at least 184 native fish species (Appendix A), counting all described and known but undescribed fishes of which we are aware. Uncertainty in the total number of fish taxa stems from the occurrence of cryptic species, some of which have been recognized (i.e., four undescribed species related to the holiday darter, Appendix A) and likely others that have not. The fauna includes at least 33 species that are endemic to the Alabama River system, approximately 18% of the native fauna. Despite the high level of endemism, many fishes are widespread within the system; for example, at least 96 species (52%) natively occurred in each of the Coosa, Tallapoosa, and Alabama sub-systems.

Alien species compose approximately 10% of the fish fauna in the Alabama River system, numbering about 19 species that may be established (Appendix A). Five species (threadfin shad, common carp, grass carp, fathead minnow, and redbreast sunfish) are widespread in the system. The redbreast sunfish occurs so commonly that the species' status as alien is questionable. Other alien species generally are restricted to narrower ranges within the basin (Appendix A), occurring as a result of local introductions that were either accidental (e.g., from bait buckets or aquaculture facilities) or to support sport fisheries (e.g., three salmonid species, restricted to cool headwaters). The red shiner is an exception, occurring widely in the upper Coosa system (Appendix A). The red shiner has been present in the upper Coosa since at least the early 1970s, and presently is one of the few cyprinids persisting in the bypassed Coosa River channel below Weiss Dam (Irwin et al. 2001). Through the 1990s, red shiners have spread up the Coosa system and threaten to reduce populations of the three native *Cyprinella* species (including the federally threatened blue shiner) through displacement and hybridization (N. M. Burkhead, unpublished data).

Extensive physical and hydrologic alteration has contributed to a relatively high level of imperilment of fishes of the Alabama River system. Ten fish species are federally listed under the Endangered Species Act, and 28 additional species are considered imperiled (i.e., 3 endangered, 9 threatened, and 16 vulnerable; Appendix A). Periodic assessments of conservation status show a steady increase in fishes considered endangered and vulnerable, with the pattern for the Alabama River system fauna largely paralleling that for the southeastern United States (Figure 2). Faunal imperilment in the Alabama River system reflects three of the major effects of multiple main-stem dams: (1) decline of diadromous and migratory species; (2) species loss in flow-modified riverine fragments downstream from dams; and (3) population isolation in tributary river systems, where populations are subject to habitat degradation (e.g., from urban development). We present evidence for each of these effects below.

Diadromous and migratory fauna.—Dams have substantially restricted the ranges of three of the four diadromous fishes native to the Alabama River system. The American eel persists in the Alabama and Cahaba rivers (Pierson et al. 1989; Mettee et al. 1996), and commonly occurs in the tailwater shoals of the downstream-most dams on the Coosa and Tallapoosa rivers (Mettee et al. 1996). Jordan (1877) reported the presence of eels in the upper Coosa system in north Georgia, prior to construction of the first hydropower dams on the Coosa and Tallapoosa (in 1914 and 1924, respectively). In systems with unobstructed passage, the American eel commonly migrates inland thousands of kilometers and inhabits a wide range of stream sizes and habitats (Helfman et al. 1997). Thus, although we lack additional predam survey data, we hypothesize that eels likely migrated throughout the Alabama River system prior to large dam construction.

The dams on the Alabama River main stem have restricted Gulf sturgeon to the portion of the river downstream from Claiborne Lock and Dam (USFWS and GSMFC 1995). Historically, the species migrated from the Gulf of Mexico upstream to the fall line in the Cahaba River (Pierson et al. 1989). Alabama shad similarly migrated from the Gulf to and above the fall line in the Coosa and Cahaba

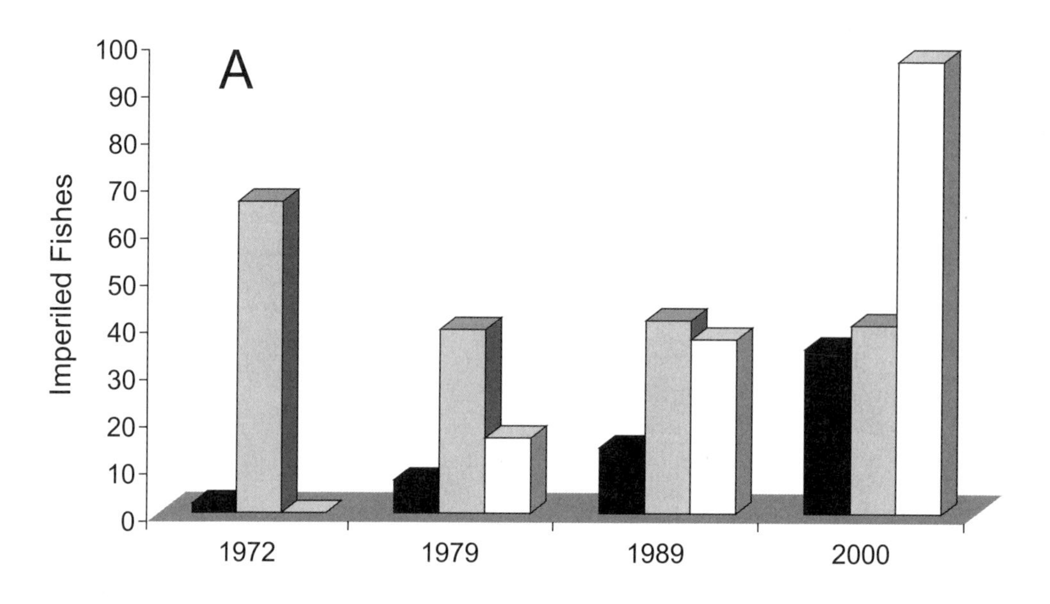

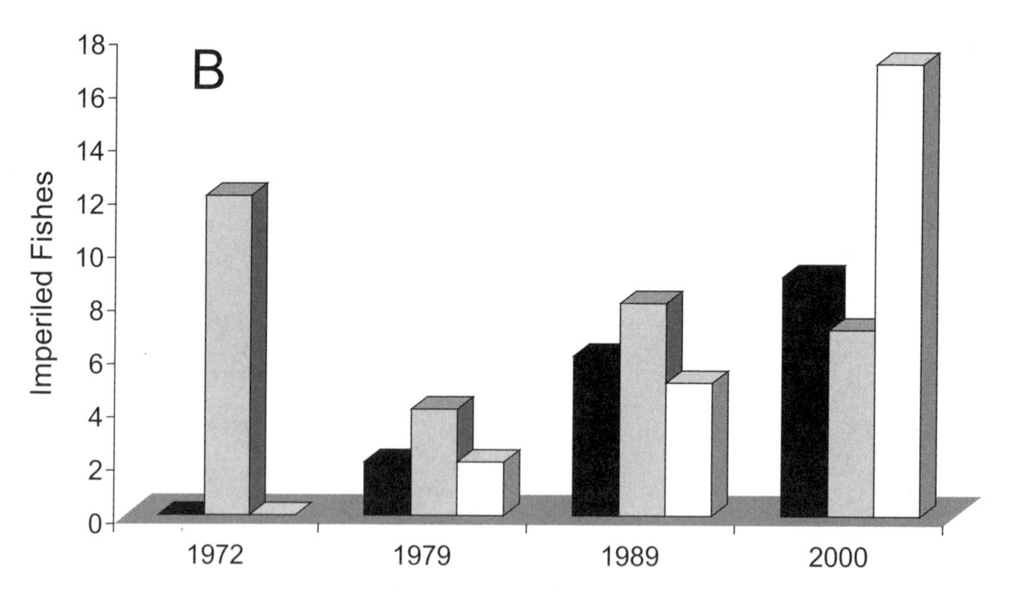

Figure 2.—Comparison of conservation statuses (black is endangered, gray is threatened, and white is vulnerable) based on classification by the American Fisheries Society for the A: southeastern United States and B: Alabama River system (derived from Miller 1972; Deacon et al. 1979; Williams et al. 1989; and Warren et al. 2000).

rivers, respectively, prior to lock and dam construction on the Alabama River (Mettee et al. 1996). This anadromous species is likely extirpated from the Cahaba River (Pierson et al. 1989), and Mettee and O'Neil (2003) report only five individuals collected from the Alabama River in the last 25 years, all downstream from Claiborne and Millers Ferry lock and dams. We hypothesize that the Gulf sturgeon and Alabama shad are extirpated from the lower, flow-regulated reaches of the Coosa and Tallapoosa rivers. Although we know of no records for Alabama shad in the upstream portions of the Coosa or Tallapoosa systems, Burkhead et al. (1997) hypothesized that the Alabama shad also is extirpated from the upper Coosa system in north Georgia. Their hypothesis is based on the observation that in Atlantic Slope drainages (e.g., the James River), the American shad migrated as far inland as the Blue Ridge province.

The Gulf striped bass, the fourth diadromous fish native to the Alabama River system, remains widespread although dams have altered migrations and population sustainability. Historically, striped bass migrated from the Gulf upstream at least to the fall line and supported popular sport fisheries in the tailwaters of the downstream-most dams on the Coosa and Tallapoosa rivers (Anonymous 1950; J. Hornsby, Alabama Department of Conservation and Natural Resources, personal communication). We lack predam records of striped bass in the more upstream portions of the Coosa and Tallapoosa systems, and thus we do not know how far upstream of the fall line native striped bass may have migrated. In any case, the main-stem dams on the Alabama River (completed between 1963 and 1972) largely blocked striped bass migration from the Gulf. The species presently occurs widely in the system; however, as a result of stocking into impoundments. Stocked populations have been of both Gulf and Atlantic origin. Lack of sufficient riverine conditions inhibits natural reproduction by populations in most of the Coosa impoundments, although stocked striped bass are known to spawn in the free-flowing portions of the Oostanaula river system upstream from Weiss Reservoir (W. Davin, Berry College, Rome Georgia, personal communication). Thus, the range of the striped bass may have increased in the Alabama River system compared to the historical range; however, most populations likely are not self-sustaining and may represent a mixture of subspecies.

River system fragmentation has also curtailed migration by resident large-river fishes, resulting in declines in at least three native species. The southeastern blue sucker is only common within the Alabama River system in the lower Alabama River, where Mettee et al. (1996) report spawning aggregations at the base of Millers Ferry Dam and postspawning movements downstream past Claiborne Dam. We hypothesize that this species historically occurred widely in larger rivers of the Alabama River system. A single record from the upper Coosa River (Scott 1950) documents occurrence above the fall line, and a number of records exist for the lower portions of the Cahaba and Tallapoosa rivers (Pierson et al. 1989; Mettee et al. 1996). Similarly, the Alabama sturgeon, which is endemic to the Mobile River drainage, historically occurred in the main channels of the Alabama, lower Cahaba, and lower Tallapoosa rivers (Burke and Ramsey 1995). An 1898 report by the U.S. Commission of Fish and Fisheries documents a large (19,000 kg, about 20,000 fish), if brief, commercial catch of Alabama sturgeon (Mayden and Kuhajda 1996). The species has continued to be encountered by fishermen in the lower portion of the Alabama River, but with decreasing frequency from the 1980s to present (Burke and Ramsey 1995; USFWS 2000). The Alabama sturgeon was federally listed as endangered in 2000, having essentially disappeared in most of its range (USFWS 2000). Overfishing and loss and fragmentation of riverine habitat as a result of dam construction are the primary suspected causes of the sturgeon's decline (Williams and Clemmer 1991; Burke and Ramsey 1995; Mayden and Kuhajda 1996; USFWS 2000). The Alabama sturgeon presently is known to persist in the Alabama River main stem downstream from Millers Ferry and Claiborne lock and dams, and in the Cahaba River (B. Kuhajda, University of Alabama, personal communication).

Finally, the lake sturgeon historically occurred in the upper Coosa River system (Scott 1950; Burke and Ramsey 1995), representing a disjunct population from those in the Mississippi, Great Lakes, and Hudson Bay drainages. Older residents of north Georgia report catching large sturgeon in the Etowah and Coosa rivers from the 1930s to 1970s, including an 86-lb individual taken with a pitchfork at the base of a low-head dam on the Etowah main stem in 1948. The last known record dates to 1980, when an individual was taken from a periodically flooded pond adjacent to the Oostanaula River. The lake sturgeon is now presumed extirpated from the Alabama River system by the Georgia Department of Natural Resources (GDNR), which has initiated a reintroduction program using individuals of Wisconsin origin.

Fish Assemblages in Flow-Modified River Segments.—The segments of the Alabama River system regulated by upstream hydropower dams all have experienced species losses and assemblage simplification. Most of the flow-regulated sections lack records for 30% or more of the minnow and/or darter species presumed native to these reaches (Figure 3). Sunfish and bass species generally show less evidence of faunal loss in flow-regulated segments (Figure 3). An exception is the short reach of the Coosa River that remains unimpounded downstream from Jordan Dam, where all three groups exhibit low percentages of native species (Figure 3). The percent of native minnow and darter species persisting in the regulated portions of the Tallapoosa appear higher than in the Coosa system; however, this variation is not obviously related to length of the segment, length of time regulated, or flow regime characteristics. The two regulated Tallapoosa segments include the most recent and among the earliest segments to be regulated, and the lowest and highest base flow provisions (Table 1).

Applying the jackknife estimator of species richness to presence-absence data from replicated PAE samples for four reaches (Coosa below Jordan Dam, Coosa below Weiss Dam, Tallapoosa below Harris Dam and Tallapoosa below Thurlow Dam) did not suggest that low species richness observations in these reaches resulted from low species detectability. Ratio of observed to estimated species richness exceeded 83% for all species groups at all sites except for centrarchid richness in the Tallapoosa downstream from Harris Dam (4 years of repeated samples; observed: estimated richness = 57%). Additionally, in all cases except the Coosa below Jordan, other sampling efforts (Table 2) obtained records for as many or more additional species as were indicated as unobserved in PAE sampling. For the Coosa below Jordan (where our richness estimates are entirely based on PAE sampling), the jackknife estimates suggest presence of only two additional minnow species (9% of native richness) and one additional centrarchid (8% of native richness). It is impossible to estimate the presence of species that are not vulnerable to any sampling. However, the fact that most of the native sunfish and bass, minnow, and darter fauna have been recorded in at least three of the unregulated portions of these systems (Figure 3), coupled with failure of replicated samples to suggest low species detectability in total sampling efforts, supported the hypothesis that fish assemblages in flow-regulated reaches have experienced species losses, particularly of river-dependent species.

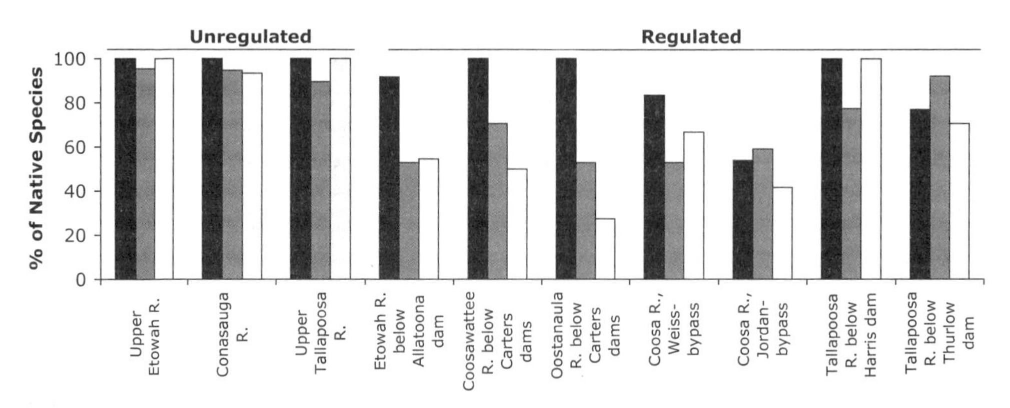

Figure 3.—Estimated percentages of native species persisting in three unregulated and seven flow-regulated segments of the Alabama River system, for three families: Centrarchidae (black bars), Cyprinidae (gray bars), and Percidae (white bars). Identities of species with known occurrences in the flow-regulated reaches are indicated in Appendix A.

Analysis of Suttkus and Gunning's long-term fish survey data suggested that the main stem of the Alabama River has also experienced dramatic changes in fish assemblage richness and composition. Species richness in collections, pooled for all sites, declined significantly over the survey period, from a high of 96 species in 1964 to a low of 34 species in 2000 (Figure 4). This decline preceded, and continued after, the change in 1983 from nighttime to daytime sampling. Groups accounting for most of the decline were percids (mostly darters), catfishes, minnows such as the "Pine Hills chub" *Macrhybopsis* sp. *cf. aestivalis,* and fluvial shiner and a group of diadromous and euryhaline species, including the Alabama shad, bay anchovy, American eel, striped mullet, southern flounder, and the hogchoker (Figure 5). Among darters, the crystal darter exhibited the strongest decline. Other darters showing marked declines were the naked sand darter, the river darter, and the saddleback darter.

Catfishes showed an abrupt change in abundance and occurrence from high during the first half of the survey (1964–1984) to low during the second half of the survey (1985–2000). Coincident with this change was the change in sampling time from evening to daylight hours. Since most catfishes are nocturnal, the most parsimonious explanation for the decline is that they were underrepresented in daylight samples due to inactivity. However, among the catfishes were five species of madtoms (the black madtom, tadpole madtom, speckled madtom, frecklebelly madtom, and freckled madtom), which were collected in the first few years of the survey (12 years prior to the start of daytime

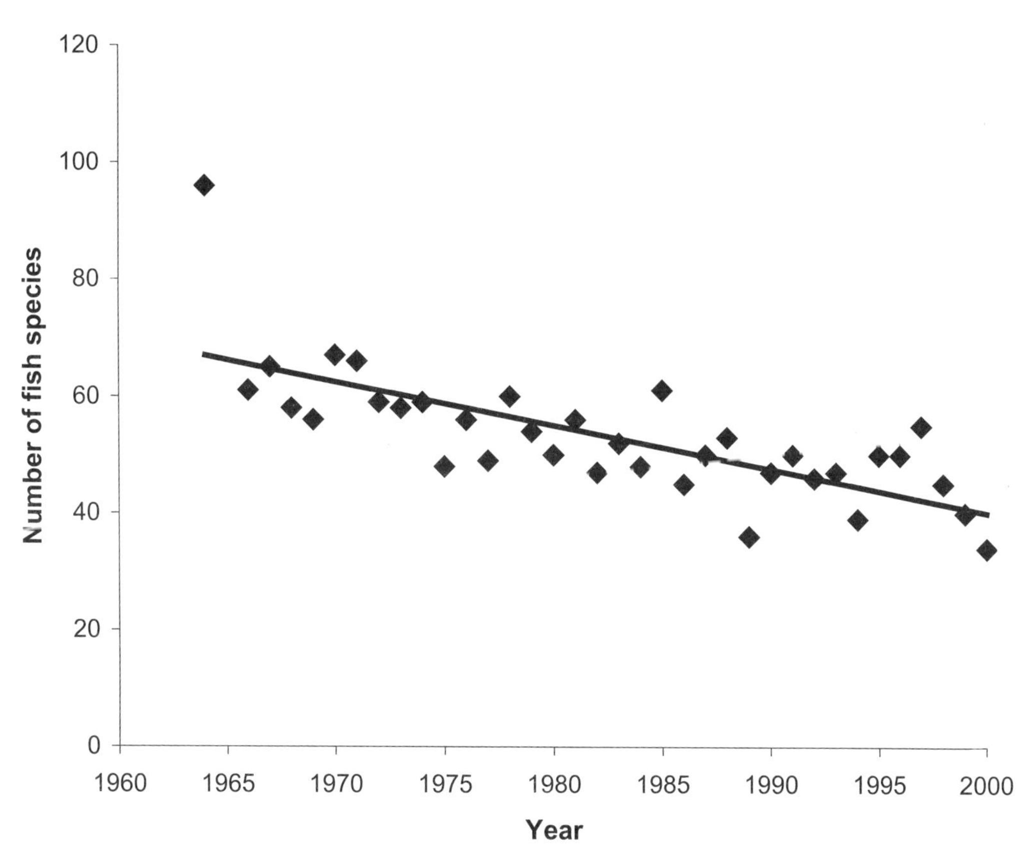

Figure 4.—Decline in total number of species collected across years in the long-term fish monitoring survey of the Alabama River main stem by R. D. Suttkus and G. E. Gunning. The coefficient of determination (r^2) for the plotted trend is 0.64.

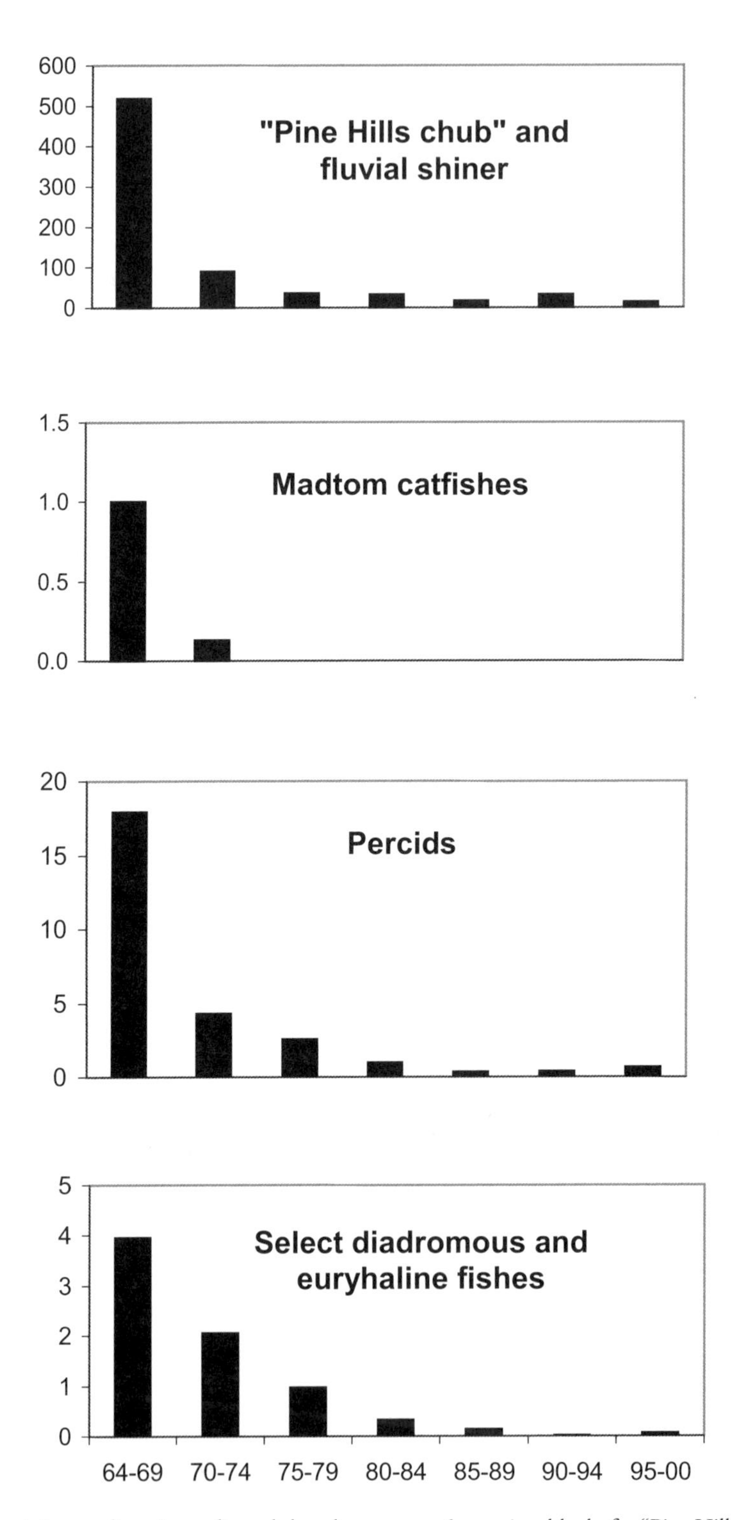

Figure 5.—Trends in sampling-time-adjusted abundance across 6-year time blocks for "Pine Hills chub" and fluvial shiner, madtom catfishes, percids, and select diadromous and euryhaline fishes collected in the long-term fish monitoring survey of the Alabama River main stem by R. D. Suttkus and G. E. Gunning.

collecting), but not after this time (Figure 5). Declines in species such as the speckled chub, fluvial shiner, frecklebelly madtom, crystal darter, and saddleback darter are attributed to the reduction in gravel bars and change from lotic to more lentic conditions in impounded portions of the survey area.

Fauna of major tributaries.—The free-flowing and unregulated portions of the ACT harbor the remaining populations of 8 of the 10 federally listed species in the system (i.e., excepting only the two listed sturgeon species) along with a large portion of the native fish fauna. For example, the Conasauga River system retains at least 71 of 77 native fishes (Walters 1997), and the upper Etowah River system holds at least 68 of 74 native fishes. Together, these two Coosa River headwater systems contain four darter species and one minnow species that are federally listed, with the remaining small-bodied protected fishes occurring in the Coosawattee system, the Cahaba system, and in the case of the pygmy sculpin, a single spring in the Coosa system (Williams 1968; Appendix A).

At least five of the federally listed darters and minnows have had their ranges restricted and fragmented as a result of main-stem impoundments. The amber darter persists in disjunct populations in the upper Conasauga and Etowah rivers, separated by Allatoona Reservoir and the flow-regulated segments of the Etowah and Oostanaula rivers (Figure 6). The goldline darter persists in populations in the upper Coosawattee and Cahaba rivers, separated by eight main-stem dams (Figure 6). The blue shiner similarly persists in fragmented populations separated by main-stem impoundments and sections of flow-regulated rivers in the upper Coosa (USFWS 1992; Figure 6). Allatoona Reservoir on the Etowah River has inundated and fragmented tributary habitats occupied by the Cherokee darter (Bauer et al. 1995) and truncated the downstream range of the Etowah darter. All of these species are hypothesized or known to have occurred more widely in main-stem shoals (or downstream portions of tributaries) before these were impounded or subjected to the effects of flow-regulation.

Discussion

The Alabama River system contains one of the most diverse temperate fish assemblages known, the full extent of which remains under discovery. Intensive faunal study continues to uncover cryptic species that have diverged from and are similar to known species and that often are narrowly distributed (Burkhead and Jelks 2000). Although the lake sturgeon has disappeared from the system, there are no recorded extinctions of Alabama River system fishes. This contrasts with the state of the system's large-river molluscan fauna; at least 32 species and 4 genera of freshwater snails are extinct as a result of the damming of the Coosa River shoals (Bogan et al. 1995; Neves et al. 1997). Of course, it is possible that fish species not discovered by science have gone extinct with the damming and fragmentation of the system. Further, if the lake sturgeon of the upper Coosa represents a unique, endemic taxon, following the pattern of other Mobile River system endemics (e.g., Alabama sturgeon) that have diverged from sister taxa in the Mississippi system, then the loss of the lake sturgeon from the Alabama River system is an extinction. Preventing future extinctions will require managing the system differently than has occurred over the last century, with direct intent to protect and restore the ecological integrity of the river system.

Conserving the fish fauna of the Alabama River system will require addressing the detrimental effects of 16 main-stem dams on native aquatic biota, preventing the spread of alien species to the greatest extent practicable, and managing future land use changes to minimize stream and river degradation. The large river main stems have been transformed from a heterogeneous continuum of flowing water habitats, to a series of slowly flowing, deep-water impoundments interspersed with fragments of unimpounded river having altered flow regimes. The major free-flowing tributary systems retain much of the system's fish fauna, but are isolated from each other and subject to effects of human population growth and increasing societal demands for water supply. In this context, the actions most essential to long-term species conservation are restoring large river habitat for fishes, including migratory species, and protecting hydrologic regimes and water quality in the free-flowing tributary systems.

Substantial potential for restoring populations of migratory, large-river fishes such as Alabama stur-

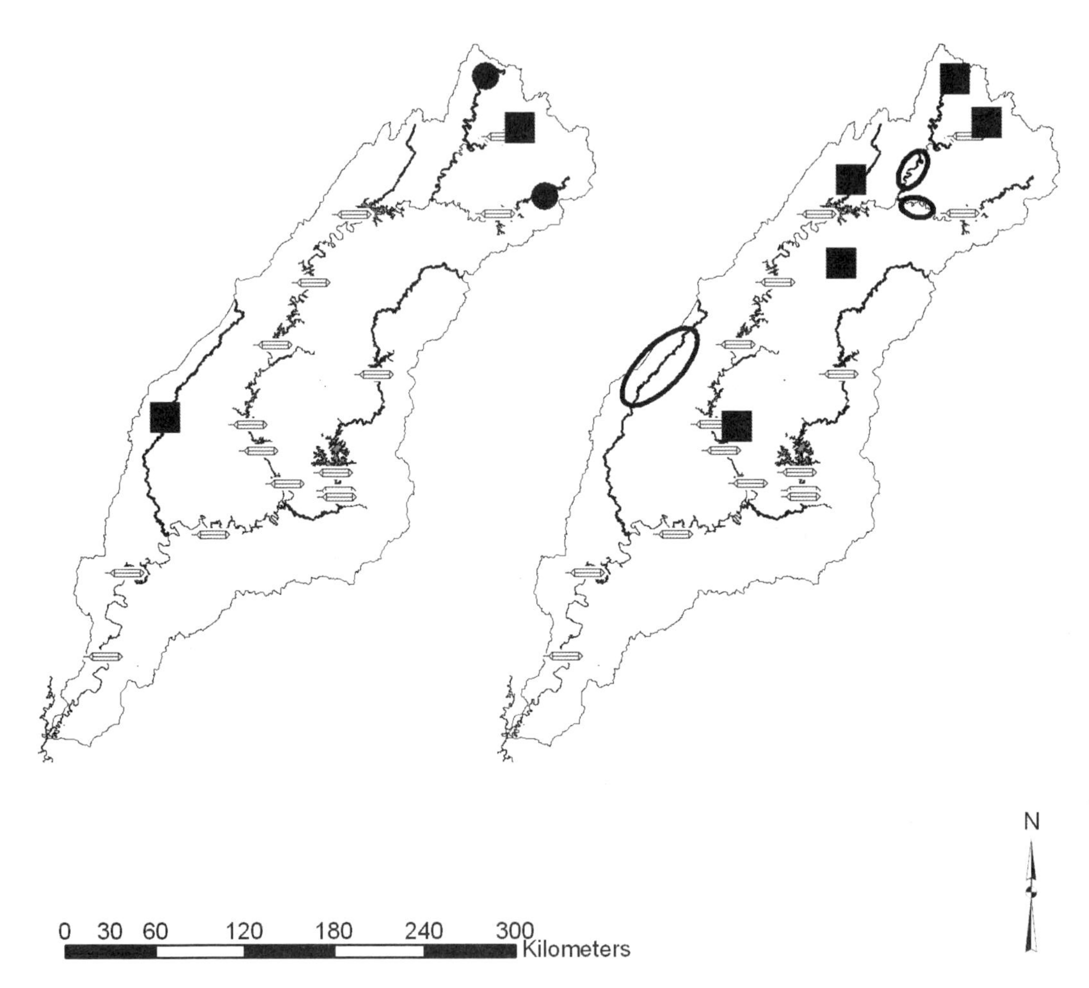

Figure 6.—Locations of extant populations of three federally protected fishes in the Alabama River system: amber darter (solid circles, left map), the goldline darter (solid squares, left map), and the blue shiner (solid squares, right map). Ovals on right map indicate reaches historically containing blue shiners. Open bars represent main-stem dams.

geon, Gulf sturgeon, Alabama shad, and southeastern blue sucker entails modifying the two downstream-most dams on the Alabama River. Enhancing fish passage at Claiborne and Millers Ferry lock and dams could restore connectivity between the lower Alabama River and the Cahaba River, encompassing over 400 km of riverine habitat from the Gulf to the fall line. Some passage occurs under present conditions; for example, southeastern blue suckers are able to swim upstream past Claiborne Dam when the dam is periodically inundated during high flows (M. F. Mettee, Alabama Geological Survey, unpublished data and personal communication). However, this represents a narrow opportunity for passage, and the success or frequency of migration by other fishes (e.g. through the locks), is unknown. Regulatory agencies and conservation groups are interested in improving fish migration success. The USACE has partnered with the World Wildlife Fund, the U.S. Fish and Wildlife Service (USFWS), and Alabama Department of Conservation and Natural Resources (ADCNR), under Section 1135 of the Water Resources Development Act, to explore options for facilitating fish passage at Claiborne Lock and Dam. Feasibility level designs have identified options, including construction of a fish lift or a vertical slot fishway to facilitate passage; funding for implementation has not been identified (M. J. Eubanks, USACE, Mobile District, personal communication). Modifying lock operations to facilitate fish passage at Millers Ferry is also being explored by the USACE and USFWS (Carl Couret, USFWS, Daphne, Alabama, and M. J. Eubanks, personal communications). Because little is known concerning the movements of most migratory fishes in the lower Alabama River system, efforts to enhance passage at dams should be accompanied by research to estimate migratory patterns and population responses of target fishes. Restoring continuous, free-flowing riverine habitat by removing the two downstream-most dams in the system may ultimately be necessary for population recovery of fishes that evolved in and require flowing-water habitat, including the diverse small-bodied fish assemblages native to the Alabama River main stem.

Present conservation efforts for two of the system's large-river fishes, Alabama sturgeon and lake sturgeon, are focused on captive propagation and fish reintroduction. Efforts by ADCNR and USFWS to propagate the Alabama sturgeon have been limited by the species' rarity; only five fish had been captured between 1997 and 2000 to use as broodstock, three of which have died in captivity (USFWS 2000). Recovery of this species obviously remains highly tentative unless the factors limiting natural reproduction and survival of Alabama sturgeon can be identified and addressed, most likely through restoration of large river habitat. The GDNR has initiated a reintroduction of lake sturgeon in the Coosa River system in Georgia, beginning with the release (in 2002) of 1,100 fingerlings produced from Wisconsin stock. The success of this program will not be known for years, but clearly depends on the availability of suitable habitat for sturgeon spawning, feeding, and overwintering, and of adequate water quality. The upper Coosa River system (upstream from Weiss Reservoir) could provide 190 km of interconnected, unimpounded riverine habitat to sturgeon and other riverine fishes, but only if the flow-regulated sections below Allatoona Dam and Carters Dam can be managed to support riverine biota.

Restoring faunal integrity in the flow-regulated river segments of the Alabama River system depends in part on changing hydropower dam operations to ameliorate detrimental effects on biota. These sections retain natural instream habitat structure (e.g., alternating shoals and pools), but are subjected to the unnatural flow regimes imposed by the upstream dams. The Federal Energy Regulatory Commission licenses operation of the nonfederal dams and requires periodic license renewal (typically at 30–50 year intervals). Although an infrequent occurrence, relicensing provides opportunities to make scientifically based changes in dam operations to enhance conditions for downstream aquatic biota. Most commonly, operational changes involve providing higher base flows to sustain aquatic habitats during periods of nongeneration, which can benefit riverine fauna. For example, increasing base flows at Jordan Dam on the Coosa River has resulted in higher fish species richness (Peyton and Irwin 1997) and higher abundances of the endangered snail *Tulatoma magnifica* (Christman et al. 1995). Similarly, increasing the base flow at Thurlow Dam on the Tallapoosa River to about 25% of mean annual flow has been followed by increases in riverine-dependent fishes downstream from the dam (Travnichek et al. 1995).

Low flows are not the only aspects of flow regimes below dams that limit biota downstream (Poff et al. 1997; Richter et al. 1997). Hydropeaking fluctuations (Cushman 1985; Bain et al. 1988; Freeman et al. 2001) and alteration of flood levels and seasonality, and of thermal and sediment regimes, also degrade habitat quality for native river biota (Sparks 1995; Collier et al. 1996; Stanford et al. 1996). Restoration of multiple aspects of natural flow regimes will be necessary to restore ecological integrity in flow-regulated rivers, but the interesting questions remain of how much and what types of restoration are necessary to conserve native biota. Clearly, full restoration of natural flow regimes in regulated rivers is not compatible with other management objectives such as hydropeaking and flood control. An adaptive management approach (Walters 1986) to modifying dam operations will thus be essential for evaluating the relative benefits to river biota of differing hydrologic changes (e.g., increasing base flows, dampening hydropeaking fluctuations, restoring seasonal flow differences; Irwin and Freeman 2002). The USACE dams, including Allatoona and Carters dams, are not subject to licensing but would also benefit from an adaptive approach to improving flow regimes. Alleviating flow-related limitations in the flow-regulated portions of the Etowah and Coosawattee rivers could substantially expand the habitat available to the imperiled riverine fauna (such as the blue shiner and amber darter), as well as the reintroduced lake sturgeon and striped bass.

Conserving the fishes of the Alabama River system will strongly depend on protecting populations in the remaining free-flowing and unregulated portions of the system that retain high proportions of the native fauna. The system has been, and continues to be, degraded by land-use activities that alter runoff and inputs of sediment, nutrients, and contaminants to the rivers. Water quality degradation in the Cahaba River as a result of wastewater discharge from multiple municipal treatment plants, and from surface mining, has been implicated in the extirpation of the blue shiner and in substantial range reductions of the goldline darter and Cahaba shiner (Mayden and Kuhajda 1989; USFWS 1990, 1992). Urban development in the Birmingham, Alabama area also has been linked to loss of fish species (Shepard et al. 1994) and reduced abundances of sensitive minnows and darters in the Cahaba River system (Onorato et al. 1998, 2000). In the Etowah River system, which is affected by development emanating from the Atlanta, Georgia area, the rapid conversion of farmland to urban and suburban developments are major threats to the amber darter, Cherokee darter, and Etowah darter (USFWS 1994; Burkhead et al. 1997) and is leading to loss of endemic fishes, minnows, darters, and sculpins in areas with urban land cover as low as 10% (Walters 2002).

Increasing demand for water supply is a direct consequence of human population growth and places additional strain on aquatic systems. Although not as disruptive of flow regimes as dams built for hydropower production, water supply reservoirs degrade and fragment habitat for stream-dependent fishes. New reservoirs are being planned for those portions of the system that yet maintain high water quality, also ensuring overlap with refugia for species eliminated from degraded streams. Quantifying how much water can be removed from a system for water supply, and how much fragmentation a stream system can sustain, without leading to losses of stream species, is critical to water resource development that conserves native aquatic biota.

The challenges of conserving the fish fauna of the Alabama River system are large, involving both societal and scientific questions, and reflect similar aquatic conservation and river management problems throughout the southeastern United States. The increase in proportion of the Alabama River system's fish fauna recognized as imperiled over the last 25 years mirrors fish imperilment for the region (Figure 2). Although the fauna has largely survived over a century of changes to the rivers and the watershed, a substantial portion of Alabama River system fishes (i.e., at least some 38 species) is now imperiled with extinction. Further, river and landscape alteration in the system, as regionally, have reached unprecedented levels of spatial extent and intensity. Whereas, in the past, many species likely had refugia in undeveloped or less disturbed portions of systems, the combination of damming and urban growth now leaves few subsystems unaffected. Unless future river and land use management strategies in the Alabama River system specifically address conservation and restoration of flowing water systems and their biota, fish species extinctions appear inevitable.

Acknowledgments

We are indebted to those who have come before us and described the natural history of the Alabama River system, those who continue to study its fauna and ecological workings, and those who are striving to conserve its biological integrity. This chapter benefited substantially from comments on an earlier draft by Robert Hughes and two anonymous reviewers. Bernard Kuhajda, Jim Williams, Patrick O'Neil, Carter Gilbert, and David Neely graciously provided information on taxonomy and distribution for a number of taxa, and we appreciate their time.

References

Anonymous. 1950. The striped bass invasion: saltwater sport provided anglers many miles upstream from the sea. Alabama Conservation 22:14–15.

Bain, M. B., J. T. Finn, and H. E. Booke. 1985. A

quantitative method for sampling riverine microhabitats by electrofishing. North American Journal of Fisheries Management 5:489–493.

Bain, M. B., J. T. Finn, and H. E. Booke. 1988. Streamflow regulation and fish community structure. Ecology 69:382–392.

Bauer, B. H., D. A. Etnier, and N. M. Burkhead. 1995. *Etheostoma* (*Ulocentra*) *scotti* (Osteichthys: Percidae), a new darter from the Etowah River system in Georgia. Bulletin of the Alabama Museum of Natural History 17:1–16.

Bogan, A. E., J. M. Pierson, and P. Hartfield. 1995. Decline in the freshwater gastropod fauna in the Mobile Bay basin. Pages 249–252 *in* E. T. LaRoe, G. S. Farris, C. E. Puckett, P. D. Doran, and M. J. Mac, editors. Our living resources: a report to the nation on the distribution, abundance, and health of U.S. plants, animals, and ecosystems. U.S. Department of the Interior, National Biological Service, Washington D.C.

Boschung, H. T. 1961. An annotated list of the fishes of the Coosa River system in Alabama. American Midland Naturalist 66(2):257–285.

Boschung, H. T. 1992. Catalog of freshwater and marine fishes of Alabama. Bulletin of the Alabama Museum of Natural History 14:1–266.

Boschung, H. T., Jr., and R. L. Mayden. 2004. Fishes of Alabama. Smithsonian Books, Washington, D.C.

Bowen, Z. H., M. C. Freeman, and K. D. Bovee. 1998. Evaluation of generalized habitat criteria for assessing impacts of altered flow regimes on warmwater fishes. Transactions of the American Fisheries Society 127:455–468.

Buckley, J. P. 1995. Navigation and hydropower related changes in the fish community of the Alabama River: analysis of an historical, long-term dataset. Master's thesis. Tulane University, New Orleans, Louisiana.

Buckley, J. P., and H. L. Bart, Jr. 1996. Navigation and fish changes in the Alabama River fish community over a 20-year interval. Southeastern Fishes Council Proceedings 33:10.

Burke, J. S., and J. S. Ramsey. 1995. Present and recent historic habitat of the Alabama sturgeon, *Scaphirhynchus suttkusi* Williams and Clemmer, in the Mobile basin. Bulletin of the Alabama Museum of Natural History 17:17–24.

Burkhead, N. M. and H. L. Jelks. 2000. Diversity, levels of imperilment, and cryptic fishes in the southeastern United States. Pages 30–32 *in* R.A. Abell, D. M. Olson, E. Dinerstein, P. T. Hurley, J. T. Diggs, W. Eichbaum, S. Walters, W. Wettengel, T. Allnut, C. J. Loucks, and P. Hedao, editors. Freshwater ecoregions of North America: a conservation assessment. Island Press, Washington, D.C.

Burkhead, N. M., S. J. Walsh, B. J. Freeman, and J. D. Williams. 1997. Status and restoration of the Etowah River, an imperiled southern Appalachian ecosystem. Pages 375–444 *in* G. W. Benz and D. E. Collins, editors. Aquatic fauna in peril: the southeastern perspective. Southeast Aquatic Research Institute, Special Publication 1, Decatur, Georgia.

Burr, B. M., and R. L. Mayden. 1999. A new species of *Cycleptus* (Cypriniformes: Catostomidae) from Gulf Slope drainages of Alabama, Mississippi, and Louisiana, with a review of the distribution, biology and conservation status of the genus. Bulletin of the Alabama Museum of Natural History 20:19–57.

Christman, S. P., F. G. Thompson, and E. L. Raiser. 1995. *Tulatoma magnifica* (Conrad) (Gastropoda: Viviparidae) status and biology in the Coosa River below Jordan Dam, Alabama. Report to Alabama Power Company, Birmingham.

Collier, M., R. H. Webb, and J. C. Schmidt. 1996. Dams and rivers: primer on the downstream effects of dams. U.S. Geological Survey, Circular 1126, Tucson, Arizona.

Cushman, R. M. 1985. Review of ecological effects of rapidly varying flows downstream from hydroelectric facilities. North American Journal of Fisheries Management 5:330–339.

Deacon, J. E., G. Kobetich, J. D. Williams, and S. Contreras. 1979. Fishes of North America endangered, threatened, or of special concern: 1979. Fisheries 4(2):30–44.

Etnier, D. A., and W. C. Starnes. 1993. The fishes of Tennessee. The University of Tennessee Press, Knoxville.

Freeman, B. J. 1998. Survey of threatened and endangered fishes in the Oostanaula, Coosawattee and Etowah rivers. Report to U.S. Fish and Wildlife Service, Daphne, Alabama.

Freeman, M. C., Z. H. Bowen, K. D. Bovee, and E. R. Irwin. 2001. Flow and habitat effects on juvenile fish abundance in natural and altered flow regimes. Ecological Applications 11:179–190.

Fuller, P. L., L. G. Nico, and J. D. Williams. 1999. Nonindigenous fishes introduced into inland waters of the United States. American Fisheries

Society, Special Publication 27, Bethesda, Maryland.

Helfman, G. S., B. B. Collette, and D. E. Facey. 1997. The diversity of fishes. Blackwell Scientific Publications, Malden, Massachusetts.

Irwin, E. R., D. Buckmeier, A. Cunha, K. Kleiner, and T. Wetzel. 1999. Nursery habitat and larval fish community relations in riverine ecosystems. Final Report to Alabama Department of Conservation and Natural Resources, Dingell-Johnson Project F-40–29, Montgomery.

Irwin, E. R., and M. C. Freeman. 2002. Proposal for adaptive management to conserve biotic integrity in a regulated segment of the Tallapoosa River, Alabama, U.S.A. Conservation Biology 16:1212–1222.

Irwin, E. R., M. C. Freeman, A. Belcher, and K. Kleiner. 2001. Survey of shallow-water fish communities in the Dead River and Terrapin Creek. Report to Alabama Power Company, Birmingham.

Jackson, H. H., III. 1995. Rivers of history: life on the Coosa, Tallapoosa, Cahaba and Alabama. The University of Alabama Press, Tuscaloosa.

Jenkins, R. E., and N. M. Burkhead. 1994. Freshwater fishes of Virginia. American Fisheries Society, Bethesda, Maryland.

Jordan, D. S. 1877. A partial synopsis of the fishes of upper Georgia; with supplementary papers on fishes of Tennessee, Kentucky, and Indiana. Annals of the New York Lyceum of Natural History 11:307–377.

Leigh, D. S. 1994. Mercury contamination and floodplain sedimentation from former gold mines in north Georgia. Water Resources Bulletin 30:739–748.

Lydeard, C., and R. L. Mayden. 1995. A diverse and endangered aquatic ecosystem of the southeastern United States. Conservation Biology 9:800–805.

Mayden, R. L. and B. R. Kuhajda. 1989. Systematics of *Notropis cahabae*, a new cyprinid fish endemic to the Cahaba River of the Mobile basin. Bulletin of the Alabama Museum of Natural History 9:1–16.

Mayden, R. L., and B. R. Kuhajda. 1996. Systematics, taxonomy, and conservation status of the endangered Alabama sturgeon, *Scaphirhynchus suttkusi*, Williams and Clemmer (Actinopterygii, Acipenseridae). Copeia 1996:241–273.

Mettee, M. F., and P. E. O'Neil. 2003. Status of Alabama shad and skipjack herring in Gulf of Mexico drainages. Pages 157–170 *in* K. E. Limburg and J. R. Waldman, editors. Biodiversity, status, and conservation of the world's shads. American Fisheries Society, Symposium 35, Bethesda, Maryland.

Mettee, M. F., P. E. O'Neil, and J. M. Pierson. 1996. Fishes of Alabama and the Mobile basin. Oxmoor House, Inc., Birmingham, Alabama.

Miller, R. R. 1972. Threatened freshwater fishes of the United States. Transactions of the American Fisheries Society 101:239–252.

Neves, R. J., A. E. Bogan, J. D. Williams, S. A. Ahlstedt, and P. W. Hartfield. 1997. Status of aquatic mollusks in the southeastern United States: a downward spiral of diversity. Pages 43–85 *in* G. W. Benz and D. E. Collins, editors. Aquatic fauna in peril: the southeastern perspective. Southeast Aquatic Research Institute, Special Publication 1, Decatur, Georgia.

Onorato, D., K. R. Marion, and R. A. Angus. 1998. Longitudinal variations in the ichthyofaunal assemblages of the upper Cahaba River: possible effects of urbanization in a watershed. Journal of Freshwater Ecology 13:139–154.

Onorato, D., K. R. Marion, and R. A. Angus. 2000. Historical changes in the ichthyofaunal assemblages of the upper Cahaba River in Alabama associated with extensive urban development in the watershed. Journal of Freshwater Ecology 15:47–63.

Peyton, J. S., and E. R. Irwin. 1997. Temporal dynamics of a shallow water fish assemblage in a shoals reach of the Coosa River. Report to Alabama Power Company, Birmingham.

Pierson, J. M., W. M. Howell, R. A. Stiles, M. F. Mettee, P. E. O'Neil, R. D. Suttkus, and J. S. Ramsey. 1989. Fishes of the Cahaba River system in Alabama. Geological Survey of Alabama, Bulletin 134, Tuscaloosa.

Poff, N. L., J. D. Allan, M. B. Bain, J. R. Karr, K. L. Prestegaard, B. D. Richter, R. E. Sparks, and J. C. Stromberg. 1997. The natural flow regime. Bioscience 47:769–784.

Richter, B. D., J. V. Baumgartner, R. Wigington, and D. P. Braun. 1997. How much water does a river need? Freshwater Biology 37:231–249.

Scott, D. C. 1950. Sampling fish populations in the Coosa River, Alabama. Transactions of the American Fisheries Society 80:28–40.

Shepard, T. E., P. E. O'Neil, S. W. MacGregor, and S. C. Harris. 1994. Water-quality and biomonitoring studies in the upper Cahaba River drainage of

Alabama. Geological Survey of Alabama, Bulletin 160, Tuscaloosa.

Smith-Vaniz, W. F. 1968. Freshwater fishes of Alabama. Auburn University, Agricultural Experiment Station, Auburn, Alabama.

Sparks, R. E. 1995. Need for ecosystem management of large rivers and their floodplains. Bioscience 45:168–182.

Stanford, J. A., J. V. Ward, W. J. Liss, C. A. Frissell, R. N. Williams, J. A. Lichatowich, and C. C. Coutant. 1996. A general protocol for restoration of regulated rivers. Regulated Rivers: Research and Management 12:391–413.

Stewig, J. D. 2001. Assessment of the largemouth bass and spotted bass populations in the 'dead' river section of the Coosa River, Alabama. M. S. thesis. Auburn University, Auburn, Alabama.

Suttkus, R. D., R. M. Bailey, and H. L. Bart, Jr. 1994a. Three new species of *Etheostoma*, subgenus *Ulocentra*, from the Gulf coastal plain of southeastern United States. Tulane Studies in Zoology and Botany 29:97–126.

Suttkus, R. D., and D. A. Etnier. 1991. *Etheostoma tallapoosae* and *E. brevirostrum*, two new darters, subgenus *Ulocentra*, from the Alabama River drainage. Tulane Studies in Zoology and Botany 28:1–24.

Suttkus, R. D., B. A. Thompson, and H. L. Bart, Jr. 1994b. Two new darters, *Percina* (*Cottogaster*), from the southeastern United States, with a review of the subgenus. Occasional Papers Tulane Museum of Natural History 4:1–46.

Thompson, B. A. 1997a. *Percina suttkusi*, a new species of logperch (subgenus *Percina*) from Louisiana, Mississippi, and Alabama (Perciformes, Percidae, Etheostomatini). Occasional Papers of the Museum of Natural History, Louisiana State University 72:1–26.

Thompson, B. A. 1997b. *Percina kathae*, a new logperch endemic to the Mobile basin in Mississippi, Alabama, Georgia, and Tennessee (Percidae, Etheostomatini). Occasional Papers of the Museum of Natural History, Louisiana State University 73:1–35.

Travnichek, V. H., M. B. Bain, and M. J. Maceina. 1995. Recovery of a warmwater fish assemblage after initiation of a minimum-flow release downstream from a hydroelectric dam. Transactions of the American Fisheries Society 124:836–844.

Travnichek, V. H., and M. J. Maceina. 1994. Comparison of flow regulation effects on fish assemblages in shallow and deep water habitats in the Tallapoosa River, Alabama. Journal of Freshwater Ecology 9:207–216.

USACE (U.S. Army Corps of Engineers). 1998. Water allocation for the Alabama-Coosa-Tallapoosa (ACT) River basin, Alabama and Georgia. Draft Environmental Impact Statement. USACE, Mobile District, Mobile, Alabama.

USFWS (U.S. Fish and Wildlife Service). 1990. Endangered and threatened wildlife and plants; endangered status determined for the fish Cahaba shiner (*Notropis cahabae*). Federal Register 55(25 October 1990):42961–42966.

USFWS (U.S. Fish and Wildlife Service). 1992. Endangered and threatened wildlife and plants; threatened status for two fish, the goldline darter (*Percina aurolineata*) and the blue shiner (*Cyprinella caerulea*). Federal Register 57(22 April 1992):14786–14790.

USFWS (U.S. Fish and Wildlife Service). 1994. Endangered and threatened wildlife and plants; determination of threatened status for the Cherokee darter and endangered status for the Etowah darter. Federal Register 59(20 December 1994):65505–65512.

USFWS (U.S. Fish and Wildlife Service). 2000. Endangered and threatened wildlife and plants; final rule to list the Alabama sturgeon as endangered. Federal Register 65(5 May 2000):26438–26461.

USFWS and GSMFC (U.S. Fish and Wildlife Service and Gulf States Marine Fisheries Commission). 1995. USFWS and GSMFC, Gulf sturgeon recovery plan, Atlanta.

Walters, C. 1986. Adaptive management of renewable resources. Macmillan Publishing Company, New York.

Walters, D. M. 1997. The distribution, status, and ecology of the fishes of the Conasauga River system. Master's thesis. University of Georgia, Athens.

Walters, D. M. 2002. Influence of geomorphology and urban land cover on stream fish assemblages in the Etowah River basin, Georgia. Doctoral dissertation, University of Georgia, Athens.

Ward, A. K., G. M. Ward, and S. C. Harris. 1992. Water quality and biological communities of the Mobile River drainage, eastern Gulf of Mexico region. Pages 278–304 *in* C. D. Becker, and D. A. Neitzel, editors. Water quality in North American river systems. Battelle Press, Columbus, Ohio.

Warren, M. L., Jr., B. M. Burr, S. J. Walsh, H. L.

Bart, Jr., R. C. Cashner, D. A. Etnier, B. J. Freeman, B. R. Kuhajda, R. L. Mayden, H. W. Robison, S. T. Ross, and W. C. Starnes. 2000. Diversity, distribution, and conservation status of the native freshwater fishes of the southern United States. Fisheries 25(10):7–29.

Wharton, C. H. 1978. The natural environments of Georgia. Georgia Department of Natural Resources, Atlanta.

Williams, B. K., J. D. Nichols, and M. J. Conroy. 2002. Analysis and management of animal populations. Academic Press, San Diego, California.

Williams, J. D. 1965. Studies on the fishes of the Tallapoosa River system, Alabama and Georgia. Master's thesis. University of Alabama, Tuscaloosa.

Williams, J. D. 1968. A new species of sculpin, *Cottus pygmaeus*, from a spring in the Alabama River basin. Copeia 1968:334–342.

Williams, J. D., and G. H. Clemmer. 1991. *Scaphirhynchus suttkusi*, a new sturgeon (Pisces: Acipenseridae) from the Mobile basin of Alabama and Mississippi. Bulletin of the Alabama Museum of Natural History 10:17–31.

Williams, J. E., J. E. Johnson, D. A. Hendrickson, S. Contreras-Balderas, J. D. Williams, M. Navarro-Mendoza, D. E. McAllister, and J. E. Deacon. 1989. Fishes of North America endangered, threatened, or of special concern: 1989. Fisheries 14(6):2–20.

Wood, R. M., and R. L. Mayden. 1993. Systematics of the *Etheostoma jordani* species group (Teleostei: Percidae), with descriptions of three new species. Bulletin of the Alabama Museum of Natural History 16:31–46.

Appendix A.—Conservation and indigenous status of the 203 fishes (freshwater, diadromous, and common marine invaders) currently known from the Alabama River system. The statuses of federally listed fishes are in boldface; other conservation designations are based on Warren et al. (2000): CS = currently stable, V = vulnerable, T = threatened, E = endangered. Occurrence symbols: native = N; native and presumably extirpated = Ext; endemic to the Alabama River system = N^E (or Ext^E); probably native = PN; marine invader = M; introduced = I; probably introduced = PI. River system abbreviations are Cona = Conasauga; Cosaw = Coosawattee; Etow = Etowah; Oost = Oostanaula; Coosa = Coosa; Talla = Tallapoosa; Cah = Cahaba; Ala = Alabama. Species occurring in flow-regulated, main-stem reaches are indicated by bold type (Figure 1, reaches 1, 2, 3, 4, 6) and/or are underlined (reaches 5 and 7). See text for data sources.

Binomen	Common name	Status	Cona	Cosaw	Etow	Oost	Coosa	Talla	Caha	Ala
Petromyzontidae	**Lampreys (3)**									
Ichthyomyzon castaneus	chestnut lamprey	CS	N	**N**	**N**	**N**	N	N	N	N
I. gagei	southern brook lamprey	CS	N	**N**	N	N	N	**N**	N	N
Lampetra aepyptera	least brook lamprey	CS	N	N	N	N	N	N	N	N
Acipenseridae	**Sturgeons (3)**									
Acipenser fulvescens	lake sturgeon	T			Ext	Ext	Ext			
A. oxyrinchus desotoi	Gulf sturgeon	**T**					Ext[a]	Ext[a]	Ext	N
Scaphirhynchus suttkusi	Alabama sturgeon	**E**					Ext[a]	Ext	N	N
Polyodontidae	**Paddlefishes (1)**									
Polyodon spathula	paddlefish	V					N	N	N	N
Lepisosteidae	**Gars (3)**									
Lepisosteus oculatus	spotted gar	CS					**N**	N	N	N
L. osseus	longnose gar	CS	N	**N**	**N**	**N**	**N**	N	N	N
Atractosteus spatula	alligator gar	V								N
Amiidae	**Bowfins (1)**									
Amia calva	bowfin	CS					N	N	N	N
Anguillidae	**Freshwater eels (1)**									
Anguilla rostrata	American eel	CS	Ext[a]	Ext[a]	Ext	Ext[a]	N	N	N	N
Engraulidae	**Anchovies (1)**									
Anchoa mitchilli	bay anchovy									M
Clupeidae	**Shads (4)**									
Alosa alabamae	Alabama shad	V					Ext	Ext[a]	Ext	N
A. chrysochloris	skipjack herring	CS			N	N	N	N	N	N
Dorosoma cepedianum	gizzard shad	CS	N	**N**	**N**	**N**	**N**	N	N	N
D. petenense	threadfin shad	CS		I	I	**I**	**I**	I	I	I
Hiodontidae	**Mooneyes (1)**									
Hiodon tergisus	mooneye	CS	N	**N**	**N**	N	N	N	N	N
Cyprinidae	**Minnows (59)**									
Campostoma oligolepis	largescale stoneroller	CS	N	**N**	**N**	**N**	**N**	**N**	N	N
C. pauciradii	bluefin stoneroller	CS			N			N		

Appendix A.—Continued.

Binomen	Common name	Status	Cona	Cosaw	Etow	Oost	Coosa	Talla	Caha	Ala
Carassius auratus	goldfish	CS	I	**I**			**I**	I		
Ctenopharyngodon idella	grass carp	CS	I		**I**	**I**	I	I	I	
Cyprinella caerulea	blue shiner	**T**	N[E]	Ext[E]	Ext[E]	Ext[E]	N[E]		Ext[E]	
C. callistia	Alabama shiner	CS	N	**N**	**N**	**N**	**N**	**N**	N	N
C. gibbsi	Tallapoosa shiner	CS						**N**[E]		
C. lutrensis	red shiner	CS	I	**I**	**I**	**I**	**I**			
C. trichroistia	tricolor shiner	CS	N	**N**	N	N	**N**		N	N
C. venusta	blacktail shiner	CS	N	**N**	**N**	**N**	**N**	**N**	N	N
Cyprinus carpio	common carp	CS	I	**I**	**I**	**I**	**I**	**I**	I	I
Hemitremia flammea	flame chub	V					N			
Hybognathus hayi	cypress minnow	CS						N	N	N
H. nuchalis	Mississippi silvery minnow	CS						N	N	N
Hybopsis lineapunctata	lined chub	V	N[E]	N[E]	Ext[E]	Ext[E]	N[E]	**N**[E]		
H. winchelli	clear chub	CS					N	N	N	N
Hybopsis sp. cf. *winchelli*	"Etowah chub"	CS			N[E]					
Luxilus chrysocephalus	striped shiner	CS	N	**N**	N	**N**	N	**N**	N	N
L. zonistius	bandfin shiner	CS		N	N			**N**		
Lythrurus atrapiculus	blacktip shiner	CS						N		
L. bellus	pretty shiner	CS					N	N	N	N
L. lirus	mountain shiner	CS	N	N	**N**	N	**N**		N	
L. roseipinnis	cherryfin shiner	CS								N
Macrhybopsis sp. *cf. aestivalis*	"Fall line chub"	V	N[E]	**N**[E]	N[E]	Ext[E, a]	N[E]	**N**[E]	N[E]	
Macrhybopsis sp. *cf. aestivalis*	"Pine Hills chub"	CS						N	N	N
Macrhybopsis storeriana	silver chub	CS		**N**	**N**	**N**	N	N	N	N
Nocomis leptocephalus	bluehead chub	CS		N	N	Ext	N	**N**	N	N
N. micropogon	river chub	CS		N	N	Ext				
Notemigonus crysoleucas	golden shiner	CS	N	**N**	N	N	**N**	N	N	N
Notropis ammophilus	orangefin shiner	CS					N	N	N	N
N. asperifrons	burrhead shiner	CS	N	N	N	N	N	N	N	N
N. atherinoides	emerald shiner	CS			Ext	**N**	N	**N**	N	N
N. baileyi	rough shiner	CS					N	**N**	N	N
N. buccatus	silverjaw minnow	CS			PN		N	N	N	N
N. cahabae	Cahaba shiner	**E**							N	
N. candidus	silverside shiner	CS					N	N	N	N
N. chalybaeus	ironcolor shiner	V								N

Appendix A.—Continued.

Binomen	Common name	Status	Cona	Cosaw	Etow	Oost	Coosa	Talla	Caha	Ala
N. chrosomus	rainbow shiner	C	N	N	N	N	N		N	N
N. edwardraneyi	fluvial shiner	CS					N	N	N	N
N. longirostris	longnose shiner	CS			N					N
N. lutipinnis	yellowfin shiner	CS			PN		PN			
N. maculatus	taillight shiner	CS								N
N. petersoni	coastal shiner	CS								N
N. stilbius	silverstripe shiner	CS	N	**N**	**N**	**N**	**N**	**N**	N	N
N. texanus	weed shiner	CS					N	**N**	N	N
N. uranoscopus	skygazer shiner	CS					N[E]	N[E]	N[E]	N[E]
N. volucellus	mimic shiner	CS	Ext				N	N	N	
N. xaenocephalus	Coosa shiner	CS					N	N	N	N
Opsopoeodus emiliae	pugnose minnow	CS	N[E]	**N**[E]	**N**[E]	N[E]	N[E]	N[E]		
Phenacobius catostomus	riffle minnow	CS	N	**N**	**N**	**N**	**N**	**N**	N	N
Pimephales notatus	bluntnose minnow	CS					N	N	N	N
P. promelas	fathead minnow	CS	I	**I**	I	I	I	I	I	I
P. vigilax	bullhead minnow	CS	N	**N**	**N**	**N**	**N**	**N**	N	N
Pteronotropis hypselopterus	sailfin shiner	CS								N
P. signipinnis	flagfin shiner	CS								N
P. welaka	bluenose shiner	V							N	N
Rhinichthys atratulus	eastern blacknose dace	CS	N	N	N	N	N			
Semotilus atromaculatus	creek chub	CS	N	N	N	N	N	N	N	N
S. thoreauianus	Dixie chub	CS					N	N	N	N
Catostomidae	**Suckers (16)**									
Carpiodes cyprinis	quillback	CS			**N**	Ext[a]	N	N	N	N
C. velifer	highfin carpsucker	CS				Ext[a]	N	N	N	N
Catostomus commersonii	white sucker	CS	PI							
Cycleptus meridionalis	southeastern blue sucker	V			Ext[a]	Ext[a]	N	N	N	N
Erimyzon oblongus	creek chubsucker	CS					N	N	N	N
E. sucetta	lake chubsucker	CS					N	N	N	N
E. tenuis	sharpfin chubsucker	CS					N	N	N	N
Hypentelium etowanum	Alabama hog sucker	CS	N	**N**	**N**	**N**	**N**	**N**	N	N
H. nigricans	northern hog sucker	CS	PI							
Ictiobus bubalus	smallmouth buffalo	CS	N	**N**	**N**	**N**	**N**	N	N	N
I. cyprinellus	bigmouth buffalo	CS							I	
Minytrema melanops	spotted sucker	CS	N	**N**	N	**N**	**N**	N	N	N
Moxostoma carinatum	river redhorse	CS	N	**N**	Ext[a]	**N**	**N**	**N**	N	N

Appendix A.—Continued.

Binomen	Common name	Status	Cona	Cosaw	Etow	Oost	Coosa	Talla	Caha	Ala
M. duquesnei	black redhorse	CS	N	**N**	**N**	**N**	$\underline{\mathbf{N}}$	**N**	N	N
M. erythrurum	golden redhorse	CS	N	**N**	**N**	**N**	**N**	$\underline{N}$	N	N
M. poecilurum	blacktail redhorse	CS	N	**N**	**N**	**N**	$\underline{\mathbf{N}}$	$\underline{\mathbf{N}}$	N	N
Ictaluridae	**Bullhead catfishes** (14)									
Ameiurus brunneus	snail bullhead	V			PN					
A. catus	white catfish	CS					I	I		
A. melas	black bullhead	CS	N	**N**	N	N	N	N	N	N
A. natalis	yellow bullhead	CS	N	N	N	N	N	$\underline{N}$	N	N
A. nebulosus	brown bullhead	CS	N	N	N	N	**N**	N	N	N
Ictalurus furcatus	blue catfish	CS	N	N	**N**	**N**	**N**	$\underline{N}$	N	N
I. punctatus	channel catfish	CS	N	**N**	**N**	**N**	$\underline{\mathbf{N}}$	$\underline{\mathbf{N}}$	N	N
Noturus funebris	black madtom	CS					N	N	N	N
N. gyrinus	tadpole madtom	CS					N	N	N	N
N. leptacanthus	speckled madtom	CS	N	**N**	N	N	N	$\underline{\mathbf{N}}$	N	N
N. munitus	frecklebelly madtom	T							N	N
Noturus sp. *cf. munitus*	"Coosa madtom"	T	N^E		N^E					
N. nocturnus	freckled madtom	CS			Ext	Ext[a]	Ext[a]	N	N	N
Pylodictis olivaris	flathead catfish	CS	N	**N**	**N**	N	$\underline{\mathbf{N}}$	$\underline{\mathbf{N}}$	N	N
Esocidae	**Pikes** (3)									
Esox americanus	redfin pickerel	CS					N	N	N	N
E. masquinongy	muskellunge	CS						I		
E. niger	chain pickerel	CS	N	Ext[a]	Ext	Ext[a]	N	$\underline{N}$	N	N
Salmonidae	**Trouts and allies** (3)									
Oncorhynchus mykiss	rainbow trout	CS	I	I	I	I	I	I	I	
Salmo trutta	brown trout	CS	I	I	I	I	I			
Salvelinus fontinalis	brook trout	CS	I	I			I			
Aphredoderidae	**Pirate perch** (1)									
Aphredoderus sayanus	pirate perch	CS					N	N	N	N
Amblyopsidae	**Cavefishes** (1)									
Typhlichthys subterraneus	southern cavefish	V					N			
Belonidae	**Needlefish** (1)									
Strongylura marina	Atlantic needlefish	CS					M	M	M	M
Fundulidae	**Topminnows** (6)									
Fundulus bifax	stippled studfish	V					N^E	N^E		
F. dispar	starhead topminnow	CS							N	N
F. notatus	blackstripe topminnow	CS								N

Appendix A.—Continued.

Binomen	Common name	Status	Cona	Cosaw	Etow	Oost	Coosa	Talla	Caha	Ala
F. nottii	bayou topminnow	CS								N
F. olivaceus	blackspotted topminnow	CS	N	N	N	N	N	N	N	N
F. stellifer	southern studfish	CS	N	N	N	N	N		Ext	N
Poeciliidae	**Livebearers (3)**									
Gambusia affinis	western mosquitofish	CS	N	N	N	N	N	N	N	N
G. holbrooki	eastern mosquitofish	CS	I		PI					
Heterandria formosa	least killifish	CS								N
Atherinopsidae	**Silversides (2)**									
Labidesthes sicculus	brook silverside	CS						N	N	N
Menidia beryllina	inland silverside	CS								N
Cottidae	**Sculpins (5)**									
Cottus sp. *cf. bairdii*	"smokey sculpin"	CS	N	N	N	N	N			
Cottus sp.	Tallapoosa sculpin	CS						N^{E}		
Cottus carolinae infernatus	Alabama banded sculpin	*CS*				N	N	N	N	N
C. carolinae zopherus	Coosa banded sculpin	CS	N^{E}	N^{E}	N^{E}					
C. paulus	pygmy sculpin	T					N^{E}			
Moronidae	**Temperate basses (3)**									
Morone chrysops	white bass	CS	I		I		I	I	I	I
M. mississippiensis	yellow bass	*CS*		I			I		I	
M. saxatilis	striped bass	CS	N	N	N	N	N	N	N	N
Elassomatidae	**Pygmy sunfishes (2)**									
Elassoma evergladei	Everglades pygmy sunfish	CS								N
E. zonatum	banded pygmy sunfish	CS					N	N	N	N
Centrarchidae	**Sunfishes (16)**									
Ambloplites ariommus	shadow bass	CS	N	N	N	N	N	N	N	N
Centrarchus macropterus	flier	CS					N	N	N	N
Lepomis auritus	redbreast sunfish	CS	PI	PI	PI	PI	PI	PI		
L. cyanellus	green sunfish	CS	N	N	N	N	N	N	N	N
L. gulosus	warmouth	CS	N	N	N	N	N	N	N	N
L. humilis	orangespotted sunfish	*CS*					I	I		I
L. macrochirus	bluegill	CS	N	N	N	N	N	N	N	N
L. marginatus	dollar sunfish	CS							N	N
L. megalotis	longear sunfish	CS	N	N	N	N	N	N	N	N
L. microlophus	redear sunfish	CS	N	N	N	N	N	N	N	N
L. miniatus	redspotted sunfish	CS	N	N	N	N	N	N	N	N

Appendix A.—Continued.

Binomen	Common name	Status	Cona	Cosaw	Etow	Oost	Coosa	Talla	Caha	Ala
Micropterus coosae	redeye bass	CS	N	**N**	**N**	**N**	**N**	**N**	N	
M. punctulatus	spotted bass	CS	N	**N**	**N**	**N**	N	**N**	N	N
M. salmoides	largemouth bass	CS	N	**N**	**N**	**N**	**N**	**N**	N	N
Pomoxis annularis	white crappie	CS	N	**N**	**N**	**N**	**N**	**N**	N	N
P. nigromaculatus	black crappie	CS	N	**N**	**N**	**N**	**N**	**N**	N	N
Percidae	**Perches (46)**									
Ammocrypta beanii	naked sand darter	CS						N	N	N
A. meridiana	southern sand darter	CS						N	N	N
Crystallaria asprella	crystal darter	V					N	N	N	N
Etheostoma artesiae	redspot darter	CS					N	N	N	N
E. brevirostrum	holiday darter	T					N[E]			
E. sp. *cf. brevirostrum*	"Conasauga snubnose darter"	T	N[E]							
E. sp. *cf. brevirostrum*	"Coosawattee snubnose darter"	E		N[E]						
E. sp. *cf. brevirostrum*	"Amicalola snubnose darter"	T			N[E]					
E. sp. *cf. brevirostrum*	"Etowah snubnose darter"	E			N[E]					
E. chlorosoma	bluntnose darter	CS						N	N	N
E. chuckwachatte	lipstick darter	V						**N[E]**		
E. coosae	Coosa darter	CS	N[E]	N[E]	**N[E]**	N[E]	N[E]			
E. davisoni	Choctawhatchee darter	CS						N		
E. ditrema	coldwater darter (Nominal)	T	N[E]	Ext[E, a]	Ext[E]	Ext[E]	N[E]			
E. sp. *cf. ditrema*	Middle Coosa + Coldwater Spr.	T					N[E]			
E. etowahae	Etowah darter	**E**			N[E]					
E. fusiforme	swamp darter	CS								N
E. histrio	harlequin darter	CS						N	N	N
E. jordani	greenbreast darter	CS	N[E]	**N[E]**	**N[E]**	**N[E]**	**N[E]**	N[E]	N[E]	N[E]
E. nigrum	johnny darter	CS						N	N	N
E. parvipinne	goldstripe darter	CS					N	N	N	N
E. proeliare	cypress darter	CS								N
E. ramseyi	Alabama darter	CS					N[E]		N[E]	N[E]
E. rupestre	rock darter	CS	N	N	N	N	N	N	N	N
E. scotti	Cherokee darter	**T**			N[E]					
E. stigmaeum	speckled darter	CS	N	N	**N**	N	**N**	**N**	N	N
E. swaini	Gulf darter	CS					N	N	N	N

Appendix A.—Continued.

Binomen	Common name	Status	Cona	Cosaw	Etow	Oost	Coosa	Talla	Caha	Ala
E. tallapoosae	Tallapoosa darter	CS						**N^E**		
E. trisella	trispot darter	E	N^E	**N^E**	Ext^E	N^E	Ext^E			
E. zonifer	backwater darter	CS						N	N	N
Percina antesella	amber darter	**E**	N^E	$Ext^{E,a}$	N^E	$Ext^{E,a}$				
P. aurolineata	goldline darter	**T**		**N^E**					N	
P. brevicauda	coal darter	T			Ext^a	Ext^a	N	N	N	
P. jenkinsi	Conasauga logperch	**E**	N^E							
P. kathae	Mobile logperch	CS	N	**N**	**N**	**N**	$\underline{N}$	$\underline{N}$	N	N
P. lenticula	freckled darter	T	N	Ext^a	N	N	N	$\underline{N}$	N	N
Percina sp.	Coosa bridled darter	V	N^E	Ext^E	N^E					
Percina sp.	muscadine darter	V						**N^E**		
P. maculata	blackside darter	CS						N	N	N
P. nigrofasciata	blackbanded darter	CS	N	**N**	**N**	**N**	$\underline{N}$	$\underline{N}$	N	N
P. palmaris	bronze darter	CS	N^E	**N^E**	**N^E**	N^E	**N^E**	**N^E**		
P. sciera	dusky darter	CS								N
P. shumardi	river darter	CS	N	Ext^a	Ext^a	N	$\underline{N}$	$\underline{N}$	N	N
P. vigil	saddleback darter	CS					Ext^a	$\underline{N}$	N	N
P. suttkusi	Gulf logperch	CS								N
Sander vitreus	walleye	CS	**N**	**N**	**N**	**N**	N	$\underline{N}$	N	N
Sciaenidae	**Drums (1)**									
Aplodinotus grunniens	freshwater drum	CS	N	**N**	**N**	**N**	N	$\underline{N}$	N	N
Mugilidae	**Mullets (1)**									
Mugil cephalus	striped mullet	CS					M		M	M
Paralichthydae	**Lefteye flounders (1)**									
Paralichthys lethostigma	southern flounder	CS								M
Achiridae	**American soles (1)**									
Trinectes maculatus	hogchocker	CS								M
Total Alabama River system endemics			15	14	18	10	16	12	5	3
Total extirpated			2	7	12	14	5	3	4	0
Total native ($N + N^E + Ext + Ext^E + M + PN$)			77	79	95	83	125	125	125	137
Total introduced			13	10	10	8	14	11	8	5

[a] Status as native and extirpated is hypothesized based on occurrences in adjacent reaches and availability of appropriate habitat.

American Fisheries Society Symposium 45:587–602, 2005

Historical Changes in the Rio das Velhas Fish Fauna—Brazil

Carlos Bernardo M. Alves* and Paulo S. Pompeu

Projeto Manuelzão – Universidade Federal de Minas Gerais, Avenida Alfredo Balena, 190 /10.012 Belo Horizonte (MG) Brasil 30130-100

Abstract.—The Rio das Velhas is a tributary of the Rio São Francisco, one of Brazil's largest rivers. It is the Rio São Francisco's second most important tributary in water volume (mean annual discharge of 631 m^3/s), with a drainage area of 27,867 km^2, length of 761 km, and mean width of 38 m. Like many other rivers around the world, it became heavily polluted in the 1900s. The Rio das Velhas is the most polluted river of Minas Gerais state because the basin contains approximately 4.5 million people. Unlike other Brazilian rivers, its fish fauna was studied from 1850 to 1856. Fifty-five fish species were recorded; 20 of them were first described at that time, when there were previously no more than 40 known species in the entire São Francisco basin. Recent fish collections, approximately 150 years later, indicate 107 fish species, but some may be locally extinct. There are good prospects of rehabilitating this fauna because of the connectivity of the Rio das Velhas with the São Francisco main stem, its well-preserved tributaries, and increased investments in sewage treatment.

Introduction

The neotropical biogeographic area is the world's richest fish species region (around 8,000 species, Schaefer 1998), but also one of the least known (Menezes 1996). Despite having one of the richest fish faunas in the world, there are few published studies regarding the past richness, distribution, and ecology of Brazilian fishes. Many of these studies are associated with modifications resulting from dam construction; for example, in the Paraná basin, compartmentalization and flow regulation significantly altered fish populations, especially of migratory species (Agostinho and Júlio, Jr. 1999).

In Brazil, untreated sewage effluents, deforestation, mining, dam construction, siltation, introduction of nonnative species, and water diversions contributed to rapid declines in fish species richness and altered spatial distributions (Agostinho and Zalewski 1996). This scenario is common to many areas of the country and is worst in highly industrialized or urbanized areas.

The aim of this chapter is to evaluate the changes in the fish assemblage of the Rio das Velhas basin in the past 150 years and associate those changes with environmental disturbances.

Methods

Study Area

The Rio das Velhas, located in central Minas Gerais state (Figure 1), is one of the most important tributaries of the Rio São Francisco, one of Brazil's largest rivers. It is the second most important tributary in water volume (mean annual discharge of 631 m^3/s), with a basin area of 27,867 km^2, length of 761 km, mean width of 38 m (CETEC 1983; PLANVASF 1986), and maximum width of 400 m (Sílvia Magalhães, Projeto Manuelzão, personal communication). The Rio das Velhas basin has the

* Corresponding author: curimata@uai.com.br

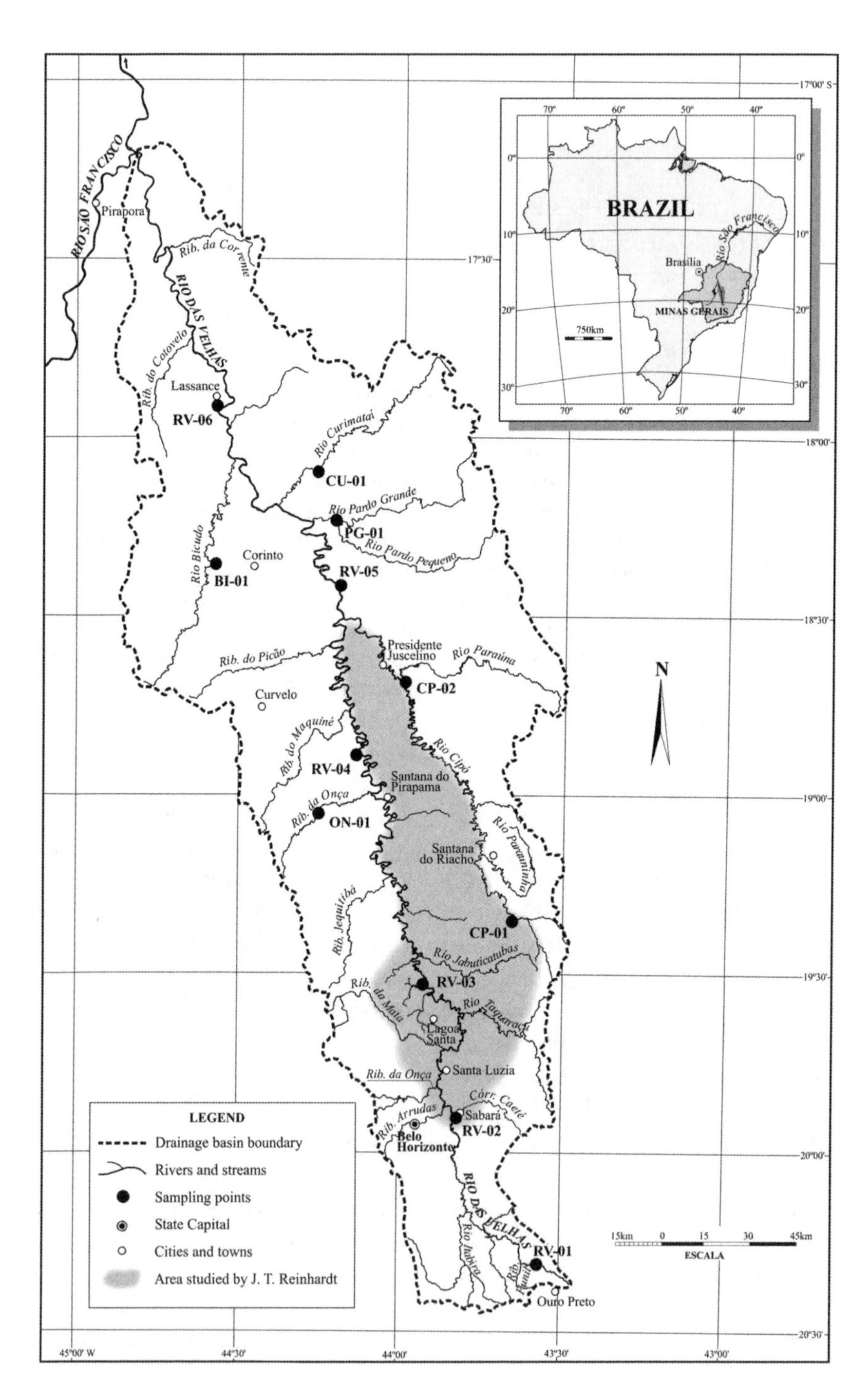

Figure 1.—Rio das Velhas basin, present sampling stations, and relative position in Minas Gerais state and Brazil. The shaded area indicates the probable collecting area of J. T. Reinhardt in the 19th century. (modified from Alves and Pompeu 2001)

largest metropolitan region, the highest gross domestic product, the largest human population, and the longest river course in the Sao Francisco basin. Its headwaters occur at 1,520 m above sea level and its confluence with the Rio São Francisco at an altitude of 478 m. Mean annual precipitation ranges from 110 to 160 cm (www.codevasf.gov.br). There are no available data on river depths, although it was navigable up to Sabará, near the Metropolitan Region of Belo Horizonte (MRBH[1]), in the 19th century (Burton 1977). Presently, many stretches are heavily silted, allowing one to wade across the river.

The Rio das Velhas' headwaters are located in a transition zone between the Atlantic rainforest and cerrado, which is the typical savanna-like vegetation of central Brazil. Both biomes are identified as world diversity hotspots because they have exceptional concentrations of endemic species undergoing exceptional loss of habitat (Myers et al. 2000). Below its headwaters to its mouth, Rio das Velhas flows only through cerrado. Another vegetation formation occurs near high elevation headwaters: the campos rupestres (literally, rock fields), a type of shrubby montane savanna, such as seen in the Rio Cipó, one of its most important tributaries. This formation is very rich in floral and faunal species, with high diversity and endemism (Costa et al. 1998).

Like many other rivers around the world, the Rio das Velhas became heavily polluted in the 20th century. The Rio das Velhas is the most polluted river of Minas Gerais state, partly because the basin has a total human population of 4.5 million people (IBGE 2000), and domestic sewage and industrial wastes of the MRBH are only partially treated (Table 1). The sewage of 3.2 million people is collected, but only 27.5% of it is primarily treated (www.copasa.com.br). New sewage treatment plants (STPs) and wastewater conveyances are being built to increase the rate of sewage treatment.

Fish Sampling

Johannes T. Reinhardt collected fish in two trips to Brazil, between 1850 and 1856. All specimens were sent to the Zoological Museum of Copenhagen University (Denmark). The sampling methods were not formally described, but Reinhardt did travel to many locations with local fishermen, and they brought him any different, rare, or interesting fish species. Reinhardt assigned Christian F. Lütken the rights to publish the monograph describing his collections (Lütken 1875). Not only the fishes, but also all field annotations, preliminary studies, and drawings were donated to Lütken. The expected sampling area of Reinhardt's collections is drawn in Figure 1 as a polygon formed by each cited collection location in Lütken's book. In the introduction of his monograph, the author pointed out the importance of the material:

> "... that was the first time ichthyological material was collected in the South American continent, as a result of a long stay of a naturalist in a single area, which permitted the local freshwater fish to be the subject of a specific study."

Present sampling stations include six sites on the Rio das Velhas main stem and other six sites on the five tributaries. The mean distance between Rio das Velhas collecting sites was 59 km (ranging from 27 to 95 km). Each tributary had only one sampling site, except Rio Cipó with two sites because of its greater length. All sites were selected depending on access and distance between upstream and downstream sites (main channel) in order to represent upper, medium, and lower river stretches. One tributary (Rio Cipó) and three main stem stations approximate Reinhardt's collection locations, justifying their selection for temporal comparisons. Except for Rio Cipó, which was sampled five times, every site was visited two or three times since 1999 to represent both dry and wet seasons.

Fish were caught with gill nets (20 m long, with 3–16 cm stretch measure mesh), seines (5 m long, 1 mm mesh), cast nets (3 cm stretch measure mesh), and kick nets (1 mm mesh). Gill nets were fished in the water column for 14 h overnight. Seines were used in shallow areas or littoral zones, kick nets were employed in near-shore aquatic macrophytes (both shorelines) and in riffles, and cast nets were used in habitats too deep to wade. The three latter methods were employed for 1–3 h. They were used only quali-

[1] The Metropolitan Region of Belo Horizonte is formed by 34 counties located around the Minas Gerais state capital, representing a total area of approximately 9,500 km^2.

Table 1.—Upriver to downriver water quality of the Rio das Velhas in 1999 (from Alves et al. 2000).

River kilo-meter	Coordinates	Dissolved oxygen (mg/L)	Biochemical oxygen demand (mg/L)	Chemical oxygen demand (mg/L)	Total dissolved solids (mg/L)	Total suspended solids (mg/L)	Volatile suspended solids (mg/L)	Ammonia (mg/L)	Chlorophyll-a (mg/m3)
20	43°34'39"W20°18'43"S	–	–	–	–	–	–	–	–
78	43°47'24"W20°05'17"S	8.0	–	–	2,355.0	11.0	–	1.2	–
114	43°48'52"W19°53'37"S	5.19	9.0	73.6	–	–	–	–	–
204[a]	43°54'39"W19°33'36"S	0.0–2.2	3.6–42.0	9.5–210.4	139.7–657.1	16.4–467.0	15.3–1,608.0	0.0–12.2	2.0–9.6
299	44°01'10"W19°14'07"S	1.0–3.4	2.0–17.9	9.9–42.2	129.3–2,081.0	2.3–1,721.0	3.7–269.0	0.0–11.0	2.4–15.5
373	44°02'14"W19°00'37"S	1.6–9.5	0.3–7.3	7.2–15.2	127.1–956.1	0.7–381.0	0.6–90.0	0.0–6.9	1.3–40.2
400	44°07'13"W18°57'07"S	0.2–12.7	0.5–11.5	5.6–40.0	111.1–941.0	0.3–416.0	6.7– ,192.0	0.0–2.9	3.5–44.3
454	44°09'04"W18°48'27"S	0.6	3.1	6.4	416.0	68.7	585.7	1.8	–
493	44°11'33"W18°40'15"S	2.9–10.6	–	–	–	–	–	–	–

[a] Approximately 40 km upstream of this point, the Rio das Velhas receives the sewage of the Metropolitan Region of Belo Horizonte (MRBH).

tatively to provide a more complete species richness list. Site lengths were 50–100 m, depending on water depth and velocity.

Results

Present samples in the basin have produced 107 species, 81 of which occurred in Reinhardt's study area (Figure 1). Lütken (2001) described 55 species in the Rio das Velhas basin, 46 were found in the main stem, and 24 in tributaries near Lagoa Santa and the Rio Cipó. Three levels of comparison with Lütken were performed: (1) all data together, (2) only the Rio das Velhas main stem, and (3) only the Rio Cipó (Figure 2). Rio Cipó had been chosen for analysis because it was the most cited tributary in the past work. Presently, it is one of the most preserved rivers and has a national park in its headwaters. For all data together, 34 species were reported by both Lütken and us, 21 species were reported only by Lütken, and 47 species were collected only by us. In the Rio das Velhas, 16 species were collected in both our studies, 27 species were reported only by Lütken, and 28 species were collected only by us. In the Rio Cipó, 13 species were collected in both studies, 11 species had occurred only in Lütken's report, and 48 species occurred only in our collections. We have added 26 fish species to Lütken's list, the great majority consisting of small-sized fishes with adults less than 10 cm long (Figure 3).

Ecological attributes have been reported for 20 extinct species in Reinhardt's study area (Table 2). Only a few of these attributes appear relevant to the species' extinctions. For example, the Siluriformes represent less than 35% of the Rio das Velhas basin fish fauna, but 70% of the locally extinct fish species.

Discussion

Surveys of fish species richness provide information for analysis of spatial-temporal and community structure patterns, assessment of biological integrity, and conservation of biodiversity (Cao et al. 2001). The number of samples, site size, and distance between sites is critical to an accurate assessment, since conclusions depend on species richness and fish assemblage composition. Cao et al. (2001) suggested the evaluation of sampling sufficiency based on the relationship between the proportion of total richness and the similarity among replicate samples. Hughes et al. (2002) calculated the optimal site distance for electrofishing Oregon rivers. For tropical

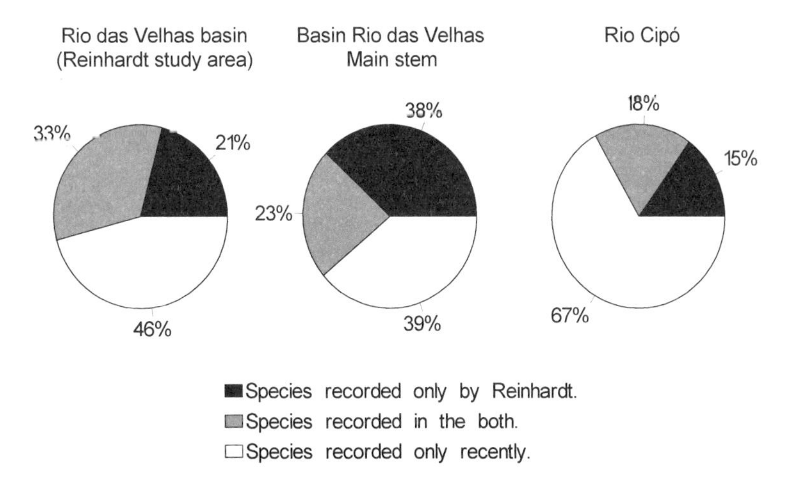

Figure 2.—Temporal and spatial comparisons of fish assemblages of the Rio das Velhas, considering the whole area studied by Reinhardt, the main stem, and Rio Cipó separately, showing the common and exclusive species between his and recent studies. (adapted from Alves and Pompeu 2001)

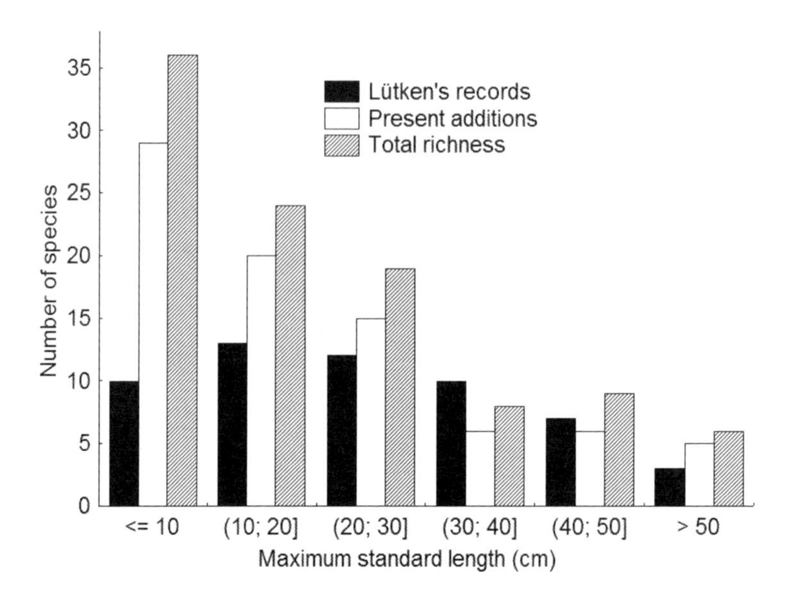

Figure 3.—Past records of fish species, recent additions to the Rio das Velhas basin, and total richness, by size-classes. (Adapted from Alves and Pompeu 2001).

waters, especially for Brazilian rivers, sampling effort studies are rare, despite their importance. Penczak et al. (1998) provided information regarding fishing effort and use of different fishing gears in the Rio Paraná basin. They demonstrated that different fishing gears play a complementary role in fish inventories and showed statistically significant differences between riversides at the same site. Climate and river size peculiarities support the need for further studies of this type to provide accurate and precise data in tropical fish surveys. For the Rio das Velhas, site selection was based on accessibility and river regions represented (upper, middle, and lower reaches), and the number of samples at each site was intended to characterize rainy and dry seasons. We used different kinds of fishing gears to maximize species richness assessments.

Costa et al. (1998) estimated fish species richness for the main river basins in Minas Gerais: São Francisco (170 species), High Paraná basin (120), Doce (77), Paraíba do Sul (59), Jequitinhonha (36), and Mucuri (44). Recent studies with new species descriptions and new occurrence records allow a more accurate estimation of Rio São Francisco' richness (176 species) within Minas Gerais boundaries (Alves et al. 1998). Because of the lack of collections in many portions of the São Francisco basin, such as small headwater streams, the whole basin may support 250–300 fish species. To support this hypothesis, we have registered 107 species in the Rio das Velhas; seven of which are new to science (*Hisonotus* sp.1, *Hisonotus* sp.2, *Planaltina* sp., *Bunocephalus* sp.1, *Bunocephalus* sp.2, *Rineloricaria* sp., and *Harttia* sp.). Voucher specimens are deposited at Museu de Zoologia da Universidade de São Paulo (MZUSP) (Appendix A).

Following Lütken's work, a number of new species records were added to the São Francisco basin list. The greatest number of species added to Lütken's listed species from the Rio das Velhas were particularly significant in the tributaries, where 48 were recorded for the first time. On the other hand, in the upper Rio das Velhas, only 28 species are newly recorded. In general, the different sampling techniques may account for the number of species added by the recent studies. Because the great majority of fish listed by Lütken were caught by fishermen, smaller species certainly were underestimated. Fishermen are always interested in fish that command good market prices or that are large enough to interest human consumers. Twenty-six species, measuring less than 10 cm total length, or 37 if we con-

Table 2.—Ecological attributes of fishes extinct in Reinhardt's study area. Bold font = extinct species in the Rio das Velhas basin.

Species[a]	Maximum length (cm)[b]	Main food Items[c]	Migratory Behavior[c]	Endemic[d]
Characiformes				
Family Characidae				
Brycon orthotaenia	40.1	fruits, insects	yes	yes
Serrapinnus piaba	4.6	plankton	no	yes
Hasemania nana	2.6	insects	no	yes
Roeboides xenodon	11.5	insects, fish scales	no	yes
Siluriformes				
Family Doradidae				
Franciscodoras marmoratus	26.2	?	no	yes
Family Auchenipteridae				
Glanidium albescens	12.4	?	no	yes
Trachelyopterus galeatus	15.7	insects, fishes, plants	no	no
Family Pimelodidae				
Bagropsis reinhardti	31.4	?	?	yes
Conorhynchos conirostris	73.4	mollusks	yes	yes
Rhamdiopsis microcephala	9.8	insects	no	yes
Pimelodella vittata	9.2	insects	no	yes
Pseudopimelodus charus	26.2	fishes	no	yes
Family Trichomycteridae				
Stegophilus insidiosus	5.0	fish mucous	no	yes
Trichomycterus brasiliensis	7.9	insects	no	no
Family Loricariidae				
Hypostomus alatus	40.6	algae	no	yes
Hypostomus francisci	28.8	algae	no	yes
Hypostomus lima	21.0	algae	no	yes
Rineloricaria lima	18.3	algae	no	?
Perciformes				
Family Sciaenidae				
Pachyurus francisci	39.3	fishes	no	yes
Pachyurus squamipennis	40.6	fishes	no	yes

[a] = according to Britski (2001); [b] = according to Lütken (1875, 2001); [c] = according to Alves *et al.* (1998); [d] = according to Reis et al. (2003).

sider species less than 20 cm, have been added to the first historical list. Comparing the size-class distribution between the past and present studies (Figure 3), smaller species currently represent the greater number of species. The only added species greater than 50 cm was the common carp *Cyprinus carpio*, which is a nonnative species.

Although Lütken's data can only be used qualitatively, since there were no measures of abundance and sampling efforts differed, this was a unique opportunity to evaluate the modifications that have occurred in the past 150 years. Many activities over this time, such as mining, agriculture, industrialization, urbanization, and population growth, likely produced direct and indirect negative effects on the fish fauna. All these activities altered chemical or physical habitats or both. For example, Pompeu and Alves (2003) demonstrated the local extinction of 70% of the original fish species of Lagoa Santa, a shallow permanent lake once connected to Rio das Velhas by a small stream. Among the factors that caused this drastic decrease were blockage from the main stem, introduction of nonnative species, changes in water level, elimination of littoral (Cyperaceae) and submerged (Characeae) vegetation, organic pollution, and siltation.

Among the 20 species not recorded recently in Reinhardt's area, 7 were collected outside it in the

Rio das Velhas basin (pirapitinga *Brycon orthotaenia*, piaba[2] *Serrapinnus piaba*, cascudos *Hypostomus alatus* and *Hypostomus francisci*, corvina *Pachyurus sguamipennis*, piaba *Roeboides xenodon*, and cangati *Trachelyopterus galeatus*), so they have been only locally extirpated. Thirteen species are apparently absent from the basin (mandi-bagre *Bagropsis reinhardti*, pirá *Conorhynchos conirostris*, mandi-serrudo *Franciscodoras marmoratus*, peixe-dourado *Glanidium albescens*, piaba *Hasemania nana*, cascudo *Hypostomus lima*, corvina *Pachyurus francisci*, mandi-chorão *Pimelodella vittata*, bagre-sapo *Pseudopimelodus charus*, candiru *Stegophilus insidiosus*, cambeva *Trichomycterus brasiliensis* bagrinho *Rhamdiopsis microcephala*, and cascudo-barbado *Rineloricaria lima*). The second of them, *C. conirostris*, plays a significant role in Rio São Francisco commercial fisheries. According to Reinhardt's observations, *C. conirostris* used to swim up-river in Rio das Velhas annually, from February to March. This movement could correspond to its reproductive migration to spawning grounds in the Rio das Velhas tributaries. The populations of this species seem to fluctuate significantly in time. After many years of being practically absent from commercial fisheries, the species became abundant following large 1996–1997 floods in the Rio São Francisco basin. In 1998, juveniles (ranging from 11.2 to 13.3 cm standard length) were caught in the Rio São Francisco near Rio Carinhanha (personal observation).

Among the species locally extinct, most are catfishes and armored catfishes. Generally, these are benthic species, living among rocks and gravel (Burgess 1989). The sedimentation that transformed the channel into a shallow river and eliminated navigability near MRBH also altered substrate composition. The major causes of this process were the mines in the Rio das Velhas headwaters, deposition of organic sediments from MRBH, vegetation clearing, sediment runoff from agriculture, and uncontrolled urbanization. Where there was once a diverse bottom of cobble, gravel, and sand, today there is a homogenized sandy substrate, eliminating many bottom feeders. Many catfishes are herbivorous or iliophagous, grazing algae and organic matter attached to rocks.

Burton (1977) described the abundance and sizes of fish species of the Rio das Velhas and the Rio São Francisco in 1867. He felt that the fishery had a greater economic value than mining. Today, fishes such as surubim *Pseudoplatystoma corruscans*, which once reached 100 kg or more, are rarely caught larger than 40 kg; most of them are 10–25 kg. Surubim is the most important commercial fish of the São Francisco basin (Godinho et al. 1997). Radio telemetry studies, with fish collected and marked in Rio São Francisco, have recently shown surubim migrates 200 km into the Rio das Velhas (Alexandre Godinho, Universidade Federal de Minas Gerais, personal communication). These data reinforce the importance of the Rio das Velhas to the São Francisco basin.

Nevertheless, there are regular fish kills in the Rio das Velhas, mainly in the beginning of the rainy season (Alves et al. 2000). Heavy organic discharges from the MRBH accumulate in the river bottom during the dry season. At the beginning of the summer rainy season, this material is suspended at the same time water temperatures are greatest. Rapid decomposition of the organic matter depletes dissolved oxygen, causing frequent fish kills. Annual kills of adults and burying of the eggs and larvae of those that survive leads to annual decreases of resident fish populations and those migrating from the Rio São Francisco. The consequences of this heavy pollution affect all river biota. Recent studies showed how the MRBH changes the expected water quality, fish, and benthic richness in the main stem (Alves and Pompeu 2001; Pompeu et al., in press).

A monitoring program for the Rio das Velhas basin has begun, analyzing chemical, physical and biological parameters (microbiological, phytoplankton, zooplankton, macroinvertebrates, and fish) at 37 sites located in the main stem, tributaries, and MRBH. The objective of this biomonitoring program is to establish the basin's general features and to locate reference sites for assessing the degree of perturbation (Hughes 1995), thereby facilitating empirical comparisons over space and time. The Rio das Velhas basin includes sites ranging from quite undisturbed to heavily polluted and highly

[2] Many characins, subfamily Tetragonopterinae are generically referred to as "piaba."

channelized reaches, so we are optimistic about assessing current and potential conditions.

Despite substantial environmental degradation and fish composition changes, we found an interesting case of culture conservation. Lütken reported that mandi-amarelo *Pimelodus maculatus* was frequently caught with adult wasps in its stomach contents. He was curious about how fish could eat so many adults of a free flying insect. The answer was that local fishermen used to put wasp nests inside their fish-traps, with the adults remaining inside the trap, attracting fishes (Lütken 1875, 2001). In 1999, we discovered a fisherman using the same fishing strategy, almost 150 years after Reinhardt recorded his observations (personal observation). Examining the contents of his trap, we found three specimens of bagre-sapo *Cephalosilurus fowleri*. Both fish species are catfish.

The present condition of the upper Rio das Velhas, which is worse than other reaches in the São Francisco basin, can be changed in the near future. There are 12 sewage treatment plants already installed within the MRBH, and others are projected. The two larger ones (STP Arrudas and STP Onça) have treatment capacities of 4,500 L/s and 3,600 L/s, respectively. The other positive features of the das Velhas basin are its direct connection with the Rio São Francisco, with no natural or artificial barriers, as well as the quality of its tributaries, which support 75% of its present fish species. The natural purification capacity in the Rio das Velhas must also be stressed because the lower river does not seem to be altered by the sewage effluents of MRBH.

Acknowledgments

We are grateful to Bob Hughes, who invited us to write this chapter and for his editorial suggestions. Reviews by Hugo Godinho, Gilmar Santos, and Phil Kaufmann helped clarify the manuscript. We thank Sílvia Magalhães for her help on the map and for geographical information about the basin. We also acknowledge the logistical and financial support of Projeto Manuelzão para Revitalização da Bacia do Rio das Velhas (UFMG), Fundo-Fundep de apoio acadêmico 1999, and Fundação O Boticário de Proteção à Natureza (grant No. 0472002).

References

Agostinho, A. A., and H. F. Júlio, Jr. 1999. Peixes da bacia do alto rio Paraná. Pages 374–400 *in* R. H. Lowe-McConnell, editor. Estudos ecológicos de comunidades de peixes tropicais. Editora da Universidade de São Paulo, São Paulo, Brazil.

Agostinho, A. A., and M. Zalewski. 1996. A planície alagável do alto rio Paraná: importância e preservação. Editora da Universidade Estadual de Maringá, Maringá, Brazil.

Alves, C. B. M., C. A. M. Estanislau, M. A. R. Araújo, M. V. Polignano, and P. S. Pompeu. 2000. Projeto S.O.S. Rio das Velhas: estudo das possíveis causas das mortandades de peixes na sub-bacia. Technical Report. Projeto Manuelzão/IEF, Belo Horizonte, Brazil.

Alves, C. B. M., and P. S. Pompeu. 2001. A fauna de peixes da bacia do Rio das Velhas no final do século XX. Pages 165–187 *in* C. B. M. Alves and P. S. Pompeu, editors. Peixes do Rio das Velhas: passado e presente. Editora Segrac, Belo Horizonte, Brazil.

Alves, C. B. M., F. Vieira, P. S. Pompeu, and P. R. Fonseca. 1998. Estudos de ictiofauna. Pages 1–154 *in* Plano diretor dos recursos hídricos das bacias de afluentes do Rio São Francisco em Minas Gerais. Governo do Estado de Minas Gerais (Technical report), Belo Horizonte, Brazil.

Britski, H. A. 2001. Sobre a obra Velhas-Flodens fiske [Peixes do Rio das Velhas]. Pages 15–22 *in* C. B. M. Alves and P. S. Pompeu, editors. Peixes do Rio das Velhas: passado e presente. Editora Segrac, Belo Horizonte, Brazil.

Britski, H. A., Y. Sato, and A. B. S. Rosa. 1988. Manual de identificação de peixes da região de Três Marias (com chave de identificação para os peixes da bacia do São Francisco). Brasília, Câmara dos Deputados/ Companhia do Desenvolvimento do Vale do São Francisco, Brasília, Brazil.

Burgess, W. E. 1989. An atlas of freshwater and marine catfishes. Tropical Fish Hobbyist Publications, Neptune City, New Jersey.

Burton, R. F. 1977. Viagem de canoa de Sabará ao Oceano Atlântico. Belo Horizonte, Editora Itatiaia, Belo Horizonte, Brazil.

Cao, Y., D. P. Larsen, and R. M. Hughes. 2001. Evaluating sampling sufficiency in fish assemblage surveys: a similarity-based approach. Canadian Journal of Fisheries and Aquatic Sciences 58:1782–1793.

CETEC. 1983. Diagnóstico ambiental do estado de Minas Gerais. Fundação Centro Tecnológico de Minas Gerais, Série de Publicações Técnicas/ SPT-010:1–158, Belo Horizonte, Brazil.

Costa, C. M. R., G. Herrmann, C. S. Martins, L. V. Lins, and I. R. Lamas. 1998. Biodiversidade em Minas Gerais: um atlas para sua conservação. Fundação Biodiversitas, Belo Horizonte, Brazil.

Godinho, H. P., M. O. T. Miranda, A. L. Godinho, and J. E. Santos. 1997. Pesca e biologia do surubim *Pseudoplatystoma coruscans* no Rio São Francisco. Pages 27–42 *in* M. O. T. Miranda, editor. Surubim. Instituto Brasileiro do Meio Ambiente e dos Recursos Naturais Renováveis, Brasília, Brazil.

Hughes, R. M. 1995. Defining acceptable biological status by comparing with reference conditions. Pages 31–47 *in* W. S. Davis and T. P. Simon, editors. Biological assessment and criteria: tools for water resource planning and decision making. Lewis Press, Boca Raton, Florida.

Hughes, R. M., P. R. Kaufmann, A. T. Herlihy, S. S. Intelmann, S. C. Corbett, M. C. Arbogast, and R. C. Hjort. 2002. Electrofishing distance needed to estimate fish species richness in raftable Oregon rivers. North American Journal of Fisheries Management 22:1229–1240.

IBGE (Instituto Brasileiro de Geografia e Estatística). 2000. Sinopse preliminar do censo demográfico. Instituto Brasileiro de Geografia e Estatística, Rio de Janeiro, Brazil.

Lütken, C. F. 1875. Velhas-Flodens fiske. Et bidrag til Brasiliens ichthyologi. Elfter Professor J. Reinhardt indsamlinger og optegnelser. Kon Dank Vidensk Selsk Skrift (Kjoebenhavn) 12:122–252.

Lütken, C. F. 2001. Peixes do Rio das Velhas: uma contribuição para a ictiologia do Brasil. Pages 23–164 *in* C. B. M. Alves and P. S. Pompeu, editors. Peixes do Rio das Velhas: passado e presente. Editora Segrac, Belo Horizonte, Brazil.

Machado, A.B.M., G.A.B. Fonseca, R.B. Machado, L.M.S. Aguiar, and L. V. Lins (eds.). 1998. Livro vermelho das espécies ameaçadas de extinção da fauna de Minas Gerais. Fundação Biodiversitas, Belo Horizonte, Brazil.

Menezes, N. A. 1996. Methods for assessing freshwater fish diversity. Pages 289–295 *in* C. E. M. Bicudo and N. A. Menezes, editors. Biodiversity in Brazil. Conselho Nacional de Desenvolvimento Científico e Tecnológico, São Paulo, Brazil.

MMA (Ministério do Meio Ambiente). 2004. Lista Nacional das Espécies de Invertebrados Aquáticos e Peixes Ameaçados de Extinção. Diário Oficial da União, 102:136-142.

Myers, N., R. A., C. G. Mittermeier, G. A. B. Fonseca, and J. Kent. 2000. Biodiversity hotspots for conservation priorities. Nature (London) 403:853–858.

Penczak, T., L. C. Gomes, L. M. Bini, and A. A. Agostinho. 1998. Importance of qualitative inventory sampling using electric fishing and nets in a large tropical river (Brazil). Hydrobiologia 389:89–100.

PLANVASF. 1986. Plano diretor para o desenvolvimento do Vale do São Francisco. Companhia de Desenvolvimento do Vale do São Francisco, Brasília, Brazil.

Pompeu, P. S., and C. B. M. Alves. 2003. Local fish extinction in a small tropical lake in Brazil. Neotropical Ichthyology 1(2):133–135.

Pompeu P. S., C. B. M. Alves, and M. Callisto. In Press. The effects of urbanization on biodiversity and water quality in the Rio das Velhas basin, Brazil. In L. R. Brown, R. H. Gray, R. M. Hughes, and M. R. Meadow, editors. Effects of urbanization on stream ecosystems. American Fisheries Society, Symposium 47, Bethesda, Maryland.

Reis, R. E., S. O. Kullander, and C. J. Ferraris, Jr. 2003. Check list of the freshwater fishes of South and Central Americanérica. Editora da Pontifícia Universidade Católica do Rio Grande do Sul, Porto Alegre, Brazil.

Schaefer, S. A. 1998. Conflict and resolution: impact of new taxa on phylogenetic studies of the neotropical cascudinhos (Siluroidei: Loricariidae). Pages 375–400 *in* L. R. Malabarba, R. E. Reis, R. P. Vari, Z. M. S. Lucena and C. A. S. Lucena, editors. Phylogeny and classification of neotropical fishes. Editora da Pontifícia Universidade Católica do Rio Grande do Sul, Porto Alegre, Brazil.

Appendix A.—Fish species of Rio das Velhas basin

	Species (Scientific names) [1]	Common names [2]	Native /alien/ locally extinct	Endemism / Conservation status [3]	MRSL (cm)	Voucher specimens
1.	*Acestrorhynchus lacustris* (Lütken 1875)	Peixe-cachoro	native		22.4	MZUSP 73745, 73755, 73775, , 73779
2.	*Anchoviella vaillanti* (Steindachner 1908)	Sardinha	native		5.4	–
3.	*Apareiodon hasemani* (Eigenmann 1916)	Canivete	native	endemic	4.7	–
4.	*A. ibitiensis* (Amaral Campos 1944)	Canivete	native		8.2	MZUSP 73687, 73813
5.	*A. piracicabae* (Eigenmann 1907)	Canivete	native		11.6	MZUSP 73681, 73804
6.	*Apteronotus brasiliensis* (Reinhardt 1852)		native		20.3	MZUSP 73674
7.	*Astyanax bimaculatus* (Linnaeus 1758)	Lambari-do-rabo-amarelo	native		12.7	MZUSP 73658, 73719, 73747, 73764
8.	*A. eigenmanniorum* (Cope 1894)	Lambari	native		9.2	MZUSP 73702, 73717, 73721
9.	*A. fasciatus* (Cuvier 1819)	Lambari-do-rabo-vermelho	native		15.5	MZUSP 73710, 73746, 73792
10.	*A. scabripinnis* (Jenyns 1842)	Lambari	native		9.3	MZUSP 73714
11.	*A. taeniatus* (Jenyns 1842)	Lambari	native		10.0	MZUSP 73827
12.	*Astyanax* sp.	Lambari	native		7.2	–
13.	*Bergiaria westermanni* (Lütken 1874)	Mandi	native	endemic	21.8	MZUSP 73806
14.	*Brycon nattereri* (Günther 1864)	Pirapetinga	native	threatened	14.7	–
15.	*B. orthotaenia* (Günther 1864)	Matrinchã	native	endemic	28.5	MZUSP 73836
16.	*Bryconamericus stramineus* (Eigenmann 1908)	Piaba	native		4.3	MZUSP 73680, 73696
17.	*Bryconops affinis* (Günther 1864)	Piaba	native		8.4	MZUSP 73679, 73791
18.	*Bunocephalus* spN.1		native	endemic	4.9	–
19.	*Bunocephalus* spN.2		native	endemic	3.7	MZUSP 73800
20.	*Callichthys callichthys* (Linnaeus 1758)	Caborja, tamoatá	native		11.4	MZUSP 73729
21.	*Cephalosilurus fowleri* (Haseman 1911)	Bagre-sapo	native	endemic	28.0	MZUSP 73667, 73756, 73815
22.	*Cetopsorhamdia iheringi* (Schubart & Gomes 1959)	Bagrinho	native		7.8	MZUSP 73676, 73695
23.	*Characidium fasciatum* (Reinhardt 1867)	Mocinha	native		8.3	MZUSP 73715, 73790

Appendix A.—Continued.

Species (Scientific names) [1]	Common names [2]	Native /alien/ locally extinct	Endemism / Conservation status [3]	MRSL (cm)	Voucher specimens
24. *C. lagosantensis* (Travassos 1947)	Mocinha	native	endemic and threatened	2.7	MZUSP 73708, 73797
25. *C. zebra* (Eigenmann 1909)	Mocinha	native		5.4	MZUSP 73666, 73689, 73700, 73751, 73795, 73814
26. *Cichla* cf. *monoculus*	Tucunaré	alien		31.0	MZUSP 73767, 73772
27. *Cichlasoma facetum* (Jenyns 1842)	Cará-preto	native		6.3	MZUSP 73726
28. *C. sanctifranciscense* (Kullander 1983)	Cará-preto	native		10.0	MZUSP 73760
29. *Crenicichla lacustris* (Castelnau 1855)		native		4.3	–
30. *Curimatella lepidura* (Eigenmann & Eigenmann 1889)	Manjuba	native	endemic	7.6	–
31. *Cyphocharax gilbert* (Quoy & Gaimard 1824)	Sagüiru	native		9.3	MZUSP 73728, 73732
32. *Cyprinus carpio* (Linnaeus 1758)	Carpa	alien		70.0	–
33. *Duopalatinus emarginatus* (Valenciennes 1840)	Mandi-açu	native	endemic	23.4	MZUSP 73738, 73812, 73819
34. *Eigenmannia virescens* (Valenciennes 1842)	Peixe-espada	native		36.0	MZUSP 73660, 73739, 73757, 73765, 73802
35. *Geophagus brasiliensis* (Quoy & Gaimard 1824)	Cará	native		15.0	–
36. *Gymnotus carapo* (Linnaeus 1758)	Sarapó, tuvira	native		24.2	MZUSP 73787
37. *Harttia leiopleura* (Oyakawa 1993)	Cascudinho	native	endemic	6.1	MZUSP 73712
38. *Harttia* sp.N	Cascudinho	native	endemic	11.5	MZUSP 73692
39. *Hemigrammus gracilis* (Lütken 1875)	Piaba	native		2.4	–
40. *H. marginatus* (Ellis 1911)	Piaba	native		2.7	–
41. *Hemipsilichthys* cf. *mutuca* (Oliveira & Oyakawa 1999)	Cascudinho	native	endemic	3.6	MZUSP 73698
42. *Hisonotus* sp.N1	Cascudinho	native	endemic	3.3	MZUSP 73682, 73694, 73793
43. *Hisonotus* sp.N2	Cascudinho	native	endemic	3.4	MZUSP 73707, 73794
44. *Homodiaetus* sp.N		native	endemic	3.6	MZUSP 73693
45. *Hoplias lacerdae* (Miranda Ribeiro 1908)	Trairão	alien		57.5	MZUSP 73655, 73735, 73837, 73839, 73842

Appendix A.—Continued.

Species (Scientific names) [1]	Common names [2]	Native /alien/ locally extinct	Endemism / Conservation status [3]	MRSL (cm)	Voucher specimens
46. *H. malabaricus* (Bloch 1794)	Traíra	native		33.8	MZUSP 73651, 73838
47. *Hoplosternum littorale* (Hancock 1828)	Chegante, tamoatá	alien		20.0	MZUSP 73730, 73743, 73744, 73766, 73770
48. *Hyphessobrycon santae*	Piaba	native	endemic	-	MZUSP 73683
49. *H. alatus* (Castelnau 1855)	Cascudo	native	endemic	28.6	MZUSP 73823, 73832
50. *Hypostomus commersoni* (Valenciennes 1836)	Cascudo	native		11.1	MZUSP 73759, 73816
51. *H. francisci* (Lütken 1874)	Cascudo	native	endemic	20.3	MZUSP 73724
52. *H. garmani* (Regan 1904)	Cascudo	native	endemic	13.2	MZUSP 73665, 73810
53. *H. macrops* (Eigenmann & Eigenmann 1888)	Cascudo	native	endemic	22.2	MZUSP 73664, 73736, 73780
54. *H. margaritifer* (Regan 1908)	Cascudo	native		7.0	MZUSP 73668
55. *Hypostomus* sp. (cited in Britski et al., 1988)	Cascudo	native	endemic	19.3	MZUSP 73754, 73811
56. *Hypostomus* spp.	Cascudo	native		38.0	MZUSP 73688, 73737, 73798, 73809
57. *Hysteronotus megalostomus* (Eigenmann 1911)	Piaba	native	endemic	3.7	MZUSP 73705
58. *Imparfinis minutus* (Lütken 1874)	Bagrinho	native	endemic	4.9	MZUSP 73678, 73704, 73796
59. *Leporellus vittatus* (Valenciennes 1849)	Piancó, piau-rola	native		24.0	MZUSP 73652, 73725, 73734
60. *Leporinus amblyrhynchus* Garavello & Britski 1987	Timburé	native		13.5	MZUSP 73653, 73672
61. *L. marcgravii* (Lütken 1875)	Timburé	native	endemic	9.9	MZUSP 73657, 73706, 73784
62. *L. obtusidens* (Valenciennes 1836)	Piau-verdadeiro	native		42.5	MZUSP 73671, 73847
63. *L. piau* (Fowler 1941)	Piau-gordura	native	endemic	16.2	MZUSP 73778
64. *L. reinhardti* (Lütken 1875)	Piau-três-pintas	native	endemic	22.6	MZUSP 73781, 73818
65. *L. taeniatus* (Lütken 1875)	Piau-jejo	native	endemic	21.4	MZUSP 73663, 73691, 73777

Appendix A.—Continued.

Species (Scientific names) [1]	Common names [2]	Native /alien/ locally extinct	Endemism / Conservation status [3]	MRSL (cm)	Voucher specimens
66. *Lophiosilurus alexandri* (Steindachner 1876)	Pacamã	native	endemic	25.5	MZUSP 73817, 73835
67. *Moenkhausia costae* (Steindachner 1907)	Piaba	native		6.3	MZUSP 73825
68. *M. sanctaefilomenae* (Steindachner 1907)	Piaba	native		5.3	MZUSP 73686
69. *Myleus micans* (Lütken 1875)	Pacu	native	endemic	17.2	MZUSP 73673, 73684, 73805
70. *Neoplecostomus franciscoensis* (Langeani 1990)	Cascudinho	native	endemic	6.7	MZUSP 73713
71. *Oreochromis niloticus* (Linnaeus 1758)	Tilápia	alien		22.2	MZUSP 73733, 73826
72. *Orthospinus franciscensis* (Eigenmann 1914)	Piaba	native	endemic	5.2	MZUSP 73763
73. *Pachyurus squamipennis* (Agassiz 1831)	Corvina	native	endemic	23.0	MZUSP 73807
74. *Pamphorichthys hollandi* (Henn 1916)	Barrigudinho	native		2.3	–
75. *Parodon hilarii* (Reinhardt 1867)	Canivete	native	endemic	11.4	MZUSP 73720, 73776
76. *Phalloceros caudimaculatus* (Hensel 1868)	Barrigudinho	native		3.2	MZUSP 73711
77. *Phenacogaster franciscoensis* (Eigenmann 1911)	Piaba	native	endemic	3.2	MZUSP 73661, 73799
78. *Phenacorhamdia somnians* (Mees 1974)	Bagrinho	native		5.6	MZUSP 73662, 73675, 73685, 73703, 73789
79. *Piabina argentea* (Reinhardt 1867)	Piaba	native		7.7	MZUSP 73654, 73690, 73697, 73753, 73786
80. *Pimelodella lateristriga* (Lichtenstein 1823)	Mandi-chorão, mandizinho	native		9.2	MZUSP 73669, 73752
81. *Pimelodus fur* (Lütken 1874)	Mandi-prata, mandi-branco	native		18.8	MZUSP 73740, 73742, 73821, 73824
82. *P. maculatus* (Lacepède 1803)	Mandi-amarelo	native		34.0	MZUSP 73782, 73844, 73848
83. *Pimelodus* sp. (cited in Britski et al., 1988)	Mandi	native	endemic	17.5	MZUSP 73820, 73828, 73830
84. *Planaltina* sp.N	Piaba	native	endemic	3.2	MZUSP 73709, 73785

Appendix A.—Continued.

Species (Scientific names) [1]	Common names [2]	Native /alien/ locally extinct	Endemism / Conservation status [3]	MRSL (cm)	Voucher specimens
85. *Poecilia reticulata* (Peters 1859)	Barrgigudinho	alien		2.7	MZUSP 73718
86. *Prochilodus argenteus* (Spix & Agassiz 1829)	Curimatá-pacu	native	endemic	39.8	MZUSP 73849
87. *P. costatus* (Valenciennes 1850)	Curimbatá-pioa	native	endemic	44.0	MZUSP 73822, 73843, 73845
88. *Psellogrammus kennedyi* (Eigenmann 1903)	Piaba	native		5.7	MZUSP 73727
89. *Pseudoplatystoma corruscans* (Spix & Agassiz 1829)	Surubim	native		>110.0	–
90. *Pygocentrus piraya* (Cuvier 1819)	Piranha	native	endemic	14.9	MZUSP 73846
91. *Rhamdia quelen* (Quoy & Gaimard 1824)	Bagre	native		29.5	MZUSP 73659, 73716, 73723, 73769
92. *Rhinelepis aspera* (Spix & Agassiz 1829)	Cascudo-preto	native		46.9	MZUSP 73831
93. *Rineloricaria* sp.N	Cascudo	native	endemic	15.0	MZUSP 73783, 73829
94. *Roeboides xenodon* (Reinhardt 1851)		native	endemic	8.0	MZUSP 73758, 73803
95. *Salminus brasiliensis* (Cuvier 1816)	Dourado	native		58.5	MZUSP 73808, 73833
96. *S. hilarii* (Valenciennes 1850)	Tabarana, dourado-branco	native		28.3	MZUSP 73749, 73834
97. *Schizodon knerii* (Steindachner 1875)	Piau-branco	native	endemic	27.7	MZUSP 73762, 73841
98. *Serrapinnus heterodon* (Eigenmann 1915)	Piaba	native		3.4	MZUSP 73677, 73701, 73750, 73774, 73788
99. *S. piaba* (Lütken 1875)	Piaba	native	endemic	2.3	–
100. *Serrasalmus brandtii* (Lütken 1875)	Pirambeba	native	endemic	16.9	MZUSP 73761
101. *Steindachnerina corumbae* (Pavanelli & Britski 1999)	Sagüiru	native		10.3	MZUSP 73656, 73699, 73722, MZUSP 73731
102. *S. elegans* (Steindachner 1875)	Sagüiru	native		12.2	MZUSP 73748, 73801
103. *Sternopygus macrurus* (Bloch & Schneider 1801)	Sarapó	native		33.1	MZUSP 73670
104. *Tilapia rendalli* (Boulenger 1897)	Tilápia	alien		20.1	MZUSP 73768, 73771, 73773, 73840

Appendix A.—Continued.

Species (Scientific names) [1]	Common names [2]	Native /alien/ locally extinct	Endemism / Conservation status [3]	MRSL (cm)	Voucher specimens
105. *Trachelyopterus galeatus* (Lütken 1874)	Cangati	native			MZUSP 73741
106. *Trichomycterus reinhardti* (Eigenmann 1917)	Cambeva	native	endemic		–
107. *Triportheus guentheri* (Garman 1890)	Piaba-facão	native	endemic	8.7	–
108. *Bagropsis reinhardti* (Lütken 1874)	Mandi-bagre	locally extinct	endemic		–
109. *Conorhynchos conirostris* (Valenciennes 1840)	Pirá	locally extinct	endemic and threatened		–
110. *Franciscodoras marmoratus* (Lütken 1874)	Mandi-serrudo	locally extinct	endemic		–
111. *Glanidium albescens* (Lütken 1874)	Peixe-dourado	locally extinct	endemic		–
112. *Hasemania nana* (Lütken 1875)	Piaba	locally extinct	endemic		–
113. *Hypostomus lima* (Lütken 1874)	Cascudo	locally extinct	endemic		–
114. *Pachyurus francisci* (Cuvier 1830)	Corvina	locally extinct	endemic		–
115. *Pimelodella vittata* (Lütken 1874)	Mandi-chorão	locally extinct			–
116. *Pseudopimelodus charus* (Valenciennes 1840)	Bagre-sapo	locally extinct	endemic		–
117. *Rhamdiopsis microcephala* (Lütken 1874)	Bagrinho	locally extinct			–
118. *Rineloricaria lima* (Kner 1853)	Cascudo-barbado	locally extinct	endemic		–
119. *Stegophilus insidiosus* (Reinhardt 1859	Candiru	locally extinct	endemic		–
120. *Trichomycterus brasiliensis* (Lütken 1874	Cambeva	locally extinct			–

[1] According to Reis et al. (2003);

[2] As cited locally or at TrÍs Marias region (Britski et al., 1988)

[3] Endemism checked in Fishbase (www.fishbase.org) and Catalog of Fishes (www.calacademy.org/reserach/ichthyology/catalog). Conservation status refers to any category of threat mentioned in Machado et al. (1998), and MMA 2004).MRSL = maximum recorded standard length in present studies in Rio das Velhas basin.

American Fisheries Society Symposium 45:603–612, 2005

Historical Changes in Large River Fish Assemblages of the Americas: A Synthesis

Robert M. Hughes*
Department of Fisheries and Wildlife, Oregon State University, Corvallis, Oregon 97331, USA

John N. Rinne
Rocky Mountain Research Station, U.S. Forest Service, Flagstaff, Arizona 86001, USA

Bob Calamusso
Tonto National Forest, U.S. Forest Service, Phoenix, Arizona 85006, USA

Abstract.—The objective of this synthesis is to summarize patterns in historical changes in the fish assemblages of selected large American rivers, to document causes for those changes, and to suggest rehabilitation measures. Although not a statistically representative sample of large rivers, the book chapters indicated that physical and biological stressors usually had a greater impact on fish assemblages than chemical stressors (where point sources were treated). In particular, flow and channel regulation combined with alien species were key factors affecting large river fish assemblages. And these factors were most pronounced for southwestern U.S. rivers. We hope that such information will aid interested citizens and government agencies in river rehabilitation and river protection or conservation. There will never be a better time to do so.

Introduction

The river reaches reported on in this book were all unwadeable with decades to centuries of fish assemblage data available. In addition, the chapter authors had studied the rivers for decades themselves and were very familiar with them and their fishes. The rivers represent a broad range of geographic settings and ecological conditions. Therefore, generalizations gleaned from the studies offer useful insights about fish assemblage ecology in large rivers, in particular the anthropogenic stressors that appeared to alter their fish assemblages. Because the rivers reported on in the chapters were not randomly selected, they cannot be deemed representative of all American rivers. Therefore, we cannot generalize to all rivers. Nonetheless, some patterns are evident (Table 1) and they may be useful for hypothesis generation, for supporting common riverine conceptual models, and for suggesting river rehabilitation tactics.

Pacific and Arctic draining rivers show a wide range in disturbance by humans. The rivers flowing to the Arctic Ocean (Slave, Mackenzie) have been minimally disturbed by humans, chiefly through fishing, and their fish assemblages remain intact. Of the three Pacific rivers, only the Willamette retains some semblance of its natural fish assemblage despite moderate flow and channel alterations. Both the Snake and Sacramento have experienced far greater flow and channel alterations, water pollution, species extirpations, and reaches dominated by alien species.

All the southwestern rivers demonstrate fundamental alterations in their flows and fish assemblages. Naturally depauperate and endemic faunas have been largely extirpated and replaced by alien

* Corresponding author: hughes.bob@epa.gov

species. In four cases, (Gila, lower Colorado, Grande, Nazas) flows have been sufficiently diverted such that they often no longer reach the sea or their confluences with the larger rivers. These rivers also rarely experience the peak flows they once did.

The mid-American rivers exhibit a wide range in disturbance, but not as great as the western rivers. The Platte, Missouri, Ohio, and Wisconsin have experienced the greatest flow and channel alterations, and they have the greatest proportions of listed and extirpated species (15–34%). But, they are still dominated by native species, unlike the southwestern rivers. The Red, upper Mississippi, Patoka and Wabash rivers have no high dams and, therefore, less altered flow regimes. Consequently, listed or extirpated species represent only 0–10% of their faunas; however, their fish assemblages have been degraded by water pollution and flow and channel alterations.

The north Atlantic draining rivers show minimal alteration of their fish faunas despite centuries of settlement and use. Again, the dams are not as high and therefore the flow regulations are not as severe, but all three have experienced reductions in the ranges and abundances of diadromous species.

The subtropical rivers also diverge in degree of disturbance and fish assemblage response. Despite massive channel alteration, the Kissimmee has lost only one species and has none listed as threatened or endangered. On the other hand, the Alabama River, with several major dams, has 22% of its fish fauna listed or extirpated. Although lacking dams, the das Velhas has lost 12–19% of its fish species because of untreated waste water and sediments. Presumably, like the Ohio and Willamette rivers, fish assemblages in the das Velhas can be expected to improve once municipal waste treatment is implemented. These broad-scale geographic patterns among large rivers are similar to those reported at the scale of entire six-digit hydrologic units for the United States (NatureServe 2005; Figure 1). However, in several cases, the declines in riverine fish faunas reported in our book appear more serious than those in Figure 1 because we describe declines and presence/absence.

As Williams et al. (1989), Abell et al. (2000), and Karr et al. (2000) reported, fish assemblages of large American rivers appear most at risk in the southwestern and southeastern United States and northern Mexico, where endemism is highest. However, our failure to find additional chapter authors from Latin America limits our confidence in that conclusion. Alien species appear most detrimental in western U.S. rivers, which have naturally depauperate fish faunas, and in rivers with fundamentally altered flow regimes, as proposed by Moyle and Light (1996). Naturalized flow regimes may limit these species to some degree in small systems like the Verde and Gila, but it will be extremely difficult to implement such changes in the larger rivers.

The species most affected by human disturbances in the studied rivers are fluvial specialists; diadromous, anadromous, or potamo-dromous species; and habitat specialists. Naturalized flow regimes and dam removals are recommended wherever possible to help rehabilitate riverine habitats and those species requiring them. Of course, that would require viewing large rivers as something more than simply barge corridors and sources of hydropower and municipal and irrigation water.

Perhaps one approach is to reexamine the objective of the Clean Water Act, "to restore and maintain the chemical, physical, and biological integrity of the Nation's waters." Marked improvements in chemical integrity have resulted from major national programs in waste water treatment beginning in the 1970s in North America and Europe. These programs should be extended to areas lacking them and expanded to include diffuse pollution derived from land use, because water pollution remains a limiting factor in many waters. In the past 20 years, biological integrity, or at least assessment and management of biological condition, has received increased attention in the United States (Southerland and Stribling 1994; Barbour et al. 2000; Hughes et al. 2000; Karr et al. 2000). Similarly, the European Union recently established the Water Framework Directive to assess and improve the ecological status of surface waters by focusing on biological indicators (Moog and Chovanec 2000; Johnson 2001). Biological assessments and criteria also should be continued and more widely implemented. Although many indi-

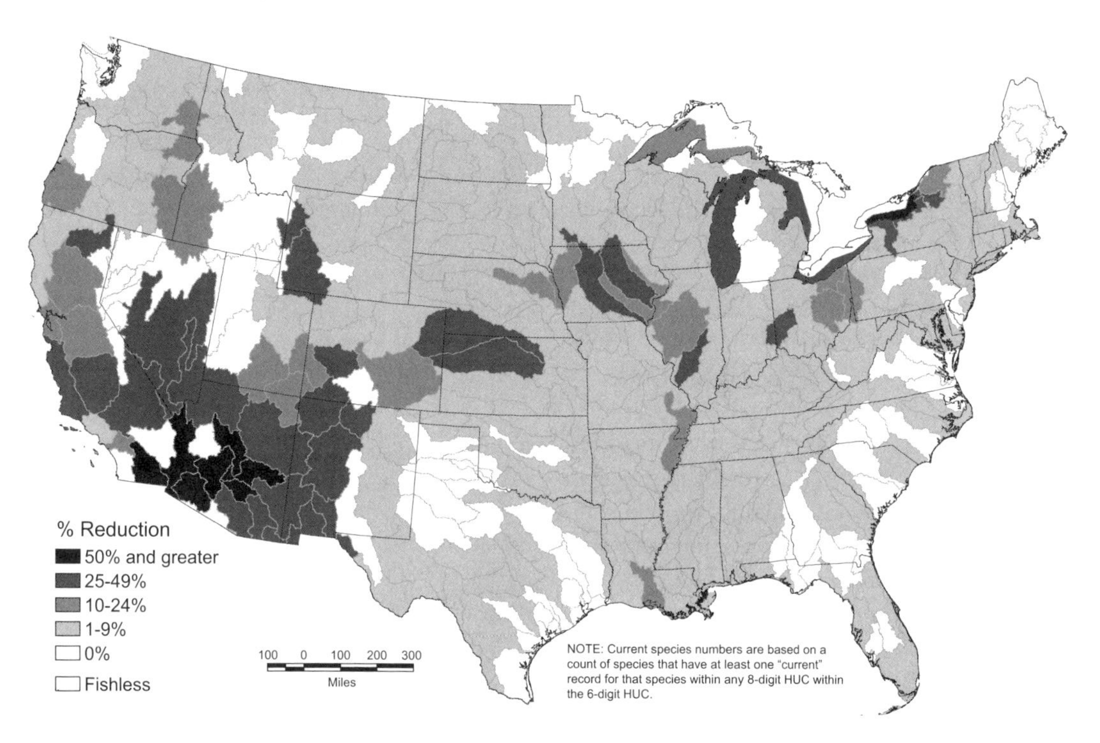

Figure 1.—Percent reductions in species richness of native fish within a six-digit hydrologic unit (from NatureServe 2005)

ces of biotic integrity (IBIs) include an alien species metric, the effects of alien species on native species should be included in those assessments (Fuller et al. 1999), as should evaluations of conditions aiding alien dominance (Marchetti et al. 2004a, 2004b). Also native fish passage up- and down-river around dams and reservoirs should be maximized while minimizing or eliminating intentional or accidental fish introductions.

However, the physical component of ecological condition has received much less attention, despite its recognition in the Clean Water Act, Water Framework Directive, and previous conceptual models (Karr and Dudley 1981; Karr et al. 1986). In particular, flow regimes and flood pulses were deemed master variables governing the ecological integrity of large rivers (Welcomme 1979; Junk et al. 1989; Poff et al. 1997). Flow regulation may be the most pervasive anthropogenic change on rivers world-wide (Stanford et al. 1996), and certainly appears to be the case in the American rivers examined in this book. Thus, if we are to see marked improvements in the ecological condition of American rivers, interested citizens and governments must direct their attention to naturalizing flow regimes and river channels where those are the limiting factors. Dam removal is an increasingly common practice for rehabilitating rivers and reducing liability, but dams also slow invasions by alien fishes (AFS 2005). Where dams and canals are not yet entirely dominant riverscape features, such as in the Amazon and Parana basins (Pringle et al. 2000), they should be avoided like the ecological plague that they appear to be in North America and Europe.

The history of Latin America rivers follows that of North American and European rivers, but with a lag of several decades (C. B. M. Alves, Federal University of Minas Gerais, Belo Horizonte). Temperate countries have been effectively treating their sewage and industrial wastes since the 1970s. Latin American countries are just beginning to install point source treatment plants. A similar lag occurs regarding dams and fish passage. Dam building was intense in Europe and North America from the 1930s–1960s, versus the 1970s–1980s in Latin America. Fish passage received attention for some North American dams in the 1960s, but has been ignored in many others, and is being reconsidered as dams are relicensed. Fish passage did not begin being seriously considered in Brazil until the 1990s. The rates of deliberate alien fish introductions have slowed in most areas of temperate countries, but are still increasing in Brazil. Most fish species identifications were well established in temperate countries by the 1950s, although new species are still being identified through reclassification. However in Brazil, nearly every extensive field study still yields undescribed species.

It is unlikely that many rivers will be returned to their natural states, which is how the National Research Council (NRC) (1992) defines restoration. A more reasonable goal is to naturalize or rehabilitate rivers so that they exhibit more natural processes and structures. In doing so, it is useful to employ a leitbild (target vision) or natural reference condition approach, even if that vision or condition is not attained (Hughes 1994; Schmutz et al. 2000). Such an approach employs historical biological and abiotic data, data from minimally disturbed reference sites, and ecological dose–response models to estimate reference conditions for various river types.

Gore and Shields (1995), Graf (2001), Poff et al. (1997), and Stanford et al. (1996) offer several recommendations for improving the physical condition of large American rivers. Take a basin and riverscape perspective to emphasize the movements of water, sediments, nutrients, wood debris, and biota along the river continuum and between rivers and their floodplains. Manage towards the dynamic nature of rivers and a naturally variable flow regime (including magnitude, frequency, duration, timing, and rate of change) versus an equilibrium or average state. Reregulate dam storage and releases to simulate natural peak and low flows. Reduce fragmentation by operating or removing dams to rehabilitate physical conditions. Let the river do the work in developing habitat heterogeneity instead of focusing on site-specific engineering projects. Give physical integrity equal emphasis as biological and chemical integrity. Include a number of flow and channel measures in all river-monitoring and management programs. Increase complexity in channel types, backwaters, floodplains, vegetation, large wood debris, substrate, and depths. Protect those rivers or river reaches that remain in minimally disturbed condition. Discourage floodplain and riparian settlements by humans through removal of flood insurance programs, irrigation subsidies, levees, and revetments. Improve information transfer among river ecologists, hydrologists, managers, policy makers, and the general public.

The chapter authors in this book frequently lamented the varying sampling methods and study designs implemented on the rivers of interest through time. One reason agencies have had greater success in monitoring and regulating water chemistry than biology was the development of standard methods for analyzing water and waste water in 1905. The methods are continually calibrated and updated and are now in their 20th edition and online (APHA et al. 2004). We in the fish assemblage monitoring profession are 100 years behind. However, there is growing recognition of the need for alternative and calibrated standard methods in national monitoring programs (Meador et al. 1993; Peck et al., in press) and in a proposed book by the American Fisheries Society (Scott Bonar, University of Arizona, Tucson, personal communication).

Our book demonstrates the value of taking broad spatial and temporal perspectives in studies of large rivers, as suggested by Minshall (1988). Large rivers and many of their fish species require study at scales of hundreds to thousands of kilometers and over decades to centuries to detect and understand current fish assemblage structures, and the effects humans have had on them. Comparative studies, such as depicted in this book, also demonstrate that different ecoregional settings affect the degree to which similar anthropogenic alterations affect fish assemblages.

Although water quality has improved greatly since the 1970s in progressive countries as a result of point source controls, diffuse pollution by nutrients and sediments remain a serious concern. We have been particularly unsuccessful in improving the physical and biological condition of our rivers because of our unwillingness to control biological pollution by alien species and physical pollution by altered flow regimes and channelization. These conditions will not improve substantially without the fundamental social changes that drive our economic systems, population growth, and ethics. Consequently, the mandates of the Clean Water Act and the European Union's Water Framework Directive for ecological integrity remain illusory in the short term. However, humans are capable of remarkable and rapid cultural changes (Nash 1989), so there is still room for realistic hope.

Acknowledgments

We thank Julie Frandsen, Debby Lehman, and the chapter authors of this book for their cooperation, diligence, and patience. This manuscript was partially funded by the USEPA through a National Center for Environmental Research (NCER) STAR Program grant R-829498–01 to Oregon State University. It benefited from reviews by Art Benke, Mike Meador, Tad Penczak, and LeRoy Poff.

References

Abell, R. A., D. M. Olson, E. Dinerstein, P. T. Hurley, J. T. Diggs, W. Eichbaum, S. Walters, W. Wettengel, T. Allnutt, C. J. Loucks, and P. Hedao. 2000. Freshwater ecoregions of North America: a conservation assessment. Island Press, Washington, D.C.

AFS (American Fisheries Society). 2005. Study report on dam removal for the AFS Resource Policy Committee. Available: http://www.fisheries.org/html/Public_Affairs/Policy_Statements. (February 2005).

APHA, AWWA, and WEF 2004. Standard methods for the examination of water and wastewater. 20th edition. American Public Health Association, American Water Works Association, and Water Environment Federation, New York.

Barbour, M. T., W. F. Swietlik, S. K. Jackson, D. L. Courtemanch, S. P. Davies, and C. O. Yoder. 2000. Measuring the attainment of biological integrity in the USA: a critical element of ecological integrity. Hydrobiologia 423:453–464.

Fuller, P. L., L. G. Nico, and J. D. Williams. 1999. Nonindigenous fishes introduced into inland waters of the United States. American Fisheries Society, Bethesda, Maryland.

Gore, J. A., and F. D. Shields, Jr. 1995. Can large rivers be restored? BioScience 45:142–152.

Graf, W. L. 2001. Damage control: restoring the physical integrity of America's rivers. Annals of the Association of American Geographers 91:1–27.

Hughes, R. M. 1994. Defining acceptable biological status by comparing with reference conditions. Pages 31–47 *in* W. S. Davis and T. P. Simon, editors. Biological assessment and criteria: tools for water resource planning and decision making. Lewis, Boca Raton, Florida.

Hughes, R. M., S. G. Paulsen, and J. L. Stoddard. 2000. EMAP-Surface Waters: a multiassemblage, probability survey of ecological integrity in the U.S.A. Hydrobiologia 423:429–443.

Johnson, R. K. 2001. Defining reference condition and setting class boundaries in ecological monitoring and assessment. Department of Environmental Assessment. Swedish University of Agricultural Sciences. Uppsala, Sweden.

Junk, W. J., P. B. Bayley, and R. E. Sparks. 1989. The flood pulse concept in river-floodplain systems. Pages 110–127 *in* D. P. Dodge, editor. Proceedings of the international large river symposium (LARS). Canadian Special Publication of Fisheries and Aquatic Sciences 106, Department of Fisheries and Oceans, Ottawa.

Karr, J. R., J. D. Allan, and A. C. Benke. 2000. River conservation in the United States and Canada. Pages 3–39 *in* P. J. Boon, B. R. Davies, and G. E. Petts, editors. Global perspectives on river conservation: science, policy, and practice. Wiley, New York.

Karr, J. R., and D. R. Dudley. 1981. Ecological perspective on water quality goals. Environmental Management 5:55–68.

Karr, J. R., K. D. Fausch, P. L. Angermeier, P. R. Yant, and I. J. Schlosser. 1986. Assessing biological integrity in running waters: a method and its rationale. Illinois Natural History Survey Special Publication 5. Champaign, Illinois.

Marchetti, M. P., T. Light, P. B. Moyle, and J. H. Viers. 2004a. Fish invasions in California watersheds: testing hypotheses using landscape patterns. Ecological Applications 14:1507–1525.

Marchetti, M. P., P. B. Moyle, and R. Levine. 2004b. Alien fishes in California watersheds: characteristics of successful and failed invaders. Ecological Applications 14:587–596.

Meador, M. R., T. F. Cuffney, and M. E. Gurtz. 1993. Methods for sampling fish communities as part of the National Water-Quality Assessment Program. U.S. Geological Survey, Open-file report 93–104, Raleigh, North Carolina.

Minshall, G. W. 1988. Stream ecosystem theory: a global perspective. Journal of the North American Benthological Society 7:263–288.

Moog, O., and A. Chovanec. 2000. Assessing the ecological integrity of rivers: walking the line among ecological, political and administrative interests. Hydrobiologia 422:99–109.

Moyle, P. B., and T. Light. 1996. Biological invasions of fresh water: empirical rules and assembly theory. Biological Conservation 78:149–161.

Nash, R. F. 1989. The rights of nature: a history of environmental ethics. University of Wisconsin Press, Madison.

National Research Council (NRC). 1992. Restoration of aquatic ecosystems: science, technology, and public policy. National Research Council. National Academy Press, Washington, D.C.

NatureServe. 2005. Fish faunal intactness indicator map. State of the environmental report. U.S. Environmental Protection Agency, Research Triangle Park, North Carolina.

Peck, D. V., D. Averill, J. M. Lazorchak, and D. J. Klemm. In press. Western pilot study field operations manual for non-wadeable rivers and streams. U. S. Environmental Protection Agency, Washington, D.C.

Poff, N. L., J. D. Allan, M. B. Bain, J. R. Karr, K. L. Prestegaard, B. D. Richter, R. E. Sparks, and J. C. Stromberg. 1997. The natural flow regime. BioScience 47:769–784.

Pringle, C. M., F. N. Scatena, P. Paaby-Hansen, and M. Nunez-Ferrera. 2000. River conservation in Latin America and the Caribbean. Pages 41–77 *in* P. J. Boon, B. R. Davies, and G. E. Petts, editors. Global perspectives on river conservation: science, policy, and practice. Wiley, New York.

Schmutz, S., M. Kaufmann, B. Vogel, M. Jungwirth, and S. Muhar. 2000. A multi-level concept for fish-based, river-type-specific assessment of ecological integrity. Hydrobiologia 423:279–289.

Southerland, M. T., and J. B. Stribling. 1994. Status of biological criteria development and implementation. Pages 81–96 *in* W. S. Davis and T. P. Simon, editors. Biological assessment and criteria: tools for water resource planning and decision making. Lewis, Boca Raton, Florida.

Stanford, J. A., J. V. Ward, W. J. Liss, C. A. Frissell, R. N. Williams, J. A. Lichatowich, and C. C. Coutant. 1996. A general protocol for restoration of regulated rivers. Regulated Rivers Research and Management 12:391–413.

Welcomme, R. L. 1979. Fisheries ecology of floodplain rivers. Longman, London.

Williams, J. E., J. E. Johnson, D. A. Hendrickson, S. Contreras-Balderas, J. D. Williams, M. Navarro-Mendoza, D. E. McAllister, and J. E. Deacon. 1989. Fishes of North America endangered, threatened, or of special concern: 1989. Fisheries 14(6):2–20.

Appendix A.—Major anthropogenic alterations in selected rivers and their effects on fish assemblages.

River	Major anthropogenic alterations	Fish assemblage and impact
Mackenzie	reduced subsistence fishing	38 native freshwater or anadromous species; minor local depletions
Great Slave	distant flow regulation; fishing	30 native freshwater or anadromous species; minor local depletions
Snake	flow regulation and diversion; channel alteration; water pollution	of 26 native species, 3 anadromous species are extirpated; 50% of sites sampled had degraded fish assemblages; 3 alien species are common or abundant, and dominant in lower reaches
Willamette	channel alteration, flow regulation; waste water treatment	of 28 native species, 1 is endemic and 4 are threatened or endangered (3 anadromous); 3 alien species occasionally captured; increased native and intolerant species; reduced % tolerant and % alien species
Sacramento	flow regulation, water pollution, and alien species	of 28 native species, 17 are endemic and 14 are extirpated, endangered, threatened, or vulnerable (6 habitat specialists; 8 anadromous); 43 alien species are stable or increasing and dominant
Virgin-Moapa	flow regulation and diversion; alien species	of 10 native species, 4 are endangered and 2 are declining in range (habitat specialists); 11 alien species are common or abundant and dominant
Verde	flow regulation and alien species	of 12 native species, 6 are extirpated (fluvial specialists); lower reaches are dominated by 6 common alien species
Gila	flow regulation and diversion; alien species	of 15 native species, 11 are endangered, threatened, or vulnerable (fluvial and habitat specialists); 5 are declining in range and numbers; 13 alien species common and dominant
Upper Colorado	flow regulation; alien species	of 8 native species, 4 are endangered (fluvial specialists); 9 alien species common and dominant

Appendix A.—Continued.

River	Major anthropogenic alterations	Fish assemblage and impact
Lower Colorado	flow regulation and diversion; alien species	all 9 endemic species uncommon, endangered, or extirpated (fluvial and habitat specialists); 26 alien species common and dominant
Grande	flow regulation and diversion	of 45 native species, 10 are extirpated, 2 are extinct, and most others are declining in range and numbers (fluvial and habitat specialists); 21 alien species common or dominant
Nazas	flow regulation and diversion, water pollution and alien species	11 of 14 species endemic; 7 species threatened or endangered (habitat and fluvial specialists); 6 alien species common or dominant; 100% of sites with degraded fish assemblages
Platte	flow regulation and diversion	of 76 native species, 26 are vulnerable because of declining ranges (fluvial and habitat specialists); 2 species extirpated; 13 alien species common
Missouri	flow regulation and channel alterations	of 107 native species, 23 are decreasing, and 16 of those are endangered or vulnerable (fluvial and habitat specialists); 11 alien species common
Red	channel and flow alterations	of 55 native species, 3 are extirpated (fluvial specialists; potamodromous); 1 alien species common
Upper Mississippi	channel alterations	of 99 species 2 decreasing (habitat specialists); increase in 1 turbidity-tolerant species; 1 alien species common
Wisconsin	flow regulation, water pollution	25 of 109 native species endangered, threatened, or vulnerable (habitat and fluvial specialists); decreased range for 7 species; 2 alien species common
Wabash	channel alterations, water pollution and flow regulation	of 158 native species, 7 are extirpated and 9 are endangered (habitat and fluvial specialists); visual predators are uncommon; 5 aliens common

Appendix A.—Continued.

River	Major anthropogenic alterations	Fish assemblage and impact
Patoka	channel alterations, water pollution and flow regulation	6 of 80 native species extirpated (habitat and fluvial specialists); sensitive minnows and darters less abundant; 61% of the channel with degraded fish assemblages; 3 alien species common
Ohio rivers	waste water treatment; flow regulation	129 native species; 2 alien species common; increased range and abundance of intolerant species (habitat and fluvial specialists); decreased abundance of tolerant species
Ohio	channel and flow alterations and wastewater treatment	out of 116 native species, 39 are vulnerable, threatened, or endangered; 35 increased, including 5 sensitive species (habitat and fluvial specialists), while 3 tolerant species declined; 6 alien species common
Susquehanna	flow regulation	out of 85 freshwater or diadromous species, anadromous species markedly reduced; two species extirpated (habitat specialists); 2 alien species common
Hudson	channel alteration, water pollution, fishing and alien species	out of 80 freshwater or diadromous species, 2 species are declining, and 1 is endan gered (anadromous); 7 aliens are common
Saint John	channel and flow alterations, water pollution	33 native species, 7 with restricted ranges (anadromous and fluvial specialists); 1 alien species common
Kissimmee	channel alteration and alien species	of 47 native species, 5 have been reduced and 1 extirpated; (habitat specialists); 4 sensitive species declined in abundance; 2 tolerant air breathing species increased in abundance; 7 alien species common

Appendix A.—Continued.

River	Major anthropogenic alterations	Fish assemblage and impact
Alabama	flow regulation	of 184 native species and 29 endemics, 38 are vulnerable, threatened, or endangered, chiefly darters and diadromous species; 2 species extirpated; 14 aliens stable or common
das Velhas	water pollution and excess sediments	of 107 native species, 13 are extirpated and 7 have reduced ranges (chiefly habitat specialists); 2 aliens common